Machine Learning and Optimization Techniques for Automotive Cyber-Physical Systems

Vipin Kumar Kukkala • Sudeep Pasricha
Editors

Machine Learning and Optimization Techniques for Automotive Cyber-Physical Systems

 Springer

Editors
Vipin Kumar Kukkala
NVIDIA
Santa Clara, CA, USA

Sudeep Pasricha
Colorado State University
Fort Collins, CO, USA

ISBN 978-3-031-28018-4   ISBN 978-3-031-28016-0   (eBook)
https://doi.org/10.1007/978-3-031-28016-0

This Springer imprint is published by the registered company Springer Nature Switzerland AG
The registered company address is: Gewerbestrasse 11, 6330 Cham, Switzerland

Preface

Modern vehicles are examples of complex cyber-physical systems (CPS) with tens to hundreds of interconnected Electronic Control Units (ECUs) that manage various vehicular subsystems. The ECU functionalities range from simple window control to highly complex Advanced Driver Assistance System (ADAS) applications such as adaptive cruise control, lane-keeping assist, and collision avoidance. With the shift toward autonomous driving and ever-increasing connectivity with external systems, connected and autonomous vehicles (CAVs) have emerged as one of the most complex automotive CPS. The modern CAV ecosystem is characterized by increased ECU count, greater software complexity, and highly complex heterogeneous vehicular networks (within and outside the vehicle). Moreover, the aggressive attempts of automakers to make vehicles fully autonomous have led to the adoption of artificial intelligence (AI)-based techniques for advanced perception and control.

These paradigm shifts have resulted in increased overall complexity and have severe performance and safety implications on the automotive CPS. This book identifies and explores the most challenging issues (listed below) in designing a safe, secure, and robust automotive CPS.

- **Safety**: The increased complexity of automotive CPS has resulted in significant overhead on resource-constrained ECUs, which can lead to missing real-time deadlines. Missing deadlines for safety-critical automotive applications can be catastrophic, and this problem will be further aggravated in the case of future autonomous vehicles. Thus, designing efficient techniques to guarantee real-time performance (i.e., ensure no deadline and timing constraints are violated) is vital to ensure the safety of automotive CPS.
- **Security**: The increased connectivity of modern vehicles has made them highly vulnerable to various sophisticated cyber-attacks. However, imposing security mechanisms on the resource-constrained ECUs can result in additional computation and communication overhead, potentially leading to further missed deadlines. Therefore, it is crucial to design lightweight security mechanisms that are tailored toward automotive systems. Moreover, the increased complexity of automotive cyber-attacks is a growing concern, and this requires developing

advanced intrusion detection systems (IDS) that can detect these sophisticated cyber-attacks.

- **Robustness**: The adoption of various machine learning and deep learning techniques for the perception and control of vehicular subsystems has resulted in high reliance on data from external systems and sensors. However, small uncertainties in the input data can lead to undesired outcomes from these perception and control systems, which can degrade the performance of the vehicle and potentially have catastrophic consequences. Hence, designing robust perception techniques and control algorithms is essential to automotive CPS design.

In summary, designing a safe, secure, and robust automotive CPS while meeting all timing and deadline constraints is not a trivial task.

To address these challenges, this book discusses emerging machine learning and optimization algorithm-based solutions for real-time scheduling (Part I), security-aware design (Part II), intrusion detection systems (Part III), robust perception (Part IV), and robust control (Part V). The brief outline of the book with different parts and the chapters within them is summarized below.

1. **Real-Time Scheduling**: Since automotive CPS are highly resource-constrained and handle various safety-critical functions, efficient scheduling is critical to enable safe and reliable vehicles. Hence, the first part of the book focuses on real-time scheduling techniques in automotive CPS.

   - Chapter "Reliable Real-Time Message Scheduling in Automotive Cyber-Physical Systems" presents a novel real-time message scheduling framework that utilizes both design time and runtime scheduling to mitigate the impact of jitter in time-triggered automotive CPS.
   - Chapter "Evolvement of Scheduling Theories for Autonomous Vehicles" describes three real-time scheduling techniques at the task, resource, and network levels to address different scheduling challenges in automotive CPS.
   - Chapter "Distributed Coordination and Centralized Scheduling for Automobiles at Intersections" discusses distributed and centralized real-time scheduling approaches for solving the problem of intersection management in connected and autonomous vehicles.

2. **Security-Aware Design**: To protect vehicles from devastating cyber-attacks, it is crucial to include security as one of the fundamental requirements when designing automotive CPS. Therefore, the second part of the book focuses on security-aware design of automotive CPS.

   - Chapter "Security Aware Design of Time-Critical Automotive Cyber-Physical Systems" presents a novel security framework that combines design time schedule optimization with runtime key management to improve the security of time-critical automotive CPS.
   - Chapter "Secure by Design Autonomous Emergency Braking Systems in Accordance with ISO 21434" discusses security-aware design using ISO

21434 by considering an autonomous emergency braking system as a case study and evaluates the impact of various adversarial actions.

- Chapter "Resource Aware Synthesis of Automotive Security Primitives" develops methods for security-aware automotive CPS design by leveraging adaptive lightweight attack detection and mitigation schemes.
- Chapter "Gradient-Free Adversarial Attacks on 3D Point Clouds from LiDAR Sensors" presents an approach using evolutionary algorithms to design novel attacks on LiDAR scans used in autonomous driving.
- Chapter "Internet of Vehicles-Security and Research Roadmap" describes various challenges in designing a secure automotive system in the context of the Internet of Vehicles (IoV) and presents a research roadmap to address the challenges.

3. **Intrusion Detection Systems**: An intrusion detection system (IDS) often acts as the last line of defense in cyber-attacks. Traditionally, firewalls and rule-based systems are employed to detect cyber-attacks. However, they are not effective in detecting sophisticated cyber-attacks due to the lack of ability to learn the complex dependencies in vehicular network data. Hence, the third part of the book focuses on machine learning-based IDS that encompasses monitoring and attack detection at an in-vehicle network level and vehicular ad hoc network (VANET) level.

- Chapter "Protecting Automotive Controller Area Network: A Review on Intrusion Detection Methods Using Machine Learning Algorithms" describes the evolution of in-vehicle networks and presents a comprehensive review of machine learning-based intrusion detection approaches in Controller Area Network (CAN)-based automotive CPS.
- Chapter "Real-Time Intrusion Detection in Automotive Cyber-Physical Systems with Recurrent Autoencoders" proposes a novel lightweight IDS that utilizes Gated Recurrent Unit (GRU)-based recurrent autoencoder networks to detect cyber-attacks in automotive CPS.
- Chapter "Stacked LSTMs Based Anomaly Detection in Time-Critical Automotive Networks" develops a novel anomaly detection framework that uses stacked Long Short-Term Memory (LSTM) with an attention mechanism and One-Class Support Vector Machine (OCSVM) to detect various cyber-attacks in automotive CPS.
- Chapter "Deep AI for Anomaly Detection in Automotive Cyber-Physical Systems" presents an anomaly detection framework that uses temporal convolutional neural attention (TCNA) network to detect cyber-attacks in automotive CPS.
- Chapter "Physical Layer Intrusion Detection and Localization on CAN Bus" explores different machine learning techniques to detect cyber-attacks and identify the ECU transmitting malicious messages using voltage characteristics of CAN signals.
- Chapter "Spatiotemporal Information Based Intrusion Detection Systems for In-Vehicle Networks" develops a convolutional LSTM-based IDS that uses

spatiotemporal information of CAN data frames to detect cyber-attacks in automotive CPS.

- Chapter "In-Vehicle ECU Identification and Intrusion Detection from Electrical Signaling" describes a linear regression and support vector machine (SVM)-based IDS that relies on electrical signaling to detect cyber-attacks and identify the source ECU in CAN-FD-based automotive CPS.
- Chapter "Machine Learning for Security Resiliency in Connected Vehicle Applications" presents a machine learning-based solution for real-time resiliency in CAV applications against cyber-attacks on perception inputs.

4. **Robust Perception**: Many safety-critical functions in today's vehicles rely on data from different sensors and external systems, and employ various machine learning models to perceive the environment. The fourth part of the book discusses robust perception techniques in automotive CPS.

- Chapter "Object Detection in Autonomous Cyber-Physical Vehicle Platforms: Status and Open Challenges" provides a comprehensive review of state-of-the-art perception models used for object detection and discusses the open challenges in integrating them into autonomous vehicles.
- Chapter "Scene-Graph Embedding for Robust Autonomous Vehicle Perception" develops a novel spatiotemporal graph learning approach based on scene-graph representation to improve perception performance of automotive CPS.
- Chapter "Sensing Optimization in Automotive Platforms" explores different optimization techniques to synthesize heterogeneous sensor placement and orientation to enhance perception in modern semi-autonomous vehicles.
- Chapter "Unsupervised Random Forest Learning for Traffic Scenario Categorization" introduces a random forest-based unsupervised learning approach to detect patterns in traffic scenarios that are used for efficient validation of perception models designed for autonomous driving.
- Chapter "Development of Computer Vision Models for Drivable Region Detection in Snow Occluded Lane Lines" presents a CNN-based model that is used to identify the drivable region in snowy road conditions when lane lines are occluded by snow.
- Chapter "Machine Learning Based Perception Architecture Design for Semi-autonomous Vehicles" describes a machine learning-based automated perception architecture exploration framework to generate robust, vehicle-specific perception solutions.

5. **Robust Control**: It is vital to design robust control strategies to ensure safe and real-time performance in vehicles. To address these challenges, the fifth part of the book focuses on robust control techniques in automotive CPS.

- Chapter "Predictive Control During Acceleration Events to Improve Fuel Economy" presents a predictive control strategy for hybrid electric vehicles that dynamically allocates the torque between the engine and electric motor by predicting the acceleration behavior.

- Chapter "Learning-Based Social Coordination to Improve Safety and Robustness of Cooperative Autonomous Vehicles in Mixed Traffic" describes a multi-agent reinforcement learning (MARL)-based control approach that enables social coordination between autonomous and human-driven vehicles to improve road safety.
- Chapter "Evaluation of Autonomous Vehicle Control Strategies Using Resilience Engineering" employs resiliency engineering (RE) and resilience assessment grid (RAG) to enhance performance and asses the operational robustness of controllers in an autonomous vehicle.
- Chapter "Safety-Assured Design and Adaptation of Connected and Autonomous Vehicles" discusses different challenges in the design and operations of CAVs that use neural network-based components for advanced perception and control, and proposes solutions for improving their safety.
- Chapter "Identifying and Assessing Research Gaps for Energy Efficient Control of Electrified Autonomous Vehicle Eco-Driving" presents a review of vehicle control systems that enables eco-driving in autonomous vehicles and discusses critical research gaps and initial studies that addressed this problem.

We hope this book provides a comprehensive review and useful information on the recent advances in safety, real-time scheduling, security, perception, and control approaches for emerging automotive cyber-physical systems.

Santa Clara, CA, USA   Vipin Kumar Kukkala
Fort Collins, CO, USA   Sudeep Pasricha
December 17, 2022

Acknowledgments

This book would not be possible without the contributions of many researchers and experts in the field of cyber-physical systems, machine learning, automotive, and embedded systems. We would like to gratefully acknowledge the contributions of Wanli Chang (University of York), Nan Chen (University of York), Shuai Zhao (University of York), Xiaotian Dai (University of York), Yi-Ting Lin (National Taiwan University), Chung-Wei Lin (National Taiwan University), Iris Hui-Ru Jiang (National Taiwan University), Changliu Liu (Carnegie Mellon University), Adriana Berdich (Universitatea Politehnica Timisoara), Bogdan Groza (Universitatea Politehnica Timisoara), Pal-Stefan Murvay (Universitatea Politehnica Timisoara), Soumyajit Dey (Indian Institute of Technology Kharagpur), Ipsita Koley (Indian Institute of Technology Kharagpur), Sunandan Adhikary (Indian Institute of Technology Kharagpur), Jan Urfei (UL Method Park GmbH), Fedor Smirnov (Robert Bosch GmbH), Andreas Weichslgartner (Cariad SE), Stefan Wildermann (Friedrich-Alexander-Universität Erlangen-Nürnberg), Arunmozhi Manimuthu (Nanyang Technological University), Tu Ngo (Nanyang Technological University), Anupam Chattopadhyay (Nanyang Technological University), Jia Zhou (Peng Cheng Laboratory), Weizhe Zhang (Harbin Institute of Technology), Guoqi Xie (Hunan University), Renfa Li (Hunan University), Keqin Li (State University of New York, New Paltz), Sooryaa Vignesh Thiruloga (Colorado State University), Xiangxue Li (East China Normal University), Yue Bao (CATARC), Xintian Hou (CATARC), Srivalli Boddupalli (University of Florida), Richard Owoputi (University of Florida), Chengwei Duan (University of Florida), Tashfique Choudhury (University of Florida), Sandip Ray (University of Florida), Abhishek Balasubramaniam (Colorado State University), Shih-Yuan Yu (University of California, Irvine), Arnav Vaibhav Malawade (University of California, Irvine), Mohammad Abdullah Al Faruque (University of California, Irvine), Joydeep Dey (Colorado State University), Friedrich Kruber (Technische Hochschule Ingolstadt), Jonas Wurst (Technische Hochschule Ingolstadt), Michael Botsch (Technische Hochschule Ingolstadt), Samarjit Chakraborty (University of North Carolina, Chapel Hill), Rodolfo Valiente (University of Central Florida), Behrad Toghi (Honda Research Institute), Mahdi Razzaghpour (University of Central Florida), Ramtin Pedarsani (University of

California, Santa Barbara), Yaser P. Fallah (University of Central Florida), Xin Chen (University of Dayton), Jiameng Fan (Boston University), Chao Huang (University of Liverpool), Ruochen Jiao (Northwestern University), Wenchao Li (Boston University), Xiangguo Liu (Northwestern University), Yixuan Wang (Northwestern University), Zhilu Wang (Northwestern University), Weichao Zhou (Boston University), Qi Zhu (Northwestern University), Thomas Bradley (Colorado State University), Samantha White (Colorado State University), Aaron Rabinowitz (Colorado State University), Chon Chia Ang (Colorado State University), David Trinko (Colorado State University), Zachary D. Asher (Western Michigan University), Parth Kadav (Western Michigan University), Sachin Sharma (Western Michigan University), Farhang Motallebi Araghi (Western Michigan University), Johan Fanas Rojas (Western Michigan University), and Richard T. Meyer (Western Michigan University).

This work was partially supported by the National Science Foundation (NSF) grants CCF-1302693, CCF-1813370, and CNS-2132385. Any opinions, findings, conclusions, or recommendations presented in this book are those of the authors and do not necessarily reflect the views of the National Science Foundation.

Contents

Part I
Real-Time Scheduling

Reliable Real-Time Message Scheduling in Automotive Cyber-Physical Systems

Vipin Kumar Kukkala, Thomas Bradley, and Sudeep Pasricha

1 Introduction

Modern vehicles have several processing elements called Electronic Control Units (ECUs) that control different functionalities in a vehicle. ECUs run various automotive applications, such as anti-lock braking control, collison avoidance, lane-keeping assist, and adaptive cruise control [50]. These automotive applications are hard real-time in nature, meaning they have strict timing (deadline) and latency constraints [1]. The ECUs are distributed across the vehicle and communicate with each other by exchanging messages. These messages can be classified as either *time-triggered* or *event-triggered*. Time-triggered messages are periodic messages, typically generated from safety-critical software applications. On the other hand, event-triggered messages are generated asynchronously due to the occurrence of an event. Event triggered messages are typically low priority and consist of maintenance and diagnostic messages.

The diverse nature of communication in automotive systems requires different network protocols to support them. The Controller Area Network (CAN) protocol is one of the most popular and widely used communication protocols in automotive systems [40]. CAN is a lightweight, low cost, broadcast communication protocol, and support message priorities and error handling [3, 4]. CAN communication

V. K. Kukkala (✉)
NVIDIA, Santa Clara, CA, USA
e-mail: vipin.kukkala@colostate.edu

T. Bradley
Department of Systems Engineering, Colorado State University, Fort Collins, CO, USA
e-mail: thomas.bradley@colostate.edu

S. Pasricha
Colorado State University, Fort Collins, CO, USA
e-mail: sudeep.pasricha@colostate.edu

© The Author(s), under exclusive license to Springer Nature Switzerland AG 2023
V. K. Kukkala, S. Pasricha (eds.), *Machine Learning and Optimization Techniques for Automotive Cyber-Physical Systems*, https://doi.org/10.1007/978-3-031-28016-0_1

3

supports transmission rates of up to 1 Mbps and a maximum payload of 8 bytes [2]. In a CAN based system, when multiple ECUs are trying to transmit messages on the bus simultaneously, the message with the lowest CAN message ID wins the arbitration and gets access to the bus first, while all the other messages wait till the next arbitration event. Some of the other commonly used in-vehicle network protocols include Local Interconnect Network (LIN), FlexRay, Media Oriented Systems Transport (MOST), and Ethernet [9].

The onset of state-of-the-art x-by-wire automotive applications (throttle-by-wire, steer-by-wire, etc.) has led to an increase in the complexity of automotive applications [8]. This has resulted in a demand for an efficient, reliable, and deterministic in-vehicle communication protocol to satisfy the deadline constraints of all time-critical applications, while still being able to meet the high bandwidth requirements of these applications [12]. This is difficult to achieve using the industry de facto standard CAN bus, as it lacks time determinism and suffers from limited bandwidth (a maximum transmission rate of only 1 Mbps, which is insufficient for many high-bandwidth vehicular applications such as pedestrian detection and lane tracking). Moreover, the event-triggered nature of the CAN makes it harder to adapt for state-of-the-art high bandwidth demanding safety-critical applications. FlexRay emerged as an alternative communication protocol that overcomes the above-mentioned limitations of the CAN protocol and offers added flexibility, higher data rates (at least $10\times$ higher compared to CAN [7]), better time determinism, and support for both time-triggered and event-triggered transmissions. As a result, it is deployed in many state-of-the-art vehicles that implement demanding applications such as Audi A4's electronic stabilization control [10], Volvo XC 90's VDDM [11], etc.

The high complexity of the embedded systems in modern-day vehicles led to many challenges that threaten the reliability [12, 27, 38, 39], security [41–45], and real-time control [46–49] of modern-day automotive systems. In this chapter, we specifically focus on one of the key reliability challenges in time-triggered transmissions known as jitter. Jitter is the stochastic delay-induced deviation from the actual periodicity of a message. At a high level, jitter can be classified into two types: *(i) bounded (deterministic) jitter* and *(ii) unbounded (random) jitter*. The former is a periodic variation that is caused by the systematic occurrences of certain events in the system (such as queuing of messages, clock jitter, etc.) whose peak-to-peak value is bounded. Moreover, due to its deterministic nature, this type of jitter can be easily predicted based on observations. The latter is an unpredictable timing noise whose peak-to-peak values are unbounded. Some of the causes of random jitter include thermal noise in an electrical circuit (resulting in delayed task executions or message transmissions) and external disturbances. Unlike deterministic jitter, such random jitter is hard to predict based on system design and simple observations.

In this study, we address the problem of random jitter in automotive systems as it can have a significant impact on the performance and safety of the vehicle. Specifically, we focus on one of the most important sources of random jitter: delay in the execution of tasks in ECUs. Failure to effectively handle jitter-induced messages from such tasks can severely affect system performance and be catastrophic. For instance, when the airbag deployment signal from the impact sensor to the inflation module gets delayed due to jitter, it can be fatal to the vehicle occupants. We conjecture that jitter handling must be incorporated from the early design phase, while designing schedules for time-critical automotive applications. At the same time, unexpected jitter variations at runtime must also be carefully handled. Hence, there is a need for an effective jitter handling approach that can be applied when designing and enforcing the schedules for time-critical automotive applications.

In this chapter, we present a novel message scheduling framework called JAMS-SG that was first introduced in [39] to handle both jitter-affected time-triggered and high-priority event-triggered messages in an automotive communication system. Our framework is demonstrated for the FlexRay protocol, but it is protocol agnostic and can be extended to other time-triggered protocols with minimal changes. JAMS-SG combines design time schedule optimization with a runtime jitter handling mechanism, to minimize the impact of jitter in the FlexRay-based automotive network.

Our novel contributions in this chapter can be summarized as follows:

- We developed a hybrid heuristic to achieve jitter-aware frame packing (packing of different signals from an ECU into messages) for the FlexRay protocol;
- We developed a heuristic approach for the synthesis of jitter-aware design time schedules for FlexRay-based automotive systems;
- We introduced a runtime scheduler that opportunistically packs the jitter-affected time-triggered and high-priority event-triggered messages in the FlexRay static segment slots;
- We compared our JAMS-SG framework with the best-known prior works in the area and demonstrated its effectiveness and scalability.

The rest of this chapter is organized as follows. Section 2 presents an overview of the FlexRay protocol, and Sect. 3 presents the related works that address the problem of message scheduling in FlexRay-based systems. We define the problem statement by introducing the system and jitter models, heuristics used in this work, and important definitions and assumptions in Sect. 4. In Sect. 5, we discuss our proposed JAMS-SG framework in detail. We discuss the experimental setup and the results from our simulation-based analysis in Sect. 6 and conclude with a summary of our work in Sect. 7.

2 FlexRay Overview

FlexRay is a high-speed serial in-vehicle network protocol designed for x-by-wire automotive applications. It supports both time-triggered and event-triggered transmissions. The overview of the FlexRay protocol is illustrated in Fig. 1. According to the FlexRay specification [5], a communication cycle is one complete instance of a communication structure that repeats periodically (e.g., every 5 ms). Each communication cycle in FlexRay consists of a mandatory static segment, an optional dynamic segment, an optional symbol window, and a mandatory network idle time block.

The static segment in FlexRay consists of multiple equal-sized slots called static segment slots that are used to transmit time-triggered messages. The static segment employs a Time Division Multiple Access (TDMA) media access scheme for the transmission of time-triggered messages, which results in a repetition of the schedule periodically. In this TDMA scheme, each ECU is assigned one or more static segment slots and cycle numbers during which its messages can be transmitted on the FlexRay bus. This ensures time determinism, which guarantees message delivery. Each static segment slot transmits one FlexRay frame, which consists of three segments: header, payload, and trailer. The *header* segment is 5-bytes long and consists of status bits, frame ID (FID), payload length, header cyclic redundancy check (CRC), and cycle count. The *payload* segment can be up to 127 words (254 Bytes) long and consists of actual data that has to be transmitted. Lastly, the *trailer* segment consists of three 8-bit CRC fields to detect errors.

The dynamic segment in FlexRay consists of variable-sized slots called dynamic segment slots that are used to transmit event-triggered and low-priority messages. A dynamic segment slot consists of a variable number of minislots (as shown in Fig. 1), where each minislot is one microtick (usually 1 μs) long. The dynamic segment employs a Flexible Time Division Multiple Access (FTDMA) media access scheme where ECUs are assigned minislots according to their priorities. If an ECU is selected to transmit a message, then it is assigned the required number of minislots depending on the size of the FlexRay frame, and hence the length of a dynamic segment slot can vary in the dynamic segment (as shown in Fig. 1). During a message transmission, all the other ECUs have to wait until the one that is transmitting finishes. If an ECU chooses not to transmit, then that ECU is assigned only one minislot and the next ECU is assigned the subsequent minislot. The symbol window (SW) is used for network maintenance and signaling for the starting of the communication cycle, while the network idle time (NIT) is used to maintain synchronization between ECUs.

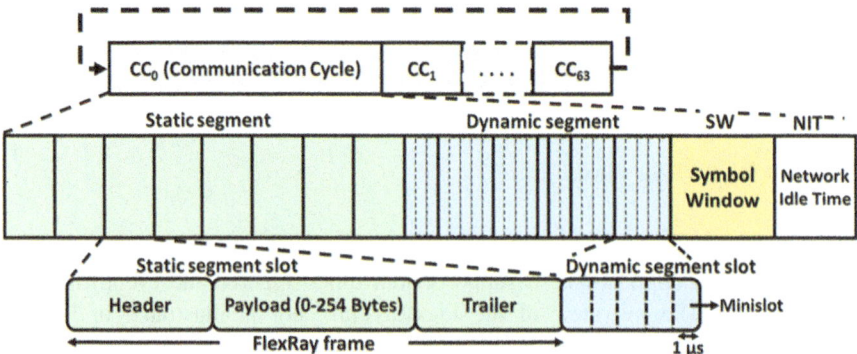

Fig. 1 Overview of the FlexRay protocol

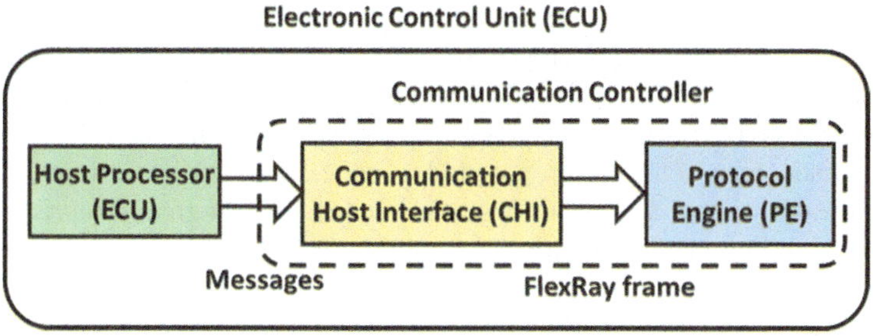

Fig. 2 Message generation and transmission in ECUs

Every automotive ECU consists of two major components: a host processor and a communication controller. The host *processor* is responsible for running automotive applications, while the *communication controller* acts as the interface between the host processor and the communication network. In a FlexRay-based ECU, the communication controller consists of a communication host interface (CHI) and a protocol engine (PE), as illustrated in Fig. 2. The CHI handles the message data generated by the host processor and sends the qualified FlexRay frames to the PE, which transmits the frames on a physical FlexRay bus (as shown in Fig. 2). Each FlexRay frame has a unique frame ID (FID) that is equal to the slot ID in which the frame is transmitted [5]. A FlexRay frame is considered to be "qualified" when the message data is available at the CHI before the beginning of the allocated static segment slot. Otherwise, a special frame called NULL frame is sent by setting a bit in the header segment of the FlexRay frame and setting all the data bytes in the payload to zero.

Jitter is one of the major reasons for the delay in the availability of message data at the CHI. Hence in this chapter, we focus on a novel frame packing and scheduling framework to overcome the delays and performance losses due to jitter in time-critical automotive systems.

3 Related Work

The message scheduling works for the FlexRay-based systems can be classified into two groups: *(i)* time-triggered and *(ii)* event triggered message scheduling. These works synthesize message schedules by optimizing various parameters, such as bandwidth, number of allocated static segment slots, response time, and end-to-end latency while ensuring all timing constraints are satisfied.

A common and crucial step prior to message scheduling is frame packing. Frame packing refers to the process of packing multiple signals into messages, to maximize the bandwidth utilization of the network [6]. In [7], an Integer Linear Programming (ILP) formulation was proposed to solve the frame packing problem, which requires multiple iterations with ILP to find an optimal solution. In [13], a Constraint Logic

Programming (CLP) formulation and heuristic were presented for reliability-aware frame packing. However, this approach could require multiple retransmissions of the packed frames to meet reliability requirements. In [14], the frame packing problem is treated as a one-dimensional allocation problem and an ILP formulation and a heuristic approach were proposed. A genetic algorithm based frame packing approach was proposed for CAN-FD systems in [15]. A fast greedy heuristic-based frame packing approach was proposed in [27]. The above-mentioned techniques either focus on optimizing bandwidth utilization or minimizing the time taken to generate a frame packing solution. *However, none of the above-mentioned works focus on generating a jitter-aware frame packing solution. Our proposed frame packing technique in this chapter uses a hybrid heuristic approach to generate a near-optimal set of messages that together make the system more resilient to jitter-induced uncertainties.*

In the case of hard real-time automotive cyber-physical systems, most parameters such as period, worst-case execution time, and deadline are known at design time. This facilitates the synthesis of highly optimized design-time schedules that are deployed during runtime to minimize the unpredictability in the system. Many works have addressed the issue of design-time scheduling of the static segment of FlexRay. One of the main objectives in these works is to minimize the number of static segment slots allocated to ensure future extensibility of the system while maximizing bandwidth utilization. An ILP-based approach is proposed in [16] to minimize the number of allocated static segment slots by considering task and message scheduling. This work was later extended in [18] by including support for multiple real-time operating systems and using ILP reduction techniques. The message scheduling problem was transformed into a two dimensional bin-packing problem in [17], and an ILP formulation and a heuristic approach were proposed for minimizing the number of allocated static segment slots. A CLP and ILP formulation for jointly solving the problem of task and message scheduling in FlexRay systems was proposed in [20] and [21], respectively. In [23], a set of algorithms was proposed to enable scheduling of event-triggered messages in time-triggered communication slots using a virtual communication layer. A few other works solve the same problem with heuristics and variants of ILP and CLP [13, 19, 22, 24, 25]. Some recent works such as [33] and [34] combine schedulability analysis and control theory, and were able to achieve fewer FlexRay static segment slots compared to many of the above-mentioned prior works. Additionally, there are works that focus on scheduling time-triggered systems using other network protocols in [28, 29, 30]. *However, the above-mentioned works focus on developing scheduling algorithms without incorporating the idea of jitter which makes them unreliable for use in real-time scenarios where jitter is prevalent and can significantly impact scheduling decisions.*

Jitter in FlexRay-based systems has been largely ignored, and there is limited literature on this topic. In [7], the authors proposed a jitter minimization technique using an ILP formulation. In [26], the message frequency is increased for the messages that are likely to be affected by jitter, to minimize the message response time. However, in both [7] and [26], it is assumed that the jitter value and number of messages that are affected by jitter are known at design time, which is unlikely in real-world scenarios. Moreover, [26] introduces significant load on the in-vehicle network due to multiple retransmissions, which is not a desired quality for resource constrained automotive systems. A jitter-aware message scheduling technique called JAMS was proposed in [27] that uses both design time and runtime schedulers to opportunistically pack jitter-affected messages in the system. However, the non-jitter-aware frame packing in [27] results in sub-optimal packing of signals into messages leading to increased message response times in the presence of jitter. Moreover, [27] considers a simple jitter model, which makes the evaluation process less efficient. In [31], an iterative design time scheduling algorithm was proposed to minimize the impact of jitter on mixed-criticality time-triggered messages. However, [31] does not effectively handle unpredictabilities due to random jitter at runtime. As random jitter can affect any message in the system, there is a need for a jitter handling mechanism that can handle jitter more comprehensively at the signal and message level, at both design time and runtime. *In this chapter, we introduce a realistic jitter model and propose a holistic message scheduling framework that achieves a jitter-aware frame packing and combines the design time schedule optimization with an improvised runtime jitter handling, to minimize the impact of jitter in FlexRay based systems. We extensively evaluate the proposed JAMS-SG framework to demonstrate its effectiveness and scalability.*

4 Problem Definition

4.1 System Model

In this study, we consider a generic automotive system model with multiple ECUs that run different time-critical automotive applications and are connected using a FlexRay bus. Executing an application may result in the generation of signal data at an ECU, which may be required for another application running at a different ECU. A signal can be a control pulse or raw data value. These signals are packed into messages by a process called frame-packing and transmitted as FlexRay frames on the bus. As discussed earlier, there are two types of applications in a typical automotive system: *(i)* time-triggered, and *(ii)* event-triggered.

Moreover, every ECU or node in the system can send both time-triggered and event-triggered messages. In a typical FlexRay system, time-triggered messages are transmitted in the static segment slots of the FlexRay while event-triggered messages are transmitted in the dynamic segment slots. However, in this work, in addition to the time-triggered messages, we facilitate the transmission of high priority event-triggered messages in the static segment of the FlexRay (details in Sect. 5.2). Hence, in this work, we focus on the challenging problem of scheduling time-triggered messages and high priority event-triggered messages in the static segment of FlexRay. We ignore the scheduling of low-priority (and typically low-frequency) event-triggered messages in the FlexRay dynamic segment, which is a much trivial problem and has a negligible impact on vehicle safety. Lastly, the terms ECU and node are used interchangeably henceforth.

4.2 Jitter Model

As discussed earlier, jitter is defined as the delay-induced deviation from the actual periodicity of the message. There are two types of jitter: *(i) bounded* or *deterministic jitter* and *(ii) unbounded* or *random jitter*. In this work, we primarily focus on random jitter, as it is hard to predict and can have a significant impact on the safety and performance of the vehicle. Our goal is to mitigate the effect of random jitter on task execution and message transmission delays. Moreover, we assume that both time-triggered and event-triggered messages are susceptible to such random jitter. However, we do not consider the impact of random jitter on low priority event-triggered messages, as such messages have minimal impact on the safety and performance of the vehicle.

Random jitter is also known as Gaussian jitter because it follows a normal distribution because of the central limit theorem [35]. In this work, we devise a specific jitter model for each signal in the system based on the signal priority and signal period. Signals with a period of less than or equal to 40 ms are treated as high priority signals and other signals are considered as low priority signals. In this work, we also model the mean jitter associated with high-priority signals as (*signal_period*/5) and (*signal_period*/4) for low priority signals. These mean jitter values are customizable and can be tuned based on the designer requirements and system specifications. A similar but more simplistic model is presented in [32] which does not consider the mixed criticality nature of the automotive applications. In a normal distribution representing jitter values (on x-axis) and number of occurrences (on y-axis), the jitter values in the tail region far from the mean occur less frequently than the values close to the mean. Hence, in this work, we mainly focus on mitigating the effect of mean jitter value associated with each signal and ensure that there are no missed deadlines.

4.3 Hybrid SA+GRASP Heuristic

A hybrid heuristic combines two or more heuristics that combine the advantages of individual heuristics and minimizes each other's disadvantages. In this study, we propose a hybrid heuristic by combining simulated annealing (SA) and greedy randomized adaptive search procedure (GRASP). Similar attempts were made in the past to combine SA and GRASP and build a hybrid heuristic in [36] and [37]. However, these efforts do not focus on the automotive domain, and they do not optimize the search space or perform tuning of hyperparameters. Our proposed

SA + GRASP hybrid heuristic aims to improve the computation speed, design space search capability, and solution optimality. Moreover, as our proposed framework uses the baseline model from JAMS [27], and the proposed SA + GRASP hybrid heuristic. Hence, we name our framework JAMS-SG where S and G represents SA and GRASP, respectively.

4.3.1 Simulated Annealing

Simulated Annealing (SA) metaheuristic is inspired from the annealing technique in metallurgy. It models the physical process of heating and controlled cooling of a material to strengthen and reduce defects. SA can effectively approximate the global optimum in very large discrete solution spaces.

There are five steps in any SA problem formulation (shown in Fig. 3): *(i) initial solution, (ii) initial temperature, (iii) random perturbations, (iv) acceptance probability,* and *(v) annealing schedule.* The SA is an iterative process, that begins by taking the initial solution and initial temperature as the inputs and tries to achieve a better solution at the end of every iteration. The temperature is progressively decreased from an initial positive value until a stopping condition is met (e.g., until temperature > 0). Each iteration constructs a new solution after making random perturbations to the current solution. If the new solution (objective function value) after making perturbations is better than the previous solution, then the new solution is accepted. Otherwise, the new solution is accepted probabilistically. SA uses an acceptance probability function, that takes the difference between the objective function values of the new and previous solutions and the current temperature of the system as the inputs, and computes the acceptance probability value. The new solution is accepted when the acceptance probability value is greater than a randomly generated number between 0 and 1. Otherwise, the new solution is discarded. In the initial stages, the SA tries to accept even a relatively poor solution when the system temperature is high. As the SA progresses, i.e., when the system temperature is lower, SA will favor accepting only those new solutions that are very close to the new solution. It is important to note that when the temperature reaches 0, SA behaves like a pure greedy algorithm. Lastly, at the end of each iteration, the temperature of the system is updated using an annealing schedule, which is responsible for the controlled cooling of the system.

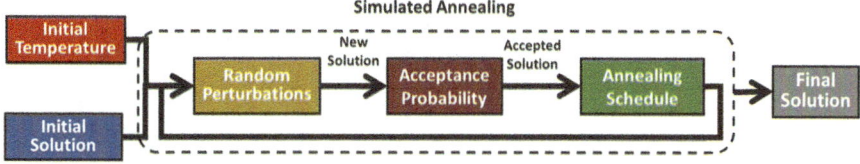

Fig. 3 Various steps involved in SA

SA is highly versatile and can deal with highly non-linear solution spaces. It is also good at dealing with arbitrary systems and cost functions while statistically guarantying an approximate global optimum. However, SA can take a very long time to converge to a good solution. Moreover, the optimality of the SA solution is heavily dependent on the selected hyperparameters, such as initial temperature, annealing schedule, and acceptance function.

4.3.2 Greedy Randomized Adaptive Search Procedure

The greedy randomized adaptive search procedure (GRASP) is a multi-start meta-heuristic. GRASP repeatedly sample stochastically greedy solutions and use an adaptive local search to refine them to a local optimum. At the end, the best of the local optima is chosen as the final solution.

The two key steps in GRASP (illustrated in Fig. 4) are: *(i) the greedy randomized construction phase* that tries to build a feasible solution and *(ii) the local search phase* that tries to explore a defined neighborhood for a local optimum. The best of the local optima is chosen as the final solution at the end. The two important aspects of the greedy randomized construction phase are its *greedy aspect* and *probabilistic aspect*. The greedy aspect involves generating a Restricted Candidate List (RCL), which consists of the best elements that will improve the partial solution (solution within the greedy randomized construction phase). The probabilistic aspect involves the random selection of an element from the RCL, to be incorporated into the partial solution. However, the solutions generated during the greedy randomized construction phase are not necessarily optimal. Hence, the local search phase tries to improve the constructed solution by iteratively using destroy and repair mechanisms, which are used to perturb the current solution and reconstruct a new solution, respectively. They help in searching for the local optimum within a defined neighborhood. Lastly, when an improved solution is found, then the best solution is updated.

GRASP is simple to construct and can be used for large optimization problems. However, as GRASP uses a greedy algorithm to evaluate the quality of the solution,

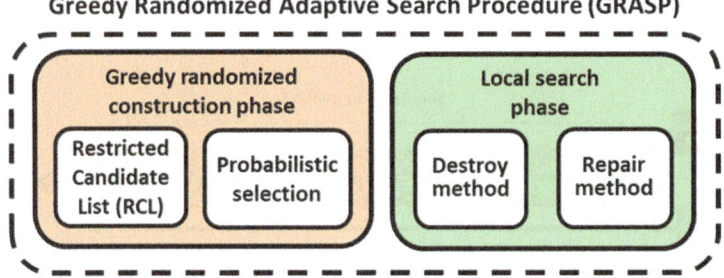

Fig. 4 Key steps involved in GRASP

it can get stuck at a local optima. Moreover, GRASP might restart at the same solution multiple times leading to re-discovering of the same local solution.

4.3.3 Hybrid Heuristic Formulation

To overcome the above-mentioned individual limitations of SA and GRASP, we propose a hybrid heuristic that combines both of them. The proposed hybrid heuristic uses SA to explore the large solution space and GRASP to find an improved local solution within a smaller neighborhood around the solution obtained from SA. In particular, the greedy construction phase of GRASP is used to make perturbations in the SA and the local search phase is used to explore the neighborhood to find a better solution. Our proposed hybrid heuristic is discussed in detail in Sect. 5.1.

4.4 Inputs and Definitions

We consider an automotive system with the following inputs:

- N denotes the set of nodes, where $N = \{1, 2, 3, \ldots, N\}$;
- For each node $n \in N$, $S^n = \{ s_1^n, s_2^n, \ldots, s_{K_n}^n \}$ represents the set of signals transmitted from that node and K_n represents the maximum number of signals in node n;
- Every signal $s_i^n \in S^n$, $(i = 1, 2 \ldots, K_n)$ is characterized by the tuple $\{ \overline{p}_i^n, \overline{d}_i^n, \overline{b}_i^n, \overline{\gamma}_i^n \}$, where $\overline{p}_i^n, \overline{d}_i^n, \overline{b}_i^n$ and $\overline{\gamma}_i^n$ denote the period, deadline, data size (in bytes), and mean jitter of the signal s_i^n, respectively;
- After frame-packing, every node maintains a set of messages $M^n = \{ m_1^n, m_2^n, \ldots, m_{R_n}^n \}$ in which every message $m_j^n \in M^n$, $(j = 1, 2, \ldots, R_n)$ (where R_n denotes the maximum number of messages in node n) is characterized by the tuple $\{ a_j^n, p_j^n, d_j^n, b_j^n, \mu_j^n \}$, where $a_j^n, p_j^n, d_j^n, b_j^n$ and μ_j^n represent the arrival time, period, deadline, data size (in bytes), and mean jitter of the message m_j^n, respectively.

In this chapter, we assume the following definitions:

- *Slot number or Slot identifier (slot ID)*: A number used to identify a specific slot within a communication cycle;
- *Cycle number*: A number used to identify a particular communication cycle in the FlexRay schedule;
- To transmit a message m_j^n on the FlexRay bus, it needs to be allocated a slot ID $sl \in \{1, 2, \ldots, N_{ss}\}$ and a base cycle number $bc \in \{0, 1, \ldots, C_{fx}\}$ where N_{ss} and C_{fx} are the total number of static segment slots in a cycle and the total number of cycles, respectively. This allocation is referred to as *message-to-slot assignment*;
- If a message m_j^n is assigned to a particular slot and a cycle, then the source node n of that message is allocated ownership of that slot. This is known as *node-to-slot assignment*.

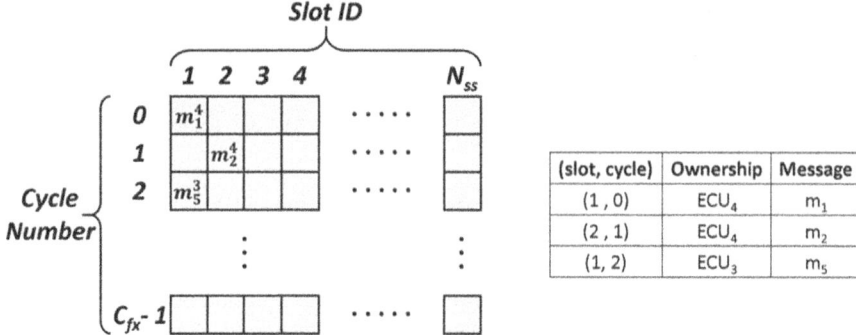

Fig. 5 Illustration of an example FlexRay 3.0.1 schedule (on the left) with slot IDs and cycle numbers; and the message-to-slot and node-to-slot allocation are shown in the table on the right

All the above-mentioned definitions are illustrated in Fig. 5 with an example FlexRay 3.0.1 schedule. In the example, message (m_1) is allocated a slot ID = 1 and cycle number = 0, which is the *message-to-slot assignment*. This implies that the source node (ECU_4) sending the message (m_1) is allocated ownership of the slot, which is the *node-to-slot assignment*.

Thus, for the above inputs, the goal of our work is to satisfy deadline constraints for time-triggered and high-priority event-triggered messages sent over the FlexRay bus. This is achieved by enabling jitter resilience during communication, which includes: *(i)* performing jitter-aware frame packing, and design time scheduling (message-to-slot assignment, node-to-slot assignment) for the time-triggered messages without violating any deadlines, and *(ii)* effectively handle jitter-affected time-triggered messages and high priority event-triggered messages at runtime and minimize the impact of jitter.

5   JAMS-SG Framework Overview

Our proposed JAMS-SG framework aims to enable jitter-aware scheduling of time-triggered messages and collocating high-priority event-triggered messages in the static segment of a FlexRay-based automotive system. The overview of our proposed JAMS-SG framework is illustrated in Fig. 6. At a high level, the JAMS-SG framework consists of design-time and runtime steps. At design time, JAMS-SG uses the proposed hybrid heuristic approach (SA + GRASP) to achieve jitter-aware frame packing of time-triggered messages and a feasible design-time schedule. At runtime, JAMS-SG handles both jitter-affected time-triggered and high-priority event-triggered messages using a multi-level feedback queue (MLFQ). The output of MLFQ and the design time schedule are given as the inputs to a runtime scheduler that opportunistically packs these jitter-affected messages into the already allocated

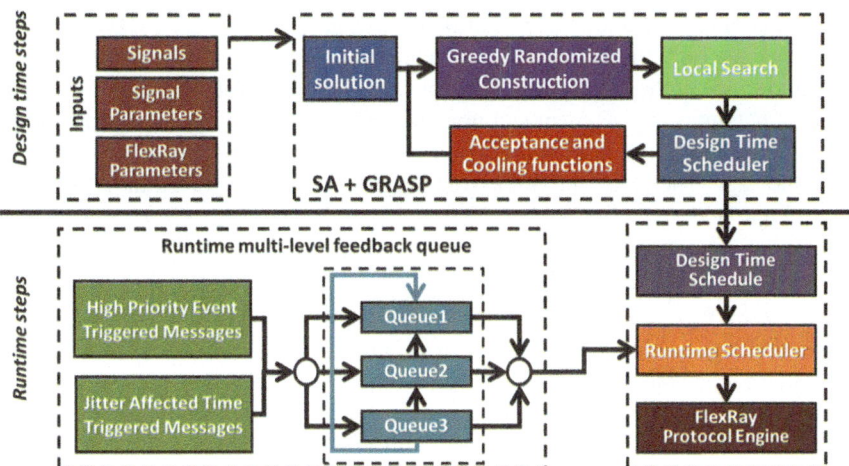

Fig. 6 Overview of our proposed JAMS-SG framework

FlexRay slots based on the available slack. Each of these steps is discussed in detail in the subsequent subsections.

5.1 Jitter-Aware Design Time Frame Packing

Frame packing refers to packing multiple periodic signals in a node into messages. This step is crucial to maximize bandwidth utilization, as it improves system performance and enhances the extensibility of the system by utilizing fewer slots than without frame packing. However, frame packing with a goal of just maximizing the bandwidth utilization can result in sub-optimal results at runtime. For instance, if one of the signals packed in the message is not available before the start of the message's allocated static segment slot due to jitter-induced delays, the entire message will be delayed and the CHI will transmit a NULL frame due to the lack of availability of complete message data. Once all of the signals in the message are available, the message can be transmitted in the next allocated static segment slot. This delayed transfer will result in increased response time of the messages and can potentially lead to missed deadlines, which can have catastrophic consequences for safety-critical applications. Thus, it is essential to have a jitter-aware frame packing technique that co-optimizes bandwidth utilization and mitigates the impact of jitter.

In this work, we define the following four necessary conditions (shown in (1)–(4)) that govern how signals can be packed into the same message. The first necessary condition is the source node condition and is expressed in (1).

$$m_k^n = \left\{ s_i^{n_1}, s_j^{n_2} \right\} \quad iff\ src\left(s_i^{n_1} \right) == src\left(s_j^{n_2} \right) \quad i, j \in [1, K_n] \ \& \ i \neq j \qquad (1)$$

The source node condition states that if two signals $(s_i^{n_1}, s_j^{n_2})$ are packed into the same message (m_k^n) they should belong to the same source node ($src()$ in (1) returns the source node of the signal). This is because the FlexRay protocol specification [5] dictates that any static segment slot in a given FlexRay cycle can be assigned to at most one node, which prevents the packing of signals from different nodes into the same message. Therefore, the frame packing problem can be solved independently for different nodes.

$$m_k^n = \left\{ s_i^n, s_j^n \right\} \quad iff \quad \overline{p}_i^n == \overline{p}_j^n \; i, j \in [1, K_n] \; \& \; i \neq j \tag{2}$$

The periodicity condition in (2) states that, only the signals with the same periods should be packed into one message. This is done primarily to minimize the retransmissions of the message frames, which in turn reduces the number of allocated static segment slots, leading to efficient bandwidth utilization. For instance, consider a scenario where two signals with periods 5 and 15 ms are packed into the same message. The resulting message will have a period of 5 ms and will retransmit the old value of 15 ms signal twice before sending the updated signal value. Therefore, packing signals with different periods leads to inefficient bandwidth utilization.

$$\sum\nolimits_{i \in sigIDs(m_k^n)} \overline{b}_i^n \leq B_{slot} \tag{3}$$

The payload condition in (3) states that, the sum of all signal sizes packed in a message ($sigIDs()$ returns the set of IDs of the signals packed in the message) should not exceed the maximum payload (B_{slot}) of the FlexRay static segment slot.

$$ResponseTime\left(m_k^n\right) \leq d_k^n \qquad \forall n, k = 1, 2, \ldots, R_n \tag{4}$$

Lastly, the deadline condition in (4) states that, the set of messages generated from frame packing should result in a feasible schedule, i.e., the response time (end-to-end latency) for all the messages should not exceed their deadline. In addition, JAMS-SG allows the system designers to specify additional timing constraints for signals, such as latency, worst-case response time, etc., which will be treated as additional constraints to the problem. The timing constraints are further discussed in Sect. 5.1.3.

Given the above-mentioned constraints, the goal of our proposed jitter-aware frame packing technique is to maximize the laxity of each resulting message while minimizing the total number of FlexRay frames. In other words, we prioritize packing of signals with similar jitter profile into the same messages. This is because of the following reason.

Consider an example scenario shown in Fig. 7, where four different signals are packed into messages in three different ways. For simplicity, we considered packing a maximum of two signals per message in this example. This constraint is not enforced anywhere else in the framework. In case (i), signals with similar jitter profile $(\overline{\gamma}_i^n)$ are packed together $((s_1^n), s_2^n)$ and $(s_3^n, s_4^n))$ which resulted in two

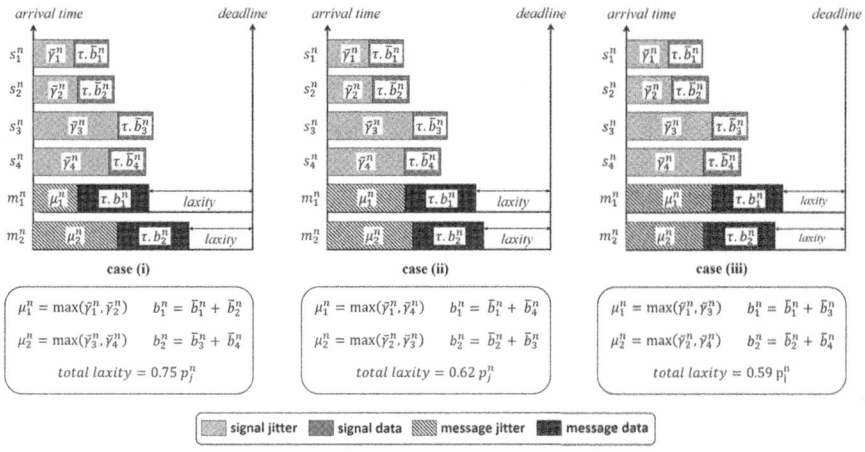

Fig. 7 Motivation example for selection of objective function for frame packing

messages (m_1^n, m_2^n) with their effective mean jitter value (μ_k^n) close to the individual signal jitter values. This results in maximizing the laxity of the messages, which is important as it provides more opportunities to pack and transmit other jitter-affected messages and also to better cope with unpredictable runtime jitter. On the other hand, in cases (ii) and (iii), the signals with very different jitter profiles are packed together resulting in lower laxity values. This makes the messages less resilient to random jitter compared to the frame packing in case (i). It is important to note that, packing one signal per message would maximize the laxity but would lead to very inefficient bandwidth utilization. Thus, to avoid packing one signal per message, we formulated a weighted objective function that achieves jitter-aware frame packing while effectively minimizing the total number of messages in the system.

In this work, we use a weighted harmonic sum of average laxity of the messages in each node as the objective function to achieve jitter-aware frame packing (shown in (5)). Laxity of any message is defined as the difference between the message deadline and sum of mean jitter value of the message and time required to transmit the message payload. The laxity of any message (m_j^n) can be computed using (6). The mean jitter value of a message (shown in (7)) is the maximum mean jitter value of all signals packed in that message. In this study, our proposed hybrid heuristic (SA + GRASP) aims to minimize the objective function (in (5)) while satisfying all of the constraints mentioned above ((1)–(4)). The parameters R_n, $\overline{b}_i^n$ and τ represent, the number of messages in the node n, data size of signal i in node n and time taken to transmit 1 byte of data on FlexRay bus, respectively.

$$minimize \sum_{n=1}^{N} \left(w * \frac{R_n}{\sum_{j=1}^{R_n} laxity_j^n} \right) \qquad (5)$$

$$laxity_j^n = d_j^n - \left(\mu_j^n + \tau. \sum_{i \in sigIDs\left(m_j^n\right)} \overline{b}_i^n \right) \qquad (6)$$

$$\mu_j^n = \max \left(\left\{ \overline{\gamma}_i^n \mid i \in sigIDs\left(m_j^n\right) \right\} \right) \qquad (7)$$

Our proposed JAMS-SG framework uses SA to explore the overall solution space and GRASP to create and refine new solutions at every iteration. The solution here refers to the signal to message packing for all the nodes in the system. To the best of our knowledge, this is the first work in this area that attempts to achieve a jitter-aware frame packing. The pseudo-code of our proposed hybrid heuristic is shown in Algorithm 1.

Algorithm 1: SA + GRASP Based Frame Packing

Inputs: Set of nodes (N), Set of time-triggered signals in each node (S^n), GRASP parameters (α, β), temperature (T), cooling rate (C_r)

1: **Initialize**: *cur_sol, prev_sol, best_sol* ← **initial_solution**(S^n)
2: **for** each *iteration* **until** *max_iterations* **do**
3:      δ = **random_int**(1, N)
4:      λ = **random_selection**(N, δ)
5:      *gr_sol* ← **greedy_randomized_construction**(α, λ, cur_sol)
6:      *ls_sol* ← **local_search**(β, N, gr_sol)
7:      *cur_sol* ← **choose_solution**(*gr_sol, ls_sol*)
8:      **if feasibility**(*cur_sol*) **then**
9:          P_{acc} = **acceptance_probability**(*cur_sol, prev_sol, T*)
10:         **if** P_{acc} > **random**(0,1) **then**
11:             *prev_sol* ← *cur_sol*
12:             **if** Φ(*cur_sol*) < Φ(*best_sol*) **then**
13:                 *best_sol* ← *cur_sol*
14:             **end if**
15:         **end if**
16:     **end if**
17:     $T *= C_r$
18: **end for**
19: *output_sol* ← *best_sol*
20: *output_schedule* ← **design_time_schedule**(*output_sol*)
Output: Set of messages in each node and a feasible design time schedule

---

The inputs to Algorithm 1 are: a set of nodes (N), set of time-triggered signals for each node (S^n), GRASP control parameters (α – which is the RCL threshold discussed later in Sect. 5.1.1, β – which is the destroy-repair threshold discussed later in Sect. 5.1.2), and the SA hyper-parameters: temperature (T) and cooling rate (C_r). The algorithm begins by initializing the current (*cur_sol*), previous (*prev_sol*), and best (*best_sol*) solutions with one signal per message (one-to-one frame packing) in step **1**. This acts as the initial solution for the SA. All solutions described in this pseudo code are data structures that have information about the signal to message packing for each node (*solution* ← [M_1, M_2 ... M_N]). Each element in the list corresponds to a frame-packing configuration for a particular node. At the beginning of each iteration in steps **3**, **4**, the algorithm selects a random number (δ) of nodes (λ) to be perturbed. In step **5**, a new solution (*gr_sol*) is constructed from the current solution (*cur_sol*) using the greedy randomized construction phase of GRASP. This new solution (*gr_sol*) is given as the input to the local search phase of GRASP in step **6** to search for a local optimum solution (*ls_sol*) within the defined neighborhood. The better of the two solutions (*gr_sol*, *ls_sol*), i.e., the solution that results in a minimal objective function value is chosen as the current solution (*cur_sol*) in step **7**. In step **8**, the *feasibility*() function is used to check for the feasibility of the chosen solution. The *feasibility*() function returns *true* when there are no missed deadlines for any message in the given solution; otherwise, it returns *false*. When a solution is feasible, the decision of accepting or discarding it is dependent on the probability of acceptance (P_{acc}). This value is computed in step **9** using the *acceptance_probability*() function, which takes the current system temperature (T), and both current and previous solutions as the input. If the current solution is accepted, then the previous solution is assigned the current solution and the best solution is also updated if the current solution has lower objective function value compared to the best solution (steps **10–16**). The function Φ() is used to compute the objective function value of the solution. Additionally, at the end of each iteration, the annealing schedule performs a controlled cooldown of the system, as shown in step **17**. At the end of *max_iterations*, the best solution is chosen as the output solution (*output_sol*) and a design-time schedule is synthesized using it (steps **19**, **20**). Thus, Algorithm 1 outputs a jitter-aware frame packing solution and a feasible design time schedule.

It is important to note that the resulting output messages will have the same period as the signal period, and their deadline will be equal to the lowest signal deadline packed in that message. In this work, we assume that all the messages have deadlines equal to their periods.

5.1.1 Greedy Randomized Construction

In this subsection, we discuss the greedy randomized construction phase of GRASP that is used to perturb the solution in SA, in step **5** of Algorithm 1.

Algorithm 2: greedy_randomized_construction $(\alpha, \lambda, cur\_sol)$

Inputs: RCL threshold (α), perturbed nodes (λ), and *cur_sol*
1:     **function** *greedy_construct*($\alpha, \hat{S}^n, partial\_sol$)
2:         **while** $\hat{S}^n \neq \{\ \}$ **do**
3:             $s = $ **random_selection**($\hat{S}^n$, 1)
4:             $\Omega \leftarrow$ **feasible_frame_ids**($s, partial\_sol$)
5:             **if** $\Omega \neq \{\ \}$ **then**
6:                 $C_{fids} \leftarrow$ **cost**(s, Ω)
7:                 $C_{min} = $ **min**(C_{fids}); $C_{max} = $ **max**(C_{fids})
8:                 $RCL \leftarrow \{fid \in \Omega \mid C_{fids}(fid) \leq C_{min} + \alpha*(C_{max} - C_{min})\}$
9:                 $chosen_fid = $ **random_select**(RCL, 1)
10:                **assign_fid**($s, chosen\_fid, partial\_sol$);
11:             **end if**
12:             **Remove** s from $\hat{S}^n$
13:         **end while**
14:         **return** *partial_sol*
15:     **end function**
16:     **Initialize**: *greedy_randomized_sol* $\leftarrow cur\_sol$
17:     **for** each node n in λ **do**
18:         $\rho = $ **random_int**(1, **length**(S^n))
19:         $\hat{S}^n = $ **random_selection**(S^n, ρ)
20:         *greedy_randomized_sol*(n) $\leftarrow$ **greedy_construct**($\alpha, \hat{S}^n, cur\_sol(n)$)
21:     **end for**
22:     **return** *greedy_randomized_sol*

Output: greedy randomized constructed solution; $\forall$ nodes $n \in \lambda$

The pseudo-code of the greedy randomized construction phase is shown in Algorithm 2. The inputs to Algorithm 2 are: RCL threshold (α; discussed in more detail below), set of nodes whose frame packing will be perturbed (λ), and the solution that needs to be perturbed (*cur_sol*). The Algorithm 2 begins by assigning the current solution (*cur_sol*) to the greedy randomized solution (*greedy_randomized_sol*) in step **16**. For each node n whose current frame packing needs to be perturbed, the algorithm selects a random number (ρ) of signals ($\hat{S}^n$) in that node and tries to greedily construct a new solution using the *greedy_construct*() function, as shown in steps **17–21**. In step **20**, the newly generated solution for the node n is updated in the *greedy_randomized_sol*(n) function. At the end, the final solution for all nodes is returned to Algorithm 1 in step **22**.

The function *greedy_construct*() in steps **1–15** takes the RCL threshold (α), set of signals whose frame packing will be changed ($\hat{S}^n$), and the current frame packing (*partial_sol*) as the inputs and tries to assign the signal to a new message. The addition of signals to new messages happens in a greedy manner, which tries to minimize the objective function value. Until the set $\hat{S}^n$ is empty, in every iteration a signal is randomly chosen from ($\hat{S}^n$) (step **3**), and a list of feasible frames (Ω) to which the signal (s) can be packed into is generated (step **4**). For a frame ID (*fid*) to be feasible, it must satisfy all three necessary conditions mentioned in (1)–(3). If there exist no feasible frame IDs for the signal (*s*), its frame packing configuration is left unchanged. Otherwise, the individual cost of adding that signal to each frame is computed using the *cost*() in step **6**. In step **7**, the minimum cost (C_{min}) and maximum cost (C_{max}) are computed, which are used in generating the restricted candidate list (*RCL*) in step **8**. The *RCL* consists of the feasible frame IDs whose associated cost of adding the signal is within the interval [C_{min}, $C_{min} + \alpha*(C_{max} - C_{min})$]. This is the greedy aspect of the algorithm. The quality of RCL depends on the RCL threshold (α) (where $0 \leq \alpha \leq 1$), which controls the amount of randomness and greediness in the algorithm. For instance, when $\alpha = 0$, the algorithm exhibits a pure greedy behavior and when $\alpha = 1$, the algorithm exhibits a purely random behavior. In step **9**, a random frame ID (*chosen_fid*) is selected from the *RCL*, and the signal *s* is assigned to that frame ID in step **10**. Furthermore, after an attempt to change the frame packing for signal *s*, it is removed from $\hat{S}^n$ in step **12**. The function terminates in step **14** when all the signals in $\hat{S}^n$ are explored and returns the perturbed solution (*partial_sol*) as the output.

5.1.2 Local Search

This is the second phase of the GRASP metaheuristic, invoked in step **6** of Algorithm 1. It iteratively explores the defined neighborhood around the greedy randomized constructed solution to look for a local optimum. This is accomplished using *destroy and repair* mechanisms that try to randomly remove a part of the solution and reconstruct it. In this study, we define *neighborhood* as the set of solutions that are generated by randomly changing the frame packing of β number of signals. The parameter β is known as the destroy-repair threshold and it controls the amount of destroy and repair operations in each iteration of the local search phase. Algorithm 3 illustrates the pseudo-code for the local search.

Algorithm 3: local_search (β, N, *gr_sol*)

Inputs: Destroy-repair threshold (β), set of nodes (N), and greedy randomized constructed solution (*gr_sol*)

1:    **Initialize**: *interm_sol, new_sol* ← *gr_sol*
2:    **for** each *ls_iteration* **until** *max_ls_iterations* **do**
3:    | η = **random_selection**(N, 1)
4:    | $\hat{S}_{ls}^{\eta}$ = **random_selection**(S^η, β)
5:    | *new_sol*(η) = **greedy_construct** (α, $\hat{S}_{ls}^{\eta}$, *interm_sol*(η))
6:    | **if** Φ(*new_sol*) < Φ(*interm_sol*) **then**
7:    | | *interm_sol* ← *new_sol*
8:    | **end if**
9:    **end for**
10:   **return** *interm_sol*

Output: Local optimum with in the defined neighborhood- if there exists one; Otherwise, the same solution as greedy_randomized_construction( ).

The inputs to the local search in Algorithm 3 are: destroy-repair threshold (β), set of nodes (N), and greedy randomized constructed solution (gr_sol). The local search begins by initializing the intermediate solution ($interm\_sol$) and new solution (new_sol) with the greedy randomized constructed solution (in step **1**). The destroy mechanism randomly selects a node (η) and changes the frame packing for β random signals ($\hat{S}_{ls}^{\eta}$) belonging to a random node (η) in steps **3, 4**. A new solution (new_sol) is reconstructed using the $greedy\_construct()$ function in step **5** (repair mechanism), and it is accepted if the new solution (new_sol) resulted in a smaller objective function value compared to the prior solution ($interm\_solution$) (steps **6–8**). At the end of $max\_ls\_iterations$, the algorithm returns the final local search solution ($interm\_sol$) in step **10**.

5.1.3   Design Time Scheduling

In this subsection, we present the jitter-aware design time scheduling heuristic that is invoked in step **20** of Algorithm 1. The heuristic takes the frame packing solution of the system as input and generates a design-time schedule. The design time schedule consists of: *(i)* message-to-slot allocation, where slot ID and cycle numbers are assigned to messages, and *(ii)* node-to-slot allocation, where their source nodes are assigned slot IDs. In this work, we design this schedule with the goal to allocate the messages as early as possible to minimize the response time of messages. In addition, we also try to minimize the number of allocated static segment slots for effective bandwidth utilization while ensuring no deadline constraints are violated. Moreover, we take advantage of cycle multiplexing in FlexRay 3.0.1, where multiple nodes can be assigned slots with the same slot ID in different communication cycles. This helps to maximize the static segment utilization while using only a minimal number of slots [17]. We add jitter awareness to the scheduling framework by considering the previously computed mean jitter of the message (μ_j^n). Additionally, we introduce a control parameter called coefficient of jitter resilience (σ) that dictates the resiliency of the design time schedule to jitter. The parameter σ is a non-negative real number that dictates how resilient the schedule is for jitter. For instance, when $\sigma = 0$, it reflects a special case called zero-jitter (ZJ) scheduling. However, in real-time systems, ZJ scheduling is not encouraged as it has no resilience to jitter. On the other hand, having a higher value for σ results in longer response times and leads to potentially missing message deadlines. Hence, it is crucial to choose an appropriate value of σ that provides sufficient jitter resilience while not resulting in longer response times and missed deadlines. In this work, we empirically set the value of σ as 0.8. In addition, we consider the concept of message repetition and slot ID utilization in a FlexRay system. For any time-triggered message in FlexRay, message repetition (rm_j^n) is defined as the ratio of message period to the cycle time of the FlexRay, and can be computed using (8).

$$rm_j^n = \frac{p_j^n}{C_{fx}} \tag{8}$$

The message repetition is an integer value as the FlexRay cycle time is chosen to be the greatest common divisor of all the message periods in the system. Moreover, any time-triggered message that is assigned a particular slot ID will end up using $\left(1/rm_j^n\right)$ of the available slots within that slot ID.

The pseudo-code of our proposed jitter-aware design time schedule considering the above-mentioned metrics is shown in Algorithm 4. The heuristic begins by taking the set of time-triggered messages in the system (M) and FlexRay parameters as the inputs and initializes all slot utilizations to zero. In addition, we define a slot cycle list (*SCL*) to keep track of the list of available cycles in a particular slot ID, and each element in it is initialized with a list [0, 1, . . . ,63] as $C_{fx} = 64$ (Sect. 4.4). After the initialization in step **1**, all the time-triggered messages (M) in the system are sorted in increasing order of message periods (step **2**). For each time-triggered message (m_j^n) in the system, we begin the search for slot ID and cycle number allocation from the computed *slot* and *cyc* in steps **4, 5**. The calculations for the initial slot ID and cycle number are based on the message parameters: arrival time (a_j^n), mean jitter value (μ_j^n), and the design parameters: coefficient of jitter resilience (σ), static segment slot duration (t_{ds}), and cycle time (t_{dc}). The computed *slot* and *cyc* are subjected to checks for three constraints in steps **7–9**: *(i)* arrival time constraint (*constraint 1*) – checks if the current slot (*slot, cyc*) begins after the arrival time plus the effective jitter ($\sigma * \mu_j^n$) of the message; *(ii)* allocation constraint (*constraint 2*) – checks if the current slot is not allocated to any other message; and *(iii)* utilization constraint (*constraint 3*) – checks if the slot ID (*slot*) utilization is below 100% after adding the current message. If all these constraints are satisfied (step **10**) and the finish time of the (*slot, cyc*) exceeds the message deadline (step **11**), the algorithm terminates with no feasible solution for the given input message set (M). Otherwise, the feasibility of allocating the current *slot* and *cyc* to the message is checked using *sc_allocation()*. In step **14**, the function returns a binary variable indicating feasibility (*feasible*) and a list of cycles (*cyc_list*) that can be allocated to the message. If the current *slot* and *cyc* are feasible, they are allocated as slot ID and base cycle, respectively, for the current message. Additionally, other cycles in the *cyc_list* are allocated to the message, and the ownership of the allocated slot ID and cycles are assigned to the message and its source node (steps **15–17**). The *SCL* for the allocated slot ID (*slot*) is updated by removing the allocated cycles (*cyc_list*) in step **18**, and the search for allocation of slot ID and cycle number for the next message is initiated. If the computed slot ID (*slot*) and cycle number (*cyc*) fail to meet any of the three constraints mentioned in steps **7–9**, the slot ID and (if needed) the cycle number is incremented accordingly (steps **23, 24**). The algorithm terminates successfully when all the messages in the system are allocated slot and cycle numbers.

Algorithm 4: design_time_schedule (*solution*)

Inputs: Set of all time-triggered messages in the system (*M*), FlexRay parameters (N_{ss}, C_{fx}, B_{slot}, t_{ds}, t_{dc}), and coefficient of jitter resilience (σ)

1:    **Initialize**: all slot utilizations $\leftarrow$ 0; $SCL = [SC_1,..,SC_{62}]$; $SC_x = [0,..,63]$

2:    **Sort** *M* in the increasing order of message periods

3:    **for** each message m_j^n in *M* **do**

4:        $cyc = \lfloor (a_j^n + \sigma * \mu_j^n)/t_{dc} \rfloor$

5:        $slot = \lceil (a_j^n + (\sigma * \mu_j^n) - (cyc * t_{dc}))/t_{dc} \rceil + 1$

6:        **while** m_j^n is **not** allocated **do**

7:           $constraint1 = (\textbf{start}(slot, cyc) \geq a_j^n + \sigma * \mu_j^n)$

8:           $constraint2 = ((slot, cyc)$ is **not** allocated to any message$)$

9:           $constraint3 = (\textbf{slot\_util}(slot) + 1/rm_j^n \leq 1)$

10:        **if** $constraint1, constraint2, constraint3$ are all **True then**

11:           **if start**$(slot, cyc) + t_{ds} > d_j^n$ **then**

12:              **exit**(*"No feasible solution"*)

13:           **else**

14:              $feasible, cyc\_list = \textbf{sc\_allocation}(m_j^n, slot, cyc, SC_{slot})$

15:              **if** *feasible* **then**

16:                 $slot_{m_j^n} \leftarrow slot; cyc_{m_j^n} \leftarrow cyc; cyc\_list_{m_j^n} \leftarrow cyc\_list$

17:                 **Assign** ownership$(slot_{m_j^n}, cyc\_list_{m_j^n}) \rightarrow \textbf{src}(m_j^n)$

18:                 **Remove** elements in cyc_list from SC_{slot}

19:                 m_j^n allocated $\leftarrow$ **True**; **break**( )

20:              **end if**

21:           **end if**

22:        **end if**

23:        $slot \mathrel{+}= 1;$

24:        **if** $slot > N_{ss}$ **then** $slot = 1; cyc \mathrel{+}= 1$

25:       **end while**

26: **end for**

Output: Message-to-slot assignment ($slot_{m_j^n}$, $cyc_{m_j^n}$) for each time-triggered message m_j^n, and slot ownership of each node *n*.

Algorithm 5 shows the pseudo-code for *sc_allocation*() function, which checks for the feasibility of allocating the slot ID and base cycle to the current message. The inputs to Algorithm 5 are: current message (m_j^n), slot ID (*slot*), base cycle (*cyc*), and *SCL* corresponding to the slot ID (SC_{slot}). The function begins by initializing a feasibility flag (*feasible*) to zero and cycle list (*cyc_list*) with an empty list and then computes the minimum number of instances (*num_instances*) of the message in C_{fx} cycles (in step **1**). From steps **2–14**, the function tries to find a feasible cycle number for each instance of the message. The search begins by initializing the feasible cycle exists flag (*fc_exists*) to zero and computing the first cycle ($k = 0$) under consideration (step **3**). In steps **4–11**, the function tries to find a cycle before the message deadline (i.e., rm_j^n- 1 cycles) by checking three different conditions (steps **5–7**): *(i)* allocation condition – checks if the cycle number in the current slot ID is unallocated; *(ii)* arrival time condition – checks if the slot begins after the message arrival time (a_j^n) and effective jitter $(\sigma * \mu_j^n)$ and; *(iii)* deadline constraint – checks if the finish time of the slot is before the deadline of the $(k+1)^{th}$ instance. If all the three conditions are satisfied, the cycle number is added to the *cyc_list*, *fc_exists* is changed to 1 (in steps **8–10**), and the search for the next instance begins. When all the message instances are allocated a feasible cycle, the function returns *feasible* as 1, and the list of allocated cycles (*cyc_list*). Otherwise, the algorithm determines that the current slot ID and cycle number are infeasible for allocating to the current message.

Algorithm 5: sc_allocation (m_j^n, *slot*, *cyc*, SC_{slot})

Inputs: current message ($\boldsymbol{m_j^n}$), slot ID (*slot*), base cycle (*cyc*), and SCL corresponding to slot ID (SC_{slot})

1:   **Initialize:** *feasible* = 0; *cyc_list* = [ ]; k = 0; *num_instances* = $\left\lfloor \frac{C_{fx}}{\lfloor rm_j^n \rfloor} \right\rfloor$;

2:   **while** k < *num_instances* **do**

3:      | *fc_exists* = 0; *test_cyc* = *cyc* + k* rm_j^n

4:      | **for** i from 0 to $(rm_j^n$- 1) **do**

5:      | *condition1* = ((*test_cyc+i*) **in** SC_{slot})

6:      | *condition2* = $(a_j^n + \sigma * \mu_j^n) \leq$ **start**(*slot*, *test_cyc* + i)

7:      | *condition3* = **start**(*slot*, *test_cyc* + i) + $t_{ds} \leq (k+1)*d_j^n$

8:      | **if** *conditions 1, 2, 3,* and *4* are all **True then**

9:      | | **Append** *cyc_list* ← (*test_cyc* + i); *fc_exists* = 1; **break( )**

10:    | **end if**

11:   | **end for**

12:   | **if** *fc_exists* == 0 **then break( )**

13:   | k += 1

14: **end while**

15: *feasible* = 1 **if length**(*cyc_list*) == *num_instances*; **else** *feasible* = 0

16: **return** *feasible*, *cyc_list*

Output: feasibility flag (*feasible*) and list of communication cycles allocated to message ($\boldsymbol{m_j^n}$) for the given *slot* and *cyc*.

5.1.4 Acceptance and Cooling Functions

The acceptance function is used to probabilistically accept the solution generated in every iteration (step **9** in Algorithm 1). The probability of acceptance of a new solution is computed using (9). The term ΔE is the difference between the objective function value of the current solution (*cur_sol*) and the previous solution (*prev_sol*), as shown in (10). This is analogous to the energy difference between the new and previous states in SA. Lastly, the cooling function shown in (11) defines the controlled cooling of the system, where C_r is the cooling rate.

$$P_{acceptance} = e^{-\left(\frac{\Delta E}{T}\right)} \tag{9}$$

$$E = \Phi\,(cur\_sol) - \Phi\,(prev\_sol) \tag{10}$$

$$Temperature\,(T) = C_r * T \tag{11}$$

5.2 Runtime Multi-level Feedback Queue

The schedule generated by the design time scheduler will only guarantee latencies for time-triggered messages when the runtime jitter experienced by the messages does not exceed their effective jitter ($\sigma * \mu_j^n$) value. However, at runtime, various internal and external disturbances may interfere with the normal operation of the FlexRay bus and might result in additional, larger jitter. Thus, it is important to handle a multitude of jitter values during runtime to minimize the impact of random jitter. In this work, we focus on handling jitter at runtime using a runtime scheduler that re-schedules jitter-affected time-triggered messages using the design-time generated schedule and the output of the Multi-Level Feedback Queue (MLFQ; discussed next) as the inputs. Moreover, in this work, we allow the transmission of high-priority event-triggered messages within the static segment of FlexRay. Consider an example scenario where a high-priority event-triggered message arrives just after the beginning of the dynamic segment, and there is a low-priority event-triggered message that is already being transmitted and ends up taking the entire duration of the dynamic segment due to its large message size. In this scenario, the high-priority event-triggered message has to wait until the beginning of the dynamic segment in the next communication cycle to start transmission if there are no other higher-priority messages. This could result in a missed deadline. Hence, we facilitate the transmission of high-priority event-triggered messages in the static segment of FlexRay by treating them similar to jitter-affected time-triggered messages within the MLFQ but with a priority lower than time-triggered messages during the runtime scheduling. This facilitates the easy rescheduling of high-priority event-triggered messages in the static segment of the FlexRay.

The MLFQ typically consists of two or more queues that have different priorities and are capable of exchanging messages between different levels of the queues using feedback connections (as shown earlier in Fig. 6). The number of queues in an MLFQ defines the number of levels, and each level queue can have a different prioritization scheme and scheduling policy. Moreover, the MLFQ attempts to resolve the issues associated with the traditional scheduling schemes (such as first come first serve (FCFS) and shortest job first (SJF)) by minimizing inefficient turnaround times for the messages and preventing message starvation.

In this work, we considered an MLFQ consisting of three level queues (as shown in Fig. 6), with queue 1 (Q1) having the highest priority, followed by queue 2 (Q2) and queue 3 (Q3), with lower priorities. In addition to prioritization between different level queues, we set priorities between different types of messages and within the messages of the same type. In this work, we prioritize time-triggered messages over event-triggered messages. Moreover, within the time-triggered messages, we compute static priorities using a Rate Monotonic (RM) policy to prioritize messages with a high frequency of occurrence. In case of a tie, priorities are resolved using a First Come First Serve (FCFS) strategy. In this work, we assume that the event-triggered messages inherit the priority of their generating node. In cases of multiple event-triggered messages from the same node, an Earliest Deadline First (EDF) scheme is employed to prioritize messages. These static priorities of the messages are used to reorder the messages in the queues and promote messages to upper-level queues. In addition, the MLFQ takes input from two separate buffers that are used to handle jitter-affected time-triggered messages and high-priority event-triggered messages.

The operation flow of the MLFQ is depicted in the flowchart in Fig. 8. It begins by checking the time-triggered (TT) message buffer for jitter-affected messages. If a TT message is available, the *load TT message* function is executed. The *load TT message* checks for a vacancy in the queues in the order Q1, Q2, and Q3 and stores the TT message in the first available queue. If the TT message buffer is empty and an event-triggered (ET) message is available in the ET message buffer, the *load ET*

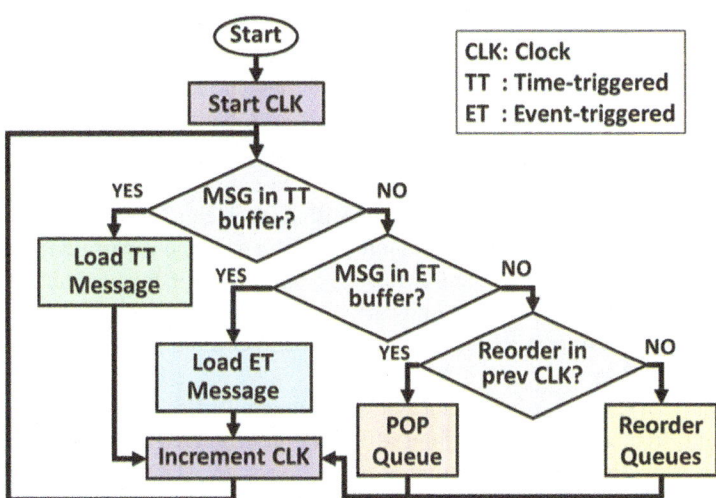

Fig. 8 The operation flow of the runtime MLFQ

Fig. 9 Packing of jitter-affected messages at runtime

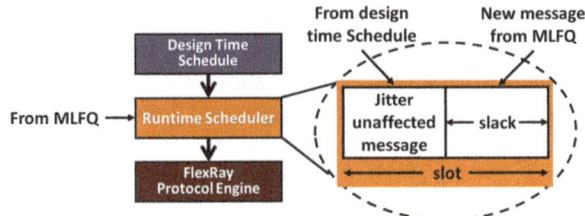

message function is executed. The *load ET message* checks for a vacancy in the queues in the order Q2, Q3, and Q1 and stores the ET message in the first available queue. In either of the cases, when all three queues are full, the message is stored in the corresponding buffer, and the same function is executed in the next clock cycle. Whenever there are no messages available in both the buffers and the reorder queue function is not executed in the preceding clock cycle, messages in the queues are reordered. The reordering happens in the order of their priorities by executing the *reorder queues* function. Otherwise, the queues are checked in the order Q1, Q2, and Q3 by running the *POP queue* function based on the conditions discussed in the following subsection.

5.3 Runtime Scheduler

We introduce a runtime scheduler that takes the output of the MLFQ and the design-time generated schedule as inputs. The runtime scheduler computes the available slack using the design-time generated schedule and stores this information, which is used at runtime to pack jitter-affected messages in the FlexRay static segment slots opportunistically. If there is a jitter-affected message in the MLFQ, the runtime scheduler checks the ownership of the next incoming slot. If the incoming slot is owned by the source node of the jitter-affected message in the MLFQ, the runtime scheduler checks for the available slack in the incoming slot. If there is non-zero slack in the incoming slot, the jitter-affected message is collocated with the jitter-unaffected message, as shown in Fig. 9. The entire jitter-affected message is rescheduled in the incoming slot if there is sufficient slack to accommodate the full jitter-affected message. Otherwise, the jitter-affected message is partitioned into two parts. The size of the first part is equal to the available slack in the incoming slot, and the remaining is the size of the second part. The second part of the message remains in the queue and is transmitted in the next feasible incoming slot by bumping up its priority.

Similarly, whenever a high-priority event-triggered message is available in the MLFQ, it is treated similar to the jitter-affected time-triggered message, except with a lower priority than the regular time-triggered message. The same steps discussed above are used to schedule the ET message at runtime.

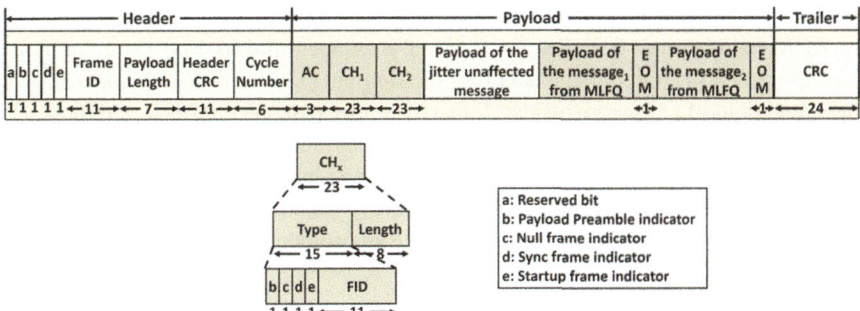

Fig. 10 The updated frame format of the FlexRay frame using the proposed segmentation and addressing scheme. The parts of the frame highlighted in a darker shade represent our modifications

This method of scheduling messages at runtime can result in packing two different message data into the payload segment of one FlexRay frame, which, unfortunately, leads to two major challenges. Firstly, there is a need for a mechanism at the receiver node to decode the payload segment correctly and distinguish between the two messages. Secondly, the implicit addressing scheme of FlexRay is lost because of combining two different messages, and the receiving nodes will not be able to identify to which specific node the message is meant.

To overcome the above-mentioned challenges, we propose a custom segmentation and addressing scheme to distinguish multiple messages that are packed in the same frame. In this scheme, we introduce one additional segment in the payload segment of the FlexRay frame that is common to multiple jitter-affected messages packed in that frame. We also introduce two more segments for each jitter-affected message packed into the frame. An illustration of our proposed segmentation and addressing scheme with two jitter-affected messages being collocated with a jitter-unaffected message is shown in Fig. 10. The first common segment is called append counter (AC), which is also the first segment in the payload. AC is a 3-bit field indicating the number of different jitter-affected messages that are packed in the current frame. In this work, we support partial message transmission to fully utilize the available bandwidth in allocated static segment slots. The second segment is called the custom header (CH), which is private for each jitter-affected message in the FlexRay frame. Every CH segment further consists of a *type* field and a *length* field. The *type* is a 15-bit field that specifies different message types (defined in [5]). The *type* field consists of one bit each for the payload preamble indicator, null frame indicator, sync frame indicator, and startup frame indicator, and an 11-bit frame ID (FID) field for specifying the FID of the jitter-affected message. The *length* field is 8-bit long and specifies the data length of the jitter-affected message in bytes. The *length* field in the custom header and the payload length field in the frame header are used to find the start bit of the jitter-affected message in the payload segment. The third segment we introduced in the payload is called the end of message (EOM), which is a 1-bit segment that is private to each jitter-affected message. The EOM field is 1 when the entire message is transmitted; otherwise, the EOM field is set to 0, indicating a partial message transmission. The remaining message data transmitted in the next feasible slot will have the remaining data size in the *length* field of its custom header. If there is more than one jitter-affected message packed in the FlexRay frame, the headers of all the messages are at the beginning of the payload segment. This gives the receiver node information about all the jitter-affected messages that are packed in the FlexRay frame. Moreover, it is important to note that the regular operation of the FlexRay protocol is not altered in any way by implementing these changes.

6   Experiments

6.1   *Experimental Setup*

To evaluate the effectiveness of the proposed JAMS-SG framework, we first compare it against two variants of the same framework: JAMS-SA and JAMS-ACC-SG. The first variant JAMS-SA uses a simulated annealing (SA) approach with no GRASP-based local search. The motivation for implementing JAMS-SA is to study the importance of local search. In JAMS-SA, the solution is subjected to random perturbations, and a new solution is created every iteration with the randomly chosen signals having an equal likelihood of grouping or splitting. The second variant JAMS-ACC-SG (accelerated SA + GRASP), behaves similar to JAMS-SA in the beginning, but then it switches to a JAMS-SG behavior (i.e., including GRASP-based local search) when the temperature of the system is sufficiently low. Based on empirical analysis, in this work, we set the threshold temperature to be 30% of the initial temperature. The motivation for an accelerated version is to save the computation time spent looking for the local optimum initially and only perform the local search after a reasonable solution is achieved. The performance of these three techniques (JAMS-SA, JAMS-SG, and JAMS-ACC-SG) is discussed in detail in Sect. 6.2. In addition, we perform a series of experiments with different weight values to determine the optimal weight parameters for the best variant of our framework.

Subsequently, we compare the best variant of our framework with various prior works: Optimal Message Scheduling with Jitter minimization (OMSC-JM [26]), Optimal Message Scheduling with FID minimization (OMSC-FM [26]), Policy-based Message Scheduling (PMSC [23]), and JAMS-greedy [27]. OMSC-JM [26] and OMSC-FM [26] use an ILP-based frame packing technique from [7] and change the message repetition to minimize the effect of jitter. OMSC-JM tries to minimize the effect of jitter by allocating more slots and performing more frequent message transmissions, while OMSC-FM aims to minimize the number of allocated slots. PMSC [23] uses a priority-based preemptive runtime scheduler that uses the message arrival times and priorities to schedule messages using heuristics.

JAMS-greedy [27] uses a greedy frame packing approach to generate the set of messages and uses a heuristic-based scheduler to synthesize design time schedules. In addition, JAMS-greedy also supports a runtime scheduler to reschedule jitter-affected messages similar to JAMS-SG. However, JAMS-greedy lacks the ability to send multiple jitter-affected messages in one FlexRay frame and also does not support partial message transmission. We also implemented a genetic algorithm (GA) based frame packing approach for FlexRay-based systems using the frame packing technique proposed for CAN-FD in [15]. We further adapted the scheduling policy proposed in [27] and combined it with [15] (hence the name JAMS-GA) to compare it with our framework. All experiments conducted with these prior works are discussed in detail in the following subsections.

To evaluate our proposed framework with its variants and against prior work, we derived a set of test cases using automotive network data extracted from a real-world 2016 Chevrolet Camaro vehicle that we have access to. In this study, for all our experiments, we considered a FlexRay 3.0.1 based system with the following network parameters: cycle duration of 5 ms (t_{dc}) with 62 static segment slots (N_{ss}), with a slot size of 42 bytes (B_{slot}) and 64 communication cycles (C_{fx}). Moreover, each experiment was run for 1000 iterations with an initial temperature $= 10,000$ and the cooling rate (C_r) set to 0.993. We chose the RCL threshold parameter (α) of GRASP 0.4, which resulted in a relatively near greedy solution in the presence of a relatively large variance. The destroy-repair parameter (β) is set to 2, which helped avoid exploring a larger neighborhood around the greedy randomized constructed solution. We randomly sampled jitter values as a function of the message period to modify the arrival times of randomly selected messages originating from a set of randomly selected jitter-affected nodes. Moreover, as discussed earlier, we chose the coefficient of jitter resilience (σ) $= 0.8$. To account for the overhead of MLFQ operations, we model additional message latency as a function of the static priority of the message (derived using the RM scheme), message data size, and the queue it is in. All the simulations are run on an Intel Core i7 3.6GHz server with 16 GB RAM.

6.2 Comparison of JAMS-SG Variants

In this subsection, we compare the proposed JAMS-SG framework with the two other variants, JAMS-SA and JAMS-ACC-SG. A series of experiments were conducted by changing the weight in the objective function (in (5)). The results were analyzed under four different scenarios: *(i)* zero, *(ii)* low, *(iii)* medium, and *(iv)* high jitter. Under zero jitter, none of the messages in the system are affected by jitter. Hence, their arrival times remain unchanged. Under the next three different jitter scenarios, the arrival times of randomly selected time-triggered messages originating from a randomly selected set of jitter-affected nodes are modified as a function of the message period. In low, medium, and high jitter scenarios, the randomly chosen messages are subjected to jitter values equal to $p_j^n/8$, $p_j^n/5$ and $p_j^n/4$, respectively (where p_j^n represents the period of message j belonging to node n). We considered a real-world automotive case study (discussed in Sect. 6.1) consisting of 19 ECUs and 248 signals to evaluate all the variants.

Figures 11a–d show the average response time of all the messages in the system for the three variants with different objective function weights under zero, low, medium, and high jitter conditions, respectively. The error bars on top of each bar represent the minimum and maximum of the average response time, and the number on top of the bar represents the number of missed deadlines. From Fig. 11a–d, across all weight values and jitter scenarios, it is evident that JAMS-SG has superior performance in response time compared to JAMS-SA and JAMS-ACC-SG in most of the cases. Most importantly, JAMS-SG never misses any deadline for any weight value and jitter scenario. This is because JAMS-SG is able to find a better jitter-aware frame-packing solution from the beginning due to its more effective GRASP-based optimization. This early exploration using GRASP helps achieve a solution that efficiently balances between minimizing the number of FlexRay messages and maximizing the laxity of the messages. Due to the lack of local search mechanisms, JAMS-SA fails to find a comparable solution. JAMS-ACC-SG suffers similarly to JAMS-SA until the local search process is initiated. But, when the local search process begins, the system temperature is already low. This forces the system to only accept the better solutions, as the acceptance probability function outputs a smaller probability value in case of a relatively bad solution. This often results in getting stuck at a local minimum, leading to subpar results. From Fig. 11, it can also be observed that as the weight value increases, the number of missed deadlines decreases across the frameworks and under different jitter scenarios. This is because of the increasing emphasis on minimizing the number of FlexRay frames in all three frameworks, resulting in fewer frames to be scheduled, which simplifies the problem. Moreover, it can be observed from Fig. 11 that choosing a very high or a very low weight value makes the system heavily biased towards optimizing the number of FlexRay frames or the laxity. To avoid this bias, we select an intermediate weight value of 2. Henceforth, all the other comparisons are made against JAMS-SG (the best variant in our analysis) with weight $(w) = 2$.

6.3   Response Time Analysis

In this subsection, we present a response time analysis by comparing JAMS-SG with message scheduling frameworks from prior works. We consider the same vehicle test case as used in the previous subsection. To induce jitter in simulations, we randomly select messages from the randomly chosen nodes, and their arrival times are delayed. Moreover, we consider an equal probability of being selected to be impacted by jitter for all nodes and the messages.

Figures 12a–c show the average response time of the messages under low, medium, and high jitter scenarios (using the configurations discussed in the previous

Fig. 11 Average response
time of all signals for
different objective function
weights (with the number of
missed deadlines shown on
the top of each bar) under (**a**)
zero, (**b**) low, (**c**) medium,
and (**d**) high jitter conditions;
for JAMS-SA, JAMS-SG, and
JAMS-ACC-SG

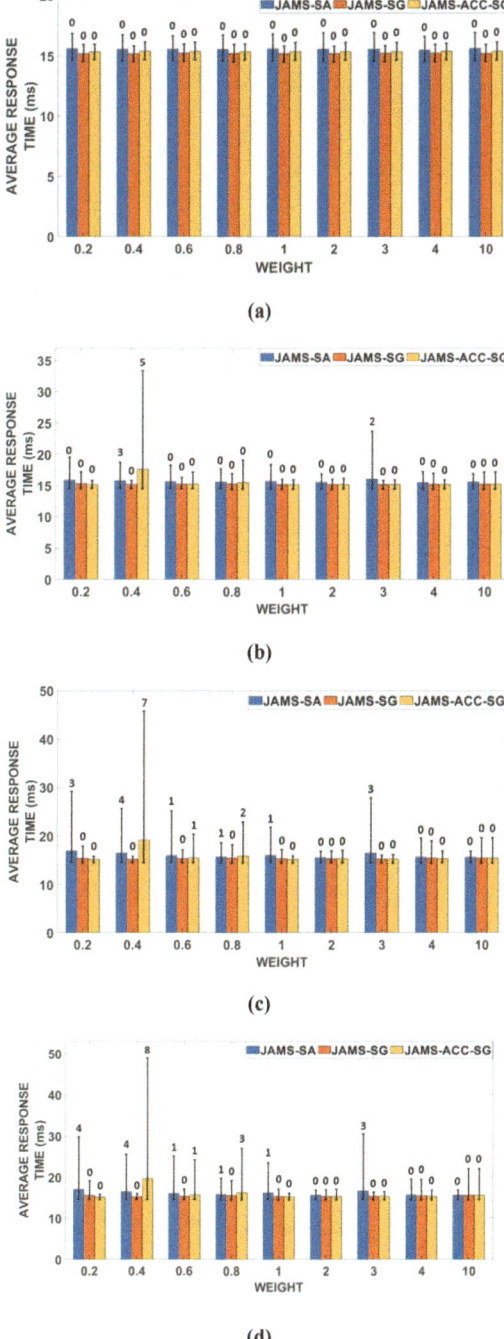

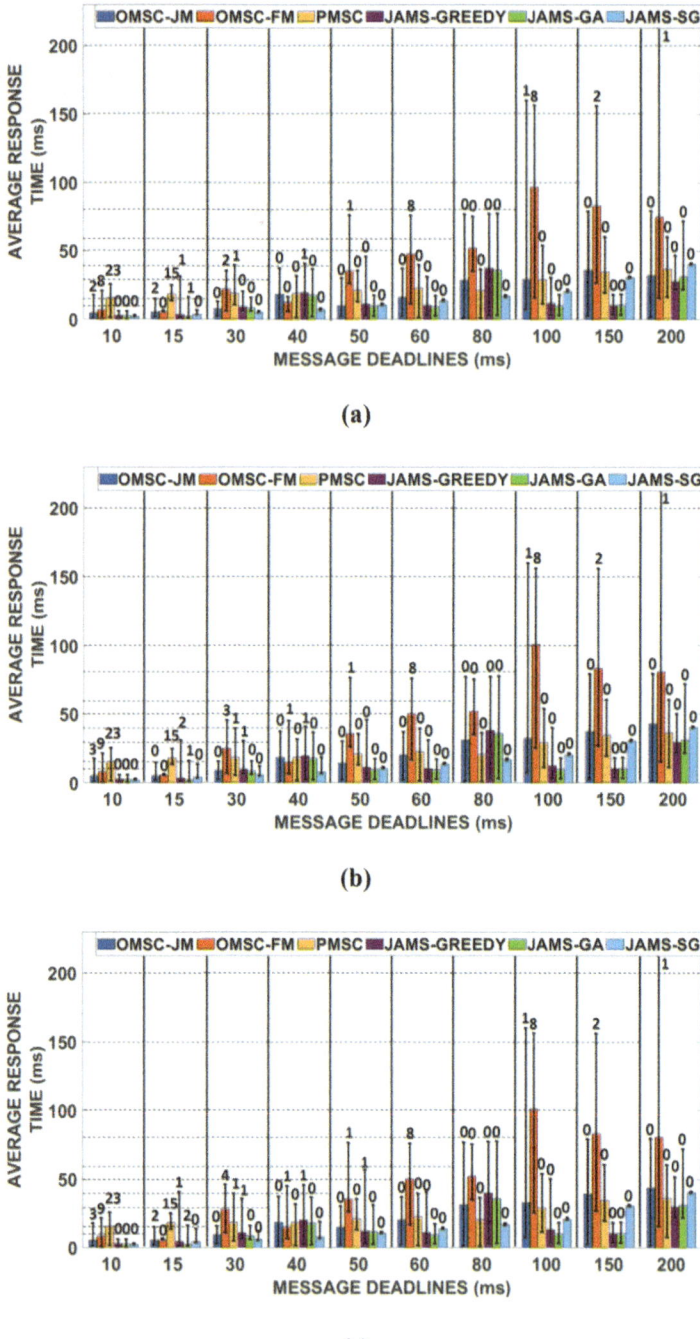

Fig. 12 Message deadlines vs. average response time (with the number of missed deadlines shown on the top of each bar) under (**a**) low, (**b**) medium, and (**c**) high jitter conditions; for the comparison frameworks (OMSC-JM [26], OMSC-FM [26], PMSC [23], JAMS-GREEDY [27], JAMS-GA [15, 27]), and our proposed JAMS-SG framework

subsection). The error bar on each bar represents the minimum and maximum average response time of the messages achieved, and the number on top of each bar represents the number of missed deadlines. The response time results are clustered into groups based on the message deadlines (on the x-axis), and the dashed horizontal line represents the deadlines. It can be seen that using OMSC-FM results in high response times under all jitter scenarios. This is because of the high emphasis on minimizing the number of static segment slots, which resulted in poor jitter resilience. On the other hand, OMSC-JM performs relatively better as it allocates extra slots for message transmission. However, it still has issues handling random jitter during runtime, especially for high-priority messages. In the PMSC technique, jitter has a strong impact on the high-priority messages because of the frame packing approach used in it. PMSC aims to use the entire static segment slot by packing the signals that are larger than the slot size and uses EDF-based preemption at the beginning of each slot. In a scenario where a high-priority message arrival gets delayed due to jitter, the node has to wait for the next transmitting slot to preempt existing transmissions of low-priority messages when using PMSC. This additional delay due to jitter and scheduling constraints can result in missed deadlines. Additionally, JAMS-greedy and JAMS-GA result in suboptimal frame packing that focuses on minimizing the number of FlexRay frames. Moreover, JAMS-greedy and JAMS-GA are relatively jitter resilient compared to other prior works due to the runtime scheduler in these works. However, these frameworks start missing deadlines when there is high jitter. It is evident that under all three jitter scenarios, JAMS-SG outperforms all the other prior works with no deadline misses. JAMS-SG achieves this by finding a balanced solution that results in optimal frame packing and jitter resilience. Moreover, the support for partial message transmission helps JAMS-SG to meet the deadline constraints under different jitter scenarios.

6.4 Sensitivity Analysis

In this subsection, we analyze the impact of the jitter on a specific subset of messages and study the behavior of the system. The same test case considered in the previous subsection is used, and the results are compared with the prior works mentioned in the previous subsection.

Figure 13a–c illustrates the message deadline vs. average response time plots under low, medium, and high jitter, for the case where only the high-priority messages (messages with a deadline ≤40 ms) are subjected to jitter. The high-priority messages affected by jitter are randomly chosen from a randomly chosen set of nodes. It can be observed that the impact of jitter results in higher response times and deadline misses for the high-priority messages in most of the prior works. In particular, for OMSC-JM and OMSC-FM, some of the low-priority messages suffer from very long response times and deadline misses. However, JAMS-SG not only results in minimal response times for most cases but also in *no deadline misses*, which is a crucial requirement for time-critical automotive cyber-physical systems.

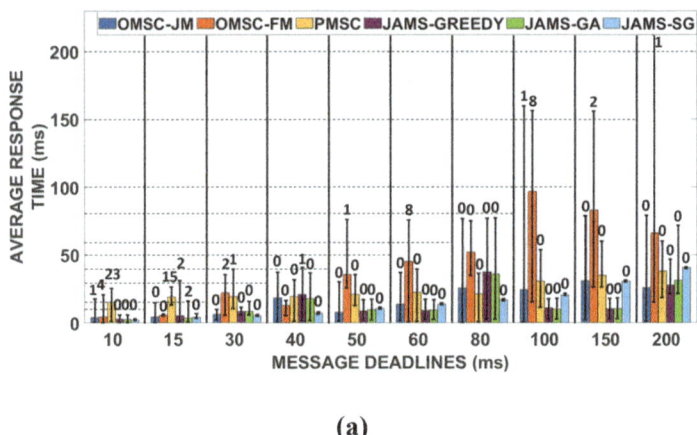

(a)

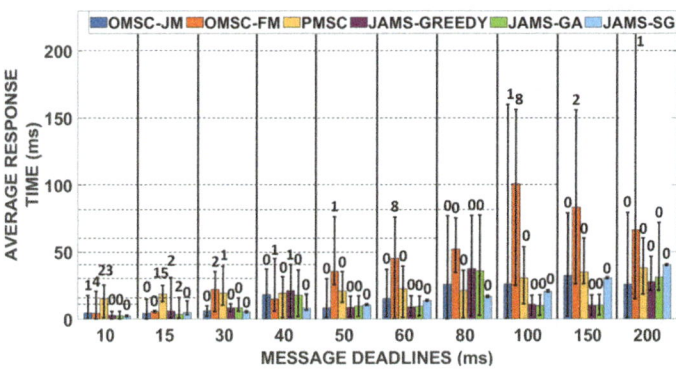

(b)

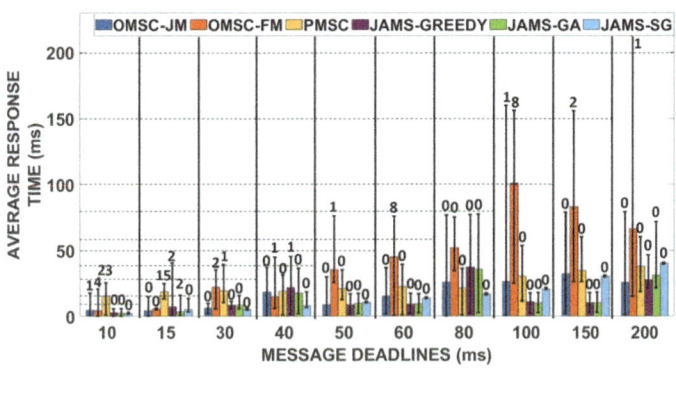

(c)

Fig. 13 Message deadlines vs. average response time with jitter affecting high-priority messages only; (with the number of missed deadlines on the top of each bar) under (**a**) low, (**b**) medium, and (**c**) high jitter conditions; for the comparison frameworks (OMSC-JM [26], OMSC-FM[26], PMSC [23], JAMS-GREEDY[27], JAMS-GA [15, 27]), and JAMS-SG

The message deadline vs. average response time plots for three jitter scenarios where only the low-priority messages (messages with deadline >40 ms) are subjected to jitter is shown in Fig. 14. It is clear that almost all of the frameworks except JAMS-SG fail to meet the deadline constraint for specific scenarios.

Thus, from Figs. 12, 13, and 14, it is evident that JAMS-SG can handle a wide variety of jitter patterns and can still meet the deadline constraints for all the messages in the system.

6.5   Scalability Analysis

To evaluate the scalability of JAMS-SG, we analyzed the performance of our JAMS-SG and comparison works under various system configurations using the test cases with varying combinations of the number of nodes and number of signals. The average response times of all the messages for the high jitter scenario for different system configurations is illustrated in Fig. 15. In the system configuration is represented as $\{p, q\}$ (on the x-axis), where p denotes the number of nodes and q is the number of signals.

The number on the top of each bar represents the number of signals that missed the deadline in that configuration. For larger test cases, some of the prior works (OMSC-JM, OMSC-FM) failed to result in a feasible solution within the 24-h time limit. It can be observed that even with increasing system size, JAMS-SG is able to meet all message deadlines for every configuration. On the other hand, the prior works suffer from multiple deadline misses due to the lack of jitter awareness. Among them, OMSC-FM seems to perform particularly poorly compared to all other works because of its heavy emphasis on minimizing the number of allocated slots, resulting in a minimal number of available slots but a lack of jitter resilience.

Lastly, Table 1 shows the time taken (at design time) for JAMS-SG and prior works to generate the best solution for the different system configurations. It is clear that JAMS-SG is able to achieve a superior solution (jitter resilient frame packing and schedule) with no deadline misses under 20 min, even for the largest test case configuration. Thus, our proposed JAMS-SG framework is highly scalable across various system complexities and jitter profiles. Moreover, unlike the frameworks proposed in prior works, JAMS-SG has no missed deadlines for all the test cases.

7   Conclusion

In this chapter, we presented a novel message scheduling framework called JAMS-SG that utilizes both design time and runtime scheduling to mitigate the effect of jitter in time-triggered automotive systems. At design time, our framework uses a hybrid heuristic (SA + GRASP) for generating jitter-aware frame packing and synthesizing a design time schedule. At runtime, our framework effectively handles

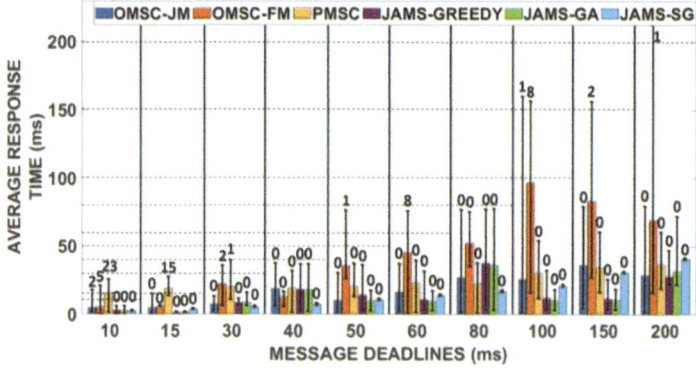

(a)

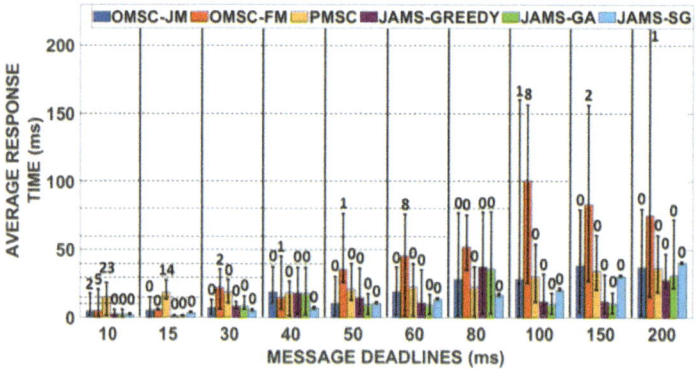

(b)

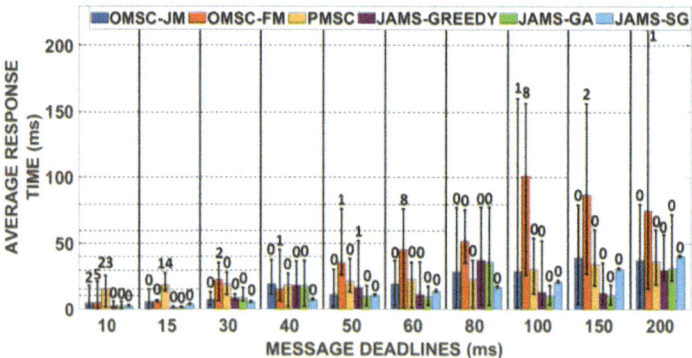

(c)

Fig. 14 Message deadlines vs. average response time with jitter affecting low-priority messages only; (with the number of missed deadlines on the top of each bar) under (**a**) low, (**b**) medium, and (**c**) high jitter conditions; for the comparison frameworks (OMSC-JM [26], OMSC-FM [26], PMSC [23], JAMS-GREEDY [27], JAMS-GA [15, 27]), and JAMS-SG

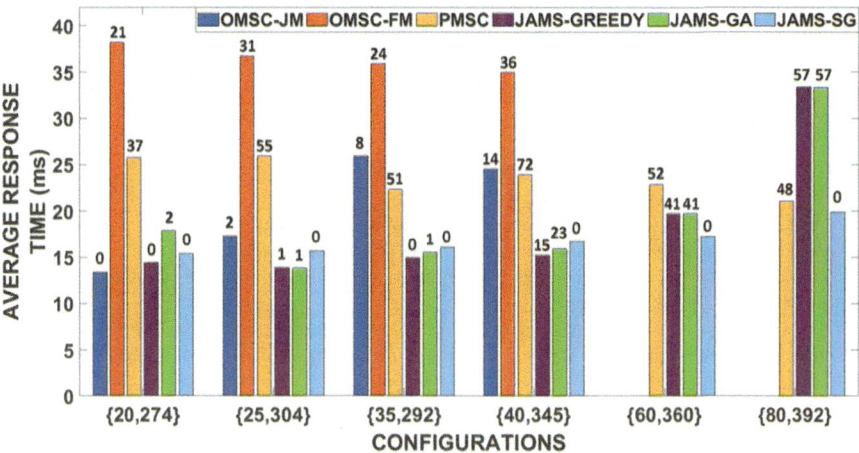

Fig. 15 Average response time for different system configurations (with the number of missed deadlines on the top of each bar) under high jitter; OMSC-JM [26], OMSC-FM [26], PMSC [23], JAMS-GREEDY [27], JAMS-GA [15, 27], and our JAMS-SG framework

Table 1 Time taken to generate the solution (in seconds) for different configurations: OMSC-JM [26], OMSC-FM [26], PMSC [23], JAMS-Greedy [27], JAMS-GA and our JAMS-SG framework

	Configurations					
	{20,274}	{25,304}	{35,292}	{40,345}	{60,360}	{80,392}
OMSC-JM	2700.54	2700.89	2700.94	2700.93	–	–
OMSC-FM	2700.49	2700.85	1116.09	2700.95	–	–
PMSC	0.72	1.89	1.947	2.58	1.31	1.65
JAMS-GREEDY	0.71	1.811	2.52	3.48	3.01	5.93
JAMS-GA	66.05	142.6	152.15	178.75	122.08	157.97
JAMS-SG	94.15	134.96	298.62	449.71	576.02	1149.80

both jitter-affected time-triggered and high-priority event-triggered messages using the proposed MLFQs and a runtime scheduler. We also devise a custom frame format to solve the addressing and segmentation challenges associated with packing multiple jitter-affected messages in the same frame. We compared our JAMS-SG framework with the best-known prior works in the area under varying jitter conditions. Our experimental analysis indicates that JAMS-SG is able to achieve significantly lower response times for most cases and, more importantly, no deadline misses. Moreover, our experiments also show that JAMS-SG is highly scalable and outperforms the best-known prior works for various system sizes and under different jitter scenarios. This makes our proposed JAMS-SG framework a promising approach to cope with random jitter in emerging automotive cyber-physical systems.

References

1. Davare, A., Zhu, Q., Di Natale, M., Pinello, C., Kanajan, S., Sangiovanni-Vincentelli, A.: Period optimization for hard real-time distributed automotive systems. In: Proceeding of the DAC (2007)
2. CAN Specifications version 2.0, Robert Bosch gmbh (1991)
3. SAE Automotive Engineering International (2016)
4. Davis, I.R., Burns, A., Bril, R.J., Lukkien, J.J.: Controller Area Network (CAN) schedulability analysis: Refuted, revisited and revised. In: Real-Time Systems (2007)
5. FlexRay. FlexRay Communications System Protocol Specification, ver.3.0.1 [Online]. Available: http://www.flexray.com
6. Saket, R., Navet, N.: Frame packing algorithms for automotive applications. In: JEC (2006)
7. Schmidt, K., Schmidt, E.G.: Message scheduling for the FlexRay protocol: The static segment. In: IEEE TVT (2009)
8. Kukkala, V., Tunnell, J., Pasricha, S., Bradley, T.: Advanced driver assistance systems: A path toward autonomous vehicles. IEEE Consum Electron 7(5) (2018)
9. Bello, L.L., Mariani, R., Mubeen, S., Saponara, S.: Recent advances and trends in on-board embedded and networked automotive systems. IEEE Trans. Industr Inform 15(2) (2019)
10. Audi A5 Driver assistance systems [Online]. Available: https://www.audi-mediacenter.com/en/the-new-audi-a5-and-audi-s5-coupe-6269/driver-assistance-systems-6281
11. Fleiss, M., Müller, T.M., Nilsson, M., Carlsson, J.: Volvo powertrain integration into complete vehicle. ATZ Worldw 118(3) (2016)
12. Kukkala, V.K., Bradley, T., Pasricha, S.: Priority-based multi-level monitoring of signal integrity in a distributed powertrain control system. In: Proceeding IFAC (2015)
13. Tanasa, B., Bordoloi, U.D., Eles, P., Peng, Z.: Reliability-aware frame packing for the static segment of FlexRay. In: Proceeding of the EMSOFT (2011)
14. Kang, M., Park, K., Jeong, M.K.: Frame packing for minimizing the bandwidth consumption of the FlexRay static segment. In: IEEE TIE (2013)
15. Ding, S., Huang, R., Kurachi, R., Zeng, G.: A genetic algorithm for minimizing bandwidth utilization by packing CAN-FD frame. In: Proceeding of the ICESS (2016)
16. Zeng, H., Zheng, W., Di Natale, M., Ghosal, A., Giusto, P., Sangiovanni-Vincentelli, A.: Scheduling the FlexRay bus using optimization techniques. In: Proceeding of the DAC (2009)
17. Lukasiewycz, M., Glaß, M., Teich, J., Milbredt, P., Flexray schedule optimization of the static segment. In: Proceeding CODES+ISSS (2009)
18. Zeng, H., Di Natale, M., Ghosal, A., Sangiovanni-Vincentelli, A.: Schedule optimization of time-triggered systems communicating over the FlexRay static segment. In: IEEE TII (2011)
19. Grenier, M., Havet, L., Navet, N.: Configuring the communication on FlexRay- the case of the static segment. In: Proceeding of the ERTS (2008)
20. Sun, Z., Li, H., Yao, M., Li, N.: Scheduling optimization techniques for FlexRay using constraint-programming. In: Proceeding of the CPSCom (2010)
21. Lukasiewycz, M., Schneider, R., Goswami, D., Chakraborty, S.: Modular scheduling of distributed heterogeneous time-triggered automotive systems. In: ASP-DAC (2012)
22. Goswami, D., Lukasiewycz, M., Schneider, R., Chakraborty, S.: Time-triggered implementations of mixed-criticality automotive software. In: Proceeding of the DATE (2012)
23. Mundhenk, P., Sagstetter, F., Steinhorst, S., Lukasiewycz, M., Chakraborty, S.: Policy-based message scheduling using FlexRay. In: Proceeding of the CODES+ISSS (2014)
24. Tanasa, B., Bordoloi, U.D., Eles, P., Peng, Z.: Scheduling for fault-tolerant communication on the static segment of FlexRay. In Proceeding of the RTSS (2010)
25. Lange, R., Vasques, F., Portugal, P., de Oliveira, R.S.: Guaranteeing real-time message deadlines in the FlexRay static segment using a on-line scheduling approach. In: WFCS (2014)
26. Schmidt, K., Schmidt, E.G.: Optimal message scheduling for the static segment of FlexRay. In: Proceeding of the VTC (2010)

27. Kukkala, V.K., Pasricha, S., Bradley, T.: JAMS: Jitter-aware message scheduling for FlexRay automotive networks. In: Proceeding NOCS (2017)
28. Zhang, L., Goswami, D., Schneider, R., Chakraborty, S.: Task-and network-level schedule co-synthesis of Ethernet-based time-triggered systems. In: Proceeding of the. ASP-DAC (2014)
29. Sagstetter, F., Andalam, S., Waszecki, P., Lukasiewycz, M., Stähle, H., Chakraborty, S., Knoll, A.: Schedule integration framework for time-triggered automotive architectures. In: Proceeding of the DAC (2014)
30. Craciunas, S.S., Oliver, R.S.: Combined task-and network-level scheduling for distributed time-triggered systems. In: Real-Time Systems (2016)
31. Novak, A., Sucha, P., Hanzalek, Z.: Efficient algorithm for jitter minimization in time-triggered periodic mixed-criticality message scheduling problem. In: Proceeding of the RTNS (2016)
32. Minaeva, A., Akesson, B., Hanzálek, Z., Dasari, D.: Time-triggered co-scheduling of computation and communication with jitter requirements. In: IEEE Transactions on Computers (2018)
33. Maldonado, L., Chang, W., Roy, D., Annaswamy, A., Goswami, D., Chakraborty, S.: Exploiting system dynamics for resource-efficient automotive CPS design. In: Proceeding of the DATE (2019)
34. Roy, D., Chang, W., Mitter, S.K., Chakraborty, S.: Tighter dimensioning of heterogeneous multi-resource autonomous CPS with control performance guarantees. In: Proceeding of the DAC (2019)
35. Yamaguchi, T.J., Ichiyama, K., Hou, H. X., Ishida, M.: A Robust method for identifying a deterministic jitter model in a total jitter distribution. In: Proceeding of the ITC (2009)
36. Witkowski, T., Antczak, P., Antczak, A.: Solving the flexible open-job shop scheduling problem with GRASP and simulated annealing. In: AICI (2010)
37. Liu, L., Mu, H., Yang, J.: Simulated annealing based GRASP for Pareto-optimal dissimilar paths problem. In: SC (2016)
38. Kukkala, V.K., Bradley, T., Pasricha, S.: Uncertainty analysis and propagation for an auxiliary power module. In: Proceeding of IEEE Transportation Electrification Conference (TEC) (2017)
39. Kukkala, V.K., Pasricha, S., Bradley, T.: JAMS-SG: A framework for jitter-aware message scheduling for time-triggered automotive networks. ACM Trans. Des. Autom. Electron. Syst. **24**(6) (2019)
40. DiDomenico, G.C., Bair, J., Kukkala, V.K., Tunnell, J., Peyfuss, M., Kraus, M., Ax, J., Lazarri, J., Munin, M., Cooke, C., Christensen, E.: Colorado State University EcoCAR 3 Final Technical Report. In: SAE World Congress Experience (WCX) (2019)
41. Kukkala, V.K., Pasricha, S., Bradley, T.: SEDAN: Security-aware design of time-critical automotive networks. IEEE Trans. Veh. Technol. **69**(8) (2020)
42. Kukkala, V.K., Thiruloga, S.V., Pasricha, S.: INDRA: Intrusion detection using recurrent autoencoders in automotive embedded systems. IEEE Trans. Comput-Aided Design Integr. Circ. Syst. **39**(11) (2020)
43. Kukkala, V.K., Thiruloga, S.V., Pasricha, S.: LATTE: LSTM self-attention based anomaly detection in embedded automotive platforms. ACM Trans. Embed. Comput. Syst. **20**(5 s): Article 67 (2021)
44. Thiruloga, S.V., Kukkala, V.K., Pasricha, S.: TENET: Temporal CNN with attention for anomaly detection in automotive cyber-physical systems. In: Proceeding of IEEE/ACM Asia & South Pacific Design Automation Conference (ASPDAC) (2022)
45. Kukkala, V.K., Thiruloga, S.V., Pasricha, S.: Roadmap for cybersecurity in autonomous vehicles. In: IEEE Consumer Electronics Magazine (CEM) (2022)
46. Tunnell, J., Asher, Z., Pasricha, S., Bradley, T.H.: Towards improving vehicle fuel economy with ADAS. SAE Int. J. Connect. Autom. Veh. **1**(2) (2018)
47. Tunnell, J., Asher, Z., Pasricha, S., Bradley, T.H.: Towards improving vehicle fuel economy with ADAS. In: Proceeding of SAE World Congress Experience (WCX) (2018)
48. Asher, Z., Tunnell, J., Baker, D.A., Fitzgerald, R.J., Banaei-Kashani, F., Pasricha, S., Bradley, T.H.: Enabling prediction for optimal fuel economy vehicle control. In: Proceeding of SAE World Congress Experience (WCX) (2018)

49. Dey, J., Taylor, W., Pasricha, S.: VESPA: A framework for optimizing heterogeneous sensor placement and orientation for autonomous vehicles. IEEE Consum. Electron. Mag. **10**(2) (2021)
50. Kukkala, V.K., Tunnell, J., Pasricha, S., Bradley, T.: Advanced driver assistance systems: A path towards autonomous vehicles. In Proceeding of IEEE Consumer Electronics Magazine (CEM) (2018)

Evolvement of Scheduling Theories for Autonomous Vehicles

Wanli Chang, Nan Chen, Shuai Zhao, and Xiaotian Dai

1 Introduction

There is a clear trend in the automotive industry towards autonomous vehicles which brings a series of new requirements for real-time scheduling, due to the evolving complexity. First, in the scheduling of real-time autonomous systems, scheduling theories for simple task models and uniprocessors have been well established, but multiprocessor systems are increasingly being employed and dependencies between tasks need to be considered [10]. Many existing works use a single recurrent event or time-triggered DAG tasks to model functional dependencies in a system [7, 8, 26, 46, 59, 60]. For example, a complete automotive task chain from on to control is described in [59] and converted to a single periodic DAG task. In addition, to avoid migration and cache-related preemption overhead, a non-preemptive global scheduling scheme is often deployed [15, 59]. That is, the nodes of a DAG are scheduled globally on all cores and preemption is not allowed during the execution of a node [47].

Figure 1 provides an example DAG which contains eight nodes with a set of edges. A node indicates a computation unit that must be executed sequentially and a directed edge describes the execution dependency of two nodes (e.g., node v_5

W. Chang (✉)
Hunan University, Changsha, China

Huawei Technologies, Shenzhen, China

N. Chen · X. Dai
University of York, York, UK
e-mail: nc952@york.ac.uk; xiaotian.dai@york.ac.uk

S. Zhao
Sun Yat-sen University, Guangzhou, China
e-mail: zhaosh56@mail.sysu.edu.cn

© The Author(s), under exclusive license to Springer Nature Switzerland AG 2023
V. K. Kukkala, S. Pasricha (eds.), *Machine Learning and Optimization Techniques for Automotive Cyber-Physical Systems*, https://doi.org/10.1007/978-3-031-28016-0_2

Fig. 1 An example DAG

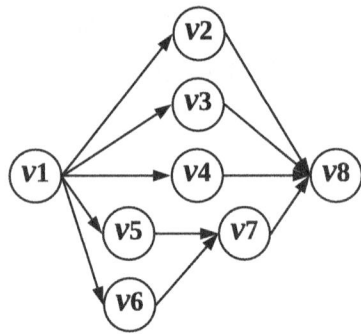

and v_7). When there are adequate cores in the system, nodes with no dependency e.g., node v_2, v_3 and v_4 can be executed in parallel. However, when the number of paralleled nodes is bigger than the number of cores available, the priority ordering between nodes becomes an issue which can impose non-negligible effects to the makespan (i.e. the execution between the start of the first node and finish time of the last node) of a DAG. In the mean time, the Worst-Case Response Time (WCRT) analysis in [33, 39] are pessimistic which can result in low system schedulability. Hence, a fine-grained scheduling policy and a less pessimistic WCRT bound are necessary.

Second, the increasing demand of autonomous systems to realize both complex functionality and high performance with limited resources necessitates extensive resource sharing. For example, to facilitate partially or fully automated driving, the AUTOSAR Classic standard (which implements static task configuration with resource isolation) is evolving to AUTOSAR Adaptive with dynamic resource sharing on multiprocessor architectures [4]. Resources sharing is referred as sharing data structures, special memory locations, and code segments, which need to be accessed in a mutually exclusive fashion. Consequently, the increasing applications of shared resources in the autonomous systems can cause blocking due to contention, while conventional requirements of timing predictability and reliability still need to be satisfied. That is, the deadlines of tasks must be met while failures during task executions must be resolved.

Satisfying both timing and reliability requirements is particularly hard. Several multiprocessor resource sharing protocols have been proposed to bound and minimize blocking time, including MSRP [27] and MrsP [14]. However, reliability has not been accounted for, which is imperative in safety-critical scenarios like autonomous systems. The common fault-tolerance methods are based on redundancy, and they may be directly applied to shared resources by scheduling repeated task executions and resource accesses a sufficient number of times to get the correct output. However, this leads to severe resource contention and undermines system schedulability. Therefore, a solution for guaranteeing both reliability and schedulability for autonomous systems with the presence of shared resource is required.

Third, on communication, Ethernet as a data link layer protocol has evolved from standard computer networks to applications of in-vehicle communication (e.g., deterministic real-time Ethernet [55]). In the emerging safety-critical systems such as highly automated vehicles, a large volume of messages with mixed types need to be transmitted on the same infrastructure, which requires deterministic and predictable timing to guarantee safety. Traditional real-time networks use non-standard Ethernet to enable high-bandwidth deterministic communication, which prohibits connectivity between different protocols and components from different vendors, as well as increases uncertainty and difficulty in timing and hazard analysis.

TSN proposed as an IEEE standard, offers an interoperable and flexible deterministic Ethernet-based solution [36]. It is widely considered as the network solution for future automobiles. The IEEE 802.1 TSN standard includes a wide range of subsets, in which one of the most important protocols is the 802.1Qbv [20, 35, 63]. The IEEE 802.1Qbv supports time-aware shaper (TAS) using TDMA (time-division multiple access)-scheduled queues to access the egress port—controlled by a gate switching logic that is driven by a synchronized global timer and a look-up scheduling table.

Control loops are often involved in the safety-critical systems, where guarantees are required on both timing of communication and control performances (measured by settling time). In general, short sampling periods enable the potential to achieve good control performance with frequent interactions between the controller and the plant. The state-of-the-art network scheduling techniques for TSN (e.g., [5, 41, 63]) cannot be directly applied, as they consider neither the hard real-time constraints on network packets nor the control performance of the system. Therefore, an integrated solution of network scheduling and controller co-design for TSN is essential for autonomous in-vehicle communications from the CPS perspective.

1.1 Organization

In this chapter, we present three interconnected fundamental works along the above directions: the real-time scheduling for DAGs on multiprocessor architectures; the reliable resource sharing in autonomous systems; and real-time scheduling and controller co-design for TSN. The rest of the chapter is organized as follows:

- **Section 2** provides the background knowledge and related research outputs of the work presented in the following sections.
- **Section 3** introduces a CPC model based on the work-conserving schedule and the classic analysis, alongside a priority ordering algorithm.
- **Section 4** presents the first fault-tolerant solution for multiprocessor MCS with shared resources. The solution contains a system execution model that is compatible with an arbitrary number of criticality levels, and a protocol, namely Multiprocessor Stack Resource Protocol Fault Tolerance (MSRP-FT) which aims to address faults during critical sections while minimizing blocking time.
- **Section 5** presents the first integrated solution of network scheduling and controller co-desig for TSN 802.1Qbv. Specifically, the first FPS approach for TSN is demonstrated. Moreover, a finer-grained analysis for the above scheduling

approach at the frame level is also included. Based on FPS and the analysis, we formulate a co-design optimization problem to decide the sampling periods and poles of real-time controllers.

- **Section 6** concludes the contents of this chapter.

2   Background

In this section, we provide the background information and related literature to motivate the research output demonstrated in the following sections. First, Sect. 2.1 reviews the work in scheduling and analysis of DAG tasks. Second, work related to fault-tolerance, resource sharing, and MCS is reviewed in Sect. 2.2. Last, relevant literature on the scheduling of TSN network is presented in Sect. 2.3.

2.1   Scheduling and Analyzing DAG Tasks in Autonomous Vehicles

The majority of the existing work on scheduling DAG tasks assumes a *work-conserving* scheduler [39]. A scheduling algorithm is said to be work-conserving if it never idles a processor when there exists pending workload. A generic bound that captures the worst-case response time of tasks scheduled globally with any work-conserving method is provided in Graham [28]. This analysis is later formalized in Melani [39] and Fonseca [25] for DAG tasks. The analysis of a single DAG task is given in Eq. (1). Notation τ_x denotes a DAG task with index x, R_x denotes the response time of τ_x, L_x denotes the length of the longest path in the DAG, W_x gives the sum of Worst-Case Execution Time (WCETs) of all nodes in the DAG, and m denotes the number of cores.

$$R_x = L_x + \left\lceil \frac{1}{m}(W_x - L_x) \right\rceil \tag{1}$$

In this analysis, the worst-case response time of a DAG task τ_x is upper bounded by the length of the critical path and the intra-task interference imposed by the non-critical nodes of τ_x itself. However, this analysis assumes the critical path can be delayed by all the concurrent nodes, which is pessimistic for scheduling methods with an explicit execution order known a priori [33, 39].

2.1.1   The State-of-the-Art in DAG Scheduling and Analysis

For homogeneous multiprocessors with a global scheme, existing scheduling (and their analysing) methods aim at reducing the makespan and tightening the worst-case analytical bound. They can be classified as either slice-based [17, 29] or node-based [18, 33]. The slice-based schedule enforces node-level preemption and divides each node into a number of small computation units (e.g., units with a WCET of one in Chang [17]). By doing so, the slice-based methods can improve node-level parallelism but to achieve an improvement the number of preemptions and migrations need to be controlled.

The node-based methods provide a more generic solution by producing an explicit node execution order, based on heuristics derived from either the *spatial* (e.g., number of successors of a node [37] and topological order of nodes [33]) or the *temporal* (execution time of nodes [18, 54, 59]) characteristics of the DAG. Below we describe two most recent node-based methods.

In Chen et al. [18], an non-preemptive scheduling method is proposed for a single periodic DAG, which always executes the ready node with the longest WCET to improve parallelism. Chen [18] prevents anomalies from occurring when nodes are executing less than their WCETs, which can lead to an execution order different from the schedule. This is achieved by guaranteeing nodes are executed in the same order as the offline simulation. However, without considering inter-node dependencies, this schedule cannot minimize the delay on the completion of DAG.

In He et al. [33], a new response time analysis is presented, which dominates the traditional bound in Graham [28] and Melani [39] when an explicit node execution order is known a priori. That is, a node v_j can only incur a delay from the concurrent nodes that are scheduled prior to v_j. Then, a scheduling method is proposed that always executes: (i) the critical path first; and (ii) the immediate interference nodes first (nodes that can cause the most immediate delay on the currently-examined path). The novelty in He [33] is considering both topology and path length in a DAG, and provides the state-of-the-art analysis against which our approach is compared. However, the method in He [33] schedules concurrent nodes based on the length of their longest complete path (a path from the source to the sink node), i.e., nodes in the longest complete path first. This heuristic is not dependency-aware, which reduces the level of parallelism that can be exploited, and hence, lengthen the finish time of a DAG task.

2.2   Real-Time Scheduling for Reliable Autonomous Driving

In this subsection, the background information and related work about real-time scheduling of reliable autonomous system are provided. More specifically, Sect. 2.2.1 introduces common faults and solutions in the embedded systems, Sect. 2.2.2 presents the research in the field of resources sharing protocols. Section 2.2.3 demonstrates the research output related to MCS.

2.2.1   Fault Tolerance

Faults in modern embedded systems can be broadly categorized as *permanent* or *transient* faults. Transient faults affect the functionality of systems for a short period of time, where permanent faults happen repeatedly and cannot be easily recovered from. Some software faults (bugs) are caused by erroneous program design, are permanent faults, and cannot be recovered by re-starting the operation [58]. Other software errors can be transient faults caused by unexpected interference among threads, and may be resolved by restarting the program [40]. Transient hardware faults can occur due to issues such as power supply fluctuations or electromagnetic interference which happen increasingly more frequently due to the decrease in transistor size and operating voltage [32]. Permanent hardware faults are the result of hardware damage or wear, and cannot be dealt with until the faulty component is replaced. In this chapter, we focus on transient faults which can be recovered by retrying the operation.

Three mainstream redundancy techniques are widely adopted in the literature to tolerate faults: *re-execution* [1], *checkpointing* [19], and *replication* [45]. The *re-execution* approach saves task status at the beginning and detects faults at the end. Once a fault is detected, the roll-back technique is applied and the whole task is re-executed. The *checkpointing* technique introduces additional checkpoints in a task and normally divides task execution into a set of uniform segments. Each small segment is tested for faults, and when a fault is detected the system rolls back to the most recent checkpoint and only re-executes the faulty segment. With replication, each task is replicated to several copies. The task and its replicas are released simultaneously and execute in parallel. When an execution finishes without incurring faults, the others are discarded.

Generally, fault detection mechanisms focus on analyzing the outputs of an execution. For example, in a lockstep dual-core architecture [50] or Triple Modular Redundancy architecture [4], multiple identical cores execute the same code and the system applies a majority vote to find the faulty component. Acceptance tests are often applied at the checkpoint to determine the correctness of an operation by checking a set of conditions that are expected to be met if the program has executed correctly [44]. In contrast, another type of fault-detection mechanism focuses on detecting the stimulus of the fault instead of the computation results. For example, acoustic wave detectors are adopted in the hardware architecture [56] to detect particle strikes that can result in transient faults during computation. Instead of using built-in hardware to detect faults, the Argus approach [38] uses detection equipment to monitor the variations of the circuits. Detailed descriptions and comparisons of such type of detecting mechanisms are included in [57].

2.2.2 Resource Sharing

Resource sharing in multicore real-time systems has been extensively studied in the past few decades with numerous resource sharing protocols available [2, 14, 27]. A comprehensive survey can be found in [11]. Here we describe the Multiprocessor Stack Resource Protocol (MSRP) [27].

The MSRP is a First-In-First-Out (FIFO) spin-based resource sharing protocol developed for fully-partitioned systems. In MSRP, each global resource (i.e., shared between cores) is associated with a FIFO queue. A task requesting a global resource

Fig. 2 The AMC model

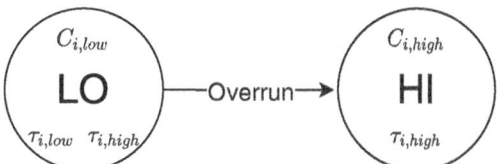

is placed in the FIFO queue and busy-waits (spins) non-preemptively until it moves to the head of the queue, at which point it will be granted the resource. The task then keeps executing non-preemptively until it releases the resource. For a local resource (i.e., shared in one core), a priority ceiling is applied, which equals the highest priority of tasks that request the resource. A task raises its priority to the ceiling during the entire access to the local resource.

When contending for shared resources, tasks will incur additional waiting time (i.e. blocking) due to mutually exclusive executions. The blocking effects incurred by tasks for accessing shared resources under MSRP can be classified as *spin delay* and *arrival blocking* [62]. With shared resources, a task can incur spin delay either *directly* or *indirectly*. *Direct spin delay* occurs when a task is being blocked directly for accessing a shared resource by other resource accesses issued from remote cores. In this case, the task is added at the tail of the FIFO queue and spin-waits until it is granted the resource. A task incurs *indirect spin delay* when it is preempted by a local higher priority task, which in turn is blocked directly from accessing a resource. *Arrival blocking* occurs when a task is released but is then immediately blocked by a local low priority task which is running non-preemptively (resp. with a higher resource ceiling) for accessing a global (resp. local) resource.

Resource sharing protocols define rules for accessing shared resources and bound the blocking delay [11]. However, they are not developed with a particular focus on system reliability, in which a resource request has to be potentially executed multiple times sequentially to tolerate faults. Hence, the additional blocking time imposed for addressing faults cannot be effectively minimized by these protocols. Based on the above, this chapter focuses on fault-tolerance for shared resources in MCS and aims to reduce the additional blocking from tolerating faults.

2.2.3 Mixed Criticality System

Baruah et al. [6] propose an Adaptive Mixed Criticality (AMC) model which is widely regarded as the most effective approach within Fixed-Priority Preemptive Scheduling [34]. The AMC model has two system modes (LO and HI) for the system that has tasks with two criticality levels (i.e., $\mathcal{L} \in low, high$). As shown in Fig. 2, the system starts in LO mode and all tasks are allowed to execute up to $C_{i,low}$. If a task overruns these budgets, the system upgrades to the HI mode (a mode switch), in which high-criticality tasks are allowed to execute with a larger budget $C_{i,high}$ and low-criticality tasks are suspended. The AMC model assumes system can monitor the running time of tasks and can be extended to have an arbitrary number of system modes according to the number of criticality levels in the system. Later on, concerning the quality of service (QoS) of low-criticality tasks after a mode switch, instead of dropping tasks brutally, many research [13, 30] propose mechanisms for MCS to degrade low-criticality tasks gracefully.

With the presence of faults, Pathan [42] proposes a mixed-criticality fault-tolerant algorithm called FTMC for systems with two criticality levels. In FTMC, the system would transit from a low-criticality mode to a high-criticality mode if any overrun happens or the number of transient faults incurred in the system exceeds a predefined threshold. Chen et al. [19] propose an online fault-tolerant MCS scheduling framework called the FTS-RHS. The framework applies the checkpointing recovery schemes which outperforms re-execution in scheduling. In addition, the DVFS techniques have been applied in MCS in [9] to provide systems with precise real-time and energy-efficient scheduling. Safari et al. [45] further extend the research topic by including the consideration of energy consumption in fault-tolerant MCS and propose a LETR-MC scheme for a system with two criticality levels.

With shared resources, Burns [12] applies the Original Priority Ceiling Protocol (OPCP) to the MCS on a uni-processor platform with two criticality levels. When the system transits to the high-criticality mode, low-criticality resource holders which are computing with the ceiling priority are suspended. They can continue to execute by inheriting the execution budget of their next release. Zhao et al. [61] extend the Priority Ceiling Protocol (PCP) [48] to HLC-PCP (Highest-Locker Criticality, Priority-Ceiling Protocol) to manage resource sharing in the MCS under AMC scheme. Han et al. [31] migrate the MSRP to the MCS and develop a criticality-aware utilization bound. However, none of the above works consider the presence of both shared resources and faults.

2.3 Real-Time TSN Scheduling for Automotive CPS

Time-sensitive networking is an enabler for Ethernet-based communication services that were not originally built to support hard real-time guarantees, such as OPC Unified Architecture (OPC-UA)[1] and Distributed Data Service (DDS).[2] The objective of TSN is to reduce the worst-case end-to-end latency for critical traffics. Here we briefly discuss the IEEE 802.1Qbv TSN (referred to as Qbv in the following text). A diagrammatic view of a Qbv-enabled switch is depicted in Fig. 3. From the figure, it can be seen that a Qbv TSN switch consists of the following major components:

---

[1] https://opcfoundation.org/about/opc-technologies/opc-ua/.

[2] https://www.omg.org/spec/DDS/1.4/PDF.

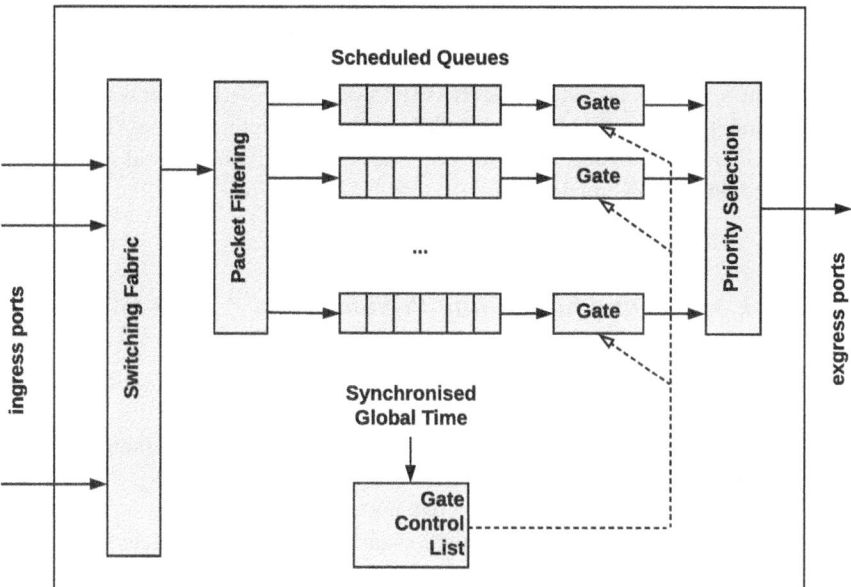

Fig. 3 An overview structure of a 802.1Qbv-capable TSN switch

- **Scheduled FIFO queues**: In a Qbv-enabled TSN switch, there are eight independent time-divided FIFO queues which are controlled by transmission gates.
  The incoming traffic is filtered by the packet filtering unit which sends a packet
  to its designated queue. This information is encoded as Class of Service (CoS) in
  the priority code point (PCP) header in the Ethernet frame.
- **Gate control list (GCL)**: The GCL can trigger gate-open and gate-close events
  periodically with a gate control cycle. The time granularity between events can be
  as low as $1ns$ depending on the specific implementation. The schedule is located
  in a GCL look-up table that is distributively configured to each TSN node. If
  multiple gates are opened at the same time, the policy in the priority selection
  unit will determine which queue is forwarded to the egress port first.
- **Time synchronization**: To allow time-divided transmission that is distributed
  through the network, a timer is globally synchronized with all the switches in
  the same network using precision time protocols (PTPs), e.g., IEEE 802.1AS or
  IEEE 802.1AS-Rev.

The mechanisms of Qbv TSN improve the flexibility in terms of traffic schedule and control. It enables interoperability between standard-compliant industrial
devices thus allowing open data exchange. It also removes the need for physical
separation of critical and non-critical communication networks. However, in a
different aspect these introduce increased design complexity that needs to be
elaborately handled.

3   Scheduling of DAGs on Multiprocessor Architectures

The content of this section is organized as follows. Section 3.1 presents the system and task model. Section 3.2 presents the CPC model that captures the two key factors of the DAG structure. Finally, Sect. 3.3 describes the scheduling algorithm for DAG tasks, based on the CPC model.

3.1   Task Model and Scheduling Preliminaries

A DAG task τ_x is defined by $\{T_x, D_x, \mathcal{G}_x = (V_x, E_x)\}$, with T_x denoting its minimum inter-arrival time, D_x gives a constrained relative deadline, i.e., $D_x \le T_x$, and $\mathcal{G}_x$ is a graph defining the set of activities forming the task. The graph is defined as $\mathcal{G}_x = (V_x, E_x)$ where V_x denotes the set of nodes and $E_x \subseteq (V_x \times V_x)$ gives the set of directed edges connecting any two nodes. Each node $v_{x,j} \in V_x$ represents a computation unit that must be executed sequentially and is characterized by its Worst-Case Execution Time (WCET), $C_{x,j}$. For simplicity, the subscript of the DAG task (i.e., x for τ_x) is omitted when the system has only one DAG task.

For any two nodes v_j and v_k connected by a directed edge $((v_j, v_k) \in E)$, v_k can start execution only if v_j has finished its execution. That is, v_j is a *predecessor* of v_k, whereas v_k is a *successor* of v_j. A node v_j has at least one predecessor $pre(v_j)$ and at least one successor $suc(v_j)$, formally defined as $pre(v_j) = \{v_k \in V \mid (v_k, v_j) \in E\}$ and $suc(v_j) = \{v_k \in V \mid (v_j, v_k) \in E\}$, respectively. Nodes that are either *directly* or *transitively* predecessors and successors of a node v_j are termed as its ancestors $anc(v_j)$ and descendants $des(v_j)$ respectively. A node v_j with $pred(v_j) = \varnothing$ or $succ(v_j) = \varnothing$ is referred to as the *source* v_{src} or *sink* v_{sink} respectively. Without loss of generality, we assume each DAG has one source and one sink node. Nodes that can execute concurrently with v_j are given by $C(v_j) = \{v_k \mid v_k \notin (anc(v_j) \cup des(v_j)), \forall v_k \in V\}$ [33].

A DAG task has the following fundamental features. First, a path $\lambda_a = \{v_s, \cdots, v_e\}$ is a node sequence in V and follows $(v_k, v_{k+1}) \in E, \forall v_k \in \lambda_a \backslash v_e$. The set of paths in V is defined as Λ_V. A *local path* is a sub-path within the task and as such does not feature both the source v_{src} and the sink v_{sink}. A *complete path* features both. Function $len(\lambda_a) = \sum_{\forall v_k \in \lambda_a} C_k$ gives the length of λ_a. Second, the longest complete path is referred to as the *critical path* λ^*, and its length is denoted by L, where $L = \max\{len(\lambda_a), \forall \lambda_a \in \Lambda_V\}$. Nodes in λ^* are referred to as the *critical nodes*. Other nodes are referred to as *non-critical nodes*, denoted as $\overline{V} = V \backslash \lambda^*$. Finally, the workload W is the sum of a task's WCETs, i.e. $W = \sum_{\forall v_k \in V} C_k$. The workload of all non-critical nodes is referred to as the *non-critical workload*.

3.2   Concurrent Provider and Consumer Model

Equation (1) indicates that minimizing the delay from non-critical nodes to the critical path (i.e., $\frac{1}{m}(W - L)$) effectively reduces makespan of the DAG. Achieving this requires the complete knowledge of the topology (i.e., the dependency and parallelism of each node) of a DAG so that the potential delay of the critical path can be identified. To support this the CPC model is presented to fully exploit node dependency and parallelism.

The CPC model has two key stages. First, the critical path is divided into a set of consecutive sub-paths based on the potential delay it can incur. Second, for each sub-path, the CPC model identifies the non-critical nodes that can 1) execute in parallel with the sub-path and 2) delay the start of the next sub-path, based on precedence constraints.

The intuition of the CPC model is: when the critical path is executing, it utilizes just one core so that the non-critical ones can execute in parallel on the remaining $(m - 1)$ cores. The time allowed for executing non-critical nodes in parallel is termed as the *capacity*, which is the length of the critical path. Note that non-critical nodes that utilize this capacity to execute cannot cause any delay to the critical path. The sub-paths in the critical path are termed *capacity providers* Θ^* and all non-critical nodes are *capacity consumers* Θ. For each provider $\theta_i^* \in \Theta^*$, it has a set of consumers $F(\theta_i^*)$ that can execute using θ_i^*'s capacity as well as delay the next provider θ_{i+1}^* in the critical path.

Algorithm 1 presents a two-step process for constructing the CPC model of an input DAG G with its critical path λ^*. Starting from the head node in λ^*, capacity providers are formed by analyzing node dependency between the critical path and non-critical nodes (Line 3-9). For a provider θ_i^*, its nodes should execute consecutively without delay from non-critical nodes in terms of dependency. That is, each node in θ_i^*, other than the head node (Line 5), only has one predecessor which is the previous node in θ_i^*.

Then, for each $\theta_i^* \in \Theta^*$, its consumers $F(\theta_i^*)$ are identified as the nodes that (1) can execute concurrently with θ_i^*, and (2) can delay the start of θ_{i+1}^* (i.e., $anc(\theta_{i+1}^*) \cap \overline{V}$ in Line 12). Accordingly, nodes in $F(\theta_i^*)$ that finish later than θ_i^* will delay the start of θ_{i+1}^* (if it exists). By doing so, the CPC model provides detailed knowledge of the potential delay caused by non-critical nodes on the critical path.

Furthermore, given an arbitrary DAG structure, a consumer $v_j \in F(\theta_i^*)$ can start earlier than, synchronous with, or later than the start of θ_i^*. For synchronous and late-released consumers, they will only utilize the capacity of θ_i^*. However, an early-released consumer can execute concurrently with certain previous providers, and therefore interfere with their consumers and impose an indirect delay to those providers. For a provider θ_i^*, $G(\theta_i^*)$ (in line 13) denotes the nodes that belong to the consumer groups of later providers, but which can execute in parallel (in terms of topology) with θ_i^*.

Algorithm 1: $CPC(\mathcal{G}, \lambda^*)$: CPC model construction

---

    Inputs   : $\{\mathcal{G} = (V, E)\}$
    Outputs   : Θ^*, $F(\theta_i^*)$, $G(\theta_i^*)$, $\forall \theta_j^* \in \Theta^*$

    Parameters : λ^*, $\overline{V} = V \backslash \lambda^*$

1 $\Theta^* = \varnothing$;
2 **for** each $v_j \in \lambda^*$, in topological order **do**
3 | $\theta_i^* = \{v_j\}$; $\lambda^* = \lambda^* \backslash v_j$;
4 | **while** $pre(v_{j+1}) = \{v_j\}$ **do**
5 | | $\theta_i^* = \theta_i^* \cup \{v_{j+1}\}$; $\lambda^* = \lambda^* \backslash v_j$;
6 | **end**
7 | $\Theta^* = \Theta^* \cup \theta_i^*$;
8 **end**
9 **for** each $\theta_i^* \in \Theta^*$, in topological order **do**
10 | $F(\theta_i^*) = anc(\theta_{i+1}^*) \cap \overline{V}$;
11 | $G(\theta_i^*) = \bigcup_{v_j \in F(\theta_i^*)} \{C(v_j) \cap \overline{V}\}$;
12 | $\overline{V} = \overline{V} \backslash F(\theta_i^*)$;
13 **end**
14 **return** Θ^*, $F(\theta_i^*)$, $G(\theta_i^*)$, $\forall \theta_i^* \in \Theta^*$

With the CPC model, a DAG is transformed into a set of capacity providers and consumers, with a time complexity of $O(|V| + |E|)$. The CPC model provides complete knowledge of both direct and indirect delays from non-critical nodes on the critical path. For each provider θ_i^*, nodes in $F(\theta_i^*)$ can utilize a capacity of $len(\theta_i^*)$ on each of $m - 1$ cores to execute in parallel while incurring potential delay from $G(\theta_i^*)$.

We now formally define the parallel and interfering workload of a capacity provider. Let $f(\cdot)$ denote the finish time of a provider θ_i^* or a consumer node v_j, $L_i = len(\theta_i^*)$ gives the length of θ_i^* and $W_i = L_i + \sum_{v_k \in F(\theta_i^*)} \{C_k\} + \sum_{v_k \in G(\theta_i^*)} \{C_k\}$ gives the total workload of θ_i^*, $F(\theta_i^*)$ and $G(\theta_i^*)$. We formally define the terms *parallel* and *interfering* workload of a provider θ_i^*. Note, $W \leq \sum_{\theta_i^* \in \Theta} W_i$ as a consumer can be accounted for more than once if it can execute concurrently with multiple providers.

Definition 1 (Parallel Workload of θ_i^*) The parallel workload α_i of θ_i^* is the workload in $W_i - L_i$ that can execute before the time instant $f(\theta_i^*)$.

For a node v_j in $F(\theta_i^*) \cup G(\theta_i^*)$, it contributes to α_i if either $f(v_j) \leq f(\theta_i^*)$ or $f(v_j) - C_j < f(\theta_i^*)$. The former case (i.e., $f(v_j) \leq f(\theta_i^*)$) indicates v_j is finished before the finish of θ_i^* and cannot cause any delay, whereas $f(v_j) - C_j < f(\theta_i^*)$ means v_j can partially execute in parallel with θ_i^* so that its delay on θ_{i+1}^* is less than C_j.

Definition 2 (Interfering Workload of θ_i^*) The interfering workload of θ_i^* is the workload in $W_i - L_i$ that executes after the time instant $f(\theta_i^*)$. For a provider θ_i^*, its interfering workload is $W_i - L_i - \alpha_i$.

With Definitions 1 and 2, Lemma 1 follows.

Lemma 1 *For providers θ_i^* and θ_{i+1}^*, the workload in W_i that can delay the start of θ_{i+1}^* is at most $W_i - L_i - \alpha_i$.*

Proof Based on the CPC model, the start of θ_{i+1}^* depends on the finish of both θ_i^* and $F(\theta_i^*)$, which is $\max\{f(\theta_i^*), \max_{v_j \in F(\theta_i^*)} f(v_j)\}$. By Definition 1, α_i will not cause any delay as it always finishes before $f(\theta_i^*)$, and hence, the lemma follows. Note that although $G(\theta_i^*)$ cannot delay θ_{i+1}^* directly, it can delay on nodes in $F(\theta_i^*)$, and in turn, causes an indirect delay to θ_{i+1}^*.   □

3.3 DAG Scheduling: A Parallelism and Dependency Exploited Method

Based on the CPC model, a scheduling method is then presented to maximize node parallelism. This is achieved by a rule-based priority assignment, in which three rules are developed to statically assign a priority to each node in the DAG. Firstly to always execute the critical path first (Sect. 3.3.1), and then two rules (Sect. 3.3.2) to maximize parallelism and minimize the delay to the critical path.

The entire presented approach has general applicability to DAGs with any topology (unlike, e.g., [25], which assumes nested fork-join DAGs only). It assumes a homogeneous architecture, however, it is not restricted by the number of processors.

3.3.1 The "Critical Path First" Execution (CPFE)

In the CPC model, the critical path is conceptually modelled as a set of capacity providers. Arguably, each complete path can be seen as the providers, which offers the time interval of its path length for other nodes to execute in parallel. However, the critical path provides the maximum capacity and hence, enables the maximized total parallel workload (denoted as $\alpha = \sum_{\theta_i^* \in \Theta^*} \alpha_i$). This provides the foundation to minimize the interfering workload on the complete critical path.

Theorem 1 *For a schedule S with CPFE and a schedule S' that prioritizes a random complete path over the critical path, the total parallel workload of providers in S is always equal to or higher than that of S', i.e., $\alpha \geq \alpha'$.*

Proof The change from S to S' leads to two effects: (1) a reduction on the length of the provider path, and (2) an increase on length of one consumer path. Below we prove both effects cannot increase the parallel workload after the change.

First, suppose the length of provider θ_i^* is shortened by Δ after the change from S to S', the same reduction applies on its finish time, i.e., $f'(\theta_i^*) = f(\theta_i^*) - \Delta$. Because nodes in θ_i^* are shortened, the finish time $f(v_j)$ of a consumer node $v_j \in F(\theta_i^*) \cup G(\theta_i^*)$ can also be reduced by a value from Δ/m (i.e., a reduction on v_j's interference, if all the shortened nodes in θ_i^* belong to $C(v_j)$) to Δ (if all such nodes belong to $pre(v_j)$) [28, 39]. By definition 1, a consumer $v_j \in F(\theta_i^*) \cup G(\theta_i^*)$ can contribute to the α_i if $f(v_j) \le f(\theta_i^*)$ or $f(v_j) - C_j \le f(\theta_i^*)$. Therefore, α_i cannot increase in S', as the reduction on $f(\theta_i^*)$ (i.e., Δ) is always equal or higher than that of $f(v_j)$ (i.e., Δ/m or Δ).

Second, let L and L' denote the length of the provider path under S and S' (with $L \ge L'$), respectively. The time for non-critical nodes to execute in parallel with the provider path is L' on each of $m - 1$ cores under S'. Thus, a consumer path with its length increased from L' to L directly leads to an increase of $(L - L')$ in the interfering workload, as at most L' in the consumer can execute in parallel with the provider.

Therefore, both effects cannot increase the parallel workload after the change from S to S', and hence, $\alpha \ge \alpha'$. $\square$

Rule 1. $\forall v_j \in \Theta^*, \forall v_k \in \Theta \Rightarrow p_j > p_k$.

Theorem 1 leads to the first assignment rule that assigns critical nodes with the highest priority, in which p_j denotes the priority of node v_j. With Rule 1, the maximum parallel capacity is guaranteed so that an immediate reduction (i.e., α) on the interfering workload of λ^* can be obtained.

3.3.2 Exploiting Parallelism and Node Dependency

With CPFE, the next objective is to maximize the parallelism of non-critical nodes and reduce the delay on the completion of the critical path. Based on the CPC model, each provider θ_i^* is associated with $F(\theta_i^*)$ and $G(\theta_i^*)$. For $v_j \in G(\theta_i^*)$, it can execute before $F(\theta_i^*)$ and use the capacity of θ_i^* to execute, if assigned with a high priority. Under this case, v_j can (1) delay the finish of $F(\theta_i^*)$ and the start of θ_{i+1}^*, and (2) waste the capacity of its own provider. A similar observation is also obtained in [33], which avoids this delay by the heuristic of early interference node first.

Rule 2. $\forall \theta_i^*, \theta_l^* \in \Theta^* : i < l \Rightarrow \min\limits_{v_j \in F(\theta_i^*)} p_j > \max\limits_{v_k \in F(\theta_l^*)} p_k$.

Therefore, the second assignment rule is derived to specify the priority between consumer groups of each provider. For any two adjacent providers θ_i^* and θ_{i+1}^*, the priority of any consumer in $F(\theta_i^*)$ is higher than that of all consumers in $F(\theta_{i+1}^*)$. With Rule 2, the delay from $G(\theta_i^*)$ on $F(\theta_i^*)$ (and hence θ_{i+1}^*) can be minimized, because all nodes in $G(\theta_i^*)$ belong to consumers of following providers and are always assigned with a lower priority than nodes in $F(\theta_i^*)$.

We now schedule the consumer nodes in each $F(\theta_i^*)$. In [33], concurrent nodes with the same earliness (in terms of the time they become ready during the execution of the critical path) are ordered by the length of their longest complete path (i.e., from v_{src} to v_{sink}). However, based on the CPC model, a complete path can be divided into several local paths, each of these local paths belong to the consumer group of different providers. For local paths in $F(\theta_i^*)$, the order of their lengths can

---

Algorithm 2: $EA(\Theta^*, \Theta)$: priority assignment

   Inputs   : Θ^*, Θ
   Parameters : p, p^{max}
   Initialize   : $p = p^{max}$, $\forall v_j \in \Theta^* \cup \Theta$, $p_j = -1$
1 `/* Assignment Rule 1. */`
2 $\forall v_j \in \Theta^*, p_j = p$; $p = p - 1$;
3 `/* Assignment Rule 2. */`
4 **for** each $\theta_i^* \in \Theta^*$, in topological order **do**
5   **while** $F(\theta_i^*) \neq \varnothing$ **do**
6   `/* Find the longest local path in `$F(\theta_i^*)$`. */`
7         $v_e, v_j \in F(\theta_i^*)$:
8         $v_e = \underset{v_e}{\text{argmax}}\{l_e(F(\theta_i^*))|suc(v_e) = \varnothing\}$;
9         $\lambda_{v_e} = v_e \cup \lambda_{v_j}$, $\underset{v_j}{\text{argmax}}\{l_j(F(\theta_i^*))|\forall v_j \in pre(v_e)\}$;
10   **if** $|pre(v_j)| > 1, \exists v_j \in \lambda_{v_e}$ **then**
11         $\{\Theta^{*\prime}, \Theta'\} = CPC(F(\theta_i^*), \lambda_{v_e})$;
12         $EA(\Theta^{*\prime}, \Theta')$;
13   **break**;
14   **else**
15   `/* Assignment Rule 3. */`
16         $\forall v_j \in \lambda_{v_e}, p_j = p$; $p = p - 1$;
17         $F(\theta_i^*) = F(\theta_i^*) \setminus \lambda_{v_e}$;
18   **end**
19   **end**
20 **end**

---

be the exact opposite to that of their complete paths. Therefore, this approach can lead to a prolonged finish of $F(\theta_i^*)$.

In the constructed schedule, we guarantee a longer local path is always assigned with a higher priority in a dependency-aware manner. This derives the final assignment rule, as given below. Notation $l_j(F(\theta_i^*))$ denotes the length of the longest local path in $F(\theta_i^*)$ that includes v_j. This length can be computed by traversing $anc(v_j) \cup des(v_j)$ in $F(\theta_i^*)$ [33]. With Rules 1-3 applied to the example DAG, it finally leads to the best-case schedule with a makespan of 13.

Rule 3*. $v_j, v_k \in F(\theta_i^*) : l_j(F(\theta_i^*)) > l_k(F(\theta_i^*)) \Rightarrow p_j > p_k$

However, simply applying Rule 3 to each $F(\theta_i^*)$ is not sufficient. Given a complex DAG structure, every $F(\theta_i^*)$ can form a smaller DAG $\mathcal{G}'$, and hence, an inner nested CPC model with the longest path in $F(\theta_i^*)$ is the provider. Furthermore, this procedure can be recursively applied to keep constructing inner CPC models for each consumer group in a nested CPC model, until all local paths in a consumer group are fully independent. For each inner nested CPC model, Rules 1 and 2 should be applied for maximized capacity and minimized delay of each consumer group, whereas Rule 3 is only applied to independent paths in a consumer group for maximized parallelism (and hence, the star mark on Rule 3). This enables complete awareness of inter-node dependency and guarantees the longest path first in each nested CPC model.

Algorithm 2 provides the complete approach of the rule-based priority assignment. The method starts from the outer-most CPC model $(CPC(\mathcal{G}, \lambda^*))$, and assigns all provider nodes with the highest priority based on Rule 1 (Line 2). By Rule 2, the algorithm starts from the earliest $F(\theta_i^*)$ (Line 4) and finds the longest local path λ_{v_e} in $F(\theta_i^*)$ (Line 8-9). If there exists dependency between nodes in λ_{v_e} and $F(\theta_i^*)\backslash\lambda_{v_e}$ (Line 9), $F(\theta_i^*)$ is further constructed as an inner CPC model with the assignment algorithm applied recursively (Line 11-12). This resolves the detected dependency by dividing λ_{v_e} into a set of providers. Otherwise, λ_{v_e} is an independent local path so that priority is assigned to its nodes based on Rule 3. The algorithm then continues with $F(\theta_i^*)\backslash\lambda_{v_e}$. The process continues until all nodes in V are assigned with a priority.

The time complexity of Algorithm 2 is quadratic. At most, $|V| + |E|$ calls to Algorithm 1 are invoked to construct the inner CPC models (Line 11), which examines each node and edge in the DAG. Mutually exclusively, Lines 16-17 assign each node with a priority value. Given that the time complexity of Algorithm 1 is $O(|V| + |E|)$, we have the time complexity $O((|V| + |E|)^2)$ for Algorithm 2. Although Algorithm 2 is recursive, this result holds as a node assigned with a priority will be removed from further iterations (Line 17), i.e., each node (edge) is processed only once.

With the CPC model and the schedule, the complete process for scheduling a DAG consists of three phases: (i) transferring the DAG to CPC; (ii) statically assigning a priority to each node by the rule-based priority assignment, and (iii) executing the DAG by a fixed-priority scheduler. With the input DAG known a priori, phases (i) and (ii) can be performed offline so that the scheduling cost at run-time is effectively reduced to that of the traditional fixed-priority system.

4   Reliable Resource Sharing in Reliable Autonomous Driving

The contents of this section is organized as follows. Section 4.1 describes the system and task model assumed in this section. Section 4.2 presents a fault-tolerance solution for MCS with shared resources, which includes a system execution model and a protocol MSRP-FT for faults which occur during critical sections.

4.1   System and Task Model

This section consider a fully partitioned system containing z identical cores (m_1 to m_z) and a set of sporadic tasks (Γ) that are scheduled by the Fixed Priority Preemptive Scheduling (FPPS) scheme. For generality, the system has tasks with $\mathcal{N}$ criticality levels which are defined by the system engineer according to their importance, denoted as $\mathcal{L} \in \{A, B, \ldots, \mathcal{N}\}$ in which A is the lowest criticality and $\mathcal{N}$ is the highest. Tasks being allocated to higher criticality levels implies a severe consequence for overall system performance if their execution in some way fails. Each task τ_i is defined by a 6-tuple $\{T_i, D_i, pri_i, m_i, l_i, \overrightarrow{C_i}\}$, including its minimum release period T_i, constrained deadline D_i (with $D_i \leq T_i$), priority pri_i, designated core m_i, criticality $l_i \in \mathcal{L}$, and a set of Worst-Case Execution Times (WCET) $\overrightarrow{C_i} = \{C_{i,A}, C_{i,B}, \ldots, C_{i,N}\}$ without accessing shared resources. The verification is more conservative for a higher criticality level [6], hence $C_{i,A} \leq C_{i,B} \leq \ldots \leq C_{i,N}$. The task τ_i with criticality l_i can execute up to C_{i,l_i} from its $\overrightarrow{C_i}$.

Within the system, there also exists a set of resources $\mathcal{R}$, each of which may be accessed by all tasks in the system in a mutually exclusive fashion by executing the *critical section* associated with the resource. Each shared resource r^x is defined by two notations: $\overrightarrow{c_i^x}$ and N_i^x, in which $\overrightarrow{c_i^x} = \{c_{i,A}^x, c_{i,B}^x, \ldots, c_{i,N}^x\}$ denotes the set of worst-case computation time τ_i needed to execute r^k with different levels of criticality, and N_i^x gives the number of requests from τ_i in one release. In this section the execution budgets of different segments of the same task (e.g. $C_{i,A}$ and $c_{i,A}^x$) increase or decrease simultaneously with the transition of system modes (see Sect. 4.2). However, to ease the presentation, the notation c^x is used to denote the worst-case time for executing r^x by all requesting tasks with any criticality level. Nested resource sharing is not considered in this section, i.e., a task can only hold one resource at a time, but can be directly supported by group locks [62].

Transient faults which can be resolved by redundancy approaches (e.g. re-execution and replication) in this section. Each fault can only affect one task at a time and the acceptance test is applied as the fault-detection technique.

4.2 A Fault-Tolerant Solution for MCS with Shared Resources

In this section, we present a new fault-tolerant solution for generic MCS that have two or more criticality levels with shared resources, to handle both task overruns and transient faults. First, we introduce a new fault-aware system model for MCS. The system model distinguishes faults occurring in normal and critical sections, which enables different fault-tolerance schemes to be implemented. Then, based on MSRP, a novel fault-tolerance multiprocessor resource sharing protocol is presented for handling faults in critical sections, which reduces the blocking time incurred for tolerating faults and guaranteeing the reliability of the system.

4.2.1 The Fault-Tolerance System Model

To handle task overruns and faults which occur during both normal and critical sections of a MCS, a fault-tolerant system model based on the extension of the AMC model [6] is introduced. Figure 4 illustrates the execution flow of the system and tasks in the model.

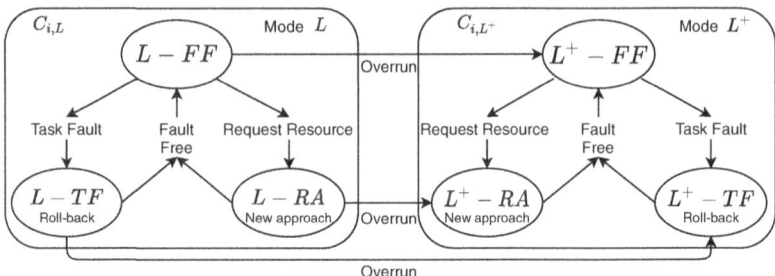

Fig. 4 The fault-tolerance system model

During a task's execution, faults can occur either in a normal or a critical section. The former is called a *task fault* and the latter a *resource fault* in this section. In the presented model, different fault-tolerant techniques are adopted to tolerate these two types of fault. The fault detection and tolerance techniques for normal and critical sections are presented in Sect. 4.2.2 and 4.2.3.

As shown in Fig. 4, each task has three execution states under a system mode (say L): fault-free (*L-FF*), task-fault (*L-TF*) and resource-access (*L-RA*). They are allowed to execute up to an execution budget $C_{i,L}$. A task executing in state *L-FF* is executing a normal section without incurring any faults. Once a fault occurs in a normal section, the task moves to state *L-TF*, at which the fault will be resolved. If a task requests a resource, it moves to state *L-RA* directly, where the fault-tolerance procedure for critical sections will be activated immediately, guaranteeing a fault-free resource access (see Sect. 4.2.3). The task moves back to state *L-FF* from *L-TF* or *L-RA* if the fault is resolved or the resource access is finished, respectively.

The system advances to the next system mode L^+ if any task in mode L overruns its budget. When an overrun occurs, tasks with criticality $l_i \geq L^+$ that are running in states *L-FF*, *L-TF* and *L-RA* will move directly to L^+-*FF*, L^+-*TF* and L^+-*RA* respectively with elevated execution budgets C_{i,L^+} and other tasks are dropped. By doing so, each overrun can bring the system to the next mode. However, there is an exception for tasks with criticality $l_i < L^+$ running in the state *L-RA* while executing with a shared resource, they are allowed to be dropped after finishing the underway critical section for the consideration of data integrity [31]. Moreover, mode changes can go in the reverse direction, when the system has less computation pressure it will resume suspended tasks and start in the lowest mode. Details of this will not be addressed here due to space constraints.

4.2.2 Fault-Tolerance of Normal Sections

In this section, we focus on transient faults which can be resolved by redundancy approaches. However, in systems with shared resources, detecting faults at the end of a task and re-executing the whole task to resolve a transient fault can lead to

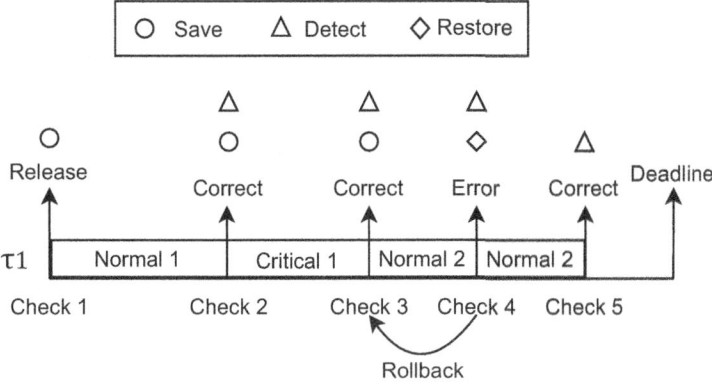

Fig. 5 Fault-tolerance in normal sections

substantial blocking time and the risk of transferring incorrect data to other tasks. To minimize the blocking time and provide reliable resource sharing, we apply different fault-tolerance approaches to handle faults that occur in normal and critical sections. This is achieved by not only inserting checkpoints at the start and end of each task but also introducing additional checkpoints around each critical section of the task. By doing so, the task execution is divided into a set of normal and critical sections. The acceptance test is assumed to be applied as the fault detection technique at each checkpoint.

In the presented fault-tolerance approach, the purposes of the checkpoints are slightly different, and so their operations vary. As shown in Fig. 5, a checkpoint (e.g. Check 1) will be set at the beginning of a task to perform a *Save* operation which involves storing the current architectural state of the system, including register files, counter values and etc. For fault-tolerance in normal sections, each checkpoint will operate a *Detect* operation to detect faults after the execution of each normal segment. If no faults are detected (e.g. at Check 2) the checkpoint will perform the *Save* operation. Otherwise, if a fault is detected (e.g. at Check 4) the task will roll back to the most recent checkpoint and perform the *Restore* operation which restores the previous data and re-performs the execution. This process repeats until the normal section is executed without any fault. Each re-attempt requires an additional *Detect* operation (e.g. at Check 5). However, for the end of the last execution segment, the *Save* operation is not needed at the checkpoint.

4.2.3   Fault-Tolerance of Critical Sections by MSRP-FT

For faults occurring in critical sections, the presented model utilizes a novel fault-tolerance multiprocessor resource sharing protocol, called MSRP-FT, in which tasks waiting for a resource can assist the resource holder to execute the associated critical section in parallel to address potential faults. The objective is to reduce

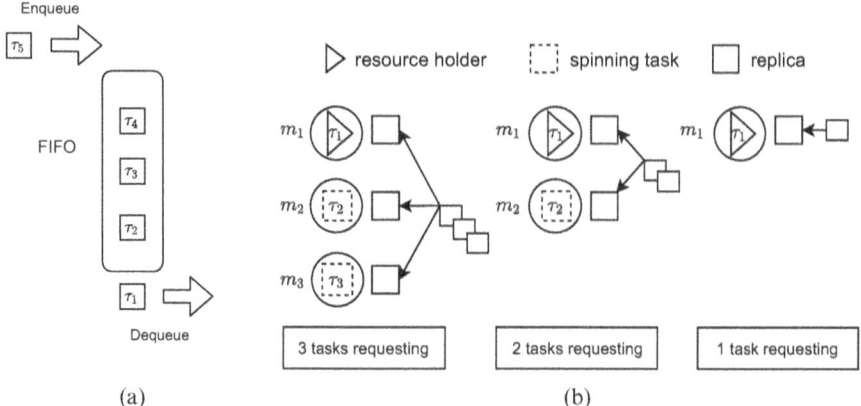

Fig. 6 Fault tolerance in critical section. (**a**) An example of a FIFO queue. (**b**) Replicas allocation based on the number of tasks in the queue

the additional blocking time caused by resolving faults in critical sections via re-executions. The mentioned MSRP-FT is introduced with the following steps.

4.2.3.1 Allocation of Replicas

Figure 6 demonstrates an example of the implementation of MSRP-FT, which is based on the resource sharing protocol MSRP. According to MSRP [27], tasks are inserted into a FIFO queue when they request a global resource. The task at the head of the queue (e.g. τ_1 in the figure) is granted the resource, other tasks spin on their own cores while checking the lock non-preemptively. With MSRP-FT, tasks are also placed at the FIFO queue when requesting shared resources. The task at the head of the FIFO queue will access the shared resource and the code segment to be executed by the head task and the internal states (e.g. variables) of the resource are replicated to a number according to the number of tasks in the FIFO queue as shown in Fig. 6b. It is worth noting that the access to the resource is always performed by the head task which obeys the mutually exclusive principle of shared resources and will not incur a race condition. Afterwards, replicas are stored in the local memory of each core and each task in the FIFO queue (including the head task) executes a replica on their host cores in parallel and updates the results on the local replica independently. If there is only one task in the FIFO queue, the head task has to execute the critical section by itself.

4.2.3.2 Submission of Replicas

Each execution of the replica is tested for faults on different cores. As shown in Fig. 7b, if a replica finishes without incurring any fault (e.g. on core m_3), it will

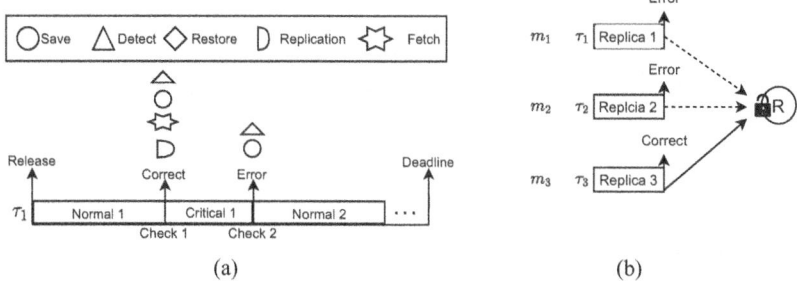

Fig. 7 Fault-tolerance in a critical section. (**a**) Operations of checkpoints around a critical section. (**b**) Submission of execution results

obtain the lock and update the shared resource with its local variables. If two overlapping requests to acquire the lock arrive, one task will commit the result and another will have no effect on the resource. The update of the resource is assumed to be conducted with an atomic action which once performed no other action can interleave with it, hence, race conditions are avoided. Once the resource is updated, other tasks are signaled to abandon the computation. In contrast, if all the resource-accessing tasks fail to obtain the correct result, they roll back and re-execute the replica until the correct result is successfully submitted. With a successful commit by any task in the FIFO queue, the head task (i.e., τ_1) is removed from the queue and continues its execution. The same procedure then repeats for the next head task within the FIFO queue.

Figure 7a shows the operations performed at the checkpoints around the critical section of τ_1. The checkpoint at the start of the critical section (e.g. Check 1) first performs *Detect* and *Save* operations to detect for faults and save the results of the execution of the previous segment, which is the same as mentioned above. It also applies *Fetch* and *Replicate* operations to fetch and replicate the corresponding operation and the shared resources to the spinning cores. A *Detect* operation is performed after the execution of the replica. Although the replica incurs faults, τ_3 already updated the result and a *Save* operation is performed to save the architectural states of the system and τ_1 continues its execution.

4.2.3.3 Working example

To clarify the implementation of the above fault-tolerance approach, the detailed execution procedure of the example stated above under two different fault-tolerance approaches is presented in Fig. 8. Figure 8a assumes that each critical section is checked for faults and any detected fault is tolerated directly by the roll-back and re-execution approach. As shown in Fig. 8a, τ_1, τ_2 and τ_3 request for a shared resource concurrently at $t = 1$. According to MSRP, τ_1 ranks first in the FIFO queue so it is granted with the resource and starts to execute its critical section immediately.

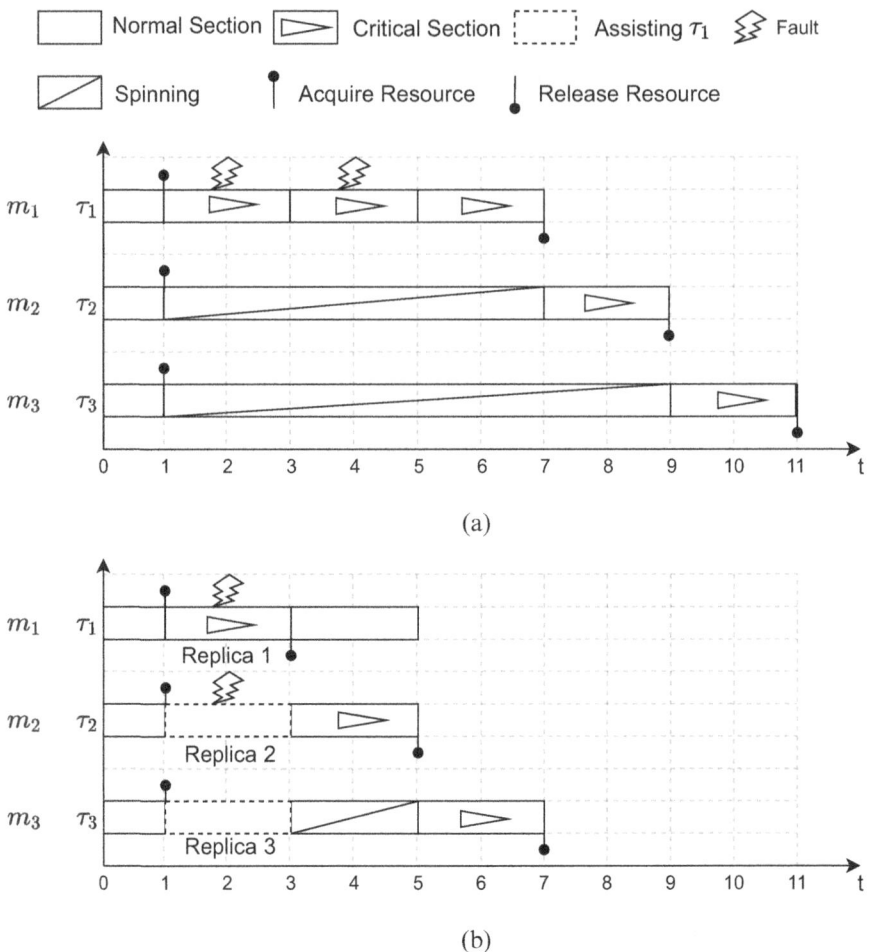

Fig. 8 A comparison between two fault-tolerance approaches under the same checkpoints setting. (**a**) Fault tolerance by simple segment re-execution. (**b**) The presented fault-tolerance method

Other tasks (τ_2 and τ_3) spin on their own cores and wait for the resource. However, τ_1 incurs two faults consecutively and re-executes its critical section twice. It finally releases the resource and leaves the FIFO queue at $t = 7$. τ_2 then becomes the head of the queue, which acquires the resource and starts its critical section from then.

With the application of the presented fault-tolerant approach, as shown in the Fig. 8b, the cores of τ_2 and τ_3 are utilized to execute $\tau_1's$ critical section in parallel instead of spinning. Although only one piece of the replica (i.e., Replica 3) is executed without faults, τ_1 can still continue its execution at $t = 3$. The chief principle of the fault-tolerant approach for critical sections is to replace wasted cycles of the spinning tasks in the FIFO queue to provide the reliability guarantee for each critical section in a single access, in pursuance of reducing the time spent on fault-tolerance and resource contention. For local resources, each task has to execute by itself as there exists no spinning tasks on remote cores.

4.2.3.4 Implementation and Run-Time Overhead

The implementation of the above approach requires the hardware architecture to have individual cache memory or dedicated memory space for each core to store replicas during the execution of the MSRP-FT, where most commercially off-the-shelf (COTS) architectures can satisfy. From the software aspect, a global scheduler will be adopted to communicate with tasks on different cores. For example, the scheduler will signal tasks to assist the head task (i.e. the resource holder) to execute the replicas in parallel. Once a successful result is submitted, the scheduler will signal other tasks to abandon the execution on replicas. Threads control methods such as *wait()* and *notify()* can be used to construct the above communication logic.

The feasibility of a task executing operations on behalf of other tasks has been validated in [51], in which once a task is preempted while spinning in the FIFO queue, the task behind it can acquire the lock first and execute the operation on behalf of the preempted task. Burns and Wellings [14] also briefly describes how the associated computations of the preempted task holder can be executed by the spinning tasks in parallel on different cores, but a detailed system design and implementation execution framework are not provided. Although the presented fault-tolerance approach is developed within a different context and serves a different purpose, that of reducing blocking time caused by resource faults, the above work has provided sufficient evidence towards the applicability and practicability of the presented approach.

Moreover, the setting of checkpoints can bring additional overheads in terms of execution time. However, there is a clear trade-off between the number of checkpoints being set and the final schedulability benefits of the presented approach. If the task has intensive resource requests (i.e. contains voluminous critical sections), the engineer can set fewer checkpoints in a flexible manner so that a balanced result can still be achieved between the time spent for each checkpoint and the advantage brought by the approach presented.

Finally, the presented fault-tolerance method can also be applied to other FIFO spin-based resource sharing protocol, e.g. MrsP [14]. The choice of MSRP made in this chapter is due to its non-preemptive spinning feature, which provides a strong guarantee to the resource-accessing and helping process. Under MSRP, spinning tasks are prevented from being preempted while assisting the resource holder, and hence avoids prolonging the helping process as well as over-complicated execution scenarios.

5 Real-Time TSN Scheduling for Automotive CPS

In this section, we present the frame-level FPS method for TSN scheduling and analysis. We present an overview in Sect. 5.1. Followed by scheduling of TSN with FPS in Sect. 5.2 and deferred queue in Sect. 5.3. A corresponding schedulability analysis is given in Sect. 5.4. Finally, the network and control co-design is formulated in Sect. 5.5 by period and control poles assignment.

5.1 Overview of Traffic Scheduling of TSN

In this section, we present an integrated solution that solves the controller-network co-design problem. Scheduling on a single TSN switch is considered and can be extended to the entire network. As we focus on the scheduling aspect, it is assumed the network communication is ideal: (i) the depth of the queues is sufficient, i.e., no traffic overflows; (ii) the channel is error-free and has a constant transmission rate. These ease the analysis and helps to understand the nature of the problem. Relaxing them in practice needs limited modifications and will be discussed in the future. The network is subjected to two basic traffic types: scheduled and unscheduled traffic, depending on a certain level of quality-of-service (QoS) is required or not. In this section, we focus on scheduled traffic and leave unscheduled traffic be transmitted using residual bandwidth with best effort.

TSN provides time synchronization and time-division transmission, which enables global scheduling through GCLs [63]. Although the schedule of TSN can be designed by hand, it soon becomes impractical as the network turns complex and more packets are added to the network. In this section, we specify the scheduling policy adopted for TSN while control systems are considered. The presented schedule minimizes the blocking of packets (including ones sent by control tasks), to improve schedulability and control performance. We then introduce a fine-grained response time analysis that bounds the worst-case latency of packets in a single Qbv switch. Below we first discuss the system model.

System Model The system contains N periodic packets[3] $\Gamma = \{\tau_1, \tau_2, \ldots, \tau_N\}$, including both control (Γ_c) and non-control packets (Γ_{nc}) sent by tasks from the application. Each packet τ_i is modelled as a 7-tuple $\{L_i, C_i, T_i, D_i, P_i, R_i, \Lambda_i\}$, representing the worst-case length of the packet L_i, transmission time C_i, period T_i, deadline D_i, priority P_i, worst-case latency R_i and the set of frames Λ_i in each release, respectively. Frames are transmitted in a non-preemptive fashion. A global

---

[3] Continuously released periodic packets will form a flow. For simplicity, we use these two terms interchangeably.

packet transmission rate v is applied to all packets, thus $C_i = L_i/v$ for τ_i. Each control packet is assigned with an implicit deadline i.e., $D_i = T_i$. To provide a more general network model for the system, the non-control packets can have arbitrary deadlines without any constraint imposed. As a consequence, at a given time instant there could be several instances of a non-control packet waiting for transmission in the switch. The priorities of all packets are assigned according to the deadline monotonic algorithm ($P_i > P_j$ if $D_i < D_j$), and each packet has a unique priority. In addition, the Maximum Transmission Unit (MTU) is considered, denoted as M, which defines the maximum data size allowed in a single transmission. For the ease of presentation, we denote M as the transmission time for sending data with a size equal to one MTU. Thus, each packet could be divided into a set of successive frames, i.e., $\Lambda_i = \{\lambda_i^1, \lambda_i^2, \ldots, \lambda_i^m\}$, with $m = \lceil L_i/M \rceil$. For a given frame λ_i^j, it inherits the analytical properties of τ_i (i.e., T_i, D_i and P_i), and has its own data length, L_i^j, and transmission time, C_i^j.

5.2   Scheduling Network Packets in TSN

In a typical Qbv switch, the network packets are queued by their arriving time (i.e., FIFO queuing) and are transmitted non-preemptively [35]. Traditionally, the synthesis of GCL schedule is performed using Satisfiability Modulo Theories (SMT) [20, 41] or Integer Linear Programming (ILP) [5]. The defined end-to-end latency imposes zero-jitter, however, with significantly reduced solution space. The scheduling in TSN networks with Quality-of-Service (QoS) requirements can be either performed at the queue level [63] or packet level [43]. With the queue-level scheduling, each FIFO queue in the Qbv switch is assigned with a priority, and packets in a queue with a higher priority are always transmitted first. However, as packets in each queue are transmitted strictly in a FIFO order, packets under the queue-level scheduling approach can incur substantial blocking, where packets with a tighter deadline but at the end of a queue cannot be favored. That is, with the queue-level scheduling, packets with different deadlines in the same FIFO queue are treated equally without concerning individual temporal requirements. For control systems, such a scheduling is not appropriate, as the delay for transmitting control packets can introduce significant impact on the control performance of the system. Thus, the packet-level (more precisely, the frame-level) scheduling is adopted to provide a finer-grained schedule, where each packet (and its frames) is scheduled strictly by its priority.

However, even with the packet-level scheduling, packets can still incur additional delay due to the FIFO queuing, as the actual transmission largely depends on the arriving time of the packets. In the worst-case, a late-arrived packet with a high priority can be blocked by all the released packets with lower priorities. To minimize the delay due to FIFO queuing, an alternative is to perform the scheduling off-line (i.e., prior to execution), with the complete knowledge of all packets in the system.[4] The offline scheduling can be performed by assuming all packets are arrived at the same time, with a packets transmission order obtained based on their priorities. If packets have different arrival times during run-time, a simple mechanism that defers the queuing of the early-arrived low-priority frames can be adopted, to maintain the queuing order obtained from the offline FPS-NP without imposing extra latency to packet transmission (see Sect. 5.3 for deferred queuing). By maintaining the offline packets transmission order during run-time, the blocking time of each packet during transmission can be minimized to one frame only, i.e., identical to the classic non-preemptive fixed-priority scheduling (FPS-NP) [23].

Based on the above discussion, to provide a fine-grained schedule and to minimize the delay due to the queuing problems, the scheduling adopted in this section is conducted before runtime on the frames of each packet in one hyper-period, with the scheduling decisions encoded into the GCL. Once a schedule is obtained, the frames can be statically allocated to the FIFO queues according to the schedule while the scheduling decisions can be mapped to the GCL to control the gates of all queues to achieve the desired execution order. To this end, the scheduling on TSN can be successfully mapped to the traditional FPS-NP, in which each packet is scheduled strictly by its priority and can be blocked maximum once during the entire transmission.

With the described scheduling approach, we avoid the packets queuing problem and can achieve the minimized delay for all packets, in the context of a Qbv switch. This is crucial for control systems as the resulting control performance can be affected by transmission delay for the control packets. To our best knowledge, this is one of the earliest work targeting at control systems in which the timeliness and performance are sensitive to the transmission delay of certain critical (i.e., control and non-control) packets. For the non-control packets, meeting their timing requirements is essential for guaranteeing the system correctness, whereas minimizing transmission delay of the time-triggered control packets are essential crucial for control performance.

For unscheduled packet flows that do not have a temporal requirements, the traffics can be scheduled using residual bandwidth left by the critical traffics with time-aware shapers [52, 53] and queue partitioning. Supporting such flows has been well-described by the above work, and will not be re-presented in this section. Targeting at such systems, a complete scheduling solution is presented that minimizes the transmission delay for all packets, in the context of the TSN Qbv switch. Last but not least, different from [20], our approach makes no assumption on the isolation of incoming packets and the construction of the GCL, e.g., isolating certain queues for a specific packet type, to provide a more general approach for using TSN in control systems.

---

[4] Such an approach is feasible as the packets are deterministic i.e., the packets sent by each task are known a prior with periodic release.

5.3   Deferred Queue

As described in Sect. 5.2, for packets with different arriving times, a mechanism is required to delay the queuing of the early-arrived low priority packets so that the minimized blocking can be guaranteed. To achieve this, a *deferred queue* with priority ordering is introduced into the Qbv switch, which is integrated into the packet filtering unit (see Fig. 3) for holding early-arrived packets temporarily, until they can be added into the scheduled queues with a correct order.

Assuming simultaneous release for all packets at the start of the system, the offline FPS-NP schedule can produce a well-planned transmission order for all packet instances released in one hyperperiod, in which each packet (a set of successive frames) is scheduled strictly based on priority. For this schedule, the blocking of each packet is minimized, as in the worst case, the ready packet with the highest priority can start transmitting after the currently transmitting frame of a low priority packet has completed. During run-time, this offline scheduling order is encoded into the priority filtering unit, which provides a reference of the expect order for incoming packets.

For each incoming packets, the priority filter examines whether this packets arrives by the expected order, i.e., all its previous packets with a higher priority have arrived. If so, this packet is dispatched to the scheduled queues immediately, at which it will be select to transmit by GCL. Otherwise (i.e., certain previous high priority packets haven't arrived yet), this packet is hold by the priority filter until (a) the missing packets arrives or (b) the scheduled queues are empty and this packet has the highest priority among all the deferred packets.

Note that the condition (b) can lead to a transmission order different from the expected one, as certain packets can be transmitted before a late-arriving higher priority packet. However, this does not introduce extra delay and can help increasing the throughput. With the deferred queuing, it is possible that all scheduled queues are empty while some packets are stored in the priority filter. Under this situation, the priority filter selects the packet at the head of the queue (i.e., with the highest priority) and send its frames into the scheduled queue for a direct transmission one by one, until a higher priority packet arrives. This guarantees that the transmission never stops as long as there exist waiting packets (either in the priority filter or the scheduled queues). In addition, for the late arriving high priority packet, its blocking is still at most C_i^j, where it can be transmitted directly after the currently-transmitting frame.

5.4   Worst-Case Response Time Analysis

With the scheduling in TSN mapped to the traditional FPS-NP, the worst-case response time for transmitting a packet in a single Qbv switch can be obtained, which bounds the time duration from when the packet enters into the switch to when the packet is transmitted. Due to the different deadline constraints of the control and non-control packets (i.e., implicit and arbitrary deadlines respectively), different analysis techniques are applied for each packet type. However, as both control and non-control packets are scheduled strictly by the FPS-NP, the basic philosophy for analyzing both types of packets is similar to that in [23], but with modifications and improvements in order to reflect the unique features of the Qbv switch and to support the analysis at the frame level.

The response time equation of a packet τ_i is given in the following equation for both control and non-control packets:

$$
R_i = \max_{\forall \lambda_i^j \in \Lambda_i}
\begin{cases}
R_i^j(0), & \text{if } \tau_i \in \Gamma_c \\
\max_{n=0\ldots\left\lceil \frac{t_i+J_i}{T_i} \right\rceil - 1} (R_i^j(n)), & \text{if } \tau_i \in \Gamma_{nc}
\end{cases}
\tag{2}
$$

In Eq. (2), $R_i^j(n)$ denotes the response time for transmitting the nth instance of frame λ_i^j in τ_i's busy period t_i, and J_i denotes the queuing time, i.e., the time window from when the first frame of τ_i reaches the Qbv Switch, until when the last frame is queued. $\left\lceil \frac{t_i+J_i}{T_i} \right\rceil$ gives the total number of times that a non-control packet can be sent within its busy period [23].

The analysis of a control packet is relatively straight forward, as at any given time, there can only exist one instance of a control packet in the system i.e., implicit deadlines. Thus, the worst-case response time of a control packet can be safely bounded by computing the maximum response time of all its frames.[5] However, for a non-control packet, multiple instances of each of its frames can co-exist due to the arbitrary deadline. Thus, the response time of a frame (with an arbitrary deadline) must be obtained by computing the maximum response time of all its instances within the busy period t_i.

Similar to [23], the busy period of a non-control packet is computed by Eq. (3), where B_i gives the worst-case blocking that τ_i can experience due to transmitting a low priority frame and $hep(i)$ refers to all indices of packets that have equal or higher priorities than P_i, including i. The recursive calculation can starts with $t_i = B_i + C_i$, and is guaranteed to converge [23], given that the total utilization for packets in $hep(i)$ is less than 1, i.e., $\sum_{\forall j \in hep(i)} (C_j / T_j) \leq 1$. We later decompose B_i in Eq. (6).

$$
t_i = B_i + \sum_{\forall k \in hep(i)} \left\lceil \frac{t_i + J_k}{T_k} \right\rceil C_k
\tag{3}
$$

---

[5] From Eq. (2), the response time of a packet equals to the response time of its last frame in each transmission, which takes into account the delay for transmitting the previous frames in one transmission.

The response time of a frame is bounded by Eq. (4), in which J_i^j denotes the time to en-queue frame λ_i^j, W_i^j gives the maximum queuing delay that λ_i^j can incur in a FIFO queue before it is selected to be transmitted and C_i^j denotes its transmission time. The time for queuing λ_i^j into a FIFO queue also contains the enqueue time of frames of τ_i that are prior to λ_i^j in one transmission. In addition, for the non-control frames, $n \cdot T_i$ is subtracted as this is the arrival time of its nth instance, relative to the start of the busy period. Note, for control frames, n is always 0.

$$R_i^j(n) = \sum_{q \in [1,j]} J_i^q + W_i^j(n) + C_i^j - n \cdot T_i \tag{4}$$

Equation (5) gives the queuing delay W_i^j of frame λ_i^j, where $hp(i)$ returns a set of packets with a priority strictly higher than P_i. This equation is also applicable to either control or non-control frames, with $n = 0$ for all control frames. Figure 9 provides an example illustrating the worst-case delay of the third ($n = 2$) instance of the second frame (i.e., $j = 2$) in packet τ_i. As shown in the figure, in the worst case, the frame (in bold) has to wait for five types of other frames to transmit before it can start, which are mapped to four types of delay, as follows. In the worst case, a frame can incur four sources of delay when waiting in a FIFO queue: (i) the blocking caused by a low-priority frame that is currently transmitting i.e., B_i; (ii) the delay by τ_i's frames prior to λ_i^j (with potential existence of multiple instances); (iii) the delay by previous instances of λ_i^j and the frames after λ_i^j in each τ_i's instance sent before λ_i^j; and (iv) the interference from the frames of each packet with a higher priority than P_i. Note that (iii) accounts for the delay cause by both the previous instances of λ_i^j itself and the frames after λ_i^j in previous instances. These delays are captured by the equation respectively.

$$W_i^j(n) = B_i + (n + 1) \cdot \sum_{q \in [1,j-1]} C_i^q + n \cdot \sum_{q \in [j,|\Lambda_i|]} C_i^q$$

$$+ \sum_{\forall \lambda_k^q \in \Lambda_k, \forall k \in hp(i)} \left\lceil \frac{W_i^j(n) + J_k^q}{T_k} \right\rceil C_k^q \tag{5}$$

Finally, B_i is given by Eq. (6), where $lp(i)$ returns the packets with a priority lower than P_i. The maximum blocking time that τ_i (and any of its frames) can incur is the longest transmission time among the frames of all the lower priority packets.

$$B_i = \max_{\forall \lambda_k^q \in \Lambda_k, \forall k \in lp(i)} (C_k^q) \tag{6}$$

Equations (2)–(6) summarises the response time analysis for bounding the worst-case transmission latency (i.e., the response time) of packets in a Qbv

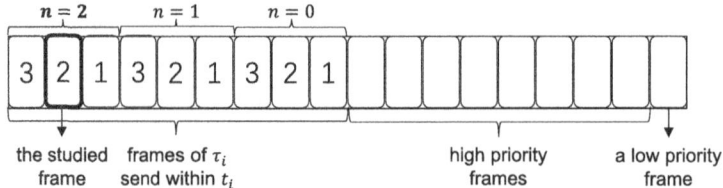

Fig. 9 The worst-case delay of a frame, which is caused by a low priority frame, high priority packets, instances of τ_i's frames prior to λ_i^j, previous instances of λ_i^j and previous instance of τ_i's frames after λ_i^j

switch for time-critical control systems. The analysis considers both implicit and arbitrary deadlines for different packet types and is fine-grained, which provides the worst-case transmission latency of each frame. Arguably, by intuition, a trivial modification that treats each frame as an independent task can be applied in an existing packet-level analysis (e.g., the one in [24]), to support the analysis at the frame-level. However, additional techniques are still required to guarantee the correct transmission order between frames that belong to the same packet and instance so that the transmission time of each individual frame can be obtained. This is achieved in our analysis by Eq. (5), which carefully examines the transmission order of different types of frames (including the ones in τ_i) and provides a tighter upper bound compared to a packet-level analysis.

The presented analysis and scheduling techniques for a single switch can be extended to support the network topology level with multiple switches and end-nodes. For the presented method, it can be implemented in each switch. For a given switch, the presented schedule takes all packets that will go through this switch and then produced a static schedule. In addition, the deferred queue is applied in each switch to handle the case in which low priority packets arrive earlier than expected. To compute the end-to-end worst-case transmission time of a packet τ_i that travels through more than one switches, the input packets of each of the switches should be given and the worst-case delay of τ_i in each switch can be effectively upper bounded by summing the worst-case delay it can incur in each switch by the above analysis.

However, with only one switch, the worst-case delay of a packet can be bounded by considering all the input packets with a synchronous release at the begin of the system. This assumption, however, may not hold in the scenario of multiple switches, in which the actual arrival time of a packet at a given switch depends on the delay it incurs at the previous switches. Thus, the analysing approach above would contain certain degree of pessimism as not all the input packets in a switch will cause a delay on τ_i, depending on their arrival times.

5.5   Controller Synthesis and Period Allocation

For a safety-critical autonomous system, for example, a self-driving car, the control functions are crucial and should always be a major concern. Further to the introduced scheduling and analysis that guarantee the timing of control packets, a well-designed controller is also required, in order to satisfy the control performance requirement and even maximize it under the schedulability constraint of the network.

Most real-time controllers targeting settling time (which will be formally defined later in this section) can run at different frequencies [3, 21, 22]. In the TSN context, this rate is bounded by (i) the maximum transmission capability; (ii) the lowest control performance requirement. Hence, there exists an optimized operational point that would produce acceptable network schedulability with maximized control performance.

5.5.1   Control Model

For a linear-time-invariant (LTI) controlled plant, its system dynamics can be described using the following differential equations:

$$\dot{x}(t) = Ax(t) + Bu(t), y(t) \quad = Hx(t) \tag{7}$$

in which A, B and H are system matrices that represent the system physical properties; $x(t)$ is the system state(s); $y(t)$ is the system output(s) and $u(t)$ is the control input(s). Assuming the sampling time is T_s and the sensor-to-actuator delay is within one sampling period, at discrete time instant k, the system dynamics evolve with the following equations:

$$x(k+1) = A_d x(k) + B_d u(k-1), y(k) \quad = Hx(k) \tag{8}$$

where $u(-1) = 0$ for $k = 0$ and

$$A_d = e^{A \cdot T_s}, B_d = \int_0^{T_s} e^{A\tau} d\tau \cdot B \tag{9}$$

To further simplify the equation, define an augmented variable z as: $z(k) = \begin{bmatrix} x(k) & u(k-1) \end{bmatrix}^T$, and substitute $x(k), u(k)$ with $z(k)$ in Eq. (8):

$$z(k+1) = \begin{bmatrix} A_d & B_d \\ 0 & 0 \end{bmatrix} z(k) + \begin{bmatrix} 0 \\ 1 \end{bmatrix} u(k) \tag{10}$$

Assuming a full state-feedback controller is used, the control input $u(k)$ is calculated by:

$$u(k) = -Kz(k) + Fr(k) \tag{11}$$

where K is the feedback gain, F is the feedforward gain and $r(k)$ is the reference. By combining Eqs. (10) and (11), the system equation therefore becomes:

$$z(k+1) = \underbrace{(A_d - B_d K)}_{A_{cl}} z(k) + B_d Fr(k) \tag{12}$$

To satisfy control stability, all the eigenvalues of the closed loop dynamic matrix, i.e. A_{cl} in Eq. (12), have to be inside the unit circle. The exact value of A_d and B_d is dependent on the sampling period T_s as seen from Eq. (9), which is equal to the period of the control packet, T_i. This control model will be used through the rest of this section.

5.5.2 Problem Definition

We use settling time (t_s) as the index of quality-of-control (QoC), which is widely used in control engineering as a compulsory design requirement [16]. Settling time is defined as the time duration from when a control system is subjected to a disturbance to when it enters steady-state, i.e., the current output has reached and stays within 5% deviation of the targeted output. There is an upper bound requirement on the settling time, e.g., the settling time of a control system should not be longer than 0.5 seconds.

Finding an optimal period is crucial for (i) guaranteeing the performance of the controller itself; and (ii) ensuring enough residual time slots for non-control-related packets so they can also meet their deadlines. Based on the aforementioned objectives and constraints, the period assignment problem can be solved as an optimization problem, which is formulated as follows:

$$\text{minimize} \quad \mathcal{J} = \sum_{\mathbb{D}} w_j \cdot t^*_{s,j}$$

$$\text{subject to} \quad R_i \leq D_i, t_{s,j} \leq t^+_{s,j} |u_j(k)| \leq u_{max}, T_i = n \cdot t_{gcd}, n \in N^+ \tag{13}$$

$$\text{where} \quad i \in \Gamma, \quad j \in \Gamma_c$$

where $w_j \in (0, 1]$ is the weight (i.e., relative importance) of the corresponding control task and $\sum w_j = 1$; $t^*_{s,j} \in [0, 1]$ is the normalized settling time of the jth controller; $\mathbb{D}$ represents the solution space of all poles that can ensure control stability; $t_{s,j}$ is the settling time of the jth controller, and $t^+_{s,j}$ is the maximum allowed settling time; $u_j(k)$ is input at discrete instance k, which is constrained by u_{max} as the maximum input threshold; The last constraint defines the time-granularity of a feasible period. To benefit from harmonic periods and to reduce the size of the GCL table, each T_i must be an integer multiple of t_{gcd}, the greatest

---

Algorithm 3: Periods and control poles assignment

---

1 **Input:** $\Gamma = \{\Gamma_c, \Gamma_{nc}\}$
2 **Output:** schedulability, S^*
3 **Initialise:** feasible and best solutions: $\mathbb{S}^f = \emptyset$, $S^* = \emptyset$
    /* construct candidate solutions: */
4 formulate the solution space: $\mathbb{S} = \{S_1, S_2, \ldots, S_n\}$.
    /* explore each candidate: */
5 **for** S_k *in* $\mathbb{S}$ **do**
6   **if** $RTA\_schedulability(\Gamma^k)$ *is True* **then**
7   **for** j *in* Γ_c^k **do**
8              $\{t_{s,j}, u_j\} = \text{pso\_find\_control\_parameters}(T_j)$
9   **end**
10   **if** $\forall j$ *in* Γ_c^k: $t_{s,j} \le t_{s,j}^+$ *and* $|u_j| \le u_{max}$ **then**
11              $\mathcal{J}_k = \sum w_j \cdot t_{s,j}$
12              $S_k \to \mathbb{S}^f$
13   **end**
14   **end**
15 **end**
    /* find the best candidate solution: */
16 **for** S_k *in* $\mathbb{S}^f$ **do**
17   **if** $\mathcal{J}_k < \mathcal{J}^*$ **then**
18          $S^* = S_k$
19   **end**
20 **end**
    /* return feasibility: */
21 **if** S^* *is not* $\emptyset$ **then**
22   return (*feasible*, S^*)
23 **else**
24   return (*infeasible*, $\emptyset$)
25 **end**

---

common divisor of all the packet periods. This is in accordance with common practice.

5.5.3 Solving the Network and Control Co-Design Problem

In a typical control application, while the periods of non-control-related packets are inflexible, the control-related packets often have adjustable periods. This additional flexibility allows fine tuning of controller periods to achieve the best overall performance (defined as in Eq. (13)). To solve the defined problem, a controller's period and its corresponding parameters under that period both have to be decided. These two steps are dependent on each other but can be decomposed into two sub-problems, i.e., the optimization process needs to (i) find the feasible periods that can satisfy schedulability constraints; (ii) find the controller parameters under the feasible periods that would satisfy control stability and minimal performance requirement, and on top of that, maximize the control performance as much as possible.

For the first problem, due to the existence of harmonic periods and that the number of control tasks is often small, the search space is manageable and thus can be solved through exhaustive search. For larger scale problems, heuristic methods can be used instead to find the feasible period configurations.

For the second problem, as pole placement for the minimum settling time under input constraints is a non-convex and non-linear problem, the solution space cannot be searched easily. We use Particle Swarm Optimization (PSO) to find the optimal controller parameters (by pole placement [16]) under certain sampling period that can minimize the settling time, while given the control performance and input saturation as constraints. PSO is a population-based optimization approach for iterative improvement of candidate solutions given a non-linear non-convex objective function and a metric of quality [49].

The optimization process is given in Algorithm 3. The solution space is first formulated in Line 4. The schedulability is then tested (Line 6) to obtain potential period configurations, and under each period configuration, the optimal poles of each control task can be found through PSO (Line 8). To speedup the process, the optimal poles under the feasible range of periods can be obtained in advance. The identified configuration is appended into the feasible solutions provided that the minimum control performance and the input constraints are both satisfied (Line 10-13). Finally, the best candidate that has the minimum $\mathcal{J}$ is selected from all the feasible solutions (Line 16-20). No feasible solution is found if $S^* = \emptyset$, in which case the algorithm fails to find a solution that satisfies all the constraints.

6 Conclusion

This chapter introduces the state-of-the-art techniques which cover three major directions of scheduling and analyzing autonomous systems. The presented solutions range from DAG task scheduling, and reliable resource sharing, to in-vehicle TSN networking. The goal is to provide autonomous systems with high-performance hard real-time scheduling, reliable resource sharing, and deterministic networking scheduling.

For scheduling and analyzing DAG tasks in autonomous systems, a CPC model is constructed to capture the two key factors of a DAG structure: dependency and parallelism. Then, a rule-based scheduling method is presented which maximizes node parallelism to improve the schedulability of single DAG tasks.

To provide reliable resource sharing in multiprocessor mixed-criticality systems, this chapter describes a fault-tolerance solution for multiprocessor MCS with shared resources. The presented system execution model and fault-tolerance resource sharing protocol reduces the blocking time imposed by guaranteeing reliable resource sharing.

To provide hard real-time guarantee for network, we introduced a network scheduling model using non-preemptive fixed-priority scheduling (FPS-NP) and the mapping of the schedule into the TSN gate control list. The schedulability of the network is discussed using non-preemptive response-time analysis with the consideration of multi frames and unconstrained deadlines. An optimization method is also proposed that could find the feasible solution with maximized overall quality of control constrained by network schedulability.

References

1. Al-bayati, Z., Caplan, J., Meyer, B.H., Zeng, H.: A four-mode model for efficient fault-tolerant mixed-criticality systems. In: IEEE Design, Automation & Test in Europe Conference & Exhibition (DATE) (2016)
2. Alfranseder, M., Deubzer, M., Justus, B., Mottok, J., Siemers, C.: An efficient spin-lock based multi-core resource sharing protocol. In: IEEE International Performance Computing and Communications Conference (IPCCC) (2014)
3. Arzén, K.-E., Cervin, A., Eker, J., Sha, L.: An introduction to control and scheduling co-design. In: Proceedings of the 39th IEEE Conference on Decision and Control, vol. 5, pp. 4865–4870. IEEE, Piscataway (2000)
4. Baleani, M., Ferrari, A., Mangeruca, L., Sangiovanni-Vincentelli, A., Peri, M., Pezzini, S.: Fault-tolerant platforms for automotive safety-critical applications. In: Proceedings of the 2003 International Conference on Compilers, Architecture and Synthesis for Embedded Systems, pp. 170–177 (2003)
5. Bansal, B.: Divide-and-conquer scheduling for time-sensitive networks. Master's Thesis, University of Stuttgart (2018)
6. Baruah, S.K., Burns, A., Davis, R.I.: Response-time analysis for mixed criticality systems. In: IEEE Real-Time Systems Symposium (RTSS) (2011)
7. Baruah, S., Bonifaci, V., Marchetti-Spaccamela, A., Stougie, L., Wiese, A.: A generalized parallel task model for recurrent real-time processes. In: Real-Time Systems Symposium, pp. 63–72 (2012)
8. Becker, M., Dasari, D., Mubeen, Behnam, S.M., Nolte, T.: Synthesizing job-level dependencies for automotive multi-rate effect chains. In: International Conference on Embedded and Real-Time Computing Systems and Applications, pp. 159–169 (2016)
9. Bhuiyan, A., Sruti, S., Guo, Z., Yang, K.: Precise scheduling of mixed-criticality tasks by varying processor speed. In: Proceedings of the 27th International Conference on Real-Time Networks and Systems, pp. 123–132 (2019)
10. Bhuiyan, A., Yang, K., Arefin, S., Saifullah, A., Guan, N., Guo, Z.: Mixed-criticality real-time scheduling of gang task systems. Real-Time Syst. **57**(3), 268–301 (2021)
11. Brandenburg, B.B.: Multiprocessor real-time locking protocols: a systematic review (2019). arXiv:1909.09600
12. Burns, A.: The application of the original priority ceiling protocol to mixed criticality systems. In: Proceedings of ReTiMiCS, RTCSA (2013)
13. Burns, A., Baruah, S.: Towards a more practical model for mixed criticality systems. In: Workshop on Mixed-Criticality Systems (colocated with RTSS) (2013)
14. Burns, A., Wellings, A.J.: A schedulability compatible multiprocessor resource sharing protocol–mrsp. In: IEEE Euromicro Conference on Real-Time Systems (ECRTS). IEEE, Piscataway (2013)
15. Buttazzo, G., Cervin, A.: Comparative assessment and evaluation of jitter control methods. In: Conference on Real-Time and Network Systems, pp. 163–172 (2007)

16. Chang, W., Chakraborty, S.: Resource-aware automotive control systems design: a cyber-physical systems approach. Found. Trends Electron. Des. Autom. **10**(4), 249–369 (2016)
17. Chang, S., Zhao, X., Liu, Z., Deng, Q.: Real-time scheduling and analysis of parallel tasks on heterogeneous multi-cores. J. Syst. Architect. **105**, 101704 (2020)
18. Chen, P., Liu, W., Jiang, X., He, Q., Guan, N.: Timing-anomaly free dynamic scheduling of conditional DAG tasks on multi-core systems. ACM Trans. Embed. Comput. Syst. **18**(5), 1–19 (2019)
19. Chen, G., Guan, N., Huang, K., Yi, W.: Fault-tolerant real-time tasks scheduling with dynamic fault handling. J. Syst. Architect. **102**, 101688 (2020)
20. Craciunas, S.S., Oliver, R.S., Chmelík, M., Steiner, W.: Scheduling real-time communication in IEEE 802.1 Qbv time sensitive networks. In: Proceedings of the 24th International Conference on Real-Time Networks and Systems, pp. 183–192. ACM, New York (2016)
21. Dai, X., Burns, A.: Period adaptation of real-time control tasks with fixed-priority scheduling in cyber-physical systems. J. Syst. Architect. **103**, 101691 (2020)
22. Dai, X., Chang, W., Zhao, S., Burns, A.: A dual-mode strategy for performance-maximisation and resource-efficient cps design. ACM Trans. Embed. Comput. Syst. **18**(5s), 85 (2019)
23. Davis, R.I., Kollmann, S., Pollex, V., Slomka, F.: Controller area network (CAN) schedulability analysis with FIFO queues. In 2011 23rd Euromicro Conference on Real-Time Systems, pp. 45–56. IEEE, Piscataway (2011)
24. Davis, R.I., Kollmann, S., Pollex, V., Slomka, F.: Schedulability analysis for Controller Area Network (CAN) with FIFO queues priority queues and gateways. Real-Time Syst. **49**(1), 73–116 (2013)
25. Fonseca, J., Nelissen, G., Nélis, V.: Improved response time analysis of sporadic DAG tasks for global FP scheduling. In: International Conference on Real-Time Networks and Systems, pp. 28–37 (2017)
26. Forget, J., Boniol, F., Grolleau, E., Lesens, D., Pagetti, C.: Scheduling dependent periodic tasks without synchronization mechanisms. In: Real-Time and Embedded Technology and Applications Symposium, pp. 301–310 (2010)
27. Gai, P., Lipari, G., Di Natale, M.: Minimizing memory utilization of real-time task sets in single and multi-processor systems-on-a-chip. In: IEEE Real-Time Systems Symposium (RTSS) (2001)
28. Graham, R.L.: Bounds on multiprocessing timing anomalies. J. Appl. Math. **17**(2), 416–429 (1969)
29. Guan, F., Qiao, J., Han, Y.: DAG-fluid: a real-time scheduling algorithm for DAGs. IEEE Trans. Comput. **70**, 471–482 (2020)
30. Guo, Z., Yang, K., Vaidhun, S., Arefin, S., Das, S.K., Xiong, H.: Uniprocessor mixed-criticality scheduling with graceful degradation by completion rate. In: 2018 IEEE Real-Time Systems Symposium (RTSS), pp. 373–383. IEEE, Piscataway (2018)
31. Han, J.-J., Tao, X., Zhu, D., Yang, L.T.: Resource sharing in multicore mixed-criticality systems: utilization bound and blocking overhead. IEEE Trans. Parallel Distrib. Syst. **28**, 3626–3641 (2017)
32. Haque, M.A., Aydin, H., Zhu, D.: Real-time scheduling under fault bursts with multiple recovery strategy. In: IEEE Real-Time and Embedded Technology and Applications Symposium (RTAS) (2014)
33. He, Q., Jiang, X., Guan, N., Guo, Z.: Intra-task priority assignment in real-time scheduling of DAG tasks on multi-cores. IEEE Trans. Parallel Distrib. Syst. **30**(10), 2283–2295 (2019)
34. Huang H.-M., Gill, C., Lu, C.: Implementation and evaluation of mixed-criticality scheduling approaches for sporadic tasks. ACM Trans. Embed. Comput. Syst. **13**(4s), 1–25 (2014)
35. IEEE 802.1 Task Group: Standard for local and metropolitan area networks – bridges and bridged networks - amendment 25: enhancements for scheduled traffic. Standard, IEEE (2016)
36. Kehrer, S., Kleineberg, O., Heffernan, D.: A comparison of fault-tolerance concepts for IEEE 802.1 time sensitive networks (TSN). In: Proceedings of the 2014 IEEE Emerging Technology and Factory Automation (ETFA), pp. 1–8. IEEE, Piscataway (2014)

37. Lin, H., Li, M.-F., Jia, C.-F., Liu, J.-N., An, H: Degree-of-node task scheduling of fine-grained parallel programs on heterogeneous systems. J. Comput. Sci. Technol. **34**(5), 1096–1108 (2019)
38. Meixner, A., Bauer, M.E., Sorin, D.: Argus: low-cost, comprehensive error detection in simple cores. In: IEEE/ACM International Symposium on Microarchitecture (MICRO) (2007)
39. Melani, A., Bertogna, M., Bonifaci, V., Marchetti-Spaccamela, A., Buttazzo, G.C.: Response-time analysis of conditional DAG tasks in multiprocessor systems. In: Euromicro Conference on Real-Time Systems, pp. 211–221 (2015)
40. Musuvathi, M., Qadeer, S., Ball, T., Basler, G., Nainar, P.A., Neamtiu, I.: Finding and reproducing heisenbugs in concurrent programs. In OSDI'08: Proceedings of the 8th USENIX Conference on Operating Systems Design and Implementation (2008)
41. Oliver, R.S., Craciunas, S.S., Steiner, W.: IEEE 802.1 Qbv gate control list synthesis using array theory encoding. In: 2018 IEEE Real-Time and Embedded Technology and Applications Symposium (RTAS), pp. 13–24. IEEE, Piscataway (2018)
42. Pathan, R.M.: Fault-tolerant and real-time scheduling for mixed-criticality systems. Real-Time Syst. **50**, 509–547 (2014)
43. Piro, G., Grieco, L.A., Boggia, G., Fortuna, R., Camarda, P.: Two-level downlink scheduling for real-time multimedia services in LTE networks. IEEE Trans. Multimedia **13**(5), 1052–1065 (2011)
44. Punnekkat, S., Burns, A., Davis, R.I.: Analysis of checkpointing for real-time systems. Real-Time Syst. **20**, 83–102 (2001)
45. Safari, S., Ansari, M., Ershadi, G., Hessabi, S.: On the scheduling of energy-aware fault-tolerant mixed-criticality multicore systems with service guarantee exploration. IEEE Trans. Parallel Distrib. Syst. **30**, 2338–2354 (2019)
46. Saidi, S.E., Pernet, N., Sorel, Y.: Automatic parallelization of multi-rate fmi-based co-simulation on multi-core. In: Symposium on Theory of Modeling and Simulation , p. 5 (2017)
47. Serrano, M.A., Melani, A., Bertogna, M., Quiñones, E.: Response-time analysis of DAG tasks under fixed priority scheduling with limited preemptions. In: Design, Automation & Test in Europe Conference & Exhibition, pp. 1066–1071 (2016)
48. Sha, L., Rajkumar, R., Lehoczky, J.P.: Priority inheritance protocols: an approach to real-time synchronization. IEEE Trans. Comput. **39**, 1175–1185 (1990)
49. Shi, Y., et al.: Particle swarm optimization: developments, applications and resources. In: Proceedings of the 2001 Congress on Evolutionary Computation, vol. 1, pp. 81–86. IEEE, Piscataway (2001)
50. Spainhower, L., Gregg, T.A.: IBM s/390 parallel enterprise server G5 fault tolerance: a historical perspective. IBM J. Res. Develop. **43**, 863–873 (1999)
51. Takada, H., Sakamura, K.: A novel approach to multiprogrammed multiprocessor synchronization for real-time kernels. In: IEEE Proceedings Real-Time Systems Symposium (RTSS) (1997)
52. Thangamuthu, S., Concer, N., Cuijpers, P.J.L., Lukkien, J.J.: Analysis of ethernet-switch traffic shapers for in-vehicle networking applications. In 2015 Design, Automation & Test in Europe Conference & Exhibition (DATE), pp. 55–60. IEEE, Piscataway (2015)
53. Thiele, D., Ernst, R., Diemer, J.: Formal worst-case timing analysis of ethernet tsn's time-aware and peristaltic shapers. In: 2015 IEEE Vehicular Networking Conference (VNC), pp. 251–258. IEEE, Piscataway (2015)
54. Topcuoglu, H., Hariri, S., Wu, M.-Y.: Performance-effective and low-complexity task scheduling for heterogeneous computing. IEEE Trans. Parallel Distrib. Syst. **13**(3), 260–274 (2002)
55. Tsai, T.-Y., Chung, Y.-L., Tsai, Z.: Introduction to packet scheduling algorithms for communication networks. In: Communications and Networking. IntechOpen, London (2010)
56. Upasani, G., Vera, X., González, A.: Setting an error detection infrastructure with low cost acoustic wave detectors. In: IEEE Annual International Symposium on Computer Architecture (ISCA) (2012)

57. Upasani, G., Vera, X., González, A.: Avoiding core's due SDC via acoustic wave detectors and tailored error containment and recovery. In: ACM/IEEE International Symposium on Computer Architecture (ISCA) (2014)
58. Vaidyanathan, K., Trivedi, K.S.: Extended classification of software faults based on aging. In: Fast Abstract, International Symposium on Software Reliability Engineering, Hong Kong. Citeseer (2001)
59. Verucchi, M., Theile, M., Caccamo, M., Bertogna, M: Latency-aware generation of single-rate DAGs from multi-rate task sets. In: Real-Time and Embedded Technology and Applications Symposium, pp. 226–238 (2020)
60. Vincentelli, A.S., Giusto, P., Pinello, C., Zheng, W., Natale, M.D.: Optimizing end-to-end latencies by adaptation of the activation events in distributed automotive systems. In: Real Time and Embedded Technology and Applications Symposium, pp. 293–302 (2007)
61. Zhao, Q., Gu, Z., Zeng, H.: HLC-PCP: aresource synchronization protocol for certifiable mixed criticality scheduling. IEEE Embed. Syst. Lett. **6**, 8–11 (2013)
62. Zhao, S., Garrido, J., Burns, A., Wellings, A.J.: New schedulability analysis for MRSP. In: IEEE Embedded and Real-Time Computing Systems and Applications (ERTCSA) (2017)
63. Zhao, L., Pop, P., Craciunas, S.S.: Worst-case latency analysis for IEEE 802.1 Qbv time sensitive networks using network calculus. IEEE Access **6**, 41803–41815 (2018)

Distributed Coordination and Centralized Scheduling for Automobiles at Intersections

Yi-Ting Lin, Chung-Wei Lin, Iris Hui-Ru Jiang, and Changliu Liu

1 Introduction

The fundamental goal of vehicles is to perform transportation tasks between sources and destinations safely and efficiently. The conflicts between vehicles occur when the corresponding vehicles intend to pass through a location at the same time, and intersections are one of the most common conflicting scenarios. Traditionally, traffic lights, stop signs, and priorities defined by traffic rules can be applied to resolve conflicts at intersections. As the technology advances, connected and autonomous vehicles (CAVs) provide a revolutionary solution at intersections, where:

- **Connectivity** provides sufficient information between vehicles and/or roadside units so that a safe and efficient passing order of vehicles can be decided.
- **Autonomy** provides precise control so that the decided passing order of vehicles can be performed.

Y.-T. Lin
Graduate Institute of Electronics Engineering, National Taiwan University, Taipei, Taiwan
e-mail: f07943102@ntu.edu.tw

C.-W. Lin (✉)
Department of Computer Science and Information Engineering, National Taiwan University,
Taipei, Taiwan
e-mail: cwlin@csie.ntu.edu.tw

I. H.-R. Jiang
Department of Electrical Engineering, National Taiwan University, Taipei, Taiwan
e-mail: huirujiang@ntu.edu.tw

C. Liu
Robotics Institute, Carnegie Mellon University, Pittsburgh, PA, USA
e-mail: cliu6@andrew.cmu.edu

© The Author(s), under exclusive license to Springer Nature Switzerland AG 2023   81
V. K. Kukkala, S. Pasricha (eds.), *Machine Learning and Optimization Techniques for
Automotive Cyber-Physical Systems*, https://doi.org/10.1007/978-3-031-28016-0_3

In this chapter, we consider connected and autonomous vehicles at intersections and introduce approaches solving the problem of intersection management, also known as the problem of conflict resolution in a more general perspective. The approaches are categorized into two categories:

- **Distributed coordination**, where vehicles coordinate and then decide a passing order of vehicles separately.
- **Centralized scheduling**, where a centralized unit, called intersection manager, decides a passing order of vehicles and provides the instructions to vehicles.

No matter an approach is distributed or centralized, also no matter from the perspective of an individual vehicle or the overall transportation system, the approach should provide the following properties:

- **Feasibility**. The decided passing order of vehicles and the corresponding trajectories, including spatial and temporal constraints, much be physically achievable by the vehicles.
- **Safety** (collision-freeness). The deciding passing order must resolve the conflict for each pair of vehicles which intend to pass through a same location. Here, we define a conflict zone, and the safety requirement is that there is at most one vehicle occupying a conflict zone at the same time.
- **Liveness** (deadlock-freeness). The deciding passing order must not lead to a deadlock, i.e., an infinite waiting between multiple vehicles.
- **Stability**. The passing order must be stable along with the time line.
- **Efficiency**. The passing order should try to optimize the traffic efficiency or minimize delays of vehicles, i.e., allow vehicles to pass through intersections as soon as possible.
- **Real-Time Decision**. The passing order should be decided in real time without delaying vehicles due to waiting the decision or the corresponding instructions.

The chapter is organized as follows. Sections 2 and 3 present our distributed coordination and centralized scheduling approaches, respectively. Section 4 provides a summary.

2 Distributed Coordination

In this section, we present a distributed coordination approach for the problem of intersection management. The approach does not require a centralized intersection manager. There are three steps to implement the approach:

1. Each vehicle broadcasts its estimated time intervals to occupy the corresponding conflict zones.
2. Given the broadcast information, all vehicles reach a consensus of the passing order by solving a conflict graph locally.

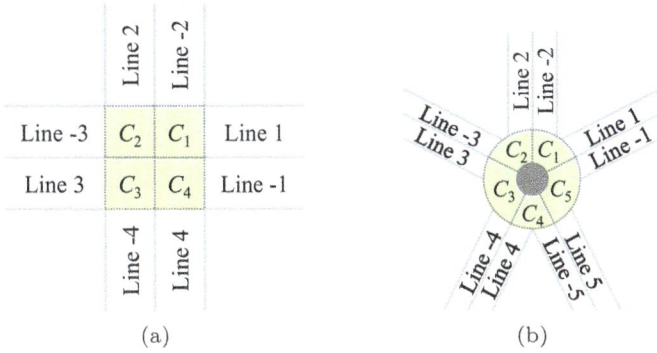

Fig. 1 Example environments: (**a**) intersection and (**b**) roundabout

3. Each vehicle adjusts its speed profile according to the passing order and updates its estimated time intervals to occupy the corresponding conflict zones.

We assume that the communication has no delay or packet loss. Figure 1 illustrates two example environments: a (real) intersection and a roundabout. A conflict zone is formulated when the extensions of two incoming lanes intersect with each other.

The rest of the discussion is organized as follows: Sect. 2.1 formulates the problem. Section 2.2 introduces the distributed approach and its theoretical guarantees. Section 2.3 provides simulation results, and Sect. 2.4 concludes the discussion.

2.1 Problem Formulation

A conflict zone is formulated when the extensions of two incoming lanes intersect with each other. The conflict zones are denoted by $C_1, C_2, \ldots, C_L$, where L is the total number of conflict zones. There are N vehicles, indexed from 1 to N, intend to pass through an intersection. The intention (the target lane after passing through the intersection) of vehicle i is G_i. The state of vehicle i at time t is denoted as $x_i(t)$. The system state at time step t is denoted as $x(t) := [x_1(t); x_2(t); \ldots; x_N(t)]$.

The system objective is to ensure that the intentions of the vehicles are satisfied efficiently and maintain the system safety. The safety constraint requires that the minimum distance between any two vehicles is larger than or equal to a threshold $d_{\min}$, e.g.,

$$\mathcal{X} := \{x \mid d(x_i, x_j) \geq d_{\min}, \ \forall i, j, i \neq j\}, \tag{1}$$

where the function d measures the minimum distance between two vehicles i and j.

In a distributed setting, each individual vehicle only has a local view and local information, i.e., vehicle i only considers the vehicles in its neighborhood $\mathcal{N}_i$. Moreover, the other vehicles' states in the safety constraint are not directly

accessible, so they need to be estimated. The navigation problem for vehicle i can
be formulated as the following optimization problem:

$$\min_{x_i} \ J(x_i, G_i), \tag{2a}$$

$$\text{s.t. } \dot{x}_i(t) \in \Gamma(x_i(t)), \tag{2b}$$

$$d(x_i(t), \hat{x}_j^i(t)) \geq d_{\min}, \ \forall j \in \mathcal{N}_i, \tag{2c}$$

where J is the objective, and $\hat{x}_j^i(t)$ is the estimation of $x_j(t)$ made by vehicle i.
Equation (2b) is the feasibility constraint to ensure that there is a low level controller
to track the trajectory, e.g.,

$$\Gamma(x_i) := \{\dot{x}_i \mid \exists u_i, \ \dot{x}_i = f(x_i, u_i)\}, \tag{3}$$

where $\dot{x}_i = dx_i/dt$, u_i is the vehicle control input (wheel angle and throttle torque),
and f describes the vehicle dynamics. It is assumed that all vehicles are equipped
with perfect controllers that can execute the planned trajectories without any error
if the planned trajectory is feasible.

In current design of autonomous vehicles, $\hat{x}_j^i$ is estimated based on local
sensors [1, 2]. In order to account for uncertainties in the estimation, the behaviors
of autonomous vehicles tend to be conservative. As a result, all vehicles may decide
to slow down to yield, which is very inefficient. The behaviors of "connected"
and autonomous vehicles can be less conservative due to more information and
more accurate estimation. From the system level, less conservative behaviors imply
smaller delay and larger throughput. Before we dive deep into the distributed
coordination solution, we first introduce several assumptions and notation.

2.1.1 Assumption on Fixed Paths

We assume that each vehicle follows a fixed path, and Eq. (2) only optimizes for the
speed profile along the path. Let x_i^* be the optimal trajectory of vehicle i that does
not consider the collision avoidance constraint, e.g.,

$$x_i^* = \arg \min_{\dot{x}_i(t) \in \Gamma(x_i(t))} J(x_i, G_i). \tag{4}$$

Hence, the path of vehicle i is fixed along x_i^*, and the vehicle only adjusts its speed
profile to meet the collision avoidance constraint. This assumption is reasonable
since vehicles are usually not allowed to change lanes at intersections. In the
following discussion, let $x_i^*(s)$ be the distance s parameterized path for vehicle i.
The speed profile for vehicle i is denoted as $s_i(t)$ which is a mapping from time to
the distance along the path. Then, $x_i^*(s_i(t))$ is the trajectory.

We say that vehicle i passes through the conflict zone C_l if there exists $s \in \mathbb{R}^+$ such that $\mathcal{B}_i(x_i^*(s)) \cap C_l \neq \emptyset$, where $\mathcal{B}_i$ denotes the area occupied by vehicle i at state $x_i^*(s)$. Define the segment on path x_i^* that intersects with the conflict zone C_l as $\mathcal{L}_{i,l} := \{s \mid \mathcal{B}_i(x_i^*(s)) \cap C_l \neq \emptyset\}$. Hence, $\mathcal{L}_{i,l} = \emptyset$ if and only if vehicle i does not pass through the conflict zone C_l. Denote the set of indices of conflict zones that vehicle i passes through as $\mathcal{A}_i := \{l \mid \mathcal{L}_{i,l} \neq \emptyset\}$. Then, two vehicles i and j pass through a same conflict zone if and only if $\mathcal{A}_i \cap \mathcal{A}_j \neq \emptyset$.

2.1.2 Notations of Discrete States

In addition to the continuous vehicle state x_i, to better describe the vehicle behaviors at intersections, we define a discrete state $\mathcal{S}_i$ for vehicle i, where

- $\mathcal{S}_i = IL$ if vehicle i is on an incoming lane, and it is not the first vehicle on the lane.
- $\mathcal{S}_i = FIL$ if vehicle i is on an incoming lane, and it is the first vehicle on the lane.
- $\mathcal{S}_i = I$ if vehicle i is at the intersection.
- $\mathcal{S}_i = OL$ if vehicle i is on an outgoing lane.

Vehicle i may enter the control area with $\mathcal{S}_i = IL$ or FIL. $\mathcal{S}_i$ can transit from IL to FIL, from FIL to I, and from I to OL, i.e., becoming the first vehicle on an incoming lane, entering the intersection, and leaving the intersection, respectively. It can leave the control area when $\mathcal{S}_i = OL$. For any vehicle i such that $\mathcal{S}_i = IL$ or OL, its front vehicle is denoted $\mathcal{F}_i$.

2.2 Distributed Coordination Approach

The key insight here is that communication can help the ego vehicle to better determine the constraint in Eq. (2c). Indeed, instead of estimating others' trajectories $\hat{x}_j^i$, what really matters to the ego vehicle is the time that other vehicles occupy the conflict zones. We design the communication protocol to be that each vehicle should broadcast the following two types of information:

- The estimated times to occupy the conflict zones once the vehicle enters a control area of the intersection, e.g., the shaded area in Fig. 1.
- The basic information such as the vehicle ID, the current state (position, heading, speed, and $\mathcal{S}_i$), and the time stayed in the control area.

Based on the broadcast information, the vehicles will seek a consensus on the passing order and compute desired time slots to pass through the conflict zones, which are then taken as temporal constraints on the vehicles' trajectories. This naturally breaks the problem into two parts as shown in Fig. 2a:

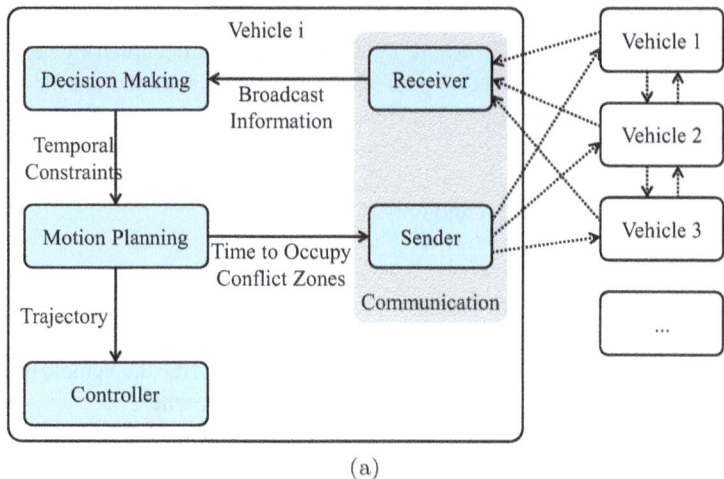

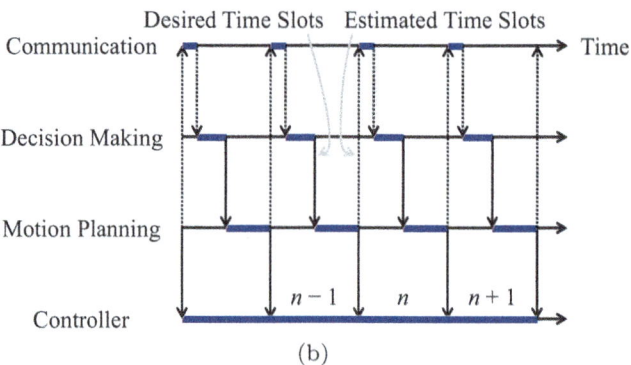

Fig. 2 Architecture of the conflict resolution mechanism. (**a**) System block diagram. (**b**) Time flow of execution

1. Decision making: determination of passing order and hence temporal constraints.
2. Motion planning: computation of trajectory.

The time flow and the coordination among different modules are shown in Fig. 2b. It is assumed that all vehicles are synchronized. At time step $n - 1$, the estimated time interval $[\mathbb{T}_{j,l}^{in,n-1}, \mathbb{T}_{j,l}^{out,n-1}]$ for vehicle j to occupy C_l is broadcast for all j and l. At time step n, vehicle i evaluates all information received from other vehicles and computes the desired time slots to pass through the conflict zones in the decision maker, i.e., $[T_{i,l}^{in,n}, T_{i,l}^{out,n}]$ for all l, which are then sent to the motion planner as temporal constraints. After motion planning, the planned trajectory is sent to the controller for execution and the estimated time slots to occupy the conflict zones given the new trajectory, i.e., $[\mathbb{T}_{i,l}^{in,n}, \mathbb{T}_{i,l}^{out,n}]$ for all l, are broadcast to other vehicles.

The decomposition of decision making and motion planning can also be adopted in centralized intersection management, where the manager takes the responsibility

---

Algorithm 1 The decision making algorithm for vehicle i for computing the temporal constraints $[T_{i,l}^{in,n}, T_{i,l}^{out,n}]$, $\forall l$ at time step n given information $[\mathbb{T}_{j,l}^{in,n-1}, \mathbb{T}_{j,l}^{out,n-1}]$, $\forall j, l$

Initialize, $n = 0$
while $\mathcal{S}_i \in \{IL, FIL, I\}$ **do**
    Receive other's information $\mathbb{T}_{j,l}^{in,n-1}, \mathbb{T}_{j,l}^{out,n-1}$
    Initialize $\mathcal{Y}_i = \emptyset$, $T_{i,l}^{in,n} = -\infty$, $T_{i,l}^{out,n} = \infty$
    if $\mathcal{S}_i = IL$ **then**
        i yields its front vehicle ($\mathcal{Y}_i = \{\mathcal{F}_i\}$)
    end if
    for j that has spatial conflicts with i ($j \in \mathcal{U}_i$) **do**
        if j has a temporal advantage over i ($j \in \mathcal{V}_i$) **then**
            if $\nexists\text{Tie}(i, j)$ or j has priority over i **then**
                i yields j ($\mathcal{Y}_i = \mathcal{Y}_i \cup \{j\}$)
            end if
        end if
        if i has a temporal advantage over j ($i \in \mathcal{V}_j$) **then**
            if $\exists\text{Tie}(j, i)$ and j has priority over i **then**
                i yields j ($\mathcal{Y}_i = \mathcal{Y}_i \cup \{j\}$)
            end if
        end if
    end for
    for j that i yields ($j \in \mathcal{Y}_i$) **do**
        for C_l that both i and j traverse ($l \in \mathcal{A}_i \cap \mathcal{A}_j$) **do**
            $T_{i,l}^{in,n} = \max\{T_{i,l}^{in,n}, \mathbb{T}_{j,l}^{out,n-1} + \Delta_{\mathcal{S}_i}\}$
        end for
    end for
    $n = n + 1$
end while

---

of decision making, and the vehicles takes the responsibility of motion planning [3]. We discuss the decision making in Sect. 2.2.1 and the motion planning in Sect. 2.2.2.

2.2.1 Decision Making

At time step n, vehicle i needs to compute the desired time interval $[T_{i,l}^{in,n}, T_{i,l}^{out,n}]$ to pass through the conflict zones given the broadcast information $[\mathbb{T}_{j,l}^{in,n-1}, \mathbb{T}_{j,l}^{out,n-1}]$ for all j and l. The basic strategy is that whoever arrives first in a conflict zone goes first.[1] However, this strategy may create deadlocks when one vehicle arrives earlier in one conflict zone, while the other vehicle arrives earlier in another conflict zone. As a result, a tie breaking mechanism is needed. Here, we first discuss a

---

[1] Note that this is different from the strategies discussed in [4], which only considers the arrival time at the intersection.

general methodology to deal with distributed coordination with multiple conflict zones, which is summarized in Algorithm 1.

If $\mathcal{S}_i = IL$, it is physically "constrained" by its front vehicle and should yield all vehicles that its front vehicle yields. The decisions when $\mathcal{S}_i = FIL$ or I are the most important as conflicts usually come among vehicles in these two states. When $\mathcal{S}_i = OL$, the vehicle no longer needs to compute the desired time interval. However, its information should be broadcast in order for the proceeding vehicles to follow the lane safely. In the following discussion, we focus on vehicle i with $\mathcal{S}_i = FIL$ or I.

2.2.1.1 Spatial Conflict

We say that there is a spacial conflict between vehicles i and j if and only if their paths pass through a same conflict zone. Consider the scenario shown in Fig. 3a, where nine vehicles locate in a six-way intersection. The shaded area denotes the six conflict zones. By adding edges between any pair of vehicles that have spatial conflicts, we formulate an undirected graph as shown in Fig. 3b, where every vertex represents one vehicle. Whenever there is an edge between two vehicles, we need to decide which vehicle goes first. In other words, the undirected graph needs to be transformed into a directed graph as shown in Fig. 3d such that the passing order is decided by the topological order. Denote the set of vehicles that have spacial conflicts with vehicle i as

$$\mathcal{U}_i := \{j \mid \mathcal{S}_j = FIL \text{ or } I, \mathcal{A}_i \cap \mathcal{A}_j \neq \emptyset\}. \tag{5}$$

Recall that $\mathcal{A}_i$ denotes the set of indices of conflict zones that vehicle i passes through. Hence, $\mathcal{A}_i \cap \mathcal{A}_j \neq \emptyset$ means that vehicle j passes through one or more conflict zones that vehicle i also needs to pass through. The graph in Fig. 3b is denoted as $\mathcal{U} := \cup_i \cup_{j \in \mathcal{U}_i} (i, j)$, where (i, j) represents an edge between i and j. There is an undirected edge between any i and j such that $j \in \mathcal{U}_i$. In literature, this graph is identified as a conflict graph [5]. Finding the optimal passing order regarding the conflict graph is NP-hard. The approach presented here is a heuristic approach which finds one feasible passing order in linear time.

2.2.1.2 Temporal Advantage

At time step n, we say that vehicle $i \in \mathcal{U}_j$ has a temporal advantage over vehicle j, if one of the following conditions holds:

- $\mathcal{S}_i = I, \mathcal{S}_j = FIL$ and vehicle j leaves some conflict zones later than vehicle i enters, i.e.,

$$\exists l \in \mathcal{A}_i \cap \mathcal{A}_j, \mathbb{T}_{j,l}^{out,n-1} > \mathbb{T}_{i,l}^{in,n-1}. \tag{6}$$

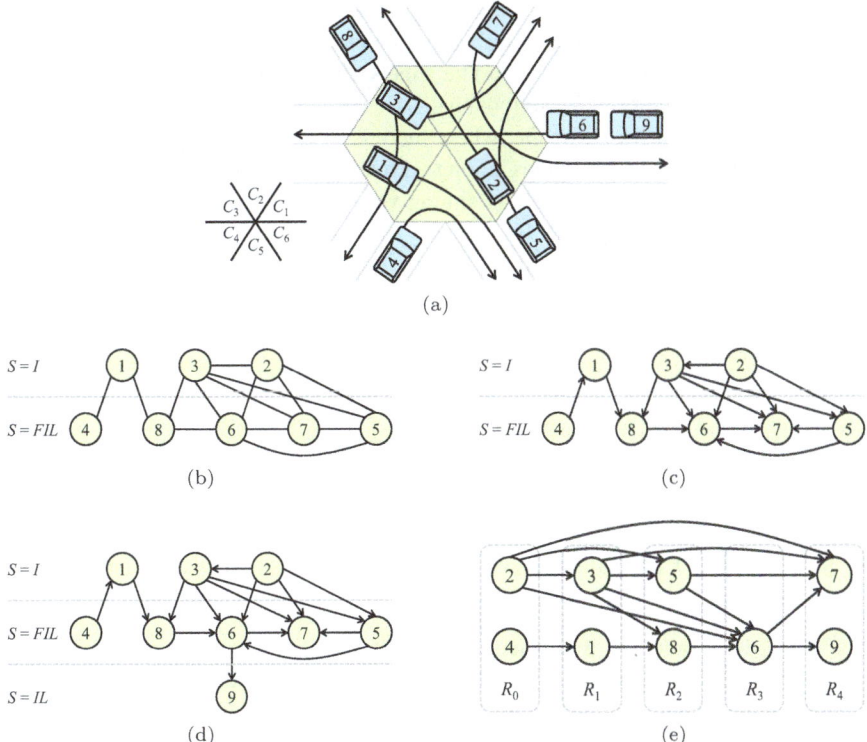

Fig. 3 The conflict graphs. (**a**) The scenario: nine vehicles in a six-way intersection. (**b**) Graph of spacial conflicts $\mathcal{U}$ at one time step. Edge $(i, j) \in \mathcal{U}$ implies that vehicle j has spacial conflicts with vehicle i at some conflict zone. (**c**) Graph of temporal advantages $\mathcal{V}$ at one time step. Edge $(i, j) \in \mathcal{V}$ implies that vehicle i has temporal advantages over vehicle j. (**d**) Graph of passing order $\mathcal{Y}$ at one time step. Edge $(i, j) \in \mathcal{Y}$ implies that vehicle j yields vehicle i. (**e**) Convergence of the graph $\mathcal{Y}$ through time. The passing order converges at time 0 for leaf vertices in $\mathcal{R}_0$, then for vertices with depth 1 in $\mathcal{R}_1$ at time 1, and so on

- $\mathcal{S}_i = FIL, \mathcal{S}_j = I$ and vehicle i leaves all conflict zones earlier than vehicle j enters, i.e.,

$$\forall l \in \mathcal{A}_i \cap \mathcal{A}_j, \mathbb{T}_{i,l}^{out,n-1} \leq \mathbb{T}_{j,l}^{in,n-1}. \tag{7}$$

- $\mathcal{S}_i = \mathcal{S}_j = FIL$ or I and vehicle i enters some conflict zones earlier than vehicle j, i.e.,

$$\exists l \in \mathcal{A}_i \cap \mathcal{A}_j, \mathbb{T}_{i,l}^{in,n-1} \leq \mathbb{T}_{j,l}^{in,n-1}. \tag{8}$$

According to the above definitions, for vehicles i and j with different discrete states, either i or j should have a temporal advantage over the other. If vehicles

i and j have the same discrete state, it is possible that both i and j have temporal advantages over the other. Denote the set of vehicles that have temporal advantages over vehicle j at time step n as $\mathcal{V}_j^n$. The superscript n in the following discussion is ignored for simplicity. It is obvious that $\mathcal{V}_j \subset \mathcal{U}_j$; and $\mathcal{V} := \cup_j \cup_{i \in \mathcal{V}_j} (i, j)$ is a directed graph as shown in Fig. 3c, where there is a directed edge from any $i \in \mathcal{V}_j$ to any j. However, there are cycles among vertices with the same discrete state, e.g., between vertices 6 and 7, as well as among vertices with different discrete states, e.g., among vertices 6, 7, and 3. If vehicles yield each other according to the graph, there are deadlocks. We will introduce a tie breaking mechanism to avoid these deadlocks.

2.2.1.3   Tie Breaking

For any vehicle i and vehicle $j \in \mathcal{V}_i$, it is called a *tie* if:

- $\mathcal{S}_i = \mathcal{S}_j$ and there exists a sequence of vehicles $\{q_m\}_1^M$ with $q_1 = i, q_M = j$, $M \geq 2$ and $\mathcal{S}_{q_m} = \mathcal{S}_i$ for all m such that $q_m \in \mathcal{V}_{q_{m+1}}$ for $m = 1, 2, \ldots, M - 1$.
- $\mathcal{S}_i = I, \mathcal{S}_j = FIL$ and there exists a sequence of vehicles $\{q_m\}_1^M$ with $q_1 = i$, $q_M = j$ and $M \geq 2$ such that $q_m \in \mathcal{V}_{q_{m+1}}$ for $m = 1, 2, \ldots, M - 1$.

Let $\text{Tie}(i, j)$ denote all these sequences. The relationship in a tie is neither symmetric nor exclusive, i.e., $\exists \text{Tie}(i, j)$ neither implies $\exists \text{Tie}(j, i)$ nor $\nexists \text{Tie}(j, i)$. For example, in Fig. 3c, there is a tie from vertex 5 to vertex 6 via the sequence $\{5, 7, 6\}$, but there is not a tie from vertex 6 to vertex 5 since $5 \notin \mathcal{V}_6$. There is a tie from vertex 2 to vertex 3 via the sequence $\{2, 3\}$ and a tie from vertex 3 to vertex 2 via the sequence $\{3, 2\}$.

We assume that each vehicle has a unique priority score P. For example, the priority score of a fire truck is higher than that of a passenger vehicle. We say that vehicle i has *priority* over vehicle j if there exists a sequence in $\text{Tie}(i, j)$ such that $P(i) > P(k)$ for all $k \neq i$ in the sequence. The basic principles are: (1) vehicles already in the intersection should always have priority over vehicles on the incoming lanes; (2) for vehicles in the same discrete state, the order implied by the priority score should not change over time. If vehicle i has priority over vehicle $j \in \mathcal{V}_i$, instead of i yielding j, vehicle j should yield vehicle i, although vehicle j has a temporal advantage. For example, in Fig. 3, we identify the score P with the vehicle index. Since vertex 5 has priority in the sequence $\{5, 7, 6\}$, the edge from 5 to 6 is reversed in Fig. 3d. Since there is a tie between vertex 2 and vertex 6, the edge from 2 to 6 is also reversed in Fig. 3d.

2.2.1.4   Passing Sequence

After tie breaking, all those remaining edges for vehicle j represent the set of vehicles that vehicle j decides to yield at time step n, which is denoted by $\mathcal{Y}_j^n$. The superscript n in the following discussion is ignored for simplicity. Indeed,

$\mathcal{Y} := \cup_j \cup_{i \in \mathcal{Y}_j} (i, j)$ is a directed graph as shown in Fig. 3d, which encodes the order for the vehicles to pass through the intersection. Note that it is not necessary for vehicle i to construct the whole graphs $\mathcal{U}$ and $\mathcal{V}$ to determine $\mathcal{Y}_i$. For example, vehicle 4 in Fig. 3 only needs to compute $\mathcal{U}_4$ and $\mathcal{V}_4$ locally to determine that $\mathcal{Y}_4 = \emptyset$. Those local decisions form the passing sequence globally. In the extreme case, the passing order follows the order specified by the priority scores. If all vehicles agree on the above tie breaking mechanism, they can solve the conflicts even if the vehicles plan and control their motions differently.

According to Algorithm 1, if $\mathcal{S}_j = IL$, the vehicle j yields its front vehicle, i.e., $\mathcal{Y}_j = \{\mathcal{F}_j\}$, as shown by vehicle 9 in Fig. 3d. If vehicle j decides to yield vehicle i, then for all $l \in \mathcal{A}_i \cap \mathcal{A}_j$, we set

$$T_{j,l}^{in,n} \geq \mathbb{T}_{i,l}^{out,n-1} + \Delta_{\mathcal{S}_j}, \tag{9}$$

where $\Delta_{\mathcal{S}_j}$ is a margin to increase the robustness of the algorithm, which is chosen such that $\Delta_{IL} > \Delta_{FIL} > \Delta_I$. Δ_{IL} is chosen to be larger than Δ_{FIL} to ensure the leading vehicles have temporal advantages over vehicles on the middle of other lanes. For example, vehicle 7 has a temporal advantage over vehicle 9 in Fig. 3d. Similarly, Δ_{FIL} is chosen to be larger than Δ_I.

2.2.2 Motion Planning under Temporal Constraints

At time step n, given the temporal constraint $[T_{i,l}^{in,n}, T_{i,l}^{out,n}]$ specified by the decision maker, the problem in Eq. (2) for vehicle i can be rewritten as:

$$\min_{s_i} \quad J(x_i^*(s_i), G_i) \tag{10a}$$

$$\text{s.t.} \quad \frac{\partial x_i^*(s_i)}{\partial s_i} \dot{s}_i \in \Gamma(x_i^*(s_i)), \tag{10b}$$

$$s_i(t) \notin \mathcal{L}_{i,l}, \; \forall t \notin [T_{i,l}^{in,n}, T_{i,l}^{out,n}], \; \forall l, \tag{10c}$$

where $s_i(t)$ is the speed profile that needs to be optimized. Equation (10c) specifies that the vehicle should only enter the conflict zone C_l in the time interval $[T_{i,l}^{in,n}, T_{i,l}^{out,n}]$. For simplicity, the constraint for vehicle following is omitted in presentation (but included in problem solution).

A method to efficiently solve the problem in Eq. (10) via temporal optimization is discussed in [6]. Here, we assume that vehicles can take unbounded deceleration, which is reasonable when vehicle speeds are low. Considering $T_{i,l}^{out,n} = \infty$ by Algorithm 1, there is always a solution of problem in Eq. (10). In the worst case, vehicle i just stops immediately. In practice, the vehicles do not necessarily need to take unbounded deceleration as this will be demonstrated in Sect. 2.3, since the conflicts are resolved before they enter the intersection. The feasibility of the problem in Eq. (10) under bounded deceleration is left as future work.

Given the optimal solution s_i^* of the problem in Eq. (10), the expected time slot $[\mathbb{T}_{i,l}^{in,n}, \mathbb{T}_{i,l}^{out,n}]$ for vehicle i to occupy the conflict zone C_l is computed as:

$$\mathbb{T}_{i,l}^{in,n} := \min_{s_i^*(t)\in\mathcal{L}_{i,l}} t \geq T_{i,l}^{in,n}, \quad \mathbb{T}_{i,l}^{out,n} := \max_{s_i^*(t)\in\mathcal{L}_{i,l}} t \leq T_{i,l}^{out,n} \tag{11}$$

If $\mathcal{L}_{i,l} = \emptyset$, then $\mathbb{T}_{i,l}^{in,n} := \infty$ and $\mathbb{T}_{i,l}^{out,n} := -\infty$. If vehicle i has entered or left C_l, then $\mathbb{T}_{i,l}^{in,n}$ and $\mathbb{T}_{i,l}^{out,n}$ are chosen as the time that it entered or left C_l, respectively.

2.2.3 Theoretical Guarantees

Here, we introduce the theoretical results to show that the proposed strategy solves the conflicts safely and efficiently in real time. The physical feasibility of the trajectories is verified in the motion planning part. Proposition 1 ensures that the passing order is completely determined. Proposition 2 states that there is no deadlock for any pair of vehicles that pass through a same conflict zone at every time step. Proposition 3 shows that a stable consensus on conflict-resolution can be reached in finite time steps. The proofs can be found in [7].

Proposition 1 (Completeness) *For any j that has spacial conflicts with i, at least one statement is true: "i yields j" or "j yields i". In other words, $j \in \mathcal{U}_i$ implies $j \in \mathcal{Y}_i$ or $i \in \mathcal{Y}_j$.*

Proposition 2 (Deadlock-Freeness) *There is no cycle in the directed graph $\mathcal{Y}$ of passing order.*

Proposition 3 (Finite Time Convergence) *If $\mathcal{S}_i$ and $\mathcal{U}_i$ remain the same for all i for more than N time steps, then $\mathcal{Y}_i^n$ and $[\mathbb{T}_{i,l}^{in,n}, \mathbb{T}_{i,l}^{out,n}]$ converge in at most N steps to $\mathcal{Y}_i^*$ and $[\mathbb{T}_{i,l}^{in*}, \mathbb{T}_{i,l}^{out*}]$ such that*

$$\mathbb{T}_{i,l}^{in*} \geq \mathbb{T}_{j,l}^{out*} + \Delta_{\mathcal{S}_i}, \ \forall l, \ \forall j \in \mathcal{Y}_i^* \tag{12}$$

Proposition 3 implies that if the sampling time is short enough compared with the time needed between two transitions of $\mathcal{S}_i$'s, the system can still reach consensus when $\mathcal{S}_i$'s are changing. Nonetheless, after a transition of some $\mathcal{S}_i$, the system needs several steps to settle down. The consistency of the passing orders $\mathcal{Y}^n$ considering those transitions is more intricate to prove, which is left as future work. Indeed, the consistency is demonstrated in simulation.

2.3 Simulation Results

In this section, we illustrate the performance of the proposed distributed conflict resolution mechanism through extensive traffic simulations. The sampling time in the system is chosen to be $dt = 0.1$ s. The robustness margins are chosen as $\Delta_{IL} = 0.5$ s, $\Delta_{FIL} = 0.3$ s and $\Delta_I = 0.1$ s. The priority score P for a vehicle is chosen to be the time that the vehicle stays in the control area. If there is a tie, then the vehicle with smaller ID has the priority. The cost function of the vehicle penalizes (1) the deviation from a target speed, (2) the magnitude of acceleration or deceleration, (3) the magnitude of jerk, and 4) the time spent in every conflict zone. The target speed varies for different vehicles.

The simulation environment is a narrow four-way intersection as shown in Fig. 1a. There is only one incoming lane and one outgoing lane in every direction. Four conflict zones are identified. The control area is the whole graph. For any $i \neq j$, there is a path from lane i to lane $-j$, so there are 12 different paths. Right turn paths only go through one conflict zone. Straight paths go through two conflict zones. Left turn paths go through all four conflict zones (a vehicle is treated as a 2D object instead of a point). In the following discussion, a microscopic case study is presented first followed by the result of macroscopic traffic simulation.

2.3.1 Microscopic Case Study

In the case study, there are four vehicles. The conditions of the vehicles (target speed, current lane, target lane, and time to enter the control area) are shown in Table 1. The paths and the executed trajectories are shown in the time-augmented space in Fig. 4a. The planned speed profiles in different time steps are shown in Fig. 4b. The left most speed profile in every subplot is the traffic-free speed profile and the others are the replanned speed profiles given the temporal constraints. Figure 5 shows the expected time intervals (the colored thick bars) for the vehicles to occupy the conflict zones. The thin vertical line indicates the current time.

In this case, vehicles 1, 2 and 3 enter the control area at the same time. According to the traffic-free speed profiles, there are temporal conflicts between vehicle 1 and vehicle 3 in conflict zones 1 and 2, and between vehicle 2 and vehicle 3 in all conflict zones. Since vehicle 2 has a temporal advantage over vehicle 3, vehicle 3 yields vehicle 2. Similarly, vehicle 1 yields vehicle 3. It takes two time steps to resolve the conflicts.

At 0.6 s, vehicle 4 enters, which creates new conflicts. The system settles down after 3 time steps as shown in Fig. 5, which verifies Proposition 3. The planned

Table 1 Conditions in the case study

Vehicle ID	Target speed (m/s)	From	To	Enter time (s)
1	10	Lane 1	Lane -3	0.2
2	12.5	Lane 2	Lane -1	0.2
3	10.75	Lane 3	Lane -2	0.2
4	17.75	Lane 4	Lane -2	0.6

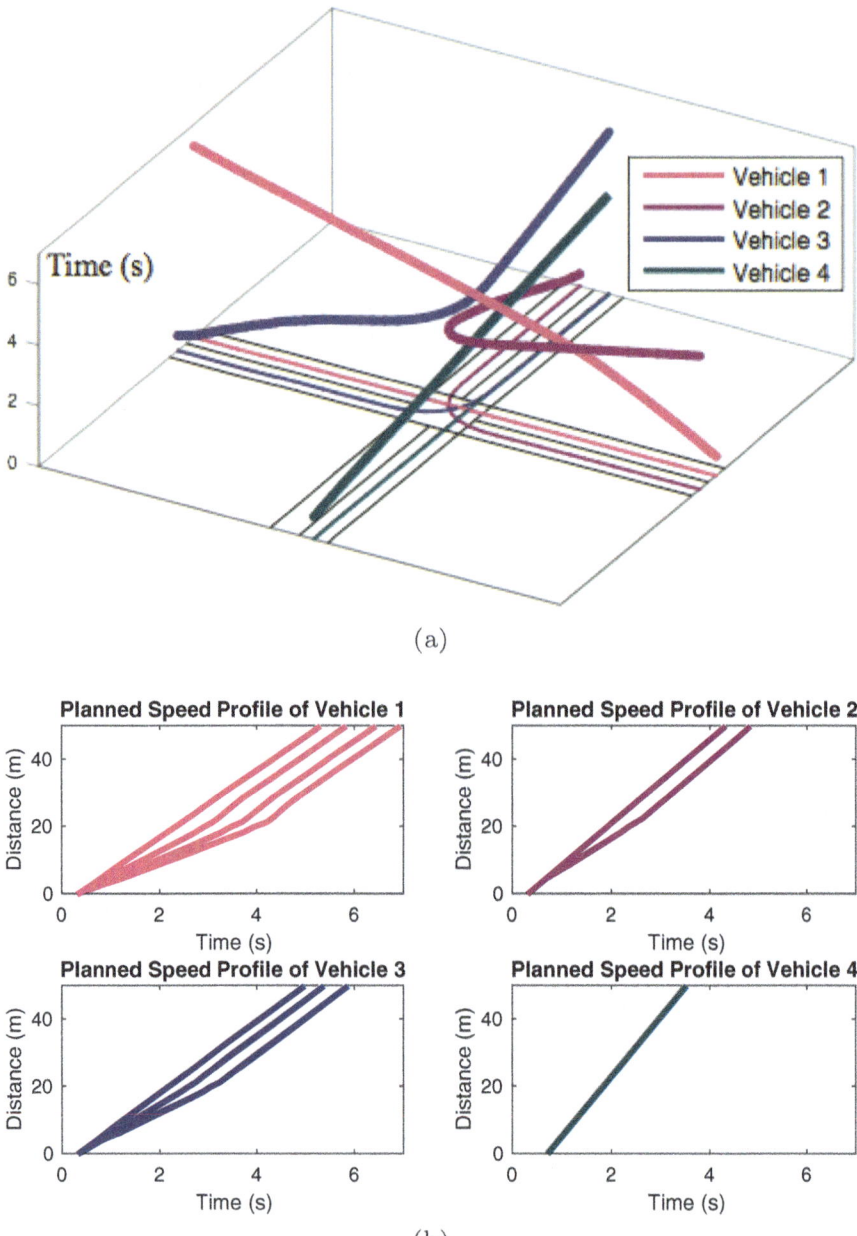

Fig. 4 Speed profiles and trajectories in the case study. (**a**) Executed trajectories in the time-augmented space. (**b**) Planned speed profiles in different time steps

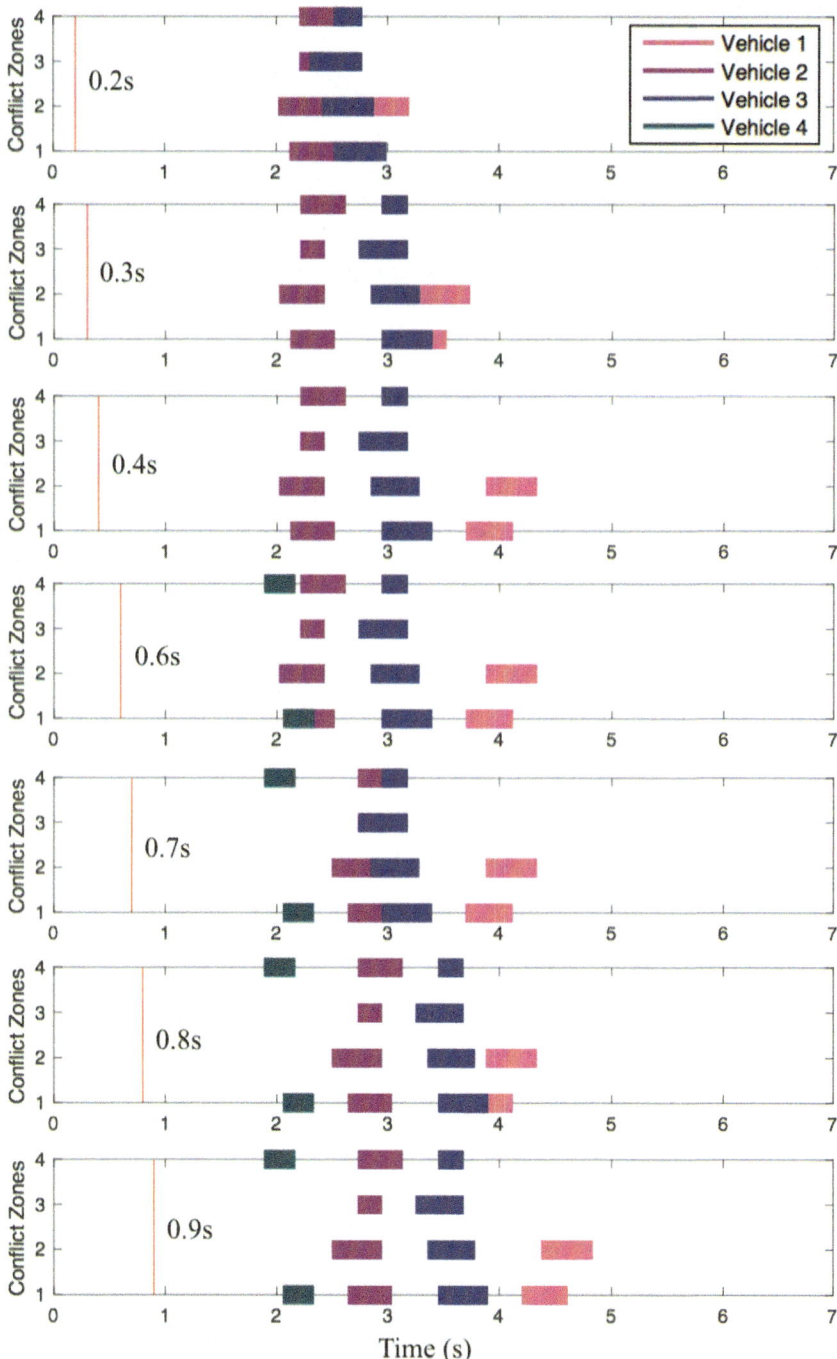

Fig. 5 Conflict resolution in the case study. The scenario in 0.5*s* is omitted since it is the same as the scenario in 0.4*s*

speed profiles change accordingly as shown in Fig. 4b. The right most speed profile
in each subplot is the executed speed profile.

2.3.2 Macroscopic Traffic Simulation

2.3.2.1 Traffic

In the macroscopic traffic simulation, the traffic is generated at every incoming lane
by a Poisson distribution where the density λ is chosen to be 0.5, 0.25 and 0.1,
which implies that on average, vehicles arrive every 2, 4, and 10 s. Two groups of
traffic are generated:

- Group 1 (G1): 50% of vehicles go straight, 25% turn right and 25% turn left.
- Group 2 (G2): all vehicles go straight.

The second group is introduced to create a relatively fair comparison among
performances under distributed strategies and performances under traffic lights.
Since we don't have left turn lane or left turn light, when a vehicle wants to turn left,
it will block all the vehicles behind, thus significantly increase the delay time. The
desired longitudinal speed v_i^r of the vehicle i follows from a uniform distribution
from 7.5 to 15 m/s.

2.3.2.2 Comparison

The proposed mechanism is compared against other mechanisms as listed below.

- Case 1 (3D): 3D intersection such as overpass without connectivity. In this case,
  there is no conflict among vehicles at the intersection. Since the delay is only
  caused by car following, the simulation result provides a lower bound for the
  delay time and an upper bound for the throughput.
- Case 2 (NC): unmanaged 2D intersection without connectivity. Vehicles are able
  to see vehicles from other directions when approaching the intersection. Then
  vehicles' strategy is: if there is no other vehicles from other directions or other
  vehicles are too far from the intersection (i.e., there is no temporal conflict even if
  the other vehicle accelerates with maximum acceleration), cross the intersection
  without stop; if there are other vehicles from other directions that are close to the
  intersection, stop and "first stop first go". The delay time in this case is upper
  bounded by the delay time in the case of a four-way-stop intersection.
- Case 3 (TL-5): 2D intersection with traffic light that changes every 5 s without
  connectivity. For example, the traffic light for the horizontal direction (lane 1 and
  lane 3) is green from 0 to 5 s and red from 5 to 10 s while the traffic light for the
  vertical direction (lane 2 and lane 4) is red from 0 to 5 s and green from 5 to 10 s.
- Case 4 (TL-10): 2D intersection with traffic light that changes every 10 s without
  connectivity.

- Case 5 (MP-IP): 2D intersection with the maximum progression intersection protocol (MP-IP) [4]. Vehicles broadcast their intentions and estimated time slots to occupy the conflict zones. Conflicting vehicles can make concurrent progress inside the intersection, though low priority vehicles need to yield high priority vehicles, i.e., entering the conflict zones after the high priority vehicles leave. In the simulation, the priority is determined by the priority score P.
- Case 6 (AMP-IP): 2D intersection with the advanced maximum progression intersection protocol (AMP-IP) [4]. In addition to MP-IP, the lower priority vehicles are allowed to cross and clear the conflict zone before the earliest possible arrival of the higher-priority vehicle to that conflict zone.

In Cases 1 to 4, there is no communication among vehicles and the vehicles are equipped with adaptive cruise control for car following. In Cases 5 and 6, vehicles communicate with one another. The two protocols only determine the passing order, not the vehicle trajectories. In the simulation, the vehicles under the two cases adopt the motion planning algorithm discussed in the previous section. The temporal constraints are determined by Eq. (9) according to the passing order. To create a fair comparison, the adaptive cruise control algorithm is integrated into the motion planning algorithm. At each time step, the output of the adaptive cruise control module will be treated as an upper bound on vehicle's acceleration, which is added to the optimization Eq. (10). In the following discussion, we analyze: (1) the average delay time and (2) the throughput in certain time horizon.

2.3.2.3   Average Delay

The delay time of a vehicle is computed as the difference between the actual time and the traffic-free time for the vehicle to travel cross the control area as shown in Fig. 4b. The average delay (mean ± standard deviation) of all vehicles traveled in the control area in 10min under different mechanisms are shown in Table 2. The proposed strategy always outperforms other mechanisms except for the case with 3D intersection which provides a theatrical lower bound of this problem. When the traffic density is low, the performances of Case 2 (without communication) and Cases 5 and 6 (with communication) are similar to the performance of the proposed method, which outperforms the cases with traffic lights. When the traffic density goes up, the performance of Case 2 gets worse dramatically as it almost functions as a stop sign mechanism. The proposed method still outperforms the cases with traffic lights (Cases 3 and 4) since it is more flexible. For example, in the proposed mechanism, four simultaneous right turns are allowed, while in the traffic light case, at most two simultaneous right turns can be tolerated.

The proposed method always outperforms Cases 5 and 6. Though more parallelism inside the intersection area (i.e., allowing more vehicles to cross the intersection at the same time) has been introduced in these two cased compared to Case 2, the rigidity of the priority queue (which does not adjust in real time) limits their performances. For example, consider the case study in Sect. 2.3.1. Since

Table 2 The delay time for traffic in 10 min

	λ	Case 1: 3D	Case 2: NC	Case 3: TL-5	Case 4: TL-10
G1	0.5	0.5 ± 0.8 s	135.6 ± 78.6 s	53.2 ± 30.6 s	57.8 ± 34.4 s
	0.25	0.2 ± 0.4 s	2.9 ± 2.8 s	3.2 ± 2.6 s	5.2 ± 3.8 s
	0.1	0.1 ± 0.2 s	0.4 ± 0.7 s	2.1 ± 2.0 s	3.8 ± 3.9 s
G2	0.5	0.5 ± 0.8 s	134.8 ± 81.0 s	22.0 ± 14.1 s	29.3 ± 17.6 s
	0.25	0.2 ± 0.6 s	9.2 ± 9.3 s	2.8 ± 2.3 s	4.4 ± 3.9 s
	0.1	0.1 ± 0.4 s	0.5 ± 0.7 s	1.9 ± 1.8 s	3.9 ± 3.9 s

	λ	Case 5: MP-IP	Case 6: AMP-IP	Proposed	
G1	0.5	31.2 ± 19.7 s	20.5 ± 13.2 s	11.4 ± 7.0 s	
	0.25	1.9 ± 1.7 s	1.2 ± 1.2 s	0.5 ± 0.7 s	
	0.1	0.4 ± 0.6 s	0.3 ± 0.6 s	0.2 ± 0.3 s	
	0.5	8.8 ± 6.4 s	6.3 ± 4.7 s	4.3 ± 3.3 s	
G2	0.25	2.5 ± 2.9 s	2.2 ± 2.9 s	2.1 ± 2.7 s	
	0.1	0.3 ± 0.5 s	0.3 ± 0.5 s	0.3 ± 0.5 s	

vehicle 4 arrives later than others, it has to wait for others according to MP-IP in Case 5. Even with AMP-IP in Case 6, vehicle 4 wouldn't be able to cut in front of vehicle 2, since it does not leave conflict zone 1 before vehicle 2 enters. Hence high-speed vehicles in Cases 5 and 6 experience larger delay compared to those in the proposed method, where they can cut into the queue only causing other vehicles to slow down slightly. Moreover, the average delay goes up from 8.8 to 52.4 s in Case 5 with "straight only" traffic $\lambda = 0.5$ if the motion planning algorithm is replaced with only adaptive cruise control (ACC). Since the travel time in the intersection is not penalized in ACC, vehicles tend to stop right before the intersection and consequently take longer time to traverse the intersection (as their acceleration is bounded) than they do when they optimize their speed profiles to slow down before approaching the intersection and then speed up to pass the intersection at full speed. Hence the efficiency of the proposed algorithm benefits from both the decision making module (determination of efficient passing order) and the motion planning module (temporal optimization) as well as their integration.

2.3.2.4 Throughput

The throughput is computed as the number of vehicles that cross the control area in a given time slot. The throughput in 10 min in all scenarios are shown in Table 3. When the traffic density is high, Case 2 reduces to the case with stop signs. Hence the throughput is roughly upper bounded by $10 \cdot 60/\delta$ where δ is the average time in seconds that is required for a single vehicle to cross the intersection. In the

Table 3 The traffic throughput (# of vehicles) in 10 min

	λ	Case 1: 3D	Case 2: NC	Case 3: TL-5	Case 4: TL-10
G1	0.5	1170	660	984	965
	0.25	590	589	589	587
	0.1	230	230	230	228
G2	0.5	1206	641	1121	1091
	0.25	599	597	595	589
	0.1	245	245	245	245
	λ	Case 5: MP-IP	Case 6: AMP-IP	Proposed	
G1	0.5	1023	1099	1139	
	0.25	590	590	590	
	0.1	230	230	230	
G2	0.5	1166	1186	1199	
	0.25	599	599	599	
	0.1	245	245	245	

simulation, $\delta \approx 1$. Hence the throughput in Case 2 is around 600 when $\lambda = 0.5$, which is much smaller than that in other cases. However, in the proposed method, the throughput almost doubles, which is higher than those in Cases 3 to 6 with traffic light or existing V2V intersection protocols, and is very close to that in Case 1 where the intersection is 3D, thus verifies the effectiveness of the proposed method.

2.4   Conclusion

This section discuss a communication-enabled distributed coordination strategy for connected and autonomous vehicles to navigate at intersections. Based on the received information, a vehicle computes a set of vehicles that it needs to yield and the desired time slots to pass the conflict zones in a decision maker. Then, it computes a desired speed profile according to the desired time slots in a motion planner and broadcasts the estimated times to occupy the conflict zones. The aggregation of these local decisions forms a global solution to a multi-vehicle navigation problem. In the simulation, it is shown that the proposed mechanism has smaller average delay and larger throughput than the comparative cases.

Although the fixed-path assumption and the discrete partitioning of the conflict zone simplifies our problem, they may potentially exclude some feasible conflict resolution strategy that can be achieved by adjusting the vehicle paths. These non-fixed-path strategies are studied in [8, 9]. A thorough analysis and comparison among all these strategies will be left for future work.

3 Centralized Scheduling

In this section, we present a centralized scheduling approach for the problem of intersection management. As shown in Fig. 6, a centralized unit installed in the roadside unit, called intersection manager, decides the passing order of the vehicles periodically. For each period, the intersection manager receives the information from vehicles within its communication range. Based on the received information, the intersection manager computes a time window to each vehicle at each conflict zone on the trajectory of the vehicle. After that, the intersection manager broadcasts these results and prepares for the next period.

The rest of the discussion is organized as follows: Sect. 3.1 presents our timing conflict graph model and problem formulation. Section 3.2 demonstrates our resource conflict model and verification approach. Section 3.3 describes our scheduling algorithm based on cycle removal. Section 3.4 discusses lane merging, a special case of intersection management. Section 3.5 provides experimental results, and Sect. 3.6 concludes the discussion.

3.1 Problem Formulation

In this section, we introduce our graph-based model and formulate the centralized intersection management problem. The notation is summarized in Table 4.

Conflict Zone Same as the definition in Sect. 2, a conflict zone is the crossing location of two trajectories, and two vehicles cannot be at (occupy) the same conflict zone at the same time. There are n conflict zones, $\Xi_1, \Xi_2, \ldots, \Xi_n$, in the intersection. This model allows us to consider different granularities of an intersection, as shown in Fig. 7.

Vehicle Each vehicle has a fixed route—it fixes its source lane, destination lane, and trajectory, and it does not change lanes before and after the intersection. Two

Fig. 6 A centralized unit installed in the roadside unit, called intersection manager, decides the passing order of the vehicles

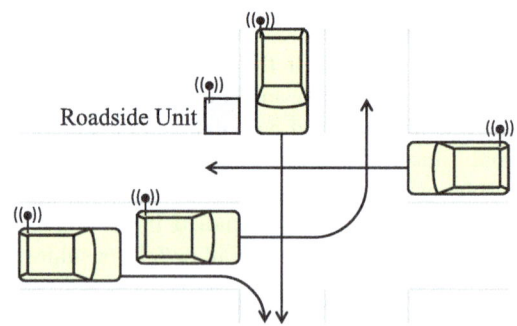

Table 4 The notation

Index	i, i'	The index of a vehicle
	j, j', j''	The index of a conflict zone
	(i, j)	The index of a vertex
	k	The index of an edge
Given (Input)	G	A timing conflict graph
	m	The number of vehicles
	n	The number of conflict zones
	Δ_i	The i-th vehicle
	Ξ_j	The j-th conflict zone
	$v_{i,j}$	The (i, j)-th vertex
	e_k	The k-th edge
	a_i	The earliest arrival time of Δ_i
	$p_{i,j}$	The vertex passing time of $v_{i,j}$
	w_k	The edge waiting time of e_k
Output	G'	An acyclic timing conflict graph
	$s_{i,j}$	The vertex entering time of $v_{i,j}$

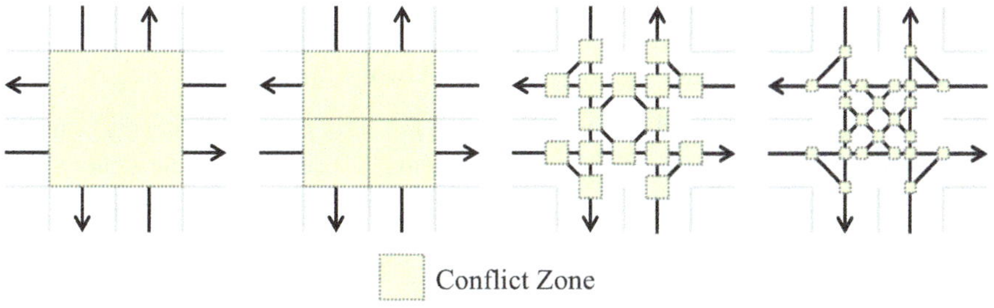

Conflict Zone

Fig. 7 The model allows us to consider different granularities of an intersection. The intersection can be modeled by 1, 4, 16, and 24 conflict zone(s), and much more alternatives are possible

vehicles have a potential *conflict* at zone Ξ_j if and only if Ξ_j is on the both trajectories.

Timing Conflict Graph A directed timing conflict graph $G = (V, E)$ is constructed by the following rules:

- There is a *vertex* $v_{i,j}$ if and only if Ξ_j is on the trajectory of Δ_i.
- There is a *Type-1 edge* $(v_{i,j}, v_{i,j'})$ if and only if the next conflict zone of Ξ_j on the trajectory of Δ_i is $\Xi_{j'}$.
- There is a *Type-2 edge* $(v_{i,j}, v_{i',j})$ if and only if Δ_i and $\Delta_{i'}$, on the same source lane and with the order where Δ_i is in front of $\Delta_{i'}$, have a conflict at Ξ_j.
- There are two *Type-3 edges* $(v_{i,j}, v_{i',j})$ and $(v_{i',j}, v_{i,j})$ if and only if Δ_i and $\Delta_{i'}$, on different source lanes, have a conflict at Ξ_j.

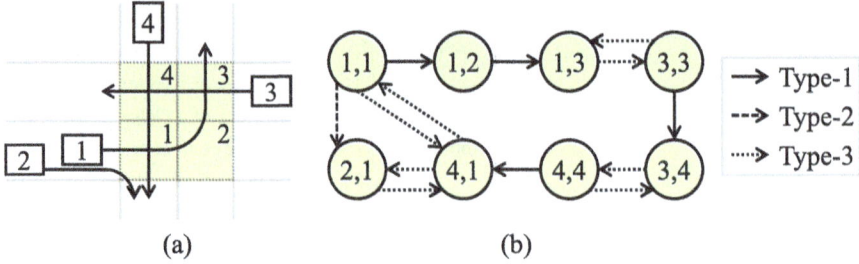

Fig. 8 (**a**) An example and (**b**) its timing conflict graph

Note that the vertex set is a subset of the Cartesian product of the sets of vehicles and conflict zones. An example and its timing conflict graph are shown in Fig. 8a, b, respectively.

Earliest Arrival Time Each vehicle Δ_i is associated with a_i, the earliest arrival time for Δ_i to arrive at the first conflict zone on its trajectory, without being delayed by any other vehicle (i.e., no vehicle is in front of Δ_i before the intersection). It can be either computed or provided by Δ_i or computed by the intersection manager.

Edge Waiting Time Each edge $e_k = (v_{i,j}, v_{i',j'})$ is associated with w_k, the waiting time "length" from Δ_i leaving Ξ_j to $\Delta_{i'}$ entering $\Xi_{j'}$, without being delayed by any other vehicle. For a Type-1 edge e_k (where $i = i'$), w_k is the time from Δ_i leaving Ξ_j to Δ_i entering $\Xi_{j'}$; for a Type-2 or Type-3 edge e_k (where $j = j'$), w_k is the time from Δ_i leaving Ξ_j to $\Delta_{i'}$ entering Ξ_j. In practice, the waiting time of a Type-2 edge e_k is smaller than that of a Type-3 edge $e_{k'}$ as vehicles from the same source lane can perform better in vehicle-following.

Vertex Passing Time Edge vertex $v_{i,j}$ is associated with $p_{i,j}$, the time "length" for Δ_i from entering Ξ_j to leaving Ξ_j.

Vertex Entering Time Each vertex $v_{i,j}$ is associated with $s_{i,j}$, the time for Δ_i to enter Ξ_j, which implies that the earliest time for Δ_i to leave Ξ_j is $s_{i,j} + p_{i,j}$. If a timing conflict graph G' is *acyclic*, the vertex entering time of each vertex is assigned as follows:[2]

- As the graph is acyclic, the assignment can follow a topological order. If there are multiple options, a Type-1 edge has a higher priority than a Type-2 or Type-3 edge.
- If $v_{i,j}$ is the first conflict zone on the trajectory of Δ_i,

---

[2] As there is dependency between vehicles, the vertex entering time of each vertex cannot be given as an input.

$$s_{i,j} = \max \left\{ a_i, \max_{k|e_k=(v_{i',j'},v_{i,j})\in G'} \left\{ s_{i',j'} + p_{i',j'} + w_k \right\}, \right.$$

$$\left. \max_{k'|e_k=(v_{i',j'},v_{i,j})\in G',e_{k'}=(v_{i',j'},v_{i',j''})\in G',e_k\neq e_{k'}} \left\{ s_{i',j''} - w_{k'} + w_k \right\} \right\}$$

(13)

Note that $j = j'$ is always true in this case. The last maximum term is to make sure that $\Delta_{i'}$ leaves $\Xi_{j'}$ for $\Xi_{j''}$ so that Δ_i can enter Ξ_j. For easier understanding, we can also set the intersection-entering point of each source lane as a conflict zone so that it is the first conflict zone of the trajectory of each vehicle from the source lane.

- Otherwise,

$$s_{i,j} = \max \left\{ \max_{k|e_k=(v_{i',j'},v_{i,j})\in G'} \left\{ s_{i',j'} + p_{i',j'} + w_k \right\}, \right.$$

$$\left. \max_{k'|e_k=(v_{i',j'},v_{i,j})\in G',e_{k'}=(v_{i',j'},v_{i',j''})\in G',e_k\neq e_{k'}} \left\{ s_{i',j''} - w_{k'} + w_k \right\} \right\}$$

(14)

Note that either $i = i'$ or $j = j'$ is always true in this case. If $i = i'$, the last maximum term is not needed.

Problem Formulation Given a conflict graph G, the earliest arrival time a_i of each vehicle Δ_i, the edge waiting time w_k of each edge e_k, and the vertex passing time $p_{i,j}$ of each vertex $v_{i,j}$, the problem is to

1. Compute an acyclic subgraph G' of G, where

   - For each vertex v_i in G, v_i is also in G',
   - For each Type-1 edge e_k in G, e_k is also in G',
   - For each Type-2 edge e_k in G, e_k is also in G',[3] and
   - For each pair of vertices $v_{i,j}$ and $v_{i',j}$ in G, there exists a path either from $v_{i,j}$ to $v_{i',j}$ or from $v_{i',j}$ to $v_{i,j}$ in G',

2. Guarantee no deadlock,
3. Assign the vertex entering time $s_{i,j}$ of each vertex $v_{i,j}$ (as the paragraph above), and
4. Minimize

$$\max_{v_{i,j}} \left(s_{i,j} + p_{i,j} \right),$$

(15)

which is the total time needed for all vehicles to go through the intersection.

---

[3] We do not consider overtaking in this section; otherwise, we can relax the constraint to potentially change Type-2 edges.

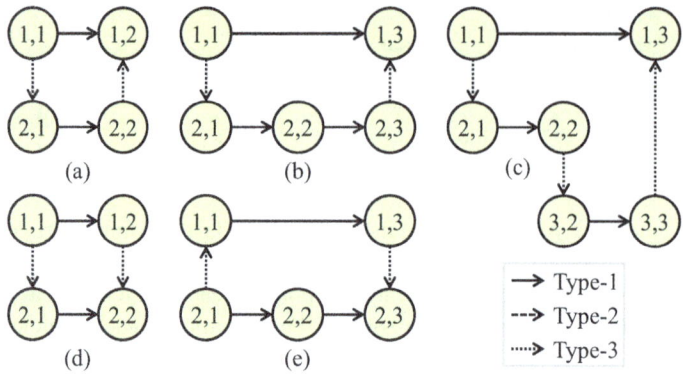

Fig. 9 (**a**)–(**c**) Examples with deadlocks and (**d**)–(**e**) examples without deadlocks

The item 1 is the safety (*collision-freeness*) property to guarantee an order for vehicles having a conflict. The item 3 follows the order to schedule vehicles, and the item 4 is the objective function. The item 2 is the liveness (*deadlock-freeness*) property. To this point, we have not detailed how to guarantee no deadlock—it will be demonstrated in the following section.

3.2 Deadlock-Freeness Verification

In this section, we will demonstrate a graph-based verification approach which can guarantee deadlock-freeness. A tailored Petri net [10] can also verify the deadlock-freeness. The verification can serve as a routine for the scheduling in Sect. 3.3 to verify deadlock-freeness for G'.

Having no cycle in G' or G does not guarantee deadlock-freeness.[4] Some examples are shown in Fig. 9.[5] All of them have no cycle in G', but Fig. 9a–c have deadlocks, and Fig. 9d–e are deadlock-free. In Fig. 9a, Δ_1 needs to enter Ξ_2 after Δ_2. However, Δ_2 even cannot enter Ξ_1(also, Ξ_2) because it is waiting Δ_1 to leave Ξ_1. That causes a deadlock. Similarly, there are deadlocks in Fig. 9b, c. On the contrary, in Fig. 9d, there is no deadlock as Δ_1 enters both Ξ_1 and Ξ_2 before Δ_2. In Fig. 9e, even if Δ_2 enters Ξ_1 first, Δ_2 can enter Ξ_2 after that so that Δ_1 is able to enter Ξ_1 (after Δ_2) and Ξ_3 (before Δ_2) without a deadlock.

---

[4] This is the reason that we need the item 2 in the problem formulation.

[5] To demonstrate the examples concisely, the examples in Fig. 9 are not associated with any intersection modeling in Fig. 7.

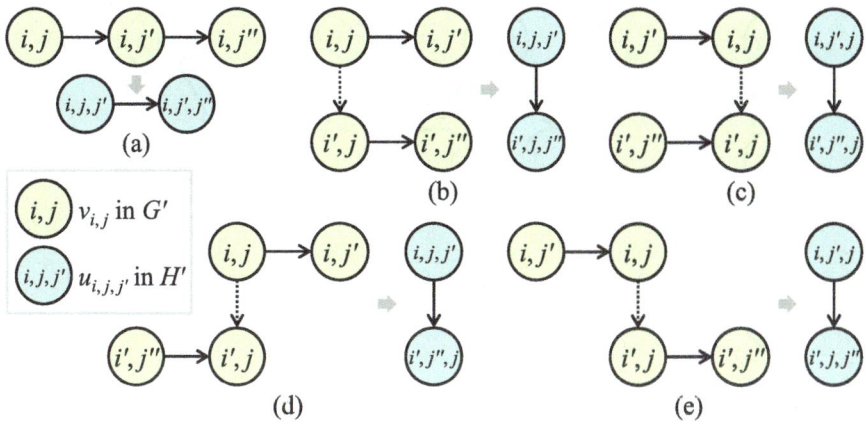

Fig. 10 The construction rules of resource conflict graphs

As illustrated above, having no cycle in G' cannot verify that there is no deadlock. Therefore, we introduce *resource conflict graphs* as follows:

Resource Conflict Graph The directed resource conflict graph H' of G' is constructed by the following rules:

- There is a *vertex* $u_{i,j,j'}$ if and only if there is a Type-1 edge $(v_{i,j}, v_{i,j'})$ in G'.
- If there are edges $(v_{i,j}, v_{i,j'})$ and $(v_{i,j'}, v_{i,j''})$ in G', then there is an edge $(u_{i,j,j'}, u_{i,j',j''})$ in H' (illustrated in Fig. 10a).
- If there are edges $(v_{i,j}, v_{i',j})$, $(v_{i,j}, v_{i,j'})$, and $(v_{i',j}, v_{i',j''})$ in G', then there is an edge $(u_{i,j,j'}, u_{i',j,j''})$ in H' (illustrated in Fig. 10b).
- If there are edges $(v_{i,j}, v_{i',j})$, $(v_{i,j}, v_{i,j'})$, and $(v_{i',j''}, v_{i',j})$ in G', then there is an edge $(u_{i,j',j}, u_{i',j'',j})$ in H' (illustrated in Fig. 10c).
- If there are edges $(v_{i,j}, v_{i',j})$, $(v_{i,j}, v_{i,j'})$, and $(v_{i',j''}, v_{i',j})$ in G', then there is an edge $(u_{i,j,j'}, u_{i',j'',j})$ in H' (illustrated in Fig. 10d).
- If there are edges $(v_{i,j}, v_{i',j})$, $(v_{i,j'}, v_{i,j})$, and $(v_{i',j}, v_{i',j''})$ in G', then there is an edge $(u_{i,j',j}, u_{i',j,j''})$ in H' (illustrated in Fig. 10e).

The general concept of the last four rules is that, if there is an edge $(v_{i,j}, v_{i',j})$ in G', then there is an edge from each vertex (which corresponds to an edge in G') involving $v_{i,j}$ to each vertex (which corresponds to an edge in G') involving $v_{i',j}$ in H'. It implies that, if Δ_i enters Ξ_j before $\Delta_{i'}$ enters Ξ_j, then Δ_i must leave Ξ_j before $\Delta_{i'}$ enters Ξ_j. The resource conflict graphs of the examples in Fig. 9 are shown in Fig. 11. We can observe that they are cyclic in Fig. 11a–c, while they are acyclic in Fig. 11d–e.

Theorem 4 H' *is cyclic if and only if* G' *has a deadlock.*

Proof From left-hand side (LHS) to right-hand side (RHS): If there is a cycle in H', we assume the cycle as $((i_0, j_0, j_0'), (i_1, j_1, j_1'), \ldots, (i_k, j_k, j_k'), \ldots, (i_l, j_l, j_l'))$,

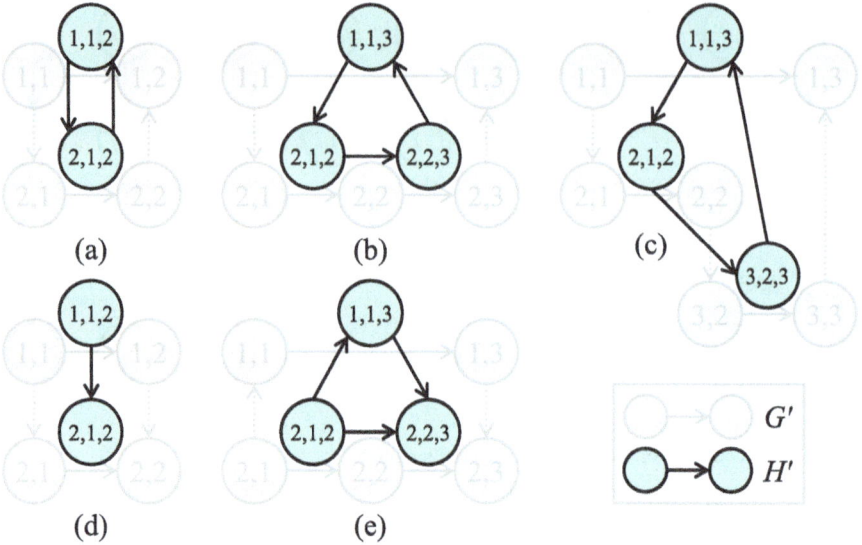

Fig. 11 The resource conflict graphs of the examples in Fig. 9

where $(i_l, j_l, j_l') = (i_0, j_0, j_0')$. By the construction rules of H', for any pair of (i_k, j_k, j_k') and $(i_{k+1}, j_{k+1}, j_{k+1}')$, at least one equality of $j_k = j_{k+1}$, $j_k = j_{k+1}'$, $j_k' = j_{k+1}$, and $j_k' = j_{k+1}'$ is true. Assume that it is equal to j^* in the true equality. By the definition of a conflict zone (that two vehicles cannot be at the same conflict zone at the same time), Δ_{i_k} must leave Ξ_{j^*} before $\Delta_{i_{k+1}}$ enters Ξ_{j^*}. This means that (i_k, j_k, j_k') blocks $(i_{k+1}, j_{k+1}, j_{k+1}')$, and thus, considering $0 \le k \le l - 1$, the cycle forms a deadlock.

From RHS to LHS: If there is a deadlock, without loss of generality, we assume that Δ_i cannot move from Ξ_j to $\Xi_{j'}$. The conditions that Δ_i cannot move from Ξ_j to $\Xi_{j'}$ include[6] (1) Δ_i cannot move from another conflict zone $\Xi_{j''}$ to Ξ_j, (2) another vehicle $\Delta_{i'}$ scheduled to enter Ξ_j earlier cannot enter Ξ_j, (3) another vehicle $\Delta_{i'}$ scheduled to leave Ξ_j earlier cannot leave Ξ_j, (4) another vehicle $\Delta_{i'}$ scheduled to enter $\Xi_{j'}$ earlier cannot enter $\Xi_{j'}$, and (5) another vehicle $\Delta_{i'}$ scheduled to leave $\Xi_{j'}$ earlier cannot leave $\Xi_{j'}$. By the construction rules of H', each of the conditions constructs an edge to vertex (i, j, j') in H'. Repeating applying the same conditions, those edges must form a cycle[7] since the numbers of vehicles and conflict zones are finite.   □

---

[6] If all of the conditions are false, then Δ_i can move from Ξ_j to $\Xi_{j'}$. A similar claim is not true for G', so having no cycle in G' cannot guarantee deadlock-freeness.

[7] Though it may not go back to (i, j, j').

By Theorem 4, H' is acyclic if and only if G' has no deadlock (deadlock-freeness). Note that we construct H' from G'. After the construction, we do not need G' in the verification.

3.3 Centralized Scheduling Approach

In this section, we develop a cycle removal algorithm based on the graph model in Sect. 3.1 and the verification approaches in Sect. 3.2.

A greedy strategy, a *First-Come-First-Serve* approach, can be adopted here to schedule the vehicles based on their earliest arrival times. However, this approach ignores the interactions between vehicles and conflict zones, and thus possibly leads to extra waiting time. To address this problem, with the graph-based model and the verification approaches, we can decide the passing order for vehicles to go through the intersection safely and efficiently by removing all cycles in the graph.

The most common method to detect and remove cycles in a directed graph is the Depth-First Search (DFS) algorithm [11]. There is a cycle in a graph only if a *back edge*, which is an edge from a vertex to itself or its ancestors, is found during the DFS traversal of the graph. Then, the method can remove any edge in the cycle to avoid having cycle in the graph. However, without optimization objective, the DFS method may not remove "good" edges to perform optimization. Furthermore, to decide a passing order, we cannot remove some edges because of the safety property (item 1) in our problem formulation, and thus the direct use of a DFS method is not feasible. On the other hand, the minimum feedback arc set problem, a special case of our problem, is NP-hard [12] and has not known to be approximable within a constant [13].

Our objective is to minimize the total time needed for all vehicles to go through the intersection, equivalent to the leaving time of the last vehicle. To remove cycles while considering the edge costs, finding a minimum spanning tree (MST) of the graph can be a potential solution, and one approach is the Kruskal's algorithm [14]. The Kruskal's algorithm repeatedly chooses a minimum-cost edge which does not form any cycle with those already-chosen edges. Kruskal also proposed the backward version of the original one, and it repeatedly removes a maximum-cost edge whose removal does not disconnect the graph. Inspired by this method, we do intend to remove the edge which results in the largest delay to the objective. This can remove cycles and benefit the objective minimization at the same time.

Based on our graph model, we develop a cycle removal algorithm. First, we compute the vertex entering time of each vertex without considering Type-3 edges. Next, the costs of Type-3 edges are estimated by their impacts on the objective. Then, we remove a Type-3 edge which has the largest cost from the graph. The impact of $(v_{i,j}, v_{i',j})$ on the objective is measured by considering $(v_{i,j}, v_{i',j})$ when recomputing the vertex entering time of each vertex. Repeating those steps, we can remove cycles and compute the vertex entering time of each vertex in the graph. It should be noted that, sometimes, we cannot remove an edge because of the last constraint of the item 1 in the problem formulation. In this case, we divide the problem into sub-problems and solve the sub-problems.

3.3.1 Definitions

We first provide some definitions which will be used in our algorithm as follows.

Edge State There are four possible states for an edge:

- An edge is **ON** if it has been decided to be kept (in G'). By the item 1 in the problem formulation, a Type-1 or Type-2 edge is always **ON**. When discussing the graph G', we only consider **ON** edges.
- An edge is **OFF** if it has been decided to be removed.
- An edge is **UNDECIDED** if it is going to be decided in the current sub-problem.
- An edge is **DONTCARE** if it is not considered in the current sub-problem.

Vertex State There are three possible states for a vertex:

- A vertex is **BLACK** if its vertex entering time has been scheduled. If $v_{i,j}$ is **BLACK**, then each edge $e_k = (v_{i,j}, v_{i',j'})$ or $(v_{i',j'}, v_{i,j})$ must be **ON** or **OFF**. On the other hand, if $e_k = (v_{i,j}, v_{i',j'})$ is **ON**, then $v_{i,j}$ must be **BLACK**.
- A vertex is **GRAY** if its vertex entering time can still be influenced by Type-3 edges. If $v_{i,j}$ is **GRAY**, then for each Type-1 or Type-2 edge $e_k = (v_{i',j'}, v_{i,j})$, $v_{i',j'}$ must be **BLACK**. When we remove edges, we only estimate the cost of an edge $e_k = (v_{i,j}, v_{i',j'})$, where at least one of $v_{i,j}$ and $v_{i',j'}$ is **GRAY**.
- A vertex is **WHITE** if its vertex entering time can be influenced by any type of edges.

Vertex Slack The vertex slack is the maximum time which can be delayed at the vertex without increasing the objective. We consider **ON** edges only. Similar to the computation of the vertex entering time, if G' is acyclic, we follow a reverse topological order and compute the vertex slack of each vertex $v_{i,j}$ as follows:

- If Ξ_j is the last conflict zone on the trajectory of Δ_i,

$$
slack\,[v_{i,j}] = \min \left\{ \max_{v_{i',j'} \in G'} \left(s_{i',j'} + p_{i',j'} \right) - \left(s_{i,j} + p_{i,j} \right), \right.
$$
$$
\left. \min_{k|e_k=(v_{i,j},v_{i',j'})\in G'} \left\{ slack\,[v_{i',j'}] \right\} \right\} \tag{16}
$$

- Otherwise,

$$
slack\,[v_{i,j}] = \min_{k|e_k=(v_{i,j},v_{i',j'})\in G'} \left\{ slack\,[v_{i',j'}] \right\} \tag{17}
$$

Algorithm 2 Cycle-removal-based scheduling

Input: G
Output: G'
1: Initialization;
2: **for** each vertex $v_{i,j} \in V$ **do**
3:    $state[v_{i,j}] \leftarrow$ WHITE;
4:    $slack[v_{i,j}] \leftarrow \infty$;
5: **end for**
6: **for** each edge $e_k \in E$ **do**
7:    **if** e_k is a *Type-3 edge* **then**
8:        $state[e_k] \leftarrow$ UNDECIDED;
9:    **else**
10:        $state[e_k] \leftarrow$ ON;
11:    **end if**
12: **end for**
13: Update-Time-Slack (G);
14: Remove-Type-3-Edges $(G, 0, m)$;
15: Output the resultant graph as G';

Edge Cost The edge cost of a Type-3 edge is the delay time of the objective caused by this edge if we keep it. For the edge $e_k = (v_{i,j}, v_{i',j})$, $(s_{i,j} + p_{i,j} + w_k)$ and $s_{i',j}$ are the vertex entering times of $v_{i',j}$ with and without considering e_k, respectively. If the delay time caused by e_k is larger than the slack of $v_{i',j}$, the objective will increase if we keep e_k. The edge cost of a Type-3 edge $e_k = (v_{i,j}, v_{i',j})$ is defined as follows:

$$cost\ [e_k] = (s_{i,j} + p_{i,j} + w_k) - s_{i',j} - slack\ [v_{i',j}] \tag{18}$$

Note that, although the edge cost may be negative, the objective will never decrease.

3.3.2 Cycle-Removal-Based Scheduling

To solve the cycle removal problem, we follow the steps listed in Algorithm 2. First, based on the problem formulation in Sect. 3.1, Type-1 and Type-2 edges must be included in G'. Thus, we set the states of all Type-1 and Type-2 edges to **ON** and the states of all Type-3 edges to **UNDECIDED**.

Next, we apply Algorithm 3 to compute the vertex entering times and slacks of vertices. At this moment, the graph G' contains only Type-1 and Type-2 edges and thus is an acyclic graph. According to its topological order, we compute the vertex entering time of each vertex by Eqs. (13) and (14) and the leaving time of the last vehicle. We also compute the slack of each vertex according to reverse topological order by Eqs. (16) and (17).

Then, we decide which edges to be removed by Algorithm 4. First, in the process of *Find-Leaders*, a vertex $v_{i,j}$ is identified as a *leader vertex* if Δ_i is the first vehicle of its source lane and Ξ_j the first conflict zone on the trajectory of Δ_i.

Algorithm 3 Update-time-slack

Input: G

  1: Initialization;
  2: Topological-Sort (G)
  3: **for** each vertex $v_{i,j}$ in topological order **do**
  4:     Compute $s_{i,j}$ by Eq. (13) or (14);
  5: **end for**
  6: $maxLeavingTime \leftarrow \max_{v_{i,j}}(s_{i,j} + p_{i,j})$;
  7: **for** each vertex $v_{i,j}$ in reverse topological order **do**
  8:     $slack[v_{i,j}] \leftarrow maxLeavingTime - s_{i,j} - p_{i,j}$;
  9:     **for** each edge $e_k = (v_{i,j}, v_{i',j'}) \in E$ **do**
10:        $slack[v_{i,j}] \leftarrow \min\{slack[v_{i,j}], slack[v_{i',j'}]\}$;
11:     **end for**
12: **end for**

Second, an **UNDECIDED** edge $e_k = (v_{i,j}, v_{i',j})$, i.e., a Type-3 edge, is identified as a *candidate edge* if $v_{i,j}$ or $v_{i',j}$ is a leader vertex. Third, we compute the edge cost of each candidate edge by Eq. (18). Fourth, we try to remove Type-3 edges in descending order of edge cost. Removing edge $e_k = (v_{i,j}, v_{i',j})$ means its reverse edge $e_{k'} = (v_{i',j}, v_{i,j})$ must be included in G' and cannot be removed. As a result, we temporarily set the state of e_k to **OFF** and $e_{k'}$ to **ON**. Then, we verify deadlock-freeness for the current G' by the verification approaches in Sect. 3.2. If G' is not deadlock-free, we recover e_k and remove $e_{k'}$ by exchanging their states and verify deadlock-freeness for G' again. If G' is deadlock-free after we decide the states of e_k and $e_{k'}$, we update the states of related vertices, identify newly set **GRAY** vertices as *leader vertices*, and recompute vertex entering times and slacks. Then, we perform the same process to the next highest cost edge.

However, sometimes G' may have a deadlock no matter we remove either e_k or $e_{k'}$. The reason is that the previous assignments of edges conflict with the decision of choosing e_k or $e_{k'}$. Backtracking the already removed edges is a solution for resolving the dilemma. Unfortunately, the backtracking suffers from a long runtime of finding a valid assignment. Therefore, instead of backtracking the removed edges, we divide the original problem to sub-problems. We partition all vehicles into two parts according to the ascending order of their earliest arrival times. The first part contains vehicles ordered before i_{end}, the second part contains the rest. Consider each pair of vehicles Δ_i and $\Delta_{i'}$, where Δ_i is in the first part, while $\Delta_{i'}$ the second. If Ξ_j is a common conflict zone on the trajectories of Δ_i and $\Delta_{i'}$, we assume that Δ_i will pass zone Ξ_j before $\Delta_{i'}$. The assumption implies the state of edge $(v_{i,j}, v_{i',j})$ is **ON** and the state of edge $(v_{i',j}, v_{i,j})$ is **OFF**. Therefore, when solving the sub-problem associated with the first part, we consider only Type-3 edges in between two vehicles belonging to the first part. For both Δ_i and $\Delta_{i'}$ in the first part, the state of their Type-3 edge $(v_{i,j}, v_{i',j})$ is set to **UNDECIDED** (to be decided in the current sub-problem); for both Δ_i and $\Delta_{i'}$ in the second part, the state of their Type-3 edge $(v_{i,j}, v_{i',j})$ is set to **DONTCARE** (ignored in the current sub-problem). After we have solved the sub-problem associated with the first part, we turn to the sub-problem

---

Algorithm 4 Remove-type-3-edges

Input: G, i_{start}, i_{end}
1: Initialization
2: **for** each vertex $v_{i,j} \in V$ **do**
3:     **if** $order[\Delta_i] \geq i_{start}$ **then**
4:         $state[v_{i,j}] \leftarrow$ WHITE;
5:     **end if**
6: **end for**
7: **for** each Type-3 edge $e_k = (v_{i,j}, v_{i',j}) \in E$ **do**
8:     **case** 1: $order[\Delta_i], order[\Delta_{i'}] < i_{start}$ **do**   do nothing;
9:     **case** 2: $i_{end} \leq order[\Delta_i], order[\Delta_{i'}]$ **do** $state[e_k] \leftarrow$ DONTCARE;
10:     **case** 3: $i_{start} \leq order[\Delta_i], order[\Delta_{i'}] < i_{end}$ **do** $state[e_k] \leftarrow$ UNDECIDED;
11:     **case** 4: $i_{start} \leq order[\Delta_i] < i_{end} \leq order[\Delta_{i'}]$ **do** $state[e_k] \leftarrow$ ON;
12:     **case** 5: $i_{start} \leq order[\Delta_{i'}] < i_{end} \leq order[\Delta_i]$ **do** $state[e_k] \leftarrow$ OFF;
13: **end for**
14: $f_{fail} \leftarrow$ FALSE;
15: $LeaderVertices \leftarrow$ Find-Leaders (i_{start}, i_{end});
16: **while** $LeaderVertices \neq \emptyset$ **do**
17:     $CandidateEdges \leftarrow$ Find-Candidates ($LeaderVertices$);
18:     **for** each edge $e_k = (v_{i,j}, v_{i',j})$ in $CandidateEdges$ **do**
19:         $cost[e_k] \leftarrow s_{i,j} + p_{i,j} + w_k - s_{i',j} - slack[v_{i',j}]$;
20:     **end for**
21:     $e_{max} \leftarrow$ Find-Max-Cost-Edge ($CandidateEdges$);
22:     $e_{max'} \leftarrow (v_{i',j}, v_{i,j})$ when $e_{max} = (v_{i,j}, v_{i',j})$;
23:     $state[e_{max}] \leftarrow$ OFF;
24:     $state[e_{max'}] \leftarrow$ ON;
25:     **if** VerifyGraph (G) is FALSE **then**
26:         $state[e_{max}] \leftarrow$ ON;
27:         $state[e_{max'}] \leftarrow$ OFF;
28:         **if** VerifyGraph (G) is FALSE **then**
29:             $f_{fail} \leftarrow$ TRUE; break;
30:         **end if**
31:     **end if**
32:     $LeaderVertices \leftarrow$ Update-Leaders ($LeaderVertices$);
33:     Update-Time-Slack (G);
34: **end while**
35: **if** f_{fail} is TRUE **then**
36:     $i_{mid} \leftarrow \frac{1}{2}(i_{start} + i_{end})$;
37:     Remove-Type-3-Edges (G, i_{start}, i_{mid});
38:     Remove-Type-3-Edges (G, i_{mid}, i_{end});
39: **end if**

---

associated with the second part based on the result derived in previously solved sub-problems. We set the Type-3 edges within the second part to **UNDECIDED** and keep those in the first part unchanged. These procedure is repeated until there are no **UNDECIDED** edges and all the vertices are **BLACK**.Finally, we obtain an acyclic graph G' and schedule the vertex entering time of each vertex in G' by Eqs. (13) and (14).

Theorem 5 *Our scheduling algorithm always finds a feasible solution.*

Proof A feasible solution should satisfy items (i) and (ii) in the problem formulation. Type-1 and Type-2 edges must be included in G', and they do not generate cycles or deadlocks. For Type-3 edges between any pair of $v_{i,j}$ and $v_{i',j}$, only one edge (either $e_k = (v_{i,j}, v_{i',j})$ or $e_{k'} = (v_{i',j}, v_{i,j})$) is selected by our algorithm. If we cannot determine all Type-3 edges at a time, the original problem of all vehicles is recursively divided into two sub-problems according to the ascending order of their earliest arrival times. For Type-3 edges in between two parts, we select only Type-3 edges from the first part to the second part. Hence, only Type-3 edges within one sub-problem have to be discussed. Every time, including a Type-3 edge in one sub-problem is verified by our verification approaches in Sect. 3.2 to guarantee cycle-freeness and deadlock-freeness. In the worst case of sub-problem division, each sub-problem solves only one vehicle. In this case, no Type-3 edges exist in between two vertices belonging to the same vehicle. As a result, the resultant G' is guaranteed to be acyclic and deadlock-free.   □

Theorem 6 *The time complexity of our scheduling algorithm is $O(E^2 \log V)$.*

Proof Our scheduling algorithm (Algorithm 2) contains three parts: vertex and edge state initialization, updating vertex entering times and slacks (Algorithm 3), and Type-3 edge removal (Algorithm 4). Vertex/edge state initialization can be done by graph traversal in $O(V + E)$ time. Vertex entering times and slacks can be computed in $O(V + E)$ time based on topological sort and graph traversal. For Type-3 edge removal, assume the induced subgraph for a sub-problem covers a vertex subset $V_s \subseteq V$ and an edge subset $E_s \subseteq E$. The running time of each sub-problem is dominated by the while loop and sub-problem division in Algorithm 4. The while loop examines each Type-3 edge at most once, and the verifier takes $O(V + E)$ time. Thus, the while loop takes a total of $O(E_s(V + E))$ time. The recurrence for the running time $T(V_s, E_s, V, E)$ of Algorithm 4 can be written as $T(V_s, E_s, V, E) = T\left(\frac{V_s}{2}, \alpha E_s, V, E\right) + T\left(\frac{V_s}{2}, \beta E_s, V, E\right) + O(E_s(V + E))$, where $\alpha + \beta \le 1$. In the base case, every sub-problem contains only one vehicle, and $T(1, E_s, V, E)$ takes $O(E)$ time. The overall running time $T(V, E, V, E)$ of Algorithm 4 is $O(VE + E(V + E) \log V)$. Therefore, our scheduling algorithm takes $O(E^2 \log V)$ time.

□

In practical cases, the number of vehicles near an intersection is less than 100, and the experimental results will show the efficiency applicable in real time.

3.4 A Special Case: Lane Merging

Lane merging is the process that vehicles from different incoming lanes merge into one outgoing lane and is one of the major sources causing traffic congestion and delay. For example, in a two-lane merging problem, we have two incoming lanes merging into one outgoing lane. There is no priority for each lane (i.e., no main or secondary lane), and vehicles are not allowed to overtake other vehicles during the process. For two-lane merging, the merging intersection is the sole conflict zone, the merging point is a representative point of the merging intersection. We can optimally solve the two-lane merging scenario by a dynamic programming algorithm. It decomposes the problem into a series of sub-problems to schedule the passing order for vehicles while minimizing the time needed for all vehicles to go through the merging point (equivalent to the time that the last vehicle goes through the merging point). We can extend the problem to a consecutive lane-merging scenario, which is fundamental to further generalization.

3.5   Experimental Results

We implemented the verification approach and scheduling algorithms in the C++ programming language. The experiments were run on a macOS mojave notebook with 2.3 GHz Intel CPU and 8 GB memory. The traffic is generated at every source lane by Poisson distribution where the parameter of Poisson distribution λ is set to as 0.1, 0.3, 0.5, 0.6, and 0.7. The higher λ, the higher traffic density. When $\lambda = 0.1$, the average time interval between two incoming vehicles is 10 s, while it is 2 s when $\lambda = 0.5$. The respective edge waiting time of a Type-1 edge, Type-2 edge, and Type-3 edge is 0.1, 0.2, and 0.2, respectively. The minimum time for a vehicle to pass a conflict zone is set to 1 second, which means a vehicle takes 1 s to pass a conflict zone without considering other vehicles.

3.5.1   Scheduling Effectiveness and Efficiency

In the first experiment, a four-way intersection is considered. For each direction, there is only one incoming lane and one outgoing lane. Four conflict zones are generated according to the crossing locations of four incoming lanes. Two traffic settings are generated. In the first setting, the earliest arrival time of the last vehicle is 30 s, meaning that the intersection manager is required to have a communication range covering all vehicles that will arrive in 30 s. In the second setting, the earliest arrival time of the last vehicle is 60 s. For each vehicle, the probability of going straight, taking a right turn, or taking a left turn is generated by a uniform distribution.

As listed in Tables 5 and 6, the proposed scheduling algorithm is compared with three approaches: (1) 3D-Intersection, (2) First-Come-First-Serve, and (3) Priority-Based. In the 3D-Intersection approach, vehicles do not consider the conflicts with vehicles on other lanes so that a vehicle is delayed only by vehicles on the same lane and in front of it. Thus, the 3D-Intersection approach provides a lower bound for the objective (T_L), although it may not be collision-free. The First-Come-First-Serve approach was introduced in Sect. 3.3. The distributed priority-based approach in [7] is modified to fit in our graph-based model and problem formulation. The Priority-Based approach iteratively decides the passing order of vehicles by their

Table 5 Results under different λ when the earliest arrival time of the last vehicle is 30 s where T_L, T_D, and RT are the leaving time of the last vehicle, the average delay time of all vehicles, and the runtime, respectively (all units are in second)

		3D-intersection			First-come-first-serve		
λ	m	T_L	T_D	RT	T_L	T_D	RT
0.1	11	33.40	0	0.001	33.40	0.00	0.003
0.3	34	40.70	0	0.003	50.70	5.85	0.005
0.5	58	42.40	0	0.006	82.40	19.58	0.009
0.6	66	40.50	0	0.009	90.39	24.65	0.011
0.7	77	46.10	0	0.010	90.20	23.68	0.013
		Priority-Based			Ours		
λ	m	T_L	T_D	RT	T_L	T_D	RT
0.1	11	33.40	0.00	0.009	33.40	0.00	0.002
0.3	34	44.50	3.17	0.007	40.80	2.23	0.015
0.5	58	68.20	10.62	0.013	60.40	6.91	0.057
0.6	66	70.10	12.31	0.020	68.70	13.65	0.119
0.7	77	74.90	13.44	0.024	72.80	13.46	0.174

Table 6 Results under different λ when the earliest arrival time of the last vehicle is 60 s where T_L, T_D, and RT are the leaving time of the last vehicle, the average delay time of all vehicles, and the runtime, respectively (all units are in second)

		3D-Intersection			First-Come-First-Serve		
λ	m	T_L	T_D	RT	T_L	T_D	RT
0.1	25	66.30	0	0.002	68.80	0.48	0.005
0.3	66	68.80	0	0.009	89.19	10.84	0.013
0.5	104	74.00	0	0.015	131.10	26.75	0.020
0.6	129	71.50	0	0.026	149.20	37.62	0.033
0.7	157	72.90	0	0.039	176.50	54.67	0.049
		Priority-Based			Ours		
λ	m	T_L	T_D	RT	T_L	T_D	RT
0.1	25	68.80	0.48	0.008	66.90	0.32	0.006
0.3	66	73.50	2.36	0.015	71.10	1.78	0.070
0.5	104	105.30	12.30	0.052	98.40	11.80	0.229
0.6	129	133.00	27.64	0.091	116.90	20.77	0.626
0.7	157	157.80	38.49	0.157	139.50	34.22	1.825

priorities, and the priorities may change after each iteration. In our experiment, for every 1.0 second, the priorities are updated according to the newly estimated earliest arrival times to intersection.

All approaches are evaluated by two criteria: (1) the leaving time of the last vehicle T_L and (2) the average delay time of all vehicles T_D. T_L is equivalent to the total time needed for all vehicles to go through the intersection. On the other hand, since the 3D-Intersection approach provides the lower bound of T_L, the average delay time of all vehicles T_D is computed as the average of the difference between each vehicle's leaving time and its leaving time in the 3D-Intersection solution. The average delay time of the 3D-Intersection approach itself is always 0.

We demonstrate the effectiveness and efficiency of our algorithm by changing (1) the traffic density and (2) the communication range.

Different Traffic Densities Table 5 shows the impact of traffic density on scheduling. Note that 3D-Intersection provides the lower bounds of T_L and T_D. When λ is 0.1, all approaches can achieve the optimal solution on T_L and T_D due to low traffic density. However, when λ becomes higher, the T_L and T_D of the First-Come-First-Serve approach increase rapidly, and our algorithm can always achieve better results than the First-Come-First-Serve approach. This is because our algorithm considers more vehicles and their interactions, i.e., a global view, and provides a systematic approach to optimize the objective. Only few cases, e.g., $\lambda = 0.6$ or 0.7 when the earliest arrival time of the last vehicle is 30 s, the priority-based approach achieves better T_D than ours. The main reason is that its frequent updates (every 1.0 second) on the earliest arrival times can sometimes mend the lack of a global view. If the update is not fast enough, its effectiveness will decline.

Different Communication Ranges The communication range of an intersection manager is an important factor. To show the flexibility of communication ranges of our algorithm, we compare Table 5 with 6 to observe the results generated by different communication ranges under same λ. In Table 5, the communication range of the intersection manager covers all vehicles that will arrive in 30 s. In Table 6, the communication range of the intersection manager covers all vehicles that will arrive in 60 s. As the communication range becomes twice larger, the T_L of our algorithm also becomes approximately twice larger, which means different communication ranges do not affect the solution quality of our algorithm.

Overall, the proposed scheduling algorithm always achieves better solutions than the First-Come-First-Serve approach under different scenarios. Our algorithm is sufficiently efficient for real-time use even when the number of vehicles reaches 100, which can be completed in around 1 second.[8] As the number of vehicles exceeds 100, the runtime grows up. However, the number of vehicles in an intersection will not exceed 100 in most cases. Even if the number of vehicles is large, we can still split the traffic and schedule the front vehicles first because it is impossible for 100 vehicles to go through the intersection in 1 second.

3.5.2 Modeling Expressiveness

In the second experiment, we show the expressiveness and generality of our modeling for different granularities of an intersection. The four-way intersection is modeled by 1 (like the previous work [7]), 4, and 16 conflict zone(s) as shown in Fig. 7. As shown in Table 7, when λ is low, different granularities of an intersection lead to near-optimal solutions because of few conflicts between vehicles. However,

---

[8] It is believed that an intersection manager has much better computational capability than a current vehicle.

Table 7 Results of the proposed algorithm under different numbers of conflict zones, where T_L, T_D, and RT are the leaving time of the last vehicle, the average delay time of all vehicles, and the runtime, respectively (all units are in second)

λ	m	1 Conflict Zone			4 Conflict Zones			16 Conflict Zones		
		T_L	T_D	RT	T_L	T_D	RT	T_L	T_D	RT
0.1	11	33.40	0.09	0.004	33.40	0.00	0.002	34.50	0.72	0.004
0.3	34	50.30	6.29	0.016	40.80	2.23	0.014	44.20	3.05	0.020
0.5	58	77.70	16.87	0.092	60.40	6.91	0.057	51.40	5.13	0.103
0.6	66	89.00	24.03	0.188	68.70	13.65	0.119	55.80	6.34	0.134
0.7	77	100.70	28.84	0.284	72.80	13.46	0.174	64.20	9.37	0.769

when λ becomes higher, the intersection modeled by 4 conflict zones always has better solutions than that modeled by 1 conflict zone. Similarly, intersection modeled by 16 conflict zones has better solution than those modeled by 1 and 4 conflict zone(s) in most cases. The finer granularity of an intersection, the more delicate intersection modeling and solution space, and thus the better scheduling results. It should be mentioned that we provide general modeling, scheduling, and verification for intersection management, and they can further assist intersection designers (i.e., governments or city planners) to design intersections (e.g., the number of conflict zones, the passing speed, the safety gap, the communication range, etc.).

3.6 Conclusion

In this section, we propose a timing conflict graph model for centralized intersection management. The model is very general and applicable to different granularities of intersections and other conflicting scenarios. We devise a resource conflict graph for formally verifying deadlock-freeness. Based on the graph-based models, we develop a cycle removal algorithm to schedule vehicles to go through the intersection safely (without collisions) and efficiently without deadlocks. The algorithm is sufficiently efficient to consider more conflict zones and more vehicles in real time. Experimental results demonstrate the expressiveness of the proposed model and the effectiveness and efficiency of the proposed algorithm.

4 Summary

In this chapter, we consider connected and autonomous vehicles at intersections and introduce distributed and centralized approaches solving the problem of intersection management. The approaches provide feasibility, safety (collision-freeness), liveness (deadlock-freeness), stability, efficiency, and real-time decision. Distributed and centralized approaches have their own advantages and disadvantages. We believe that they are suitable for different intersections. For example, a distributed approach for a small intersection; a centralized approach for a large intersection. The trade-offs between different factors and properties should be handled to match the real-world scenarios.

References

1. Liu, C., Tomizuka, M.: Enabling safe freeway driving for automated vehicles. In: 2016 American Control Conference, pp. 3461–3467. IEEE, Piscataway (2016)
2. Liu, C., Chen, J., Nguyen, T.-D., Tomizuka, M.: The robustly-safe automated driving system for enhanced active safety. Technical report, SAE Technical Paper (2017)
3. Dresner, K., Stone, P.: Multiagent traffic management: an improved intersection control mechanism. In: Proceedings of the Fourth International Joint Conference on Autonomous Agents and Multiagent Systems, pp. 471–477. ACM, New York (2005)
4. Azimi, S., Bhatia, G., Rajkumar, R., Mudalige, P.: Reliable intersection protocols using vehicular networks. In: Proceedings of the ACM/IEEE 4th International Conference on Cyber-Physical Systems (ICCPS'13), pp. 1–10. ACM, New York (2013). https://doi.org/10.1145/2502524.2502526
5. Pandit, K., Ghosal, D., Zhang, H.M., Chuah, C.-N.: Adaptive traffic signal control with vehicular ad hoc networks. IEEE Trans. Veh. Technol. **62**(4), 1459–1471 (2013)
6. Liu, C., Zhan, W., Tomizuka, M.: Speed profile planning in dynamic environments via temporal optimization. In: 2017 IEEE Intelligent Vehicles Symposium (IV), pp. 154–159. IEEE, Piscataway (2017)
7. Liu, C., Lin, C.-W., Shiraishi, S., Tomizuka, M.: Distributed conflict resolution for connected autonomous vehicles. IEEE Trans. Intell. Veh. **3**(1), 18–29 (2017)
8. Zhou, H., Liu, C.: Distributed motion coordination using convex feasible set based model predictive control. In: International Conference on Robotics and Automation (ICRA 2021) (2021)
9. An, J., Giordano, G., Liu, C.: Flexible MPC-based conflict resolution using online adaptive ADMM. In: European Control Conference (ECC 2021) (2021)
10. Peterson, J.L.: Petri nets. ACM Comput. Surv. **9**(3), 223–252 (1977)
11. Kleinberg, J., Tardos, E.: Algorithm Design. Pearson, Addison Wesley, London (2006)
12. Karp, R.M.: Reducibility among combinatorial problems. In: Complexity of Computer Computations. The IBM Research Symposia Series, pp. 85–103 (1972)
13. Kann, V.: On the approximability of np-complete optimization problems. Ph.D. Thesis, Department of Numerical Analysis and Computing Science, Royal Institute of Technology, Stockholm (1992)
14. Kruskal, J.B.: On the shortest spanning subtree of a graph and the traveling salesman problem. Am. Math. Soc. **7**(1), 48–50 (1956)

Part II
Security-Aware Design

Security-Aware Design of Time-Critical Automotive Cyber-Physical Systems

Vipin Kumar Kukkala, Thomas Bradley, and Sudeep Pasricha

1 Introduction

Today's vehicles are complex cyber-physical systems with tens of interconnected Electronic Control Units (ECUs) that control various subsystems in the vehicle. The introduction of Advanced Driver Assistance Systems (ADAS) in vehicles to support the goals of autonomy has resulted in an increase in the number of ECUs, which in turn has increased the complexity of the in-vehicle network that connects the ECUs. Moreover, state-of-the-art ADAS relies on information from various external systems using advanced communication protocols such as vehicle-to-vehicle (V2V) and vehicle-to-infrastructure (V2I) [1]. These advances increased the complexity of automotive systems, which introduced several other challenges related to reliability [2–6], real-time performance [7–10] and security [11–15] of automotive systems. In this chapter, we focus on improving security in automotive systems. The increased connectivity of today's vehicles has made them highly vulnerable to various sophisticated cyber-attacks. Therefore, ensuring the security of automotive systems is a crucial concern and will become further crucial as connected and autonomous vehicles (CAVs) become more ubiquitous.

The most commonly seen cyber-attacks on vehicles include masquerade, replay, and denial of service (DoS) attacks [16]. In a *masquerade* attack, the attacker

V. K. Kukkala (✉)
NVIDIA, Santa Clara, CA, USA
e-mail: vipin.kukkala@colostate.edu

T. Bradley
Department of Systems Engineering, Colorado State University, Fort Collins, CO, USA
e-mail: thomas.bradley@colostate.edu

S. Pasricha
Colorado State University, Fort Collins, CO, USA
e-mail: sudeep.pasricha@colostate.edu

© The Author(s), under exclusive license to Springer Nature Switzerland AG 2023
V. K. Kukkala, S. Pasricha (eds.), *Machine Learning and Optimization Techniques for Automotive Cyber-Physical Systems*, https://doi.org/10.1007/978-3-031-28016-0_4

121

pretends to be an existing ECU in the system. In a *replay* attack, the attacker eavesdrops on the in-vehicle network, captures valid messages transmitted by other ECUs, and sends them on the network in the future. In a *DoS* attack, the attacker ECU floods the in-vehicle network with random messages, thereby preventing the normal operation of valid ECUs. Most of these attacks require access to the in-vehicle network, which can be acquired either physically (e.g., using on-board diagnostics (OBD-II)) or remotely (e.g., using LTE or Bluetooth). Various real-world approaches to gaining access to the in-vehicle network and taking control of the vehicle by sending malicious messages are discussed in detail in [17–20].

Traditional in-vehicle network protocols, such as controller area network (CAN), FlexRay, etc., fail to address key security concerns such as confidentiality, authentication, and authorization as they do not have any inherent security features. Thus, additional security mechanisms (e.g., encryption-decryption) must be implemented in ECUs to prevent unauthorized access to the in-vehicle network. The two most widely used encryption techniques are - *symmetric key encryption* and *asymmetric key encryption*. The former uses the same key for encryption and decryption operations, while the latter uses a public-private key pair that has a strong mathematical relation. Both mechanisms incur computational overhead on the ECUs, which may catastrophically delay the execution of real-time automotive tasks and message transfers, e.g., a delay in the messages from impact sensors to airbag deployment systems could lead to severe injuries for vehicle occupants. Thus, it is highly crucial to carefully introduce security mechanisms in the vehicles.

The individual ECU utilizations of a FlexRay-based automotive system consisting of four ECUs running 12 different hard real-time automotive applications (each with multiple tasks) is illustrated in Fig. 1. Each ECU has a real-time utilization due to the execution of real-time automotive tasks (*RT Util*) and a security utilization because of the execution of security operations (*Sec Util*). The numbers on top of each bar show the number of applications that miss their deadlines when executed on the corresponding ECU. Along the x-axis, the *no security mechanism* case has no security mechanism implemented (hence it only has the real-time utilization), while in the *unoptimized security mechanism* case, all the ECUs employ AES-256 for encryption and decryption of messages. In the latter case, it can be seen that the total utilization for ECUs 3 and 4 (sum of real-time and security task utilizations) exceeds 100% (represented by the red dotted line) because of the overhead of security-specific encryption/decryption task executions, resulting in missed deadlines for four applications. Lastly, the *optimized security mechanism* case represents our goal in this work, to integrate all required security mechanisms while keeping utilization of all ECUs below 100%, without any deadline violations.

In this chapter, we present a novel security framework called *SEDAN*, which was first introduced in [11]. *SEDAN* is a lightweight (minimal overhead on the ECUs) security framework that aims to maximize the overall security of the automotive system without violating real-time deadline constraints and per-message security constraints. Moreover, the *SEDAN* framework employs symmetric key cryptography as it is less computationally intense compared to the asymmetric key cryptography to enhance the security of the vehicle. Our novel contributions in *SEDAN* are:

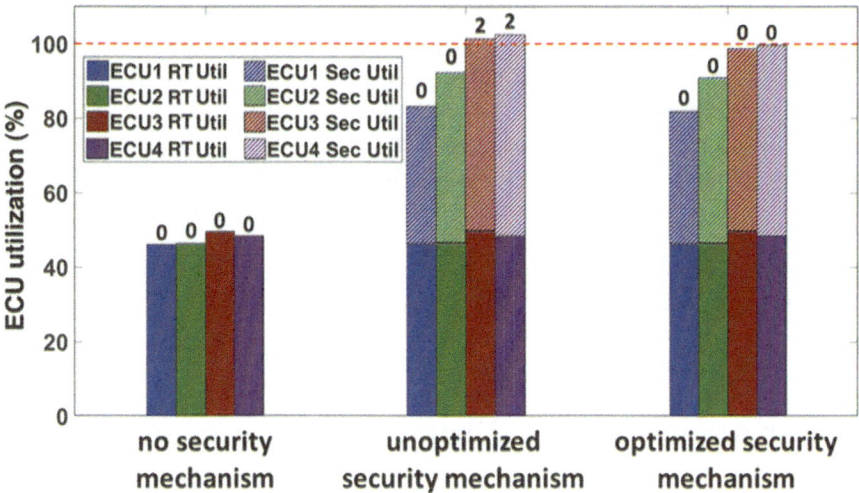

Fig. 1 Motivation for a lightweight (low overhead) vehicular security framework. The number on top of each bar indicates the number of missed real-time application deadlines

- We introduced a novel quantitative methodology to derive the security requirements for various messages in an automotive system based on ISO 26262
  standard and formulated a new metric to quantify the overall security of a system;
- We devised a heuristic-based key management technique to provide adequate
  security for various message types and ensure that the utilization of all ECUs is
  below 100%;
- We developed an approach for the joint exploration and synthesis of message
  schedules and security characteristics in TDMA-based automotive systems and
  also proposed a technique to efficiently map tasks to ECUs while meeting real-
  time message deadlines and ECU utilization goals;
- We extracted network traffic and ECU execution data from a real-world vehicle
  (2016 Chevrolet Camaro) and compared *SEDAN* with [21], the best-known prior
  work in the area, to demonstrate the effectiveness and scalability of *SEDAN*.

2 Related Work

Security in automotive systems was not a primary concern until recently. The first
full vehicle hack in 2010 [17] highlighted the need for concrete security measures in
automotive systems. In [17], the researchers had physical access to the vehicle and
were able to control various systems in the vehicle by injecting custom messages
into the CAN bus. Moreover, they reverse-engineered a subset of the ECUs and
were able to update the firmware on those ECUs by sending custom CAN messages.
Later in [18], they were able to perform the same attacks remotely. In [19], the

researchers hacked the radio in a 2014 Jeep Cherokee and were able to control the vehicle remotely. They used the telematics system in the radio to send remote messages to the vehicle, which were injected into the CAN bus to take control of various vehicular subsystems. In [20], the authors recently developed a Trojan app that was executed on a smartphone connected to the vehicle infotainment system via Bluetooth. They used this app to send custom CAN messages into the in-vehicle network. All these attacks have raised serious concerns about security in automotive systems.

Since the traditional in-vehicle network protocols do not provide any security features, it is hard to prevent unauthorized access to the in-vehicle network. However, one of the popular solutions in the literature to prevent unauthorized access is authenticating the sender ECU using message authentication codes (MACs). Several works, such as [22–28], advocate the use of MACs to improve security in automotive systems. In [23], a mixed integer linear programming (MILP) formulation was proposed to minimize the overhead for MAC computation and end-to-end application latency in a CAN-based system. Moreover, the authors in [23] use the same MAC for a group of ECUs. In [24], the authors extended [23] to minimize the security risks associated with grouping different ECUs. An authentication protocol called LCAP was presented in [26] to encrypt messages that utilized hash functions to generate hashed MACs to authenticate ECUs. In [27], an RC4 encryption-based authentication was implemented to improve security in CAN-based systems. Another lightweight authentication scheme based on PRESENT [29] was introduced in [28] and evaluated on FPGAs. However, cryptanalysts have demonstrated successful attacks on both RC4 and PRESENT. In [30], a technique based on obfuscating CAN message identifiers (IDs) was presented to protect a fleet of vehicles. *However, all the above-mentioned techniques are designed for event-triggered protocols (such as CAN) and do not apply to more scalable and sophisticated time-triggered protocols.*

A lightweight authentication technique is proposed in [22] that uses cipher-based MACs that are generated using the ECU local time stamp and a secret key. However, this technique requires strong synchronization between the ECUs, and any uncertainty can result in a full system failure. In [31], a device-level technique is presented, which uses an enhanced network interface (NI) to authenticate ECUs in the system by using hardware-based security modules (HSMs). In [25], FPGAs are employed as co-processors for ECUs to handle all security tasks that are implemented based on the TESLA [32] protocol. However, the techniques in [25, 31] require additional compute resources and many modifications to the existing automotive systems, which is not very practical and cost-efficient. In [33], the authors proposed a virtual local network (VLAN) based solution for improving security in Ethernet-based automotive systems. They introduced an integer linear programming (ILP) model to minimize message routing times and authenticate the messages by making multiple message transmissions on different routes. However, this technique results in inefficient bandwidth utilization and poor scalability. A co-design framework is introduced in [34] to improve message response times while meeting security concerns. However, only a small subset of messages are considered for encryption to guarantee control performance, which makes it impractical for safety-critical automotive systems and also exposes the system to various vulnerabilities.

A time delayed release of keys approach (adapted from the TESLA protocol [32]) is proposed in [21], in conjunction with simulated annealing based heuristic to minimize the end-to-end latency of messages by co-optimizing task allocation and message scheduling. This is one of the very few holistic frameworks that integrate the concept of security with real-time system design from the beginning of the system design phase. This work is extended in [35] by including V2V communication, using dedicated short-range communication (DSRC). In [36], a lightweight authentication technique for vehicles called LASAN is proposed, which uses the Kerberos protocol. The authors extended this work in [37] by presenting a comprehensive analysis of LASAN and compared with the TESLA [32] protocol. Though the LASAN technique demonstrated superior performance over other works, it has stringent requirements for a trusted centralized ECU, which creates a single point of failure. A security mechanism using different authentication methods was proposed for real-time systems in [38]. A group-based security service model is presented in [39] that tries to maximize the combined security of the system. However, as the model does not consider the timing constraints, it cannot be implemented in time-critical automotive systems.

An intrusion detection system (IDS) based on principal component analysis (PCA) is proposed in [40]. An IDS that detects the presence of an attacker by monitoring the increased transmission rates of the messages is proposed in [41]. In [42], the usage of reactive runtime enforcers called safety guards is proposed to detect the discrepancies between the input data from sensors and the output of the controllers. In [43], a challenge-response authentication approach was proposed to detect the presence of attackers. However, this technique requires prior and proprietary information about the sensors to function correctly.

All the above-mentioned prior works for securing time-triggered systems have various limitations: (i) they do not consider the utilization overhead on ECUs and latency overhead on messages due to the implemented security mechanisms, which results in over-optimistic results; (ii) they utilize only one key size for all messages, which that does not account for the heterogeneous security goals in real-time systems; (iii) they do not consider precedence constraints between tasks and messages, and; (iv) they consider homogenous single core ECUs which do not accurately represent today's vehicles. In this chapter, we present the *SEDAN* framework that addresses these limitations of the prior works. Moreover, *SEDAN* improves the security in vehicles with time-triggered networks while satisfying all security, utilization, and message timing constraints. We demonstrate it for the FlexRay protocol, but it can be easily extended to other time-triggered protocols, e.g., TTEthernet.

3 Problem Definition

3.1 System and Application Model

In this subsection, we present the automotive system model that was considered in *SEDAN*, where multiple ECUs execute different time-critical applications and are connected using a FlexRay-based network, as shown in Fig. 2. Each ECU has of two major components: a host processor (HP) and a communication controller (CC). The HP primarily runs the automotive and security applications, whereas a CC acts as an interface between the HP and the in-vehicle network (in this case, FlexRay bus) and is responsible for packing message data into frames, sending and receiving messages, and filtering out unwanted messages. Moreover, *SEDAN* considers heterogeneous HPs with different numbers of cores, which aligns with the state-of-the-art. It is important to note that the heterogeneity in this work is limited to varying the number of homogeneous cores per HP (i.e., multicore parallelism).

Each automotive application consists of dependent and independent tasks that are mapped to different ECUs and executed in the corresponding HPs. If two dependent tasks are mapped to the same ECU, they exchange information using shared memory. Otherwise, the tasks communicate with each other by exchanging messages over the FlexRay bus. A message contains one or multiple signals that are generated as a result of task execution on the ECU. The Signals are packed into messages by the HP and are sent to the CC to transmit as FlexRay frames on the bus. Automotive applications can be categorized into one of two types: *(i)* time-triggered (periodic) or *(ii)* event-triggered (aperiodic). Most safety-critical applications, e.g., collision avoidance, lane keep assist, anti-lock braking, etc., are

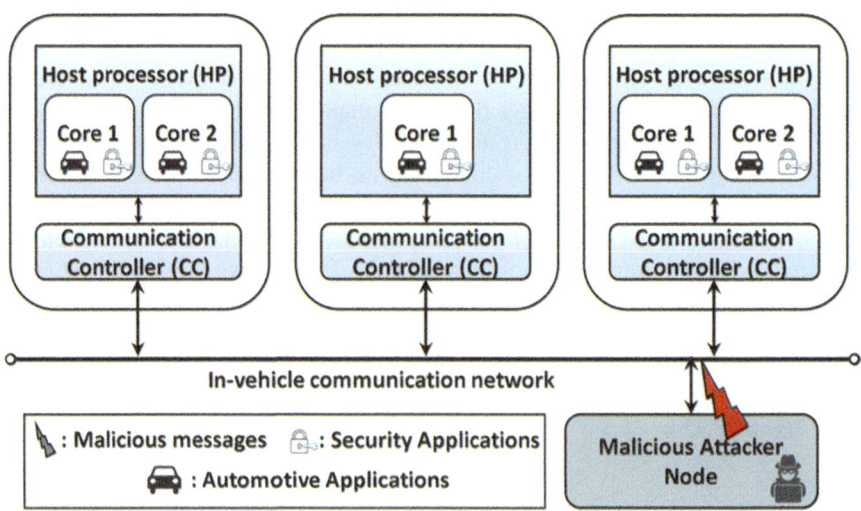

Fig. 2 Overview of the automotive system model used in *SEDAN*

time-triggered and generate time-triggered messages. Event-triggered messages are generated by maintenance and diagnostic applications. Moreover, much like real-time applications across other domains, the execution characteristics of time-critical automotive applications are known at design time. In this work, we focus on time-triggered applications as they significantly impact system performance and vehicle safety. Additionally, time-triggered messages generated by these applications have strict timing and deadline constraints. Thus, it is vital to optimize the security of the time-triggered messages while ensuring that no real-time deadline constraints are violated. In this work, we adapt various state-of-the-art standards, namely, Advanced Encryption Standard (AES) with key sizes 128,192 and 256 bits and evaluate Rivet-Shamir-Adleman (RSA) with key sizes 512, 1024, 2048, and 4096 bits, and Elliptic Curve Cryptography (ECC) with key sizes 256 and 384 bits to improve system security.

3.2 FlexRay Communication Protocol

FlexRay is an in-vehicle network protocol designed to support high-speed real-time complex automotive applications such as drive-by-wire applications. It supports both time-triggered and event-triggered transmissions and offers a data rate of up to 10 Mbps. The structure of the FlexRay protocol is illustrated in Fig. 3. A communication cycle is one complete instance of a communication structure that repeats periodically. Each communication cycle (also known as *cycle*) consists of a mandatory static segment, optional dynamic segment, optional symbol window, and mandatory network idle time. The static segment consists of multiple equally sized time slots that are used to send time-triggered messages. Each static segment slot consists of a header, payload (up to 254 bytes), and trailer segments. The TDMA media access scheme is employed in FlexRay static segment, where each ECU is assigned a particular static segment slot and a cycle number to transmit messages.

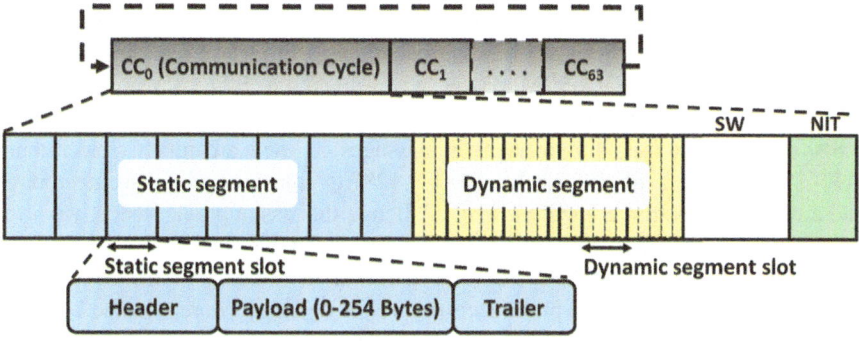

Fig. 3 Structure of the FlexRay protocol

On the other hand, the dynamic segment consists of variable-sized dynamic segment slots that are used to send event-triggered messages. Moreover, the dynamic segment employs a Flexible-TDMA media access scheme where the highest priority ECU gets access to the bus. The symbol window segment is used for signaling the start of the first communication cycle and network maintenance. Lastly, the network idle time segment helps with maintaining inter-ECU synchronization.

3.3 Attack Models

In this work, we focus on protecting the vehicle from masquerade and replay attacks as they are the most common, hard to detect, and can have a severe impact. The increased connectivity of modern vehicles with the external environment has created multiple pathways (attack vectors) to gain access to the in-vehicle network and ECUs. An attacker can choose a variety of attack vectors to gain access to the in-vehicle network and masquerade as an existing ECU or replay valid message transmissions to achieve malicious goals. In this study, we considered the most common and practical attack vectors in vehicles, which include connecting to the OBD-II port, connecting to systems that communicate with the external systems (such as infotainment systems), probe-based snooping on the vehicle bus, and replacing an existing ECU. Our framework can still be effective even when the attacker gains access to the in-vehicle network via other attack vectors.

3.4 Security Model

In this work, we focus on achieving the following key security objectives in vehicles: *(i)* confidentiality of message data and *(ii)* the authentication of ECUs. Meeting these objectives is crucial as it can help prevent masquerade and replay attacks. *Confidentiality* refers to the practice of protecting information from unauthorized ECUs, whereas *authentication* refers to the process of correctly identifying an ECU. In this study, we employ AES to achieve confidentiality by encrypting message data using a shared secret key. Moreover, we evaluate the choice of using RSA and ECC for setting up shared secret keys. However, it should be noted that neither RSA nor ECC is used for encrypting messages as they are much slower than AES. While AES with 128-bit keys (AES-128) is considered very secure today, the advent of quantum computing may challenge this assumption. Hence, we also consider AES-192 and AES-256. As each ECU can have messages of various criticalities, every ECU in the system can run all three variants of AES. Section 4.6 discusses the complete encryption/decryption flow in detail. Moreover, the key size for encrypting/decrypting messages is assigned based on the security requirements of a message, which is discussed in detail in Sect. 4.3.

3.5 Definitions

The *SEDAN* system model has the following inputs:

- Set of heterogeneous (1 or 2 core) ECUs $N = \{1, 2, \ldots, N\}$;
- Set of automotive applications $A = \{1, 2, \ldots, \lambda\}$ and set of tasks in the system $T = \{T_1 \cup T_2 \ldots \cup T_\lambda\}$, where T_a is the set of tasks in an application $a \in A$;
- Each task in T has a unique task ID $T_{ID} = \{1, 2, \ldots, G\}$;
- After task allocation, each task t is represented as $t_{q,n}$ where $q \in T_{ID}$ is the task ID, and $n \in N$ is the ECU to which the task t is mapped;
- Every task t is characterized by the 4-tuple $\{\tilde{a}_{q,n}, \tilde{p}_{q,n}, \tilde{d}_{q,n}, \tilde{e}_{q,n}\}$, where $\tilde{a}_{q,n}$, $\tilde{p}_{q,n}$, $\tilde{d}_{q,n}$, and $\tilde{e}_{q,n}$ represent the arrival time, period, deadline, and execution time of the task, respectively;
- For each ECU $n \in N$, $S_n = \{s_{1,n}, s_{2,n} \ldots, s_{K_n,n}\}$ is the set of signals transmitted from the ECU; K_n is the total number of signals in n;
- Every signal $s_{i,n} \in S_n$, $(i = 1, 2 \ldots, K_n)$ is characterized by the 4-tuple $\{\bar{a}_{i,n}, \bar{p}_{i,n}, \bar{b}_{i,n}, \bar{d}_{i,n}\}$, where $\bar{a}_{i,n}$, $\bar{p}_{i,n}$, $\bar{b}_{i,n}$, and $\bar{d}_{i,n}$ are the arrival time, period, deadline, and data size (in bytes) of signal $s_{i,n}$ respectively;
- After frame packing, each ECU has a set of messages $M_n = \{m_{1,n}, m_{2,n}, \ldots, m_{R_n,n}\}$, where R_n is the total number of messages in n;
- Every message $m_{j,n} \in M_n$, $(j = 1, 2, \ldots, R_n)$ is characterized by the 5-tuple $\{a_{j,n}, p_{j,n}, d_{j,n}, b_{j,n}, \Delta_{j,n}, \psi_{j,n}\}$ where $a_{j,n}$, $p_{j,n}$, $d_{j,n}$, $b_{j,n}$, $\Delta_{j,n}$, and $\psi_{j,n}$ are the arrival time, period, deadline, data size (in bytes), and minimum security requirement of the message $m_{j,n}$ (see Sect. 4.3), respectively. $\psi_{j,n}$ is a binary variable that has a value $= 1$ when the security constraints of the message are satisfied. Otherwise $\psi_{j,n} = 0$;

Problem Objective: In this work, we focus on maximizing security (aggregate security value, described in Sect. 4.4) while synthesizing a design time schedule for time-triggered tasks and messages that satisfy three types of constraints: *(i)* real-time timing and deadline constraints for tasks and messages in all applications; *(ii)* minimum security constraints for each message in the system, *(iii)* ensure no ECU utilization exceeds 100%.

4 *SEDAN* Framework: Overview

A high-level overview of the *SEDAN* framework is illustrated in Fig. 4, with all the design time steps in gray boxes and the runtime steps in green boxes. The steps involved in the *SEDAN* framework can be mainly classified into two categories: *(i)* security operations that improve the security of the system and *(ii)* real-time operations that satisfy the application's real-time performance objectives. At *design time*, *SEDAN* begins by allocating tasks to available ECUs in the system and generates the set of signals needed for inter-task communication. These signals are packed into messages using a frame packing approach, and security requirements

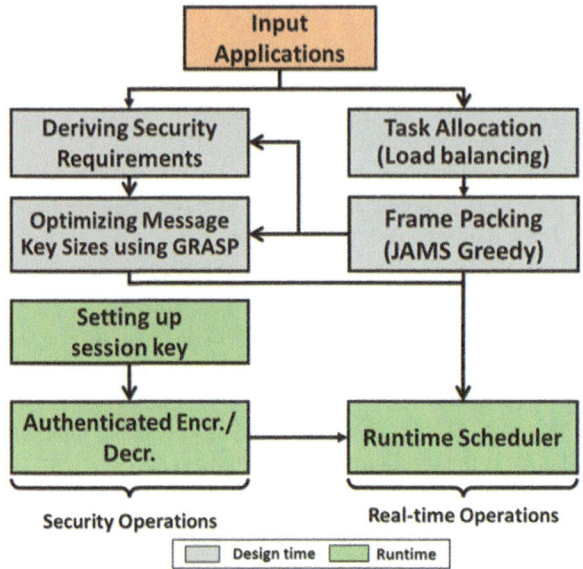

Fig. 4 Overview of the *SEDAN* framework

are derived for each message. The size of the keys used for encryption and decryption of the messages are optimized using a greedy randomized adaptive search procedure (GRASP) metaheuristic. At *runtime*, *SEDAN* starts with setting up the session keys, which will be used for generating keys used for authenticated encryption and decryption of messages. Lastly, a runtime scheduler schedules messages at runtime by using the previously generated keys and the optimal design time schedule. Each of these steps is discussed in detail in the subsequent subsections.

4.1 Task Allocation

This is the first step of the *SEDAN* framework and occurs at design time. The main goal of this step is to quickly allocate each task in the system to an available ECU that results in uniform real-time utilization across ECUs. This makes the *load-balancing task allocation scheme* a good choice for this step. Moreover, if there are some tasks that need to be allocated to certain ECUs, e.g., due to being in close proximity to sensors or actuators that they use heavily (or exclusively), we pre-allocate those tasks and do not include them in the set of mappable tasks for allocation.

For any task (t_q), the real-time utilization of the task ($\widetilde{U}_{t_q}$) is defined as the ratio of execution time $(\tilde{e}_q)$ and the period $(\tilde{p}_q)$ of the task, as shown in (1). The real-time utilization of any given ECU ($\widetilde{U}_n$) is the sum of the real-time utilizations of the tasks ($\widetilde{U}_{tq,n}$) allocated to that ECU, and is computed using (2):

$$\widetilde{U}_{t_q} = \frac{\tilde{e}_q}{\tilde{p}_q} \tag{1}$$

$$\widetilde{U}_n = \sum_{q=1}^{G_n} \left(\widetilde{U}_{tq,n} \right) \tag{2}$$

Our proposed *load-balancing task allocation* scheme begins by initializing all the ECUs' real-time utilization ($\widetilde{U}_n$) to zero and computing the real-time utilization of all the tasks ($\widetilde{U}_{t_q}$) in the system using (1). The allocation subsequently occurs in three steps: *(i)* the set of ECUs in the system is sorted in the increasing order of the ECU real-time utilization ($\widetilde{U}_n$); *(ii)* the first unallocated task in the set of tasks (T), sorted in decreasing order of real-time utilization, is selected and allocated to the least loaded ECU (i.e., ECU with the lowest utilization); and *(iii)* the task's real-time utilization ($\widetilde{U}_{t_q}$) is added to the allocated ECU's real-time utilization ($\widetilde{U}_n$). These three steps are repeated until all the unallocated tasks in T are allocated. If any task, $t \in T$, cannot be allocated to an ECU during this process, then there exists no solution for the given configuration. Otherwise, at the end of this step, each task in the system is allocated to an ECU. After the task allocation step, the set of signals S_n is generated for each ECU based on the precedence constraints of tasks in the application.

We also explored other allocation schemes that minimize the total communication volume between ECUs. However, it resulted in allocations that resulted in non-uniform load across ECUs, which violated the ECU utilization constraints after implementing security mechanisms.

4.2   Frame Packing

Frame packing is defined as the grouping of signals in each ECU into messages. This is done to maximize the bandwidth utilization of the communication bus. The set of signals generated by the task allocation step is given as the input to this step. The following conditions need to be satisfied to successfully pack the signals into messages: *(i)* for any two signals to be packed into the same message, they must originate from the same source ECU; *(ii)* signals with the same periods are packed together to avoid multiple message transmissions; and *(iii)* the total computed payload of the message is the sum of the size of the cipher generated by AES and the size of the MAC; and should not exceed the maximum possible FlexRay payload size. Because of the nature of AES, *the size of the generated cipher is independent of the key size.* However, the size of the cipher is dependent on the input size to the AES, which is the sum of signal sizes grouped in that message. Thus, the cipher size can be expressed as ⌈*sum of signal sizes in the message/*16⌉, and the size of MAC is set to the maximum of the minimum required MAC size (49 bits, explained further in Sect. 4.3; a designer can also use a value greater than 49). In this work, we adapted a fast *greedy frame packing* heuristic proposed in [2] and enhanced it by integrating the computed payload size definition to generate a set of messages for each ECU.

4.3 Deriving Security Requirements

In this subsection, we present a novel methodology used in *SEDAN* to derive security requirements for each message. We employ a risk classification scheme defined in ISO 26262 [44] known as the Automotive Safety Integrity Level (ASIL) as the basis for deriving security requirements for each message in the system. Four different ASILs: ASIL-A, ASIL-B, ASIL-C, and ASIL-D, are defined in the ISO 26262 standard to classify applications based on their risk upon failure. Applications classified as ASIL-D have the lowest failure rate limit indicating high criticality, while ASIL-A applications are less critical and subject to fewer security requirements. The underlying assumption for deriving security requirements based on ASIL groups is that the applications that demand high safety levels are more critical and need to be better protected from cyber-attacks. Hence, the higher the safety requirement, the higher the security requirement.

In this work, we define two security requirements for every message based on their ASIL classification.

The <u>first</u> requirement is the *minimum key size* required to encrypt the message depending on its ASIL group, which is as follows: ASIL-A (128 bits), ASIL-B (128 bits), ASIL-C (192 bits), and ASIL-D (256 bits). The following methodology is followed to derive ASIL groups for all messages in the system. Each application is assigned an ASIL depending on the criticality and tolerance to failure. Each task in that application inherits the same ASIL, and so do the signals generated by these tasks. When these signals are packed into messages, the highest ASIL group among the signals in that message is assigned as the ASIL group ($m_{j,n}^{AG}$) of the message. We also assign a security score $\left(m_{j,n}^{SS}\right)$ to each safety-critical message depending on its assigned key size. In this study, we consider the following score based on the key size: 128-bit key (score = 1), 192-bit key (score = 2), and 256-bit key (score = 3). Additionally, each message is assigned a weight value called *ASIL weight* $\left(m_{j,n}^{AW}\right)$. A high *ASIL weight* value indicates a high message criticality and is analogous to a Risk Priority Number (RPN) that can be calculated using Hazard Analysis and Risk Assessment (HARA) approaches [45]. Using the above-mentioned metrics, we derive a *security value* $\left(m_{j,n}^{SV}\right)$ for each message as shown in (3). Lastly, to quantitatively compare the security of different systems, we propose a metric called *Aggregate Security Value (ASV)*, which is computed using (4).

$$m_{j,n}^{SV} = m_{j,n}^{AW} * m_{j,n}^{SS} \tag{3}$$

$$\text{Aggregate Security Value } (ASV) = \frac{\sum_{n=1}^{N} \sum_{j=1}^{R_n} \left(\psi_{j,n} * m_{j,n}^{SV} \right)}{\sum_{n=1}^{N} R_n} \tag{4}$$

where $\psi_{j,n}$, and R_n are defined in Sect. 3.5. ASV is the ratio of the sum of security values of all messages in the system for which minimum security requirements are satisfied to the total number of messages in the system. ASV can be used to compare the security of various systems using the same encryption scheme. A system with a higher ASV value is more secure than a system with a lower value.

The <u>second</u> requirement is the minimum number of Message Authentication Code (MAC) bits required for a message based on the assigned ASIL group. This is derived using the failure rate limit of the ASIL group of the message. The failure rate limit is typically expressed as FIT (Failure in Time), which denotes the maximum number of acceptable failures per 1 billion hours of usage. Based on the specifications in the standard, ASIL-D has 10 FIT, ASIL-B and C have 100 FIT, and ASIL-A has 1000 FIT as their maximum limits. In other words, ASIL-D applications need less than 10^{-8} failures per hour, while ASIL-A applications can have up to 10^{-5} failures per hour. In this work, we derive the security requirements for each message in the system using the following method:

- Consider a message ($m_{j,n}$) with period ($p_{j,n}$) (in milliseconds);
- The number of transmissions of $m_{j,n}$ per second are $10^3/p_{j,n}$.
- The number of transmissions of $m_{j,n}$ per hour are $(3600*10^3)/p_{j,n}$.
- If there are k bits in the MAC field of a message, the *probability of failure due to an attacker guessing a valid MAC* (e.g., using brute-forcing or other methods) is 2^{-k} for one transmission of that message;
- Therefore, the probability of failure due to a compromised MAC for an hour-long transmission is $((3600*10^3)/p_{j,n})*2^{-k}$.
- For an ASIL-D application, the probability of failure needs to be less than 10^{-8} per hour, i.e., $((3600*10^3)/p_{j,n})*2^{-k} \leq 10^{-8}$.
- Thus, the minimum number of MAC bits ($\Delta_{j,n}$) required for the message ($m_{j,n}$) according to the ASIL-D requirement is:

$$\Delta_{j,n}(D) = k \geq \left\lceil Q + \log_2 \left(\frac{1}{p_{j,n}} \right) \right\rceil \tag{5}$$

where Q is a constant and has a value of 48.35 for ASIL-D. Similarly, the minimum number of MAC bits required ($\Delta_{j,n}$) for other ASIL groups are calculated using

(5) by using $Q = 45.04$ for ASIL-B and ASIL-C and $Q = 41.72$ for ASIL-A. The different values of Q for each ASIL group are computed based on the FIT limit of that ASIL. Thus, for an ASIL-D message, for the most stringent (smallest) period, we observed ($=1$ ms), $\Delta_{j, n}(D) = 49$ bits (thus this is used in frame packing).

4.4 Optimizing Message Key Sizes Using GRASP

This is the last step of the design time process in *SEDAN*. This step aims to assign an optimal key size for each message in the system that maximizes the ASV while meeting all the security requirements and real-time deadline constraints. Additionally, we model the overhead caused by the security tasks (i.e., encryption and decryption) in terms of the additional ECU utilization (security-induced utilization) and latency (response time) of the message. For any given message $(m_{j,n})$ that is encrypted or decrypted using a block cipher, the security-induced ECU utilization $(\overline{U}_{m\,j,n})$ due to the message is computed using (6).

$$\overline{U}_{m\,j,n} = \left(\left\lceil \frac{b_j}{b_{size}} \right\rceil * \frac{T_{encr/decr}}{p_j} \right) \tag{6}$$

where b_{size} denotes the block size in bytes, and $T_{encr/decr}$ represents the time taken to encrypt or decrypt one block of data. Since AES is the encryption algorithm used in this study, the above equation can be re-written as shown in (7).

$$\overline{U}_{m\,j,n} = \left(\left\lceil \frac{b_j}{16} \right\rceil * \frac{T_{AES(X)}}{p_j} \right) \tag{7}$$

where $T_{AES(X)}$ is the time taken to encrypt or decrypt one block (16 Bytes) of data using AES with an X-bit long key (where X can be 128, 192, or 256). The security-induced utilization of any ECU $(\overline{U}_n)$ (computed using (8)) is the sum of the security-induced utilizations of all transmitted and received messages $(\overline{U}_{m\,j})$ for that ECU. Hence, the total utilization of any ECU (U_n) is the sum of the real-time utilization $(\tilde{U}_n)$ and security-induced utilization $(\overline{U}_n)$ as shown in (9). Moreover, to avoid uncertainties and undesired latency overheads, we always ensure that the utilization of any ECU does not exceed 100%.

$$\overline{U}_n = \sum_{j=1}^{R_n} \overline{U}_{m\,j,n} \tag{8}$$

$$U_n = \tilde{U}_n + \overline{U}_n \tag{9}$$

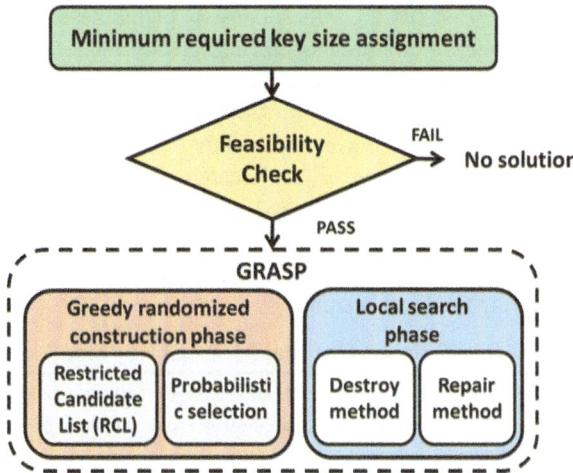

Fig. 5 Overview of the GRASP-based optimal message key size allocation step in *SEDAN*

In this study, we propose a heuristic approach to achieve this goal based on the *greedy randomized adaptive search procedure (GRASP)* metaheuristic [46]. An overview of this approach is illustrated in Fig. 5. Our proposed approach begins by taking the set of messages from the output of frame packing (Sect. 4.2) and the derived security requirements (Sect. 4.3) as inputs. An *initial solution* is generated by assigning the minimum required key sizes for all the messages based on the derived security requirements. This initial solution is subjected to a feasibility check which investigates the: *(i)* total ECU utilization (U_n) for all ECUs and *(ii)* number of missed deadlines using a design time scheduler. Moreover, we adapt the fast design time scheduling heuristic proposed in [2] to generate an optimal design time schedule. The initial solution is given to the GRASP only when there are no utilization violations at any ECU (i.e., $U_n \leq 100\%$ ∀ ECUs) and deadline misses for any message. If any of the above-mentioned conditions fail, the optimal message key size allocation step terminates, and the system does not have a feasible solution. GRASP intelligently explores various message key sizes (that are greater than or equal to the minimum key size requirement for a message) and design time schedule configurations (i.e., assigning messages and ECUs to FlexRay static segment slots) to select a solution that maximizes ASV, with no security violations, real-time deadline misses, and ECU utilization violations (i.e., no ECU utilization exceeds 100%).

The GRASP metaheuristic is an iterative process in which each iteration has two major phases: *(i) greedy randomized construction phase* that tries to build a local feasible solution and *(ii) local search phase* that tries to investigate the neighborhood for a local optimum. In the end, the best overall solution is chosen as the final solution. The greedy randomized construction phase has two key aspects- the *greedy aspect* and the *probabilistic aspect*. The greedy aspect involves generating a Restricted Candidate List (RCL), which consists of the best elements that will improve the partial solution (solution within the greedy randomized construction phase). And the probabilistic aspect involves selecting a random element from the RCL, which will be incorporated into the partial solution. It is important to note that the solutions generated during the greedy randomized construction phase are not necessarily optimal. Hence, a local search phase is used to improve the partial solution from the greedy randomized construction phase. The local search is an iterative process that uses destroy and repair mechanisms to search for local optimum within a defined neighborhood. The best solution is updated if an improved solution is found during the local search.

Algorithm 1: GRASP Based Optimal Message Key Size Assignment

Inputs: Set of nodes (N), Set of all messages (M), $init\_solution$, $max\_iterations$, RCL threshold (α), and destroy-repair threshold (β)

```
1:   best_solution ← init_solution
2:   for iteration = 1, …, max_iterations do
3:   │   current_solution ← greedy_randomized_construction(α, N, M)
4:   │   current_solution ← local_search(β, N, M, current_solution)
5:   │   if current_solution > best_solution do
6:   │   │   best_solution ← current_solution
7:   │   end if
8:   end for
```

Output: Optimal message key sizes for every message that results in maximum ASV and a feasible design time schedule with no deadline misses, security violations, and utilization of all ECUs below 100%.

Algorithm 1 presents an overview of our GRASP-based optimal message key size assignment approach. The inputs to Algorithm 1 are a set of nodes (N), a set of all the messages in the system (M), and the minimum required message key size assignment ($init\_solution$), which is the initial solution given to GRASP to reduce the search space. In addition, the tunable parameters such as maximum iterations ($max\_iterations$), RCL threshold (α), and a destroy-repair threshold (β) are given as input to GRASP to efficiently look for solutions in the search space. The algorithm starts by assigning the $init\_solution$ to the $best\_solution$ in step **1**. GRASP iteratively tries to find a better solution in steps **2–8** until $max\_iterations$ is reached. In each iteration $greedy\_randomized\_construction()$ in step **3**, generates a local feasible solution ($current\_solution$) which is updated using $local\_search()$ in step **4**. If a better solution is found at the end of the local search phase, the $best\_solution$ is updated in steps **5–7**. The output of the algorithm is an optimal message key size for every message and a feasible design time schedule with no deadline misses, no security violations, and no ECU utilization exceeding 100%. _Note_: Every solution in GRASP consists of two attributes _(i)_ key sizes for all the messages and _(ii)_ ASV of the system as a result of the key size assignment. Moreover, every solution generated by GRASP ensures that no message is allocated a key size less than the key size assigned in the initial solution, and the overall system ASV is always greater than the ASV of the initial solution.

4.4.1   Greedy Randomized Construction Phase

The greedy randomized construction phase tries to generate a feasible solution in every iteration of GRASP by increasing the key sizes of some of the non-ASIL-D messages. The goal here is to maximize the ASV of the system without any deadline, security, and ECU utilization violations. Moreover, it also ensures that no message is allocated a key size less than the key size allocated in the initial solution (minimum required key size). The solution generated by the greedy randomized construction phase will be given as the input to the local search phase for refinement.

Algorithm 2: greedy_randomized_construction (α, N, M)

Inputs: RCL threshold (α), set of nodes (N), and set of all messages (M)

1:   $\tilde{M} \leftarrow \{m \in M \mid m^{AG} \neq \text{ASIL-D}\}$

2:   **Increment** m^{SS} by 1 $\forall$ $m \in \tilde{M}$ and compute m^{SV}

3:   **Sort** $\tilde{M}$ in the increasing order of m^{SV}

4:   **while** $\tilde{M} \neq \{ \, \}$ **do**

5:   $\quad$ $SV_{min} = \min (\{m^{SV} \, \forall \, m \in \tilde{M}\})$

6:   $\quad$ $SV_{max} = \max (\{m^{SV} \, \forall \, m \in \tilde{M}\})$

7:   $\quad$ $RCL \leftarrow \{m \in \tilde{M} \mid m^{SV} \geq SV_{min} + \alpha * (SV_{max} - SV_{min})\}$

8:   $\quad$ $\bar{m} \leftarrow$ random element from RCL

9:   $\quad$ **Increment** the key size of $\bar{m}$ to the next higher key size

10: $\quad$ **if feasibility_check( )** $==$ *false* **do**

11: $\quad\quad$ **Revert** the key size of $\bar{m}$ back to its previous key size

12: $\quad\quad$ **Decrement** m^{SS} by 1 for $\bar{m}$ and compute m^{SV}

13: $\quad$ **end if**

14: $\quad$ **Remove** $\bar{m}$ from $\tilde{M}$

15: **end while**

16: *current_solution* $\leftarrow$ {**calculate_ASV( )**, message key size assignment}

Output: Local feasible solution that results in a feasible schedule with no deadline misses, security violations, and utilization of all ECUs below 100%.

Algorithm 2 shows the pseudocode of the greedy randomized construction phase where the inputs are: set of nodes (N), set of messages (M), and RCL threshold (α). A set of non-ASIL-D messages ($\tilde{M}$) is generated in step **1**. In step **2**, the security score of each message (m^{SS}) in $\tilde{M}$ is incremented by one, and the security values of the messages (m^{SV}) are updated using (3). The $\tilde{M}$ is sorted in the increasing order of m^{SV} and the ties are resolved based on the message period in step **3**. In steps **4–15**, the algorithm tries to find a local solution by incrementing key sizes for some messages that would result in no deadline, security, and ECU utilization violations. The minimum (SV_{min}) and maximum (SV_{max}) security values of messages in $\tilde{M}$ are computed in steps **5, 6** respectively. The RCL consists of messages in $\tilde{M}$, that will result in increased ASV when their key size is incremented. Hence, the messages whose security value (m^{SV}) is within the interval $[SV_{min} + \alpha (SV_{max} - SV_{min}), SV_{max}]$ are added to the RCL in step **7**. This is the greedy aspect of the greedy randomized construction step. Moreover, GRASP employs an RCL threshold ($\alpha \in [0, 1]$) to regulate the quality of the generated RCL. The threshold (α) controls the amount of greediness and randomness in the algorithm. The $\alpha = 1$ case corresponds to a pure greedy approach, while $\alpha = 0$ is equivalent to a purely random approach. A random message ($\overline{m}$) is selected from the *RCL* (probabilistic selection) in step **8**, and its key size is incremented to the next higher key size in step **9** (i.e., 128 → 192 or 192 → 256). The *feasibility_check()* in step **10**, checks for any *(i)* ECU utilization violations (i.e., any ECU utilization >100%) and *(ii)* deadline misses using the design time scheduling heuristic proposed in [2]. If any of the above-mentioned checks fail, the *feasibility_check()* will return *false* and reverts the key size of ($\overline{m}$) back to its previous key size in step **11**. Moreover, the m^{SV} of $\overline{m}$ is re-computed after decrementing the m^{SS} by one in step **12**. Otherwise, the key size increment is left unchanged. The message ($\overline{m}$) is removed from $\widetilde{M}$ and the steps **5–14** are repeated until there are no messages left in $\widetilde{M}$. Lastly, in step **16**, the current message key size assignment and the ASV of the system (using *calculate_ASV()*) are assigned to the *current_solution*. The function *calculate_ASV()* is implemented using (4).

4.4.2 Local Search Phase

The local search phase tries to iteratively improve the solution found in the greedy randomized construction phase by investigating a defined neighborhood in the search space. The local search phase achieves this by using *destroy* and *repair* methods, which remove a part of the solution and recreate a feasible solution, respectively. In this study, we define the neighborhood as the set of solutions that are generated by randomly changing key sizes for β number of messages. The parameter β is known as the destroy-repair threshold, which controls how much to destroy or repair in each iteration of the local search. These random changes in message key sizes help in recovering from suboptimal ordering (sorting in the increasing order of m^{sv}) of messages in the greedy randomized construction phase.

The pseudocode of the local search procedure is illustrated in Algorithm 3. The *destroy()* function in steps **1–4** randomly selects a message from the set of messages that are allocated a key size higher than the minimum required key size and decreases the key size to the next smaller key size. The function *min_score()* in step **2** returns the minimal security score demanded by the assigned ASIL group. The *repair()* method in steps **5–18** aims to increase the key size for β non-ASIL-D messages and computes the local solution using *local_solution()*. The *repair()* step always selects a message that results in a maximum increase in the ASV of the system (as shown in step **8**). The ties in step **8** are resolved based on the ASIL group, and if multiple messages have the same ASIL group, one message is selected at random. In steps **19–29**, the local search algorithm iteratively explores the neighborhood around the *current_solution* using *destroy()* and *repair()* to find a better solution. In each iteration, the value of β is chosen randomly from [2, β_{max}]. In steps **21–24**, the function *destroy()* is modeled as a stochastic process that is controlled by the key decrease probability (p_{kd}). Lastly, the *current_solution* is updated if a better *local_solution* is found in the repair method in steps **25–28**. In each iteration of GRASP, at the end of the local search phase, a local optimum is found if there exits one. Otherwise, the solution remains unchanged from the greedy randomized construction phase.

Algorithm 3: local_search (β, N, M, *current_solution*

Inputs: Destroy-repair threshold (β), set of nodes (N), set of all messages (M), and *current_solution*

1: **function** *destroy* (M)
2: $\quad$ $M_d = \{m \in M \mid m^{SS} > \textbf{min\_score}(m^{AG})\}$
3: $\quad$ **Decrement** the key size of a random message ($\bar{m}$) in M_d
4: **end function**
5: **function** *repair* (β, M, N)
6: $\quad$ $M_r = \{m \in M \mid m^{AG} \neq \text{ASIL-D}\}$
7: $\quad$ **while** ($\beta > 0$) or ($M_r \neq \{\ \}$) **do**
8: $\quad\quad$ $\bar{m} = \{m \in M_r \mid \Delta \text{ASV is maximum}\}$
9: $\quad\quad$ **Increment** the key size of the message ($\bar{m}$)
10: $\quad\quad$ **if feasiblity_check( )** $==$ *false* **do**
11: $\quad\quad\quad$ **Revert** the key size of ($\bar{m}$) back to the previous key size
12: $\quad\quad$ **else do**
13 $\quad\quad\quad$ $\beta = \beta - 1$
14: $\quad\quad$ **end if**
15: $\quad\quad$ **Remove** ($\bar{m}$) from M_r
16: $\quad$ **end while**
17: $\quad$ **return** {**calculate_ASV( )**, message key size assignment}
18: **end function**
19: **for** *local_iteration* = 1,…, *max_local_iterations* **do**
20: $\quad$ $\beta = $ **random_integer**(2, β_{max})
21: $\quad$ **if** $p_{kd} > $ **random**($0,1$) **do**
22: $\quad\quad$ *destroy* (M)
23: $\quad\quad$ $\beta = \beta - 1$
24: $\quad$ **end if**
25: $\quad$ *local_solution* $\leftarrow$ **repair** (β, M, N)
26: $\quad$ **if** *local_solution* $>$ *current_solution* **do**
27: $\quad\quad$ *current_solution* $\leftarrow$ *local_solution*
28: $\quad$ **end if**
29: **end for**

Output: Local optimum with in the defined neighborhood- if there exists one; Otherwise, the same solution as greedy_randomized_construction( ).

It is important to note that when the message key size is changed, the size of the output cipher and MAC (or the message size) remains unchanged. The key size only affects the time taken to encrypt/decrypt the message, which impacts the security-induced utilization of the sender and receiver ECUs. Moreover, the real-time utilization of the ECUs also remains unchanged, as the execution time of time-triggered tasks does not change with changing message key sizes.

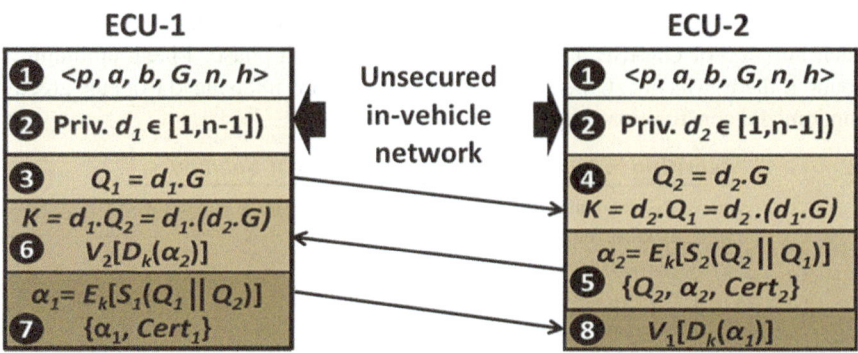

Fig. 6 Overview of steps involved in setting up a session key using the STS protocol with ECC

4.5 Setting Up Session Key

In this subsection, we discuss the first runtime step of the *SEDAN* framework. It involves settings up session keys required for generating keys that will be used for the encryption and decryption of messages. This is a crucial step in improving the security of the vehicle, as using the same key every time for encryption and decryption for the entirety of the vehicle's lifetime makes the system highly vulnerable to cyber-attacks. Hence, during runtime, we generate a new key for every session (called session key), which will be used for generating keys that will be used for encryption and decryption of messages.

A session is defined as the time duration between the start of a vehicle to turning off the vehicle. Since we use symmetric key encryption, all ECUs in the system need to have the same secret key to function properly. As traditional automotive networks do not have any inbuilt security features, exchanging the session keys between ECUs over an unsecured channel is a major challenge. In this work, we adapt the Station-to-Station (STS) key agreement protocol [47], which is based on the famous Diffie-Hellman key exchange method [45], to the automotive domain (as simple Diffie-Hellman is vulnerable to man-in-the-middle attacks), to securely transfer session keys between ECUs over an unsecured FlexRay bus. Moreover, within the STS protocol, we employ elliptic curve cryptography (ECC) as the basis for key agreement instead of RSA. This is mainly because ECC is faster and has a lower memory footprint for the same level of security compared to the RSA (as discussed in Sect. 5.2). The overview of steps involved in STS protocol with ECC for two ECU cases is illustrated in Fig. 6.

The STS approach begins with two ECUs agreeing upon a set of domain parameters that define the elliptic curve. These parameters are shown in the first step in Fig. 6, where the parameter p defines the field, a and b define the elliptic curve, G is the generator and n is its order, and h is the co-factor. Additionally, each ECU utilizes an asymmetric key pair for authentication operations (sign and verify). In the second step, each ECU generates a random private number (d_1 in

ECU 1 and d_2 in ECU 2), which is not shared with any other ECU in the system. In step 3, ECU1 performs an elliptic curve scalar multiplication (hereafter referred to as scalar multiplication) of the private number d_1 and generator G. The output Q_1 is transmitted to ECU2 over an unsecured FlexRay bus. In step 4, a similar scalar multiplication between d_2 and G is performed at ECU2, but the output Q_2 is not sent to ECU1. ECU2 then computes the common secret key K (session key) by performing the scalar multiplication of the private number d_2 and the received output Q_1. In step 5, ECU2 computes the signature (S_2 ()) of the concatenation of Q_2 and Q_1 (represented as $Q_2 \ || \ Q_1$) using its private key of the asymmetric key pair. The output signature is encrypted (E_k ()) using the computed session key from the previous step, which produces the cipher α_2. The scalar multiplication output (Q_2), output cipher (α_2), and certificate ($Cert_2$) are all transmitted to ECU1 over the unsecured FlexRay bus. The certificate is issued by a trusted certificate authority (CA), which is used to prove the ownership of a public key. The certificate consists of the public key of the owner and signature of the CA and will be programmed in the ECUs by the manufacturer. The public key of the CA is used to verify the certificate and extract the public key of the owner. In step 6, when the ECU1 receives the output of step 5 from ECU2, it performs a scalar multiplication of private number d_1 and Q_2 to produce the shared secret key K (session key). Moreover, ECU1 utilizes the key K to decrypt (D_k ()) the received cipher (α_2) and verifies ($V_1()$) the decrypted output using the public key extracted from the certificate of ECU2 ($Cert_2$). The session key K is accepted by ECU1 only when the verification is successful, implying a successful authentication of ECU2. In step 7, ECU1 computes the signature ($S_1()$) of the concatenation of Q_1 and Q_2 (represented as $Q_1 \ || \ Q_2$) using its private key of the asymmetric key pair. The resulting output is encrypted using the key K that generates the cipher (α_1), which is transmitted to ECU2 along with the certificate ($Cert_1$). Lastly, in step 8, at ECU 2, the received cipher (α_1) is decrypted using the key K, and the output is verified using the public key extracted from the certificate of ECU1 ($Cert_1$). The session key K is accepted to use for the session only when the verification is successful. Thus, all the ECUs are authenticated, and a common secret key (session key) is established at every ECU without actually exchanging the actual key over the unsecured bus. Additionally, the STS protocol uses no timestamps and provides perfect forward secrecy. Using a standard AES key schedule at every ECU, this session key is then used to generate 128-bit, 192-bit, and 256-bit keys. These resulting keys are used for encrypting and decrypting messages at runtime. Moreover, in order to avoid interference with the time-critical messages, the messages related to the security operations utilize a small number of reserved FlexRay frames. To speed up the startup process, we assume that the manufacturer pre-programs some of the session keys during manufacturing. New keys are generated continuously during the idle time of an ECU, saved in local memory, and used in future sessions. To further speedup this process, the public keys of the trusted ECUs can be pre-programmed in the ECU's tamper-proof memory, thereby avoiding the verification of the certificate, which saves both computation time and network bandwidth.

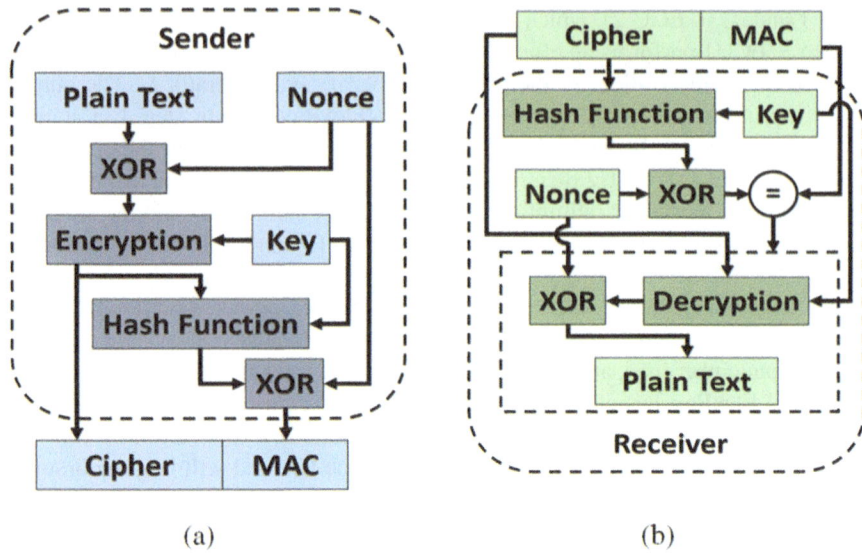

(a)   (b)

Fig. 7 (**a**) Authenticated encryption at sender ECU; (**b**) Authenticated decryption at receiver ECU

It is essential to highlight that, even if there was an attacker already in the system during the key setup phase, the attacker could not compute the secret key with the publicly available results due to the discrete logarithm problem [48]. Moreover, the common man-in-the-middle attack that breaks the standard Diffie-Hellman approach [49] fails with STS as the attacker cannot authenticate successfully.

4.6 Authenticated Encryption/Decryption

In this subsection, we discuss the various steps involved in authenticated encryption employed in *SEDAN*. Authenticated encryption refers to simultaneously providing a message with confidentiality and authenticity, which is a well-known technique in the literature. We discuss this step in detail here to highlight how *SEDAN* leverages this process to achieve a more secure runtime system. The authenticated encryption and decryption phases are illustrated in Fig. 7a, b, respectively.

The authenticated encryption at the sender ECU begins with an XOR operation between the *plain text* (message data) and a *nonce* (random number), and the result is encrypted using AES with the key size assigned to the message (as discussed in Sect. 4.4). The XOR operation with a nonce is performed to avoid generating the same cipher every time when the input data is the same for long durations. Even though protecting the system from side-channel attacks is not within the scope of this work, this simple step could be the first step in preventing information leakage. A cryptographic hash function (MD5) takes the output cipher and the key used for encryption to produce a hash, which is XORed with a nonce to generate the MAC. The output MAC size is truncated if needed and set to be at least the size computed in Sect. 4.3. The generated MAC is then transmitted with the encrypted message data in the payload section of the FlexRay frame.

The authentication decryption at the <u>receiver</u> ECU begins by authenticating the sender ECU of a received message. The received cipher and the selected key are given to the same cryptographic hash function whose result is XORed with a nonce to generate a local MAC. The authentication of the sender ECU is successful only when the local MAC matches the received MAC. Otherwise, the authentication process fails, and the received message is discarded. After successful authentication of a sender ECU, AES decryption is initiated, and the output is XORed with the nonce to extract the original message data as plain text.

As discussed in Sect. 3.3, we mainly focus on protecting the system from masquerade and replay attacks as they are the most common, hard to detect, and severely impact system safety and performance. The system is protected against masquerade or impersonation attacks by <u>authenticating the ECUs</u> in the system using the STS protocol, which establishes the session keys used for encryption and decryption only after successful authentication. The attacker fails to authenticate due to the lack of trusted certificates and cannot masquerade as a legitimate ECU. Moreover, the MAC generated in the authenticated encryption protects the system from replay attacks. During the MAC generation, it is essential to XOR the output of the hash function with the nonce as it makes the messages resilient to replay attacks. During a replay attack, the authenticity of the replayed message fails as the nonce used in computing the local MAC at the receiver is different from the nonce used in generating the received MAC at the sender. This mismatch in MAC will result in discarding the message sent by the attacker. Moreover, in the event of a man-in-the-middle attack, where the attacker tries to modify the message payload, the MAC comparison fails, resulting in protecting the <u>integrity of the messages</u>. Also, if an attacker eavesdrops on the network, the attacker would still be unable to decrypt the encrypted messages, as no keys are exchanged on the network. In this manner, we achieve <u>confidentiality of the message data</u>. Hence, using the proposed *SEDAN* framework, we were able to achieve all the security objectives, namely confidentiality, integrity, and ECU authenticity (as discussed in Sect. 3.4).

4.7   Runtime Message Scheduler

Runtime message scheduling is the last step in the *SEDAN* framework. It takes the unique values of the cipher and MAC generated in the previous step and packs them into FlexRay frames generated during the frame packing step (Sect. 4.2). Other control fields, such as the fields in the header and trailer segments that are required for the transmission of FlexRay frames, are also added by the scheduler. The runtime scheduler uses the design time generated message schedule and interacts with the FlexRay protocol engine to schedule messages on to the FlexRay bus at runtime.

5 Experiments

5.1 Experimental Setup

We evaluated the performance of our proposed *SEDAN* framework by comparing it with the best-known prior work [21]. In [21], the authors proposed a technique that uses simulated annealing to minimize the end-to-end latencies of all in-vehicle network messages and uses symmetric key encryption with the time-delayed release of keys to improve security in a vehicle system. Since [21] does not support variable key sizes, three different variants of [21] are implemented using AES encryption with fixed key sizes of 128, 192, and 256 bits, which are referred to as 'Lin et al. AES-128', 'Lin et al. AES-192', and 'Lin et al. AES-256' respectively in the experimental results. We generated several test cases based on automotive network and ECU computation data extracted from a real-world vehicle (2016 Chevrolet Camaro) that we have access to. We modeled the network and ECU computation data as directed acyclic graphs (DAGs), which were generated using TGFF [50]. We developed multiple synthetic test cases by scaling this data based on different combinations of the number of ECUs, number of applications, number of tasks in each application, and the range of periods. Moreover, we assume that the deadline for both tasks and messages are equal to their period. Lastly, we considered the FlexRay 3.0.1 [51] protocol with the following network parameters for all experiments: cycle duration of 5 ms with 62 static segment slots, with a slot size of 42 bytes, and 64 communication cycles.

5.2 Benchmarking Encryption Algorithms

To accurately capture the runtime behavior of session key generation and authenticated encryption/decryption steps, we implemented various encryption algorithms in the software. We implemented AES-CBC with key sizes of 128, 192, and 256 bits, RSA with key sizes of 512, 1024, 2048, and 4096 bits, and the ECC with key sizes of 256 and 384-bits using OpenSSL [52]. All these algorithms were executed on an ARM Cortex-A9 CPU on a ZedBoard, which has similar specifications as state-of-the-art ECUs [53, 54].

Table 1 shows the average AES encryption/decryption times with different standard key sizes for one block of data (16 Bytes) on an ARM Cortex A9 CPU. These values are used to model the latency overhead on each message due to the added security mechanisms at design time. They are also used in scheduling decisions and computing the response time of the messages. The encryption and decryption times of RSA with 512, 1024, 2048, and 4096-bit keys and ECC with 256 and 384-bit keys are also shown in Table 1. These values are considered in choosing between RSA and ECC as the cryptographic scheme in the STS protocol. The NIST recommends a key size of 2048-bits for RSA [55], while

Table 1 Execution time (ms) of AES, RSA, and ECC on ARM Cortex A9

Cryptographic scheme	Key size	Encryption / Decryption	
AES	128	0.35	
	192	0.393	
	256	0.415	
Cryptographic scheme	Key size	Public key operation	Private key operation
RSA	512	2.01	19.89
	1024	6.48	139.15
	2048	23.65	911.8
	4096	91.52	6283.2
ECC	256	59.8	17.1
	384	182.4	50.4

NSA recommends a 256-bit key size for SECRET level and a 384-bit key size for TOP SECRET level using ECC [56]. Moreover, ECC with 224, 256, and 384-bit key sizes provides similar security as RSA with 2048, 3072, and 7680 key sizes, respectively [57]. In this work, we consider the minimum key sizes based on the above-mentioned recommendations. From Table 1, it can be seen that RSA is faster for verifying signatures (operation performed using the public key) and much slower for generating signatures (operation performed using the private key). On the other hand, ECC is much faster for generating signatures while relatively slower for verifying signatures. It is important to note that the security (provided by RSA using the equivalent key size) doubles when the ECC key size is increased from 256 to 384. However, since the automotive systems are highly resource-constrained, we choose to employ ECC with a 256-bit key size (which still provides higher security than the minimum recommended key size for RSA) for cryptographic operations in the STS key agreement protocol. Moreover, the ECC execution time values are used in estimating the worst-case time required for setting up a session key, which is 0.24 s for a 256-bit key, while an equivalent RSA 2048 takes 3.72 s. Thus, it is evident that ECC is much faster than RSA for a similar level of security. Moreover, ECC can provide a similar level of security compared to RSA, with a much shorter key size. Lastly, when we profiled the MD5 hashing algorithm used in the authenticated encryption step, we observed that processing one block of data takes about 2.68 μs.

Moreover, with the increasing complexity of automotive applications, designing security mechanisms that result in minimal power consumption is crucial. Hence, we profiled the security mechanisms studied in this work and presented the power consumption results in Table 2. Other overheads, such as memory consumption, are not explicitly modeled as most modern-day ECUs have sufficient memory to store the small keys needed for secure transfers. Additionally, the designer can limit the number of pre-computed session keys that can be stored to minimize the memory overhead. Based on the results in Tables 1 and 2, it is evident that ECC has lower computation and memory overhead than RSA for the same level of security. Hence, in *SEDAN,* we authenticate the ECUs in the system and setup session keys using the

Table 2 Power consumption of AES, RSA and ECC on ARM Cortex A9

Cryptographic scheme	Key size	Encryption / Decryption	
AES (mW)	128	57.76	
	192	58.04	
	256	60.19	
Cryptographic scheme	Key size	Public key operation	Private key operation
RSA (W)	512	0.28	0.65
	1024	0.34	1.22
	2048	0.72	1.91
	4096	1.08	2.58
ECC (W)	256	0.62	0.33
	384	0.93	0.58

STS protocol using the ECC. Additionally, we use AES to encrypt and decrypt the messages in the system using the keys computed from the session key.

5.3   GRASP Parameter Selection

To get an efficient solution using the GRASP, it is essential to select the appropriate values for the threshold parameters α and β_{max}. We ran a series of simulations by changing the value of α from 0 to 1 with an increment of 0.2, and the greedy randomized construction phase was run 1000 times using different input test cases. We observed that the mean solution approached a greedy solution, while the variance approached zero as α tends to 1. On the other hand, when α is small and close to zero, the mean solution approaches a random solution with high variance. Therefore, we selected $\alpha = 0.8$, which provided a good quality solution to the local search phase that resulted in a near greedy solution in the presence of a relatively large variance.

Moreover, we observed that $\beta_{max} = 3$ provided enough randomness to look for other solutions in each iteration of the local search phase. A higher value of β_{max} could result in an exhaustive local search leading to unreasonably long computation times. Also, the minimum value of β needs to be 2 to increase the key size of at least one message when the key size is reduced in the event of a destroy operation. This prevents the generation of a solution that results in lower ASV compared to the solutions in previous iterations. Lastly, a relatively small value for $p_{kd} = 0.3$ is chosen to avoid frequent key size decrements.

5.4 Response Time Analysis

In this subsection, we present the response time analysis by comparing our proposed *SEDAN* framework with the three variants of [21]. Response time of a message is defined as the end-to-end latency, which is the aggregate of the time for encryption and MAC generation, and queuing delay at the sender ECU; transmission time on the Flexray bus, and the time for MAC verification and decryption at the receiver ECU. We evaluated our proposed *SEDAN* framework, and the comparison works using three different test cases: (1) low input load- system with 5 ECUs (3 single-core and 2 dual-core) and 77 tasks that produced 57 (time-triggered) signals; (2) medium input load- system with 12 ECUs (9 single-core and 3 dual-core) and 126 tasks with 93 signals; and (3) high input load- system with 16 ECUs (12 single-core and 4 dual-core) and 243 tasks with 196 signals. The average message response time for the low, medium, and high input load cases with their deadlines on the x-axis is illustrated in Figs. 8(a–c). The confidence interval on each bar represents the minimum and maximum average response time of messages. The dashed horizontal lines represent different message deadlines. The number on top of each bar is the number of deadlines misses.

From Figs. 8(a–c), it is clear that *SEDAN* outperforms the three variants of [21] and achieves significantly lower average response times for all the messages under all input load cases. *SEDAN* achieves this by balancing security and real-time performance goals by optimizing key sizes while meeting message security requirements and ensuring that all ECU utilizations are below 100%. This prevents the messages from experiencing additional delays on top of the latency caused by the encryption-decryption processes. Moreover, all three variants of [21] experience significant authentication delays (time taken from the transmission of the message to decryption of the message) compared to *SEDAN*, which increases the response time of the messages when using [21]. These high authentication delays in [21] are because of the time-delayed release of keys, which is employed in all three variants of [21]. Also, the periodic computation of keys in every session at each ECU in all three variants of [21] results in high ECU utilization overhead resulting in increased response time and power consumption. Lastly, the requirement of large message buffers to hold multiple messages for longer durations in [21] (due to the time-delayed release of keys) further increases power consumption and response time.

5.5 Security Analysis

Table 3 shows the number of security violations in each technique under three different input load cases (as discussed in the previous sub-section). A security violation is defined as an instance when the derived security constraints (defined in Sect. 4.3) for a message are not met. From Table 3, it can be seen that the *SEDAN* and Lin et al. AES-256 are the only techniques that do not violate any security

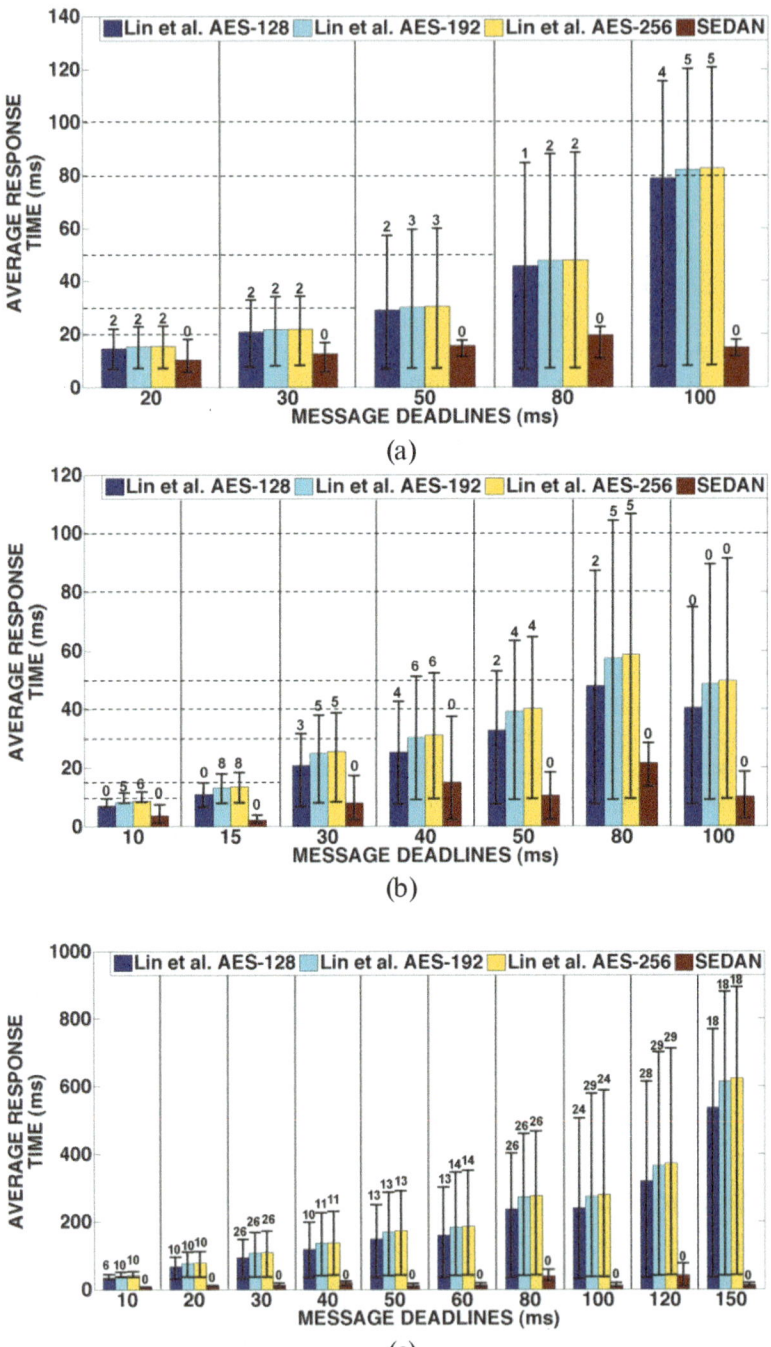

Fig. 8 Comparison of the average response time of all messages under (**a**) low; (**b**) medium, and (**c**) high input application load conditions for Lin et al. AES-128, AES-192, AES-256 [21], and *SEDAN* (with the number of missed deadlines shown on the top of the bars)

Table 3 Total number of security violations for each input load configuration

Framework	Lin et al. 128	Lin et al. 192	Lin et al. 256	SEDAN
Low load	28	12	0	0
Medium load	45	16	0	0
High load	96	31	0	0

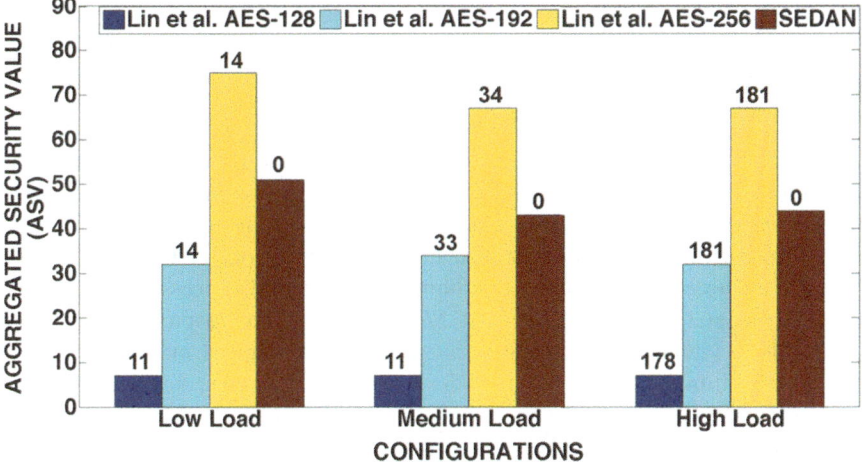

Fig. 9 Comparison of aggregate Security Value (ASV) under each input load configuration for Lin et al. AES-128, AES-192, AES-256 [21], and *SEDAN* (with the number of missed deadlines on top of bars)

requirements. However, it is essential to note that, unlike *SEDAN*, Lin et al. AES-256 has no intelligent key size assignment scheme and assigns all the messages with 256-bit keys irrespective of their ASIL group, which helps in meeting the message security requirements. But this results in increased ECU utilization, which in turn incurs additional latency overheads for messages. Moreover, unlike all three variants of [21], *SEDAN* does not exchange or release keys on an unsecured communication bus. This helps prevent an attacker from gaining knowledge about the current and previously used keys, which provides additional security to the systems. *SEDAN* also does not require frequent key computation at each ECU within a single session, as done in [21], which helps reduce utilization overheads in ECUs when *SEDAN* is employed.

Lastly, the ASV for the three input load cases, with numbers on top of each bar showing the number of messages that missed deadlines, is illustrated in Fig. 9. It can be seen that Lin et al. AES-256 achieves the highest ASV. However, this comes at the cost of multiple missed deadlines. Thus, *SEDAN* is able to satisfy minimum security requirements (i.e., all messages have at least the minimum key size required by the designer) and all real-time deadlines for all messages while providing an ASV value that is higher than that for Lin et al. AES-128 and Lin et al. AES-192.

Thus, *SEDAN* represents a promising framework that can intelligently manage the limited computing resources in vehicles while improving the overall security of the system. Moreover, from Fig. 9 and Table 3, it is evident that *SEDAN* is able to do a better job of balancing security and real-time performance goals by intelligently optimizing key sizes and accurately integrating overheads of security primitives while making task and message scheduling decisions.

6 Conclusions

In this chapter, we presented a novel security framework called *SEDAN* that combines design time schedule optimization with runtime symmetric key management to improve security in time-critical automotive systems without utilizing any additional hardware. We demonstrated the feasibility of our *SEDAN* framework by implementing cryptographic algorithms on real-world processors. Moreover, the experimental results indicate that *SEDAN* is able to reason about security overheads to intelligently adapt security primitives during the message and task scheduling, ultimately ensuring that both security and real-time constraints are met. Such a framework promises to be extremely useful as we move towards connected autonomous vehicles with large attack surfaces by *enabling security to be a first-class design objective* without sacrificing real-time performance objectives.

References

1. Kukkala, V.K., Tunnell, J., Pasricha, S., Bradley, T.: Advanced driver-assistance systems: a path toward autonomous vehicles. IEEE Consumer Electronics Magazine. **7**(5) (2018)
2. Kukkala, V.K., Pasricha, S., Bradley, T.: JAMS: Jitter-Aware Message Scheduling for FlexRay Automotive Networks. In: IEEE/ACM International Symposium on Network-on-Chip (2017)
3. DiDomenico, C., Bair, J., Kukkala, V.K., Tunnell, J., Peyfuss, M., Kraus, M., Ax, J., Lazarri, J., Munin, M., Cooke, C., Christensen, E.: Colorado State University EcoCAR 3 Final Technical Report. In: SAE World Congress Experience (WCX) (April 2019)
4. Kukkala, V.K., Bradley, T., Pasricha, S.: Priority-based Multi-level Monitoring of Signal Integrity in a Distributed Powertrain Control System. in: Proceedingsof IFAC Workshop on Engine and Powertrain Control, Simulation and Modeling (July 2015)
5. Kukkala, V.K., Bradley, T., Pasricha, S.: Uncertainty Analysis and Propagation for an Auxiliary Power Module. In Proceedings of IEEE Transportation Electrification Conference (TEC) (June 2017)
6. Kukkala, V.K., Pasricha, S., Bradley, T.: JAMS-SG: a framework for jitter-aware message scheduling for time-triggered automotive networks. ACM Transactions on Design Automation of Electronic Systems (TODAES). **24**(6) (2019)
7. Tunnell, J., Asher, Z., Pasricha, S., Bradley, T.H.: Towards improving vehicle fuel economy with ADAS. SAE International Journal of Connected and Automated Vehicles. **1**(2) (2018)
8. Tunnell, J., Asher, Z., Pasricha, S., Bradley, T.H.: Towards Improving Vehicle Fuel Economy with ADAS. In Proceedings of SAE World Congress Experience (WCX) (2018)
9. Asher, Z., Tunnell, J., Baker, D.A., Fitzgerald, R.J, Banaei-Kashani, F., Pasricha, S., Bradley, T.H.: Enabling Prediction for Optimal Fuel Economy Vehicle Control. In: Proceedings of SAE World Congress Experience (WCX) (2018)

10. Dey, J., Taylor, W., Pasricha, S.: VESPA: a framework for optimizing heterogeneous sensor placement and orientation for autonomous vehicles. IEEE Consumer Electronics Magazine (CEM). **10**(2) (2021)

11. Kukkala, V., Pasricha, S., Bradley, T.: SEDAN: Security-Aware Design of Time-Critical Automotive Networks. IEEE Transaction on Vehicular Technology (TVT). **69**(8) (2020)

12. Kukkala, V.K., Thiruloga, S.V., Pasricha, S.: INDRA: Intrusion Detection using Recurrent Autoencoders in Automotive Embedded Systems. IEEE Transactions on Computer-Aided Design of Integrated Circuits and Systems (TCAD). **39**(11) (2020)

13. Kukkala, V.K., Thiruloga, S.V., Pasricha, S.: LATTE: LSTM Self-Attention based Anomaly Detection in Embedded Automotive Platforms. ACM Transactions on Embedded Computing Systems (TECS). **20**(5s), Article 67 (2021)

14. Thiruloga, S.V., Kukkala, V.K., Pasricha, S.: TENET: Temporal CNN with Attention for Anomaly Detection in Automotive Cyber-Physical Systems. In: Proceedings of IEEE/ACM Asia & South Pacific Design Automation Conference (ASPDAC) (January 2022)

15. Kukkala, V.K., Thiruloga, S.V., Pasricha, S.: Roadmap for Cybersecurity in Autonomous Vehicles. In IEEE Consumer Electronics Magazine (CEM) (2022)

16. Studnia, I., Nicomette, V., Alata, E., Deswarte, Y., Kaâniche, M., Laarouchi, Y.: Survey of security threats and protection mechanisms in embedded automotive systems. In: IEEE Dependable Systems and Networks Workshop (2013)

17. Koscher, K., Czeskis, A., Roesner, F., Patel, S., Kohno, T., Checkoway, S., McCoy, D., Kantor, B., Anderson, D., Shacham, H., Savage, S.: Experimental security analysis of a modern automobile. In: IEEE Symposium on Security and Privacy (2010)

18. Checkoway, S., McCoy, D., Kantor, B., Anderson, D., Shacham, H., Savage, S., Koscher, K., Czeskis, A., Roesner, F., Kohno, T.: Comprehensive Experimental Analyses of Automotive Attack Surfaces. USENIX Security Symposium (2011)

19. Miller, C., Valasek, C.: Remote Exploitation of an Unaltered Passenger Vehicle. In: Black Hat USA (2015)

20. Izosimov, V., Asvestopoulos, A., Blomkvist, O., Törngren, M.. Security-Aware Development of Cyber-Physical Systems Illustrated with Automotive Case Study. In: IEEE/ACM Design, Automation & Test in Europe (2016)

21. Lin, C.W., Zhu, Q., Sangiovanni-Vincentelli, A.: Security-Aware Mapping for TDMA-Based Real-Time Distributed Systems. In: IEEE/ACM International Conference on Computer-Aided Design (2014)

22. Zalman, R., Mayer, A.: A secure but still safe and low cost automotive communication technique. In: IEEE/ACM Design Automation Conference (2014)

23. Lin, C.W., Zhu, Q., Phung, C., Sangiovanni-Vincentelli, A.: Security-Aware Mapping for CAN-Based Real-Time Distributed Automotive Systems. In: IEEE International Conference on Computer-Aided Design (2013)

24. Lin, C.W., Zhu, Q., Sangiovanni-Vincentelli, A.: Security-aware modeling and efficient mapping for CAN-based real-time distributed automotive systems. IEEE Embedded Systems Letter. **7**(1) (2015)

25. Han, G., Zeng, H., Li, Y., Dou, W.: SAFE: Security Aware FlexRay Scheduling Engine. In: IEEE/ACM Design, Automation & Test in Europe (2014)

26. Hazem, A., Fahmy, H.A.: LCAP- A Lightweight CAN Authentication Protocol for Scheduling in-Vehicle Networks. In: Embedded Security in Cars Conference (2012)

27. Chavez, M.L., Rosete, C.H., Henriquez, F.R.: Achieving Confidentiality Security Service for CAN. In: IEEE International Conference on Electronics, Communications and Computers (2005)

28. Yoshikawa, M., Sugioka, K., Nozaki, Y., Asahi, K.: Secure in-Vehicle Systems against Trojan Attacks. In: IEEE International Conference on Computers and Information Science (2015)

29. Bogdanov, A., Knudsen, L.R., Leander, G., Paar, C., Poschmann, A., Robshaw, M.J., Seurin, Y., Vikkelsoe, C.: PRESENT: An Ultra-Lightweight Block Cipher. In: International Workshop on Cryptographic Hardware and Embedded Systems (2007)

30. Lukasiewycz, M., Mundhenk, P., Steinhorst, S.: Security-aware obfuscated priority assignment for automotive CAN platforms. ACM Transactions on Design Automation on Electrical Systems. **21**(2) (2016)
31. Shreejith, S., Fahmy, S.A.: Security Aware Network Controllers for Next Generation Automotive Embedded Systems. In: IEEE/ACM Design Automation Conference (2015)
32. Perrig, A., Canetti, R., Tygar, J.D., Song, D.: The TESLA Broadcast Authentication Protocol. In: RSA Cryptobytes (2005)
33. Lin, C.W., Yu, H.: Coexistence of Safety and Security in Next-Generation Ethernet-Based Automotive Networks. In: IEEE/ACM Design Automation Conference (2016)
34. Zheng, B., Deng, P., Anguluri, R., Zhu, Q., Pasqualetti, F.: Cross-layer codesign for secure cyber-physical systems. IEEE Transactions on Computer-Aided Design of Integrated Circuits and Systems. **35**(5) (2016)
35. Lin, C.W., Zheng, B., Zhu, Q., Sangiovanni-Vincentelli, A.: Security-aware design methodology and optimization for automotive systems. ACM Trans. Des. Autom. Electron. Syst. **21**(1) (2015)
36. Mundhenk, P., Steinhorst, S., Lukasiewycz, M., Fahmy, S.A, Chakraborty, S.: Lightweight Authentication for Secure Automotive Networks. In: IEEE/ACM Design, Automation & Test in Europe (2015)
37. Mundhenk, P., Paverd, A., Mrowca, A., Steinhorst, S., Lukasiewycz, M., Fahmy, S.A., Chakraborty, S.: Security in automotive networks: lightweight authentication and authorization. ACM Trans. Des. Autom. Electron. Syst. **22**(2) (2017)
38. Xie, T., Qin, X.: Improving security for periodic tasks in embedded systems through scheduling. ACM Trans. Embed. Comput. Syst. **6**(3) (2007)
39. Lin, M., Xu, L., Yang, L.T., Qin, X., Zheng, N., Wu, Z., Qiu, M.: Static security optimization for real-time systems. IEEE Transaction on Industrial Informatics. **5**(1) (2009)
40. Liang, H., Jagielski, M., Zheng, B., Lin, C.W., Kang, E., Shiraishi, S., Nita-Rotaru, C., Zhu, Q.: Network and System Level Security in Connected Vehicle Applications. In: IEEE/ACM International Conference on Computer-Aided Design (2018)
41. Waszecki, P., Mundhenk, P., Steinhorst, S., Lukasiewycz, M., Karri, R., Chakraborty, S.: Automotive electrical and electronic architecture security via distributed in-vehicle traffic monitoring. IEEE Trans. Comput. Aided Design Integrated Circuits Syst. **36**(11) (2017)
42. Wu, M., Zeng, H., Wang, C., Yu, H.: Safety Guard: Runtime Enforcement for Safety-Critical Cyber-Physical Systems. In: IEEE/ACM Design Automation Conference (2017)
43. Dutta, R.G., Guo, X., Zhang, T., Kwiat, K., Kamhoua, C., Njilla, L., Jin, Y.: Estimation of Safe Sensor Measurements of Autonomous System under Attack. In: IEEE/ACM Design Automation Conference (2017)
44. ISO 26262: Road Vehicles- Functional Safety, ISO Standard (2011)
45. Diffie, W., Hellman, M.: New directions in cryptography. IEEE Trans. Inf. Theory. **22**(6) (1976)
46. Feo, T.A., Resende, M.G.: Greedy randomized adaptive search procedures. J. Glob. Optim. **6**(2) (1995)
47. O'Higgins, B., Diffie, W., Strawczynski, L., De Hoog, R.: Encryption and ISDN- a Natural fit. In: International Switching Symposium (1987)
48. ElGamal, T.: A public key cryptosystem and a signature scheme based on discrete logarithms. IEEE Trans. Inf. Theory. **31**(4) (1985)
49. Adrian, D., Bhargavan, K., Durumeric, Z., Gaudry, P., Green, M., Halderman, J.A., Heninger, N., Springall, D., Thomé, E., Valenta, L., VanderSloot, B.: Imperfect Forward Secrecy: how Diffie-Hellman Fails in Practice. In: ACM SIGSAC Conference on Computer and Communications Security (2015)
50. Dick, R.P., Rhodes, D.L., Wolf, W.: TGFF: Task Graphs for Free. In: IEEE/ACM International Workshop on Hardware/Software Codesign (1998)
51. FlexRay. FlexRay Communications System Protocol Specification, ver.3.0.1. [Online]. Available: http://www.flexray.com
52. OpenSSL: Cryptography and SSL/TLS toolkit [Online]. Available: http://www.openssl.org/

53. NXP, MPC5775K [Online] www.nxp.com/docs/en/data-sheet/MPC5775KDS.pdf
54. NXP, i.MX 6 [Online] www.nxp.com/docs/en/fact-sheet/IMX6SRSFS.pdf
55. Barker, E., Dang, Q.: Recommendation for Key Management: Application-Specific Key Management Guidance. NIST special publication 800–57 Part 3, Revision 1 (2015)
56. Barker, E., Chen, L., Roginsky, A., Vassilev, A., Davis, R.: Recommendation for Pair-Wise Key-Establishment Schemes Using Discrete Logarithm Cryptography. NIST Special Publication 800-56A Revision 3 (2018)
57. National Security Agency. The Case for Elliptic Curve Cryptography. www.nsa.gov/business/programs/elliptic_curve.shtml

Secure by Design Autonomous Emergency Braking Systems in Accordance with ISO 21434

Adriana Berdich and Bogdan Groza

1 Introduction

In the past century, vehicles mediated a dramatic change of our ecosystem, not only by allowing us to safely travel over great distances but also by allowing us to change the environment in which we live by deploying cutting-edge infrastructure that would have been impossible to build on human power alone. More recently, the degree of autonomy of vehicles drastically improved. This happened not only by the introduction of various driver assistance technologies, such as automatic cruise control and autonomous emergency braking systems, but also with basic self-driving capabilities that are going to be extended until fully autonomous vehicles will travel the roads. Being such an important asset and now having such an enormous potential for being controlled by malicious pieces of software, it is no surprise that vehicles become a potential cybersecurity target.

Fortunately, so far, attacks on vehicles have been only demonstrative in nature, such as the experimental analysis provided the research in [8], a now famous attack on a Jeep car [21], and more recently some remote attacks on TESLA cars [25]. As security become manifest, the stakeholders had to react with standards and regulations that facilitate the deployment of security countermeasures and the proper incident response mechanisms. The array of standards ranges from AUTOSAR requirements on cryptography layers and secure on-board communication [3] to the more recently released requirements for intrusion detection systems on Electronic Control Units (ECU) [2]. In parallel to these, standards for security evaluation and for the assessment of security incidents were released. These include the more recent

A. Berdich · B. Groza (✉)
Faculty of Automatics and Computers, Politehnica University of Timisoara, Timisoara, Romania
e-mail: adriana.berdich@aut.upt.ro; bogdan.groza@aut.upt.ro

© The Author(s), under exclusive license to Springer Nature Switzerland AG 2023   155
V. K. Kukkala, S. Pasricha (eds.), *Machine Learning and Optimization Techniques for Automotive Cyber-Physical Systems*, https://doi.org/10.1007/978-3-031-28016-0_5

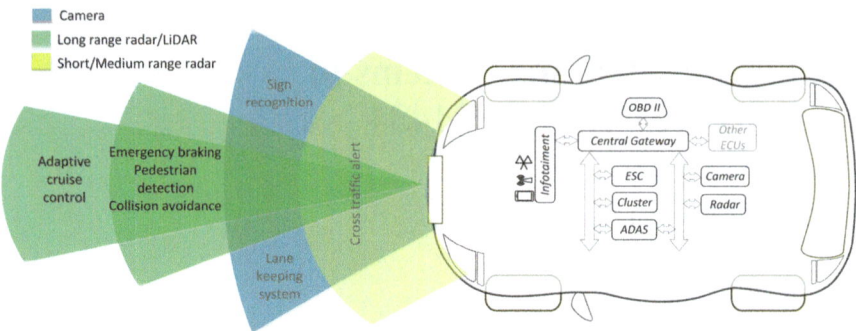

Fig. 1 Car equipped with camera and radar sensors for driver assistance or partial autonomous driving

ISO 21434 [15] containing the cybersecurity guidelines which we will follow in this work.

The introduction of these standards provides enormous help for an industry which now produces almost 80 million vehicles each year. But the problems are far from being solved since specifications inside standards do not provide the exact procedures and security mechanisms to mitigate the attacks which are up to the manufacturer. And more, no security mechanism is perfect and manufacturers have to imagine clever ways to mitigate the attacks. For this reason, we will focus on a secure-by-design Automatic Emergency Breaking (AEB) system, a system which is intended to trigger the brakes in order to avoid collisions with another vehicle or pedestrian. A car equipped with an AEB system and various other sensors is suggested in Fig. 1. Various long or short-distance radars and cameras report data to an Advanced Driver-Assistance Systems (ADAS) ECU which decides to request braking to the Electronic Stability Control (ESC) ECU. More details about this architecture will be added in a forthcoming section. To facilitate a security analysis according to ISO 21434 [15] we will need to proceed to a more in-depth evaluation at the control system level, clarifying the security mechanisms that should be put in place as well as the effects of various types of attacks.

The exposition in this chapter is structured as follows. We begin by providing some background for the reader that is unfamiliar with the AEB system and ISO 21434 in Sect. 2. Then we proceed to an in-depth analysis on adversary actions and their impact on the AEB system in Sect. 3. This is the most demanding section of our work and is extremely important since it allows us to set room for the exact security specifications for such systems. In Sect. 4 we proceed to an analysis following the ISO 21434 activities leading to specific security goals. Finally, Sect. 5 holds the conclusion of our work.

2   Background

In this section we set a brief background on the AEB system that will serve to us as a case study and on ISO 21434 which provides the guidelines for a security-aware design.

2.1   The AEB System in a Nutshell

The Automatic Emergency Braking (AEB) is one of the main ADAS functionalities designed to detect slow or stopped vehicles and pedestrians ahead and to trigger the brakes immediately. It is thus a system that can save the lives of passengers and of the traffic participants, pedestrians in particular. Other ADAS functionalities include the Adaptive Cruise Control (ACC) which is present in many older vehicles as well and the more recent Blind Spot Monitoring (BSM), Forward Collision Warning (FCW), Lane Departure Warning (LDW) and Lane Keeping Assist (LKA). As the names suggests, these systems ensure that the driver is signalled for the presence of objects on the sides (BSM), the approach toward a stationary object on the front (FCW) or the departure from the lane (LDW), eventually helping the driver by keeping the car to follow the lane (LKA). To provide a crisper case study, we will focus on the AEB system alone.

The scope of the AEB module is to prevent the accidents or to minimize the injuries resulting from such accidents by reducing the vehicle speed automatically when an obstacle, e.g., bicycle, pedestrian or sudden braking of the lead vehicle, is detected with the help of the long-range radar and front camera. The AEB system has more than a decade of use, Volvo first introduced the system in 2009. Since 2014, the European New Car Assessment Program (Euro NCAP)[1] introduced specific evaluations for the autonomous braking in the AEB City and AEB Interurban tests for low speed and high speed scenarios. The AEB feature is available in the majority of the recently released cars thus becoming an ubiquitous functionality.

The AEB module is a safety component and is part of the Forward Collision Avoidance (FCA) system. In Fig. 2 we give an overview of the AEB system function suggesting one vehicle that approaches a pedestrian. When the front camera and the front long-range radar detect the obstacle, an acoustic signal is activated and a visible warning light for the driver is displayed on the cluster. Afterwards, three braking stages follow: a first stage of slow pre-braking (partial braking), then a second stage of intensive pre-braking (partial braking) and finally the full braking stage. This image should be sufficient to understand the functionality of the AEB, the concrete system model will be detailed later.

---

[1] https://www.euroncap.com/en/vehicle-safety/safety-campaigns/2013-aeb-tests/.

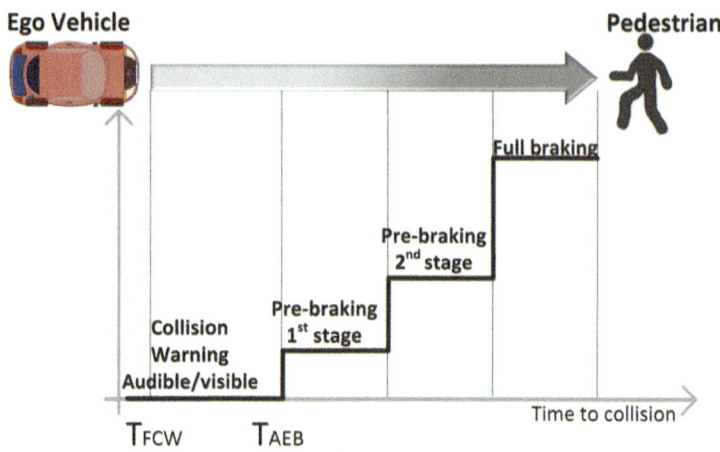

Fig. 2 Overview of the AEB system signalling and actions

2.2 Overview of ISO 21434 Activities

The automotive industry heavily relies on the V model for the development cycle of in-vehicle components. There are good reasons behind this choice, the most relevant being the rigorous interaction between the design and testing stages. It is no coincidence that a similar view can be expressed for the cybersecurity-aware design of in-vehicle components. On the left side of Fig. 3 we illustrate a generic development cycle in the automotive industry as a V model. It starts on the left branch with the stakeholder's requests which come from costumers and legislation, followed by requirement engineering, architecture design, software design, implementation and integration. On the right branch we have software testing, system testing, system validation and finally the homologation of the product. Similarly, on the right side of Fig. 3 we depict as a V-model the cybersecurity related tasks according to ISO 21434 using some of the activities outlined in Annex A of the standard. They start on the left branch with the item definition, followed by a Threat Analysis and Risk Assessment (TARA) then the definition of the cybersecurity goals and claims, the cybersecurity concept, specification and requirements and finally the integration and verification. On the right branch there are verification reports for each of these steps culminating with a final validation. In the cybersecurity related V model from Fig. 3b we highlight the first four steps, from item definition to cybersecurity specifications and requirements which are the subject of our analysis in this work. These four activities will be detailed for our AEB model in a forthcoming section.

Although it was published less than one year ago, it is worth mentioning that ISO 21434 has been already also used in several recent works. An earlier overview of ISO 21434 can be found in [18]. A whitepaper which places ISO 21434 in the context of other automotive and cybersecurity standards is also made available by BSI Group (British Standards Institution) [6]. A tool entitled ThreatGet which

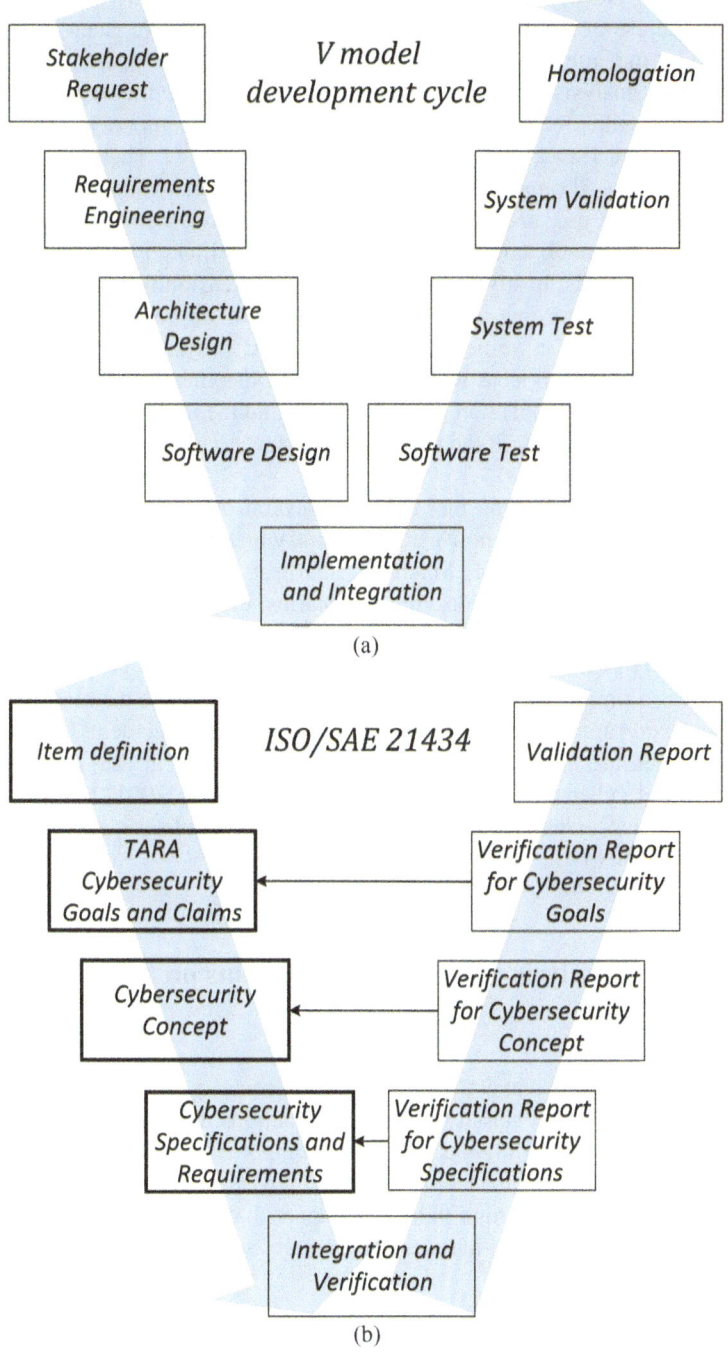

Fig. 3 Development cycle in the automotive industry as a V model (**a**) and cybersecurity related activities according to ISO 21434 expressed as a V-model (**b**)

is compliant to the ISO 21434 is proposed by the authors in [30]. The authors exemplify the use of the tool that they design for an automotive gateway ECU. As underlined by the authors [30], the analysis which ISO 21434 facilitates is an asset driven security analysis which focuses on assets to determine the impact as well as the attack path—this requires specific treatment for each component. The authors in [27] present an ISO 21434 risk assessment methodology. The risk assessment they propose is based on an offline phase, which assess the damage scenarios and asset dependencies, etc., and on an online phase which assess the risk of a reported incident. Another attack surface assessment based on ISO 21434 is presented in [26]. Their analysis is mostly focused on the attack feasibility rating. This rating along with the impact rating can be used to determine the risks according to ISO 21434. The analysis that we perform here on the AEB system is in-line with the previously mentioned works as we follow the same specifications from ISO 21434. What differs is the component on which we focus and the in-depth adversary model and protection mechanisms at the control system level which are not present in the previously mentioned works.

It is also worth mentioning that there are several other works concerned with security assessments for automotive components, which were published well before the release of the ISO 21434. Maybe the earliest is [13] which tries to drive security requirements from various threat scenarios, including ECU corruption and spoofed CAN messages, etc. An analysis centered on the use of the Body Control Module (BCM) as the critical gateway component is done in [11]. The work in [12] performs a similar risk analysis but it is centered on vehicle instrument clusters which can be corrupted to mislead the driver and thus cause accidents. Another risk and countermeasure analysis was done in [4]. The authors in [28] discuss a risk assessment and cybersecurity analysis based on ISO/IEC 27001. In [34] another cybersecurity risk assessment is discussed which accounts for several types of attacks on the CAN bus.

3   In-Depth Analysis of Adversarial Actions on AEB Control Systems

Existing research works that apply the ISO 21434 security standard, generally take a straight forward way in classifying adversary actions and their impact based on generic assumptions regarding the attack of a component. Here we will proceed to a more in-depth analysis that accounts for exact adversarial manipulations of CAN frames and we try to determine the exact impact that these actions will have on safety. This analysis is needed in order to accurately assess the risks and understand the countermeasures.

3.1 Detailed AEB System Model

In order to simulate the vehicle reaction in case of distinct attacks on the CAN bus, we will use an existing Simulink model for Autonomous Emergency Braking with Sensor Fusion from MathWorks[2] and add adversarial behaviour to the model. Other works have also used Simulink models to test and validate attacks from the CAN bus. For example, the authors in [9] use a Simulink model of the Anti-lock Braking System (ABS) developed by Mathworks to test a multilevel monitor for the isolation and detection of attacks over the senors and CAN bus for a Cyber Physical System (CPS). In [17] an ACC model from Simulink is used to validate a method for the detection and mitigation of spoofing attacks on the radars which are used by the ACC system. The authors check the integrity of the radar sensor data based on a spatio-temporal challenge-response (STCR) which transmits signals in random directions and identifies then excludes signals reflected from untrustworthy directions.

In Fig. 4 we depict the model of the AEB and the placement of six potential adversaries denoted as A1–A6. The Simulink AEB model contains two functional parts, the AEB functionality and the vehicle and environment component. The AEB functionality is also split in two ECUs: Camera/Radar ECU and AEB/ADAS ECU. The Camera ECU acquires data from the hardwired components, i.e., camera and radar, based on which it derives the information about obstacles and computes the relative distance and velocity which are transmitted to the ADAS ECU using the Private CAN bus communication (Pr-CAN). The ADAS ECU, in addition to this information received from the Camera ECU, also receives the longitudinal velocity from the ESC ECU using the Chassis CAN bus communication (C-CAN). The ADAS ECU implements an AEB controller which computes the deceleration request needed to stop the vehicle in order to avoid the collision with the obstacle, i.e., the pre-braking stages which is the AEB status. Additionally the ADAS ECU implements a speed controller (designed as a PID controller) to determine the required acceleration in order to maintain the ego velocity setpoint, i.e., the throttle. Finally, based on the AEB status and the internally computed acceleration, the ADAS ECU computes the throttle position. The deceleration request and the throttle position are transmitted over the C-CAN to the ESC ECU which controls the vehicle based on the requested commands. Additionally, in our model, the ESC ECU computes the vehicle position and trajectory represented as: XY position, XY velocity, yaw rate, yaw angle and longitudinal velocity. For simplicity, in the Simulink scenario, the curvature of the road is set as a constant. As stated, there are six adversary positions in Fig. 4, representing the six signals which can be attacked in the model.

---

[2] https://nl.mathworks.com/help/driving/ug/autonomous-emergency-braking-with-sensor-fusion.html

3.2 Adversary Model and Attack Strategies

For an accurate description of the attacks at the control system level, we need to formalize the adversarial actions. The adversary actions consist in manipulating a specific signal and we will use $y_\blacklozenge(k)$ to denote a signal at step k which is a positive integer, i.e., $k \in Z_N^+$, and $\blacklozenge$ is a placeholder to denote the six possible adversaries A1–A6 on six possible signals: deceleration (i.e., the braking stages), acceleration, ego velocity, relative distance, relative velocity and curvature (each corresponding to the 6 location points in Fig. 4), i.e., $\blacklozenge \in$ {brake, throttle, vego, rdist, rvel, curvature}. Then the value of $y_\blacklozenge(k)$ at each step will be either the legitimate signal $\overline{y}_\blacklozenge(k)$ or a signal originating from the adversary $\widetilde{y}_\blacklozenge(k)$. For example, $\overline{y}_{vego}(k)$ will be the legitimate signal for ego velocity and $\widetilde{y}_{vego}(k)$ is the adversarial signal corresponding for the ego velocity.

Most, if not all, of the existing works focusing on attacks and intrusion detection for CAN buses consider three types of attacks: replay, Denial of Service (DoS) and fuzzing attacks. These attacks can be easily formalized as follows:

1. *replay attacks*—by this attack, CAN frames are re-transmitted, possibly with a random delay, containing previously recorded signals, i.e., $\widetilde{y}_\blacklozenge(k) \leftarrow \overline{y}_\blacklozenge(i), i < k$
2. *fuzzing attacks*—are a modification attack in which random values are injected in the datafield of CAN frames, essentially meaning that the attack signal becomes a random value, i.e., $\widetilde{y}_\blacklozenge(k) \leftarrow$ rand,
3. *DoS attacks*—prohibit CAN frames from being transmitted on the bus and are specifically difficult to address since an adversary can always write high priority frames on the bus or even destroy legitimate frames with error flags or by distortions of the data-field that deem them unusable, which means that the signals are effectively lost, i.e., $\widetilde{y}_\blacklozenge(k) \leftarrow \perp$.

For all of the previous attacks, in the later model where we evaluate them, we assume a probability of occurrence, simply denoted as p, which is the probability of an adversarial signal (or CAN frame) to replace a legitimate one. Assigning a

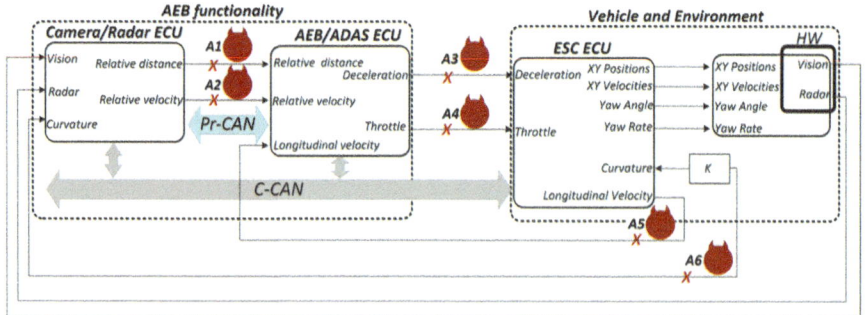

Fig. 4 The AEB model with the attach surfaces

probability to the event of an attack is in line with the practical side of the problem since adversaries usually insert manipulated frames that compete with legitimate frames on the bus. This also responds to the situation in which an intrusion detection system is in place on the controller and only some of the adversary frames will go undetected and accepted as legitimate.

Still, these three types of attacks described above are insufficient for giving a complete image over both the attacker and defense capabilities. Notably, an intrusion detection system may be into place and arithmetic attacks which take advantage of the system model and inject specific values that are expected to cause a particular behavior of the car may be an option. Such a scenario has received little or no attention at all in the research literature related to car security. A reason for which will carefully examine three flavours of stealthy attacks which were also pointed out in well known control system security paper [7]. These include three flavours of stealthy attacks: surge attacks, bias attacks and geometric attacks [7].

For this reason, let us consider that an intrusion detection system may be in place. The problem addressed by an intrusion detection system is thus to distinguish between the two values $\overline{y}_\blacklozenge(k)$ and $\widetilde{y}_\blacklozenge(k)$, $\forall k \in Z_N^+$. Since no intrusion detection system is perfect, a small false negative rate exists, i.e., some of the adversarial frames may go undetected. To introduce more specialized attacks and countermeasures, we may consider that a simple change detection algorithm stays at the core of the intrusion detection method implemented on the ECU. Such an algorithm may account for statistical distances between value and various range checks. The work in [7] dedicated to control systems security, suggests the use of cumulative sums (CUSUM) statistics over the reported value and some predicted value for the same signal. This methodology is indeed well suited for our scenario since we can infer the value of one signal from another signals available in the car. For example, using the relative velocity reported by a radar and the velocity of the car, we can compute the distance to the object and compare it with the reported one, etc. Generally speaking, having the predicted value of the signal $y'_\blacklozenge(k)$ and a bias b we can use the following recurrent sum to detect an attack [7]:

$$S_\blacklozenge(k) = \max\{0, S_\blacklozenge(k-1) + |y_\blacklozenge(k) - y'_\blacklozenge(k)| - b\}, S(0) = 0$$

Whenever the error between the predicted value and the reported value is greater than the bias b, the error is added and, when the sum reaches a signalling threshold τ, the signal will be deemed as adversarial. That is, if $S_\blacklozenge(k) > \tau$ we consider $y_\blacklozenge(k)$ to be an attack signal and if $S_\blacklozenge(k) \leq \tau$ we consider $y_\blacklozenge(k)$ to be legitimate. Having this change detection procedure in mind, an adversary can mount the following three stealthy attacks that are described in [7]:

1. *surge attacks*—are the modification attacks in which the value of the signal is set to the maximum value (or minimum value) such that it will inflict the maximum damage on the system; to remain stealthy and go undetected by the cumulative summing, the attack value at step $k + 1$ will be $y_{\blacklozenge,\max}$ only if the corresponding sum at the next step $S_\blacklozenge(k + 1) \leq \tau$ while otherwise the attack signal will stay at

$y'_\blacklozenge(k) + |\tau + b - S_\blacklozenge(k)|$ (note that in this way $|\tau + b - S_\blacklozenge(k)|$ is the maximum value that can be added to the legitimate signal such that an intrusion will not be detected).

2. *bias attacks*—are the modification attacks in which a small constant $c = \tau/n + b$ is added at each step to the attacked signal, i.e., $\widetilde{y}_\blacklozenge(k) \leftarrow y_\blacklozenge(k) + \tau/n + b$, ensuring that the attack remains undetected for n steps (this happens so since the threshold is divided over the n steps of the attack),

3. *geometric attacks*—are the modification attacks in which a small drift is added to the attacked signal in the beginning and the drift becomes increasingly larger in the next steps using a geometric expansion, i.e., $\widetilde{y}_\blacklozenge(k) \leftarrow y_\blacklozenge(k) + \beta\alpha^{n-k}$ where α is fixed and $\beta = \frac{(\tau + nb)(\alpha^{-1} - 1)}{1 - \alpha^n}$.

To sum up, in the light of these attack strategies, we are concerned with assessing the impact of two kinds of adversarial actions: those attacks that will go undetected due to the non-zero false negative rate of the in-vehicle IDS and the stealthy attacks in which adversary actions deviate by a small margin from the predicted values, thus remaining undetected. We discuss the impact of the attacks in what follows.

3.3 Attack Evaluation on the AEB Model

One of the test scenarios from the Euro NCAP car safety performance assessment programme [33] is the Car-to-Pedestrian Nearside Child test dedicated to the AEB functionality. In Fig. 5 we depict this scenario. Two vehicles are stationary on the right side of the road and there is a pedestrian nearby, crossing the road at one meter from the cars. The ego vehicle is travelling on the left lane of the road, the view of the pedestrian crossing is obstructed for the driver. The AEB system has to activate the automatic braking in order for the car to stop and avoid collision with the pedestrian.

This scenario is used in the Simulink model as well and we will analyze the adversarial impact on it. We consider attack points A1–A5 and leave A6 outside the discussion since A6 represents the curvature of the road and attacking this value will lead the vehicle outside the lane, not causing a collision with the pedestrian in front

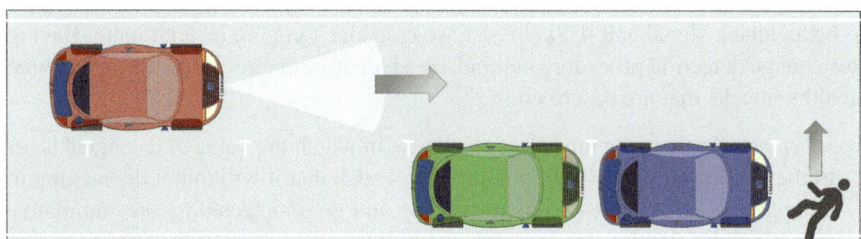

Fig. 5 Overview of the AEB scenario

which is our attack target. An attack on the curvature is relevant, but will not fit our specific use case. More, the countermeasure which we later propose will hold for manipulating data related to A6 as well.

Signal Interpretation To clarify the impact of the attacks, in Figs. 6, 7 and 8 we illustrate the signals starting from a clear scenario without adversarial interventions and then add several types of attacks to the signals. We show the plot markers for each signal in Fig. 6 which corresponds to the case when there is no adversarial intervention. We use two types of plots: (i) FCW/AEB status plots which show the status of the collision warning and deceleration/acceleration stages and (ii) velocity/distance plots which show the velocities and the distances toward the object in front. Note that each plot has distinct axes on the left and right side. For the FCW/AEB status, we plot the status signals on the left axis (marked with black) and the accelerations measured in m/s^2 on the right axis (marked with gray). The status signal from the left axis includes the signal which indicates the activation of the FCW with solid line, AEB status (braking stages) with dashed line and the signal which indicates that the ego car was stopped which is marked with a dashed-dotted line. On the right axis, we plot the deceleration with dotted line and the acceleration with a dotted line marked by circles. For the velocity/distance plots, we again plot two axes, one for the velocity and another one for the distance. The signals plotted on the left axis (marked with black) are the following: the preset velocity for the vehicle marked with dashed-dotted line, the ego velocity marked with solid line and the relative velocity marked with dashed line. The signals plotted on the right axis (marked with gray) are the following: the relative distance marked with dotted line and the headway marked with dotted-circle line.

Attack-Free Scenario A few words on the plots for the attack-free scenario may be helpful. In Fig. 6 we illustrate the signals without the adversarial interventions. As the relative distance between the ego car and the obstacle becomes lower than 15 meters (velocity/distance plot), the FCW system becomes active (FCW/AEB status) and the AEB begins the braking stage for the car. As expected, the AEB status follows the three pre-braking stages until the car is stopped. The acceleration is decreasing during braking, thus the ego velocity is decreasing from the preset velocity until the car eventually stops. The dotted line marked with circles shows the headway which is the difference between the relative distance and the length of the car (the headway is about 2 meters when the cars stops).

Replay Attacks In Fig. 7 we depict the AEB signals under a replay attack on the braking (deceleration) signal and the adversarial signal $\tilde{y}_{\mathsf{brake}}(k)$ corresponding to the deceleration. To inflict maximum damage, the replayed valued for the deceleration is the minimum value known to the intruder, i.e., $\tilde{y}_{\mathsf{brake}}(k) = 0$. We will use the following attack probabilities: $p = 0.25$, $p = 0.5$ and $p = 0.75$. In the FCW/AEB status plots from Fig. 7 it can be easily observed that the adversarial signal $\tilde{y}_{\mathsf{brake}}(k)$ corresponding to deceleration (gray dotted line) is set to zero more often when the attack probability is increasing, which means that the adversary deactivates the brake request more often. In the velocity/distance plots from the

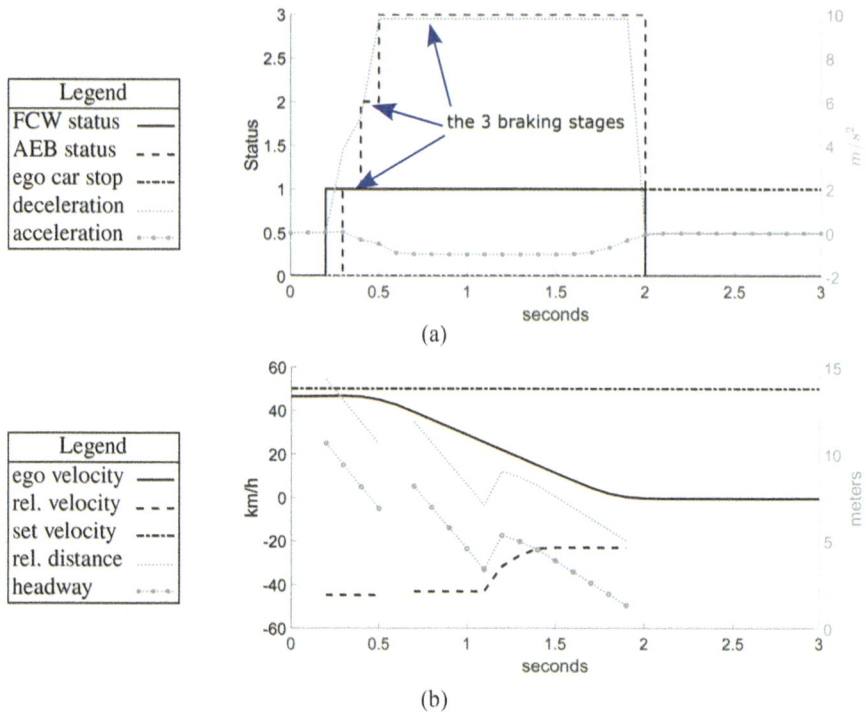

Fig. 6 Signals without adversarial intervention. (**a**) FCW/AEB status plot. (**b**) Velocity/distance plot

Fig. 7 we depict the response of the car to the attack. The ego velocity (black solid line) in Fig. 7b is 0 km/h when the headway is 0.19 meters which means that there is no impact at an attack probability of $p = 0.25$. In Fig. 7d, the ego velocity is 25.53 km/h when the headway is 0 meters which means that at an attack probability of $p = 0.5$ the impact will take place. In Fig. 7f, the ego velocity is 42.08 km/h when the headway is 0 meters which means that at an attack probability of $p = 0.75$ again there will be an impact at a considerable speed. As expected, the impact severity of the attack is increasing with the attack probability.

Fuzzing Attacks In Fig. 8 we illustrate the signals when fuzzing attacks take place, with an attack probability of $p = 0.5$, on three attack surfaces A1, A3 and A4. In Fig. 8a, b we depict the impact for a fuzzing attack on the deceleration signal. The adversarial signal $\tilde{y}_{brake}(k)$ corresponding to deceleration (gray dotted line) has random values which reduce the brake intensity or even deactivate the brake request in order to produce impact. The ego velocity (black solid line) is 50.54 km/h when the headway is 0 meters which means that when the fuzzing attack on deceleration takes place with probability of $p = 0.5$ it induces an impact at considerable speed. In Fig. 8c, d we show the fuzzing attack on acceleration. The adversarial signals

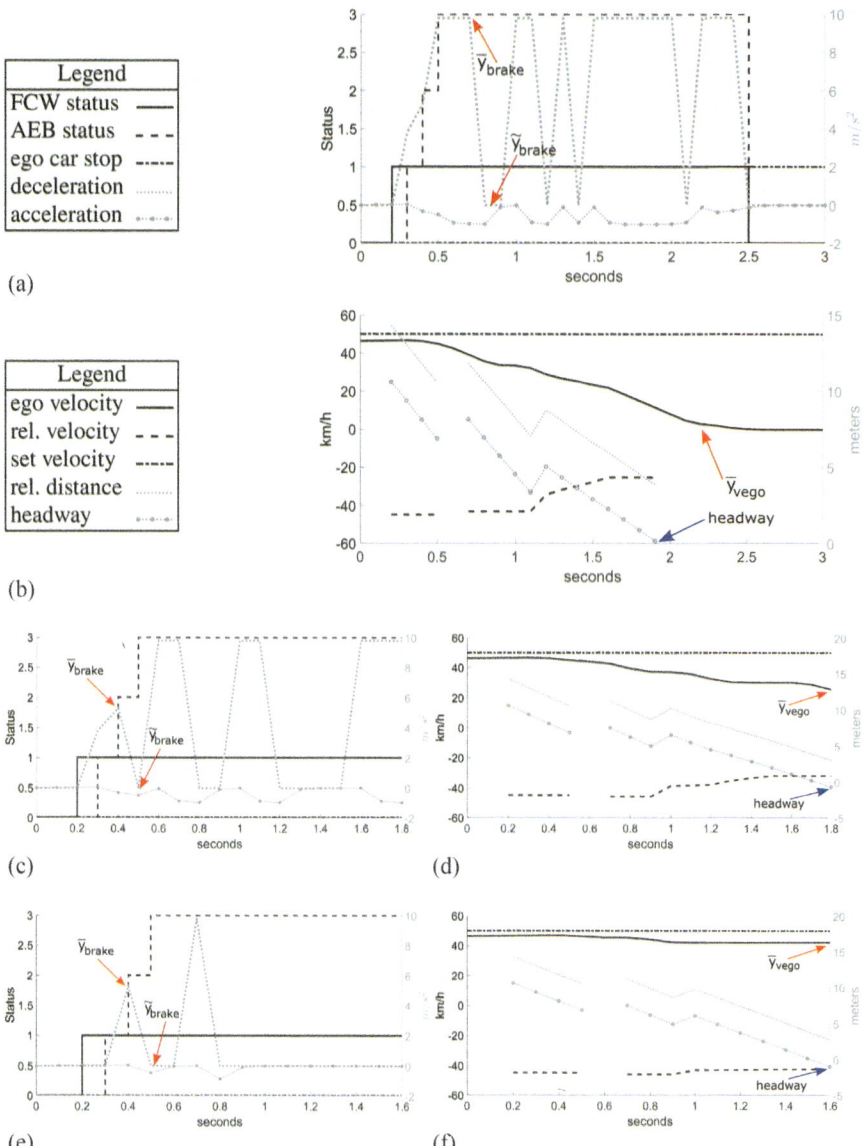

Fig. 7 Signals after a replay attack on deceleration y_{brake} with distinct attack probabilities: (**a**) $p = 0.25$—FCW/AEB status, (**b**) $p = 0.25$—velocity/distance plot (**c**) $p = 0.5$—FCW/AEB status, (**d**) $p = 0.5$—velocity/distance plot, (**e**) $p = 0.75$—FCW/AEB status and (**f**) $p = 0.75$—velocity/distance plot

$\tilde{y}_{throttle}(k)$ do not significantly impact the functionality of the AEB system. The car is stopped in time, in some cases only the FCW is activated after the car is stopped. In Fig. 8e, f we depict the impact of a fuzzing attack on the relative distance signal.

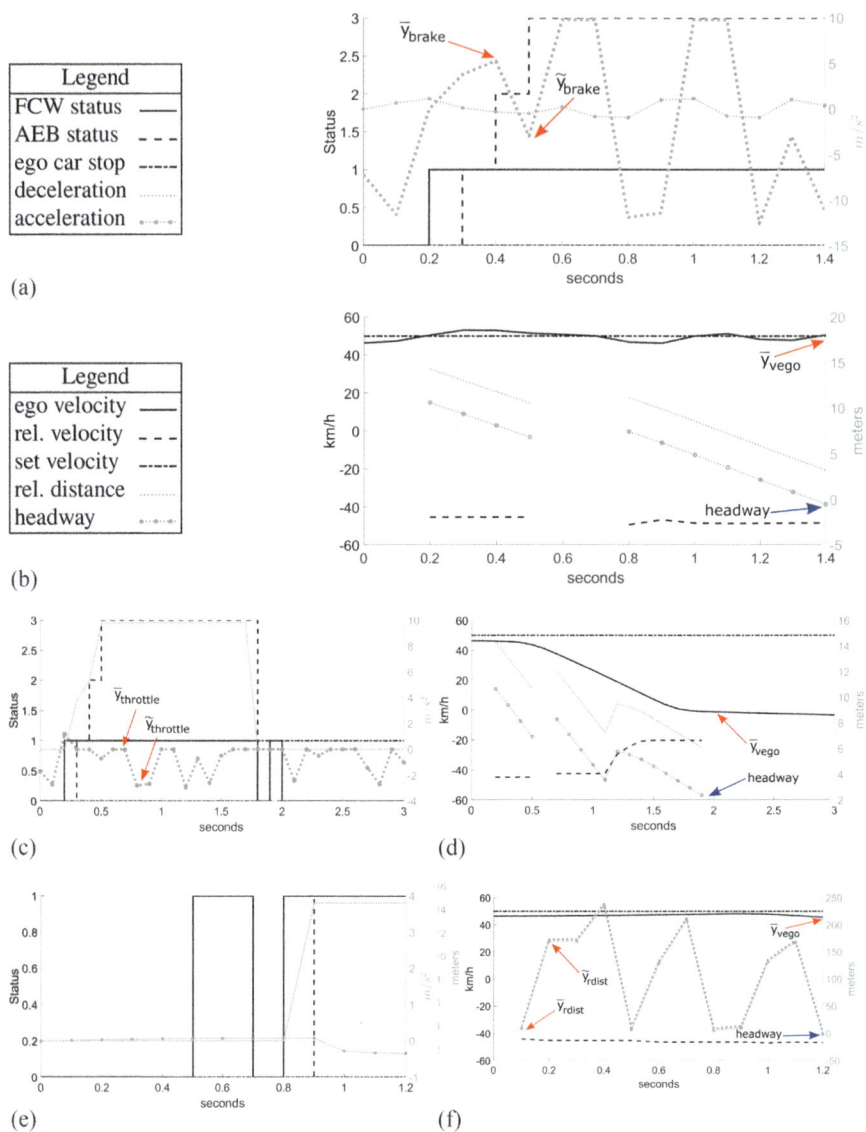

Fig. 8 Signals under fuzzing attack with attack probability of $p = 0.5$ on: (**a**) deceleration y_{brake}—FCW/AEB status, (**b**) deceleration y_{brake}—velocity/distance plot, (**c**) acceleration $y_{throttle}$—FCW/AEB status, (**d**) acceleration $y_{throttle}$—velocity/distance plot, (**e**) relative distance y_{rdist}—FCW/AEB status and (**f**) relative distance y_{rdist}—velocity/distance plot

The adversarial signal $\widetilde{y}_{rdist}(k)$ corresponding to the relative distance (gray dotted line) produces a delay in the activation of the AEB system which causes an impact at a velocity of 45.61 km/h.

Table 1 AEB results: collision velocity and distance to target in case of replay, fuzzing and DoS attacks at various success rates

Attack	Signal	p = 0.25 (0.2 for DoS)		p = 0.5		p = 0.75 (0.7 for DoS)	
		Collision veloc-ity[km/h]	Distance to target[m]	Collision veloc-ity[km/h]	Distance to target[m]	Collision veloc-ity[km/h]	Distance to target[m]
Replay	A3: decelera-tion	No coll.	0.19	25.53	0	42.08	0
Fuzzing	A3: decelera-tion	43.99	0	50.54	0	57.27	0
	A4: throttle	No coll.	2.32	No coll.	2.37	No coll.	2.26
	A1: relative distance	33.89	0	45.61	0	45.61	0
	A2: relative velocity	No coll.	1.32	No coll.	1.32	No coll.	1.32
	A5: long. velocity	No coll.	1.46	No coll.	1.46	No coll.	1.90
DoS	A3: decelera-tion	No coll.	1.32	9.01	0	32.33	0

DoS Attacks In case of DoS attacks we chose to apply the attack again only on deceleration signal. Since our simulation is running with a step of 0.1s we can simulate only the attack probabilities which are multiple of 0.1, i.e., $p = 0.2$, $p = 0.5$ and $p = 0.7$. For brevity, we omit the plots for this attack scenario and we refer the reader to the results in Table 1. The DoS attack on deceleration does not cause an impact for an attack probability $p = 0.2$, but it produce impact with a collision velocity of 9.01km/h and 32.33 km/h in case of attack probabilities $p = 0.5$ and $p = 0.75$ respectively. We use "no coll." to denote that no collision took place.

Also, in Table 1 we summarize as numerical data the collision velocity and distance to target in case of replay and fuzzing at various attack success rate, i.e., $p = 25$, $p = 0.5$ and $p = 0.75$. In case of the replay attack we apply the attack only on the deceleration signal since it has the most significant effect. The attacks are causing a collision at $p = 0.5$ and $p = 0.75$ and the collision velocity is increasing with the attack probability. In case of fuzzing attacks, as can be already observed in the previous figures, no collision happens when the adversarial signals are $\tilde{y}_{\text{throttle}}(k)$, $\tilde{y}_{\text{rvel}}(k)$ and $\tilde{y}_{\text{vego}}(k)$ corresponding to the throttle, relative distance and longitudinal velocity. On the other hand, when the fuzzing attacks are applied on deceleration and relative distance, the collision takes place at all attack probabilities with an impact velocity which increases with the attack success rate from 43.99 to 57.27 km/h in case of the deceleration and from 33.89 to 45.61 km/h in case of the relative distance.

3.4   Impact of Stealthy Attacks

We now discuss the impact of the three types of stealthy attacks: surge attacks, bias attacks and geometric attacks. The work in [7] evaluated the impact of these attacks on a control system for a chemical reactor process and we will now evaluate it on our AEB system. This type of attacks assume that a change detection mechanism is in place. Distinct from the work in [7], we do not use a state predictor to infer the next value but we can infer the value of some of the parameters from the others and use this in the cumulative sum. Namely, we estimate the relative distance from the relative velocity along with the longitudinal velocity and we also estimate the deceleration as the derivative of the longitudinal velocity. The estimated values of the signals are used in place of $y'_\blacklozenge(k)$ when computing the cumulative sum $S_\blacklozenge(k)$. The rest is similar in the change detection mechanism and in the computation of the attack values. In what follow we will demonstrate the impact of stealthy attacks on the relative distance and deceleration. The attacks on the relative distance will have little effects and will not cause a collision, while the attacks on deceleration will lead to collisions at significant speed.

Stealthy Attacks on Relative Distance   In Fig. 9 we illustrate the signals when a stealthy attack on relative distance takes place, i.e., the adversarial signal is $\widetilde{y}_{rdist}(k)$. In the FCW/AEB status plots from Fig. 9 it can be seen that the AEB functionality is not influenced by the stealthy attacks on relative distance as the car stops in time to avoid the collision. This is because the stealthy attack cannot take advantage of the larger random values of the previously demonstrated fuzzy attack (this will make the attack detectable). In the velocity/distance plots of Fig. 9 we show the adversarial signal $\widetilde{y}_{rdist}(k)$ corresponding to the relative distance and the headway which is also influenced by the attack. In Figure (b), corresponding to surge attacks, the adversarial signal $\widetilde{y}_{rdist}(k)$ sets the relative distance to 15 meters (maximum value of the relative distance in normal conditions) on several points. But still, no impact occurs. In Figure (d), corresponding to a bias attack, the adversarial signal $\widetilde{y}_{rdist}(k)$ smoothly distorts the relative distance signal placing it slightly below the real distance—still, there is no collision. In Figure (f), corresponding to a geometric attack, the adversarial signal $\widetilde{y}_{rdist}(k)$ starts from the real signal value and progressively increases in time in order to maximize the damage at the end. But again, there is no collision. Thus, in our simulation, none of the stealthy attacks on the relative distance influenced the AEB functionality in such way as to cause an accident. The effects will become more serious when the deceleration is attacked in the same way.

In terms of parameters for the stealthy attacks on relative distance, we did set the bias $b = 2$ because the maximum error between the signal computed by the ECU and the predicted signal is around 2 m. The threshold was set equal with the bias, i.e., $\tau = 2$ and the number of steps was set to $n = 25$ for the bias and geometric attacks, because our simulation has 50 steps in case when no attack takes place and

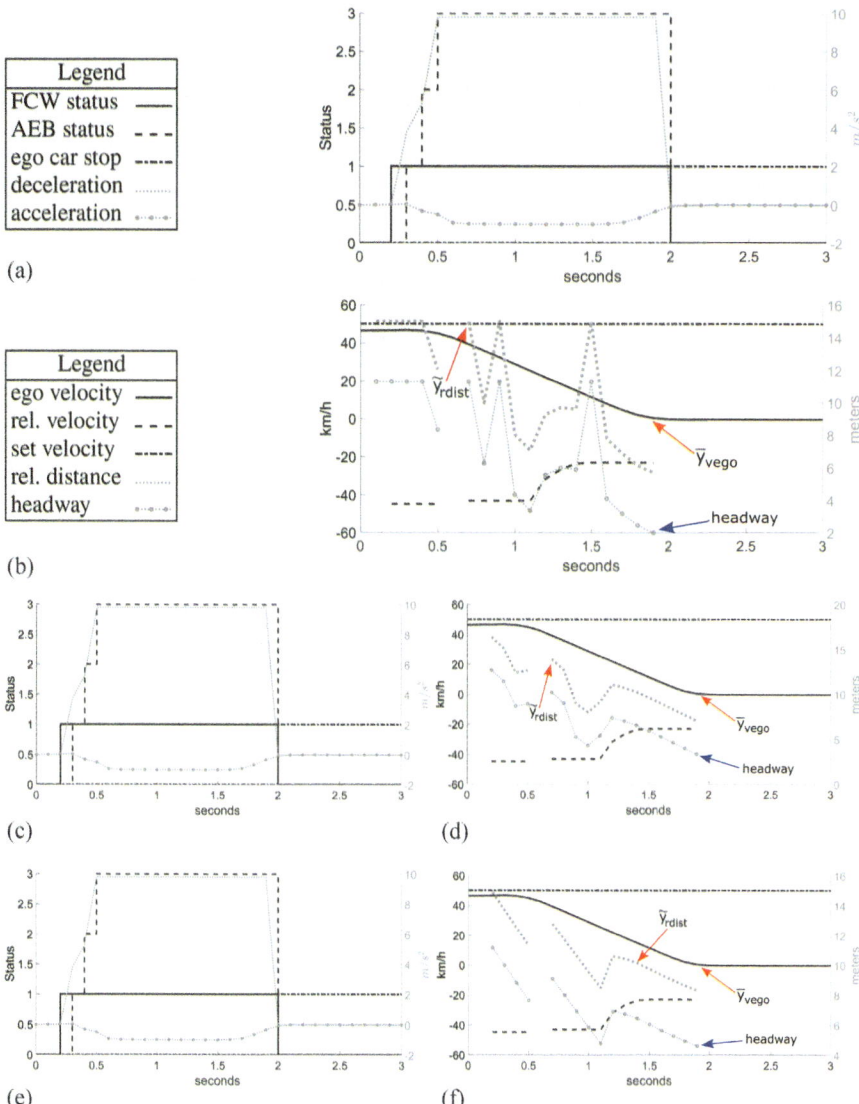

Fig. 9 Signals under stealthy attacks on relative distance y_{rdist}: (**a**) surge attack—FCW/AEB status, (**b**) surge attack—velocity/distance plot (**c**) bias attack—FCW/AEB status, (**d**) bias attack—velocity/distance plot, (**e**) geometric attack—FCW/AEB status and (**f**) geometric attack—velocity/distance plot

as such we will obtain the full attack during the first half of the simulation. For the geometric attack we set parameter $\alpha = 0.9$ to maximize the attack impact. Different parameters may yield distinct results.

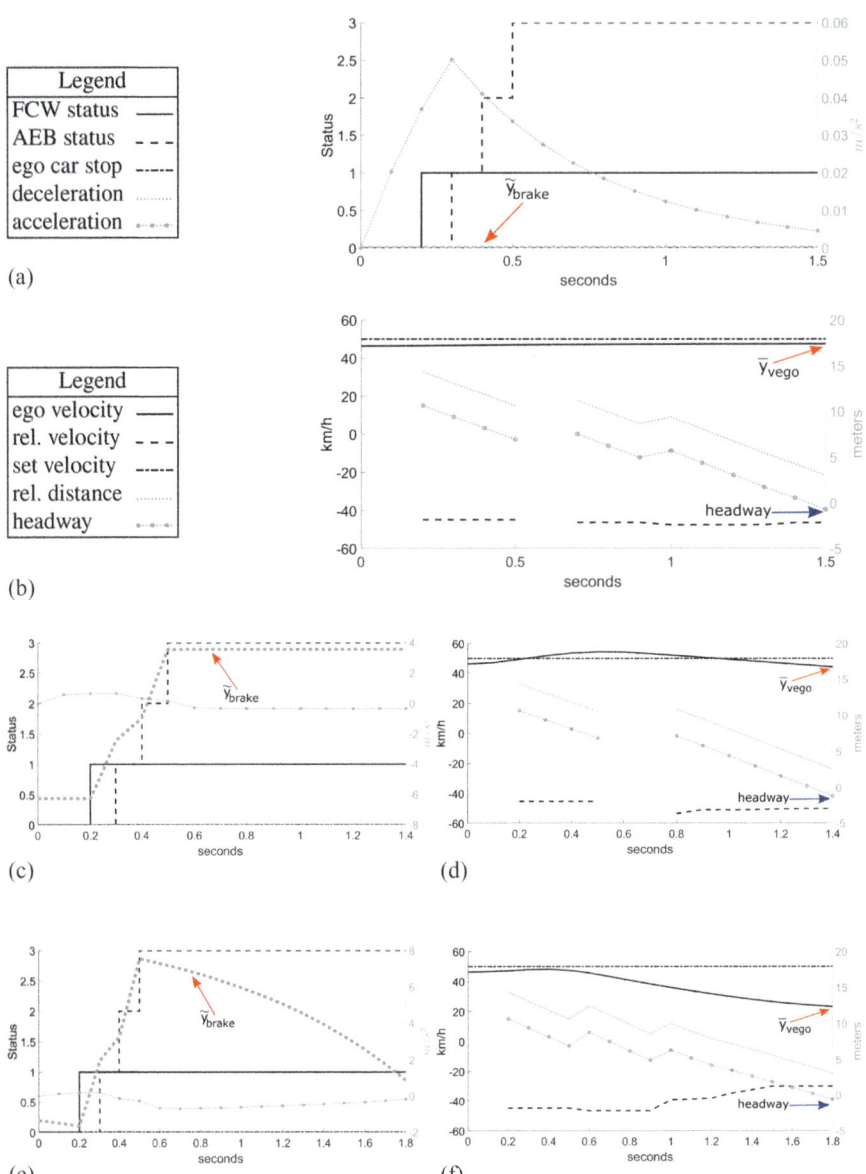

Fig. 10 Signals under stealthy attacks on deceleration y_{brake}: (**a**) surge attack—FCW/AEB status, (**b**) surge attack—velocity/distance plot (**c**) bias attack—FCW/AEB status, (**d**) bias attack—velocity/distance plot, (**e**) geometric attack—FCW/AEB status and (**f**) geometric attack—velocity/distance plot

Stealthy Attacks on Deceleration In Fig. 10 we illustrate the signals when a stealthy attack on deceleration takes place, i.e., the adversarial signal is $\tilde{y}_{brake}(k)$. Now the attack will clearly result in an impact with the pedestrian. In Figure (a) and (b) we depict the effect of a surge attack on the deceleration signal. In the FCW/AEB status plots of the figure, the adversarial signal $\tilde{y}_{brake}(k)$ is always $0\,m/s^2$ (gray dotted line) because in the attack implementation $y_{min} = 0$, the acceleration is small (close to $0\,m/s^2$) as the vehicle speed is constant and in the velocity/distance plots it can be observed that the ego velocity remains near the preset velocity, i.e., no brake request comes from the AEB controller. Even if the AEB status is correctly shown in the FCW/AEB status plot, which follows the 3 braking stages, the deceleration request signal received by the ESC ECU is corrupt, requesting no deceleration, and thus the car continues to maintain the preset velocity. Note that the AEB status parameter is internal to the AEB controller and only the deceleration is communicated to the ESC ECU for the car to decelerate (this can be easily seen in the model from Fig. 4). Therefore in the plot from Fig. 10, while the AEB status is still set to 3 (full braking) inside the AEB controller, the deceleration value is subject to a stealthy attack and is much lower, misleading the ESC controller that the car should not brake and eventually leading to a collision. This attack causes an impact at an ego velocity of 47.51 km/h (note that the headway is 0 m) which means that the surge attack on deceleration creates an impact at considerable speed. In Figures (c) and (d) we depict the effect of the bias attack on the deceleration signal. In the FCW/AEB status plot, (c), the adversarial signal $\tilde{y}_{brake}(k)$ (gray dotted line) is increasing until $4\,m/s^2$ are reached, but in normal conditions the deceleration should reach a much higher $10\,m/s^2$ (Fig. 6). This leads to a much slower braking even if the AEB status request is set to full braking. In the velocity/distance plot (iv), the ego velocity is still near the preset velocity, reaching around 44.42 km/h at the time when the headway is 0 meters, i.e., when the collision occurs. This again means that the bias attack on deceleration produces an impact at considerable speed. Finally, in Figure (e) and (f) we depict the effect of a geometric attack on the deceleration signal. In Figure (e), the adversarial signal $\tilde{y}_{brake}(k)$ (gray dotted line) is increasing until $8\,m/s^2$ are reached after which the geometric attack occurs and maximizes the damage as it abruptly decreases and the deceleration request gets near $0\,m/s^2$. In Figure (f), the effect of this attack can be observed as the ego velocity is decreasing to 23.37 km/h when the headway is 0 m, i.e., the time of collision. This is a slightly lower impact velocity compared to the other two stealthy attacks. Still, all the three stealthy attacks on deceleration had caused an impact while they remained undetected by the change detection mechanism. We also note that in this case, another protection mechanism may be put in place: as the AEB controller orders the car to decelerate and the reported velocity does not decrease according to the expectations, the AEB controller may determine that an attack takes place. However, the ESC controller will not know which of the frames are the legitimate ones, i.e., lower or higher deceleration, and cannot act to correct the issue.

In terms of parameters for the stealthy attacks on deceleration, we set the bias $b = 6$ because the maximum error between the signal computed by the ECU and

Table 2 AEB results: collision velocity and distance to target in case of stealthy attacks or deceleration and relative distance

Attack	Signal	Collision velocity[km/h]	Distance to target [m]
Surge	A1: relative distance	No coll.	2.01
Bias		No coll.	3.40
Geometric		No coll.	4.63
Surge	A3: deceleration	47.52	0
Bias		44.42	0
Geometric		23.37	0

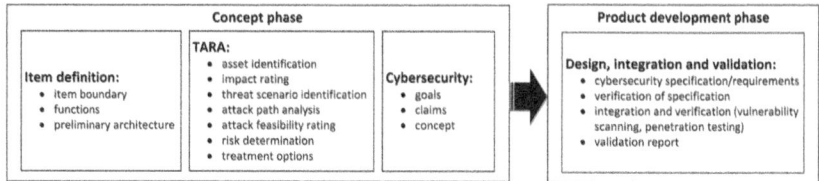

Fig. 11 Example of activities flow according to ISO 21434

the predicted signal is around $6 \, \text{m/s}^2$. The threshold was set equal with the bias, i.e., $\tau = 6$, the number of steps and parameter α were set again to $n = 25$ and $\alpha = 0.9$ respectively.

Table 2 summarizes in terms of numerical data the collision velocity and the distance to the target in case of stealthy attacks on deceleration and relative distance. The collision occurs in case of all the three attacks on deceleration signal, while no collision occurs when the stealthy attacks are applied to the relative distance as discussed previously. We use "no coll." to denote that no collision took place.

4 Secure-by-Design AEB in Accordance to ISO 21434

Benefiting from the previous attack analysis, we will now follow the steps of ISO 213434 in order to point out specific security goals and the means to assure them for the AEB system.

4.1 Overview of the ISO 21434 Cybersecurity Design Flow

For an accurate overview of the steps required by ISO 21434, we will first introduce an operational overview of the activities presented in the standard. These activities will be then detailed with respect to the AEB system that we use as an example. Figure 11 gives an overview of the activities presented in ISO 21434. These activities are also exemplified in case of a headlight system which is used as a case study inside the standard.

The security related activities start from the definition of the item which is going to be secured, the boundaries of the system in which it is incorporated, its functions and a preliminary architecture. All these form the first step of the concept phase. A more tedious step follows which consists in the TARA (Threat Analysis and Risk Assessment). This step asks us to delve into more details regarding the identification of assets that need to be protected (either logical assets such as CAN frames, or physical assets as a specific sensor), rating the impact, identifying the threat, attack path, rating the attack feasibility, determining the risk level and the treatment option. The last step of the concept phase consists in the determination of the cybersecurity goals, claims and the introduction of the concept. After a correct understanding of the security goals, claims and concept are available, the specifications and requirements will be detailed in the product development phase. Next, the specifications are reviewed and then the product is integrated and verified, e.g., by the use of penetration testing tools, etc. Then the final validation report is completed. We address the concept phase according to ISO 21434 for the AEB system in what follows.

4.2 From Item Definition to Risk Determination

Item Definition The first step of the concept phase is the item definition which includes the boundary, functions of the item and its preliminary architecture. The item boundary for the AEB system is presented in Fig. 12. It includes the interfaces

Fig. 12 In-vehicle network architecture and item boundary for the AEB system

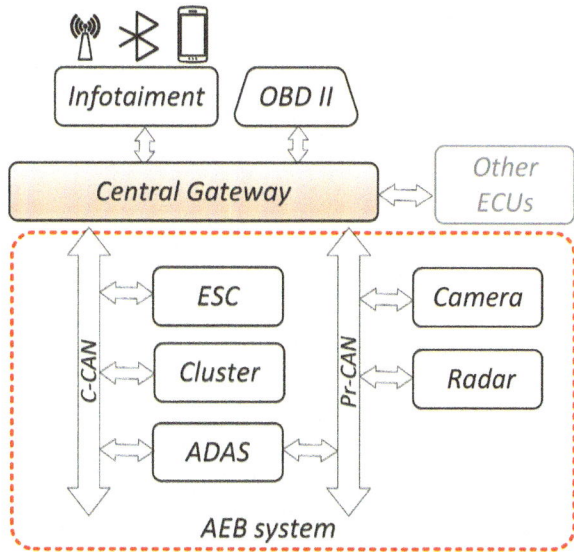

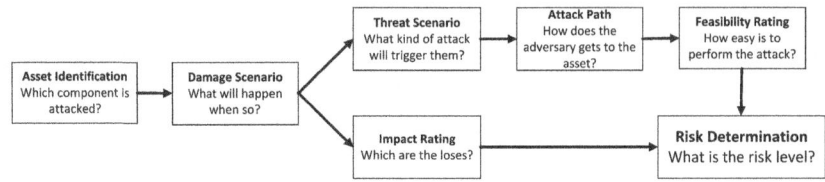

Fig. 13 Steps for risk determination according with to ISO 21434

with internal and external items, forming as such the environment in which the AEB system resides. The function of the AEB, i.e., assisting the driver in avoiding collision with front objects, has already been clarified. The preliminary architecture of the AEB is encircled in the middle and it includes two CAN buses: the Chassis CAN (C-CAN) which is used for data exchange between the ADAS ECU, the ESC ECU and the cluster and the Private CAN bus (Pr-CAN) which is used for data exchange between the ADAS ECU, Camera and Radar.

Once the item is defined, a series of activities follow starting with the identification of the asset which needs to be protected up to the determination of the risk level. In Fig. 13 we show these activities according to ISO 21434. For an easier understanding of these steps, we formulate one question for each step which summarizes its expected outcome. Next, we will address all the steps in this image for the AEB.

Asset Identification The first step consists in identifying the assets that we are going to protect. These assets include logical objects such as the firmware or the CAN frames. Due to obvious space constraints for the current work, we will focus on CAN frames alone and consider that the rest of the components, such as the firmware and the hardware components are secured, e.g., by digitally signed software updates and the appropriate tamper resistant hardware such as TPMs (Truster Platform Module).

According to ISO 21434, the CAN frames as an asset, have to respond to the three classical security objectives: Confidentiality, Integrity and Availability (CIA). Confidentiality may not be a necessary requirement since there is no need to hide the content of the frames from an adversary (there are no privacy concerns). But integrity and availability are critical. The former has to ensure that the content has not been modified or injected by an adversary, while the later must ensure that the corresponding frames are delivered on time by the ECU.

In Table 3 we detail the consequences when the two objectives, i.e., integrity (I) and availability (A), are not met for the frames which carry the signals from the AEB system. In the third column, the damage inflicted on the system is outlined. This includes the activation/deactivation of brakes, inability of the vehicle to maintain speed or estimate the time or distance toward the front object, etc. Generally, the attacks on integrity will mislead the car or the AEB system on the distance to the object or the time to collision, etc. The attacks on the availability of the signal will simply disable the corresponding functionality in the absence of required data.

Impact Rating Impact rating is divided on four distinct chapters: safety (S), financial (F), operational (O) and privacy (P). A few justification on how we select these values in Table 3 may be needed. First, our table contains only the value for the safety impact since the financial, operational and privacy impact is identical as we argue at the end of this paragraph. The unexpected deactivation of automatic braking, in case of signal integrity which complies to the adversary manipulations from the previous section, can have fatal consequences on the life of pedestrians, etc. For this reason, we consider attacks on integrity to have a severe impact due to the fatal injuries that may result from accidents. In case when the communication is lost, i.e., a DoS attack, the driver may still be warned by a visible or audible signal from the instrument cluster, showing that the functionality is not responding and as such he should increase his vigilance. A reason for which we consider that the loss of availability will have a more moderate safety impact. The same impact assessment holds for the rest of the parameters: throttle, relative distance, relative and longitudinal velocity. Not last, it is worth mentioning that based on our detailed analysis from Sects. 3.3 and 3.4, attacks on relative and longitudinal velocities will have more impact at higher speeds. The current version of ISO 21434, explicitly states that the financial impact refers to the costs of the road user. In most instances, car insurance companies cover these costs, although they may only cover the repair costs of the cars that are damaged by an inattentive driver and not the costs to repair their own car. We will consider that the financial costs should be moderate in general, although we cannot exclude that the financial impact may also run up to major, e.g., in case of impact at high velocities and the lack of the appropriate insurance. The operational impact is moderate since in case when the AEB functionality is lost, there is partial degradation of a vehicle function but the car is still fully controllable by the driver who is still able to brake. The privacy risks should be negligible.

Threat Identification This step has been refined by the specific attacks embedded in our adversary model. The threat to availability (A) is posed by the DoS attacks. When it comes to integrity (I) our model accounts for different kinds of manipulations, i.e., replay, fuzzing, surge, bias and geometric attacks. Separating these threats is relevant, because they can be addressed in different ways as we will discuss later when introducing the cybersecurity requirements, e.g., some of the attacks can be addressed by simple change detection mechanism while others require cryptographic authentication, etc.

Attack Path Analysis For the attack paths we considered the regulations concerning the approval of vehicles with regard to cybersecurity and cybersecurity management systems [1] proposed by the United Nations Economic Commission for Europe (UNECE). These regulations were recently investigated by the authors in [5]. Due to page limitations, we only use here the most significant attack surfaces, e.g., OBD (On-Board Diagnostics) connector, cellular interface, USB ports, etc.

Table 3 Asset identification and impact assignment

Asset	Obj	Damage scenario	Safety	Justification
Deceleration (CAN frame)	I	**D1.** Unexpected activation or deactivation of the brakes or jumping from one braking stage to another	severe	Can lead to accidents at high speed, see attack assessment from Sects. 3.3 and 3.4
	A	**D2.** Unable to activate AEB function	Moderate	The AEB can't activate the brakes and stop the car, driver may be warned of the lost functionality
Throttle (CAN frame)	I	**D3.** Unexpected self acceleration/ deceleration	severe	Unexpected self acceleration/deceleration may produce accidents at high speed
	A	**D4.** Car unable to self accelerate and maintain preset velocity	Moderate	Unexpected loss of throttle signal would slow down the car but the driver should eventually notice this and compensate for the correct speed, visible/audible signal may also warn the driver for loss function
Relative distance (CAN frame)	I	**D5.** AEB system is mislead on the correct distance to front obstacle, results in unexpected activation/deactivation of the AEB system	severe	Similar to the integrity (I) attack on throttle, may produce severe accidents
	A	**D6.** AEB system is unable to estimate the distance to the front object	Moderate	Similar to the availability (A) attack on throttle, may still be noticeable for the driver that can compensate
Relative velocity (CAN frame)	I	**D7.** AEB system is mislead on the correct time to collision, unexpected activation/deactivation of the AEB system	severe	Similar to the integrity (I) attack on throttle, may produce severe accidents
	A	**D8.** AEB system unable to estimate time to collision	Moderate	Similar to the availability (A) attack on throttle, may still be noticeable for the driver that can compensate
Ego velocity (CAN frame)	I	**D9.** AEB system is mislead on the correct acceleration request, results in self acceleration/deceleration	severe	Similar to the integrity (I) attack on throttle, may produce severe accidents
	A	**D10.** AEB system unable to compute the acceleration request, results in car unable to self accelerate and maintain the preset velocity	Moderate	Similar to the availability (A) attack on throttle, may still be noticeable for the driver that can compensate

Table 4 Risk determination for deceleration under the first two damage scenarios D1 and D2

Damage scenario	Threat scenario	Attack path	Feasibility rating	Risk value
D1. Unexpected activation or deactivation of the brakes or switching between braking stages	**T1**. Replay,	OBD II connector	High	5
	T2. Fuzzing,	Cellular interface	High	5
	T3. Stealthy attacks	Corrupted applications (3rd party)	Low	3
		USB port	Medium	4
		Malicious software (malware)	Low	3
		Software/hardware vulnerabilities from the development process	Low	3
		Unauthorized hardware added	Low	3
		Corrupted software update	Low	3
D2. Unable to activate AEB function	**T4**. DoS	OBD II connector	High	3
		Cellular interface	High	3
		Corrupted applications (3rd party)	Low	2
		USB port	Medium	2.5
		Malicious software (malware)	Low	2
		Software/hardware vulnerabilities from the development process	Low	2
		Unauthorized hardware added	Low	2
		Corrupted software update	Low	2

Attack Feasibility Rating According to the specifications in the Annex G of the standard [15], attack feasibility is a mixture between 5 components: required time, expertise, knowledge of the component, window of opportunity and equipment. The aggregate attack potential resulting from the scores of these 5 components ranges from very low to high. For simplicity we will not detail the score based on each of the previous 5 components, but provide some arguments for the aggregate rating that we present in Table 4. The OBD II connector has an attack path with a feasibility rating set to high since such an attack can be accomplished with low effort due to existing commercial 3-rd party OBD devices that are common. The feasibility rating in case of the cellular interface is also high, i.e., the attack path can be accomplished with low effort because the cellular interface are used for telematics and several attacks were already reported, e.g., [8, 21]. Hosted 3-rd party corrupted applications have a low feasibility rating because they require expertise and a corrupted provider while the automotive software market is well controlled.

The USB port has a medium feasibility rating since USB sticks commonly carry unwanted software. The introduction of malicious software (malware) has a low feasibility since it requires expertise from multiple experts. Also, we score a low feasibility rating for the software or hardware development which is again subject to a mature development process. Vulnerabilities may still be possible due to the high software complexity, possibly reaching 8 million lines of code for a single ECU, and numerous companies, e.g., 8–11, working on the software for a single ECU [19]. A low feasibility is associated to the addition of a new unauthorized ECU and for software updates, i.e., corrupted software stacks on the ECU which evades detection to cause the attack, since this requires multiple experts to design. Additionally, for damage scenarios D8 and D10, another attack path can be considered, i.e., the manipulation of information collected by the sensors from the environment which can be at least ranked as having a medium feasibility. There is an increasing number of works that show clever manipulations of environmental data such as traffic signs [24], traffic lights [35], road lanes [29] and distances toward objects [32, 36].

We note that the attack path can be subject to a more complex feasibility analysis as done by the authors in [27]. They consider that the feasibility of an attack path is the product of probabilities associated to each edge from the path. The ISO 21434 however does not quantify feasibility as a probability and it was the choice of the authors from [27] to associate a probability to each rating, e.g., when the risk is high $p \in [0.9, 1]$ and when the risk is medium $p \in [0.5, 0.9)$.

Risk determination. According to ISO 21434 [15], risk values are determined based on the impact rating and the attack feasibility using the following relation: $R = 1 + I \times F$, where R is the risk value, I is the impact rating and F is the feasibility rating. For the impact rating, in our calculation for the risk values we consider the maximum of the four types of impact (safety, financial, operational and privacy) which in this case is given by the safety component. This means that in case of the threats T1, T2, T3 which correspond to fuzzing, replays and stealthy attacks, the impact is major. While for a DoS the impact is only moderate. These values can be retrieved from Table 3. According to the standard, the four class impact, expressed as {*negligible*, *moderate*, *major*, *severe*}, is translated to numerical values as {0, 1, 1.5, 2}. The feasibility which is expressed as a four rank class {*very low*, *low*, *moderate*, *high*} is translated to numerical values as follows {0, 1, 1.5, 2}. Consequently, a moderate impact incurs a numerical cost $I = 1$ and the severe impact corresponds to $I = 2$. A high feasibility ranking corresponds to $F = 2$ and consequently the impact ranking is $R = 1 + 2 \times 2 = 5$. In Table 4 we depict the attack paths and the determined risk only for deceleration under the first two damage scenarios D1 and D2. For the rest of the signals and associated damages, risk determination should be done in a similar manner.

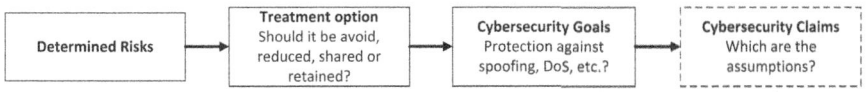

Fig. 14 Steps for product development according with to ISO 21434

4.3 From Determined Risks to Cybersecurity Goals and Concept

In Fig. 14 we illustrate the steps that follow from risk determination to integration and verification in case of cybersecurity attacks according to ISO 21434. Again, for an easier understanding of these steps, we formulate one question for each step which summarizes its expected outcome. According to ISO 21434 [15], the determined risks have to be treated in one of the following four ways: (a) avoided by not starting or continuing a specific activity, (b) reduced by using a proper security mechanism, (c) shared, for example with insurances or (d) retained. The cybersecurity goals are the result of the threat analysis and risk assessment (TARA) which we performed in the previous section. A cybersecurity goal, which results from the previous threat analysis, is a requirement to protect an assets against a threat according to ISO 21434 [15]. The cybersecurity claims must be formulated only in case of rationales for retaining or sharing the risks according to the same requirement. The claims can also include conditions for specific goals or functions for specific aspects, such as the use of a secure communication channel according to ISO 21434 [15].

Returning to the AEB system, in Table 5 we show the treatment option, cybersecurity goals and requirements for the threat scenarios on deceleration signal. For the rest of the assets (signals) in our analysis the details would be similar and we omit them for brevity.

Treatment The treatment is the same in case of all assets: to reduce the risk. It is not acceptable to share the risk with an insurance company since they may result in fatal accidents. Clearly, the risk cannot be retained either, nor avoided by cancelling the AEB functionality which would contradict the main purpose of the system. Thus, the only treatment is to reduce the security risk. For this reason, the cybersecurity claims are not needed in our table. Cybersecurity claims are needed only when the treatment option is to retain or share the risk.

Cybersecurity Goals and Requirements The decision to reduce the risk, moves us to the obvious goal to protect the signals, deceleration in particular, as outlined in Table 5, against spoofing, DoS, replay as well as against stealthy attacks.

For fuzzing on deceleration we proposed two requirements: (a) a change detection mechanism which needs to be implemented on the deceleration signal form the AEB ECU and (b) a verification procedure enforced by cryptographic security for the received data, which needs to be implemented on all ECUs in order

Table 5 Treatment option, cybersecurity goal and requirements for the threat scenarios on the deceleration signal

Threat scenario	Treatment option	Cybersecurity goal	Cybersecurity requirement	
			Description	Allocation
T1. Fuzzing on deceleration	Reducing the risk	Deceleration shall be protected against spoofing by authentication or change detection mechanism	Verify if received data comes from a valid entity	All ECUs from attack path
			Implement change detection mechanism	AEB ECU
T2. DoS on deceleration	Reducing the risk	Deceleration shall be protected against DoS attacks by detection and recover the signal	Measures to detect and recover from a denial of service attack shall be employed	AEB ECU and Instrument cluster
T3. Replay on deceleration	Reducing the risk	Deceleration shall be protected against replay attacks by authentication (including the appropriate freshness parameters, timestamp)	Authentication (include strong time parameters, timestamp)	AEB ECU
T4. Stealthy on deceleration	Reducing the risk	Deceleration shall be protected against stealthy attacks by authentication	Verify if received data comes from a valid entity	All ECUs from attack path

to check that CAN frames comes from a valid entity. The implementation of the second requirement should offer sufficient protection but it requires cryptographic capabilities that may be too expensive for some controllers and the use of a change detection mechanism may be cheaper and still provide some degree of protection. For DoS on deceleration, the cybersecurity requirement is to implement measures to detect and recover from a denial of service attack on the AEB ECU and instrument cluster. Authentication, implying the existence of strong time-variant parameters, i.e., timestamps, is needed against replay attacks. For the stealthy attacks, i.e, surge, bias and geometric, the requirement is to implement mechanisms to verify the source of the received data.

This description of the cybersecurity goals and requirements would be incomplete if we do not further give concrete suggestion on the exact mechanism that should be used for achieving these goals. Regarding intrusion detection, we have already pointed out a basic change detection mechanism. There are various other mechanisms that have been considered for intrusion detection in the literature including changes of specific parameters in heavy-duty J1939 vehicles [16], entropy analysis [20, 22] or Hamming distances possibly coupled with Bloom filters [10].

Other authors have proposed the use of precedence graphs [14], Markov models [23] or finite-state automatons [31].

Regarding authentication mechanisms that validate the source of the frames, we have to leverage the discussion toward the AUTOSAR security standard for on-board communication [3]. According to the AUTOSAR SecOC [3] standard, the communication between two ECUs needs to be secured by authentication. In order to achieve this, the messages from the sender ECU contains a Protocol Data Unit (PDU) which helds the data and the timestamp or the freshness value (CNT) which is computed internally by the sender ECU and increases in time. Based on the PDU, including the freshness value (CNT), a cipher-based message authentication code (CMAC) is computed and the sender ECU transmits the PDU, CNT and CMAC to the receiver. Subsequently, the receiver ECU checks the CNT and if the CNT is correct it is used for the CMAC verification.

The AUTOSAR SecOC [3] standard specifies three security profiles on pages 62–63 that have to use 32-bit truncated CMAC-AES for authentication. Assuming that the secret key is secure, this would lead to a probability of 2^{-32} for an adversary to inject a valid frame (this is equivalent to a false negative event). However, the situation is much worse if we consider replays, not last correlated with stealthy manipulations, since the authentication tag of replayed frames is computed with the correct key. The only way to circumvent these attacks is with the proper freshness parameters which according to AUTOSAR SecOC [3] have 8 bits in profile 1, 0 bits in profile 2 and 4 bits in profile 3. This means that there are 256, 0 and 16 possible values for the time-variant parameter which is slightly low (or non-existent). At best, assuming an 8 bit counter, the probability of an injection would be $2^{-8} = 0.3\%$. The 4-bit length for the freshness parameter is too low for serious security demands.

To illustrate the effectiveness of this layer of cryptographic protection, we chose the deceleration signal which had significant impact in case of fuzzing and stealthy attacks. We illustrate the behaviour in case of attacks with attack probability $p = 2^{-4}$ and $p = 2^{-8}$ that would result from using the corresponding freshness parameter on 4 or 8 bits. In Fig. 15 we illustrate the signals under: (a), (b) fuzzing attack with $p = 2^{-4}$, (c), (d) surge attack with $p = 2^{-4}$ and (e), (f) fuzzing attack with $p = 2^{-8}$. When the attack probability is reduced to $p = 2^{-8}$, the AEB system is not affected by the adversarial signal $\tilde{y}_{brake}(k)$. In Table 6 we show the collision velocity and distance to target in case of fuzzing, and stealthy attacks on deceleration with attack probability $p = 2^{-4}$ and $p = 2^{-8}$. In case of an attack probability of $p = 2^{-4}$, the fuzzing attack causes an impact at significant velocity, the surge and bias attack cause impact at low velocity, while in case of the geometric attack no collision takes place. In case of an attack probability of $p = 2^{-8}$ no collision takes place. Also, by comparing with the results from Sects. 3.3 and 3.4 the collision velocity is decreasing as the attack probability decreases, leading eventually to no impact when the attack probability is only $p = 2^{-8}$. The 8-bit freshness parameter is sufficient if we consider randomized injections with previously recorded frames, but it is too low for a more powerful adversary that records the order of the frames on the bus. For this reason, extending the 8-bit freshness parameter to a larger, 32 or 64-bit counter is needed, but this can only be achieved with the larger CAN-FD

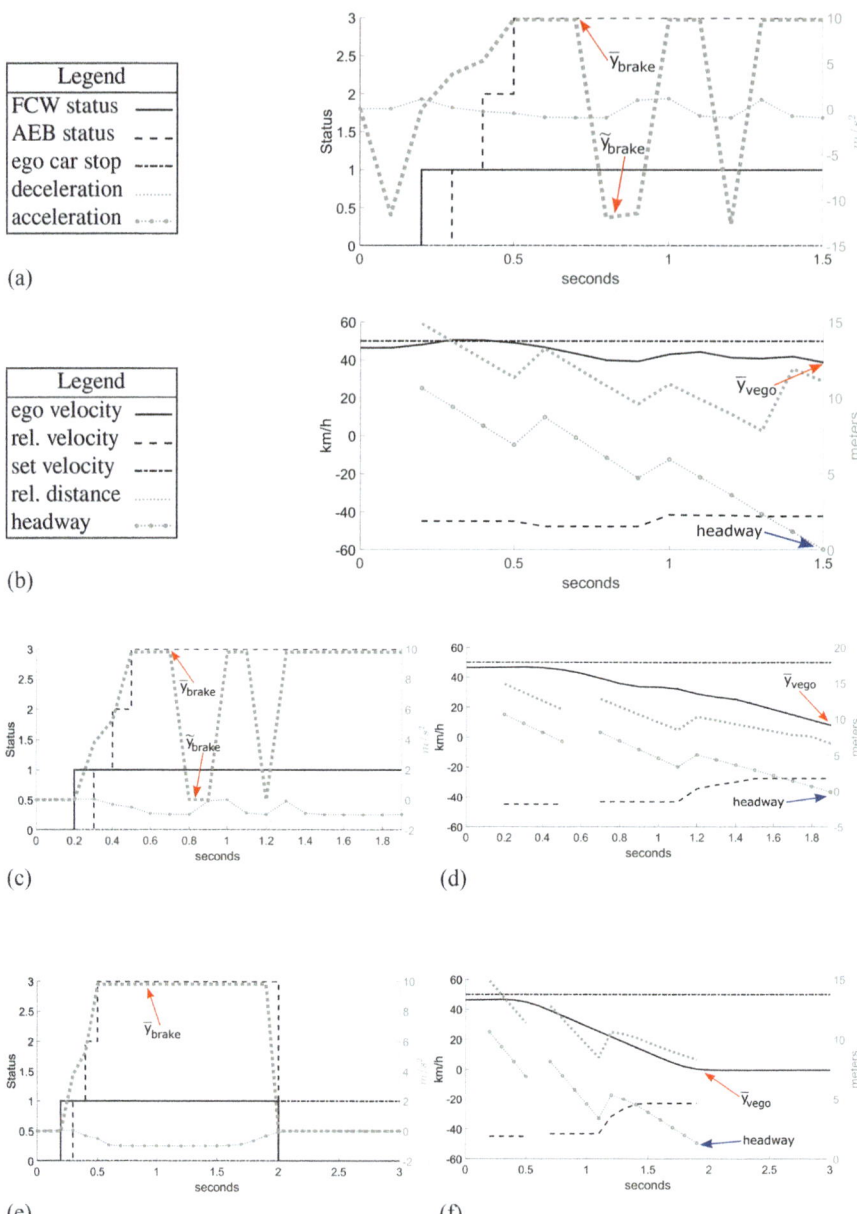

Fig. 15 Signals under fuzzing and surge attacks on deceleration y_{brake} with $p = 2^{-4}$ and $p = 2^{-8}$: (**a**) fuzzing attack with $p = 2^{-4}$—FCW/AEB status, (**b**) fuzzing attack with $p = 2^{-4}$—velocity/distance plot (**c**) surge attack with $p = 2^{-4}$—FCW/AEB status, (**d**) surge attack with $p = 2^{-4}$—velocity/distance plot, (**e**) fuzzing attack with $p = 2^{-8}$—FCW/AEB status and (**f**) fuzzing attack with $p = 2^{-8}$—velocity/distance plot

Table 6 AEB results: collision velocity and distance to target in case of fuzzing, and stealthy attacks on deceleration with attack probability $p = 2^{-4}$ and $p = 2^{-8}$

	$p = 2^{-4}$		$p = 2^{-8}$	
Attack	Collision velocity[km/h]	Distance to target[m]	Collision velocity[km/h]	Distance to to target[m]
Fuzzing	38.91	0	No coll.	1.32
Surge	7.99	0	no coll.	1.32
Bias	6.38	0	no coll.	1.32
Geometric	no coll.	0.92	no coll.	1.32

frames that have 512 bit datafields. We believe this is the only alternative for a high level of security.

A further step is the cybersecurity validation at the vehicle level which is followed by the production of the actual item or component, i.e., clauses 11 and 12 of ISO 21434 [15]. These details are out of scope for the current presentation which was focused on the cybersecure-aware design alone. Lastly, specific operation and maintenance activities, suggested in clause 13 of ISO 21434 [15], will also occur during the vehicle lifetime which may also lead to re-designs of the cybersecurity goals and claims.

5   Conclusion

Two lines of defence are advocated by our analysis. One of them is the inclusion of intrusion detection systems, such as the basic change detection outlined in our analysis. This line of defense requires a careful selection of specific parameters, i.e., thresholds and biases, which have to be the subject of careful engineering maturity testing and verification which are not fully possible in our work. The second line of defense is the adoption of cryptographic security that will ensure that each frame is authentic and, similarly important, fresh in order to remove the possibility of a replay attack. For this, using regular 64 bit CAN frames has its limit. More specifically, in accordance to AUTOSAR SecOC [3], at most 8 bits are used as freshness parameter which offers only a limited protection against replay attacks. For this reason, we believe that the adoption of CAN-FD which extends the datafield to 512 bits and allows a larger freshness parameter, such as a 64 bit timestamps as commonly available in network synchronization protocols, is the only way to ensure that freshness and thus complete source authentication is achieved.

References

1. Addendum 154 – UN regulation no. 155: uniform provisions concerning the approval of vehicles with regards to cyber security and cyber security management system (2021)
2. AUTOSAR: Specification of intrusion detection system protocol, r20–11 edition (2020)
3. AUTOSAR. Specification of secure onboard communication, r20–11 edition (2020). No. 654
4. Ben Othmane, L., Ranchal, R., Fernando, R., Bhargava, B., Bodden, E.: Incorporating attacker capabilities in risk estimation and mitigation. Comput. Secur. **51**, 41–61 (2015)
5. Brandt, T., Tamisier, T.: The future connected car–safely developed thanks to Unece Wp. 29? In: 21 Internationales Stuttgarter Symposium, pp. 461–473. Springer, Berlin (2021)
6. Brown, M.: Addressing the challenges of a sector in transformation and preparing to meet new cyber compliance requirements (ISO/SAE 21434). BSI Group (2022)
7. Cárdenas, A.A., Amin, S., Lin, Z.-S., Huang, Y.-L., Huang, C.-Y., Sastry, S.: Attacks against process control systems: risk assessment, detection, and response. In: Proceedings of the 6th ACM Symposium on Information, Computer and Communications Security, pp. 355–366 (2011)
8. Checkoway, S., McCoy, D., Kantor, B., Anderson, D., Shacham, H., Savage, S., Koscher, K., Czeskis, A., Roesner, F., Kohno, T.: Comprehensive experimental analyses of automotive attack surfaces. In: 20th USENIX Security Symposium (USENIX Security 11) (2011)
9. Gautham, S., Jayakumar, A.V., Elks, C.: Multilevel runtime security and safety monitoring for cyber physical systems using model-based engineering. In: International Conference on Computer Safety, Reliability, and Security, pp. 193–204. Springer, Berlin (2020)
10. Groza, B., Murvay, P.-S.: Efficient intrusion detection with bloom filtering in controller area networks. IEEE Trans. Inf. Forens. Secur. **14**(4), 1037–1051 (2018)
11. Groza, B., Gurban, H.-E., Murvay, P.-S.: Designing security for in-vehicle networks: a body control module (BCM) centered viewpoint. In: 2016 46th Annual IEEE/IFIP International Conference on Dependable Systems and Networks Workshop (DSN-W), pp. 176–183. IEEE, Piscataway (2016)
12. Gurban, E.H., Groza, B., Murvay, P.-S.: Risk assessment and security countermeasures for vehicular instrument clusters. In: 2018 48th Annual IEEE/IFIP International Conference on Dependable Systems and Networks Workshops (DSN-W), pp. 223–230. IEEE, Piscataway (2018)
13. Henniger, O., Apvrille, L., Fuchs, A., Roudier, Y., Ruddle, A., Weyl, B.: Security requirements for automotive on-board networks. In: 2009 9th International Conference on Intelligent Transport Systems Telecommunications,(ITST), pp. 641–646. IEEE, Piscataway (2009)
14. Islam, R., Refat, R.U.D., Yerram, S.M., Malik, H.: Graph-based intrusion detection system for controller area networks. IEEE Trans. Intell. Transp. Syst. **23**, 1727–1736 (2020)
15. ISO/IEC, ISO/SAE DIS 21434 - Road Vehicles - Cybersecurity Engineering, International Organization for Standardization, Geneva, Switzerland (2021)
16. Jichici, C., Groza, B., Ragobete, R., Murvay, P.-S., Andreica, T.: Effective intrusion detection and prevention for the commercial vehicle sae j1939 can bus. IEEE Trans. Intell. Transp. Syst. **23**, 17425–17439 (2022)
17. Kapoor, P., Vora, A., Kang, K.-D.: Detecting and mitigating spoofing attack against an automotive radar. In: 2018 IEEE 88th Vehicular Technology Conference (VTC-Fall), pp. 1–6. IEEE, Piscataway (2018)
18. Macher, G., Schmittner, C., Veledar, O., Brenner, E.: ISO/SAE DIS 21434 automotive cyber-security standard-in a nutshell. In: International Conference on Computer Safety, Reliability, and Security, pp. 123–135. Springer, Berlin (2020)
19. Mader, R., Winkler, G., Reindl, N.: Thomas amd Pandya. The car's electronic architecture in motion: the coming transformation. In: 42nd International Vienna Motor Symposium (2021)
20. Marchetti, M., Stabili, D., Guido, A., Colajanni, M.: Evaluation of anomaly detection for in-vehicle networks through information-theoretic algorithms. In: 2016 IEEE 2nd International Forum on Research and Technologies for Society and Industry Leveraging a Better Tomorrow (RTSI), pp. 1–6. IEEE, Piscataway (2016)

21. Miller, C., Valasek, C.: Remote exploitation of an unaltered passenger vehicle. Black Hat USA 2015(S 91) (2015)
22. Müter, M., Asaj, N.: Entropy-based anomaly detection for in-vehicle networks. In: 2011 IEEE Intelligent Vehicles Symposium (IV), pp. 1110–1115. IEEE, Piscataway (2011)
23. Narayanan, S.N., Mittal, S., Joshi, A.: Obd_securealert: an anomaly detection system for vehicles. In: 2016 IEEE International Conference on Smart Computing (SMARTCOMP), pp. 1–6. IEEE, Piscataway (2016)
24. Nassi, B., Mirsky, Y., Nassi, D., Ben-Netanel, R., Drokin, O., Elovici, Y.: Phantom of the ADAS: Securing Advanced Driver-Assistance Systems from Split-Second Phantom Attacks, pp. 293–308. Association for Computing Machinery, New York (2020)
25. Nie, S., Liu, L., Du, Y.: Free-fall: hacking tesla from wireless to can bus. Brief. Black Hat USA 25, 1–16 (2017)
26. Plappert, C., Zelle, D., Gadacz, H., Rieke, R., Scheuermann, D., Krauß, C.: Attack surface assessment for cybersecurity engineering in the automotive domain. In: 2021 29th Euromicro International Conference on Parallel, Distributed and Network-Based Processing (PDP), pp. 266–275. IEEE, Piscataway (2021)
27. Püllen, D., Liske, J., Katzenbeisser, S.: ISO/SAE 21434-based risk assessment of security incidents in automated road vehicles. In: International Conference on Computer Safety, Reliability, and Security, pp. 82–97. Springer, Berlin (2021)
28. Razikin, K., Soewito, B.: Cybersecurity decision support model to designing information technology security system based on risk analysis and cybersecurity framework. Egypt. Inf. J. 23, 383–404 (2022)
29. Sato, T., Shen, J., Wang, N., Jia, Y., Lin, X., Chen, Q.A.: Dirty road can attack: security of deep learning based automated lane centering under {Physical-World} attack. In: 30th USENIX Security Symposium (USENIX Security 21), pp. 3309–3326 (2021)
30. Schmittner, C., Schrammel, B., König, S.: Asset driven ISO/SAE 21434 compliant automotive cybersecurity analysis with threatget. In: European Conference on Software Process Improvement, pp. 548–563. Springer, Berlin (2021)
31. Studnia, I., Alata, E., Nicomette, V., Kaâniche, M., Laarouchi, Y.: A language-based intrusion detection approach for automotive embedded networks. Int. J. Embed. Syst. 10(1), 1–12 (2018)
32. Sun, J., Cao, Y., Chen, Q.A., Mao, Z.M.: Towards robust {LiDAR-based} perception in autonomous driving: general black-box adversarial sensor attack and countermeasures. In: 29th USENIX Security Symposium (USENIX Security 20), pp. 877–894 (2020)
33. Test protocol – AEB VRU systems, version 3.0.3. In: Vulnerable Road User (VRU) Protection. Euro NCAP (2020)
34. Wang, Y., Wang, Y., Qin, H., Ji, H., Zhang, Y., Wang, J.: A systematic risk assessment framework of automotive cybersecurity. Autom. Innov. 4(3), 253–261 (2021)
35. Yan, C., Xu, Z., Yin, Z., Ji, X., Xu, W.: Rolling colors: adversarial laser exploits against traffic light recognition. In: 31st USENIX Security Symposium (USENIX Security 22). USENIX Association, Boston (2022)
36. Zhou, C., Yan, Q., Shi, Y., Sun, L.: DoubleStar: long-range attack towards depth estimation based obstacle avoidance in autonomous systems. In: 31st USENIX Security Symposium (USENIX Security 22). USENIX Association, Boston (2022)

Resource Aware Synthesis of Automotive Security Primitives

Soumyajit Dey ⓘ, Ipsita Koley ⓘ, and Sunandan Adhikary ⓘ

1 Introduction

With the evolution of transportation systems, modern-day vehicles are no more mere mechanical systems. Contemporary automotive architectures are designed as a collection of cyber-physical control loops with an aim to provide energy-efficient performance, safety, comfort, and connected mobility features. These software-governed automotive controllers supervise a plethora of functionalities like engine control, power management, regenerative braking, lane-keeping, comfort features, etc. Examples from domains like safety would be features like Vehicle Stability Control (VSC), Anti-lock Braking System (ABS), Roll Stability Control (RSC), etc. Convenience features like Adaptive Cruise Control (ACC) are also ubiquitous in most vehicles nowadays.

Features are implemented in the form of control programs mapped to Electronic Control Units (ECUs) as real-time tasks. *Sensing* tasks process sensor measurements, *communication tasks* interface with communication hardware and send such measurements over the communication channels and *control* tasks compute the desired control input for respective actuators. An ECU may host multiple control tasks. Moreover, some control functionalities may require tasks spanning over multiple ECUs to attain some global control objective. For example, on identifying a life-threatening situation, the Central Locking System (CLS) that controls the power door locking mechanism works alongside the crash detection system to ensure occupants' safety.

With time, the number of ECUs in modern vehicles has been increasing with the quest for more features that make transportation safer and more convenient.

S. Dey (✉) · I. Koley · S. Adhikary
Department of Computer Science and Engineering, Indian Institute of Technology Kharagpur,
Kharagpur, West Bengal, India
e-mail: soumya@cse.iitkgp.ac.in; ipsitakoley@iitkgp.ac.in; mesunandan@kgpian.iitkgp.ac.in

© The Author(s), under exclusive license to Springer Nature Switzerland AG 2023   189
V. K. Kukkala, S. Pasricha (eds.), *Machine Learning and Optimization Techniques for Automotive Cyber-Physical Systems*, https://doi.org/10.1007/978-3-031-28016-0_6

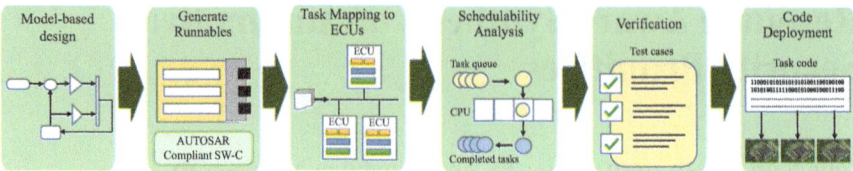

Fig. 1 Automotive software development flow

For example, BMW 7-series models have as many as 150 ECUs. Generalized bus-oriented architectures have come into the picture to realize the real-time collaboration of such a large number of ECUs. The electrical/electronic (E/E) [1] architecture of modern vehicles is divided into different functional domains (powertrain, body control, infotainment, etc). Based on the bit rate, fault-tolerance, and soft/hard real-time requirements of each domain, the intra-vehicular network [2] of a contemporary car comprises network elements with multiple lightweight protocols making it heterogeneous in nature. Example protocols include Controller Area Network (CAN) [3], FlexRay [4], Local Interconnect Network (LIN) [5], Media oriented systems transport (MOST) [6]. The E/E architecture of modern vehicles ensures the inter-operability of such heterogeneous network protocols.

The current design flow of automotive architectures (Fig. 1) in the industry is compartmentalized into model-based design, development, and standardized implementation on target platforms, each followed by testing and verification against certain specifications. The implementations of control software in ECUs are generalized by certain standard guidelines set by AUTOmotive Open System ARchitecture also known as AUTOSAR [7], which is a predominant entity built as a worldwide development partnership among automotive industries. Such standards encourage the model-based development of modularized and reusable control functionalities for automotive subsystems. This is followed by an AUTOSAR compliant conversion of these model-based designs of controllers into runnable programs. Then the control programs are implemented on the ECUs and invoked from the system level as control tasks based on the task allocations in ECUs.

Today, the design specification and implementation of automotive controllers are mostly carried out using Synchronous Reactive (SR) models such as those modeled in the Simulink and Stateflow tools [8]. TIER 1 suppliers organize the control functionalities as a hierarchy of subsystems and define them as a network of blocks in Simulink. Code implementation of each subsystem is generated as a set of functions. These sets of runnables are then standardized with AUTOSAR-guided specifications as Autosar Software Components (SWC) [8, 9]. The AUTOSAR model specifies the data, execution, and call dependencies for all the functions. All the SWCs from the TIER 1 suppliers are collected and connected to a system-level model by the OEMs and Carmakers. Using AUTOSAR tools, they map the runnables or the functions into tasks. Schedulability analysis is performed using platform-specific utilities (eg. symtavision for Infineon ECUs). Accordingly, tasks are allocated to the processors. Formal tests are conducted on the initial designs using some verification tool, like

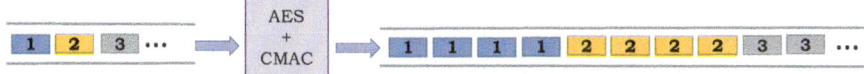

Fig. 2 Effect of CMAC/AES on CAN Traffic (numerals denote message id-s)

Simulink Test. Simulink Test facilitates test case and test suite definitions along with automation of test harness generation. Following the simulation tests, the C-code of the model is generated using Embedded Coder or Simulink Coder and implemented on the processor.

Security requirements have been an afterthought in the automotive software development flow discussed above. AUTOSAR mandates cryptographic mechanisms, like MAC (Message Authentication Code) to authenticate communications through the intra-vehicular network [10]. However, such cryptographic methods incur computation and communication overload. For example, securing an 8 byte CAN message using AES encryption and SHA-2 MAC algorithm will generate 6 CAN frames for a single CAN frame [11]. To deal with such bus load issues, AUTOSAR has suggested using truncated MAC [10]. The problem with MAC is that it can detect an external cyber attack, but fails to detect insider attacks and denial-of-service attacks like bus-off [12]. The embedded platforms where the automotive controllers are implemented are mostly of low computation power. On the other hand, cryptographic security algorithms, like MAC, incur significant computation and communication load. For example, on a 96 MHz ARM Cortex-M3-based Electronic Control Unit (ECU), some of the well-known control law computations take approximately $5 \mu s$ while a 128-bit MAC computation for a single message takes $100 \mu s$ [13]. On the other hand, if CMAC hash and AES-128 encryption algorithm are used to secure CAN frames, each CAN frame will be replaced by 4 CAN frames (Fig. 2) [11]. Imagine the load on CAN traffic if every CAN packet is secured this way. As most automotive CPSs are safety-critical with hard real-time deadlines, it naturally raises the question of how practical it is to implement cryptographic security algorithms on such embedded platforms. As an alternate solution, a number of researchers have proposed to use control-theoretic light-weight attack detectors[14] in place of periodic cryptographic security checks. These detectors are designed by exploiting the control theoretic properties of automotive CPSs [14, 15].

This naturally leads us to the problem of evolving automotive software design flows, which must consider the smooth integration of lightweight security primitives along with software controllers while maintaining verifiability, schedulability, and other platform constraints. The above also brings up the question of how such detection systems can be of practical use and in what way such existing approaches may be improved, more specifically in the automotive context. In this regard, we now discuss the major contributions to this chapter.

1. In control-theoretic light-weight attack detectors, the residue i.e., the difference between the actual sensor measurements and the estimated sensor measurements

is compared with a threshold. An alarm is raised when the residue surpasses the threshold value. The performance of the detector depends on the value of the threshold. If a lower valued threshold is selected, even noise can be considered an attack. This will lead to false alarms. On the other hand, a smartly crafted stealthy attack, like a zero-dynamic attack [15], can easily bypass a higher-valued threshold. So, the question is how to wisely compute the threshold value. In Sect. 3, we discuss some methodologies to synthesize such fixed threshold-based detectors.

2. Next, we consider a more informed attack scenario. An external attacker can snoop into intra-vehicular networks through OBD port and telematic units [16]. Widely used intra-vehicular network protocols like CAN transmits data in broadcast mode. Therefore, an attacker who has access to the intra-vehicular network can analyze transmitted data packets and design optimal attacks. We discuss in Sect. 4, how adaptive and intelligent threshold-based detectors can be designed to thwart such attack attempts.

3. While light-weight detection is an important task in the context of security-aware automotive CPS design, another important feature is what to do when an attack is detected. A number of researchers have proposed robust controller design methods to make the system robust against attacks. In this chapter, we discuss an alternative approach. In Sect. 5, we present how an intermittent MAC along with additional control logic can diminish the effect of the attack on the system.

4. Finally, in Sect. 6, we have presented how to realize some of these security-aware automotive CPS design methods in a Hardware-in-Loop (HIL) experimental setup.

2 Background and Related Work

2.1 System Model for Secure CPS

Similar to other model-based CPSs, the automotive software design life-cycle also conceptualizes modular subsystems for certain desired operating regions. For efficient and real-time control computation, a nonlinear plant $\dot{x}(t) = f(x)$ is usually linearized around such an operating point in the form of a linear time-invariant (LTI) system expressed as follows.

$$\dot{x}(t) = \Phi x(t) + \Gamma u(t) + w(t), \ y(t) = Cx(t) + v(t) \tag{1}$$

Here $x(t) \in \mathbb{R}^n$ is system state, $u(t)$ is output of the controller, $y(t) \in \mathbb{R}^m$ is system output under the influence of physical process noise $w(t) \in \mathbb{R}^n \sim \mathcal{N}(0, \Sigma_w)$ and measurement noise $v(t) \in \mathbb{R}^m \sim \mathcal{N}(0, \Sigma_v)$ at time t (w and v are independent Gaussian random variables with Σ_w and Σ_v as variance parameters). Also, Φ, Γ, and C are transition matrices, derived from the physical plant equations. Φ is known as the state transition matrix, Γ is known as the input-to-state transition matrix and C is the output matrix. The output of the plant is sensed and used for control input generation.

Depending on the plant-state characteristics and the available sampling periods in the ECU, the plant outputs are sampled periodically. For this, the control program uses the discretized versions of the above state-space equations, i.e.

$$x[k+1] = Ax[k] + Bu[k] + w[k], \ y[k+1] = Cx[k+1] + v[k] \tag{2}$$

Above equations express the k-th sampling iteration of the discretized system i.e., $t \in [hk, h(k+1)]$, where h is the chosen sampling interval. Therefore, the new transition matrices become $A = e^{\Phi h}$, $B = \int_{hk}^{h(k+1)} e^{\Gamma s} \Gamma ds$ [17]. At the controller side, this sensed plant output $y[k+1]$ is received once every sampling period. The controller needs to estimate the actual plant states using this output in order to calculate a suitable control input u to control the plant dynamics. To estimate the plant states from the sensed outputs, typically an *observer* is used.

$$\hat{x}[k+1] = A\hat{x}[k] + Bu[k] + Lr[k], \ r[k] = y[k] - C\hat{x}[k], \ u[k+1]$$
$$= -K\hat{x}[k+1] \tag{3}$$

As shown in Eq. 3 the estimated state at $(k+1)$-th iteration is denoted using $\hat{x}[k+1] \in \mathbb{R}^n$ and it is derived using a similar state-space equation like Eq. 2 along with a suitable correction $Lr[k]$ in order to track the actual state. The quantity $r[k] = y[k] - C\hat{x}[k]$ is known as system residue and it signifies the error between estimated and actual outputs. The observer gain L is designed in such a way that minimizes the residue [17]. The feedback control input $u[k+1]$ at the $(k+1)$-th sampling iteration is calculated based on the current estimated state $\hat{x}[k+1]$. We consider K as a pre-calculated optimal control gain. The control input thus calculated is then used to actuate the plant and stabilize it around the target operating point. As an example, consider Fig. 3 which demonstrates such closed-loop interaction between plant(s) and controller(s). We represent as a high-level view of a system under control where different subsystems with corresponding dynamics are modeled as plants (denoted as P_i for the i-th plant) and for each of them, suitable measurements are obtained using a set of sensors (denoted as S_i for the i-th plant). For each P_i, we have a corresponding control (denoted as c_i) and estimation task (denoted as e_i) implemented in the ECU. The communicated data and control outputs are suffixed with the plant names and tasks are suffixed with the plant indices for better understanding.

Like standard information processing systems, there are three fundamental security properties of any computer-controlled system and the information it deals with, i.e., *confidentiality, integrity and availability* (CIA). Now, there are several kinds of Man-in-the-Middle attacks that can observe and then utilize some system-specific knowledge to corrupt the communicated data to hamper its *integrity*. Insider attacks of this kind do not target *confidentiality* and can not be stopped by the

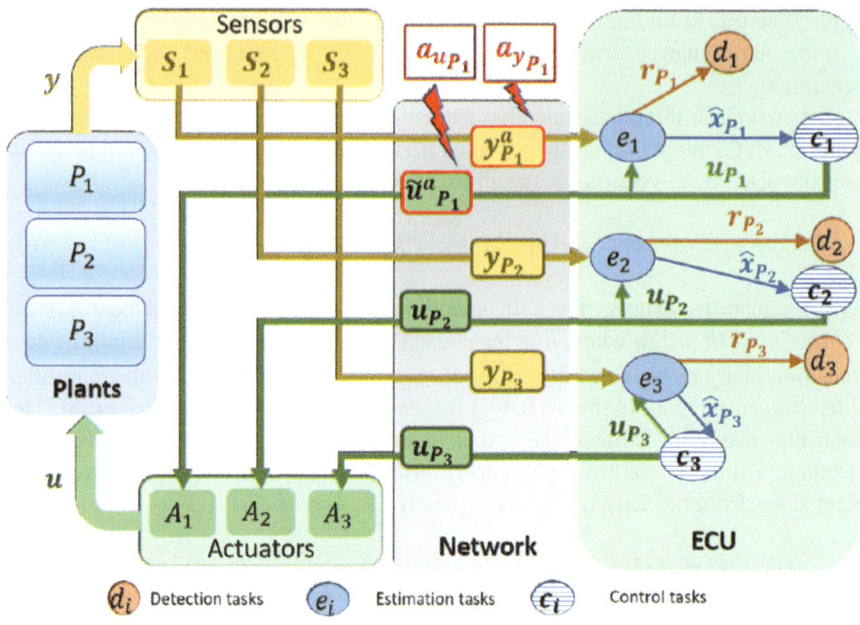

Fig. 3 Component-level overview of secure CPS under false data injection (FDI)

encryption policies mandated as per the AUTOSAR standards (truncated CMAC with 128 bit AES). Such attacks can be generalized as *False Data Injection Attacks* which corrupt the actual sensor and actuator data with a certain amount of false data. Our discussion in this chapter will primarily focus on the effects and counter-measures of False Data Injection (FDI)-type Man-in-the-middle attacks on control loops. In Fig. 3, such attacks on measurement and actuation are denoted by the variables $a_{y_{P_i}}$ and $a_{u_{P_i}}$ for a plant P_i. We assume that the ECU also runs a detection task d_i for every i-th loop in the system. We start with an example of simple threshold-based detection. Threshold-based detection tasks (Fig. 3) are designed to monitor the transmitted sensor data and flag an attack or anomaly in system output whenever the residue (or some derived statistics from it) surpasses some pre-fixed constant detector threshold Th i.e.,

$$\|r[k]\|_p > Th, \ (\|r\|_p = (\sum \|r\|^p)^{1/p}) \tag{4}$$

where p is the chosen norm. A suitable threshold value on the residue statistics can be chosen to constrain the estimation error. For a control loop, the state progression under FDI attacks can be expressed as follows.

$$x^a[k+1] = Ax^a[k] + B\tilde{u}^a[k] + w[k], \tag{5}$$

$$y^a[k+1] = Cx^a[k+1] + v[k] + a_y[k], \tag{6}$$

$$\hat{x}^a[k+1] = A\hat{x}^a[k] + Bu^a[k] + Lr^a[k], \ r^a[k] = y^a[k] - C\hat{x}^a[k] \qquad (7)$$

$$u^a[k+1] = -K\hat{x}^a[k+1], \ \tilde{u}^a[k+1] = -K\hat{x}^a[k+1] + a_u[k] \qquad (8)$$

Here $x^a[k]$, $\tilde{u}^a[k]$, $y^a[k]$ and $\hat{x}^a[k]$ are the attacked variants of the state, control input, output and estimated state vectors (from Eqs. 2 and 3) respectively at k-th iteration. By *attacked*, we mean to say under the influence of additive false data $a_u[k]$ injected on actuation, and $a_y[k]$ injected on sensor data at k-th sampling iteration. Also, $r^a[k]$ denotes the system residue under attack scenario (from Eq. 3) at k-th sampling instance. Note the difference between $\tilde{u}^a[k]$ and $u^a[k]$. The first one is the control input at k-th iteration under the influence of an additive FDI attack on actuation $(a_u[k])$ and the second one is not affected by the actuation attack but is calculated using the estimated states derived from the FDI affected $(a_y[k])$ sensor readings y^a. The estimated states are calculated in the controller side itself (i.e., not transmitted via the network under attack or not actuated via the actuator under attack). Therefore, unlike the actual plant state calculation, (where $\tilde{u}^a[k]$ is used to calculate $x^a[k+1]$), $u^a[k]$ is used for the calculation of estimated state $\hat{x}^a[k+1]$. In Fig. 3, we demonstrate such a data falsification attack on an automotive communication network. The attack vector at k-th sampling iteration is symbolically represented as $\mathcal{A}[k]^T = [a_u[k]^T, a_y[k]^T]^T$. If the attacker continues the false data injection for l sampling iterations, then the l length attack vector is expressed as follows.

$$\mathcal{A}_l = [\mathcal{A}[1] \cdots \mathcal{A}[l]] = \begin{bmatrix} a_u[1] \cdots a_u[l] \\ a_y[1] \cdots a_y[l] \end{bmatrix}$$

Since automotive control loops are highly *safety-critical*, an intelligent attacker can design the FDIs while utilizing system model knowledge with an aim to make the system states *unsafe*. To achieve this, the attacker has to compromise the sensors/actuators or the intra-vehicular communication networks (e.g. the CAN bus). In this process, the attacker might get detected as the residue-based detection tasks are always running in the ECU looking for anomalies where the residue-statistic changes undesirably or beyond a certain threshold. Therefore, the attacker also needs to design the false data in a way such that it can maintain its *stealth* while making the system eventually unsafe [15]. Considering n as the dimension of the system and $X_S \subset \mathbb{R}^n$ being the safe region of system states, the following is the criteria that an N-length *stealthy and successful FDI* attack vector $\mathcal{A}_N$ has to satisfy.

$$\|r^a[k]\|_p \leq Th \ \forall k \in [1, N] \textbf{ and } x^a[N+1] \notin X_S \qquad (9)$$

Figure 4 demonstrates such a successful attack injected into the sensor and actuation data of an Automatic Cruise Control (ACC) system. The states of the system are deviation (D) from the reference trajectory and the velocity (V) of the vehicle. The velocity (V) is considered as system output and is controlled using acceleration

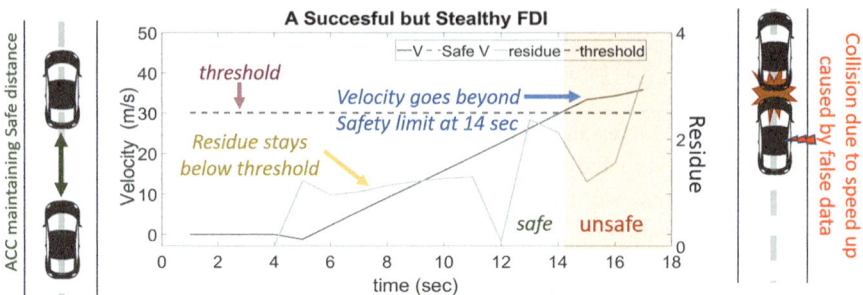

Fig. 4 A successful yet stealthy FDI

control input. The safety boundary for V is considered to be 30 m/s to maintain a safe distance from the preceding vehicle. Following are the states and transition matrices of the ACC plant sampled at every $T_s = 0.1$ s, $A = \begin{bmatrix} 1 & 0.1 \\ 0 & 1 \end{bmatrix}$, $B = \begin{bmatrix} 0.005 \\ 0.1 \end{bmatrix}$, $C = \begin{bmatrix} 0 \\ 1 \end{bmatrix}$, $D = [0]$, $x^a = \hat{x}^a = \begin{bmatrix} D \\ V \end{bmatrix}$. The controller and observer gains used to achieve the closed-loop functionalities are respectively $K = [0.9171 \ 1.6356]$ and $L = [0.8327 \ 2.5029]^T$. The system is equipped with a residue-based detector with $threshold = 2.5$. The stealthy attacker can inject falsified velocity and acceleration data communicated between the plant and controller. Figure 4 is a plot of velocity and residue under a 13 length false data injection attack vector launched on this Automatic Cruise Control system (ACC). We can see that the false data is successfully pushing the velocity of the follower vehicle beyond the safety limit, i.e., 30 m/s at 14 sec time instant but the residue still remains below $threshold$. Hence, this is a successful 13-length attack vector as per Eq. 9 as it successfully makes the system unsafe before the alarm is raised by the detector. In this chapter, we consider this generalised set of false data injection type attack, and discuss a security-aware design that can protect the system against them.

2.2 Automotive Software Tools and Standards

Automotive controllers are developed in a modularized fashion with different initial models representing different subsystems aimed to handle specific control operations. The high-level modeling is typically done using a formalism that supports hybrid specification of continuous dynamics and discrete switching logic together. In this development process, Model-Based Design (MBD) methodology is used in early stages by the TIER-I suppliers. MBD tools enable design, testing and verification to be performed in a single design platform. Stateflow/Simulink is widely used by control system designers for this purpose. Controller specifications are defined as networks of Simulink components or state-flow models that are developed and validated separately as part of a hierarchical system model [18]. Along with simulation-based testing in the system state-space, the design verifiers associated with these platforms verify the high-level design as a hybrid automaton using formal engines. This enables an integrated verification and correction of the developed control system model in the development stage itself. The code generation for the target platform is another feature integrated into these model-based design tools. The Simulink Coder is one such popularly used tool that modularizes the functionalities of each subsystem and provides the binaries for integration to a system-level model. An AUTOSAR compliant conversion of these model-based designs of controllers into runnable programs is done thereafter. These modular tasks are then cluster-wise mapped to the ECUs so that they can be invoked from the system level as control tasks based on the task allocations following the AUTOSAR standards [8, 9]. Thereafter, a thorough schedulability analysis of the collection of tasks in a given ECU core is done in order to validate this. Given the safety-critical and real-time nature of tasks, such automotive performance analysis tools need to be correct with high confidence (even if conservative). Tools like Symtavision SymTA suite (for Infineon ECUs), Inchron chronSUITE are popularly used for this purpose. These tools help in calculating the end-to-end response time for a given task mapping and verifying simulated system response given certain safety, performance criteria, and resource budget. Standard protocols like CAN, Flexray, etc., are also supported in order to analyze the communication busloads and optimize them. After rounds of tests and required design updates, the code for the final design is generated for the target ECUs. After analysing and verifying the generated code using integrated code verifiers (e.g., Polyspace [19]), the binaries are implemented in the ECU following the mapping strategy. Modern automotive ECUs follow a layered software architecture with the AUTOSAR runtime environment (RTE) interfacing with the AUTOSAR software components (SWCs). A service layer follows this application layer that interacts with ECU and Microcontroller abstraction layer (MCAL), which is equipped with complex and low-level device drivers. This facilitates multiple control features to be executed as real-time tasks while sharing the same physical platform.

The crypto stack of AUTOSAR provides an interface for Message Authentication Codes (MACs), Secure Hash Algorithms (SHA), and key-based authentication methods. Crypto service manager (CSM) [20] is the service layer module that interacts with the crypto interface (CryIf) in the ECU abstraction layer and enables communication with the cryptographic software or hardware via the crypto driver module in MCAL. Data packets are transmitted as Protocol Data Units (PDUs) and unpacked into Service Data Units (SDUs) at the receiver's end following the protocol control information (PCI). Standard AUTOSAR guidelines for Secure Onboard communication mandate the use of 128-bit AES with Cipher-based MACs while transmitting PDUs through the communication buses. This prevents unauthorized tampering of data communication but it does not ensure protection against false-data injection type insider attacks. To thwart one such powerful attack i.e., Record and Replay Attack, freshness value (FV) is introduced along with the MACs. But the use of these cryptographic authentications increases the processing and communication overheads. This is why the AUTOSAR Secure onboard communication (SecOC) directive suggests the use of truncated MACs. This might result in a lower security level, but the use of a 128-bit key size with more than 64-bit MAC is considered to provide significant security against unauthorized intrusions. The security profile for CAN communication suggests the use of 28 most significant bits from MAC (calculated using 128-bit AES with CMAC) and 4 least significant bits from the freshness value.

2.3 Related Studies

There exists a significant amount of work that had shown how an adversary can gain access to the intra-vehicular network physically or remotely [16, 21–24]. Once the access to the intra-vehicular network is gained, any ECU with safety-critical tasks can be compromised and the attack will pose like an *insider attack* effort. Since CAN is a protocol using which most of the safety-critical control messages are broadcasted, it is an ideal attack surface for an FDI attacker. The authors in [12] exploit the in-built error-handling protocol of CAN to send a victim ECU to bus-off mode using a compromised ECU. Authors in [25] take this attack strategy further by extending the bus-off period. They choose an optimal victim message ID, observe when the ECU recovers from bus-off, and re-transmit that ID to target the preceding error transmission frames, thus pushing the ECU back to bus-off. Now that the victim ECU is compromised repeatedly, the attacker can inject fabricated data packets in the CAN bus in the disguise of this victim ECU for a long enough period to make a control loop unsafe. A *denial-of-service* type attack is demonstrated in [21]. Authors show how individual brakes of a real car can be locked and communication with the engine control module, body control module can be disabled by injecting random data packets into the CAN bus. A *false data injection* attack can be inflicted in this way by crafting data packets with false speed information and injecting them into the CAN bus. A *replay attack* methodology is discussed in [26] on the keyless entry system of a vehicle. There are various other automotive attacks in the literature [27, 28]. Such intrusions can cause serious damage to the system but are hard to catch.

To combat such attacks, the integration of cryptographic schemes is proposed by researchers in the automotive domain. The use of Cipher-based Message Authentication Codes (CMAC) based on symmetric key ciphers like AES was chosen as part of Secure Hardware Extension (SHE) for automotives [29]. Since these are computationally simpler than the asymmetric approaches, they are ideal for real-time use with less computational power. But sharing of secret keys among all participating ECUs makes the intra-vehicular network prone to insider attacks. Keys being pre-programmed into the ECUs have been exploited in [21]. To prevent this, the use of Cyclic Redundancy Codes (CRC) along with CMAC suggested in [30] ensures the integrity of intra-vehicular communication.

Control-theoretic monitoring systems are also proposed to deal with power-hungry cryptographic algorithms [13]. These mechanisms offer basic safety checks on a CPS while it operates. There are statistical change detection methods like χ^2-test, Cumulative Sum (CUSUM) [31, 32], that are implemented to detect whether the system output or the system states are anomalous. The residue of the system is monitored for this purpose. If the residue statistic goes beyond a certain pre-calculated threshold, the system is found to be anomalous. Though such lightweight control-theoretic security primitives can limit the attacks, they can also be fooled [14, 15].

Since the standalone use of cryptographic algorithms to secure a CPS is not resource-friendly and control-theoretic anomaly/attack detection units are not sufficient for security either, combining both is usually suggested and is a good choice to build a resource-aware *Intrusion Detection System* (IDS) for CPSs. Authors in [33] proposed such an IDS for securing plant controller communication with reduced resources by sporadically using the cryptographic schemes with attack-resilient control-theoretic detection tasks running in the background. Such intermittent activation of cryptographic schemes is made further resource-aware by utilizing the weakly-hard design constraints of a CPS in [34]. They also explore formal methodologies to ensure that resource awareness would not compromise the safety and security of the CPS.

Another approach is to design the detection task adaptively enough to detect attacks based on the current state of the systems. Authors of [35, 36] have proposed such anomaly detectors that vary their detection thresholds. The work in [35] proposes two greedy algorithms based on formal methods to generate a set of monotonically decreasing thresholds in off-line mode. On the other hand, the authors of [36] formulate an attacker-defender game to solve the adaptive threshold selection problem. In [37], the authors show that the using windowed residue statistic with an optimally chosen threshold, one can have a better idea about the history of the states which can be useful in terms of better attack detection. The work in [38] takes this statistical analysis further toward guided learning of attacked state detection using reinforcement learning (RL) and model knowledge.

In the context of attack mitigation, [39] presented a secure state estimation problem which is further leveraged to compute attack-mitigating robust control inputs using RL. A recent work [40] presented an online attack recovery method by estimating the current system state from the latest trusted data using the checkpoint method from [41] followed by which they synthesized recovery control inputs using a linear program (LP) and formal methods. In [42], trusted hardware components are used as a high-assurance unit to increase the security of the system. As the decider unit, they proposed a side-channel analysis-based intrusion detection system. In the case of connected and automated vehicles, as discussed in [43], on detection of an attack, the system is switched to adaptive cruise control from cooperative adaptive cruise control.

There are several works that address the overall security-aware co-design perspective for automotives. Authors in [44] propose a Lightweight Authentication for Secure Automotive Networks (LASAN), which suggests optimization of the cryptographic protocols with asymmetric key encryption based on available power, compute, and communication resources. The work in [45] suggests cross-layer co-design of security framework, keeping the performance in mind along with a schedulable solution. This design space exploration solution was applied to an automotive case study to achieve a refined co-design.

3 Lightweight Attack Detection

As we discuss the security-aware design of automotive CPSs, we must keep in mind that the design should ensure the real-time requirements of the safety-critical systems. Hundreds of ECUs collaboratively work together to attain global objectives. Additional communication load due to security primitives must not hamper the real-time aspects of automotive networks. Being light-weight and handling real-time communication has been the primary motivation for designing intra-vehicular network protocols like CAN, FlexRay, etc. Most safety-critical CPSs are connected via CAN but CAN does not have any authentication scheme. It does contain a cyclic redundancy check (CRC) field, however, it can be broken via simple reverse engineering [46]. To ensure utmost security, securing every data packet using some cryptographic method seems the most promising strategy. To date, the traditional cryptographic techniques (for example, message authentication code also known as MAC along with some encryption techniques like RSA, AES, etc.) are known to provide the best security against false data injection (FDI) attacks. But, they incur computational and communication overheads which may lead some of the safety-critical tasks to miss their deadlines (refer to Fig. 2 and corresponding discussion in Sect. 1).

An alternative solution that has been widely suggested by a number of researchers in the literature to deal with the above limitations in the context of security-aware automotive CPSs is to use residue-based attack detectors [14, 15, 32]. A residue is computed as the difference between actual and estimated sensor measurements $r^a[k] = y^a[k] - C\hat{x}^a[k]$ (see Eq. 7). As explained in Sect. 2.1 either some norm of the residue or some statistical derivation of the residue [37] is compared with a threshold value Th. The detector's efficiency depends highly on the value of the threshold. The following two measures are used to quantify the detector's performance. The first one is *true positive rate* (TPR) i.e., the probability at which the detector raises an alarm when an FDI attack is taking place. The second one is *false alarm rate* (FAR) i.e., the probability at which the detector raises an alarm when no attack is taking place. An efficient detector will have higher TPR and lower FAR.

Let us consider an example of a zero-dynamic attack demonstrated on a trajectory tracking control (TTC) system (see Fig. 5). The states of TTC are deviation (D) from the reference trajectory and the velocity (V) of the vehicle, $x^a = [D\ V]^T$. The system matrices are $A = \begin{bmatrix} 1 & 0.1 \\ 0 & 1 \end{bmatrix}$, $B = [0.005\ 0.1]^T$. The system is equipped

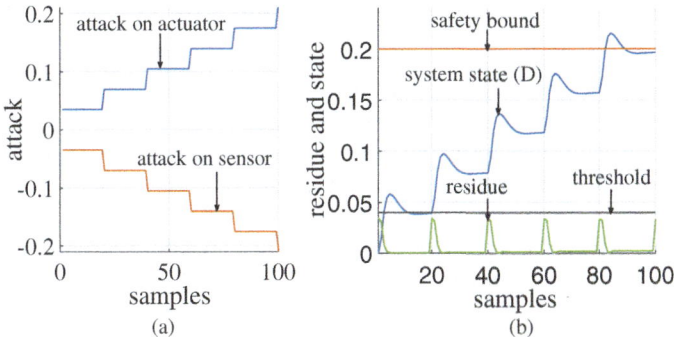

Fig. 5 Zero dynamic attack: (**a**) Attack vectors (**b**) Effect of attack on system and detector

with only distance sensor i.e. $C = [1\ 0]$. Here, the control input is the acceleration of the vehicle. The controller and observer gains of this TTC system are $K = [16.03\ 5.66]$ and $L = [1.87\ 9.65]^T$ respectively. The attacker can modify both distance measurement and acceleration data. A residue-based attack detector is also in place with a constant threshold $Th = 0.4$. The safety limit for state D has been set as 0.2 unit. Consider an FDI attacker crafted stair-case-like attack vector as given in Fig. 5a. Here, by *attack vector* we mean a sequence of false data to be injected to sensor data or actuator signal as mentioned earlier in Sect. 2.1. The intensity of the attack values is constant for a certain number of consecutive samples and then it is increased. While the actuation attacks $a_u[k]$ are positive, attacks on sensor measurements $a_y[k]$ are negative. This is because the attack on the control signal accelerates the changes in plant states and drives the system towards an unsafe region. On the other hand, the attack on sensor measurement hides the reflection of the system's drastic change in the measurements (see Eq. 9). Thus, the detector task that can only see the measurements, not the actual system states, is hoodwinked into thinking that the system is operating as desired. This smartly crafted attack can successfully make the system unsafe as can be seen in Fig. 5b. We are saying the attack is stealthy because following the successful attack criteria mentioned in Eq. 9, the residue remains below the threshold value all the time. It will never trigger an alarm to notify that an attack has taken place. This reduces the detector's TPR. One can reduce the threshold further to improve the detection rate. But, in that case, the detector will consider even small process and/or measurement noises as attacks. This increases FAR and thus reduces the detector's efficacy. Therefore, it is necessary to determine an optimal threshold for which the detector's performance is enhanced. We now discuss 2 state-of-the-art approaches that can be found in the literature for determining the optimal threshold to improve detectability as well as to reduce false alarms.

3.1   Optimal Static Threshold-Based Detector

In [37, 47], the authors considered a *stateful* [32] detection system and provided a theoretical base on how to correlate the characteristics of the detector with system dynamics. A *stateful* detector in the context of residue-based light-weight detection mechanism means the decision of the attack detector does not rely only on the current measurement; rather, a number of past measurements are also considered. Examples of such detectors include CUmulative SUm (CUSUM), windowed χ^2-test statistics-based detectors, etc. Let us summarize the idea by considering a CUSUM detector. Consider we have m sensors to measure that plant state i.e., $y[k]$, $y^a[k] \in \mathbb{R}^m$. Therefore, the residue $r[k]$, $r^a[k] \in \mathbb{R}^m$. When no attack is taking place, the mean and covariance of residue are respectively $E[r_k] = 0$ and $E[r_k r_k^T] = \Sigma$. Using the subscript i, we denote the i-th sensor as $y_i[k]$ where $i \in \{1, 2, \ldots, m\}$. Consequently, we have $r_i[k] \sim \mathcal{N}(0, \sigma_i^2)$. Here, σ_i is the i-th diagonal entry of the covariance matrix Σ. The condition for detecting a false data injection attack using CUSUM detector [48] is:

$$S_i[k] = max(0, S_i[k-1] + |r_i^a[k]| - b_i) \quad \text{if } S_i[k-1] \leq Th_i \tag{10}$$

$$= 0 \qquad\qquad\qquad\qquad\qquad \text{if } S_i[k-1] > Th_i \tag{11}$$

The test sequence S_i is initialized with 0 for all $i \in \{1, 2, \ldots, m\}$. Th_i and b_i are the threshold and bias selected for the i-th sensor. So, basically, CUSUM detector checks whether a certain sensor is under attack. When the cumulative sum sequence S_i exceeds Th_i, an alarm is triggered to raise an attack situation.

The efficacy of this detector depends on the bias b_i and the threshold Th_i. Since $|r_i^a[k]|$ is non-negative, if a sufficiently large value is not selected for b_i, the test sequence S_i may grow unboundedly. This inherent unboundedness of CUSUM may lead to false alarms. Because, due to some measurement and process noise, the value of $|r_i^a[k]|$ can be greater than 0 even if an attack is not taking place. Therefore, first, the value of b_i must be selected wisely relative to the characteristics of the residue r_i. Following this, a suitable Th_i needs to be computed to achieve a desired FAR.

The authors of [37] established a lower bound on b_i as $b_i > \bar{b} = \sigma_i \sqrt{2/\pi}$ in Theorem 1 in [37]. Once, b_i is determined, the value of the threshold Th_i has to be computed such that the false alarm rate never crosses a desired value. To do so, we define *run length* κ_i of CUSUM (Eq. 11) as the number of iterations needed to reach $S_i[k] > Th_i$ i.e.

$$\kappa_i = inf\{k \geq 1 : S_i[k] > Th_i\} \tag{12}$$

The average run length (ARL_i) is the expected value κ_i which is related to FAR as $ARL = 1/FAR$. Considering the desired FAR as FAR^*, we need to find out Th_i such that $ARL_i = 1/FAR^*$ provided $b_i > \bar{b}_i$. Authors of [37] presented a Markov chain approach for approximating ARL_i to determine the pair $< b_i, Th_i >$ such that Eq. 12 holds.

3.2   *Variable Threshold-Based Attack Detector*

A different line of approach was presented in [35] to synthesize the threshold for residue-based detectors to enhance TPR and reduce FAR. As a potential example of a targeted performance degrading attack, they consider the situation when the reference point of a controller changes due to the occurrence of some event. For example, if the driver rotates the steering wheel, the yaw rate of the vehicle needs to be changed to maintain the lateral dynamics of the vehicle. For such kinds of systems, an attacker can obstruct the vehicle from reaching the proximity of the new reference by injecting even smaller faults at the later stage of the system dynamics (when nearing the reference). From the perspective of designing a security-aware system, this brings in an interesting trade-off. Assume we want to design a static threshold-based detector where a constant threshold will be used throughout. We look into two cases. First, a lower-valued threshold is determined considering the required false data to be injected at the later phase of settling time. In this case, any process or measurement noise induced by the environmental disturbances in the system will be considered an attack. This will lead to false alarms. Second, a higher-valued threshold is selected considering the required attack amount at the earlier phase of settling time. This will help an attacker easily bypass the detector. The attacker can inject a sequence of small false data to make the system unsafe (as demonstrated in Fig. 5). Such scenarios have motivated the authors of [35] to design a variable threshold-based detector that may ensure reduced FAR while identifying even small attack efforts that may lead to potential performance degradation.

As a motivating example, we again consider the same example trajectory tracking control (TTC) system. We can see in Fig. 6a, that due to the process noise $w[k]$ and measurement noise $v[k]$ (Eq. 2), the violation in system's desired performance is negligible. This is due to the intrinsic robustness of the controller. On the other hand, we can see the system gradually becomes unstable when the system is under the influence of a smart attacker (Fig. 6a). Consider *three* such possible residue based detectors: with the smaller threshold th, the bigger threshold Th and the variable threshold curve v_{th} in Fig. 6b. The detector considers even the harmless noise as an attack when th is used, while the actual attacker could bypass the detector when

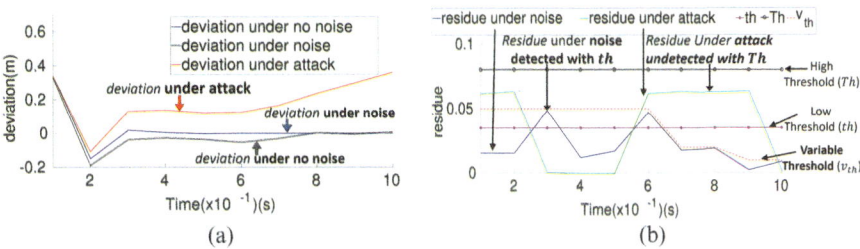

Fig. 6 Noise and attack simulation on trajectory tracking system. (**a**) Effect of noise and attack. (**b**) Static vs dynamic threshold

Th is used. However, using the variable threshold curve v_{th} (dotted black line in Fig. 6b), the attack does not remain stealthy while harmless noise is allowed to pass, reducing the false alarm rate. With this motivation, the authors of [35] presented two greedy approaches for synthesizing variable thresholds by leveraging formal methods like Satisfiability Modulo Theory (SMT) [49]. The following paragraph contains detailed explanations of those methods.

Given the closed-loop system dynamics in Eq. 2, and the reference point x_{des}, the target performance criteria *pfc* is to reach some n-dimensional closed ball (polytope) $\mathcal{B}_\epsilon(x_{des})$ with radius $\epsilon > 0$ around x_{des} (i.e., the closed region $\{x \in \mathcal{R}^n \mid \| x - x_{des} \| \leq \epsilon\}$) within a finite number of iterations, starting from an initial state $x[0] \in \mathcal{I} \subset \mathbb{R}^n$. Hence,

$$pfc : x[l] \in \mathcal{B}_\epsilon(x_{des}), \text{ where } l > 0 \text{ is the finite number of iterations.}$$

The attacker's objective would be $x[l] \notin \mathcal{B}_\epsilon(x_{des})$ after l closed-loop iterations (Eqs. 5–8). The property *pfc* captures both control performance and stability criteria. Assume the system already has some rudimentary monitoring scheme, like a range monitor for the sensor measurements, in place. Let us denote such monitoring rules as *mdc*. Authors of [35] present two counter-example guided methods to synthesize variable thresholds. They generate a stealthy attack vector i.e., a sequence of attacks that can ensure violation of *pfc* while *mdc* fails to detect it. Using this attack vector, they include a new threshold to the variable threshold set and again generate another attack vector. This step is continued until no attack vector can be generated with the current set of thresholds. We first explain the attack vector generation method using SMT [35].

The following are fed to Algorithm 1 as input: i) dynamics of the plant P, ii) the controller gain K to control the plant P, iii) estimator gain L, iv) desired performance criteria *pfc* of the closed-loop system, v) specification of existing attack monitor *mdc*, vi) set of thresholds *Th* (this is initially a null set), and vii) finite duration T for satisfying *pfc*. The system states, estimated states, and control inputs are initialized in line 2. Note that, u^a differs from $\tilde{u}^a$ by the fact that u^a is the control input before being communicated to the plant, and $\tilde{u}^a$ is the control input which is modified by the attacker and received by the plant (see Eq. 8). Consider attack is taking place at every iteration in $\{1, 2, \ldots, T\}$. At every iteration $k \in \{1, \ldots, T\}$, the variable $a_y[k]$ and $a_u[k]$ signifying false data are assigned a value non-deterministically (line 4). Following Eq. 6, the false data $a_y[k]$ is added to the measurement $y^a[k]$ which is transmitted from plant to controller (line 5). The controller computes estimated measurement $\hat{y}^a[k]$ and thereby the residue $r^a[k]$. System states $x^a[k + 1]$ and estimated states $\hat{x}^a[k + 1]$ are updated in lines 8–9. Note that since estimator and controller reside in the same embedded platform and we are only considering network-level attack on CPS, x^a is updated with $\tilde{u}^a$ while $\hat{x}^a$ is updated with u^a. Finally, control input $u^a[k + 1]$ is computed in line 10 and modified control input $\tilde{u}^a[k + 1]$ is calculated by introducing $a_u[k]$ to $u^a[k + 1]$ in line 11. This way, the closed-loop system progression for T iterations is unrolled and symbolically represented. We say that an attack is stealthy but successful

---

Algorithm 1 Attack vector synthesis [35]

Require: Plant P, controller K, observer L, Control property pfc, existing monitoring constraint
  mdc, computed threshold vector Th, attack duration T
Ensure: Attack vector $\mathcal{A}$(if it exists, otherwise NULL)
  1: **function** ATTVECSYN(P, K, L, Th, pfc, mdc, T)
  2:   $x^a[1] \leftarrow \mathcal{I}; \hat{x}^a[1] \leftarrow 0; u^a[1], \tilde{u}^a[1] \leftarrow -K\hat{x}^a[1];$ $\triangleright$ Initialization
  3:   **for** $k = 1$ to T **do**
  4:       $a_y[k], a_u[k] \leftarrow non-deterministic\_choice;$
  5:       $y^a[k] \leftarrow Cx^a[k] + D\tilde{u}_a[k] + a_y[k];$
  6:       $\hat{y}^a[k] \leftarrow C\hat{x}^a[k] + Du^a[k];$
  7:       $r^a[k] \leftarrow y^a[k] - \hat{y}^a[k];$
  8:       $x^a[k+1] \leftarrow Ax^a[k] + B\tilde{u}^a[k];$
  9:       $\hat{x}^a[k+1] \leftarrow A\hat{x}^a[k] + Bu^a[k] + Lr^a[k];$
 10:       $u^a[k+1] \leftarrow -K\hat{x}[k+1];$
 11:       $\tilde{u}^a[k+1] \leftarrow u^a[k+1] + a_u[k];$
 12:   **end for**
 13:   $\mathbb{A} \leftarrow$**assert**$((\forall Th[p] \in Th, \|r^a[p]\| < Th[p] \ \&\& \ mdc) \rightarrow pfc)$
 14:   **if** $\mathbb{A}$ is *violated* **then**
 15:       **return** $\mathcal{A} \leftarrow \begin{bmatrix} a_u[1] \cdots a_u[T] \\ a_y[1] \cdots a_y[T] \end{bmatrix};$
 16:   **else**
 17:       **return** NULL;
 18:   **end if**
 19: **end function**

---

when predicates $\| r^a[k] \| < Th[k]$ and mdc are satisfied, but pfc is violated.
Negation of this is modeled by assertion $\mathbb{A}$ in line 13. The function ATTVECSYN()
in Algorithm 1 thus non-deterministically models all possible T consecutive closed-
loop executions under stealthy attacks. After this, the assertion on the system states
and residue is given as input to an SMT tool with the assert clause. If the assertion is
violated, the algorithm gives as output a successful stealthy attack vector (line 15).
Else, it returns NULL (line 18) which signifies that the performance criteria pfc of
the system can not be violated by any stealthy attack of duration T samples. Using
this algorithm, the authors of [35] presented two greedy algorithms to synthesize a
set of variable thresholds.

3.2.1 Pivot-Based Threshold Synthesis Method

This method generates a new threshold at every iteration. The steps of the method
are demonstrated in Figs. 7, 8, 9, and 10 and discussed in detail below.

Step 1:   Initially, the threshold set is considered to be empty. The function
    ATTVECSYN() in Algorithm 1 is called with the empty threshold set and other
    parameters. If an attack vector $\mathcal{A}$ is returned, it implies that the existing monitor
    mdc fails to detect the successful attack and a new threshold for the residue-based
    detector is needed. The maximum residue generated by the current attack vector

Fig. 7 Step 1 (pivot-based threshold synthesis): from the 1-st attack vector, add 1st threshold to Th that can detect maximum residue

Fig. 8 Step 2 (pivot-based threshold synthesis): check if attack vector exists with new Th, look for new threshold on LHS of existing ones keeping monotonic decreasing order intact

Fig. 9 Step 3 (pivot-based threshold synthesis): check if attack vector exists with new Th and step 2 fails, look for new threshold on RHS of existing ones keeping monotonic decreasing order intact

Fig. 10 Step 4 (pivot-based threshold synthesis): check if attack vector exists with new Th and step 3 fails, modify an existing threshold keeping monotonic decreasing order intact

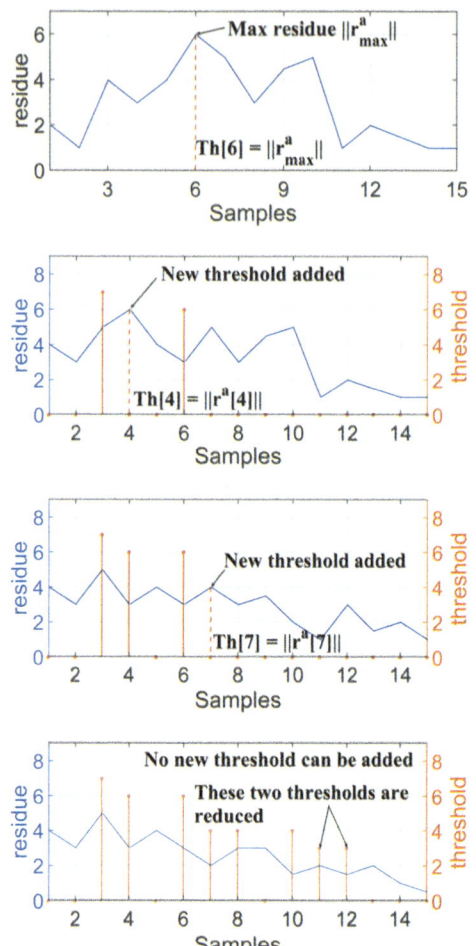

is selected as the first threshold (Fig. 7). This ensures that the new threshold will be able to detect the current attack vector.

Step 2: The function ATTVECSYN() is called again with the updated threshold set to check if any attack vector exists. If so, it implies that the current threshold set is not enough to detect all attacks, and a new threshold must be included to Th. As we aim to generate a monotonic decreasing set of thresholds, first we see if we can add a new threshold on the left-hand side of the existing ones such that the monotonic decreasing order is maintained. It is demonstrated in Fig. 8. For any of the existing thresholds $Th[p] \in Th$, we try to find out whether the current attack has produced any residue $\| r^a[k] \| \geq Th[p]$ for $k \leq p$. Multiple such candidate residues may exist. The maximum of them is considered to be the new threshold. This new threshold ensures the current attack will be detected.

Step 3:   In step 2, if any higher valued threshold can not be found on the left-
hand side of any existing threshold, the method checks if a new threshold can
be added on the right-hand side of some existing threshold (Fig. 9). For any of
the thresholds $Th[p] \in Th$, we try to find out whether the current attack has
produced any residue $\| r^a[k] \| \geq Th[p]$ after p-th instance, i.e., $k > p$. A new
threshold is added at i if only $r^a[i]$ is at least as much as $Th[k]$ for all $k > i$.
This ensures the monotonic decreasing order.

Step 4:   If no new threshold can be added following the rules in Steps 2 and 3,
then one of the existing thresholds needs to be modified to detect the current
attack vector. To do so, the proposed approach computes the difference between
existing thresholds $Th[p] \in Th$ and the corresponding residue $\| r^a[p] \|$. The
threshold $Th[i]$ is selected as a candidate if $Th[i] - \| r^a[i] \|$ is minimum
among all $Th[i] \in Th$ and the value of $Th[i]$ is comparatively reduced than
earlier. If this modification violates the monotonic decreasing property of Th, all
the $Th[p] \in Th$ for $p > i$ are reduced. This is demonstrated in Fig. 10.

Steps 2–4 are repeated until the function ATTVECSYN() returns no attack vector
with the modified threshold set. This returned threshold set Th is the final one.
Since this approach may take a longer time to converge, the authors of [35] proposed
another greedy approach that we discuss next.

3.2.2   Step-Wise Threshold Synthesis Method

While the previous approach computes a single threshold at each iteration, this
method computes a sequence of thresholds together at each step. The steps of this
method are pictorially presented in Fig. 11. Let us discuss the steps in detail.

Step 1:   Here as well, the threshold set is initialized to be empty. The function
ATTVECSYN() in Algorithm 1 is called with the empty set. If it returns an attack
vector, it implies that there is a need for a threshold to detect this attack vector.
For introducing the first sequence of thresholds, the maximum value among the
$\| r^a[i] \|$'s where $1 \leq i \leq T$ is selected, say $\| r^a[j] \|$. The first sequence
of thresholds is computed as $Th[p] = \| r^a[j] \|$, for all $p \in \{1, .., j\}$. This is
demonstrated in Fig. 11a. The name of the method is justified by the fact that the
threshold set computed using this method will always looks like steps.

Step 2:   With the updated threshold set Th, the function ATTVECSYN() is again
called to check if the new threshold is enough to detect every attack vector. If
the function returns an attack vector, it indicates the need for new thresholds.
Let $Th[i]$ be the last non-zero threshold value. To create a new step, this method
finds out maximum $\| r^a[k] \|$ for $k > i$ such that $\| r^a[k] \| \leq Th[i]$. Say, the
maximum is $\| r^a[j] \|$. The threshold set is then updated as $Th[p] = \| r^a[j] \|$
for all $i < p \leq j$ (Fig. 11b). This ensures the desired monotonic decreasing
order property of Th.

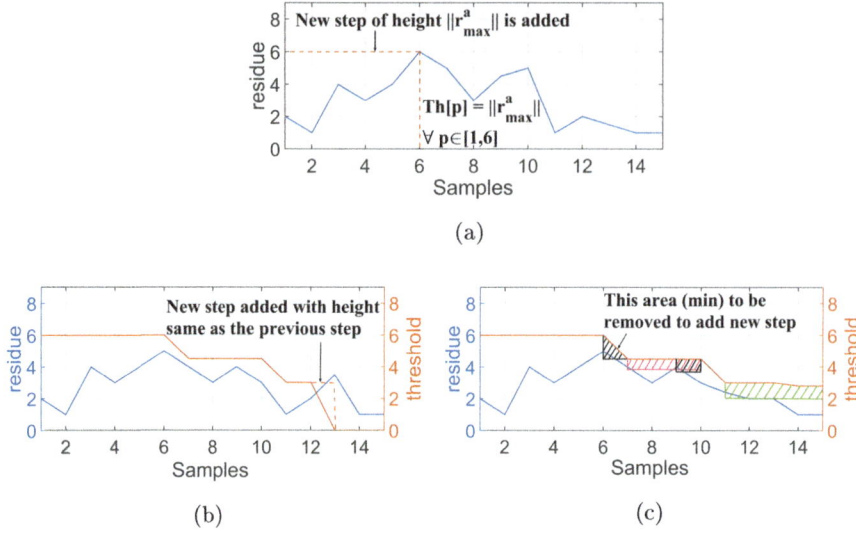

Fig. 11 Step-wise threshold synthesis. (**a**) Step 1: from 1-st attack vector, add 1st step of thresholds in Th. (**b**) Step 2: check if attack vector exists with new threshold Th, look for new step downwards to maintain the monotonic decreasing order. (**c**) Step 3: check if attack vector exists with new threshold Th, and step 2 fails, create new steps out of the old ones by keeping the monotonic decreasing order intact

Step 3: A situation may occur when no new step can be generated in the threshold set. This can happen when there is no zero element in Th. In such cases, the height of some existing steps needs to be modified to ensure that the current attack will be detected with the modified threshold set. Instead of reducing the height of an entire step, we break a portion or the whole step whichever involves minimum effort i.e., the minimum area under the threshold curve that can be removed to detect the current attack. From Fig. 11, it can be seen that the thresholds in Th create an area under the threshold curve. At each sampling instant i, an area $Area_i$ is computed as follows. Find p, $i < p \leq T$ such that for all $k > p$, $Th[k] \leq \| r^a[i] \|$ but for all $k \leq p$, $Th[k] > \| r^a[i] \|$. $Area_i$ is the segment under threshold curve Th from i-th to p-th sample. The sampling instant for which this area is minimum is selected, say that is the j-th instance. By removing $Area_j$, new step is generated as $Th[l] = \| r^a[j] \|$ for all $l \in (j, p]$. This is demonstrated in Fig. 11c.

Steps 2–3 are repeated until no new attack vector can be found upon calling ATTVECSYN() every time Th is updated. In [35], authors analyzed that the step-wise method converges must faster than the pivot-based one. Also, the step-wise method performs better in terms of FAR than the pivot-based method.

4   AI-Based Adaptive Attack Detection

The previous section brings forth some algorithmic and heuristic-based approaches to improve the detection tasks in CPSs. It introduces the concept of an adaptive detection technique. These heuristic-based approaches definitely improve the detection when compared to a fixed threshold-based detector, but it does not formalize or quantify the improvements. In an attempt to achieve so, the authors of [36] presented an attacker-defender game to guide the adaptive threshold selection problem though the proposed detector is not evaluated on any closed-loop CPS. Moreover, the optimization problem for threshold selection is solved in real-time and may cause computation overhead. Reinforcement learning is already been successfully used in the domain of estimation [50], energy efficiency [51] of safety-critical CPSs with real-time requirements. An RL-based adaptive threshold-based attack detector learns from the affected system dynamics and adaptively tunes the threshold of the residue-based anomaly detectors. The main motivation behind considering an RL-based strategy is the following challenge. The false data injected into sensors and actuators by the stealthy attacker are highly system-specific, random, and do not follow any statistical distribution. Therefore, the parameters of the adaptive attack detector cannot be directly derived from the injected false data signature. The performance of the proposed detector depends on how well it is trained against the optimal attack vectors. Thus, we also design an RL agent for mimicking the attacker's behavior during the training phase.

Let, for the discrete LTI system shown earlier in Eq. 2, the estimation error $e[k]$ be defined as $e[k] = (x[k] - \hat{x}[k])$. The Gaussian assumptions of noise and initial states $(x[0] \sim \mathcal{N}(0, \Sigma_{x[0]}))$ ensure that $e[k] \sim \mathcal{N}(0, \Sigma_e)$ (steady state covariance matrix of this estimation error is Σ_e). Therefore, the system residue $r[k] = Ce[k] + v[k]$. Being a linear function of two other independent gaussian random variables estimation error and measurement noise, the residue is also normally distributed i.e., $r \sim \mathcal{N}(0, \Sigma_r)$, where, $\Sigma_r = E[r[k]r[k]^T] - E[r[k]]E[r[k]]^T = E[(Ce[k])(Ce_=[k])^T] + E[v[k]v[k]^T] = C\Sigma_e C^T + \Sigma_v$.

As a popular detection scheme, the χ^2-test can be used on $r[k]$ to find out how anomalous $x[k]$ is i.e., whether it is affected by injected false data. This helps one understand how *bad* the estimated output is compared to the actual controlled-plant output according to the χ^2-test. Let $g[k]$ denote the χ^2-test result at k-th sample and $g[k] = \sum_{i=k-l[k]+1}^{k} r_i^T \Sigma_r^{-1} r_i$. A window size of $l[k]$ is considered during the χ^2-test at k-th sampling instance since taking historical data into account produces a more accurate estimation compared to only considering instantaneous data. Consider that there are m available sensors to sense different plant outputs (i.e., $m \leq n$, n being the dimension of the system). The degree of freedom (DOF) for this test is $m \times l[k]$. When there is no FDI attack, $g[k]$ follows χ^2 distribution with mean $ml[k]$ (Fig. 12) since $r[k]$ and $e[k]$ follows 0 mean Gaussian distribution as discussed above. If $Th[k]$ is the threshold that is chosen at k-th sampling instance, $g[k]$'s probability density function (PDF) along with its cumulative distribution function (CDF) w.r.t. $Th[k]$ can be defined as,

Fig. 12 χ^2-distribution

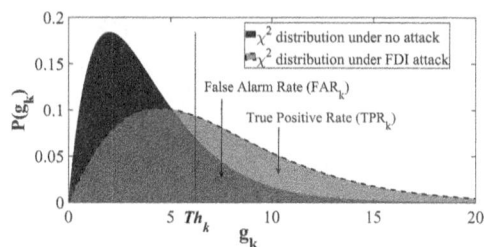

$$P(g[k]) = \frac{g[k]^{\frac{ml[k]}{2}-1}e^{-\frac{g[k]}{2}}}{2^{\frac{ml[k]}{2}}\Gamma(\frac{ml[k]}{2})} \tag{13}$$

$$P(g[k] \leq Th[k]) = \frac{\gamma(\frac{ml[k]}{2}, \frac{Th[k]}{2})}{\Gamma(\frac{ml[k]}{2})} \tag{14}$$

Here, Γ and γ are ordinary and lower incomplete gamma functions respectively [52, 53]. It is considered to be a *false alarm* when $g[k] > Th[k]$ even in the absence of an attacker. We quantify this with the *false alarm rate* (FAR), calculated with the ratio of the number of times a false alarm is raised falsely and the total number of alarms raised. We denote the FAR at k-th sample where the χ^2-test result $g[k]$ is compared with the threshold $Th[k]$ as $FAR[k]$. In Fig. 12, the black area under the solid curve and the grey area under the dashed curve represent the distribution of $g[k]$ under no attack and attack respectively. Therefore, $FAR[k]$ should be the fraction of area under the probability distribution curve of un-attacked $g[k]$ that is constrained by $g[k] > Th[k]$; thus computed as $FAR[k] = 1 - P(g[k] \leq Th[k])$.

As proven in Theorem 1 in [38], the spurious data $a^y[k]$ and $a^u[k]$ added by the attacker to the sensor, and the actuator transmissions respectively introduce non-centrality to the actual χ^2 distribution of system residue. The gray area under the dashed curve in Fig. 12 is the distribution of $g^a[k]$ obtained from the residue $r^a[k]$ (Eq. 7). The resulting $g^a[k]$ is compared to $Th[k]$ in order to flag an attack. Following Corollary 1 in [38], the variance of $g[k]$ is $\sigma[k] = 2ml[k]$ and variance of $g^a[k]$ is $\sigma^a[k] = 2(ml[k] + 2\lambda[k])$, where $\lambda[k] > 0 \implies \sigma^a[k] > \sigma[k]$. From the Theorem 1 in [38] we can see that the expected deviation of $g^a[k]$ from its mean is more than the expected deviation of $g[k]$ from $ml[k]$ which makes the distribution of $P(g^a[k])$ wider and thereby flatter (since the area under both curves is unity). Therefore, $P(g^a[k] > Th[k]) > P(g[k] > Th[k])$ as shown in Fig. 12. As the window size $l[k]$ increases, the PDF of $g^a[k]$ becomes even flatter and hence more distinguishable from the PDF of $g[k]$. So, intuitively speaking, the non-centrality of χ^2 distribution improves $TPR[k]$ i.e., attacks are more detectable for a properly chosen window size $l[k]$ parameter.

For an optimally chosen $l[k]$, the non-centrality of $g^a[k]$ is more evident and hence produces better TPR for a certain threshold $Th[k]$. We can also optimally choose a $Th[k]$ to attain the minimum possible $FAR[k]$ (for a certain window size) during the absence of an attack (i.e., when only noise is present, in PDF of $g[k]$) and change it during the attack to attain the maximum possible TPR. The main trick here is to understand that the system is under attack more often in the true positive case and reduce false alarms. Therefore, the pertinent problem becomes *how to achieve the above by suitable choice of threshold $Th[k]$ and the test parameter $l[k]$ in order to identify the attack as quickly as possible without too many false alarms.* So, given a closed-loop CPS, one needs to learn when is the system under some stealthy FDI attack and when it is running normally. The work in [38] leverages the non-centrality property of $g^a[k]$ and learns when the system is becoming affected by a successful and stealthy FDI attack. The problem of synthesizing an optimal detector at every k-th simulation step can thus be formulated as the following optimization problem.

$$J_t = \max_{l[k],Th[k]} w_1 \times TPR[k] - w_2 \times FAR[k] \text{ s.t. } FAR[k] < \epsilon, \, l[k] < l_{max} \quad (15)$$

The cost function J_t aims to minimize $FAR[k]$ and maximize $TPR[k]$ at every simulation step. Here, w_1, w_2 are respective non-negative weights assigned to TPR and FAR depending on attacked (TPR increment gets more importance) and non-attacked (FAR reduction gets more importance) situations. ϵ is the maximum allowable FAR and l_{max} is the maximum allowed χ^2 window length. At each k-th step, given the current measurement $y[k]^a$, the solution of the above optimization problem is a pair $< l[k]^*, Th[k]^* >$, where $l[k]^*$ and $Th[k]^*$ are the optimal χ^2 window length and threshold respectively that lead to maximum $TPR[k]$ and minimum $FAR[k]$ w.r.t. current measurement of the system states. But this formulation has to work for all possible FDI attacks within the sensor and actuation limits. The authors in [38] take a nice approach to ensure that the detection works even in the worst case. They learn the optimal attack possible at k-th iteration that maintains its stealth but imparts the most significant damage to the system safety. The following subsection explains how such an attacking policy can be learned.

4.1 Optimal Attack Policy Design

As we discussed in Sect. 2, the attacker's motive is to steer the system beyond the safe set X_S while trying to remain stealthy by *reducing the TPR* i.e., fooling the detector. Given the sensor measurement $y^a[k-1]$, we present this attack estimation problem as the following optimization problem.

$$J_a = \max_{a^y[k],a^u[k]} -w_1 \times TPR[k] + w_2 \times FAR[k] +$$

$$\sum_{i=0}^{\infty} (|x^a[i+1]| - X_S|)^T W_3 (|x^a[i+1]| - |X_S|) \quad (16)$$

s.t. $x_0^a, \hat{x}_0^a \in X_R$ (17)

$$u^a[k] = -K\hat{x}^a[k], \ \tilde{u}^a[k] = u^a[k] + a^u[k], \ | u^a[k] |,$$

$$| \tilde{u}^a[k] | \leq \epsilon_u \ \forall k \in [0, \infty] \tag{18}$$

$$y^a[k] = Cx^a[k] + D\tilde{u}^a[k] + v[k] + a^y[k],$$

$$| y^a[k] | \leq \epsilon_y \ \forall k \in [0, \infty] \tag{19}$$

$$r^a[k] = Cx^a[k] - C\hat{x}^a[k], \ g^a[k] \leq Th[k] \ \forall k \in [0, \infty] \tag{20}$$

$$\hat{x}^a[k+1] = A\hat{x}^a[k] + Bu^a[k] + L(Cx^a[k] - C\hat{x}^a[k]), \ \forall k \in [0, \infty] \tag{21}$$

$$x^a[k+1] = Ax^a[k] + B\tilde{u}^a[k], \ \forall k \in [0, \infty] \tag{22}$$

Here, w_1 and w_2 are weights that denote the relative priorities of the attack initiative similar to the optimal threshold cost function J_t (Eq. 15). This is because our intention is to design an optimal and stealthy FDI attack for a system equipped with the adaptive detector designed above. While an attacker tries to decrease $TPR[k]$ (and increase $FAR[k]$ simultaneously as a by-product), the detector's objective is to increase $TPR[k]$ and decrease $FAR[k]$ based on the value of $\lambda[k]$ (Eq. 15). The last component of J_a is important to establish it as the worst-case attack. It accounts for the deviation of the current system state from the safety boundary X_S using a quadratic weighted distance metric where W_3 is a diagonal matrix with relative weights signifying the safety-criticality of each dimension. The constraints in 18 and 19 ensure that the attack efforts are practical, within the allowable ranges and utilize the LTI system properties. In case of an invalid or beyond the range sensor data and control signal, their effects will be trimmed by the saturation limit and won't produce a desirable effect of the attack. An intelligent adversary's another aim is to remain stealthy, thus bypassing the detector. This is taken into account in constraint 20 while estimating the optimal attack. The constraints 21 and 22 ensure system progression following Eqs. 5–8.

4.2   The MARL Based Framework

This section discusses the Multi-Agent Reinforcement Learning (MARL) based implementation that is the methodology to build an adaptive threshold-based detection module as discussed in [38]. The goal of the adaptive detector is to detect a stealthy FDI attack before it is successful in making the system unsafe. In the introduction of Sect. 4, we have explained how changing the detection thresholds by leveraging the non-centrality of the system residue can be useful to increase TPR and reduce FAR. We utilize that notion here. The detection and attacker modules implicitly learn how the system model behaves normally and under FDI attacks by analyzing system outputs, states, residue, etc. The smart attacker module should challenge the adaptive detection module by posing the most stealthy yet effective

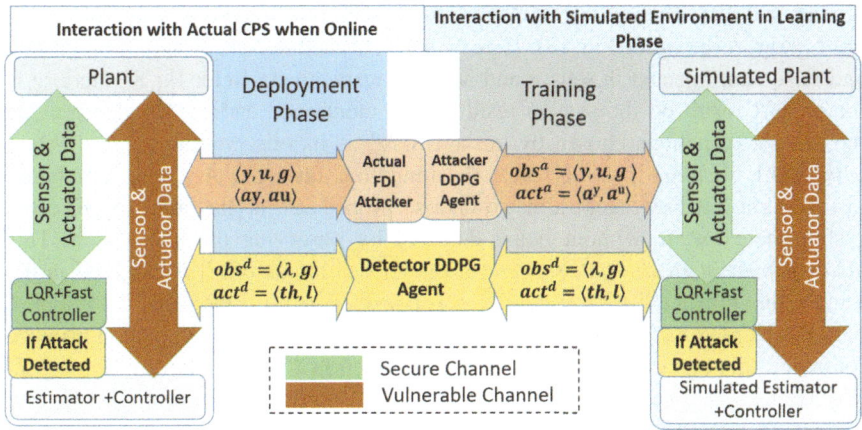

Fig. 13 The RL-based methodology [38]

FDI attacks depending on current system behavior. On the other hand, the intelligent detector module learns the best possible FDI attack on the CPS by observing the non-centrality of the χ^2-distribution of system residue and it adaptively changes the threshold to expose the attacker with a promise of increased TPR and reduced FAR. RL algorithms are capable of automatically updating their strategy by learning from prior experiences. Moreover, since labeled data for falsified system states are not available, a simulation environment for the targeted system model can be useful. This explains how integrating an MARL framework like this can be useful in the context of a security-aware CPS design. In the context of the timing overhead, it is reasonably low when we use a trained RL agent for inferencing at run time as seen in previous literature [51].

A plant-controller closed-loop system equipped with a χ^2-based detector (as shown in Fig. 3) is modeled similar to the real-world system under test. This is then used as the environment for the RL agents (both under attack and without attack situations). Individual RL agents are built as part of our methodology to act as the FDI attacker and the adaptive threshold-based detector that run simultaneously in a closed-loop with the system environment (see Fig. 13). These agents (Λ) interact with the environment by observing certain states from the environment (obs) and learn how intelligent choice of action (act) values can influence the environment towards the fulfillment of their objectives i.e., earning higher rewards (Rwd).

RL Agent For FDI Attack Estimation Following the intelligent attacker modeling in Sect. 4.1, it can be considered that the attacker has information about the system characteristics (Eq. 2) and it can manipulate the sensor and actuator data communicated between the plant and controller side. An *Attacker RL Agent* Λ^a intelligently injects false data into the system by observing the sensor data, actuator data, and the χ^2-test result on the system residue. These false data injections should be bounded by the sensor and actuator saturation limits (refer

Eqs. 18 and 19). The *actions* of Λ^a are $act^a = [a^y, a^u]$ and the *observations* are $obs^a = [y, u, g]$ (refer Fig. 13). Here y, u, a^y, a^u denote sensor and actuator data and false data injected in sensor and actuator respectively (refer Eq. 8). Here, g is the χ^2-test result on the system residue r as mentioned earlier. We also provide the last set of actions chosen by the agent Λ^a as its observation, i.e.,$obs^a[k] = [y[k], u[k], g[k], act^a[k - 1]]$ at k-th simulation instance. At every simulation instance, the attacker agent aims to choose proper $act^a[k]$ in order to make the system state unsafe without being detected by observing the above data. This *measurement of stealth and success* of the chosen attack effort is captured in the *reward* function $Rwd[k]^a(obs^a[k], act^a[k], obs^a[k + 1])$. Like usual RL policies, the agent is *rewarded* against its choice of $act^a[k]$ at every k-th simulation instance following this reward function. The reward function for Λ^a is built following J_a (Eq. 16) i.e.,

$$Rwd^a[k] = -w_1 \times TPR[k] + w_2 \times FAR[k]$$
$$+ (\mid x^a[k + 1] \mid - \mid X_S \mid)^T W_3(\mid x^a[k + 1] \mid - \mid X_S \mid) \qquad (23)$$

The notations carry the same meaning as in Eq. 16 and the index k denotes their value at k-th simulation instance. As described earlier, the two parts of $Rwd^a[k](obs^a[k], act^a[k], obs^a[k+1])$ have opposing objectives. The part $-w_1 \times TPR[k] + w_2 \times FAR[k]$ accounts for stealthiness with minimized FAR and $(\mid x^a[k + 1] \mid - \mid X_S \mid)^T W_3(\mid x^a[k + 1] \mid - \mid X_S \mid)$ accounts for the success of a chosen FDI attack action $act^a[k]$. Λ^a moves towards gaining a higher $Rwd^a[k]$ at every simulation instance by choosing an optimal action $act^a[k]$. Therefore, a learned optimal attack estimation agent would ensure that the overall return (the cumulative reward discounted over time) is maximized, which translates to the fact that the attacker agent will estimate the false data in a way such that the system under attack goes unsafe as quickly as possible without being detected. Note that this agent also gives an idea of the actual system states which we can use for secure state estimation.

RL Agent For Adaptive Detection The *Variable Threshold-based Detector Agent* Λ^d also acts on the same system environment under FDI attack as a competitor to the Attacker RL Agent. It chooses an optimal attack detection threshold $Th[k]$ and a suitable χ^2-window $l[k]$ at k-th iteration by observing the χ^2 statistics g of the system residue, current non-centrality λ of this χ^2 distribution and the previous action $act^d[k - 1]$ chosen by itself. The intuition behind its formulation is already discussed above and the authors in [38] provide rigorous mathematical proof. Depending on the observations from the attacked environment, the Detector agent chooses a χ^2-window length and threshold. Therefore, we consider the action vector $act_k^d = [Th[k], l[k]]$ and observation vector $obs^d[k] = [g[k], \lambda[k], act^d[k - 1]]$ (refer Fig. 13). The reward function $Rwd^d[k](obs^d[k], act^d[k], obs^d[k + 1])$ is designed following J_t from Eq. 15, i.e.,

$$Rwd^d[k](obs^d[k], act^d[k], obs^d[k+1]) = \begin{cases} TPR[k] & \text{when } \lambda[k] > \delta \\ -FAR[k] & \text{when } \lambda[k] \leq \delta \end{cases}$$

$$(24)$$

Here also, the variables carry a similar meaning as in Eq. 15. With a goal to increase the discounted reward over time in an episode, Λ^d chooses $act^d[k]$ at every k-th simulation instance. The structure of the reward function thus ensures that the Neural Network will be trained such that it can always choose its actions to maximize TPR and minimize FAR.

Learning Technique Here we use Deep Deterministic Policy Gradient (DDPG) algorithm [54]. Each DDPG agent consists of an *actor* neural network that deterministically chooses an action (act) by observing the states (obs) of the environment. Another Deep Q-Network(DQN) acts as a critic. In each simulation instance, the actor chooses an action ($act[k]$) by exploring the action space randomly. The transitions from $obs[k]$ to $obs[k+1]$ due to the action $act[k]$ taken are stored in the experience replay buffer along with the corresponding reward $Rwd[k]$ achieved during this transition. Note that this action was chosen based on the maximum possible return. The critic network calculates corresponding Q values in every iteration picking a random batch from the replay buffer and updates itself by the mean square loss between the calculated Q values from consecutive iterations. The actor-network policies are updated using the policy gradient over the *expected Q value return*. Figure 14 depicts the learning flow of a DDPG agent. The training algorithm finally learns the highest expected return from its experiences and then keeps updating the RL policy to output the optimal action that earns the expected maximum return. This, in turn ensures the objective functions we chose to define

Fig. 14 A DDPG RL agent with actor and critic networks [38]

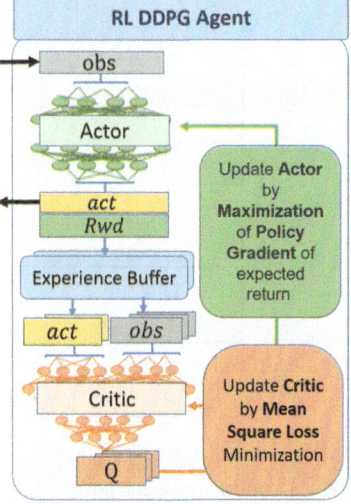

$Rwd^d[k]$ and $Rwd^a[k]$ in Eq. 15 and Eq. 16 respectively are maximized. Given a secure CPS model, we first train such DDPG agents as discussed with system-specific simulation data so that they can reprise their designated roles in the environment. The *learning process* is collaborative and competitive. The standard DDPG algorithm as in Algorithm 1 in [54] is modified according to the requirements in our case. Interested readers are encouraged to read [54] for a detailed discussion on how DDPG policy optimization works. The work in [55] is also another interesting read to know more about how Agent Environment Cycle (AEC) Games model turn-based games like our MARL setup where the Attacker and Detector compete with each other in every iteration by taking optimal action in a stochastic system environment. Without going into those implementation-specific challenges, we stick to the CPS design aspect without sidelining the main topic of discussion in this section. In the next section, we move on to describe the most plausible next action that should follow an intelligent attack detection in a secure CPS design.

5 Attack Mitigation

To complete the circle of the discussion on the security-aware design of CPS, in this section we briefly talk about system recovery steps to be taken once attack attempts are detected. The idea proposed by most researchers is to switch the system to a *secure mode* once an attack is detected. We explain this idea using Fig. 15. By secure mode, we mean an operational mode where every communication is secured via some cryptographic methods, like MAC, RSA, AES, etc. For schedulability, un-important messages can be dropped in this mode. We assume that the cryptographic methods provide utmost security and that no stealthy attacks are possible in this mode. Therefore, the attack model is to exploit the normal mode of operations

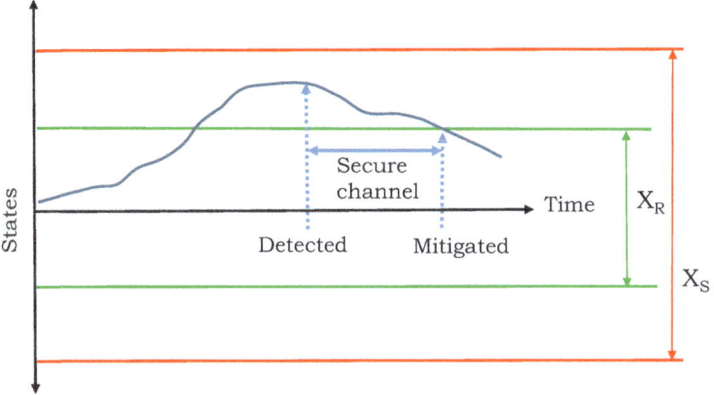

Fig. 15 Attack mitigation through secure channel

and stealthily drive the system to some unsafe region. In Fig. 15, X_R denotes the operating region (in a single dimension) where the system is expected to reside when no attack is taking place. Also, X_S denotes the safety region of the system. An attacker would try to steer the system toward an unsafe zone (see Eq. 9). Once such an attack is detected, the system will be switched to a secure channel. And, during this secure mode, the controller will mitigate the damage done by the attacker (Fig. 15) by steering the system back to its preferable operating region.

While in [34] it is suggested to simply use the available controller gain (K as mentioned in Eq. 2) for attack mitigation, the authors in [38] have suggested using additional control input along with the inputs from the usual feedback controller, following the theory of [56]. As it is shown in Fig. 15, due to an attack, the system may go beyond the preferred operating region X_R. It is desirable to bring back the system from $X_S \setminus X_R$ as early as possible. The motivation behind this is that the duration spent in secure mode must be as minimum as possible. Also, the faster the system is back to the desired operating region, the better will be its average performance. As we have already discussed in Sect. 3, the cryptographic methods incur quite a significant computational and communication load. Thus, it is infeasible to secure every communication. Releasing a secure channel at the earliest will help other control loops to use it. It can be seen in [38] that the use of additional control inputs can actually speed up the recovery process. We briefly demonstrate the idea here.

The authors in [56] have proposed an SMT-based method to pre-calculate a sequence of control inputs that take the system from $X_S \setminus X_R$ to X_R provided during this time, system state should always be retained within X_S. This means safety is guaranteed during recovery. Since $X_S \setminus X_R \in \mathcal{R}^n$, it is not possible to compute control sequence for all possible points in $X_S \setminus X_R$. As a solution to this problem, the authors in [56] proposed a region-wise control synthesis method. They divide $X_S \setminus X_R$ into such sub-regions that the control sequence computed to take the system trajectory from the center of each sub-region to X_R will also work for every other point in that sub-region, as elaborated in Fig. 16. Initially, the length of the control sequence is set as $t = 1$ and the method tries to compute safe control sequences considering the entire $X_S \setminus X_R$ as the source region. If failed, the source region is reduced to half and this process repeats until a safe control sequence can be found. If

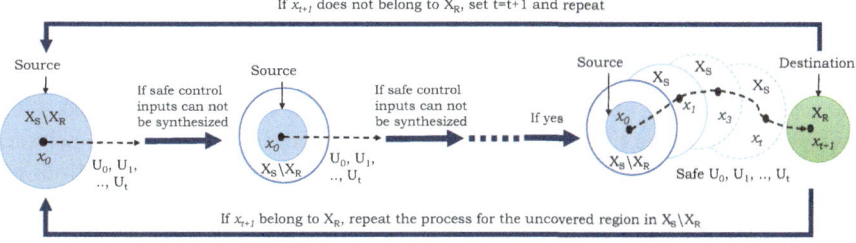

Fig. 16 Fast control action synthesis

the final safe control sequence of length t can not take the system to X_R, the length t is incremented by 1 and the process is repeated until a safe and successful control sequence is found for every sub-region in $X_S \setminus X_R$. This incremental process is also adapted in [35] such that it also ensures that the length of the final control sequence is minimum. This means this is the minimum required sequence of control inputs to bring back the system from $X_S \setminus X_R$ to X_R while safety remains intact.

The above line of work however relies on attacks getting detected in the normal mode of operations. The security model assumes that using the secure mode is costly for the entire system and hence it should be relinquished to other potential subsystems under attack as soon as possible. In that case, how do we guarantee safety from stealthy attacks in a non-probabilistic way? For that, the use of secure and normal modes of operations must be interleaved by design so that every control task switches among such modes periodically. The normal model duration should be chosen in such a way that a stealthy attack cannot drive the system to an unsafe region as verified formally in the model itself [34]. This duration should be followed by a secure mode of suitable duration which helps the system crawl back to its desired performance region [34, 38]. The sequence keeps repeating with such sporadic integrity checks in between [33] coupled with recovery control mechanisms [56]. In the next section, we present a hardware-in-loop experimental set-up on a real-time platform to demonstrate the security-aware design of an automotive CPS considering the variable threshold-based security scheme discussed in Sect. 3.2.

6 A Platform Level Example

In this section, we first discuss a security-aware control implementation on an automotive-grade ECU setup. The setup contains an Infineon Tricore AURIX TC397 ECU where software controllers are mapped. A Hardware-In-Loop (HIL) simulator (ETAS LabCar) is used to emulate the automotive plants. These plants are periodically manipulated by control tasks co-scheduled in a single core of the Infineon ECU. The plant and ECUs are connected via CAN bus, interfaced using the integrated CAN shield in the ECU. The closed-loop setup is depicted in Fig. 17. The ECU is running two control tasks for two automotive plants i.e., Trajectory Tracking Control (TTC) and Electronic Stability Program (ESP). The details of the TTC loop is described in Sect. 3. It controls the longitudinal deviation (D) of the vehicle from a desired trajectory and maintains a target Velocity (V) by changing the acceleration(acc) of the vehicle. The ESP regulates the yaw rate (γ) and sideslip (β) of the vehicle by controlling the steering angle (θ). The system transition, controller, and observer gain matrices of the ESP are taken from [38]. Both of these tasks are implemented as runnables and invoked periodically depending on the sampling period of the discretized plant model considered while designing the LQR controllers, i.e., 100 ms for TTC and 40 ms for ESP. In every sampling period, these statically scheduled tasks are invoked, and control inputs are calculated using the

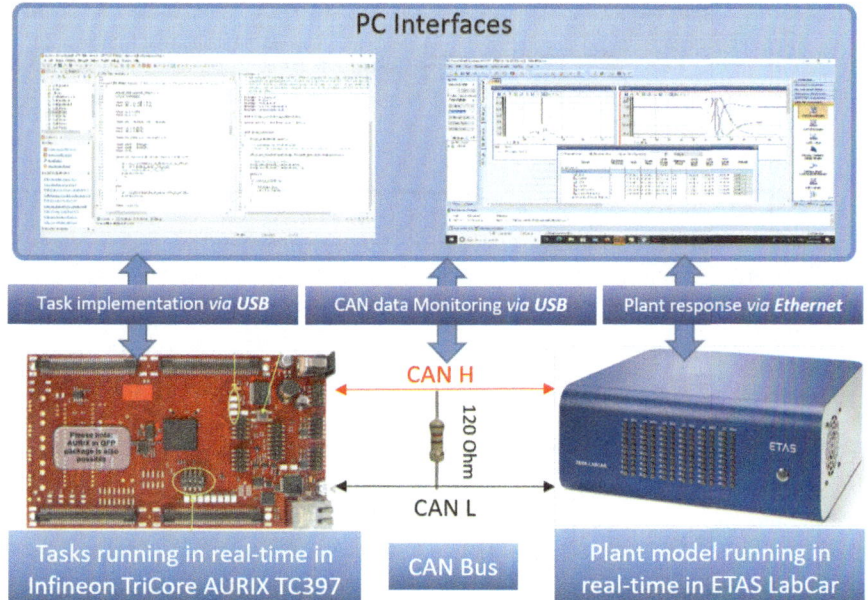

Fig. 17 Real-time automotive test bed

estimated plant states and transmitted via CAN. The control inputs actuate the plant, which emulates itself in real-time in the HIL. The plant states are estimated using estimation tasks that read the received plant output data from the sensor readings transmitted from the plant through CAN. These functions are called before the control tasks in order to supply the estimated plant state to the control tasks. The estimation tasks also run with similar periodicity as their corresponding control calculations. Note that the last calculated control input is also used in order to estimate the current state based on the sensor data. As a detection task, we consider the variable threshold-based detector. The detection task synthesizes thresholds using the pivot-based method as explained in Sect. 3.2. It uses the norm-based generalized detection scheme as mentioned in Eq. 4.

We consider an attack model where an ECU connected to the same CAN bus is compromised. Therefore, it has the capability to inject false data into the CAN transmissions. The attack model is feasible because a compromised ECU can send a real sender to bus-off mode for some interval and mimic the actual sender [12]. We emulated this insider attack scenario by running the attacker routine from a different core of the same ECU. The false data injected were synthesized for TTC using the SMT-based FDI attack synthesis method used in [34, 35]. As mentioned in our attack model in Sect. 2.1, the successful attack criteria while synthesizing the attack vectors is given by Eq. 9. The plot in Fig. 18 shows the outputs of the TTC plant under attack. This is a screenshot taken from the ETAS LabCar environment that shows a 1.3 seconds long stealthy (i.e., 1-norm of the residue under this attack always

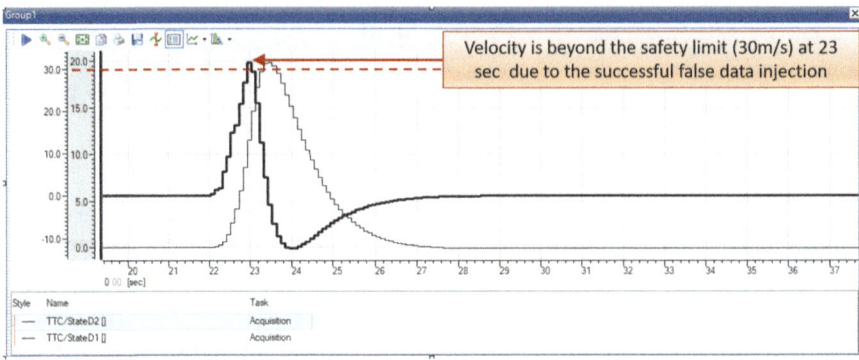

Fig. 18 Successful FDI emulation on TTC

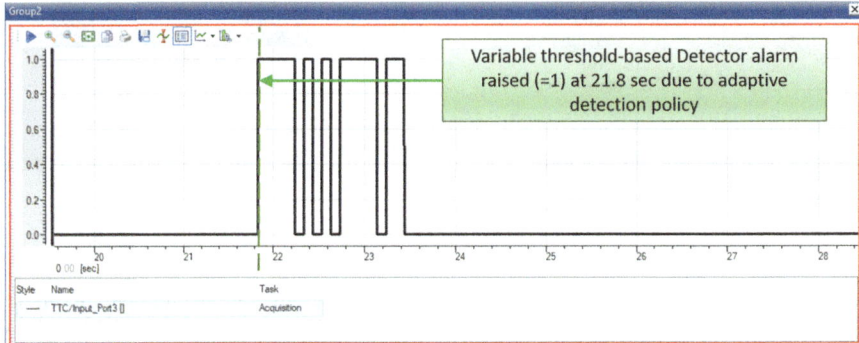

Fig. 19 Adaptive detection of attack before success for TTC

remains below a static threshold of 2.5) attack vector injection (refer Fig. 4, which shows a simulation of the same). The x-axis of Figs. 18 and 19 represents time. The two different scales in the y-axis of Fig. 18 represent two states of TTC i.e., deviation from the trajectory (D) in meters (the first scale from the right) and velocity (V) in meters/sec (the second scale from the right). Y axis of Fig. 19 denotes the detector output. 1 signifies attack detection and 0 signifies no attack scenario. The variable threshold-based detection task is implemented as mentioned earlier in this section. As we can see, both states are starting from 0 units and the 13-length FDI attack vector drives the velocity of the vehicle (the bold one) beyond the safety limit i.e., 30 m/s at 23-rd sec bypassing the static threshold-based detector (in Fig. 18). Whereas, the variable threshold-based detection task selects certain thresholds in real-time which are able to detect this attack attempt (at $\sim$ 21.8 s in Fig. 19) before the attack becomes successful or the system becomes unsafe. Validation of MARL-based detectors and mitigation strategies have been reported in [38] where the full system is simulated in Matlab. In the future, we plan to have a HIL-based validation of the same in this automotive test-bed.

7 Conclusion

This chapter discusses methods for security-aware automotive CPS design leveraging adaptive lightweight attack detection and mitigation schemes. The presented design methodology reduces the compute and communication overhead incurred by standard cryptographic methods suggested by AUTOSAR. We discuss two heuristic-based algorithms for variable threshold selection and a multi-agent reinforcement learning (MARL)-based adaptive threshold selection method in order to increase false data injection attack detectability and decrease false alarm rate in a system. The heuristic-based methods choose thresholds based on solver based vulnerability analysis. Thus, this design technique provides a guarantee that the synthesized variable threshold-based detectors will detect a false data injection attack attempt. The more scalable approach employing the adaptive detector infers a stochastically optimal threshold in order to catch a competing FDI attacker agent, which is designed to falsify the sensed and actuated data. We also discuss a formal method-based attack mitigation scheme which is activated via a secure channel once the attacks are detected. Overall they promote an end-to-end security-aware CPS design idea.

The objectives of this security-aware co-design framework targeting automotive systems had been, (i) lightweight, real-time detection of FDI attacks (ii) while maintaining the least possible false alarm rate; and (iii) guaranteeing the mitigation of the attack-effect as early as possible so that (iv) the compute and communication overhead incurred by the cryptographic schemes are reduced. We discuss the evaluation of the variable threshold-based detection technique in a real-time automotive test bed in order to demonstrate its applicability. Essential future extension of such work is to test the performance of the proposed RL-based adaptive detection and formal mitigation units in this automotive test bed.

Acknowledgments We acknowledge generous grants received from *IHUB NTIHAC Foundation—IIT Kanpur* and *Meity (Grant No. AAA.22/8/2021-CSRD-Meity)* for partially supporting this work.

References

1. Möller, D.P., Haas, R.E.: Guide to Automotive Connectivity and Cybersecurity. Springer, Berlin (2019)
2. Reif, K.: Automotive Mechatronics. Springer, Berlin (2014)
3. HPL SC: Introduction to the controller area network (CAN). Application Report SLOA101, pp. 1–17 (2002)
4. Makowitz, R., Temple, C.: Flexray-a communication network for automotive control systems. In: 2006 IEEE International Workshop on Factory Communication Systems, pp. 207–212. IEEE, Piscataway (2006)
5. Ruff, M.: Evolution of local interconnect network (LIN) solutions. In: 2003 IEEE 58th Vehicular Technology Conference, vol. 5, pp. 3382–3389. IEEE, Piscataway (2003)

6. Sumorek, A., Buczaj, M.: New elements in vehicle communication "media oriented systems transport" protocol. Teka Komisji Motoryzacji i Energetyki Rolnictwa. **12**(1), 275–279 (2012)
7. Bo, H., Hui, D., Dafang, W., Guifan, Z.: Basic concepts on AUTOSAR development. In: 2010 International Conference on Intelligent Computation Technology and Automation, vol. 1, pp. 871–873. IEEE, Piscataway (2010)
8. Deng, P., Cremona, F., Zhu, Q., Di Natale, M., Zeng, H.: A model-based synthesis flow for automotive CPS. In: Proceedings of the ACM/IEEE Sixth International Conference on Cyber-Physical Systems, pp. 198–207 (2015)
9. Chakraborty, S., Al Faruque, M.A., Chang, W., Goswami, D., Wolf, M., Zhu, Q.: Automotive cyber–physical systems: a tutorial introduction. IEEE Des. Test **33**(4), 92–108 (2016)
10. AUTOSAR: Specification of secure onboard communication. AUTOSAR CP Release **R20-11**(969), 1–28 (2017)
11. Munir, A., Koushanfar, F.: Design and analysis of secure and dependable automotive CPS: a steer-by-wire case study. IEEE Trans. Depend. Sec. Comput. **17**(4), 813–827 (2018)
12. Cho, K.T., Shin, K.G.: Error handling of in-vehicle networks makes them vulnerable. In: Proceedings of the 2016 ACM SIGSAC Conference on Computer and Communications Security, pp. 1044–1055 (2016)
13. Lesi, V., Jovanov, I., Pajic, M.: Integrating security in resource-constrained cyber-physical systems. ACM Trans. Cyber-Phys. Syst. **4**(3), 1–27 (2020)
14. Mo, Y., Sinopoli, B.: False data injection attacks in control systems. In: Preprints of the 1st Workshop on Secure Control Systems, pp. 1–6 (2010)
15. Teixeira, A., et al.: Secure control systems: a quantitative risk management approach. IEEE Control Syst. Mag. **35**(1), 24–45 (2015)
16. Checkoway, S., McCoy, D., Kantor, B., Anderson, D., Shacham, H., Savage, S., et al.: Comprehensive experimental analyses of automotive attack surfaces. In: USENIX Security Symposium, San Francisco, vol. 4, pp. 447–462 (2011)
17. Åström, K.J., Wittenmark, B.: Computer-Controlled Systems. Prentice-Hall, Hoboken (1997)
18. Becker, M., Mohamed, S., Albers, K., Chakrabarti, P., Chakraborty, S., Dasgupta, P., et al.: Timing analysis of safety-critical automotive software: the AUTOSAFE tool flow. In: 2015 Asia-Pacific Software Engineering Conference, pp. 385–392. IEEE, Piscataway (2015)
19. Boulanger, J.L.: Industrial Use of Formal Methods: Formal Verification. Wiley, Hoboken (2013)
20. AUTOSAR: Specification of crypto service manager. AUTOSAR FO Release **R22-11**(402), 1–337 (2020)
21. Koscher, K., Czeskis, A., Roesner, F., Patel, S., Kohno, T., Checkoway, S., et al.: Experimental security analysis of a modern automobile. In: 2010 IEEE Symposium on Security and Privacy, pp. 447–462. IEEE, Piscataway (2010)
22. Miller, C., Valasek, C.: A survey of remote automotive attack surfaces. Black Hat USA. **2014**, 94 (2014)
23. Miller, C., Valasek, C.: Remote exploitation of an unaltered passenger vehicle. Black Hat USA **2015**, 91 (2015)
24. Mazloom, S., Rezaeirad, M., Hunter, A., McCoy, D.: A security analysis of an in-vehicle infotainment and app platform. In: 10th {USENIX} Workshop on Offensive Technologies (2016)
25. Serag, K., Bhatia, R., Kumar, V., Celik, Z.B., Xu, D.: Exposing new vulnerabilities of error handling mechanism in CAN. In: 30th USENIX Security Symposium, pp. 4241–4258 (2021)
26. Alrabady, A.I., Mahmud, S.M.: Analysis of attacks against the security of keyless-entry systems for vehicles and suggestions for improved designs. IEEE Trans. Veh. Technol. **54**(1), 41–50 (2005)
27. Francillon, A., Danev, B., Capkun, S.: Relay attacks on passive keyless entry and start systems in modern cars. In: Proceedings of the Network and Distributed System Security Symposium. Eidgenössische Technische Hochschule Zürich, Department of Computer Science (2011)
28. Rouf, I., Miller, R.D., Mustafa, H.A., Taylor, T., Oh, S., Xu, W., et al.: Security and privacy vulnerabilities of in-car wireless networks: a tire pressure monitoring system case study. In: USENIX Security Symposium, vol. 10 (2010)

29. Dworkin M.: Recommendation for block cipher modes of operation: The CMAC mode for authentication. Special Publication (NIST SP), National Institute of Standards and Technology, Gaithersburg, MD (2016). https://doi.org/10.6028/NIST.SP.800-38B
30. Zalman, R., Mayer, A.: A secure but still safe and low cost automotive communication technique. In: Proceedings of the 51st Annual Design Automation Conference, pp. 1–5 (2014)
31. Willsky, A.S., Deyst, J.J., Crawford, B.S.: Two self-test methods applied to an inertial system problem. J. Spacecraft Rockets 12(7), 434–437 (1975)
32. Giraldo, J., Urbina, D., Cardenas, A., Valente, J., Faisal, M., Ruths, J., et al.: A survey of physics-based attack detection in cyber-physical systems. ACM Comput. Surv. 51(4), 1–36 (2018)
33. Jovanov I, et al.: Sporadic data integrity for secure state estimation. In: 2017 IEEE 56th Annual Conference on Decision and Control (CDC). IEEE, Piscataway (2017)
34. Adhikary, S., Koley, I., Ghosh, S.K., Ghosh, S., Dey, S., Mukhopadhyay, D.: Skip to secure: securing cyber-physical control loops with intentionally skipped executions. In: Proceedings of the 2020 Joint Workshop on CPS&IoT Security and Privacy, pp. 81–86 (2020)
35. Koley, I., Ghosh, S.K., Dey, S., Mukhopadhyay, D., KN, A.K., Singh, S.K., et al.: Formal synthesis of monitoring and detection systems for secure cps implementations. In: 2020 Design, Automation & Test in Europe Conference & Exhibition, pp. 314–317. IEEE, Piscataway (2020)
36. Ghafouri, A., Abbas, W., Laszka, A., Vorobeychik, Y., Koutsoukos, X.: Optimal thresholds for anomaly-based intrusion detection in dynamical environments. In: International Conference on Decision and Game Theory for Security, pp. 415–434. Springer, Berlin (2016)
37. Murguia, C., Ruths, J.: Characterization of a cusum model-based sensor attack detector. In: 2016 IEEE 55th Conference on Decision and Control, pp. 1303–1309. IEEE, Piscataway (2016)
38. Koley, I., Adhikary, S., Dey, S.: Catch me if you learn: real-time attack detection and mitigation in learning enabled CPS. In: 2021 IEEE Real-Time Systems Symposium, pp. 136–148. IEEE, Piscataway (2021)
39. Zhou, Y., Vamvoudakis, K.G., Haddad, W.M., Jiang, Z.P.: A secure control learning framework for cyber-physical systems under sensor attacks. In: 2019 American Control Conference (ACC), pp. 4280–4285. IEEE, Piscataway (2019)
40. Zhang, L., Chen, X., Kong, F., Cardenas, A.A.: Real-time attack-recovery for cyber-physical systems using linear approximations. In: 2020 IEEE Real-Time Systems Symposium, pp. 205–217. IEEE, Piscataway (2020)
41. Kong, F., Xu, M., Weimer, J., Sokolsky, O., Lee, I.: Cyber-physical system checkpointing and recovery. In: 2018 ACM/IEEE 9th International Conference on Cyber-Physical Systems, pp. 22–31. IEEE, Piscataway (2018)
42. Mohan, S., Bak, S., Betti, E., Yun, H., Sha, L., Caccamo, M.: S3A: secure system simplex architecture for enhanced security and robustness of cyber-physical systems. In: Proceedings of the 2nd ACM International Conference on High Confidence Networked Systems, pp. 65–74 (2013)
43. Zhao, C., Gill, J.S., Pisu, P., Comert, G.: Detection of false data injection attack in connected and automated vehicles via cloud-based sandboxing. IEEE Trans. Intell. Transp. Syst. 23, 9078–9088 (2021)
44. Mundhenk, P., Paverd, A., Mrowca, A., Steinhorst, S., Lukasiewycz, M., Fahmy, S.A., et al.: Security in automotive networks: lightweight authentication and authorization. ACM Trans. Des. Autom. Electron. Syst. 22(2), 1–27 (2017)
45. Zheng, B., Deng, P., Anguluri, R., Zhu, Q., Pasqualetti, F.: Cross-layer codesign for secure cyber-physical systems. IEEE Trans. Comput.-Aided Des. Integrated Circuits Syst. 35(5), 699–711 (2016)
46. Ewing, G.: Reverse-engineering a crc algorithm. https://www.cosc.canterbury.ac.nz/greg.ewing/essays/CRC-Reverse-Engineering.html. Accessed 06 Feb 2021
47. Tunga, R., Murguia, C., Ruths, J.: Tuning windowed chi-squared detectors for sensor attacks. In: 2018 Annual American Control Conference, pp. 1752–1757. IEEE, Piscataway (2018)
48. Page, E.S.: Continuous inspection schemes. Biometrika 41(1–2), 100–115 (1954)

49. Moura, L.D., Bjørner, N.: Z3: an efficient SMT solver. In: International Conference on Tools and Algorithms for the Construction and Analysis of Systems, pp. 337–340. Springer, Berlin (2008)
50. Ferdowsi, A., Challita, U., Saad, W., Mandayam, N.B.: Robust deep reinforcement learning for security and safety in autonomous vehicle systems. In: 2018 21st International Conference on Intelligent Transportation Systems, pp. 307–312. IEEE, Piscataway (2018)
51. Wang, Y., Huang, C., Zhu, Q.: Energy-efficient control adaptation with safety guarantees for learning-enabled cyber-physical systems (2020). arXiv:200806162
52. Artin, E.: The Gamma Function. Courier Dover Publications, New York (2015)
53. Jameson, G.: The incomplete gamma functions. Math. Gazette **100**(548), 298–306 (2016)
54. Lillicrap, T.P., Hunt, J.J., Pritzel, A., Heess, N., Erez, T., Tassa, Y., et al.: Continuous control with deep reinforcement learning. arXiv preprint arXiv:150902971 (2015)
55. Terry, J.K., Grammel, N., Black, B., Hari, A., Horsch, C., Santos, L.: Agent environment cycle games (2020) arXiv:200913051
56. Fan, C., Mathur, U., Mitra, S., Viswanathan, M.: Controller synthesis made real: reach-avoid specifications and linear dynamics. In: International Conference on Computer Aided Verification, pp. 347–366. Springer, Berlin (2018)

Gradient-Free Adversarial Attacks on 3D Point Clouds from LiDAR Sensors

Jan Urfei, Fedor Smirnov, Andreas Weichslgartner, and Stefan Wildermann

1 Introduction

In recent years, more and more research is being done in the area of automated driving and the first autonomous vehicles are in use already today. In California, for instance, Google subsidiary Waymo is launching its self-driving vehicles [17] in some cities and in Beijing, the company Baidu Inc. has introduced the first autonomous cabs [26]. But also for the drivers of non-autonomous cars, driving is made easier and safer by a steadily increasing number of assistance systems [13], which are envisioned to pave the road to fully autonomous cars.

SAE International has defined various stages (so-called levels) for the transition towards automated driving [27]. At the highest level, Level 5, a vehicle must be able to respond autonomously to every possible situation without the intervention of a driver. Increased levels of automation necessitate more and more information about the surroundings of the car, making it necessary to develop new sensors to improve the vehicle's perception capabilities.

J. Urfei
UL Method Park GmbH, Erlangen, Germany
e-mail: Jan.Urfei@ul.com

F. Smirnov
Robert Bosch GmbH, Renningen, Germany
e-mail: fedor.smirnov@de.bosch.com

A. Weichslgartner
CARIAD SE, Ingolstadt, Germany
e-mail: andreas.weichslgartner@cariad.technology

S. Wildermann (✉)
Friedrich-Alexander-Universität Erlangen-Nürnberg, Erlangen, Germany
e-mail: stefan.wildermann@fau.de

© The Author(s), under exclusive license to Springer Nature Switzerland AG 2023
V. K. Kukkala, S. Pasricha (eds.), *Machine Learning and Optimization Techniques for Automotive Cyber-Physical Systems*, https://doi.org/10.1007/978-3-031-28016-0_7

One of these sensors is a LiDAR system, which measures the distance to an object via the time of flight of the reflection of an emitted light beam [3]. In contrast to object detection via a camera system, a LiDAR sensor does not necessarily require perfect visibility conditions, therefore it can be used very well both during the day and at night [34]. The complete coverage of a vehicle's entire surroundings is often achieved by deploying a laser scanner that can move by 360° and mounting it on the roof of the vehicle. Alternatively, it is possible to install multiple immovable LiDAR sensors and then combining their data. For example, five LiDAR sensors were mounted on the roof of a vehicle for recording the Audi Autonomous Driving Dataset [9]. Almost all car manufacturers are expected to install such a LiDAR system in future vehicles with autonomous driving functions in addition to the existing sensor technology in order to detect three-dimensional objects and thus ensure safe automated driving [30].

Machine learning algorithms, especially deep neural networks, are becoming more and more common for the recognition and classification of 3D objects or 2D images. Hereby, their main objectives are to (a) recognize patterns—i.e., objects— in the given data and (b) classify these objects by mapping each recognized pattern onto one of multiple known classes, e.g., cars, pedestrians or bicycles. However, the inner logic of these algorithms is very difficult to interpret for humans, so that it is typically difficult or even impossible to understand how they arrive at a particular recognition/classification decision.

Due to its crucial role for the control of autonomous vehicles and the necessity to operate on data from outside the car, the LiDAR system is becoming a potential target for so-called *Adversarial Attacks*, where classification errors are induced by purposefully introducing changes to the input (see Sect. 2.3). The resulting classification errors can have negative consequences ranging from the display of erroneous warnings to accidents which could involve property damage or even the loss of human life [23, 30]. Consequently, it is of utmost importance that the classification models not only accurately classify the objects in the input data, but are also resilient to adversarial attacks.

In order to develop resilient classification algorithms, it is—in the first step— necessary to identify attack patterns to which the currently used classification algorithms are particularly vulnerable, to then—in the second step—research how the algorithms can be adapted to make them resilient to these patterns, e.g., using adversarial training techniques [10] (see Sect. 2.3.3).

In this work, we evaluate how gradient-free optimization methods can generate attack patterns in situations where none or very little information is available about the classification system. Since the implementation details of the classifying algorithms are typically kept secret by the manufacturers—and therefore, are unlikely to be available for the development of attack patterns—this attack scenario seems more realistic, so that the resulting insights are likely to prove more relevant than gradient-based techniques which require insights of the attacked machine-learning model.

1.1 Contributions

The manuscript at hand provides the following contributions to the research of adversarial attacks on neural networks used for segmentation tasks in the context of autonomous driving:

- formalization of adversarial attack for LiDAR as generic multi-objective optimization problem
- solving this optimization problem by a gradient-free optimization
- black box for adversarial attacks, works independent of used classifier
- experimental results providing insights about the suitability of evolutionary approaches and their encoding for the design of adversarial attacks on LiDAR

1.2 Outline

The remainder of this article is structured as follows. In Sect. 2, we give the background to the addressed research problem by introducing the usage and technology of LiDAR sensor data in the context of autonomous driving. We review adversarial attacks in the space of machine learning in general as well as existing attacks on LiDAR in specific. Afterwards, we specify the underlying threat model in Sect. 3 before we present our proposed approach in Sect. 4. In Sect. 5, we use LiDAR data from the real-world dataset KITTI to evaluate the proposed Evolutionary Algorithms (EAs)-based methodology to generate attack patterns. Section 6 concludes our work and outlines future research directions.

2 Background

This section presents the basic principle of LiDAR sensors, segmentation of their data as the basis for autonomous driving, as well as existing attacks on them.

2.1 Semantic Segmentation of LiDAR Data

Light Detection and Ranging (LiDAR) sensors use an optical measurement method to determine distances to objects and to locate them [20]. For this purpose, a light wave is emitted in the form of a laser beam pulse and the reflection of this light wave is detected again. Due to the constant speed of light, the distance to the reflecting object is then calculated based on the time-of-flight method. 3D LiDAR systems are increasingly propagated to be deployed for driver assistance systems. For this purpose, the sensor is mounted on a vehicle, usually together with

Fig. 1 Example of 2D segmentation (left): camera image (top) and 2D segmentation (bottom) of a data sample from [9]. Example of 3D segmentation of a LiDAR point cloud (right). Data sample is from [4]

cameras, in order to be able to provide the three-dimensional image with textures. Often, inclinometer and GPS sensors are also added in order to be able to correct the coordinates of the measured points. LiDAR sensors are gaining increasingly importance in autonomous driving, as they are able to capture high-resolution 360° 3D data and, in contrast to, e.g., cameras, are more robust to varying weather and lighting conditions.

The first fundamental step for interpreting data from LiDAR but also camera is *semantic segmentation*. A semantic segmentation represents a mapping between an object (pixel, 3D point) and a category (e.g., street, car, pedestrian, etc.). Sometimes this step is also referred as pixelwise or pointwise classification. We use both terms with the same semantic meaning in the remainder of this article. Examples of 2D and 3D semantic segmentations can be seen in Fig. 1. In the field of autonomous driving, semantic segmentations are enormously important because they can assign meaning to objects and thus form the basis for autonomous decisions. For example, semantic segmentation is often the base for object detection, object location [16], and further planning such as trajectory and motion planning [15].

Deep Learning models like deep neural networks are often used to calculate such semantic segmentations. Various networks exist, which produce different results. For example, [25] and [36] propose techniques where each individual point is assigned a class. In [11], an additional enclosing frame is computed for each object. In most cases, e.g., also for [11, 25, 36], these networks are a special kind of Convolutional Neural Networks (CNNs). The topology differs from network to network, but is also not relevant for the methodology presented in this section: It is the stated goal to consider the segmentator as a black box, so that no knowledge of how it works is required to apply the proposed adversarial attack.

2.2   Existing Attacks on LiDAR Sensors

Several attacks on perception for autonomous vehicles have already been inves-tigated. However, the study of attacks on perception based on LiDAR sensors is still very young. Attacks can happen on different levels, and thus there are also different classifications of attacks on LiDAR. For example, [5] classifies attacks as *sensor-level attacks* that directly target the manipulation of the sensors measurement scheme, *physical-world attacks* in which the environment is modified (e.g., by adding stickers on traffic signs), and *Trojan attacks* that directly target the software, particularly the neural networks. Shin et al. [28] classify the attacks according to the channels of the sensor: *Regular channel sensor attacks* correspond to attacks on the sensor level, *side-channel attacks* misuse any physical quantity to indirectly attack the sensor, whereas *transmission-channel attacks* target the sensor data when it is transmitted from the sensor to the system which processes it.

A big part of work on sensor-level attacks on LiDAR focuses on *spoofing attacks*. A LiDAR sensor works by sending a laser pulse and listening for its echo to determine the distance to an object. In a spoofing attack, the attacker sends a laser pulse with a second laser during the listening phase of the LiDAR sensor which will be interpreted as the echo from an object [24]. The attacker can thus create fake objects and obstacles or change the position of existing objects with this technique. To do this, however, the attacker must know the frequency at which the system operates, the order in which the environment is scanned, and the size of the listening window, since these parameters differ between the various systems. If this knowledge is known, the time can be determined at which a response pulse must be sent in order to make points appear at an arbitrary location from the LiDAR's point of view.

Besides spoofing, *saturation attacks* are further sensor-level attacks. Here, a powerful second laser is used to illuminate the LiDAR sensor [29]. This leads the measurement into a saturation so that the sensor becomes blind. Such attacks are consequently also called blinding attacks or jamming attacks.

Such attacks manipulate the sensor measurement. However, the goal of an attacker would not only be the mere modification of 3D points. Rather, he or she would specifically want to manipulate the vehicle's decisions in that way. As the perception and planning of autonomous vehicles strongly rely on sensors, attacks on LiDAR would have the potential to manipulate the reactive behavior of the vehicle. Examples are forcing the car into an emergency break which may harm the passengers, and freezing attacks in which the car does not move any more to cause traffic jams [5]. Meshcheryakov et al. [21] categorize such attack vectors as *adversarial machine learning attacks* when directly targeting the machine learning component of the perception system. In this context, Cao et al. [35] as well as Sun et al. [30] have shown that machine learning techniques for perception are robust to randomly generated fake points; it was not possible to effectively reach the goal of fooling the perception system when blindly spoofing points. Instead, it is necessary to search for effective patterns for spoofing fake points. As a consequence, [5] and

[30] propose *adversarial attacks* to find such patterns. The remainder of this section concentrates on such adversarial machine learning attacks, which have the aim to deceive the machine learning algorithms by applying artificial modifications to the input data.

2.3   Adversarial Attacks

The term adversarial attacks describe attacks that attempt to exploit vulnerabilities in the machine learning models [33]. While also cluster algorithms, regression models, as well as other classifiers like a support vector machine [33] are susceptible, we will focus on attacks on neural networks in the following.

The goal of an adversarial attack is to slightly alter (tamper) the input of a classifier resulting in a different outcome. These alterations are often not perceivable by the human eye [6]. The goal of neural networks is to generalize well and to be robust to small input perturbations. However, Szegedy et al. [31] showed that *adversarial examples* exist which exhibits exactly these attributes.

These examples are rooted in the non-linear nature of neural networks and the resulting *blind spots*. The non-linear layers (each layer with a non-linear activation function such as the ReLu or sigmoid function) of a neural network generalize between input and output. This is the case even for very different inputs of the same class. In the case of image classification, e.g., this could be an object seen from another side and thus has a different contour and features. If this case is not represented in the training data, the network has to abstract from the other training data to identify the object correctly anyway. This results in blind spots, because the classification is not based on any known training data but on a generalization. If this learned generalization is not correct, these blind spots can then be deliberately exploited or used to generate incorrect output [31].

Szegedy et al. [31], have thus shown for the first time that neural networks are vulnerable and that small changes in the input can achieve large changes in the output. A further step would be to check whether there are general patterns that always lead to the same change and are transferable to other networks. This will be discussed in more detail in a subsequent section. Figure 2 shows several examples from the MNIST dataset [14] where, despite a small change, the original figure is still recognizable to the human eye. The altered images were used as input to a CNN that is supposed to recognize the given number. The output of the CNN can be seen in the caption, showing that the alterations resulted in an incorrect classification.

2.3.1   Attack Goal

Adversarial attacks can be categorized into two classes *targeted* and *un-targeted attacks*. While the former tries to classify the input with a given output class, the

(a) Classified as 5   (b) Classified as 8   (c) Classified as 8   (d) Classified as 3

Fig. 2 Adversarial examples using the MNIST dataset. The left image represents the original image, the right image the corresponding attack. The number recognized by the neural network is shown in the caption. (**a**) Classified as 5. (**b**) Classified as 8. (**c**) Classified as 8. (**d**) Classified as 3

latter has the goal of any misclassification and therefore decrease the reliability of the model [33]. An untargeted attack has less conditions and is easier to perform [6].

These two categories can be also formalized [6]. If x is a valid input and x' is the altered input, y is the target class, and $c^*(x)$ the classification function, then follows for an untargeted attack:

$$c^*(x) \neq c^*(x') \tag{1}$$

We can extend Eq. (1) for a targeted attack as follows:

$$c^*(x) \neq y \quad \wedge \quad c^*(x') = y \quad \wedge \quad x \neq x' \tag{2}$$

The generation of an attack can then be seen as an optimization problem that tries to minimize the input perturbations (see for example Eq. (3), so that depending on the attack target the formula (1) or (2) applies. The p stands for the selected L_p norm.

$$\arg\min_{x'}\{||x - x'||_p\} \tag{3}$$

According to Carlini and Wagner [6], it is important to design an effective objective function with a good distance measure. However, current distance metrics are not perfectly reproducing differences experienced by the human perception. This leaves room for future research.

2.3.2 Known Information of the Attacker

Adversarial attacks can be further categorized depending on the given information of the model [33]: black-box, white-box, and gray-box attacks.

White-box attacks assume complete knowledge of the model [1]. This also includes all parameters and hyperparameters of the model [33]. Once this is known, there is an effective way to compute an adversarial example using the *fast gradient sign* method [10]. Since by white-box access the objective function is known, its derivation can be used in a back-propagation algorithm to optimize or create an attack like this.

If no exact information about the model and its parameters is available, this is referred to as a *black-box attack* [33]. Furthermore, it must then be clarified whether the attacker has the ability to make queries to the model. If the attacker is able to do this, he must at least know the structure of the input and output data in order to be able to correctly formulate the input and interpret the output for a query made to the model [33]. Often, however, the possible queries to a model are limited within a time period and it is a matter of having to make as few queries as possible [7].

An intermediate solution, but rarely used in existing literature, is a *gray-box attack*. In this, the attacker has incomplete information about the system [33].

2.3.3 Adversarial Training

In literature, many counter measurements against adversarial attacks were proposed. While a lot of these mechanisms were broken, adversarial training seems to add certain robustness. Adversarial training describes the attempt to better protect neural networks against adversarial attacks by giving the predicted data from the normal neural network to another network that uses that classification instead of the original classification from the training data [6]. The second model then learns the behavior of the first model including its already learned knowledge [6]. Adversarial training can therefore also be seen as active learning, where new training data is generated and classifications are automatically applied [10].

Furthermore, the deliberate use of adversarial examples, in conjunction with the original classifications, leads to a better ability of the network to generalize [31]. With a better ability to generalize, the network is at the same time more robust against minor changes.

Normally-trained neural networks are by themselves not protected against adversarial attacks [10]. However, if adversarial examples are included in the training process (for example by an objective function based on the *fast gradient sign* method and including the adversarial example), this leads to a noticeable improvement of the error probability of the network [10]. Adversarial training, however, still does not prevent the existence of adversarial examples [6].

2.4 Adversarial Attacks on LiDAR Point Clouds

In [35], attacks on 3D point classification are performed by modifying existing points of the point cloud and adding new points. The classification model is the PointNet deep neural network. The goal of the attack is to generate a point cloud that is as little different as possible from the original input, but causes the classifier to assign the input an incorrect class label. The authors of [35] formulate this as an optimization problem and use a gradient-based optimization algorithm to determine the adversarial point cloud.

Cao et al. [5] and Sun et al. [30] investigate adversarial machine learning on LiDAR-based point cloud classification in the context of autonomous driving. Both works have the goal of fooling the system into classifying a vehicle in close range of the attacked vehicle, as these can cause an emergency stop or a vehicle freezing. Moreover, in both works a maximum of 200 points are added to the point cloud. [5] has empirically evaluated how many points can be practically spoofed by such an attack. While this depends on the angle between the attacking laser (used for spoofing points) and the LiDAR sensor, 60 points could be generated reliably. As a consequence, both try to cause a false classification with as few additional points as possible.

Sun et al. [30] observed that LiDAR measures only few points of cars which are partially occluded, e.g., by other cars, and vehicles that are very distant. They therefore took point clouds of partially occluded and distant cars from LiDAR measurements as well as generated some artificially with a 3D renderer. Then, they added these point clouds into pristine LiDAR measurements and found out that the attack has a very high success rate of the point cloud classifiers detecting fake cars in the close vicinity.

Cao et al. [5] follow an optimization-based approach to find adversarial points by minimizing an adversarial loss function. The formalization of this function is based on the functionality of the applied machine learning algorithm. They solve the corresponding optimization problem using an Adam optimizer in conjunction with global sampling. The global sampling is needed to reduce sticking in local optima, which in a gradient-based method like the Adam optimizer can lead to the optimization converging to sub-optimal solutions. A major disadvantage is that by using such a gradient-based method, knowledge of the model used is necessary. In contrast, Sun et al. [30] treat it as a black-box problem, as this is more similar to a real scenario.

Also physical-world attacks on LiDAR-based perception systems have been investigated. In general, adversarial examples are special input patterns with the goal of leading a neural network to a wrong classification. Such adversarial examples can also be objects in the physical world scanned by the sensor. For example, Tu et al. [32] present an adversarial example that can be physically realized, e.g., by means of a 3D printer. When mounting the object on the rooftop of a car, this car will be entirely hidden from the LiDAR-based perception of other autonomous vehicles.

3   Threat Model

The state-of-the-art approach to estimate the security risk a system imposes is a Security Risk Assessment (SRA). For example, in the automotive domain, Modular Risk Assessment (MoRa) [2] is used to be compliant with ISO/SAE 21434 [19]. Recently, specific risk assessment frameworks for autonomous driving, like SARA [22], were proposed. An SRA consists of identifying assets, security goals, threat analysis, an attacker model, damage potentials, and an overall risk

Table 1 Attacker types according to [22]

Attacker	Expertise	Knowledge	Equipment
Thief	Layman	Public	Standard
Evil Mechanic	Expert	Restricted	Specialized
Organized Crime	Proficient	Sensitive	Specialized
Hacktivist	Experts	Sensitive	Multibespoke
Researchers	Experts	Public	Specialized
Foreign Government	Experts	Critical	Multibespoke

level. Classical security goals, known as the CIA triad, are confidentiality, integrity, and availability. As these three attributes are often insufficient, the authors of [22] extended the model with authenticity, non-repudiation, authorization, unlinkability, and trustworthiness. These security goals are targets of certain threats. In the classic STRIDE model [12] threats are: spoofing, tampering, repudiation, information disclosure, denial of service, elevation of privilege. They origin from a certain attacker type who has certain knowledge, expertise, and equipment/budget. Table 1 gives an overview of various attacker types and their capabilities. In the case of our work, the threat of tampering the input data of the neural network pointwise classifier would target the integrity of the segmentation. This could have safety implications, as for example, classifying a pedestrian or an e-bike rider as driveable area could have severe consequences. We assume that the attacker has public knowledge, like the researcher attacker (see Table 1), hence a black-box attack (see Sect. 2.3.2), as the intrinsics of commercial classifiers used in autonomous driving are only available to the manufacturer. Treating the neural network as a black box is more realistic and makes the attack more achievable than a white-box attack. Further, we assume that the attacker can directly manipulate the 3D point clouds that serve as inputs to the semantic segmentation neural network. The format of the 3D points can be obtained by sniffing the communication between the LiDAR sensor and the black box or by reverse engineering the interface. Besides, we have no time constraints on creating tampered input points, i.e., we target an offline attack.

4 Gradient-Free Adversarial Attacks on Semantic Segmentation of LiDAR Data

In this section, we formalize the adversarial attack on semantic segmentation of LiDAR point clouds as a multi-objective optimization problem. We summarize the main principles of optimization and give a short outline of the functionality of Evolutionary Algorithms (EAs), before presenting our gradient-free adversarial attack methodology using EA-based optimization.

4.1 Adversarial Attack as Multi-Objective Optimization Problem

Given a *3D point cloud* $X = \{x_i | x_i \in \mathbb{R}^3, i = 1, \ldots, n\}$. Each of the n 3D points $x_i \in X$ can be uniquely identified by identifier i. Moreover, let $\mathcal{X} \subseteq \{\mathbb{R}^3\}^n$ denote the domain of 3D point clouds. *Semantic segmentation* is given as a function $c : \mathcal{X} \to \mathcal{Y}$ with $\mathcal{Y} \subseteq \mathbb{Z}^n$ that maps a point cloud $X \in \mathcal{X}$ to a segmentation $Y = c(X) \in \mathcal{Y}$. Particularly, each 3D point $x_i \in X$ is assigned a corresponding class label $y_i \in Y$.

Attack Capabilities In general, a 3D LiDAR scan can be attacked by (a) modifying existing points, (b) removing points, and (c) adding points, where (a) can be realized, e.g., by spoofing attacks and (b) by blinding attacks. In the following, we concentrate on (a) by modifing existing points of the original scan.

Formally, with X being the pristine scan, we modify a subset $T \subseteq X$ of 3D points according to a transformation π. The modified points are given as $T' = \pi T$. Each of the attacked points $x_i \in T$ is modified by vector π_i, thus $x'_i = x_i + \pi_i$. The attacked 3D scan is given as $X' = X \setminus T \cup T'$ and the resulting segmentation is $Y' = c(X')$.

The degrees of freedom for the attack are thus given as (i) the selection of the set T of points to modify as well as (ii) the transformation π to apply on these points.

Attack Goals The first goal, as usual for adversarial attacks, is to minimize the perturbation inflicted by the attack. Ideally, the modification is not recognizable by the human eye. We can use an L_p norm to measure the *distance* between the two point clouds X and X'. The L_p norm is a commonly used metric for adversarial perturbation of fixed-shape data [35]. The first goal is therefore to minimize the distance between both point clouds:

$$\text{minimize} \left\| X - X' \right\|_p. \tag{4}$$

The second goal is to cause *misclassification* or *classification errors* due to the modified points. In this section, we focus on an untargeted attack. Thus, we aim at changing the label of as many 3D points as possible with our attack. With Y being the segmentation of the original scan and Y' the segmentation of the attacked scan, this objective is formulated as

$$\text{maximize} \left| \left\{ x_i \in X \cap R | y_i \in Y \wedge y'_i \in Y' \wedge y_i \neq y'_i \right\} \right|. \tag{5}$$

Here, R is a *region of interest* and only the labels of points $x_i \in R$ within this region are evaluated. In general, this region can cover the complete scan, thus containing all points $x_i \in X$. In some cases, however, it may be of interest to target a specific region of the point cloud, e.g., a region containing another car, a pedestrian, or an e-bike rider.

The attack can thus be perceived as a *multi-objective optimization problem* with the two objectives in Eqs. (4) and (5). Both objectives are conflicting, that is, an improvement in one can lead to the deterioration of the other. We thus first give an introduction to multi-objective optimization and EAs as a meta-heuristic for solving them, before presenting our gradient-free adversarial attack methodology.

4.2 Evolutionary Algorithms for Multi-Objective Optimization

An optimization problem is defined by a single objective function O or multiple objective functions O_k, $k = 1, 2, \ldots$ which represent some system properties resulting from a particular action. The set of all possible actions which can be taken constitutes the *search space*. The optimal solution is the action which is preferable to any other solution within the entire search space. The notion of preference depends on the optimization problem and, in particular, on the number of objective functions: With one minimization objective function $O(s)$, solution s_i is preferable to solution s_j iff it is at least as good w.r.t. the objective function, i.e., iff $O(s_i) \leq O(s_j)$.[1] In case of multiple objective functions, the notion of preference between problem solutions is established using the concept of *dominance*. Solution s_i is said to *weakly dominate* solution s_j iff s_i is at least as good as solution s_j w.r.t. all objective functions. Furthermore, solution s_i is said to *strongly dominate* solution s_j iff s_i weakly dominates s_j *and* s_i is better than s_j in at least one objective function. In a multi-objective optimization problem, all solutions which are not strongly dominated by any other solution in the search space are considered *Pareto-optimal* and are summarized to the so-called *Pareto front* of the optimization problem.

Brute-force enumeration of the search space is impractical for very large search spaces and/or objective functions which require prohibitively long computation times for each considered action. Typically, such problems are solved using meta-heuristic optimization approaches which only consider a small subset of the search space but cannot provide guarantees about the optimality of the found solutions. *Gradient descent* (gradient ascent in case of maximization problems) is an example of a metaheuristic optimizer. In this approach, a problem is optimized by iteratively selecting a single solution and determining its quality and the gradient of the objective function at the corresponding point of the search space. The solution in the next iteration is then selected by following the direction for which the objective value grows/decreases (in case of a maximization/minimization problem, respectively). Approaches based on gradient descent have recently gained great popularity, in particular for the training of neural networks (which can be considered as an optimization problem minimizing the loss function). While they do have many

---

[1] The phrasing in this explanation refers to a minimization problem. An explanation focused on maximization problems is omitted since minimization problems can be trivially transformed into maximization problems and vice versa.

advantages, their application is restricted to problems with (a) a single objective function which is (b) explicitly known and (c) differentiable. Since in this work, the generation of attack patterns is interpreted as a multi-objective optimization problem whose objective functions can only be accessed as a black-box, the presented approach is instead based on an EA.

Evolutionary Algorithms (EAs) are population-based iterative metaheuristic optimizers which tackle optimization problems by emulating biological evolution. The optimization problem is interpreted as an iterative adaptation of a population (a set of evaluated problem solutions), where the fitness of individuals is measured by the objective function(s). In each iteration, a group of individuals is selected as parents, with individuals with higher fitness having a higher probability of being selected. These parent individuals are then used to create new individuals by means of *crossover* and *mutation* operators which mimic the reproduction in nature. These new individuals are then used to update the population, where low-quality individuals from previous iterations are replaced with high-quality individuals generated in the current iteration, hereby improving the average fitness level of the whole population.

The individuals of the population are represented by a *phenotype* and a *genotype*. The phenotype of the individual is a representation which can be processed by the objective function(s), while the genotype is typically represented as a simple data structure which can be easily stored, deconstructed into subparts, and modified— such as a bit- or an integer string. This "genetic" representation of problem solutions makes it possible to use problem-independent recombination operators to create new, previously unknown problem solutions, while at the same time making sure that these solutions will be of similar or even better quality than their parents.

In the context of this work, the usage of EAs is mainly motivated by their independence from the objective function of the optimization problem: In contrast to, e.g., gradient-based approaches, EAs use the objective function merely as a look-up table to determine the quality of individuals without, however, requiring any knowledge of the function or relying on any assumptions about its properties such as linearity or continuity. Furthermore, operating on a population of solutions and, in particular, generating new individuals based on a diverse set of known solutions, enables EAs to direct the exploration into multiple search space directions. Compared to optimizers operating based on a single solution, EAs are, thus, typically much more effective at optimizing problems with multiple objectives. Consequently, EA are a natural fit for the problem addressed in this work. In the following, we describe how to solve the gradient-free adversarial attack described before with an EA.

4.3 *Gradient-Free Adversarial Attack Methodology*

An EA can be used to generate adversarial attacks by solving the proposed multi-objective optimization problem. Figure 3 illustrates the methodology. The

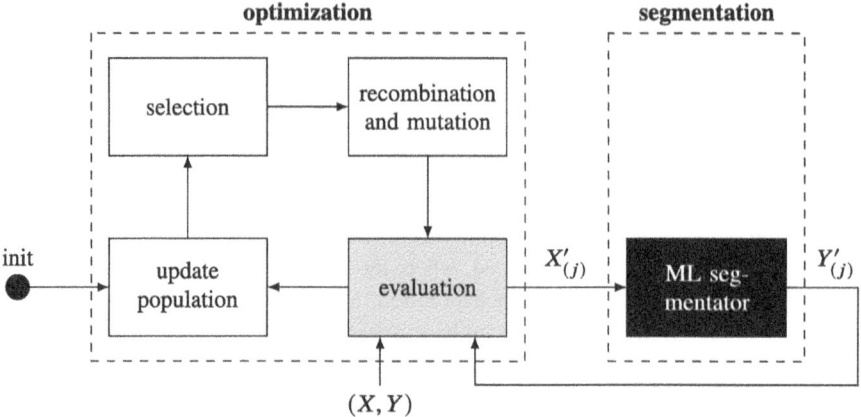

Fig. 3 Schematic illustration of gradient-free adversarial attacks on LiDAR classification using EAs

optimization part (left) consists of the EA-based optimization loop. A population of solution candidates is iteratively updated by first selecting individuals from the population for reproduction based on their fitness values. Offsprings are created by recombination of the genotype information of the selected individuals and by mutation. The evaluation step evaluates the objective functions according to Eqs. (4) and (5) for each offspring. The fitness is based on the obtained objective function values and the population is updated with the new solution candidates.

EAs as meta-heuristics can be adopted for solving arbitrary single-objective and multi-objective combinatorial optimization problem. However, it is necessary to provide a "genetic" encoding (genotype) of the search space. In case of the presented adversarial attack, the search space is defined by choosing (i) the set T of attacked 3D points, and (ii) the transformation π applied on them. Sect. 4.4 presents two options for encoding this search space.

Let j be a specific individual of the population of the EA with $T_{(j)}$ and $\pi_{(j)}$ resulting in attacked point cloud $X'_{(j)}$. The evaluation of the distance objective in Eq. (4) requires to compute the distance between $X'_{(j)}$ and the pristine point cloud X. The computation of the classification error objective in Eq. (5) is based on the class labels obtained when applying segmentation function c on $X'_{(j)}$. As EAs do not require any gradients of objective functions, we treat the machine-learning model c as a black box and simply apply it on the attacked scan to obtain the class labels $Y'_{(j)} = c(X'_{(j)})$. With this information, the EA computes the fitness of j.

The encoding of the search space is fundamental. In fact, the choice of encoding influences the specific attack capabilities on the one hand. On the other hand, it has a direct impact on the size of the search space, and with that, the convergence of the optimization process towards optimized solutions. We next present two encoding alternatives of the adversarial attack.

4.4 Encoding

There exist multiple degrees of freedom to encode the selection of attacked 3D points T and the chosen transformation π of the adversarial attack. We propose two approaches for encoding this scheme. The first encoding enables to choose a fixed amount of 3D points and to select the modification of their position. The second encoding describes the shape and position of a cuboid in space. All 3D points contained within this volume will be modified according to a transformation vector that is also encoded. Both approaches are described next.

4.4.1 Manipulation of Arbitrary 3D Points

The first alternative is a straight-forward encoding of the search space. It encodes the selection of a fixed amount of 3D points as subset of the LiDAR scan and the transformation of each point of the subset in each axis individually. The points are identified with their identifiers (denoted by ID in the following) and the concrete selection of points will be a degree of freedom during optimization. Always the same amount of points will be modified (except one point is chosen twice). By default the amount is set to 60 points, because Cao et al. [5] have empirically shown that it is possible to reliably modify 60 points in a practical spoofing attack. This value can be set as parameter for the optimization and is denoted by m in the following. Each of the selected m points can be manipulated individually and the respective transformation vectors are the second degree of freedom for the optimizer. Figure 4 visualizes the structure of the encoding. The genotype contains m fields for the selection of 3D points by their IDs, and $m \times 3$ fields that contain a transformation vector for each of the selected points. The first field encodes the modification of the first point on x-axis, the second the modification on y-axis, and the third on z-axis. The next triple belongs to the second point and so on.

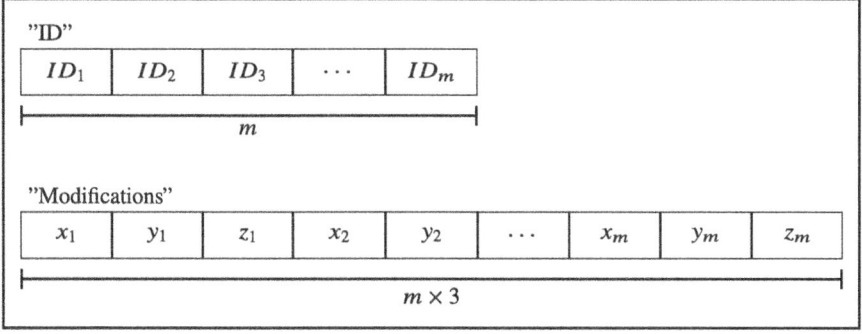

Fig. 4 Setup of one individual within the approach of manipulate individual points

When the EA is initialized before the start of the optimization loop, an initial population has to be generated by randomly setting genotype instances. For the random initialization of the ID fields, a uniform distribution over all n points of the 3D point cloud is used. Also the transformation vectors are randomly initialized based on a uniform distribution, but this time over the maximum boundaries of the space captured by the LiDAR sensor.

Every LiDAR sensor has a maximal range and a field of view, which depends on the sensor itself and its mount point at the vehicle. After applying the transformations on the selected points, it is possible that a point is outside of the space that is captured by the sensor. This means the point would not appear in the recorded scan and therefore not in the interpreted segmentation. To prevent this case, a clipping into the visible area is implemented. In a first step the (x, y)-coordinates are checked (i.e., the position in the horizontal plane) and possibly set to the borders of the visible area. The vertical field of view depends on the distance of 3D points to the sensor and the sensor parameters such as the viewing angle. Due to the physics and geometry of the perceptual field of the sensor, the greater the distance to the sensor, the greater the range of vision it covers. In a second step, the point is thus clipped into the vertical perceptual area of the LiDAR sensor. This approach is comparable to the clipping done in [6].

Additionally, it is recognized if the same point is selected multiple times. In this case, the point is modified only once. This is the only case, where less than the defined amount of points will be edited.

As mentioned above, the choice of an encoding also affects the size of the search space. For our analysis of the search space encoded by the presented genotype, only the combinations for selecting the m 3D points are in focus without considering the selection of the point transformations. This encoding chooses an amount of m points from an overall of n 3D points within the original scan. The number of possible combinations is thus given by n^m. To give a better understanding of the complexity from a practical point of view, the LiDAR scans used in our experimental evaluation contained $n = 150\,000$ 3D points. A value of $m = 60$ will yield to $150\,000^{60} \approx 3.676\,847 \times 10^{310}$ possibilities. Even though only the selection of points is considered for this complexity analysis, this alone represents an immense search space. For any optimization algorithm, it is hard to deal with such a big space and it will be hard to find good solutions.

4.4.2 Manipulation of 3D Points within Cuboid

Selection of individual points results in an immense search space. As second alternative, the points are not individually selected by their ID, but rather through their position in space. The selection of points is done via a cuboid whose size and position can be freely varied. All points of the scan that are within the volume defined by the cuboid are selected. The position as well as the size of the cuboid can be determined by the EA. The same transformation vector is applied to all points within the volume. In difference to the first approach, the amount of modified points

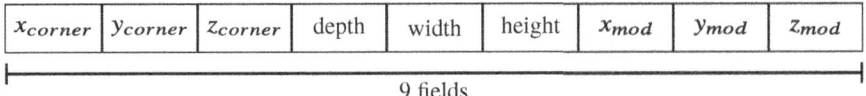

9 fields

Fig. 5 Architecture of a solution for modifications inside a volume in space

is not fixed but depends on how many points fall into the volume. This not only depends on the cuboid size but also its position because the density of points in the environment is not uniformly distributed and usually much denser directly in front of a vehicle. As illustrated in Fig. 5, the complete encoding only has nine values, which can describe an attack: the coordinate of one corner of the cuboid, its width, depth, and height, and finally the transformation vector to apply on the selected points.

While with the first encoding, the number of modified points is bounded, the selection via a cuboid volume could result in an arbitrary amount of points, ranging from a shape containing zero points to a shape containing the complete point cloud. To steer the optimization towards more practical attacks, it is thus also possible to specify constraints for feasible cuboids, e.g., that they contain a minimal number of points as well as constraints on the maximum cuboid size. Whenever offspring is obtained through recombination that violates the constraints, it is marked as infeasible and not used for advancing the population. We evaluate the impact of such constraints on the attack in Sect. 5.4.

For generating the initial population, genotype instances are randomly initialized as follows. The first values for the position are randomly selected according to a normal distribution. The remaining values are initialized within a uniform distribution. For the values which define the cuboid size, the interval borders are set to cover the maximum boundaries of the space captured by the LiDAR sensor. When constraints are defined regarding the cuboid size, an infeasible genotype instance will be initialized repeatedly until the constraint is fulfilled. The same clipping procedure as presented for the previous encoding is applied to ensure that transformed points are in the visible space of the sensor.

The analysis of the size of the search space has to consider which point combinations can be selected by different cuboids. As for the first analysis, the point transformation is not considered, even though there are less combinations than in the first encoding as all points are modified according to the same vector. So the solution space depends on the possible sizes and positions of the cuboid. The spanning corners of the cuboid (a rectangle can be described by two opposite corners) could be every pair of 3D point from the scan. With knowledge about the maximal range and resolution of the sensor, the number of points that are in the horizontal plane spanned by the x-axis and y-axis are calculated with:

$$\frac{x_{max} - x_{min}}{\Delta x} \cdot \frac{y_{max} - y_{min}}{\Delta y} \tag{6}$$

Here, x_{min} and x_{max} are the minimum and maximum measurable distances along the x-axis. For a LiDAR which is installed on the roof of a vehicle and takes a 360° measurement, x_{min} should have the same absolute distance as x_{max}. The same applies to y_{max} and y_{min}. Accordingly, $x_{max} - x_{min}$ results in the measurable interval in x direction. The resolution Δx and Δy is measured per axis and indicates how fine-grained points can be recognized. For a resolution of two centimeters for a LiDAR sensor as used to obtain the scans from our experiments, there would be 1.44×10^8 possible points in the horizontal plane described by the x-axis and y-axis. When additionally considering the vertical field of view of the sensor, i.e., the z-axis, this will be $\approx 4.32 \times 10^{11}$ possible points as an over-approximation. The amount of possible points has to be squared because the cuboid is described by two corners. This will lead to $\approx 1.866 \times 10^{23}$ different solutions (i.e., cuboids). In comparison with the first encoding, this search space is much smaller and therefore better tractable for optimization algorithms.

4.5 Evaluation

For evaluating an individual j, the first step is to decode the information encoded in the genotype. When the set of selected points and the transformation vector(s) are obtained, the transformation is applied resulting in attacked point cloud $X'_{(j)}$, i.e., the phenotype. For the evaluation of the classification error objective in Eq. (5), we defined a Region of Interest (RoI) R. In the general case, this would cover the complete space captured by the LiDAR sensor. For the evaluation, the class labels of all points would be considered. However, the RoI enables attacks on concrete objects in a point cloud like pedestrians, cyclist, or cars. The encoding will be the same as presented before, i.e., all n points of the point cloud can be selected for manipulation. The difference is in the evaluation of $X'_{(j)}$. For the evaluation of the number of classification errors acc. to Eq. (5), we only consider those points which lie in the RoI and whose class labels have changed, i.e., all points $x_i \in X \cap R$ with $y_i \neq y'_i$. In contrast, for calculating the distance metric acc. to Eq. (4), all points are included no matter whether they are inside or outside the RoI.

4.6 Implementation Details

The optimization of the attack can be divided into two larger sections in general as illustrated in Fig. 3: the optimization loop itself and the segmentation of the attacked LiDAR scan. For the optimization, the Java-based framework Opt4J [18] is used. As the framework should be able to work with any machine-learning model or even other techniques for semantic segmentation, we chose to provide segmentation as a service on a local web server, accessible through a standardized Internet Protocol (IP). The advantage of this approach is that the segmentation is encapsulated and can be exchanged easier, resulting in a high flexibility and interchangeability. The communication between the segmentation and the optimization framework is provided via HTTP, and thus works independently from the used programming language of the segmentator. Ideally, both optimizer and segmentator are recommended to be on the same machine to avoid communication becoming the bottleneck, as the EA has to access the segmentation very often during evaluation.

5 Experiment Results

5.1 Hypotheses and Focus of Experiments

In this section, the presented adversarial attack methodology is evaluated experimentally. The main goal of the experiments is to assess whether it is possible to introduce attack patterns resulting in a change of the classification results and whether the EA is an appropriate tool to achieve this. Furthermore, the different encodings are compared to investigate how they differ in effectiveness. The hypothesis and research questions to be validated by the experiments are:

- How does the employed norm influence the result?
- How do the different encodings differ in effectiveness?
- Are EAs effective to generate adversarial attack patterns?

For the following experiments, the convolutional neural network *SqueezeSegV3* [36] was used. However, the experiments can also be performed with any other semantic segmenter, since it is assumed to be a black box in which only the data formats have to match.

5.2 Run Time and Effectiveness

During the generation of the attack pattern, the time required for the classification of a 3D point cloud is the main bottle neck, since this operation has to be run for each considered pattern. The time required to generate such an attack is highly dependent on the underlying computer architecture and is essentially determined by the time required to classify a 3D point cloud. For example, on a computer with an Intel i7-8700 processor, an Nvidia GeForce RTX 2080 with 8GB of graphics memory as the graphics card, and 32GB of RAM, classifying a point cloud takes 0.12 s on average.

For all experiments presented in the following, we used an implementation of the NSGA2 algorithm [8] provided by the OPT4J framework [18] as the EA. The algorithm was used with a population size of 100 individuals, with 25 new individuals generated in each iteration. The time per iteration including classification and all further calculations took approx. 11.5 s. The following experiments have been optimized over 10,000 iterations, which thus corresponds to a total run time of about 32 h.

5.3   *Manipulation of Arbitrary 3D Points*

The first experiment evaluates the encoding presented in Sect. 4.4.1. In this experiment, we used the KITTI data set [4], which amongst others includes LiDAR point clouds recorded from a driving car. We executed the gradient-free attack methodology to generate attack patterns by modifying up to 60 3D points, with a maximal displacement of 3 m in each direction. The attack patterns in this experiment were generated as untargeted attacks, and we used and compared the L_1 and the L_2 norms for evaluating the distance objective function from Eq. (4).

Figure 6 illustrates the change of the two design objectives (number of classification errors and distance between pristine and attacked scans) over the course of the optimization when using each of the two norms. Figure 6a displays the number of classification errors, whereas the distance is depicted in Fig. 6b. The x-values correspond to the iteration of the optimization in both diagrams. The y-axis shows the values of the respective objective functions. In each case, the figure shows the number of misclassifications or the distance objective of the individual that achieved the best corresponding value up to the respective iteration. They also display the average of each objective over all non-dominated solutions found up until the respective iteration of the optimization.

The two metrics result in a similar course of the optimization, where the objective values of the generated solutions exhibit a rapid convergence towards (local) optima in the initial phase of the optimization, followed by only minor improvements in

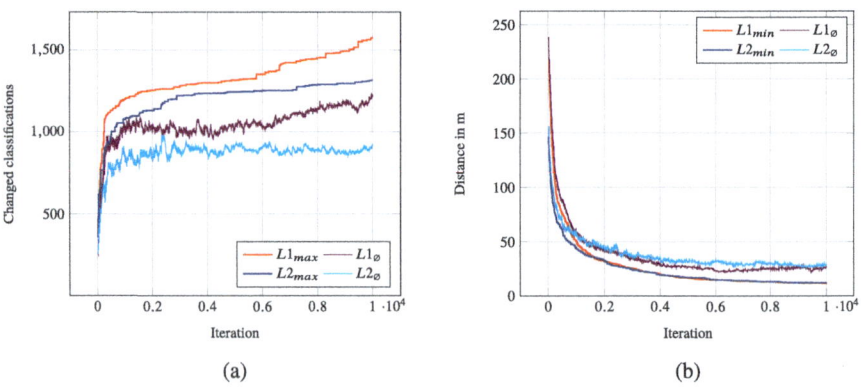

Fig. 6 Results using the encoding for arbitrary point modifications (see Sect. 4.4.1). (**a**) Maximizing the incorrect classifications. (**b**) Minimizing the perturbation

the later optimization stages. These results, thus, suggest that an EA is capable of improving attack patterns without relying on gradients of the objective functions.

The two investigated distance metrics lead to similar results w.r.t. the achieved number of classification errors, with the L_1 metric being slightly better on average. The differences between the two metrics w.r.t. the distance between pristine and attacked point cloud are even smaller. This could be caused by the fact that in this particular case, the metrics produce very similar results, since the distances in each dimension are close to 1. It is, however, important to point out that the same values of the metrics do not necessarily suggest identical modifications. For instance, a displacement by (0.5; 0.5; 0.5 m) would result in a value of 1.5 m when using the L_1 metric, while a calculation with the L_2 metric would yield 0.87 m. The choice of the distance metric, therefore, determines how much a certain modification influences the design objective quantifying the overall perturbation.

Based on the depicted results, it is also possible to draw conclusions about the functionality of the segmentation algorithm. Figure 6a shows that the best attack patterns were able to change the classification results for between 1300 and 1550 3D points. Thus, since only 60 points were modified in this experiment, it follows that the modification of a single point can have an impact on the classification results of a large number of other points. Figure 7 shows the final Pareto fronts obtained when using the L1 norm (red) and the L2 norm (blue). Each point represents the number of misclassifications and the distance of one non-dominated individual at the end of the optimization. Naturally, a larger modification of the 3D points also results in a larger number of classification errors. Figure 7 also shows that an attack pattern generated with this encoding results in 1574 and 1316 classification errors if each of the selected 60 3D points from a LiDAR scan is changed on average by $\frac{52.50\,\text{m}}{60} \approx 0.88$ m (L1-norm) and $\frac{61.65\,\text{m}}{60} \approx 1.03$ m (L2-norm), respectively.

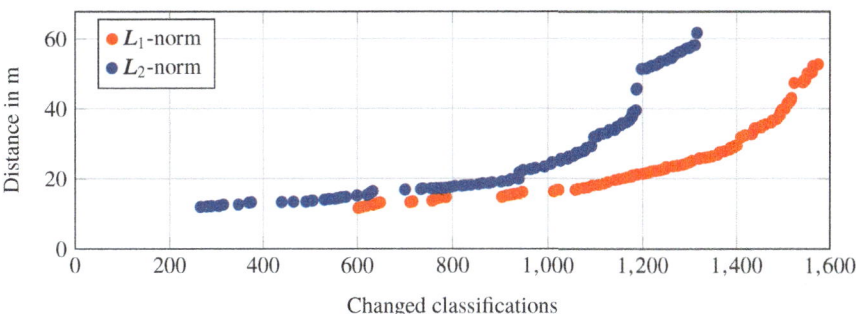

Fig. 7 Pareto fronts of the individuals at the end of the optimization using the encoding for arbitrary point modifications

5.4   *Manipulation of 3D Points Within Cuboid*

In addition, we evaluated the adversarial attack when using the cuboid-based selection of the 3D points for manipulation. To keep the results comparable, we used the same LiDAR scan as in the previous experiments. The maximum extend to which a single point could be modified w.r.t. a single dimension was also constrained to 3 m and, just like in the previous experiment, the focus was on untargeted attacks comparing the L_1 and L_2 norms for evaluation of the distance metric in Eq. (4). We evaluated an attack in which no constraint on the size of the cuboids was set, and an attack with the constraints that there are at least 30 points inside the cuboid and the maximum cuboid size is $125\,\mathrm{m}^3$. Figure 8 shows the results of the unconstrained attack, whereas Fig. 9 shows the results of the constrained attack, however, only for the L_2 norm.

The two plots in Fig. 8 show the course of two optimizations using the encoding presented in Sect. 4.4.2 without constraints on the cuboid size. The x-values correspond to the iteration of the optimization in both diagrams. The y-axis shows the values of the respective objective functions. In these plots, the (a) best (w.r.t. the number of misclassifications and the distance) individual found up until the respective iteration and (b) the average over all non-dominated solutions found so far ar shown are shown in (a) red and (b) purple, respectively, when using the L_1 metric for the distance objective. The plots illustrating the optimization using the L_2 metric highlights the best solution in blue and a light blue for the average over all non-dominated solutions. Figure 8a shows the objective function measuring the number of classification errors. The perturbation required to achieve this is shown in the diagram on the right (Fig. 8b).

It is noticeable that when only the number of classification errors is considered, the two metrics do not differ significantly from each other. Both the best solution and the average are close to each other. On average, at the end of the optimization,

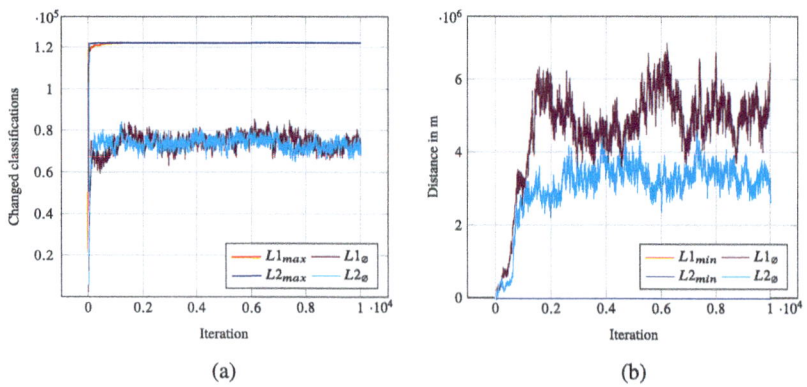

Fig. 8 Experiment results when using a cuboid to select the attacked 3D points (see Sect. 4.4.2). (**a**) Maximizing the incorrect classifications. (**b**) Minimizing the perturbation

classification errors can be introduced for approximately 75,000 points ($\approx$ 61 % of all points). The sum of all point shifts in an attack is on average 5500 km (L_1 norm) or approx. 3500 km (L_2 norm). Hereby, it is difficult to estimate the number of the 3D points that were actually modified. The best solution in both experiments in terms of distance is from the beginning a solution that hardly modifies the LiDAR-scan and therefore has a distance of close to zero. Since the corresponding individual is nearly optimal w.r.t. the distance objective, it can hardly be dominated and remains in the set of non-dominated solutions throughout the entire optimization. The average of the solutions in Fig. 8b is subject to a lot of noise. This possibly suggests an excessively large search space.

Immediately after the first iterations (for both L_1 and L_2), a non-dominated solution is found that is close to the achievable maximum regarding the number of classification errors (all points in a scan, in this case 122,526). At this point, it should be noted that the perturbation (distance) necessary to create this solution is very high influencing the average curve. Since the distance increases rapidly, although it should be minimized, and Sect. 5.3 has shown that it is correlated to the achieved number of classification errors (see Fig. 7), it can be assumed that the majority of the solutions of a population at the beginning of the optimization are oriented towards the solution with the maximum amount of misclassifications

It is also noticeable that after approximately 2000 iterations the values of the objective functions no longer improve, but roughly stagnate. Overall, such attacks do lead to a large number of classification errors (i.e., misclassifications). However, the underlying attack patterns hardly represent sensible or feasible attacks, since they require vast modifications and, given the number of classification errors, have to cover a large part of the measurable range (assuming that the ratio between number of classification errors and the number of modified points remains roughly the same as the ratio observed in Sect. 5.3 ($\approx$ 25)).

The two plots in Fig. 9 show the course of an optimization with the same parametrization as in the experiments in Fig. 8. The difference is that in this experiment the constraints on the generated cuboids were used.

The values of the objective functions are also plotted on the y-axis and the values on the x-axis correspond to the respective generation. The development of number of classification errors is similar to the previous experiment. At the beginning of the optimization, high-quality solutions are quickly found and from the 2000th iteration onward the optimization stagnates. This experiment thus confirms the previous assumption according to which not all 10,000 generations have to be calculated.

In contrast to the results illustrated in Fig. 8a, however, a much smaller number of classification errors is achieved in this experiment. At the end of the optimization, the number of classification errors is on average 22,000 ($\approx$ 17.4 % of all points), instead of 75,000. The constraints thus seem to demonstrate that the volume of the cuboid has a strong influence on the effectiveness of the attack pattern, a smaller cuboid results in a significantly smaller number of classification errors. Please note that the plot on the right illustrates the change of the average value of the distance metric throughout the experiment and not the minimal value found during the optimization (this explains the fact that the illustrated value increases in later phases

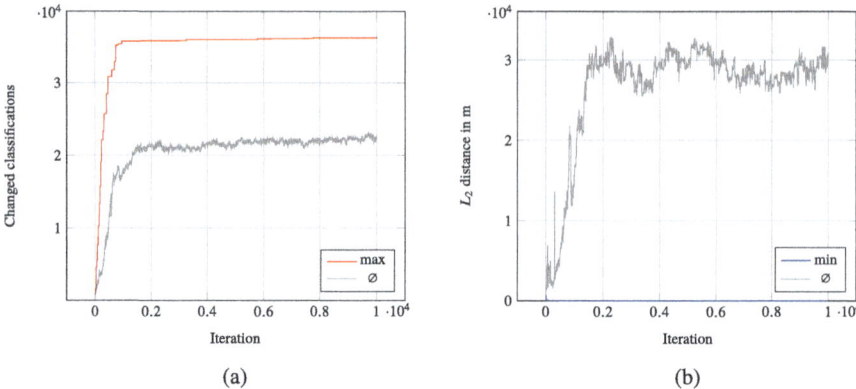

Fig. 9 Results using a cuboid to select the attacked 3D points, the L_2 distance norm, and constraint cuboid size (see Sect. 4.4.2). (**a**) Maximizing the incorrect classifications. (**b**) Minimizing the perturbation

of the experiment, in spite of the fact that the distance is being minimized). With a smaller cuboid, it is also not surprising that the average distance is significantly lower than in the experiment without constraints. In the experiment with restrictions, the average distance is about 30 km. Furthermore, since the beginning of the optimization, a solution with a very small distance is present in the population. Hereby, cuboids which do not contain at least thirty points are excluded by the constraint. Since the distance is to be minimized, this solution is optimal with regard to the distance objective function and is kept as a non-dominated solution by the EA until the end of the optimization. Compared to the results in Fig. 8b, the average shown in Fig. 9b is subject to less change from generation to generation. This could also be caused by the limited size of the cuboid, which means that fewer points fall out of or fall into to the cuboid that results from recombining parent individuals in the EA.

Overall, the introduction of constraints significantly narrows down the search space, while at the same time limiting the achievable number of classification errors. Due to smaller perturbation (distance), the resulting attack patterns would be easier to implement in reality. The usage of constraints thus makes it possible to adapt the optimization more easily to real conditions.

5.5 Attacking Class Labels of 3D Points Within an RoI

Since the results presented in Sect. 5.4 gave evidence that a restriction of the search space has a positive effect on the results, two additional experiments have been carried out where an RoI is defined. An attack only targets at changing class

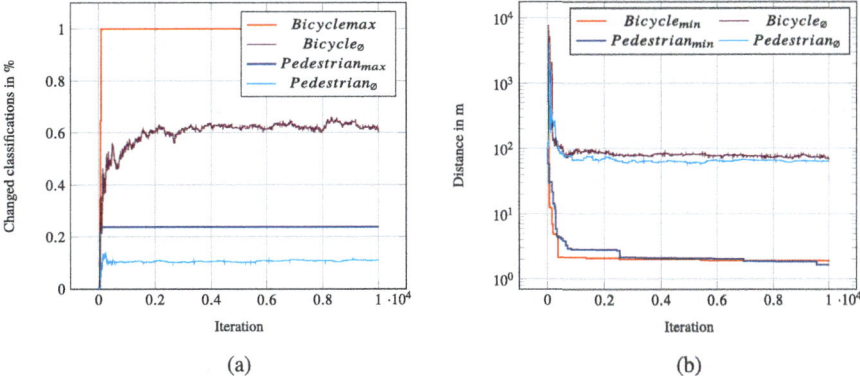

Fig. 10 Experiment results when using a predefined RoI. (**a**) Maximizing the incorrect classifications. (**b**) Minimizing the perturbation

labels within the RoI. The implementation details of the approach used in these experiments is provided in Sect. 4.5.

Similarly to the previous sections, we first detail the parameter settings we chose for the experiment. A region with a pedestrian has been selected as the focus of the attack. A second region contains a bicycle. Since both regions are significantly smaller than $125\,\mathrm{m}^3$, the maximum cuboid size has been reduced to $25\,\mathrm{m}^3$. Both a pedestrian and a bicycle can thus be covered. In these two experiments, the L_2 metric has been used. All other parameters are the same as in the Sects. 5.4 and 5.3.

Figure 10a illustrates the change of the number of classification errors over the course of the optimization. In contrast to previous plots, the values on the y-axis are not shown in absolute terms, but as a percentage of the maximum number of points within the selected region (which is 884 for the attack on the bicycle and 221 for the attack on the person). This allows the two attacks to be compared. Figure 10a shows that classification errors can be introduced for on average 15 % of all points of the region containing the person. In the case of the bicycle, this number is even higher with up to 60 %. Although these numbers of classification errors are quite high, the question whether they would be sufficient to deceive or confuse a subsequent interpreting algorithm cannot be deduced from this experiment.

Figure 10b illustrates the distances which were generated in the two experiments. In contrast to the previous experiments, the y-axis is shown logarithmically to provide a better overview of the data. The plots clearly show that the objective function gradually converges towards a minimum over the course of the optimization, which represents a significant difference compared to the experiment using a cuboid-based selection without constraints. Compared to the solutions generated in the early phase of the optimization, the solutions generated later require a perturbation (distance) which is smaller by a factor of up to 66. The best solution even only has a distance of approx. 1.6 m. Since in this variant solutions are only evaluated in a subspace of the possible solution space, it can be assumed that the EA optimizes the population in such a way that the cuboid size mainly includes the points that achieve a change in the RoI. Thus, points that result in a change outside the RoI increase the distance but at the same time do not necessarily increase the number of classification errors, enabling the EA to minimize the distance without decreasing the number of classification errors.

The average distance of these two attacks is approximately 65 m and is thus also physically possible, since, e.g., 300 classification errors can be achieved by introducing an average modification of merely 21.6 cm per point. In Fig. 11 the points which are modified and/or classified incorrectly are highlighted by color. The first image on the top left shows the original segmentation. There, the person is marked in red and the bicycle in cyan. It is clearly visible that the points that are changed (which are highlighted yellow and green) are all contiguous due to the encoding, but often only cover a part of the person or bicycle. Furthermore, often already the modification of a single point introduces classification errors in its surrounding area. This is, for instance, the case for the red points: In all but the top left image, points are highlighted in red when their positions were not modified but their class labels have changed due to the modification of other points. There are also cases where such points are not contiguous but individually spread in the scan. This is an interesting phenomenon, but such single misclassified points will likely be considered as noise and only have a minor influence on subsequent algorithms, as they only affect individual points of an object.

Table 2 provides a more detailed description of the modifications of the points. It shows the average number of classification errors and the average distance of all attack patterns at the end of the optimization. The standard deviation as well as the minimum and maximum deviations per coordinate axis were also calculated for the distance. According to this, in the region of the bicycle and the person on average 300 and 62 points are modified, respectively. In the attack on the person, it is noticeable that the points are mainly modified in the vertical direction. In the y-direction (to the left and right from the direction of view of the person in the scan), the modifications are very different across the solutions, which can be seen in the comparatively high standard deviation. Likewise, the solutions vary greatly in the height modification of the points. Only in the x-direction (direction in which the person is walking) are the points hardly modified in almost all solutions of the last generation. In the attack on the bicycle, it is mainly the height of the points that is modified. It can therefore be assumed that modifying the height of the points has a greater influence on the classification of these objects than modifications in the other directions.

5.6 Interpretation of Results

In the following, we come back to the hypotheses and research questions formulated at the beginning of this section.

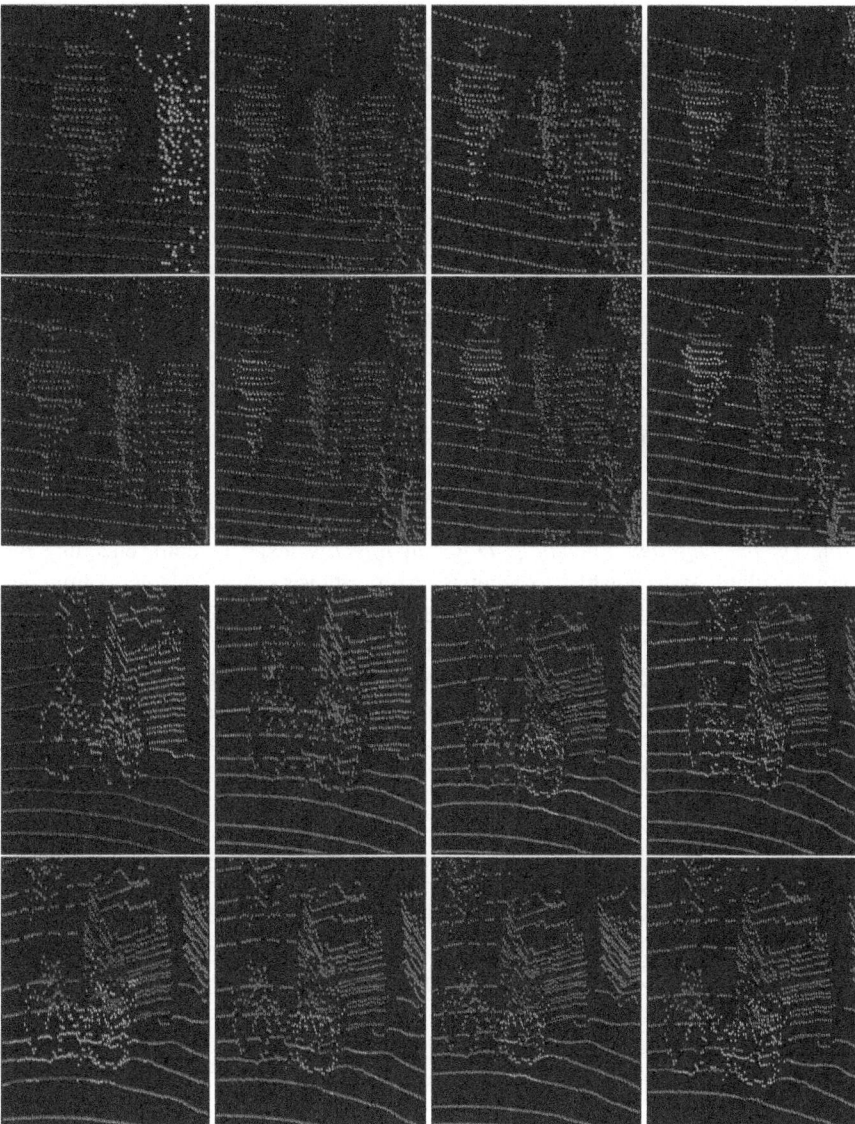

Fig. 11 Different attacks focused on the region of a person (top) and a bicycle (bottom) generated by the optimization. Marked in red are points that have not been modified but whose class has changed. Those shown in yellow have been modified in position with a change in class and for those shown in green modifying their position has not caused a classification errors. The first image shows the original segmentation

How Does the Employed Norm Influence the Result? The experimental results provide the opportunity to compare different distance metrics (L_1 and L_2). It can be observed that the two metrics result in solutions of similar quality regarding the

Table 2 Modifications of the population at the end of the optimization

Attacked Object	∅ Modifications		Point X	Y	Z
Bicycle	300.29	∅	0.035	−0.001	0.035
		Min	0.033	−0.004	−0.127
		Max	0.051	0.009	0.163
		Std	0.003	0.004	0.086
Person	61.66	∅	0.1	−0.14	0.86
		Min	−0.028	−1.206	−0.049
		Max	0.158	0.323	1.255
		Std	0.032	0.323	0.333

number of classification errors, with the L_1 metric performing slightly better for the experiment with targeted points. Comparing the changed point classifications achieved in the experiment shown in Fig. 8a, the two metrics work similarly well.

How Do the Different Encodings Differ in Effectiveness? Overall, encoding the entire available space enables to generate attack patterns spanning larger contiguous areas, compared to the approach which modifies only specific points. Within this larger area, there are then more classification errors, but the creation of these patterns also requires significantly more extensive perturbations. A different takeaway is that constraints which are used to constrain the cuboid size (see Fig. 9) or to focus misclassifications within a predefined RoI (see Fig. 10) improve the ability of the EA to generate effective attack patterns.

Are EAs Effective to Generate Adversarial Attack Patterns? The results of the experiments measuring the effectiveness of the attacks are illustrated in Figs. 6a, 8a and 9a. Hereby, successful attacks can be recognized by a positive value of the average or the maximum, since this can be interpreted as a change of at least one classification result. The results depicted in the figures, therefore, give evidence that the patterns generated by the EA result in successful attacks on the LiDAR sensor with the used segmentation network (the SqueezeSegV3 [36]). A first conclusion from the experiments is, thus, that a gradient-free approach such as an EA can be used to generate effective attack patterns.

Moreover, all experiment runs presented in this article have in common that no significant quality improvement could be observed after approximately 2000 generations. Thus, 80 % of the optimization run time could be saved without reducing the solution quality. Since EAs are heuristic optimizers, there is a possibility that they converge in local optima. This problem is usually addressed by a large number of optimization runs, which is why the time saved should be used to repeatedly start new EA runs.

In order to assess the usage of an EA as an approach for the generation of attack patterns, the results of this work have to be compared to already existing attacks in this area (see Sect. 2.4). In [30], Sun et al. were able to perform a successful attack with 60 changed points in 80% of all cases by spoofing a car into the close proximity of the vehicle. If more points are modified, the chances of success are even greater. However, Sun et al. work with object detection, which measures whether a whole object (the car) is detected. In this case, the attack is considered successful. Our work modifies and evaluates the changes of individual points that together form an object. Success is thus defined differently in our work (see Sect. 5.2). Since we do not make assumptions about a subsequent interpretive algorithm, or how objects are created from a segmented point cloud, the successes can only be compared to a limited extent. However, the number of points changed in an attack on a pedestrian is in a similar range to a successful attack from [30]. In the case of the attack on a bicycle, there are even significantly more, which means higher chances of success (see above). In [30] there are also restrictions on the horizontal angle within which points can be changed. Our work also restricts the size of the transformation vectors.

Randomly generated attack patterns have been tested in both [30] and [5], with no success. With the approach proposed in this work, where the EA was used for a structured modification of random points, it was possible to achieve a number of classification errors which significantly exceeded the number of modified points, i.e., we were able to change the class labels of a higher number of points than the number of points whose positions were manipulated.

With a predefined RoI, the average number of modified points is well within the range or close to the number used for the attacks from [30] and [5]. Thus, we consider the attack patterns generated by the presented approach as realistic for, e.g., spoofing attacks.

6 Conclusion and Future Work

6.1 Conclusion

In this work, we have proposed an approach for the design of attacks on LiDAR scans used in the area of autonomous driving. In contrast to existing approaches for similar attacks, the presented approach treats the functionality of the image recognition approach as a black box, i.e., the presented approach requires no information except the input and the output data of the machine-learning model. The approach was implemented and integrated into an EA-based optimization framework. The presented experimental results show that the approach can not only be used to generate attacks which result in misclassifications, but can additionally minimize the number of input changes required to cause a misclassification. They also demonstrate how the effectiveness of the approach varies for different ways of modifying the input, different genetic encodings, and different distance metrics.

6.2 Outlook

Beside the natural options of conducting additional experiments and investigating the usage of other optimizers and/or other classification systems, there are three main directions for possible extensions of the work at hand: (1) In order to reduce the large search space, it would be interesting to investigate approaches for the formulation of search space constraints. Such approaches would also be promising to introduce additional rules for the modifications of the input. (2) Designing additional objectives would enable more focused attacks where the goal could, for instance, be to not just maximize the number of classifications but instead to deliberately change the classification label of an object (e.g., a pedestrian) to a particular other target label (e.g. a car). (3) Finally, it would be highly interesting to investigate whether the attack patterns designed for a given LiDAR scan can also be used as an adversarial example for the attack on other LiDAR scans.

References

1. Alzantot, M., Sharma, Y., Chakraborty, S., Zhang, H., Hsieh, C.J., Srivastava, M.B.: Genattack: practical black-box attacks with gradient-free optimization. In: Proceedings of the Genetic and Evolutionary Computation Conference, Association for Computing Machinery, GECCO'19, New York, pp. 1111–1119 (2019). https://doi.org/10.1145/3321707.3321749
2. Angermeier, D., Beilke, K., Hansch, G., Eichler, J.: Modeling security risk assessments. In: 17th Escar Europe: Embedded Security in Cars (2019). https://doi.org/10.13154/294-6670
3. Bastos, D., Monteiro, P.P., Oliveira, A.S.R., Drummond, M.V.: An overview of lidar requirements and techniques for autonomous driving. In: 2021 Telecoms Conference (ConfTELE), pp. 1–6 (2021). https://doi.org/10.1109/ConfTELE50222.2021.9435580
4. Behley, J., Garbade, M., Milioto, A., Quenzel, J., Behnke, S., Stachniss, C., Gall, J.: Semantickitti: a dataset for semantic scene understanding of lidar sequences. In: Proceedings of the IEEE International Conference on Computer Vision (ICCV), pp. 9296–9306 (2019). https://doi.org/10.1109/ICCV.2019.00939
5. Cao, Y., Xiao, C., Cyr, B., Zhou, Y., Park, W., Rampazzi, S., Chen, Q.A., Fu, K., Mao, Z.M.: Adversarial sensor attack on lidar-based perception in autonomous driving. In: Proceedings of the 2019 ACM SIGSAC Conference on Computer and Communications Security, Association for Computing Machinery, New York, pp. 2267–2281 (2019). https://doi.org/10.1145/3319535.3339815
6. Carlini, N., Wagner, D.: Towards evaluating the robustness of neural networks. In: 2017 IEEE Symposium on Security and Privacy (SP) (2017), pp. 39–57. https://doi.org/10.1109/SP.2017.49
7. Chen, J., Jordan, M.I., Wainwright, M.J.: HopSkipJumpAttack: a query-efficient decision-based attack. In: 2020 IEEE Symposium on Security and Privacy (SP), pp. 1277–1294. https://doi.org/10.1109/SP40000.2020.00045
8. Deb, K.: A fast elitist non-dominated sorting genetic algorithm for multi-objective optimization: Nsga-2. IEEE Trans. Evol. Comput. **6**(2), 182–197 (2002)
9. Geyer, J., Kassahun, Y., Mahmudi, M., Ricou, X., Durgesh, R., Chung, A.S., Hauswald, L., Pham, V.H., Mühlegg, M., Dorn, S., Fernandez, T., Jänicke, M., Mirashi, S., Savani, C., Sturm, M., Vorobiov, O., Oelker, M., Garreis, S., Schuberth, P.: A2D2: audi autonomous driving dataset (2020). https://www.a2d2.audi. 2004.06320

10. Goodfellow, I.J., Shlens, J., Szegedy, C.: Explaining and harnessing adversarial examples. In: 3rd International Conference on Learning Representations, ICLR 2015, San Diego (2015)
11. He, K., Gkioxari, G., Dollár, P., Girshick, R.B.: Mask R-CNN. In: 2017 IEEE International Conference on Computer Vision (ICCV), pp. 2980–2988. https://doi.org/10.1109/ICCV.2017. 322
12. Hernan, S., Lambert, S., Ostwald, T., Shostack, A.: Threat modeling-uncover security design flaws using the stride approach. MSDN Magazine-Louisville, pp. 68–75 (2006)
13. Jiménez, F., Naranjo, J.E., Anaya, J.J., García, F., Ponz, A., Armingol, J.M.: Advanced driver assistance system for road environments to improve safety and efficiency. Transp. Res. Proc. **14**, 2245–2254 (2016)
14. LeCun, Y., Jackel, L.D., Bottou, L., Cortes, C., Denker, J.S., Drucker, H., Guyon, I., Müller, U.A., Säckinger, E., Simard, P., Vapnik, V.: Learning algorithms for classification: a comparison on handwritten digit recognition. In: Neural Networks: The Statistical Mechanics Perspective, pp. 261–276. World Scientific, Singapore
15. Li, X., Sun, Z., Cao, D., He, Z., Zhu, Q.: Real-time trajectory planning for autonomous urban driving: framework, algorithms, and verifications. IEEE/ASME Trans. Mechatron. **21**(2), 740–753 (2016). https://doi.org/10.1109/TMECH.2015.2493980
16. Li, Y., Ma, L., Zhong, Z., Liu, F., Chapman, M.A., Cao, D., Li, J.: Deep learning for lidar point clouds in autonomous driving: a review. IEEE Trans. Neural Netw. Learn. Syst. **32**(8), 3412–3432 (2020)
17. LLC W: Expanding our testing in san francisco (2021). https://blog.waymo.com/2021/02/ expanding-our-testing-in-san-francisco.html
18. Lukasiewycz, M., Glaß, M., Reimann, F., Teich, J.: Opt4j: a modular framework for meta-heuristic optimization. In: Proceedings of the 13th Annual Conference on Genetic and Evolutionary Computation. Association for Computing Machinery GECCO'11, New York, pp. 1723–1730 (2011). https://doi.org/10.1145/2001576.2001808
19. Macher, G., Schmittner, C., Veledar, O., Brenner, E.: ISO/SAE DIS 21434 Automotive Cybersecurity Standard - in a Nutshell. Springer, Berlin (2020)
20. McManamon, P.: LiDAR Technologies and Systems, 1st edn. Society of Photo-Optical Instrumentation Engineers (SPIE), Bellingham (2019) . https://doi.org/10.1117/3.2518254
21. Meshcheryakov, R., Iskhakov, A., Mamchenko, M., Romanova, M., Uvaysov, S., Amirgaliyev, Y., Gromaszek, K.: A probabilistic approach to estimating allowed SNR values for automotive lidars in smart cities under various external influences. Sensors 22(2) (2022). https://www. mdpi.com/1424-8220/22/2/609
22. Monteuuis, J.P., Boudguiga, A., Zhang, J., Labiod, H., Servel, A., Urien, P.: SARA: security automotive risk analysis method. In: Proceedings of the 4th ACM Workshop on Cyber-Physical System Security, pp. 3–14 (2018)
23. Petit, J., Shladover, S.E.: Potential cyberattacks on automated vehicles. IEEE Trans. Intell. Transp. Syst. **16**(2), 546–556 (2015). https://doi.org/10.1109/TITS.2014.2342271
24. Petit, J., Stottelaar, B., Feiri, M., Kargl, F.: Remote attacks on automated vehicles sensors: experiments on camera and lidar. In: Black Hat Europe (2015)
25. Qi, C.R., Su, H., Mo, K., Guibas, L.J.: Pointnet: deep learning on point sets for 3D classification and segmentation. In: 2017 IEEE Conference on Computer Vision and Pattern Recognition (CVPR), pp. 77–85 (2017). https://doi.org/10.1109/CVPR.2017.16
26. Reuters: Baidu, pony.ai approved for robotaxi services in beijing (2021). https://www.reuters. com/technology/baidu-ponyai-approved-robotaxi-services-beijing-2021-11-25/
27. SAE International: Taxonomy and definitions for terms related to driving automation systems for on-road motor vehicles (2021)
28. Shin, H., Son, Y., Park, Y., Kwon, Y., Kim, Y.: Sampling race: bypassing timing-based analog active sensor spoofing detection on analog-digital systems. In:10th USENIX Workshop on Offensive Technologies (WOOT'16), USENIX Association, Austin (2016). https://www. usenix.org/conference/woot16/workshop-program/presentation/shin

29. Shin, H., Kim, D., Kwon, Y., Kim, Y.: Illusion and dazzle: adversarial optical channel exploits against lidars for automotive applications. In: Cryptographic Hardware and Embedded Systems–CHES 2017, vol. 10529, pp. 445–467. Springer, Cham, (2017). https://doi.org/10. 1007/978-3-319-66787-4_22

30. Sun, J., Cao, Y., Chen, Q.A., Mao, Z.M.: Towards robust lidar-based perception in autonomous driving: general black-box adversarial sensor attack and countermeasures. In: 29th USENIX Security Symposium (USENIX Security 20). USENIX Association, pp. 877–894 (2020). https://www.usenix.org/conference/usenixsecurity20/presentation/sun

31. Szegedy, C., Zaremba, W., Sutskever, I., Bruna, J., Erhan, D., Goofellow, I., Fergus, R.: Intriguing properties of neural networks. In: International Conference on Learning Representations (ICLR), Banff, AB, Kanada (2014)

32. Tu, J., Ren, M., Manivasagam, S., Liang, M., Yang, B., Du, R., Cheng, F., Urtasun, R.: Physically realizable adversarial examples for lidar object detection (2020). CoRR abs/2004.00543. https://arxiv.org/abs/2004.00543

33. Vorobeychik, Y., Kantarcioglu, M.: Adversarial Machine Learning, 1st edn. Morgan & Claypool, San Rafael (2018). https://doi.org/10.2200/S00861ED1V01Y201806AIM039

34. Wallace, A.M., Halimi, A., Buller, G.S.: Full waveform lidar for adverse weather conditions. IEEE Trans. Veh. Technol. **69**(7), 7064–7077 (2020). https://doi.org/10.1109/TVT.2020. 2989148

35. Xiang, C., Qi, C.R., Li, B.: Generating 3D adversarial point clouds. In: 2019 IEEE/CVF Conference on Computer Vision and Pattern Recognition (CVPR), pp. 9128–9136 (2019). https://doi.org/10.1109/CVPR.2019.00935

36. Xu, C., Wu, B., Wang, Z., Zhan, W., Vajda, P., Keutzer, K., Tomizuka, M.: Squeezesegv3: spatially-adaptive convolution for efficient point-cloud segmentation. In: Computer Vision – ECCV 2020, vol. 28, pp. 1–19 (2020). https://doi.org/10.1007/978-3-030-58604-1

Internet of Vehicles: Security and Research Roadmap

Arunmozhi Manimuthu, Tu Ngo, and Anupam Chattopadhyay

1 Evolution of Automotive Connectivity

With growth of autonomy, and connectivity, the automobile industry is undergoing tremendous change. Autonomous Vehicles are being adopted in various scenarios, including public transport, due to its sophisticated user experience, value-added services, and higher efficiency. This is aided by wide-ranging communication protocols allowing a vehicle to interact with other vehicles, roadside infrastructure as well as pedestrians with handheld smart devices (Fig. 1). These includes Wi-Fi, Bluetooth, Ultra-WideBand (UWB), Near Field Communication (NFC), IEEE 802.11p (DSRC). These protocols help in data collection, storage, transfer, and live updates about the overall traffic scenario in real-time, which helps in an overall improved transportation management. In the following sections, these communication protocols are discussed in the context of IoV and their corresponding security challenges.

Cyber Security Research Centre (CYSREN), Nanyang Technological University & DESAY SV Automotive.

A. Manimuthu (✉)
Cybersecurity and Business Analytics, Operations and Information Management Department, Aston Business School, Aston University, Birmingham, UK
e-mail: m.arunmozhi@aston.ac.uk

T. Ngo
Cyber Security Research Centre (CYSREN), Nanyang Technological University, Singapore, Singapore
e-mail: anhtu.ngo@ntu.edu.sg

A. Chattopadhyay
School of Computer Science, Nanyang Technological University, Singapore, Singapore
e-mail: anupam@ntu.edu.sg

© The Author(s), under exclusive license to Springer Nature Switzerland AG 2023
V. K. Kukkala, S. Pasricha (eds.), *Machine Learning and Optimization Techniques for Automotive Cyber-Physical Systems*, https://doi.org/10.1007/978-3-031-28016-0_8

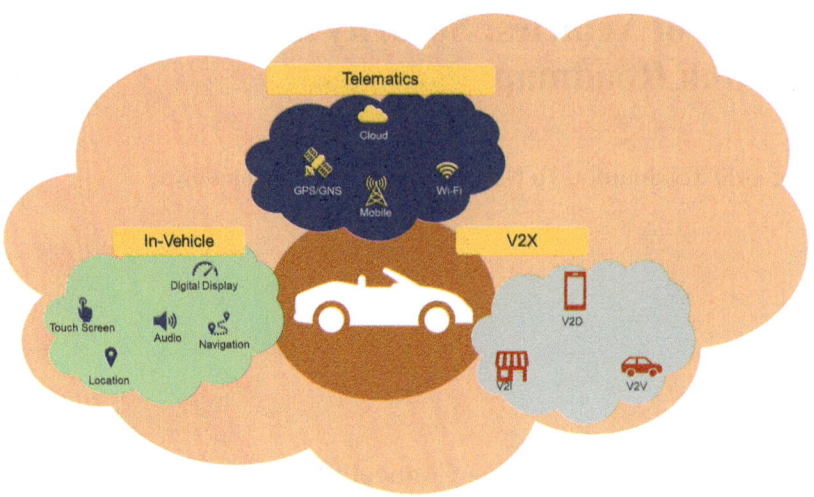

Fig. 1 IoV ecosystem

2 Autonomous Vehicle Standards

VANET (Vehicular Ad hoc NETwork) is a type of wireless ad hoc network in which automobiles serve as mobile nodes. VANETs have a standardized communication system that integrates all elements from the hardware to the application layer. Dedicated short-range communications (DSRC), wireless access in vehicle environments (WAVE), and IEEE 802.11p are the key components available in VANETs [1, 2].

- **DSRC: Dedicated Short Range Communications**—For DSRC, the Federal Communications Commission (FCC) set allocated the 75-MHz range between 5850 and 5925 GHz. The DSRC spectrum consists of 7 channels, each with a 10 MHz range [3]. From low to high, these channels are numbered 172, 174, 176, 178, 180, 182, and 184, with channel 178 serving as the master controller and the other six used for consumer services. Channel 172 is responsible for data integrity check and low bandwidth, whereas channel 184 which requires high power to operate is allocated for user's security.
- **WAVE: Wireless Access in Vehicular Environments**—The WAVE IEEE 1609 series specifies a framework for establishing V2V and V2I communications [4]. It defines safety standards and regulatory procedures for adopting wireless communication protocols with different transportation applications like remote connectivity, location sharing, etc.
- **IEEE 802.11p**—An update from IEEE 802.11b where it works between the data link and physical layers for connection and communication establishment between high-speed vehicles that requires rapid data exchange with zero or least possible latency in real-time [5]. The specifications of the physical and data link layer in VANETs ecosystem are defined in 802.11p.

2.1 IEEE 1609

The absence of high-speed connections and standardized communications interfaces between automobile manufacturers limits the capabilities of AV. The IEEE 1609 addresses these two issues by providing a suitable framework with multiple adversarial functions and operating modes at the device and operation levels [2, 6]. WAVE defines a collection of independent, standardized services and interfaces that enable safe vehicle-to-vehicle (V2V) and vehicle-to-infrastructure (V2I) connection and communication. Certain specifications and procedures form the basis for a wide range of services, including automated tolling, navigation, traffic management, and localization. IEEE 1609 series defines the V2X framework and communications procedures for secured mobile connectivity between automobiles and network service providers [7]. These protocols can be used in combination with RF equipment for remote system diagnostics.

- **IEEE P1609.0** —Framework for multi-channel DSRC/WAVE devices which helps in establishing communication with remote vehicular infrastructure.
- **IEEE 1609.1-2006—*Resource Overview*—**It describes the steps for vehicular safety services, data and device-level safety management schemes.
- **IEEE 1609.2-2006—*Security Services*—**Assists in message analysis, and formatting. It provides a suitable networking infrastructure for data and message exchange.
- **IEEE 1609.3-2007—*Networking Services*—**It provides frameworks and procedures for secured data exchange, access point management and protocol selections.
- **IEEE 1609.4 -2006- *Multi-Channel Operations*—**Exclusive for MAC and Data link layers. Assists in secured transactions and digital payments.
- **IEEE P1609.11 *Intelligent Transportation Systems (ITS)*—**It Provides security frameworks for V2I and V2V via OTA support.

5G networks are redesigned (6G) to make use of new technologies like software-defined networking (SDN) which supports faster installation, updated device drivers, software patches and network virtualized functions (NFV). This intensely linked connectivity eventually makes automobiles and connected networks more vulnerable to attacks [8–10]. SAE, ISO and NIST have conducted a series of related studies and released corresponding guidelines to identify IoV vulnerabilities and firmware upgrade challenges. As an outcome, "Information Security and Information Security Management System" and "Software Upgrade and Software Upgrade Management System" were released. These guidelines help the manufacturers, and automotive software service providers to understand their requirements and prepare them for new design updates for limiting the vulnerabilities [11]. The Cybersecurity Management System (CSMS) provides safety guidelines and a set of design procedures for vehicle manufacturers.

2.2 ISO/SAE DIS 21434 Automotive Cybersecurity Standard

The ISO and SAE are collaborating on the creation of a cybersecurity guideline for vehicular traffic. The ultimate objective of the ISO-SAE 21434 and NIST [12] are to (a) define a formal methodology in ensuring information security for in-vehicle components and systems-level applications, (b) reduce the probability of successful attack surfaces and attack vectors, and (c) provide quick response to cyber threats without affecting the normal vehicular operations. These guidelines helps to create a foundation for a standardized information security cycle in the automobile sector. This specification describes the important features for hardware or process improvement. ISO/SAE DIS 21434 [13] focuses on cyberattacks across the vehicle life cycle, including development, testing, production, installation and maintenance, and dismantling. ISO/SAE 21434 "Road Vehicle Information Security Engineering" developed jointly by ISO and SAE is primarily concerned with the vehicle's infrastructure building exclusively to protect vehicle data.

In Europe, the communication architecture and protocols are conceived and standardised by the European Telecommunications Standards Institute (ETSI) [14], through the ETSI Intelligent Transport Systems (ITS) series of standards. ETSI ITS is inspired by IEEE 1609 whereas ISO 24089 is established with the goal of proposing safe vehicular operation and data security requirements for onboard equipment. GB-T32960 is a technical standard for electric vehicle's wireless communication systems. It provides guidelines for charging, navigation and vehicle health management.

3 Network Model

Interactions among IoV elements such as vehicle, cloud, roadside units (RSU) naturally result in a multi-level information exchange that helps in safe and secured vehicular movements. It assists the motorists/passengers with relevant data for smooth transportation [15]. The IoV components act as smart objects and interact with one another in message and command exchange.

3.1 Various Connectivity Platforms

• **In-vehicle platforms**—OSEK/VDX is an open-ended architecture for distributed control units in automobiles just as Android Auto and Apple Carplay. It establishes a software architecture for automobile's process control with the goal of software portability, and promotes reusability. Development of conceptual protocols for real-time operating frameworks (OS-OSEK OS and

OSEKtime OS), connectivity within and between ECUs (OSEK COM), and networking structures and monitoring techniques are all part of OSEK/VDX.

- **Mobile platforms**—Data offered by mobiles can be used to operate automotive apps and helps with connectivity. WiFi and Bluetooth available within the mobile devices can potentially be used to interact with RSUs and other V2I facilities. Digital payments, localization and remote monitoring facilities can be enabled while the vehicle is connected to the mobile network [16]. Automotive software developers may offer conventional software testing platforms that can be tested using these devices when connected with the vehicle's internal network.

- **Internet-based platforms**—Through the Internet, apps may connect with other automobiles or with trusted third party service providers. Google Play Store and Apple Software Store, for example, offer navigational guidance, audio streaming including Android Auto and Apple CarPlay.

Several initiatives have been developed by ISO and SAE to eliminate application barriers and improve interoperability [13]. Several efforts have centered on creating generic multi-platform interface frameworks which allow various vehicular networks, components and sensors to communicate with each other. This not only promotes code compatibility between different manufacturers but also limit the possibility of attacking surfaces in real-time.

3.2   V2V Communication

Vehicle-to-vehicle (V2V) communication allows cars to communicate wirelessly about their movement, geo-location, and traffic patterns. V2V allows vehicles to multi-cast signals thereby giving them 360-degree "consciousness" of surrounding vehicles. Vehicles having the necessary software (or safety programs) can use the information from other vehicles to defend themselves against potential risks. To alert the driver, the apps and software tools use graphical, vibrations, and sound signals, or a mix of these alerts. These alerts can enable drivers to take action in order to avoid collisions [3]. These V2V signals may detect risks during traffic, topography, or weather and have a range of more than 200 m. Vehicle-to-vehicle digital communication can boost the effectiveness of safety systems and save lives [17].

3.3   V2I Communication

Vehicle-to-Infrastructure (V2I) communication is the wireless exchange of data between vehicles and road infrastructure. V2I is often wireless and bi-directional and is enabled by a system of infrastructure, firmware, and circuitry. Infrastructure components including road markings, street signs, and traffic signals can transmit information to the vehicle, and vice versa. Rich, timely information may be leveraged to provide security, accessibility, and potential impacts with the recorded and shared data from both vehicle and infrastructure ends. Greater redundancies are needed to assist AV to read and decode road laws in more simple and easy way [4]. Here are a few examples of new technologies that can help to enhance transportation, accessibility, and security.

- **Smart Road Marks**: In any driving conditions, we require pavement markers that are noticeable to both humans and computers. Pavement road signs integrate with vehicle sensors help to detect boundaries beyond the eyesight range, enhancing detection and tracking them even in the worst weather.
- **Smart Digital Signs**: In addition, we require guiding signs that are visible to both humans and vehicles in all types of road conditions. Retro-reflective signs improve readability, resulting in more precise navigation and smart decision-making. Furthermore, smart signs can be used in conjunction with traditional signage. The Dedicated short-range multi-Channel test tools are stand-alone and used for V2V and V2I connection and communication. The test tools are unbiased third-party solution that supports SAE J2735, IEEE 802.11b, IEEE 802.11p and WAVE protocols [13].

3.4   In-Vehicle Infrastructure

Internal communication between different sub-systems within the vehicle introduce considerable complexity in an AV system design, which also needs to flawlessly operate in tandem with the IoV communication modules. ECUs are typically responsible for engine control, gearbox operations, ignitions, power transmission or steering, as well as GPS and Bluetooth functions. Thus, multi-layered security and decentralized security solution must be developed and integrated with such a heterogeneous ECU environment.

4   Security

Security becomes an indigenous component in the autonomous vehicle domain. Every activities performed and the operations executed must be secured from external attackers. In recent times, numerous cyberattacks are reported with multiple data breaches, theft and ransomware issues. Vulnerabilities can be everywhere and thus, it is very essential to ensure that passenger's data and transportation is secured throughout the operation. This chapter will provide inputs on the current challenges in IoV along with the importance of automotive standards. Further, this chapter will explain few security models which helps to ensure security in real-time.

Current Challenges in IoV
Attackers pose a potential threat to car security in the IoV ecosystem. Some of the actions carried out as a result of data breaches include exposing the passenger's data,

compromising the vehicle's critical equipment (brakes, ABS, etc.) and compliance processes, and, in extreme cases, compromised access control of the vehicle. With an increase in the number of actuators and sensors in vehicles, attackers may be able to get personally identifiable information (PII) from the vehicular network, including location history, entertainment preferences, and financial data. Attackers can get unauthorized entry to automobiles through the use of remote keys, wireless key cards, and mobile applications, which have already supplanted conventional physical car keys. This is possible by eavesdropping on the communication between a smartphone or digital key cards and the vehicle. As a result of security and safety concerns, managing virtual automobile keys may be just as complex as keeping real keys. Automobiles pose a severe threat because of their numerous communication ports, which may be used as possible attack surface. Once the attacker has access to the internal AV systems, further attacks towards the safety breach follows.

The increased technological improvements has widened the security gaps. Researchers have demonstrated the ability to hack into a car (either directly or remotely) via telematics systems, the on-board diagnostics (OBD-II) port or any other open ports with least security encryptions [18, 19].

As more mobile applications for interacting with automobiles, they become a target for malicious operators. In the instance of the Nissan Leaf, for example, security experts revealed how they might access the vehicle's AC, steering control, armrests, telematics unit, and ABS. These unauthorized operations may drain the battery and leave the car inoperable in an EV scenario. Concerns have also been raised about the number of security flaws in the Android and iOS mobile hardware and software [12, 13].

4.1   CPS-IoT Design Challenges

An AV is considered as Cyber-Physical System (CPS). Usually, CPS is a sophisticated, standalone decentralized distributed system linked to a pool of computing and storage nodes. The CPS intends to recognize and understand the impact on the physical environment and evaluate the effects of any modifications on its operating condition. It will help to make automated intelligent decisions without altering the physical setting within the vehicle. CPS can remotely control, monitor, regulate, and automate various in-vehicle activities because it is a distributed decentralized closed-loop process with few customizations from every automaker.

The widespread use of Web-based components and application usages in CPS networks has bridged the gap between CPS and IoT. As a result, there are abundant application models and real-time use cases in the traditional Operational Technology (OT) domain. This includes distributed network of sensor nodes to control and regulate vehicle operations like braking, acceleration and remote diagnostics. Due to the increase in attacks and vulnerabilities, automakers are searching for ways to combine ECUs by their operation domains. However, for a variety of reasons, this simplistic and restricted analysis of threats and related defense methods are insufficient and misleading due to the absence of common SAE/ISO/NIST standards for CPS operations.

4.2 Need for Security by Design

There is a lack of rigorous security assessment, with much of it occurring too late in the design process. Furthermore, some fail-safe system tasks may pool data to the external repositories for processing. This activity can become a potential target for the attacker. The only way of creating "Security by Design" systems that will be robust, resilient and flexible in the long run is to design from the attacker's perspective focusing more on the vulnerable components and processes [20]. Third-party suppliers are extensively relied upon by automakers to offer system-level components, firmware, and physical embedded devices for their cars. However, automakers lag in setting stringent firewall criteria [21]. Thus, any component in charge of vital functions, such as acceleration, brakes, etc. must fulfill the highest security standards and updates listed by ISO and SAE. Many of these updates are given via authorized security patches and authenticated firmware updates. However, each embedded device may pose their own set of security challenges. One of the best examples of cyberattack is data breach in Jeep [22]. In the current version of JEEP, the Renesas V850ES/FJ3 chip resides between the CAN bus and the head unit. It is set up as a read-only device, collecting vehicle information via the CAN and informing the operator about regular maintenance status, diagnostics, and warnings. The study observed and reverse-engineered the V850 firmware, reconfigured it to grant read/write rights, learned the CAN message, and successfully updated the software itself from a remote location. This capability, like other vulnerabilities, required a few security researchers many weeks to create, but once identified, it could be weaponized and bundled along with other penetration testing tools and vulnerability scanners.

4.3 Design-Thinking Concepts

As digital cabin equipment providing unique, customized digital experiences for drivers and passengers, it also poses serious security concerns and information security problems. It will affect the overall customer experience thereby breaking the integrity of the whole system. There are many differences in design, execution and customization between vehicle manufacturers. Some common design-thinking concepts must be addressed while customization. These are as follows:

- No common scheme for confirming the integrity, authenticity, and reliability of firmware and software updates

- Limited security monitoring system for data flow, accessibility control or unauthorized message filtering schemes
- No common standard followed for connection and communication with cloud and infrastructure.

4.4 Cyberattacks in Recent Times

Security flaws have been identified most notably in the authentication procedure. Nissan, for example, released a mobile app that allows access to the car to obtain data such as battery charge level and range, as well as enabling climate control ahead of the journey and other functions [23]. The only authentication information necessary while pairing the smart gadget to the vehicle was the Vehicle Identification Number (VIN). The VIN is usually found at the bottom area of the windscreen, accessible from the outside. To interrupt the vehicle's normal operation, an attacker may pool the vehicle with vulnerable software such as ransomware. The Association of British Insurers (ABI) reported 160k claims in the quarterly May 2019, equivalent to around 100 million pounds or 1.5 million per day. Attackers are exploiting known loopholes in Keyless Entry Systems utilizing low-cost and easily available technologies such as software-defined radio (SDR) frequency devices.

Researchers [24] discovered 19 flaws in the Mercedes-Benz E-Class in August 2020, which might allow hackers to wirelessly open the car door and start the engine. In September 2016, the IoT botnet 'Mirai' [25] has momentarily disabled numerous high-profile services, including OVH—A SaaS-based web service provider and Dyn—a web security company via a DDoS. Thus, AVs equipped with multiple sensors and wireless communication capabilities pose a potential threat and can be easily exploitable by external attackers.

5 Security Objectives

An attack may, either intentionally or unintentionally, cause a safety threat for the vehicle's occupants. These assaults can be indirect, such as distracting the driver with alert and modifying the loudness of the multimedia unit, manipulating the throttle or steering angle. An example of such an attack is the Lexus OTA update failure. More recently, a Mobileye 630 PRO and Tesla Model X hack [26] duped the Advanced driver assistance and autopilot control triggering the brakes and steering towards moving cars. These activities are executed by spoofing the vehicle control system. As we become more reliant on authentication and real-time in-vehicle safety monitoring from things like traffic congestion, emergency signs, roadworks, or accidents, it will have an influence on human driving behavior. But over-reliance on these activities may lead to increased implications from attacks like DoS and

Distributed Denial-of-Service (DDoS) assaults [23, 27]. Spywares like ransomware might be used to construct botnets for cryptojacking or to perform DDoS attacks.

6   Importance of Automotive Standards

Most generic attack assessments frequently misguide the CPS security objectives, which may aim to bring balance between vulnerabilities, cost, and accessibility. As a result, any unsafe procedures may be operationally acceptable. These procedures operate within a regulated environment formed by the SAE/ISO standards [9] that allows for certain security protocols to be used in CPS operations.

Root of Trust (RoT), security perimeter analysis and modeling are some of the basic security by design elements that are yet to be focused in depth during CPS security testing, analysis and evaluation. Addressing individual threats in an ad-hoc and isolated fashion will not assist much in security design techniques. Therefore, security by design can help to identify the vulnerabilities and isolate any particular system without affecting the complete AV. But in reality, this tendency of generic attack studies is worsened by the lack of strong guidelines that assists the AV's road safety regulations.

7   Trust Management

Several prior studies offered techniques for enabling security, authentication, and credential verification in vehicular networks. For example, VANETs uses digital signature (DSA) during message broadcast to preserve privacy, security, and defense against DDoS threats. Furthermore, IEEE 802.11p was modified for V2I communications with a lightweight authentication that targets vehicle data security, confidentiality and privacy. Proven methods for security management for VANETs help to secure all the interconnected networking devices used for data exchanges, command executions and condition monitoring in real-time. With these features, the probability of attacks can be gradually reduced thereby ensuring the privacy preserved safe driving in a closed-loop connected environment. Researchers [25] has enhanced the location accuracy by incorporating neighborhood locations and geospatial information into account. Furthermore, researchers [28] employed portable design suits that provide security for the vehicle during location sharing and navigation. While vehicle networks are opened up to other heterogeneous connections, various security concerns arise, as described by Kaiwartya et al. [29]. These design problems are closely linked with data that compromises the location precision, authenticity, confidentiality, and in some worst cases, attackers can spoof the location with their desirable coordinates.

Many experts, regardless of field, define trust as a level of risk, or ambiguity, and it sets an expectation about how an automobile industry will perform in the future.

Trust is a multifaceted concept frequently described as the degree of probability of an individual's readiness towards any anticipated/expected outcomes from the vehicle irrespective of its operations, control and command execution. Despite adopting concepts and categories of trust from the scientific literature, there is no clear consensus on the notion of trust in communications systems, especially in the AV domain. Thus, trust becomes a critical component of security in vehicle ad hoc networks.

Security experts have presented a unique situation-based trust model, called Situation-Aware Trust (SAT). The purpose of SAT is to create a new security framework through the use of cryptographic techniques that provide trust information as well as used for dynamic key management. A real-time message content validation (RMCV) schemes were proposed to address similar trust management issues. This RMCV will provide a trust score for all forms of commands and messages shared in IoV environment. The maximum value of final trust ratings received from neighborhood devices or vehicles are used to determine the message trustworthiness. Similar trust models include Lightweight Self-Organized Trust (LSOT) which uses both security certifications and user recommendations for trust management. There are many hybrid trust management mechanisms available. One such hybrid scheme is the Beacon-based Trust Management (BTM) system which uses message beacons for cross-checking and validating the level of trust. In general, the trust management methods use public key infrastructure (PKI) and digital signatures to validate trust and provide trust scores.

7.1 Components of Trust

The concept of trust is based on the quality of interactions and quantity of data exchanged between involved entities without the need for cross verification and security checks.

- **Direct Trust**—It is characterized by an AV user's immediate observations and their level of interactions. For example, general command executions like toll ticketing, speed limit verification etc where trust level will be very high as these are monitored by the government agencies. Some security experts describe trust as "direct information obtained by the AV to analyze the RSU using predetermined metrics and evaluation methods". Although it is considered that direct trust is more important than indirect trust, the combination of both is considered when evaluating a vehicle's trust during vulnerability assessment in V2X environment.
- **Indirect Trust**— It expresses the views of AVs neighboring/trusted nodes about the target infrastructure, taking into consideration of every previous interaction with the node. This includes the history of interactions, trust levels with other AVs, etc. Some researchers analyzed the indirect observation of trust by combining the robustness and reliability of nodes. Reliability is achieved when

the connection between the AV and the RSU happens in a controlled and secured environment without worrying about any cyberattacks. Robustness is the ability of both parties to carry out the task and perform operations only based on fool-proof trust management without any bias or altercations due to external interference.

7.2 Attributes of Trust

When calculating the above-mentioned trust components, some potential trust qualities are taken into account [30].

1. **Similarity level**—The degree of similarity between any two vehicles refers to the degree of comparable digital services, network connectivity, operation execution and control methods with externals resources. This includes AVs mode of message exchange, design standards and protocol usage. In many cases, message beacons, location tracking and navigation details are used to analyse, evaluate and generate the trust score.
2. **Familiarity Index**—Familiarity expresses how smooth AV and RSU are interacting with each other. A high familiarity score indicates that the AV has extensive information about the RSU or vice versa. This characteristic is inspired by social media websites like Facebook and Instagram, where an increased familiarity index leads to increased trust which eventually leads to a smooth relationship with one another.
3. **Timeliness**— It denotes how often the interaction between two parties has happened. This includes their recent history of connections. These values are calculated by combining the present interaction instance and the time when similar interaction occurred earlier.
4. **Packet Delivery Rate**—The delivery rate is directly related to data integrity. Any unauthorized activities identified will directly be reflected in the trust management index. Thus, the packet delivery rate is kept as one of the major criteria in determining the trust level between all the participating agents in the IoV network. Any malicious activities will limit the delivery rate and affect the whole data exchange process. In this way, AVs can become more resilient to similar attacks in the future.
5. **Coordinated Operation**—This attribute prioritizes the type of service offered by the RSU rather than the mode of connectivity or operational procedures. On-demand services are exclusively offered from the external devices or infrastructure only to particular AV due to very high trust score and zero history of attacks. Similar to social media, network entities demonstrate a mutual connection establishment and share their services via decentralized multi-hop communication without the need for data integrity check, analysis, testing and evaluation.

6. **Connection Readiness**—AVs readiness to interact with infrastructure or external devices for operations or service utilization. This characteristic is critical while maintaining robustness and reliability. Thus many third party service providers offer rewards, discount coupons to encourage cooperative behavior and boost their platform usage among the AV manufacturers.

7. **Interaction time**—It is predicted that significantly longer interactions contribute to improved collaboration among parties, which paves way for increased confidentiality. This eventually leads to an increased trust level in every upcoming interaction between the two entities in the IoV environment.

8. **Interaction Frequency**—Greater the frequency of communication, the higher the opportunity to understand each entity's behavior.

7.3 Classification of Trust Management Models

Vehicle trust management is broadly classified into three categories [31–33]: Data-centric, Participant-centric, and Hybrid trust management models.

1. **Data-Centric Trust Management**—This type of trust management strategy is primarily concerned with the data credibility and authenticity of information communicated between participating entities. This information mostly consists of logged data and alerts. Because data-centric trust models analyze the authenticity and integrity of each occurrence, bottlenecks and security breaches. Due to a lack of sufficient evidence, reasonable connection time and trustability between the participants, this model will perform poorly in many cases. Thus recent research in data-centric trust management focuses on the 2 different levels of trust.

   • Data trustworthiness is determined by assigning weights to observations or incidents or activities (e.g., connection frequency, time, etc.) shared by surrounding vehicles/RSUs. Scores are based on a vehicle's connection duration, the number of instances, connection history and location vicinity. Thus, a vehicle with longer connection time or frequent data exchanges or a high level of vicinity from neighborhood vehicles/RSU will have a higher level of trust.

   • The message trust level is determined by warning, conflicts and similarity between the participating agents.

2. **Participant-Centric Trust Management**—The model category focuses on the trustworthiness of the participating vehicles by accessing each of their reliability, credibility, and neighborhood endorsements. As a result, the authenticity of the data, originality of the sender/receiver and data consistency are ensured. In this method, the efficiency of the vehicles and security levels of all the participants have to be assessed frequently to restrict any vulnerabilities. If this process is not ensured then confidentiality becomes a big question as there is no assurance that

the data is free of errors or originated from the authorized vehicle. In this method, there are 2 different levels as follows:

- Prior to picking a cluster leader, the trust score of each vehicle is evaluated by combining direct (participant themselves) and indirect (neighbor vehicle or infrastructure) trust ratings. A vehicle with a trust level greater than a set threshold (combined value from all the participants irrespective of time, connection and mode of operation) is considered reliable; otherwise, it is classified as a vulnerable entity. As per the trust score, notifications will be shared before and after data exchange to all participating members.
- To isolate the network or vehicle from connecting with a vulnerable entity, trust scores obtained from the last instance of connection and communication are used as reference points. Thus any participant will have the privilege to get access to this score. But the major drawback of this mode is the degree of trustworthiness. Since the operation is purely based on the last connection and communication instances, any history of vulnerable activities will be masked from the participants. Thus reliability becomes a big question in this method of trust evaluation.

3. **Hybrid Trust Management**—This type of trust management approach combines both data and participant-based trust assessment, i.e., the legitimacy of the transferred data along with the neighbor's endorsement towards the entities.

8 Security Model for Automotive CPS

In the IoV environment due to the emergence of high level automation, it becomes mandatory to focus on the safety and security of AVs. As the number of vulnerabilities and attack surfaces are increasing, it is very much essential for the automaker to look for smart methods and algorithms to ensure a safe operating environment for their vehicles. Irrespective of design, all the manufacturers follow NIST and SAE automation standards [9, 13] for ensuring the safety and security. Their guideline suggests series of steps in an orderly fashion: "Identify, Protect, Detect, Respond, and Recover" in order to protect the AV. Government organizations and federal agencies play their part in providing frameworks, guidelines, amendments and measures that help to maintain a safe IoV ecosystem thereby attracting more public participation.

8.1 Storage Security

Storage is one of the potential attack vectors targeted by hackers. User data and applications are the key components considered to be vulnerable and thus additional security is provided. Storage security are series of rules and settings based on a certain level of control parameters. Access to these parameters and settings will be given only to the authorized participants and trusted third parties. This helps to ensure system security where the entire AV is closely monitored for any malicious and vulnerable activities. Storage security is not alone the responsibility of AV but also all the participating entities. Each of the participating agents must ensure their storage is free from any malicious activities and vulnerabilities. Storage area network (SAN) is one of the emerging concepts to ensure storage security [8]. SAN helps with the customized high-speed storage network. It provides infrastructure support for sharing storage facilities with multiple devices. Thus many automakers use SAN for safe data storage. SAN implementation requires certain considerations:

1. Network accessibility must be provided to the authorized users alone with two levels of authorization
2. Reliability of data and signal strength during the connection and communication must be ensured all the time
3. Cyberattacks and malicious activities must be closely monitored
4. Authenticity of the participating agents must be carefully verified by high-level security schemes

Storage security and data protection is monitored using certain primary tasks according to NIST 800. These procedures are generic and followed by the SAN users with certain customization as per the consumer's requirements. These include the following:

• Encrypting the data that are highly sensitive
• Frequent software updates and security patch releases
• Removing unauthorized, unlicensed and vulnerable services from the network
• Promoting the security policies among the users and notifying them about the malicious activities
• Deploying the latest data security algorithms that ensures data privacy

By following these set of procedures, it is very much possible for the AV user to ensure data and storage security. Some of the common functions listed above help to protect the system from Ransomware attacks, and data leakages.

8.2   Computing Node Security

It refers to the security provided for the embedded devices and controllers available in the IoV environment to perform their assigned tasks without any attacks and vulnerabilities. Since different components have a different levels of security.They are categorised based on several parameters as per the NIST 800 and SAE J3016 [13]. Some of them include:

1. **Connectivity**—Depends on the protocol used for connecting the system with other RSU or IoV elements, their security is provided/updated.
2. **Level of criticality**—Device computation, performance and operation are directly related to different levels of criticality. For example, the controller circuitry responsible for braking and acceleration are highly critical when compared to the music system. Depending on the tasks and computing nature, their security levels have to be prioritized.
3. **Performance time**—In general the operating time plays a significant role in determining the computing behavior of the devices. High-speed computation is more desirable in any application execution. The attacker and the defender need to cautiously handle time.

Virus, malware, malicious instant push messages, unauthenticated software updates, and vulnerable codes are some of the key players that are actively compromising computation security. Thus, it becomes mandatory to implement suitable encryption techniques at the device level. This helps to ensure the credibility of the computing devices in every action they perform.

8.3   Communication Security

In the communication network, the motive of the attacker is to breach the security perimeter thereby gaining unauthorized access to the data. Some of the common communication network attacks include DoS, DDoS, Man-in-the-Middle, malicious code injection, privilege escalation, etc. All these attacks primarily focus on breaching confidentiality, integrity and availability (CIA triad) [34]. According to the National Security Agency (NSA)—"Measures and controls put in place to prevent unauthorized individuals from accessing information obtained from a communications network and to ensure the integrity and authenticity of such communication channel". Despite the level of security provided for the communication network, these are the predominant threats identified:

1. **Malicious code injection**—Injected code will potentially mask, modify, destroy or bypass sensitive information intended for the authorized agents, thereby worsening the entire system behavior. These infiltrated data can misguide the users about the services offered by the RSU or other infrastructure devices within their close proximity.
2. **Localization attack**—Intercepting communications containing the localization details of sensors or other IoT devices provides attackers an ability to infiltrate the entire AV system from remote locations. The importance of encrypting the location details from an intruder emerges from the fact that sensor nodes have standard operating procedures. But when connected with ECUs, it is least possible to track them among the pool of AV components. Thus, it is critical to hide the node sensor positions and their associated operations.

3. **Application centric attack**—Communication establishment differs as per the type of application used by the AV. For example, Digital payment uses NFC, Telematics uses Bluetooth and internal ECUs use CAN bus for communication. It is possible for the attacker to specifically target these application-specific communication protocols. Apple Carplay, Android Auto and Google Assisted Navigation have provided additional authentication and security functions to ensure the credibility of the communication network before the communication establishment. It is now the responsibility of the users to safely handle the application installations.

8.4   Sensor and Actuator Security

In general it is also stated as device-level security. For every operation, the command executions are carried out by the controllers and their associated sensors and actuators. Hence it is very much essential to protect these devices from any potential threats. Sensor security is directly related to the confidentiality, authenticity and integrity of the vehicle operation. Thus they become the potential target vectors for the attackers. Some of the following security concerns were focused on by the security experts in order to secure the sensors from performing their normal operations:

1. Labelling the sensors based in the type of operations
2. Operating voltage range
3. Operating time including the reverse response time
4. Criticality levels
5. Type of connectivity and communication protocol used

Some of the most common attacks with the sensors and actuators are snooping, sinkhole, jamming, tampering, Sybil, wormhole, spoofing, etc. Any attacks with these devices will have a direct and immediate impact on the users as well as the surrounding environment. Thus, the Target of Evaluation (TOE) and its security objectives (SO) depends purely on their security functional requirement (SFR) [35]. SFR holds a list of objectives with which it is possible to record, monitor and closely evaluate the activities performed by the sensors. During this process of evaluation, all anonymous activities can be easily identified. In order to protect the devices from attacks there are few hardware security guidelines proposed by SAE and ISO. These includes:

- Secure booting
- Authorized device level security patch installations
- Device level encryption

9 Privacy

Today, automobiles rolled out of factories are enabled with a huge amount of innovative technologies and sophisticated embedded devices. Consumers use their functions to ease their use cases like navigation, traffic awareness, voice assistance, RSU assistance, remote diagnostics, parking assistance, etc. Such technical advancements frequently rely on user's personally identifiable information to work efficiently and provide precise details. Several automakers, security agencies and research groups have realized the need for privacy and have already begun to develop guidelines.

ISO and SAE standards are frequently updated in order to ensure privacy for all connected devices. Auto alliance group has developed a framework for vehicle manufacturers in developing a privacy-preserving scheme at the device level. Fair Information Practice Principles (FIPPs) lays the foundation for safe privacy practices and provided regulatory laws for global automakers to practice in their automobile manufacturing processes [36]. In US, Government Accountability Office (GAO) has managed to release data and other asset privacy reports for vehicles. According to GAO, automakers must ensure privacy from the very beginning step of their manufacturing process and provide software support to their users for an extended time duration. It has also provided guidelines for vehicle owners to practice operations in a safe and secured way. Federal Trade Commission (FTC) circulated a notice to all vehicle manufacturers under the title "Careful Connections: Building Security in the Internet of Things". Notifying the user's about their device-level security, type of encryption used, license details and privacy-preserving policies are stressed in this FTC guidelines.

Regulatory policies proposed under Electronic Communications Privacy Act ("ECPA") help with the norms for anti-interception of data associated with the user's privacy. These norms are applicable for all third party applications, network service providers, connection and communication support and storage service agents. Federal Communications Act ("FCA") has enforced laws and usage policies for the network service providers while providing infrastructure support to the users. Depending on the user request and duration of the subscription, the service provider need to take full responsibility for all the activities happening within its network. The European Union General Data Protection Regulation (GDPR) derived a regulatory policy to limit the e-privacy (electronic-privacy) issues. According to GDPR, 6 laws are kept as standard measurement metrics while handling the personal information of the users [37]. These are as follows:

1. User consent
2. Contract period
3. Legal policies and compliance
4. Risk assessment duration
5. Legitimacy of ownership
6. Transparent data handling mechanisms

Methodologies adopted by the manufacturer have also updated frequently with security and software updates to retain their credibility in the consumer market.

9.1 Location Privacy

GPS and GNS play a predominant role in collecting the location details from the vehicle. Common factors identified from the location details are *Time, Position and Orientation and Asset Identification.*

- **Time**—Every user within the IoV will follow certain driving and operation patterns while using their vehicles. These include playing music, using Alexa or Siri, connecting to Bluetooth, attending calls, standard toll ticketing, digital payments, and pit stop points. All these activities can be easily identified by tracking the vehicle timing information. For example, consider the following scenario where a person buys food from the same outlet continuously for several months within the same time duration using some mode of digital payment. By analyzing the time details, it is easy for the attacker to visit the same outlet at the same time without the knowledge of the user.
- **Position and Orientation**—It is completely taken care by available sensors like GPS, GNS and LiDAR within the AV. When these components are hacked, the location coordinates can be easily obtained. Information shared in social networking websites like Facebook, WeChat, etc will hold enough details about their users including their most recent visit location and time. Another important consideration while using these sensors is the period of data storage. As these sensors operate continuously, their associated data storage facility is also supposed to be completely secured.
- **Asset Identification**—This location privacy factor is purely based on the type of application used by the user. On the basis of authorization provided by the user, applications will collect the details about the asset. With these details, desired services will be offered to the users. Example: Using Google maps for finding the nearest hotel or booking a taxi or ordering food online etc. Thus it is very much essential to understand the privacy-preserving policies offered before using any third party apps.

9.2 Data Privacy

Data privacy is different from data security. The former focus on personal privileges, preference and rights while later concentrating on protecting it from attacks. Thus rights of the individual are the top priority and in ensuring their rights government agencies, and application service providers often revise their user policies. These regulations and laws concentrate on the attacks and the level of criticality faced by different entities within the IoV. Privacy-preserving programs offered by application developers help the user to keep track of recent privacy issues and alert them with necessary precautions. Privacy policies are carefully developed by the third party service providers. Their policies must address the following things completely to the users before and after collecting data:

- Mode of data collection
- Type of storage
- Data management and usage policies
- Data sharing privileges
- Regulatory compliance
- Code of conduct and enforcement regulations

OTA security patch updates helped around 100 billion people around the globe saving about 3.5 billion USD from data breaches. Since IoV has many potential attacks vectors, it is complex to manage the data privacy without the user's support. Despite recent breakthroughs in data privacy policies, legal strategies and laws, companies and governments often infringe or undermine consumers' privacy. As a result, some claim that users have already lost the privacy battle. Some of the key questionnaires every application service provider must transparently address to their users:

- Security levels while gathering, usage, storage and sharing of data
- Clarity in building an adequate data governance regulatory foundation capable of dealing with privacy issues
- Ensuring the authenticity and credibility of the information shared with the user
- Showing transparency in data handling with its use cases
- Detailing the users about their data portability and erasure
- Maintaining legitimacy by implementing the latest government privacy regulations

Service providers are forced to practice this mode of operation due to the emerging nature of data-related attacks. Privacy Impact Assessment (PIA) is done in a systematic and regular fashion to completely access the privacy levels. The company's reputation and credibility will be at risk in the event of any adversaries. As data privacy is critical, it needs to be double-checked and carefully handled while sharing with any third parties. In the event of long-term storage, suitable security infrastructure has to be provided. Some companies have combined the privacy and compliance functions. By combining both, they tried to bring more clarity to the roles and responsibilities of all the participating entities within the IoV.

10 Security for IoV

Threats and vulnerabilities are evolving continuously along with number of sensors and embedded controllers within the vehicular surface. Thus it becomes mandatory to execute threat, vulnerability and risk analysis (TVRA). In this chapter, we have discussed TVRA model with suitable steps and procedures. It is evident from the recent attacks and threats, the attack surfaces are also need to be focused. We have listed few potential attack surfaces along with their know attacks within the IoV systems. Finally, we have discussed few crucial aspects for continuous assessment of security with some common testing methods that are in practice by the vehicle manufacturers and testers in real-time.

10.1 Threat, Vulnerability and Risk Analysis (TVRA)

According to the European Telecommunications Standards Institute (ETSI), TVRA uses the product of the likelihood of an attack and the impact of that attack on a system to identify the risk to the system [8]. TVRA identifies the assets of the system and their weaknesses, threats and threat agents that may attack the system.

This is the brief overview of TVRA model:

- *Assets* (physical, human or logical) may have *weaknesses* that may be attacked by *threats*.
- A *threat* is enacted by a *threat agent*, which may lead to *unwanted incidents*.
- A *vulnerability*, which is a combination of *weaknesses*, can be exploited by *threats*.

The TVRA method consists of these following steps:

- **Step 1**: Identify Target of Evaluation (TOE), which includes a high-level description of the main assets of TOE, TOE environment and a specification of the goal, purpose, scope of TVRA.
- **Step 2**: Identify the security objectives, which includes a high-level statement of the security aims and issues to be resolved.
- **Step 3**: Identify functional security requirements from security objectives.
- **Step 4**: Refine the high-level asset descriptions (step 1) and additional assets derived from steps 2 and 3.
- **Step 5**: Identify and classify system's vulnerabilities, threats, and unwanted incidents.
- **Step 6**: Quantify occurrence likelihood and impact of the threats.

- **Step 7**: Establish the risks.
- **Step 8**: Identify countermeasure framework, which results in a list of alternative security devices and capabilities to reduce the risk.
- **Step 9**: Analyse countermeasure cost-benefit, which helps identify the most suitable security devices and capabilities.
- **Step 10**: Specify detailed requirements for security devices and capabilities.

10.2 Security Perimeter

According to Anupam et al. [20], security perimeter aims to segregate AV into different security domains, with different threat environments, which has a direct impact on the validity of the trust model. Therefore, a design for the trust infrastructure as well as the security mechanisms is essential.

A security perimeter enables a holistic and systematic approach to AV system design and analysis. Figure 2 visualizes a three-layer design showcasing AV security perimeters. An AV security architecture is considered well-designed if it implements decent control mechanisms to reinforce the boundary of each layer so that practical security assumptions can be made to enable systematic security analysis and design.

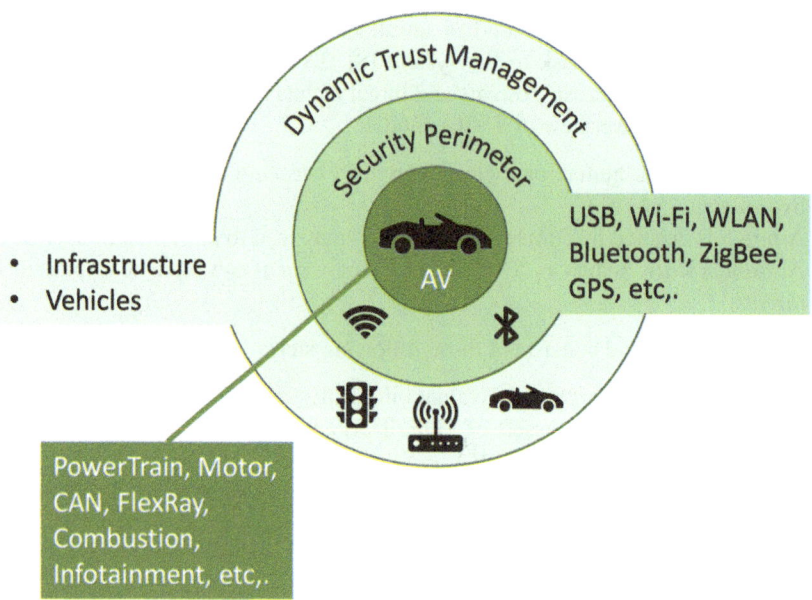

Fig. 2 Three-layer AV design

10.3   *Vehicular Attack Surfaces*

For an autonomous vehicle, there are a wide range of attack surfaces such as:

- Airbag ECU
- Remote Link Type App
- OBD II
- Bluetooth
- USB
- DSRC-Based Receiver (V2X)
- Passive Keyless Entry
- Remote Key
- TPMS (Tire Pressure Monitoring Sensor)
- ADAS System ECU
- Lighting System ECU (interior and exterior)
- Engine and Transmission ECU
- Steering and Braking ECU
- Vehicle Access System ECU

These attack surfaces above are visualized through Fig. 3.

In [39], the author stated that the number of cyber-attacks in the connected automotive industry is increased by a factor of six from 2010 to 2018. Of all the attacks:

- 21.4% are remote attacks
- 8% are keyless-entry attacks
- 5% are physical attacks via OBD port
- 8% are against OEM's mobile applications

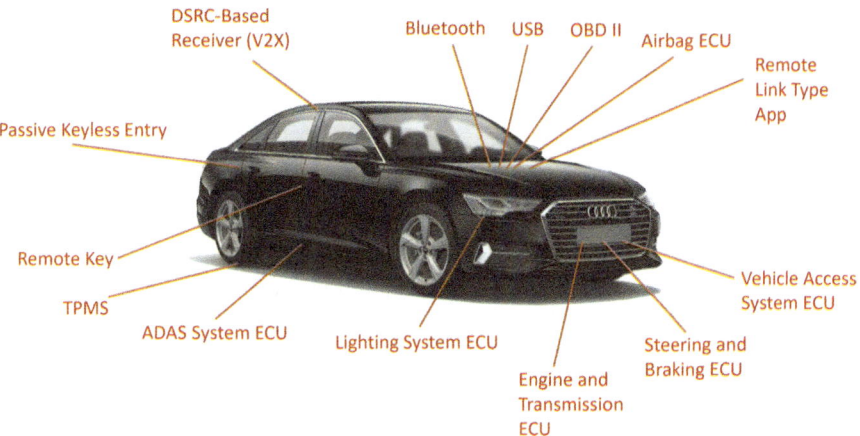

Fig. 3 AV attack surfaces [38]

In the context of black-hat attacks, 91% of the attacks are wireless.

11 Known Attacks on IoV Systems

11.1 Sensor/Actuator Attack

Today's modern vehicles often contain from 60 to more than 100 sensors [40] in various types, which lie in vehicle's sensing layer. Each of these sensors is responsible for a specific functionality such as safety, diagnostics, convenience, and environment monitoring. The sensing layer is vulnerable to both physical and remote attacks. A lot of research has been conducted to explore the vulnerability of vehicular sensors such as magnetic encoder (used in ABS) [41], inertial sensor (accelerometer, gyroscope) [42, 43], TPMS [44], GPS, LiDAR, and camera [28], etc.

Autonomous vehicles nowadays rely heavily on sensors so that the vehicle's machine learning models can make precise decision. Attacks on LiDAR and camera system has recently drawn a lot of attention from researchers since these sensors are the first components of the perception module, which is crucial for most functionalities in a typical AV system.

With the advancement of deep learning techniques for perception models, more researchers are focusing on adversarial attacks, whose aim is to fool the machine learning model. Previous camera attacks involve blinding a camera with extra sources of light, i.e., Light Amplification by Stimulated Emission of Radiation (LASER) beam or infrared LED. LiDAR attacks follow pretty similar trend to camera attacks. One of the pioneering work is from [45] where they created a fake object nearer than spoofer. Most recently, there was research into LiDAR-camera fusion system attack, in which the authors create a 3-D adversarial object that can fool either camera, LiDAR or even LiDAR-camera fusion algorithms.

11.2 Remote Wireless Network Attacks

Modern intelligent vehicles contain a great number of ECUs and various types of wireless connection protocols. These growing connectivity in vehicles leads to the increase in security vulnerability. By exploiting the connection mechanisms, attackers can compromise telematics ECUs. In 2011, Checkoway et al. [46] got access to remote code executions on a telematics ECU of a vehicle via Bluetooth and long-range wireless connection. They extracted the ECU's firmware and reverse-engineered the code. According to the authors, if attackers can pair their smartphones with Bluetooth ECU, they can send malicious codes from their smartphones to compromise the ECU.

11.3   OTA Software Update Attacks

Nowadays Over-the-air (OTA) update is becoming a standard option in automotive industry. OTA update enables upgrading of vehicle's functionalities, bug fixation in ECU's software, remotely. It is really conducive for both OEM and vehicle's owners since the software update can be applied immediately. Thus, these are a few advantages of OTA update [47]:

- **Lower cost:** nearly real-time, without requiring vehicle's owner to go to dealer for software updates.
- **Improved safety:** updates are applied immediately to improve safe functionalities.
- **Improved customer satisfaction:** hassle-free experience, can get information about the update without going to the dealer.
- **Frequent updates:** in case of more severe issues, manufacturers can just roll out the updates without requires a mass recall, which makes the updates more frequent due to reduced cost concerns.
- **Increased values:** vehicle's value can increase by maintaining consistent OTA updates, which help automakers gain more revenue.

Despite the fact that OTA update brings a lot of advantages, it requires access to in-vehicle communication network, which is highly critical. In 2020, researchers from U.S. National Highway Traffic Safety Administration (NHTSA) studied about automotive update mechanisms via both physical and OTA means. They proposed five feasible attack scenarios which may happen during OTA update process, i.e. malicious control of vehicles, denial-of-service, vehicle/contents theft, intellectual property theft/private information exfiltration and performance tuning/unauthorized feature activation. In 2012, Mulliner and Michele [48] exploited the insecurity of firmware installation process during firmware over-the-air (FOTA) update. This research shows that the downloaded firmware is vulnerable to adversarial modification by a time-of-check-to-time-of-use (TOCTTOU) attack. In 2020, Wen et al. [49] made a comprehensive analysis using OBD-II USB dongles as a new OTA attack surface.

11.4   OBD-Based Attacks

On-board Diagnostic (OBD) port is a connection port that allows people to collect information such as emission, mileage, speed and data from other vehicle's components. Since the OBD port does not usually encrypt the transmitted data, it is an open gateway for attacks to access various components of vehicles. To perform attacks, attackers have to physically access the OBD port, since it does not allow any kinds of remote access. Some devices that are attached to OBD port can transfer data between the port and a computer through wired or wireless connection. In

2019, Christensen and Dannberg successfully performed a man-in-the-middle attack through AutoPi Cloud interface [19]. Moreover, once attackers gain access to OBD port, it is shown that they can control other components such as warning light [44], window lift, airbag control system, horn [50], or even injecting codes into ECU.

12 Security Management

Security management for autonomous vehicles is a crucial aspect for continuous assessment of security challenges. Below are the two AV security incidents [20]:

1. **Threat Analysis and Risk Assessment (TARA):** there are some methods for assessment of these factors:

   - **Attack Tree Analysis:** the threat is analysed using the attack trees. The worst-case scenario and the risk can be determined using the combination of multiple potential threats.
   - **Threat, Vulnerability and Risk Analysis:** this is a standard approach in cyber-physical system, where the threat is associated with system's assets.
   - **Software Vulnerability Analysis:** this is an assessment technique for software code's vulnerability. The development environment and real-world environment of software implementation significantly differs from each other, which is the root cause for software vulnerability. The Open Web Application Security Software (OWASP) is a methodology to build secure software projects.

2. **Security testing methods**

   - **Penetration testing:** this is part of a security audit, either in a black-box or white-box setting. In black-box setting, the system details are unknown to the tester, while in white-box setting, it is assumed that the attackers have knowledge about internal details of a system.
   - **Red teaming:** this process detects network and system vulnerabilities by assuming the role of attacker.
   - **Fuzz testing:** a tremendous amount of random data passes through the system to make it crash. This kind of testing aims to test coding errors and security loopholes.
   - **Network testing:** a large number of packets bypasses through system's network, which serves as a network resilience testing.

13 IoV Security Analysis: Research Roadmap

The IoV is developed on top of an existing vehicular infrastructure that supports multiple embedded devices, networking platforms and computational device. As a result, IoV encounters certain similar security issues just as conventional vehicular networks (communication and application based threats and vulnerabilities) in real-time. Thus it is very much essential to concentrate on both traditional attacks along with other emerging network or autonomy-related threats. Guidelines provided by the NIST and SAE helps the designers to develop simulation models and testing prototypes to understand the behaviour of the vehicular system during attacks. These activities guides the developers and automakers to adapt new security mechanisms to protect the AV.

14 Security Analysis Through Simulation

Simulation modeling is used to tackle critical challenges in a safe and effective manner as IoV involves multiple players for its smooth operation. AV simulation gives a significant approach of analysis that is simple to verify, discuss, and deploy in the actual environment. Simulation modeling delivers important solutions for vulnerabilities related to all OSI layers. Depending on the Simulation software capabilities, AV design can be simulated and analyzed in a dynamic environment including the ability to clearly visualize the traffic scenario. It helps to predict and estimate the key parameters required for design, development, analysis, testing,installation, control, process, operation, and support [46, 51] Some of the key benefits of simulation are as follows:

1. **Detailed Visualization**—Simulation models may be animated in 2D/3D, making designs and ideas easier to verify, share, and interpret. By witnessing a model in operation, developer acquire faith in it and can convincingly implement findings to develop AV prototype.
2. **Increased accuracy**—A simulation model may record many more characteristics of AV than an analytical model, allowing for more accuracy and precision in predicting and analyzing the impact of any faults in the design.
3. **Insights to system dynamics**—Modeling and analysis help to understand the system dynamics in a more detailed way especially the risk prone areas.
4. **Reduced risk**- Simulation modeling allows the developer to test and explore many "what-if" possibilities in a safe and controlled environment.
5. **Save money and time**—Simulation of AV takes lesser time and incurs less money than testing and analyzing using hardware components and tools.
6. **Limits uncertainty**—Uncertainty in design, operation duration, component selections and outcomes may be easily represented in simulation models, enabling for risk measurement and the development of more sustainable and robust AV system.

15 Lightweight and Side-Channel Resistant Security Protocols

Security protocols are set of operating procedures capable of assisting the system or device in executing a series of commands and perform tasks in safe and secured environment. The ultimate goal of lightweight security protocol is to consume less storage, computing resources, and less energy to deliver improved security solutions to the system [52]. Lightweight security protocol design is also linked with the initiatives in the world of cryptography to propose smaller and faster cryptographic primitives.

Kocher et al. [53] first proposed the idea of side-channel analysis for cryptographic schemes. They observed that any physical device will leave behind certain footprints during its operation and computing procedures. *Side-channel* information such as computing time, electromagnetic radiation and power consumption can therefore lead to system-level attacks. Countermeasures to these attacks need to be incorporated at cipher-level and eventually to the security protocol.

16 Post-Quantum Cryptography (PQC)

Continuous progress in the field of quantum computing presents a threat towards current public-key cryptographic primitives, which form an integral part of many security protocols, e.g., TLS/IPSec. As a result, new cryptographic primitives are being designed and standardized, which are generally known as quantum-safe cryptography or Post quantum cryptography. Future IoV communication needs to consider quantum-safe cryptography as part of the protocol.

17 Conclusion

Safety and security for IoV will remains as an important area as the communication and connectivity related threats and vulnerabilities are continuously emerging. Even though many standards and protocols are developed and deployed, it is essential to perform TVRA. This helps to ensure privacy and data protection. The challenges discussed and the security objectives furnished will gives an overview about the attacks happened in IoV. Network models, security models and their allied trust management helps with smooth and secured IoV operation. Attack surfaces listed will helps the designer, manufacturer and developers to concentrate at the component and firmware levels. Finally, the proposed research road map will showcase the recent development in the automotive domain using simulation models, cryptography tools and methods.

References

1. Hasrouny, H., Samhat, A.E., Bassil, C., Laouiti, A.: Vanet security challenges and solutions: a survey. Veh. Commun. **7**, 7–20 (2017)
2. Lee, M., Atkison, T.: Vanet applications: past, present, and future. Veh. Commun. **28**, 100310 (2021)
3. Petit, J., Schaub, F., Feiri, M., Kargl, F.: Pseudonym schemes in vehicular networks: a survey. IEEE Commun. Surv. Tutorials **17**(1), 228–255 (2014)
4. Wang, J., Huang, Y., Feng, Z., Jiang, C., Zhang, H., Leung, V.C.: Reliable traffic density estimation in vehicular network. IEEE Trans. Veh. Technol. **67**(7), 6424–6437 (2018)
5. Wang, Y., Li, F.: Vehicular ad hoc networks. In: Guide to wireless ad hoc networks, pp. 503–525. Springer (2009)
6. Hussain, S.S., Ustun, T.S., Nsonga, P., Ali, I.: IEEE 1609 wave and IEC 61850 standard communication based integrated EV charging management in smart grids. IEEE Trans. Veh. Technol. **67**(8), 7690–7697 (2018)
7. Labertcaux, K., Hartenstein, H.: VANET: vehicular applications and inter-networking technologies. John Wiley & Sons (2009)
8. ETSI: CYBER; Methods and protocols; Part 1: Method and pro forma for Threat, Vulnerability, Risk Analysis (TVRA). IEEE Trans. Intell. Transp. Syst. [Online]. Available: https://www.etsi.org/deliver/etsi_ts/102100_102199/10216501/05.02.05_60/ts_10216501v050205p.pdf
9. IEEE Guide for Wireless Access in Vehicular Environments (WAVE) Architecture, IEEE Std 1609.0–2019 (Revision of IEEE Std 1609.0–2013), pp. 1–106, 10 April 2019, https://doi.org/10.1109/IEEESTD.2019.8686445
10. Karati, A., Islam, S.H., Biswas, G., Bhuiyan, M.Z.A., Vijayakumar, P., Karuppiah, M.: Provably secure identity-based signcryption scheme for crowdsourced industrial internet of things environments. IEEE Internet Things J. **5**(4), 2904–2914 (2017)
11. Sakiz, F., Sen, S.: A survey of attacks and detection mechanisms on intelligent transportation systems: VANETs and IoV. Ad Hoc Netw. **61**, 33–50 (2017)
12. Barrett, M.P.: Framework for improving critical infrastructure cybersecurity. National Institute of Standards and Technology, Gaithersburg, MD, USA, Tech. Rep (2018)
13. DRAFT INTERNATIONAL STANDARD ISO/SAE DIS 21434. International Organization for Standardization, Geneva, CH, Standard, Feb. (2021)
14. Virtualization, N.F.: European telecommunications standards institute (ETSI). Industry Specification Group (ISG) (2013)
15. Cheng, J., Cheng, J., Zhou, M., Liu, F., Gao, S., and Liu, C.: Routing in internet of vehicles: a review. IEEE Trans. Intell. Transp. Syst. **16**(5), 2339–2352 (2015)
16. Lin, X., Li, X.: Achieving efficient cooperative message authentication in vehicular ad hoc networks. IEEE Trans. Veh. Technol. **62**(7), 3339–3348 (2013)
17. Kim, S., Ulfarsson, G.F.: Traffic safety in an aging society: analysis of older pedestrian crashes. J. Transp. Saf. Secur. **11**(3), 323–332 (2019)
18. Engoulou, R.G., Bellaïche, M., Pierre, S., Quintero, A.: Vanet security surveys. Comput. Commun. **44**, 1–13 (2014)
19. Christensen, L., Dannberg, D.: Ethical hacking of IoT devices: OBD-II dongles (Dissertation). (2019). Retrieved from http://urn.kb.se/resolve?urn=urn:nbn:se:kth:diva-254571
20. Chattopadhyay, A., Lam, K.-Y., Tavva, Y.: Autonomous vehicle: security by design. IEEE Trans. Intell. Transp. Syst. **22**(11), 7015–7029 (2021)
21. Mejri, M.N., Ben-Othman, J., Hamdi, M.: Survey on VANET security challenges and possible cryptographic solutions. Veh. Commun. **1**(2), 53–66 (2014)
22. Agrafiotis, I., Nurse, J.R., Goldsmith, M., Creese, S., Upton, D.: A taxonomy of cyber-harms: defining the impacts of cyber-attacks and understanding how they propagate. J. Cybersecur. **4**(1), tyy006 (2018)
23. Lu, Z., Qu, G., Liu, Z.: A survey on recent advances in vehicular network security, trust, and privacy. IEEE Trans. Intell. Transp. Syst. **20**(2), 760–776 (2018)

24. Sedjelmaci, H., Senouci, S.M.: An accurate and efficient collaborative intrusion detection framework to secure vehicular networks. Comput. Electr. Eng. **43**, 33–47 (2015)
25. Antonakakis, M., April, T., Bailey, M., Bernhard, M., Bursztein, E., Cochran, J., Durumeric, Z., Halderman, J.A., Invernizzi, L., Kallitsis, M. et al.: Understanding the mirai botnet. In: 26th USENIX security symposium (USENIX Security 17), pp. 1093–1110 (2017)
26. Nie, S., Liu, L., Du, Y., Zhang, W.: Over-the-air: How we remotely compromised the gateway, BCM, and autopilot ECUS of tesla cars. Briefing, Black Hat USA (2018)
27. Ghosal, A., Conti, M.: Security issues and challenges in V2X: a survey. Comput. Netw. **169**, 107093 (2020)
28. Cao, Y., Wang, N., Xiao, C., Yang, D., Fang, J., Yang, R., Chen, Q.A., Liu, M., Li, B.: Invisible for both camera and LiDAR: security of multi-sensor fusion based perception in autonomous driving under physical-world attacks. In: 2021 IEEE symposium on security and privacy (SP), May 2021. [Online]. Available: https://doi.org/10.1109/SP40001.2021.00076
29. Kaiwartya, O., Abdullah, A.H., Cao, Y., Altameem, A., Prasad, M., Lin, C.-T., Liu, X.: Internet of vehicles: motivation, layered architecture, network model, challenges, and future aspects. IEEE Access **4**, 5356–5373 (2016)
30. Huang, D., Hong, X., Gerla, M.: Situation-aware trust architecture for vehicular networks. IEEE Commun. Mag. **48**(11), 128–135 (2010)
31. Hbaieb, A., Ayed, S., Chaari, L.: A survey of trust management in the internet of vehicles. Comput. Netw. **203**, 108558 (2022)
32. El-Sayed, H., Ignatious, H.A., Kulkarni, P., Bouktif, S.: Machine learning based trust management framework for vehicular networks. Veh. Commun. **25**, 100256 (2020)
33. Malik, N., Nanda, P., He, X., Liu, R.P.: Vehicular networks with security and trust management solutions: proposed secured message exchange via blockchain technology. Wirel. Netw. **26**(6), 4207–4226 (2020)
34. Ouchani, S. and Khaled, A.: Security assessment and hardening of autonomous vehicles. In: Risks and security of internet and systems: 15th international conference, CRiSIS 2020, Paris, France, November 4–6, 2020, Revised Selected Papers 15, pp. 365–375. Springer International Publishing (2021)
35. Liu, H., Hu, Z., Song, Y., Wang, J., Xie, X.: Vehicle-to-grid control for supplementary frequency regulation considering charging demands. IEEE Trans. Power Syst. **30**(6), 3110–3119 (2014)
36. Cate, F.H.: The failure of fair information practice principles. In: Consumer protection in the age of the 'information economy', pp. 351–388. Routledge (2016)
37. Costantini, F., Thomopoulos, N., Steibel, F., Curl, A., Lugano, G., Kováčiková, T.: Autonomous vehicles in a GDPR era: an international comparison. In: Advances in transport policy and planning, vol. 5, pp. 191–213. Elsevier (2020)
38. "Intel Automotive Research Workshops (2016). [Online]. Available: https://www.intel.com/content/dam/www/public/us/en/documents/product-briefs/automotive-security-research-workshops-summary.pdf
39. Bell, S.: 2018: A pivotal year for black hat cyber attacks on connected cars (2018) [Online]. Available: https://www.tu-auto.com/2018-a-pivotal-year-for-black-hat-cyber-attacks-on-connected-cars
40. El-Rewini, Z., Sadatsharan, K., Sugunaraj, N., Selvaraj, D.F., Plathottam, S.J., Ranganathan, P.: Cybersecurity attacks in vehicular sensors. IEEE Sensors J. **20**(22), 13752–13767 (2020)
41. Shoukry, Y., Martin, P., Tabuada, P., Srivastava, M.: Non-invasive spoofing attacks for anti-lock braking systems. In: Cryptographic hardware and embedded systems - CHES 2013, pp. 55–72. Springer, Berlin, Heidelberg [Online]. Available: https://doi.org/10.1007/978-3-642-40349-1_4
42. Tu, Y., Lin, Z., Lee, I., Hei, X.: Injected and delivered: fabricating implicit control over actuation systems by spoofing inertial sensors. In: 27th USENIX security symposium (USENIX Security 18), pp. 1545–1562. USENIX Association, Baltimore, MD (2018) [Online]. Available: https://www.usenix.org/conference/usenixsecurity18/presentation/tu

43. Trippel, T., Weisse, O., Xu, W., Honeyman, P., Fu, K.: WALNUT: Waging doubt on the integrity of MEMS accelerometers with acoustic injection attacks. In: 2017 IEEE European symposium on security and privacy (EuroS P), pp. 3–18 (2017)

44. Rouf, I., Miller, R., Mustafa, H., Taylor, T., Oh, S., Xu, W., Gruteser, M., Trappe, W., Seskar, I.: Security and privacy vulnerabilities of in-car wireless networks: a tire pressure monitoring system case study. In: Proceedings of the 19th USENIX conference on security, ser. USENIX Security' 10, p. 21. USENIX Association (2010)

45. Shin, H., Kim, D., Kwon, Y., Kim, Y.: Illusion and dazzle: adversarial optical channel exploits against lidars for automotive applications. In: Cryptographic hardware and embedded systems - CHES 2017 - 19th international conference, Taipei, Taiwan, September 25-28, 2017, Proceedings, ser. Lecture notes in computer science, W. Fischer and N. Homma, Eds., vol. 10529, pp. 445–467. Springer (2017) [Online]. Available: https://doi.org/10.1007/978-3-319-66787-4_22

46. Checkoway, S., McCoy, D., Kantor, B., Anderson, D., Shacham, H., Savage, S., Koscher, K., Czeskis, A., Roesner, F., Kohno, T.: Comprehensive experimental analyses of automotive attack surfaces. In: 20th USENIX security symposium (USENIX Security 11). San Francisco, CA: USENIX Association, Aug. (2011) [Online]. Available: https://www.usenix.org/conference/usenix-security-11/comprehensive-experimental-analyses-automotive-attack-surfaces

47. Halder, S., Ghosal, A., Conti, M.: Secure over-the-air software updates in connected vehicles: a survey. Comput. Netw. **178**, 107343 (2020) [Online]. Available: https://www.sciencedirect.com/science/article/pii/S1389128619314963

48. "Read it twice! a Mass-Storage-Based TOCTTOU attack. In: 6th USENIX workshop on offensive technologies (WOOT 12). Bellevue, WA: USENIX Association, Aug. (2012) [Online]. Available: https://www.usenix.org/conference/woot12/workshop-program/presentation/mulliner

49. Wen, H., Chen, Q.A., Lin, Z.: Plug-N-Pwned: comprehensive vulnerability analysis of OBD-II dongles as a new Over-the-Air attack surface in automotive IoT. In: 29th USENIX security symposium (USENIX Security 20), pp. 949–965. USENIX Association, Aug. (2020) [Online]. Available: https://www.usenix.org/conference/usenixsecurity20/presentation/wen

50. Koscher, K., Czeskis, A., Roesner, F., Patel, S., Kohno, T., Checkoway, S., McCoy, D., Kantor, B., Anderson, D., Shacham, H., Savage, S.: Experimental security analysis of a modern automobile. In: 2010 IEEE symposium on security and privacy, pp. 447–462 (2010)

51. Contreras-Castillo, J., Zeadally, S., Guerrero-Ibañez, J.A.: Internet of vehicles: architecture, protocols, and security. IEEE Internet Things J. **5**(5), 3701–3709 (2017)

52. Mundhenk, P., Paverd, A., Mrowca, A., Steinhorst, S., Lukasiewycz, M., Fahmy, S.A., Chakraborty, S.: Security in automotive networks: lightweight authentication and authorization. ACM Trans. Des. Autom. Electron. Syst. (TODAES) **22**(2), 1–27 (2017)

53. Kocher, P.C.: Timing attacks on implementations of Diffie-Hellman, RSA, DSS, and other systems. In: Annual international cryptology conference, pp. 104–113. Springer (1996)

Part III
Intrusion Detection Systems

Protecting Automotive Controller Area Network: A Review on Intrusion Detection Methods Using Machine Learning Algorithms

Jia Zhou, Weizhe Zhang, Guoqi Xie, Renfa Li, and Keqin Li

1 Introduction

1.1 Background and Motivation

The automotive industry is undergoing rapid changes. The in-depth integration of advanced information technology and automotive technology enables the vehicles equipped with more intelligent functions and more connections with outside. Despite a higher level of comfort, safety, efficiency and personalized experience providing for drivers, the vehicles are also exposed to negative risks brought by the new technologies. The rich connectivity with external environments also means more potential access points which can be exploited by malicious adversaries. The adversaries can further intrude the safety-critical in-vehicle network via compromising the bridge nodes. Considering that vehicle is a man-in-the-loop cyber physical system, the attacker can further gain the ability to control the physical components

J. Zhou
Department of New Networks, Peng Cheng Laboratory, Shenzhen, China
e-mail: zhoujia@hnu.edu.cn

W. Zhang (⊠)
School of Computer Science and Technology, Harbin Institute of Technology, Harbin, China

Department of New Networks, Peng Cheng Laboratory, Shenzhen, China
e-mail: wzzhang@hit.edu.cn

G. Xie · R. Li
Key Laboratory for Embedded and Network Computing of Hunan Province, College of Computer Science and Electronic Engineering, Hunan University, Changsha, China
e-mail: xgqman@hnu.edu.cn; lirenfa@hnu.edu.cn

K. Li
Department of Computer Science, State University of New York, New Paltz, NY, USA
e-mail: lik@newpaltz.edu

© The Author(s), under exclusive license to Springer Nature Switzerland AG 2023
V. K. Kukkala, S. Pasricha (eds.), *Machine Learning and Optimization Techniques for Automotive Cyber-Physical Systems*, https://doi.org/10.1007/978-3-031-28016-0_9

of automotive and manipulate its behaviors. It may result in a threat to human life or deeper security issues to the whole society. Security concern has become one of the most challenging issues for in-vehicle network which cannot be ignored.

In-vehicle network is the underlying base for the implementation of automotive functions such as driving safety, autonomous driving, intelligent in-cabin system, and body control. Accordingly, the in-vehicle network is also in the process of innovation to meet future requirements. With the rapid development of intelligence and connectivity of vehicles, the architecture of in-vehicle network is undergoing evolution from distributed model to domain model and zonal model. It is getting more complex and sophisticated, which usually comprises several networks responsible for different functions. In this chapter, we mainly focus on currently the most popular in-vehicle communication protocol Controller Area Network (CAN), which is directly responsible for the safety of vehicles. From our point of view, CAN will still bear an important role in ensuring driving safety in the future in-vehicle network. How to defend automotive CAN bus draws much attention from the public as well as academia.

CAN is capable of providing reliable and real-time communication to ensure the safety of the automotive control systems. But there is no any inherent mechanism at its birth to defend against malicious adversary. Its characteristics such as broadcast nature, plain-text transmission, lack of message authentication, and weak access control make the automotive CAN network vulnerable to cyber attack. Security schemes such as cryptographic measures are introduced in the automotive domain. Message Authentication Code (MAC), which can provide the ability to verify the data integrity as well as identify the sender seems like a good option. It is implemented based on a symmetric cryptographic mechanism, which can favor the deployment on automotive embedded systems by reducing the computational complexity. However, the extremely limited length of the CAN frame cuts the effect of the deployment of message authentication codes. For example, the maximum data payload of a data frame of the standard CAN protocol is only 8 bytes. The longer message authentication code results in a shorter payload which degrades the efficiency of the communication system, while the shorter message authentication code results in an insufficient security level. To mitigate this issue, the longer authentication tag can be transmitted via extra frames. Unfortunately, it can result in a heavier bus load which might affect the real-time performance of the system.

The intrusion detection method can be a simple but efficient solution for protecting in-vehicle network. It can monitor the network traffic and detect anomalies during the runtime of vehicles. Different from the encryption and authentication measures, intrusion detection methods do not occupy the limited bandwidth and payload of the in-vehicle network. It works based on the observation and analysis of network traffic. The intrusion detection system was firstly introduced for in-vehicle network by Hoppe et al. [12]. The authors proposed three ways to utilize features, which are the increase in the frequency of CAN frames, the observation of signal characteristics as well as the abuse of CAN identifiers to detect attacks. More schemes based on intrusion detection methods are designed since then. One way of designing the intrusion detection system is to build a physical model or pre-defined rules to detect unexpected behaviors. However, the in-depth knowledge about the system is always required for this kind of approaches. Besides, it is difficult to design a closed-loop expression to detect attacks in real cases. Machine learning (ML) is one of the most promising technologies nowadays which can also favor the solution for security concerns of in-vehicle network. ML can extract latent patterns from traffic to provide an effective and flexible solution for intrusion detection on in-vehicle network.

1.2   Contributions and Outline

In this chapter, we survey the studies which take advantage of machine learning technologies to detect intrusion for automotive CAN bus. The structure of our chapter can be seen in Fig. 1. To provide a better understanding about the application scenarios, we firstly introduce the in-vehicle network architecture and how it evolves. Next, we provide a detailed description about the intrusion detection methods exploiting ML algorithms. According to the domain knowledge used for extracting features by ML, we divide these approaches into four categories, which are semantics-based methods, literal-based methods, timing-based methods and signal characteristics-based methods respectively. Our contributions can be concluded as follows:

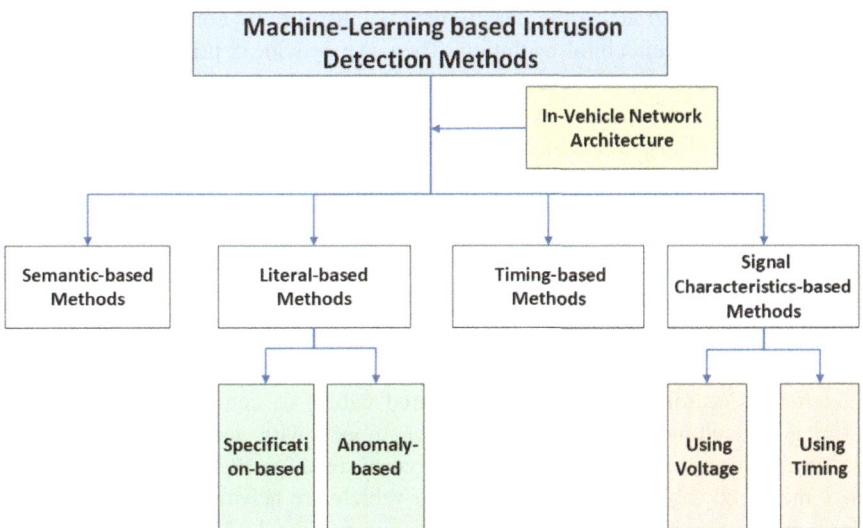

Fig. 1 Structure of the chapter

1. We provide an introduction about current and future in-vehicle network architecture. The evolution trend emphasizes the importance of CAN for driving safety and the necessity to protect it.
2. We classify the machine learning-based intrusion detection methods based on the domain knowledge exploited to extract features. The domain knowledge can be referred to those low-level characteristics in CAN such as timing characteristics or signal shapes, or the high-level characteristics such as the data payload of CAN frames or their semantic values.
3. We provide a detailed description for each category of intrusion detection methods. In each section, we firstly introduce the basic insight of how it works and discuss the disadvantages of the traditional methods. Then, we introduce the existing work based on machine learning algorithms.

The organization of this chapter is as follows: Sect. 2 provides the description about the current and future in-vehicle network architecture. Sections 3 to 6 describes the intrusion detection methods exploiting machine learning algorithms from four aspects, which are semantics-based methods, literal-based methods, timing-based methods and signal characteristics-based methods respectively. Finally, Sect. 7 concludes this chapter.

2   In-Vehicle Network Architecture

In this section, we first provide a description of the in-vehicle network architecture and how it will upgrade in the near future. We also briefly conclude the benefits brought by the architectural evolution. Then, we provide a primer on CAN and illustrate the necessity for research on protecting CAN. From our point of view, CAN will not be abandoned by the future in-vehicle network and will face more security risks. Thus, defending CAN from attacks is important for protecting vehicles no matter for the current or future in-vehicle network.

2.1   Evolution of In-Vehicle Network

The hardware of in-vehicle network mainly consists of two parts, which are the Electronic Control Units (ECUs) and wired cables to connect the ECUs. The ECU is an automotive embedded device equipped with abilities of computing, communication and control. The data and control signals of ECUs can be exchanged over the wired cables. All ECUs inside the vehicle are networked with each other through the internal communication system to form a whole. The whole system can provide the ability from sensing the driving environment to making decisions and implementing high-level automotive driving.

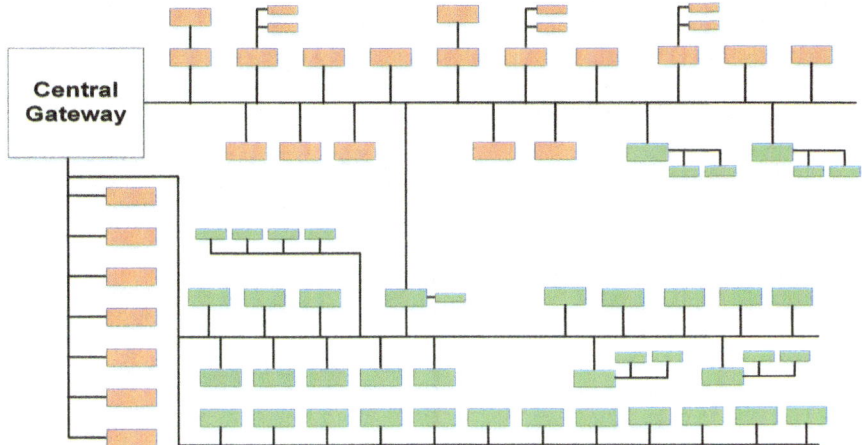

Fig. 2 Distributed in-vehicle network architecture

The traditional in-vehicle network adopts a distributed architecture (as shown in Fig. 2). All ECUs are scattered on the network and work distributively. Generally, it equips low-speed communication protocols such as CAN and LIN (Local Interconnect Network) as the backbone network. The distributed in-vehicle network enables the transition of automotive from mechanization to electronics. However, the increasing number of electronic functions and ECUs has lead to a heavy, large-scale in-vehicle network, making the wiring harness system the third-heaviest automotive component after the engine and chassis [35]. The bulky wiring harness system increases the total weight of the vehicle, resulting in higher energy consumption and cost. Besides, the increasing number of ECUs makes the in-vehicle network more complex. It could lead to a higher cost of software development as well as a higher cost of software verification and validation which might increase the risk of uncertainty.

Furthermore, the demand for automobile intelligence and the rising connections with outside are forcing the innovation of the communication architecture of in-vehicle network. Various advanced communication technologies such as 5G, WIFI, Bluetooth, and Vehicle-to-Everything (V2X) have been deployed on vehicles, which makes vehicles as a complex communication system. To realize the advanced intelligent functions of vehicles, the concept of Software Defined Vehicles (SDV) has gradually become the mainstream for automotive software development. The high integration of automotive technology and information technology increases the complexity of the intelligent connected vehicles continuously, which requires a scalable design of architecture and coordination of ECUs with higher computing power. To meet these requirements, the architecture of in-vehicle network would evolve from the traditional distributed architecture to a new generation of centralized architecture. Specifically, as shown in Fig. 3, it would gradually evolve into a

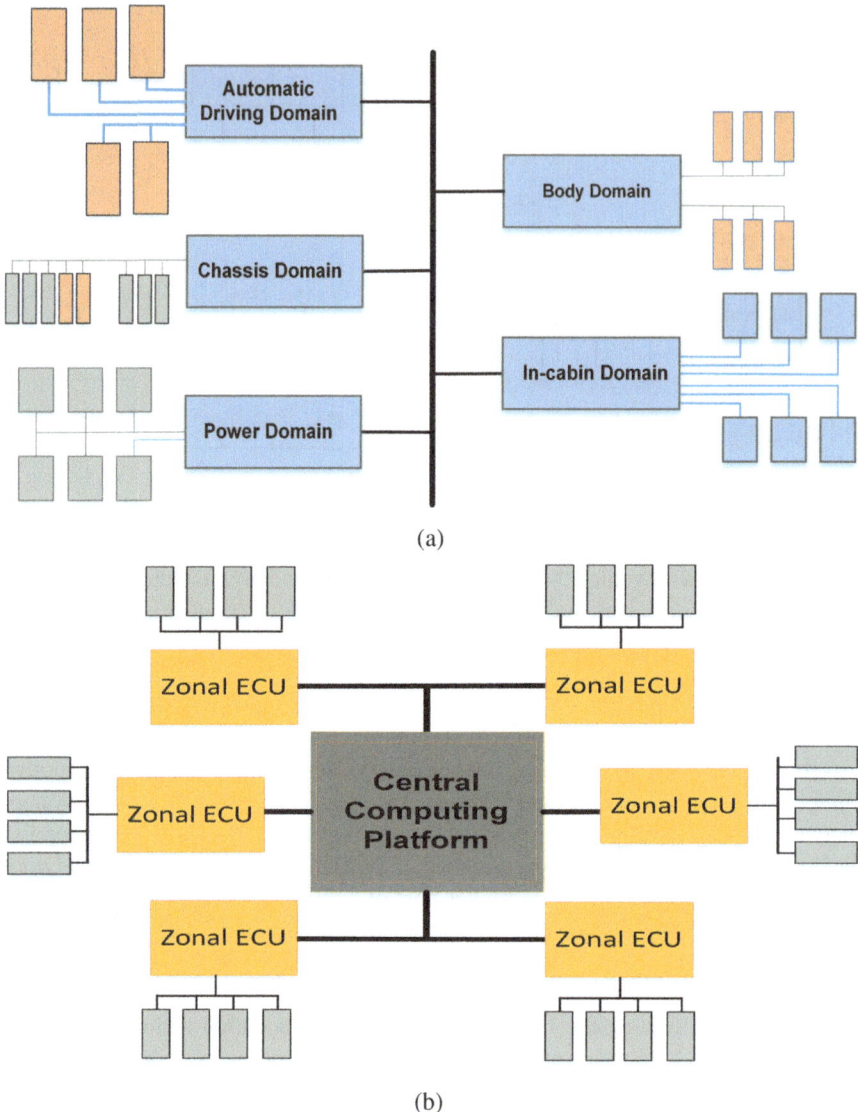

Fig. 3 New in-vehicle network architecture. (**a**) Domain architecture. (**b**) Zonal architecture

domain architecture (Fig. 3a), and further, a more centralized architecture called zonal architecture (Fig. 3b).

One common scheme of domain architecture is to divide the in-vehicle network into five different control domains according to their functions. Each domain is equipped with a Domain Control Unit (DCU) to centralize the functions and computing resources within the domain. The DCU is a higher-performance automotive microcontroller designed to solve the problem of performance bottleneck of distributed in-vehicle network. As shown in the left of Fig. 3, it consists of the power domain responsible for powertrain control and optimization, the chassis domain responsible for driving behavior (braking, steering, transmission, etc.), the body domain responsible for body and comfort control, the in-cabin domain responsible for entertainment, and the automatic driving domain responsible for assisting vehicle driving. The DCU consolidates the functionality within each domain and communicates with other DCUs via high-speed backbone network (such as Ethernet, etc.). In-domain ECUs attached to the DCU are degraded to low-level ECUs or actuators with limited computing and communication resources. Low-speed communication protocols (such as CAN, LIN, etc.) are exploited to connect the DCU with the in-domain nodes.

The zonal architecture further improves the degree of centralization by organizing a three-layer architecture. It consists of the following key components, including (1) computing resources which are a central computing platform, multiple zonal ECUs and many low-level ECUs; (2) communication resources which are high-speed backbone network (such as Ethernet, etc.) to connect the central computing platform with zone ECUs and low-speed local area network (such as CAN, LIN, etc.) to connect the zone ECUs with low-level ECUs. The hardware inside the local area network can be consolidated by the upper level zone ECU, while the hardware of zone ECUs can be further consolidated by the central computing platform. Highly consolidation of hardware resources makes it more available to separate software and hardware to achieve the concept of software-defined vehicles. It can manage the needs of more advanced and intelligent functions for future vehicles.

Currently, most car manufactures are in the stage of transition from distributed architecture to domain architecture. In general, the upgrade of the in-vehicle network architecture can bring advantages in terms of cost reduction and driving intelligence, which are listed as follows:

1. Reduction on hardware cost: Benefiting from architecture evolution, the total number of ECUs can be significantly reduced to optimize the utilization of computing resources. In addition, the layout of the wiring harness system can be optimized, lowering the total weight and hardware cost of vehicles.
2. Reduction on development and verification cost: The highly integration of hardware can favor the application of scalable software-driven framework for decoupling of hardware and software, leading to faster development cycle and lower cost of software development and verification.
3. Support for implementation of OTA: The Over the Air (OTA) technology can achieve the goal to upgrade the automotive software remotely through wireless access points of vehicles. It can provide a convenient, timely, and lower cost of recall management by cutting the necessity to bring the vehicles back. The centralized architecture with fewer ECUs and unified software architecture can reduce the verification complexity of the OTA update process.
4. Support for implementation of advanced intelligent functions: Vehicle intelligence requires the powerful hardware as well as the advanced software devel-

opment model. The application of the scalable software-driven framework, high performance computing platform and heterogeneous communication architecture which are benefited from the new in-vehicle architecture can make it possible to implement advanced functions like intelligent in-cabin system and high-level autonomous driving.

2.2   The Necessity for Protecting CAN

CAN is currently the most mature protocol with the highest market share, and has been required to be implemented on production vehicles. It is widely used in automotive network related to safety-critical functions such as automobile transmission and body control. The safety-critical information, e.g., the engine or cruise control is exchanged over the CAN bus. The data in CAN is exchanged via the unit called data frame. Its structure can be divided into five fields, including arbitration field, control field, data field, CRC (Cyclic Redundancy Check) field and ACK (Acknowledgement) field (can be seen in Fig. 4). The arbitration field bears the identifier which can be used for identifying different frames as well as competing the rights of transmitting on the bus.

Safety is always the first priority for vehicles. Despite the proportion of CAN for the in-vehicle network is getting smaller as the architecture evolves, the urgency for research on protecting CAN is even getting stronger. The reasons can be explained as follows. Firstly, CAN will not be abandoned by the future in-vehicle network due to its high efficiency and low cost. Despite many advanced technologies such as high-speed Ethernet and high performance computing devices are introduced, the lowest level network for both the domain model and zone model would still be developed as a signal-oriented communication paradigm. Such design can provide reliable and real-time data exchange to ensure the safety of vehicles. CAN is still going to play critical role in these areas, especially the networks for safety-critical functions. Secondly, the risk of in-vehicle network being attacked increases significantly. The evolution of in-vehicle network architecture is along with the trend that the number of communication technologies used in vehicles increases. That also opens more doors for attacks, resulting in higher security concerns for vehicles. The attackers can intrude on the in-vehicle network by exploiting the flaws in the hardware or software of these access points. Since CAN was originally designed to work in an isolated environment, CAN does not take any security concerns into

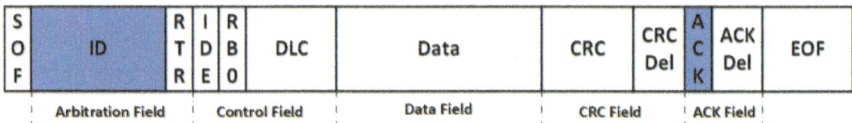

Fig. 4 The CAN data frame format

consideration [8], making CAN vulnerable to attacks. It has been demonstrated that the adversary can manipulate the vehicles' behavior after obtaining access to the safety-critical CAN bus [18]. Thus, we claim that defending CAN from attacks is critical for ensuring the safety of vehicles no matter on the current or future in-vehicle network.

3   Semantic-Based Intrusion Detection Methods

3.1   Motivation and Basic Idea

The data transmitted on the in-vehicle network has specific physical meaning for describing the current states and dynamics of the vehicle. An example of physical variables transmitted on in-vehicle CAN is listed in Table 1. For instance, the data can be explained as the speed of the engine, vehicle velocity or the state of the headlights. These data are transmitted and exchanged over the in-vehicle network to control the various functions of vehicle.

For a given dynamic of automotive system, there should be a certain correlation between data read from different sensors since they obey the same physical law. Under normal circumstance, the variable which indicates the inclination angle of the accelerator pedal should change accordingly when the driver presses the pedal. The speed of the engine and vehicle velocity would increase. In the meantime, the automotive gear would also switch in time. The different parts of the vehicle collectively respond to the act of pressing the accelerator pedal in a correlated and consistent manner. Therefore, the physical properties of vehicles can be abstracted by the physical model built from the semantic traffic. The correlation among different sensors can be exploited to detect anomaly. We assume that the attacker cannot compromise all relevant ECUs simultaneously which is plausible in real scenarios. The intrusion detection is to identify any observation which is inconsistent with expected behavior.

To detect unexpected behavior, the first priority is to construct the model for describing the relationship between variables obeying same physical laws. One

Table 1 An example of physical variables on CAN

Physical variables	
Vehicle speed	Position of steer
GPS speed	Torque of wheel
Acceleration pedal	Wheel angle
Brake pedal	Gear
Engine RPM	Coolant temperature
Fuel rate	Ambient temperature
Fuel/Air commanded equivalence	Air intake temperature
Master cylinder pressure	Boost pressure

way is to build the physical model manually based on the physical expression or experience. Cho et al. [5] proposed an anomaly detection method called Brake Anomaly Detection for the brake-by-wire system. Under normal circumstances, the behavior of the vehicle should be consistent with the driver's intent and the surrounding driving environment. The authors chose the Brush tire model [2] as the normal behavior model to characterize the frictional relationship between the tire and the ground. The attack to the brake-by-wire system can be observed by checking the consistency between the driver's input and the actual data captured from the in-vehicle network. The model also takes into account the change in the coefficient of friction of the tires under different weather and road conditions. Similarly, Ref. [10] designed a delicate ring-based architecture to organize multiple correlations by utilizing the physical model and experience. In this study, ten variables and nine nodes in total comprise the well-designed correlation ring to improve the robustness of detection while reducing the overall computation overhead.

However, these methods require in-depth understanding about the target system and expertise, which may not always be available. Researchers resort to machine learning algorithms to construct the model automatically that reflects the physical laws. It is mainly based on the insight that multiple sensors readings are directly proportional to the same physical phenomenon under normal circumstances [1]. Thus, the model can be generated from semantic traffic of in-vehicle network without the requirement for the in-depth knowledge of the control system. The machine learning algorithms to be exploited can be varied including artificial neural network [33], random forest regressor [20], deep autoencoder [11], and CNN model [13].

3.2 Machine Learning-Based Methods

Reference [20] formulated the problem to detect anomalies as a machine learning prediction problem that can be resolved by the regression model. The authors selected a set of correlated sensor data as features of the regression model based on domain knowledge and pairwise correlations firstly. The sensor signals which can be used for calculating vehicles' speed are taken as an example in this study. They included engine speed, acceleration on both longitudinal and lateral orientation, brake pedal ratio, steering angle, gear, and so on. During the training phase, the feature readings are fed into a Random Forest Regressor to train a regression model. While in the testing phase, the output values of the model can be estimated continuously based on the trained regression model. The anomaly can be flagged once the difference between the observed value and the estimated value is larger than a predefined threshold.

A more advanced learning technique for generating the physical model automatically is introduced in an intrusion detection system called context-aware intrusion detection system (CAID) [33]. CAID exploits the Bottleneck Artificial Neural Network (ANN) to develop the reference model of the automotive control system.

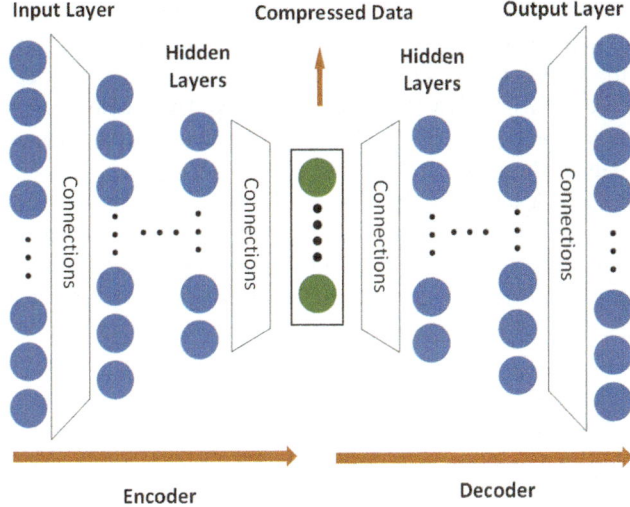

Fig. 5 Architecture of deep autoencoder neural network

The bottleneck ANN is designed as a network model in that the input and output layers are with the same number of neurons while the hidden layer is with a significantly less number of neurons. The sensor signals to describe the state of the engine control unit, such as fuel rate, absolute throttle position, engine RPM, and seven other signals, are collected to validate the performance of the proposed method. The parameters of the model can be generated in the training phase. During the testing phase, the estimated value can be obtained by reconstructing an input via the trained bottleneck ANN. CAID can detect anomalies by checking the similarity of actual readings of the sensor against the estimated values.

Reference [11] devised a deep autoencoder-based intrusion detector to extract the inherent redundancy of related sensors. The autoencoder (as shown in Fig. 5) is composed of two parts which are the encoder and the decoder respectively. Both the encoder and decoder are deep neural networks with multiple hidden layers. The aim of the encoder is to compress the input into low-dimensional features as much as possible, while the decoder aims to restore the compressed features to the original data as much as possible. By cascading the encoder and the decoder together, the autoencoder can extract the pattern of the input data. The overall process of the research [11] is as follows. Firstly, the authors selected a set of correlated data as input. The evaluation is performed on a publicly available dataset. It includes three categories of data, which are data from sensors on CAN bus, data from GPS sensors, and data from IMU sensors. Next, the deep autoencoder is adopted to learn the consistent pattern of these sensor data from the trustful training dataset. The learned consistent pattern can be expressed as the normal behavior of the automotive control system. In [11], the evaluated autoencoder network is designed with a 4-layers encoder and 4-layers decoder. The authors defined three different means to measure the error of the input against the reconstructed output. The training process of the encoder and decoder can be repeated to update the parameters of the model by minimizing the reconstruction error. Finally, anomalous behaviors can be detected by checking the reconstruction error during running. The reconstruction error shall be ranged within a predefined bound. If the reconstruction error exceeds the bound, an intrusion can be alarmed.

Reference [13] designed a framework that comprises an anomaly detection method based on Convolutional Neural Network (CNN) as well as an ensemble classifier which consists of multiple traditional machine learning algorithms. The ensemble classifier is to evaluate the effectiveness of the proposed CNN-based anomaly detection method. The proposed CNN-based method introduces a multi-stage attention Long Short-Term Memory (LSTM) model to enable the algorithm can focus on the significant parts of the data. The authors provided a comprehensive evaluation of four distinct anomaly types generated by [31] which are instant, constant, gradual drift, and bias to a publicly available dataset, and their combinations.

3.3 Summary

The semantic-based methods exploit the fact that the CAN traffic over the automotive network bears specific physical meanings for representing the dynamics or states of vehicles. Thus, these physical variables can be used to construct the abstract of the physical properties of vehicles. Machine learning algorithms can build the model automatically without requiring in-depth knowledge of the target system. The inconsistency with expected behavior can be regarded as an intrusion. Despite reducing the effort for generating the model compared to the traditional methods, the proprietary nature of CAN makes the obtainment of the specific meanings of the CAN frames a non-trivial work. It hinders the research on semantics-based methods since the specific meanings of the frames are kept confidential from the public.

4 Literal-Based Intrusion Detection Methods

4.1 Motivation and Basic Idea

There are two main limitations of semantic-based intrusion detection methods. First, it is non-trivial to obtain the semantic meaning of data from in-vehicle network. The automotive industry is not willing to disclose the detailed specification of their CAN messages considering the concerns on intellectual property and security. That is, the detailed meaning of automotive CAN messages cannot be obtained publicly. Second, the selection of input data requires domain knowledge or correlation computation. The performance of such methods on irrelevant data beyond the selected sensor has not been verified. These limitations hinder the application of semantic-based intrusion detection methods.

In this section, we introduce one more intuitive kind of method called literal-based intrusion detection method. It is unnecessary to obtain or derive the semantics of the CAN messages painstakingly. The binary streams (literal value) can be exploited directly as the input for the intrusion detection system. Firstly, the inherent correlations are extracted by analyzing the binary stream of CAN traffic. The extracted correlations can be used to characterize the normal behavior of the system or pattern of the anomalies. After building the required model for the target system, the intrusion can be reported by comparing the expected data with the observed one.

The main insight behind the literal-based intrusion detection methods can be summarized as follows. CAN is highly deterministic and predictable during operations to manage the requirements for strict real-time, and provide stable and reliable services. The stable operational patterns for CAN shall be observed in the absence of cyber attack. It has been pointed out in [9, 21] that the model of normal behavior can be established from the analysis of CAN data streams without understanding the semantics of CAN messages. Information entropy is a measurement to describe the uncertainty of a system. The more orderly and deterministic the system is, the lower the information entropy is. Reference [24] proposed the entropy-based intrusion detection methods for in-vehicle network. The entropy of the data traffic can be computed for representing the state of CAN traffic. When the entropy value deviates from the normal range, it means that there is an attack mounted on the in-vehicle network. However, the estimation of the entropy value can be affected easily by different driving scenes, which results in a high false positive rate.

Machine learning algorithms are better options for processing the binary streams of CAN messages. Generally speaking, the overall process of these methods can be concluded as two phases, which are the training phase and the testing phase as

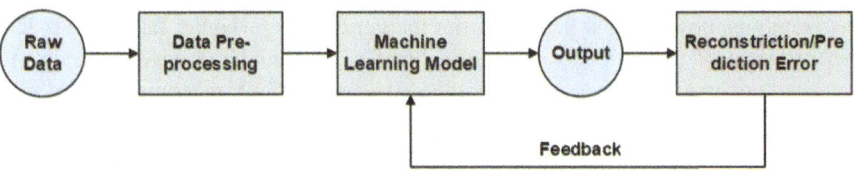

Fig. 6 The training phase

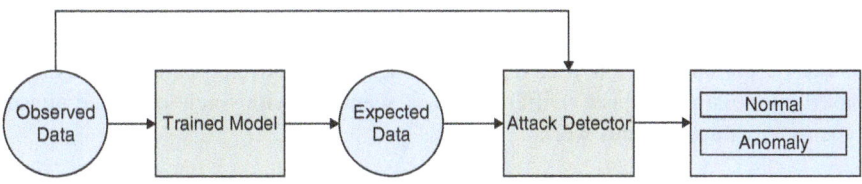

Fig. 7 The testing phase

illustrated in Figs. 6 and 7 respectively. The aim of the training phase is to develop the model for intrusion detection by extracting latent patterns from the traffic of in-vehicle network. The raw data, i.e. the literal binary data of CAN messages after pre-processing can be fed into the machine learning algorithm to train the model. The feedback process is to minimize the reconstruction/prediction error to improve the model performance. The training phase can be performed offline in a controlled environment considering it is a time-consuming task. During the testing phase, the observed data is compared with the output (expected data) by the trained model to detect anomalies. The observed data and the expected data are fed into the attack detector together to identify whether their difference exceeds a well-designed threshold.

The methods in this section can be divided into two categories according to whether the attack sample is required in the training phase, which is specification-based methods using attack samples for the training model and anomaly-based methods using normal samples for generating the model.

4.2 Specification-Based Methods

Methods in this category require labeled attack samples for training the classification model. The model can learn the patterns of the CAN traffic under attack during the training phase. The intrusion can be detected once any similar patterns are observed during the testing phase.

Xie et al. [34] proposed a generative adversarial network (GAN) based intrusion detection method, which can be shown in Fig. 8. Technically, the GAN model consists of two core components: generator (G) and discriminator (D). The basic principle of how GAN works is as follows. The generator utilizes random noise as input and tries to output synthetic data to deceive the discriminator. On the contrary, the discriminator utilizes the ground truth as input and tries to make decisions as accurately as possible that the data from the generator is whether fake or not. The performance of the generator and discriminator can thus be improved during the repeated adversarial process. In [34], the real attacked CAN messages are fed into GAN for training the intrusion detection model.

CANintelliIDS [14] is designed based on a convolutional neural network (CNN) combined with an attention-based gated recurrent unit (GRU) model. Similar to LSTM, the GRU model is suitable for solving the prediction problem of sequential data. Besides, the utilization of GRU can be helpful for improving the efficiency as well as reducing the memory consumption considering its more simplified design and fewer parameters compared to LSTM. The intrusion detection model is trained based on the attack dataset. Different attack scenarios with single or mixed attack types are evaluated in this work.

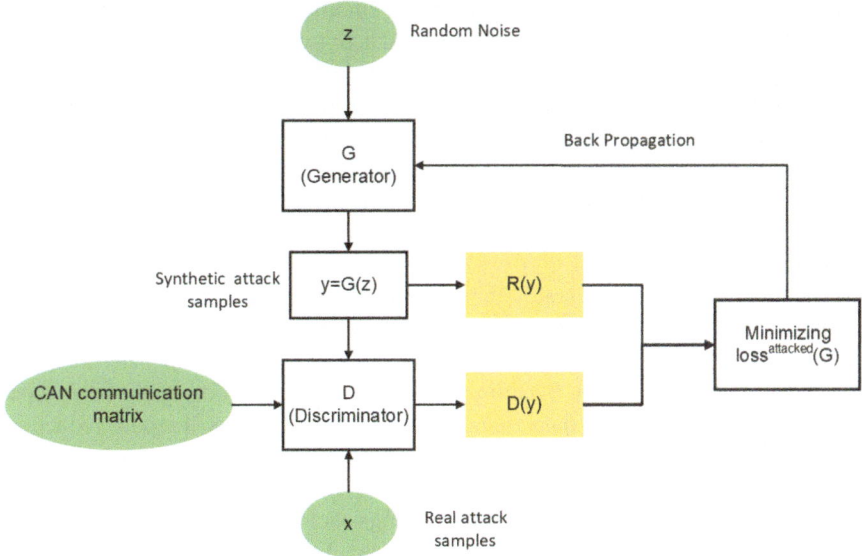

Fig. 8 Training process of GAN generator in Ref. [34]

4.3 Anomaly-Based Methods

The methods belonging to this category do NOT require the attack labeled data during the training phase. The intrusion detection model is generated from the attack-free CAN traffic under normal circumstances. If there is any deviation from the normal model is detected, an intrusion can be alarmed. Compared with specification-based methods, the performance of anomaly-based methods to detect unknown attacks is preferable.

CANnolo [21] implements LSTM as the hidden layer of the auto-encoder. The auto-encoder is used to automatically learn the normal patterns of raw CAN data without semantics. At runtime, CANnolo utilizes the trained model to reconstruct the CAN streams. The Mahalanobis distance between the reconstructed and the observed sequences is computed as an anomaly score to indicate how likely the CAN bus is under attack. Reference [29] designed an LSTM-based RNN model constituted by two non-recurrent hidden layers and two recurrent LSTM layers. To improve the accuracy of the detection model, the features on the time dimension are combined with features on the data dimension as input for LSTM neural network [39]. Besides, the multi-task LSTM framework is utilized to implement parallel computing locally as well as on the mobile edge. The mobile edge can break the limitation of onboard computing capacity.

HDAD [32] introduces the concept of hyper-dimensional computing (HDC) to detect intrusion for in-vehicle network. HDC is a novel computing paradigm that simulates the working mechanism of neuronal circuits in the human brain. It works using high-level and abstract patterns of neural activity. Firstly, the training data are encoded into hypervectors (HVs) to learn pattern. Only normal patterns are required in the training dataset. The number of dimension for HV can be set as 10,000 or larger. Then, the pattern decoder is subject to reconstruct the HV to the original data. Finally, the reconstruction error is used for determining if there is an intrusion. The authors claimed that the adoption of HDC can benefit from compact model size, reduced computation cost, and one-shot learning in contrast to deep learning-based approaches.

The research on HDC is still at the preliminary stage. To improve the efficiency of the intrusion detection model, CLAM [28] improves the process of data pre-processing to cut the dimensionality of raw CAN traffic which can favor the acceleration of computation. Specifically, READ [22] method designed for reverse engineering of automotive data frames is introduced to assist the data reduction in CLAM. READ method can analyze the traffic and extract signals that vary continuously without supervision. These extracted signals can be explained as physical signals with specific physical meanings such as vehicle speed and engine speed. In the step of data pre-processing, the signal boundaries can be determined by READ methods. Thus, instead of using the whole CAN frame as input, only the bits bounded by data pre-processing are conveyed to the intrusion detection model for improving efficiency. It should be noted that the CLAM model also does NOT need to know the semantics of CAN frames. The CLAM model consists of a 1-D Convolution Network and bi-directional LSTM with an attention mechanism. The attention mechanism can enable the model to focus on the important parts of the data.

4.4 Summary

The literal-based intrusion detection methods can automatically extract intrinsic relationships among variables and develop the intrusion detection model by ana-lyzing the binary stream of CAN frames. The semantics of frames are not required. The intrusion detection model can be trained by either the attack-free samples to generate the normal patterns of CAN frames or the attack-labeled samples to detect well-known intrusion. That is, the literal-based intrusion detection can be directly applied to CAN frames from the data link layer without knowledge of the protocol specifications of the upper layer (application layer). The protocol specifications of the application layer for automotive CAN bus are kept confidential from the public. Different specifications are defined for different car manufacturers and even different car models. From this perspective, compared to semantic-based methods, literal-based intrusion detection methods seem more attractive to both security technicians in the automotive industry as well as researchers from academia.

5 Timing-Based Intrusion Detection Methods

5.1 Motivation and Basic Idea

Considering that the vast majority of CAN frames are triggered periodically, i.e. CAN frames are queued for transmission at a fixed rate, there are some regularities of timing characteristics that can be found from CAN frames traffic. Illegal data due to unauthorized intrusion attacks can disrupt the regularities. Based on this observation, researchers propose that intrusion detection can be implemented by digging into the temporal patterns of CAN data traffic. The inconsistency with expected temporal patterns can be regarded as an anomaly. Similar to literal value-based methods, the timing-based method can also cope with the disadvantage of the proprietary nature of CAN data specifications. The traditional approach [25] builds the mathematical model to describe the timing behavior precisely of CAN frames traffic by utilizing real-time scheduling theory. However, the main downside is that it requires in-depth domain knowledge for building the model and it is hard to build a model adapted to different driving scenes.

5.2 Machine Learning-Based Methods

Tomlinson et al. [30] introduced three straightforward machine learning algorithms (Autoregressive Integrated Moving Average, Z-score, and supervised threshold) combined with time-defined windows to identify abnormal timing changes for CAN traffic. Reference [26] proposed a deep convolutional neural network (DCNN) model-based intrusion detection method. The authors designed a data pre-processing module called frame builder to convert the raw CAN traffic to the data fitted for the CNN model. Subsequently, the DCNN model learns temporal sequential patterns of raw CAN traffic automatically without hand-designed features. The CAN data with labels indicating whether normal or not is required for the training process. The Recurrent neural network (RNN) is naturally designed to cope with time sequence data. Reference [27] designed an RNN model with a 1-layer hidden layer of 100 nodes. From the evaluation results, the proposed RNN model can handle more realistic scenarios in that the period can fluctuate. The period fluctuation can often be observed in CAN traffic collected from real vehicles. It is mainly caused by the process of multiple ECUs to compete with the right of CAN bus usage. The attack samples are needed for computing the final output.

Generative Adversarial Network (GAN) is introduced in [15] to extract temporal features for modeling normal behaviors by attack-free training dataset. The authors improved the original GAN model by introducing a modified evolutionary algorithm to produce multiple generators instead of one single generator. This modification can increase the chance to obtain a better performance generator in the process of adversary game, which can mitigate the issue of instability in GAN. Since no given attack sample is used for the training model, the data collection process shall be undertaken when driving under different conditions to capture as many normal features as possible. It can be helpful for reducing false positives.

5.3 Summary

Timing-based methods build the intrusion detection model by analyzing the timing characteristics of CAN traffic automatically. As same as the literal-based methods, the semantic values of CAN traffic are NOT required for timing-based methods. The timing-based methods can effectively detect attacks that essentially change the timing behavior of CAN frames, such as denial of service (DoS) attack, suspension attack and injection attack. However, from another perspective, the performance of such methods can be significantly degraded when dealing with more sophisticated attacks which do not influence the timing characteristics. Due to the broadcast nature of CAN, the attacker can eavesdrop and learn the temporal patterns of the target frames silently and stealthy. Next, the attacker can bypass the deployed timing-based intrusion detection system by injecting malicious frames with the same identifier and similar transmission pattern as the victim.

6 Signal Characteristics-Based Intrusion Detection Methods

6.1 Motivation and Basic Idea

Another way to design an intrusion detection system is to exploit the unique hardware characteristics of automotive ECUs to generate a digital fingerprint. Specially, the tiny but measurable differences in specific characteristics (such as voltage or timing) can be obtained from the electrical signal transmitted on the bus medium. The extracted difference can then be utilized as a device fingerprint to enable authentication in CAN. The intrusion can be detected when the actual sending ECU (predicted data) of the newly received CAN frame is inconsistent with its legitimate sending ECU (expected data).

The difference in hardware is mainly due to the imperfect manufacturing processes, which results in the characteristics of unique, stable, and hard to replicate to enable higher security. It was first introduced in [23] which exploits the difference of signal characteristics in the physical layer to identify ECUs for in-vehicle network. This study has demonstrated that the signal characteristics driven by the hardware of ECUs can be unique while remaining stable within a certain range for several months. Inspired by this observation, more researches to protect the in-vehicle network by utilizing low-level signal characteristics of CAN frames are proposed.

An idea of implementing intrusion detection based on signal characteristics is to explicitly define the relationship between the collected data and the hardware characteristics of the sending ECU by establishing a model. Most of such works exploit a linear model to represent the relationship of the accumulation of derived signal characteristics over time or data samples. Viden [4] adopts voltage measurements to build the model to source the sending node. Viden measures the voltage of CAN high and CAN low respectively during the transmission of dominant bits. These measurements are gathered to derive a voltage instance containing six statistics to describe the distribution of measurements. The voltage instance can be expressed as the transient behavior of voltage of sending ECU. At last, Viden constructs a linear model called voltage profile by utilizing the continuously obtained voltage instance. The main reason why Viden can work is that the voltage instances derived from the same ECU shall be nearly equivalent. Thus, the voltage profile can be constructed as a linear model by which the sending ECU can be correctly identified.

Different from Viden, Refs. [3, 19, 38] exploited the skew in clocks of electronic devices to establish the linear model for intrusion detection. The clock skew is defined as the difference in frequency between clocks. The common insight behind these methods is based on the observation that the clock skew is nearly constant for single ECU and unique among different ECUs. Thus, the linear model which represents the timing behavior of clock can be built for detect anomalies. Deviations from the established model can be used to trigger an alarm for intrusion on in-vehicle network. For example, the sudden change of the slope of the linear model can be regarded as an indication that the attack is mounted.

6.2 Machine Learning-Based Methods

Besides the model-based methods, the problem of identifying the sending ECU for newly received CAN frames can also be regarded as a classification problem. The CAN frames from the same ECU are considered to be of the same class. If the actual class of any CAN frames (identified by the intrusion detection system) is inconsistent with its expected class (determined by the frame identifier), it indicates that the adversary performs an attack by injecting frames with falsified ID. The supervised machine learning algorithms can be used to solve such classification problem. Generally speaking, the overall process of methods belonging to this category can be summarized into three phases as shown in Fig. 9.

The first step is to preprocess the electrical CAN signal to derive the characteristics from the physical layer. The signal characteristics exploited in this phase can be varied from voltages measurements to timing characteristics, which is the same as the model-based methods. Subsequently, the statistical features in the time and/or frequency domain are extracted from the measurements. Finally, the supervised learning-based classification algorithms are adopted to generate a classifier to distinguish the attack from the normal CAN traffic.

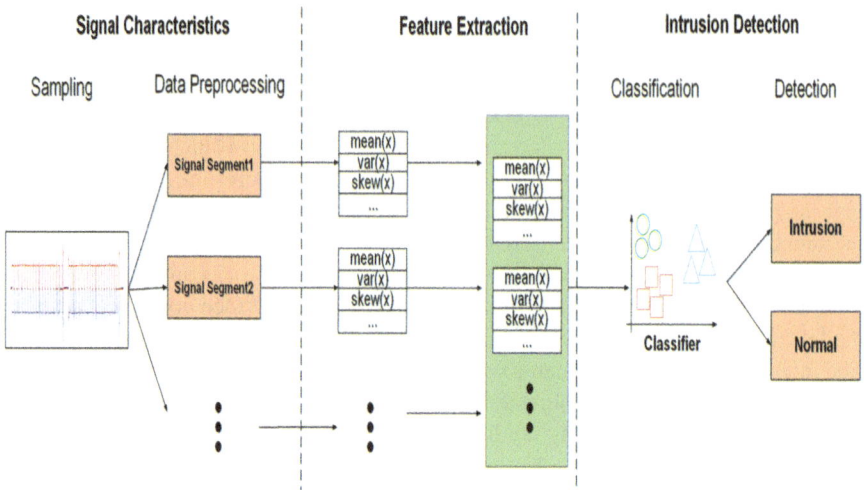

Fig. 9 Workflow of the supervised learning algorithms-based methods

6.2.1   Signal Characteristics Derivation

6.2.1.1   Using Voltage as Signal Characteristics

Choi et al. [6] proposed an approach to source the transmitting ECU by measuring the voltage of an array of the same consecutive bits. Specifically, it requires that an identical predefined bit sequence is embedded in all CAN frames transmitted on the bus. To achieve this, all data frames on the bus are set as the extended frame format with a 29-bit identifier. A predefined bit sequence which is 18-bit long is placed at the extended identifier field. Subsequently, the voltages of the pre-assigned bit sequence for every newly received CAN frame are sampled and measured. Obviously, reprogramming for all active ECUs on CAN bus is required to add the predefined bit sequence to each CAN frame. Besides, it can NOT be applied to the natural extended frames (the extended part of the identifier is already occupied). These limitations hinder its deployment on real production vehicles.

References [7, 16, 17] improved the process for extracting signal characteristics based on voltage. More specifically, SCISSION [16] and EASI [17] divide the string of consecutive dominant bits into three parts, which are the rising edge, the falling edge, and the holding edge of the dominant state part (as shown in Fig. 10). The approach adopted by VoltageIDS [7] is similar except that it only considers the 1-bit length holding edge of the dominant state part. Next, the voltage is measured and gathered separately for each part. The significant features of voltage on the rising edge and falling edge could be suppressed without such actions considering that their length is too short (resulting in much fewer samples) compared to the holding edge. By doing so, the combined features including the voltage measurements as well as the signal shape can be extracted to better represent

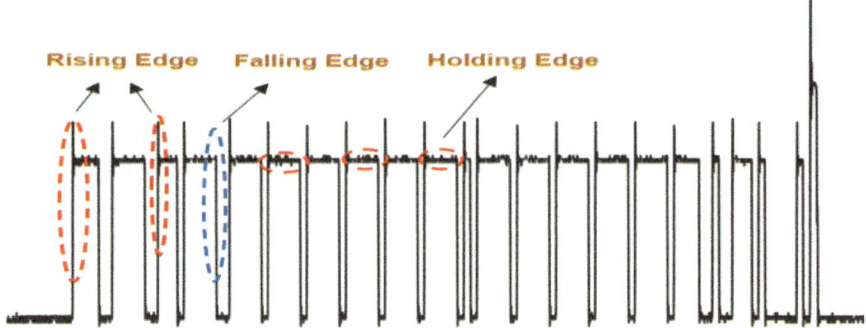

Fig. 10 An example of CAN electrical signal

the signal characteristics of the sending ECU. In addition, EASI [17] designs a low-cost solution to improve the efficiency of preprocessing the electrical signal. The authors optimized the interval for signal sampling and introduce the Random Interleaved Sampling technique, which greatly reduces the sampling rate and system resource requirements. It can favor the development on in-vehicle network.

6.2.1.2   Using Timing as Signal Characteristics

Apart from the voltage characteristics, the timing characteristics of CAN electrical signal can also be utilized to construct the intrusion detection system. Most existing works [3, 19, 36] using timing characteristics estimate the clock skew based on the periodic CAN traffic. Considering that most CAN frames are transmitted nearly periodically, the skew in the clock of the transmitter can be estimated by the difference between the expected and the actual arrival time of periodic traffic. From the observations of CAN traffic from real vehicles, the actual period of many frames can fluctuate a little wild and some frames might stop transmission for a while in real cases [19]. To mitigate these challenges, CANvas [19] improves the estimation process by introducing the concept of hyper-period. However, the dependency on periodic traffic still remains which makes it unavailable to aperiodic frames or sporadic frames.

BTMonitor [37] employs the timing characteristics of a single CAN frame to build the intrusion detection model, by which the dependency on periodic traffic can be cut. The insight behind BTMonitor is that the electrical signal length which is driven by the hardware of the transmitter can reflect the timing characteristics of sending ECU. Thus, the clock skew can be derived by measuring the signal length from a single frame, making the signal preprocessing process independent of the periodic traffic. To capture the signal which can accurately reveal the hardware characteristics of sending ECU, the signal segment in the identifier field shall be excluded from the measurement process. The reason is that multiple ECUs on the bus might initiate the transmission simultaneously and compete for the right of bus

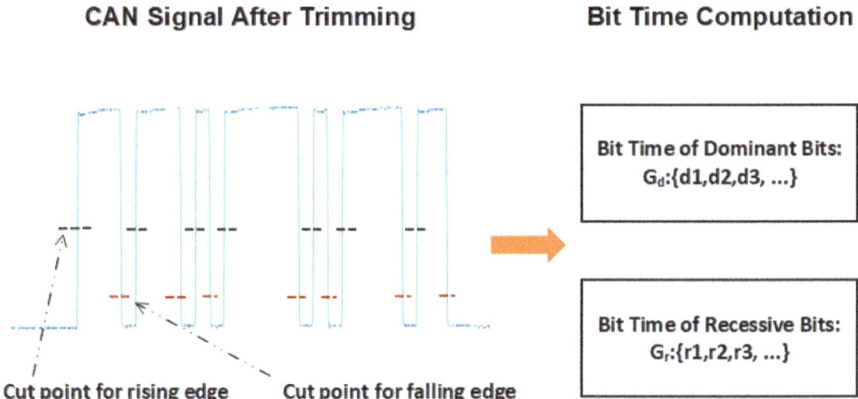

Fig. 11 Signal characteristics derivation in BTMonitor [37]

usage during the arbitration process. Thus, the signal segment in the identifier field might reflect the signal characteristics of more ECUs beyond the transmitter. For the same reason, the duration of the signal during the acknowledge field is also excluded to remove the effect on signal length by other nodes.

After the trimming, BTMonitor divides the remaining signal into different segments along the consecutive edges. These signal segments can be referred to as two categories, which are dominant bits and recessive bits. To reduce the requirement for a high sampling rate for measuring device, BTMonitor takes the rising edge as well as the falling edge into consideration. The point to divide the signal on the rising edge is different from the point on the falling edge. Finally, BTMonitor measures and computes a corresponding bit time for each signal segment. The calculated bit time of each category is gathered up to form data samples that represent the timing characteristics of sending ECU. The process is shown in Fig. 11.

6.2.2 Feature Extraction and Intrusion Detection

Once the signal characteristics are obtained, the preprocessed data is fed into the next phase to extract statistical features in the time and/or frequency domain. The extracted features can be used as device fingerprints to identify different ECUs. As an example, BTMonitor adopts eight statistical features in the time domain for each of the categories of dominant bits and recessive bits, i.e. 16 statistical features in total to represent one received data sample. The selected features are shown in Table 2. Then, the generated fingerprint is input into the classifier for intrusion detection.

During the training phase, supervised learning algorithms along with labeled samples (training datasets) are used to train the classifier. During the runtime phase, the newly derived device fingerprints are fed into the trained classifier to predict its

Table 2 Selected features in time domain by BTMonitor [37]. x represents bit time. N is the number of data

Feature	Description
Mean	$\bar{x} = \frac{1}{N} \sum_{i=1}^{N} x(i)$
Standard deviation	$\sigma = \sqrt{\frac{1}{N} \sum_{i=1}^{N} (x(i) - \bar{x})^2}$
Variance	$\sigma^2 = \frac{1}{N} \sum_{i=1}^{N} (x(i) - \bar{x})^2$
Skewness	$\gamma = \frac{1}{N} \sum_{i=1}^{N} (\frac{x(i) - \bar{x}}{\sigma})^3$
Kurtosis	$\beta = \frac{1}{N} \sum_{i=1}^{N} (\frac{x(i) - \bar{x}}{\sigma})^4 - 3$
RMS (Root mean square)	$A = \sqrt{\frac{1}{N} \sum_{i=1}^{N} x(i)^2}$
Highest value	$H = \max(x(i))$
Energy	$en = \frac{1}{N} \sum_{i=1}^{N} x(i)^2$

actual sending ECU. If the predicted sending ECU is inconsistent with the legitimate sending ECU, an intrusion is alarmed. Varied machine learning algorithms, such as Linear Support Vector Machines [6, 7, 17], Packed Decision Trees (BDT) [6, 7], Logistic Regression [16, 17, 37], Naive Bayes Classifiers [17], Neural Networks [6], etc. are used to generate classification models.

6.3 Summary

The difference in the signal can be utilized to generate the fingerprint for the ECU. The derived fingerprint can then provide the ability to authenticate the sending ECU and detect intrusion. We summarize the overall process of machine learning-based methods in three steps, which are signal characteristics derivation, feature extraction, and intrusion detection respectively. These methods can be divided into two categories based on the exploited signal characteristics, which are the methods using signal voltage and the methods using signal timing. The general process of feature extraction and intrusion detection in both categories is similar. The statistical features in the time and/or frequency domain are extracted from the extracted signal characteristic and combined as the device fingerprint. Finally, popular machine learning algorithms are utilized as the classification model to detect intrusion. The signal characteristics-based methods can provide high security for automotive CAN bus considering that the fingerprint is derived from the inherent physical characteristics and is hard to be duplicated. However, how to obtain an effective but stable fingerprint from the mutable and sensitive signal is the major challenge to be solved.

7 Conclusion

CAN is the most important communication protocol for the current in-vehicle network and aged for over 35 years. With the rapid development of connectivity and intelligence for today's vehicles, the underlying internal communication system is updated accordingly to manage the future's needs. In this chapter, we firstly take a discussion about the traditional and tomorrow in-vehicle network architecture as well as the advantages brought by the new architecture, aiming to provide a whole picture of how in-vehicle network evolves. The necessity of protecting CAN for ensuring the safety of vehicles is emphasized to motivate the research on defending techniques. Subsequently, we introduce different approaches to detect intrusion by categories based on the domain knowledge used in machine learning algorithms.

The variables with specific physical meanings in CAN can respond to a physical phenomenon in a correlated way. These observations can be exploited to detect intrusion which is detailed in semantic-based intrusion detection methods. Further studies reveal that the latent relationship can be extracted without requiring semantics of CAN frames. Literal-based intrusion detection methods provide a detailed description of how it works from two aspects according to whether the attack sample is required for training the model. Timing-based intrusion detection methods exploit the fact that most CAN traffic is triggered periodically thus the timing of CAN traffic can exhibit specific patterns. However, the main drawback is that it cannot deal with attack scenarios in which the timing characteristics are not affected. At last, signal characteristics-based intrusion detection provides a novel way of fingerprinting the ECUs by measuring the low-level characteristics of CAN electrical signals. Considering it is derived from the unique and inherent hardware characteristics, it can provide high security for in-vehicle CAN bus.

In conclusion, we survey the machine learning-based intrusion detection methods for automotive CAN bus and provide the introduction from the perspective of the exploited domain knowledge. We hope this chapter can help the interested reader to understand and grasp the status and research of machine learning-based intrusion detection methods comprehensively.

References

1. Akowuah, F., Kong, F.: Physical invariant based attack detection for autonomous vehicles: Survey, vision, and challenges. In: 2021 Fourth international conference on connected and autonomous driving (MetroCAD), pp. 31–40. IEEE, Piscataway (2021)
2. Bakker, E., Nyborg, L., Pacejka, H.B.: Tyre modelling for use in vehicle dynamics studies. SAE Trans. **96**, 190–204 (1987)
3. Cho, K.T., Shin, K.G.: Fingerprinting electronic control units for vehicle intrusion detection. In: 25th USENIX conference on security symposium (USENIX Security), pp. 911–927. USENIX Association, Berkeley (2016)
4. Cho, K., Shin, K.G.: Viden: Attacker identification on in-vehicle networks. In: 2017 ACM conference on computer and communications security (CCS), pp. 1109–1123. ACM, New York (2017)
5. Cho, K.T., Shin, K.G., Park, T.: CPS approach to checking norm operation of a brake-by-wire system. In: ACM/IEEE sixth international conference on cyber-physical systems (ICCPS), pp. 41–50. ACM, New York (2015)

6. Choi, W., Jo, H.J., Woo, S., Chun, J.Y., Park, J., Lee, D.H.: Identifying ecus using inimitable characteristics of signals in controller area networks. IEEE Trans. Veh. Technol. **67**(6), 4757–4770 (2018)

7. Choi, W., Joo, K., Jo, H.J., Park, M.C., Lee, D.H.: Voltageids: low-level communication characteristics for automotive intrusion detection system. IEEE Trans. Inf. Forens. Secur. **13**(8), 2114–2129 (2018)

8. Di Natale, M., Zeng, H., Giusto, P., Ghosal, A.: Understanding and using the controller area network communication protocol: theory and practice. Springer Science & Business Media (2012)

9. Groza, B., Murvay, P.S.: Efficient intrusion detection with bloom filtering in controller area networks. IEEE Trans. Inf. Forens. Secur. **14**(4), 1037–1051 (2018)

10. Guo, F., Wang, Z., Du, S., Li, H., Zhu, H., Pei, Q., Cao, Z., Zhao, J.: Detecting vehicle anomaly in the edge via sensor consistency and frequency characteristic. IEEE Trans. Veh. Technol. **68**(6), 5618–5628 (2019)

11. He, T., Zhang, L., Kong, F., Salekin, A.: Exploring inherent sensor redundancy for automotive anomaly detection. In: 2020 57th ACM/IEEE design automation conference (DAC), pp. 1–6. IEEE, Piscataway (2020)

12. Hoppe, T., Kiltz, S., Dittmann, J.: Security threats to automotive can networks–practical examples and selected short-term countermeasures. In: International conference on computer safety, reliability, and security (SAFECOMP), pp. 235–248. Springer, Berlin, Heidelberg (2008)

13. Javed, A.R., Usman, M., Rehman, S.U., Khan, M.U., Haghighi, M.S.: Anomaly detection in automated vehicles using multistage attention-based convolutional neural network. IEEE Trans. Intell. Transp. Syst. **22**(7), 4291–4300 (2020)

14. Javed, A.R., Ur Rehman, S., Khan, M.U., Alazab, M., Reddy, T.: CANintelliiDS: detecting in-vehicle intrusion attacks on a controller area network using CNN and attention-based GRU. IEEE Trans. Netw. Sci. Eng. **8**(2), 1456–1466 (2021)

15. Kavousi-Fard, A., Dabbaghjamanesh, M., Jin, T., Su, W., Roustaei, M.: An evolutionary deep learning-based anomaly detection model for securing vehicles. IEEE Trans. Intell. Transp. Syst. **22**(7), 4478–4486 (2020)

16. Kneib, M., Huth, C.: Scission: signal characteristic-based sender identification and intrusion detection in automotive networks. In: ACM SIGSAC conference on computer and communications security (CCS), pp. 787–800. ACM, New York (2018)

17. Kneib, M., Schell, O., Huth, C.: EASI: Edge-based sender identification on resource-constrained platforms for automotive networks. In: The 2020 network and distributed system security symposium (NDSS), pp. 1–16. ISOC, San Diego (2020)

18. Koscher, K., Czeskis, A., Roesner, F., Patel, S., Kohno, T., Checkoway, S., McCoy, D., Kantor, B., Anderson, D., Shacham, H., Savage, S.: Experimental security analysis of a modern automobile. In: IEEE symposium on security and privacy (S&P), pp. 447–462. IEEE, Piscataway (2010)

19. Kulandaivel, S., Goyal, T., Agrawal, A.K., Sekar, V.: Canvas: fast and inexpensive automotive network mapping. In: The 28th USENIX conference on security symposium (USENIX Security), pp. 389–405. USENIX Association, Berkeley

20. Li, H., Zhao, L., Juliato, M., Ahmed, S., Sastry, M.R., Yang, L.L.: Poster: intrusion detection system for in-vehicle networks using sensor correlation and integration. In: The 2017 ACM SIGSAC conference on computer and communications security (CCS), pp. 2531–2533 (2017)

21. Longari, S., Valcarcel, D.H.N., Zago, M., Carminati, M., Zanero, S.: CANnolo: an anomaly detection system based on LSTM autoencoders for controller area network. IEEE Trans. Netw. Serv. Manag. **18**(2), 1913–1924 (2020)

22. Marchetti, M., Stabili, D.: Read: reverse engineering of automotive data frames. IEEE Trans. Inf. Forens. Secur. **14**(4), 1083–1097 (2018)

23. Murvay, P.S., Groza, B.: Source identification using signal characteristics in controller area networks. IEEE Signal Process. Lett. **21**(4), 395–399 (2014)

24. Müter, M., Asaj, N.: Entropy-based anomaly detection for in-vehicle networks. In: IEEE intelligent vehicles symposium, pp. 1110–1115. IEEE, Piscataway (2011)

25. Olufowobi, H., Young, C., Zambreno, J., Bloom, G.: Saiducant: specification-based automotive intrusion detection using controller area network (CAN) timing. IEEE Trans. Veh. Technol. **69**(2), 1484–1494 (2019)

26. Song, H.M., Woo, J., Kim, H.K.: In-vehicle network intrusion detection using deep convolutional neural network. Veh. Commun. **21**, 100198 (2020)

27. Suda, H., Natsui, M., Hanyu, T.: Systematic intrusion detection technique for an in-vehicle network based on time-series feature extraction. In: 2018 IEEE 48th international symposium on multiple-valued logic (ISMVL), pp. 56–61. IEEE (2018)

28. Sun, H., Chen, M., Weng, J., Liu, Z., Geng, G.: Anomaly detection for in-vehicle network using CNN-LSTM with attention mechanism. IEEE Trans. Veh. Technol. **70**(10), 10880–10893 (2021)

29. Taylor, A., Leblanc, S., Japkowicz, N.: Anomaly detection in automobile control network data with long short-term memory networks. In: 2016 IEEE international conference on data science and advanced analytics (DSAA), pp. 130–139. IEEE (2016)

30. Tomlinson, A., Bryans, J., Shaikh, S.A., Kalutarage, H.K.: Detection of automotive can cyberattacks by identifying packet timing anomalies in time windows. In: 2018 48th Annual IEEE/IFIP international conference on dependable systems and networks workshops (DSN-W), pp. 231–238. IEEE (2018)

31. van Wyk, F., Wang, Y., Khojandi, A., Masoud, N.: Real-time sensor anomaly detection and identification in automated vehicles. IEEE Trans. Intell. Transp. Syst. **21**(3), 1264–1276 (2020). https://doi.org/10.1109/TITS.2019.2906038

32. Wang, R., Kong, F., Sudler, H., Jiao, X.: Brief industry paper: Hdad: hyperdimensional computing-based anomaly detection for automotive sensor attacks. In: 2021 IEEE 27th real-time and embedded technology and applications symposium (RTAS), pp. 461–464. IEEE (2021)

33. Wasicek, A., Pesé, M.D., Weimerskirch, A., Burakova, Y., Singh, K.: Context-aware intrusion detection in automotive control systems. In: 5th ESCAR USA conference, pp. 21–22 (2017)

34. Xie, G., Yang, L.T., Yang, Y., Luo, H., Li, R., Alazab, M.: Threat analysis for automotive can networks: a GAN model-based intrusion detection technique. IEEE Trans. Intell. Transp. Syst. **22**(7), 4467–4477 (2021)

35. Zeng, W., Khalid, M.A., Chowdhury, S.: In-vehicle networks outlook: achievements and challenges. IEEE Commun. Surv. Tutorials **18**(3), 1552–1571 (2016)

36. Zhao, Y., Xun, Y., Liu, J.: Clockids: A real-time vehicle intrusion detection system based on clock skew. IEEE Internet Things J. **9**, 15593 (2022)

37. Zhou, J., Joshi, P., Zeng, H., Li, R.: Btmonitor: bit-time-based intrusion detection and attacker identification in controller area network. ACM Trans. Embed. Comput. Syst. **18**(6), 1 (2020)

38. Zhou, J., Xie, G., Zeng, H., Zhang, W., Yang, L.T., Alazab, M., Li, R.: A model-based method for enabling source mapping and intrusion detection on proprietary can bus. IEEE Trans. Intell. Transp. Syst. (2022)

39. Zhu, K., Chen, Z., Peng, Y., Zhang, L.: Mobile edge assisted literal multi-dimensional anomaly detection of in-vehicle network using LSTM. IEEE Trans. Veh. Technol. **68**(5), 4275–4284 (2019)

Real-Time Intrusion Detection in Automotive Cyber-Physical Systems with Recurrent Autoencoders

Vipin Kumar Kukkala, Sooryaa Vignesh Thiruloga, and Sudeep Pasricha

1 Introduction

Modern-day vehicles are highly sophisticated cyber-physical systems (CPS) that consist of multiple interconnected embedded systems known as Electronic Control Units (ECUs). The ECUs run various real-time automotive applications that control different vehicular subsystem functions. Moreover, ECUs are distributed across the vehicle and communicate with each other using the in-vehicle network. In recent years, the number of ECUs being integrated into the vehicles and the complexity of software running on these ECUs has been rapidly increasing to enable various state-of-the-art Advanced Driver Assistance Systems (ADAS) features such as adaptive cruise control, lane keep assist, collision avoidance, and blind spot warning. This resulted in an increase in the complexity of the in-vehicle network over which huge volumes of automotive sensor and real-time decision data, and control directives are communicated. This increased complexity of modern-day vehicles has led to various complex challenges that pose a serious threat to the reliability [1–4], security [5–9], and real-time control of automotive systems [10–13].

V. K. Kukkala (✉)
NVIDIA, Santa Clara, CA, USA
e-mail: vipin.kukkala@colostate.edu

S. V. Thiruloga
Department of Electrical and Computer Engineering, Colorado State University, Fort Collins, CO, USA
e-mail: sooryaa@colostate.edu

S. Pasricha
Colorado State University, Fort Collins, CO, USA
e-mail: sudeep.pasricha@colostate.edu

© The Author(s), under exclusive license to Springer Nature Switzerland AG 2023
V. K. Kukkala, S. Pasricha (eds.), *Machine Learning and Optimization Techniques for Automotive Cyber-Physical Systems*, https://doi.org/10.1007/978-3-031-28016-0_10

Today's vehicles heavily rely on information from various external systems that utilize advanced communication standards such as 5G technology and Vehicle-to-X (V2X) [14] to support various ADAS functionalities. Unfortunately, this makes automotive embedded systems highly vulnerable to various cyber-attacks that can have catastrophic consequences. The cyber-attacks on vehicles discussed in [15–17] have presented different ways to gain unauthorized access to the in-vehicle network and override the vehicle controls by injecting malicious messages. With connected and autonomous vehicles (CAVs) on the horizon, these security concerns will get further aggravated and become a serious threat to the safety of future autonomous vehicles. Therefore, it is crucial to prevent unauthorized access to in-vehicle networks by external attackers to ensure the security of automotive CPS.

Traditional computer networks utilized firewalls to defend the networks from external attackers. However, no firewall is foolproof, and no network can be fully secure from attackers. Thus, there is a need for an active monitoring system that continuously monitors the network to identify malicious messages in the system. These systems are commonly referred to as intrusion detection systems (IDS). An IDS that is deployed in a vehicle can be used to continuously monitor the in-vehicle network traffic and trigger alerts when suspicious messages or known threats are detected. Thus, IDS acts as the last line of defense in automotive CPSs.

At a high level, IDSs are categorized into two types: *(i) rule-based* and *(ii) machine learning based*. Rule-based IDSs look for traces of previously observed attack signatures in the network traffic, whereas machine learning-based IDSs observe for the deviation from the learned normal system behavior to detect cyber-attacks. Rule-based IDS can have faster detection rates and very few false alarms (false positive rate) but are limited to detecting only previously observed attacks. On the other hand, machine learning-based IDS can detect both previously observed and novel attacks but can suffer from relatively slower detection times and higher false alarm rates. An efficient IDS needs to be lightweight (have minimal overhead), robust, and highly scalable. More importantly, practical IDSs need to have comprehensive attack coverage (i.e., detect both known and unknown attacks) with high detection accuracy and low false alarms, as recovering from false alarms can be costly.

Moreover, obtaining the signature of every possible attack is highly impractical and would limit us to only detecting known attacks. Hence, we believe that machine learning-based IDSs provide a more pragmatic solution to this problem. Additionally, large volumes of message data can be collected due to the ease of acquiring in-vehicle network data, which further assists the use of advanced deep learning models for detecting cyber-attacks in automotive CPS [9].

In this chapter, we present a novel IDS framework called *INDRA*, first introduced in [6], that monitors the in-vehicle network messages in a Controller Area Network (CAN) based automotive CPS to detect various cyber-attacks. During the offline phase, *INDRA* uses a deep learning-based recurrent autoencoder model to learn the normal system behavior in an unsupervised manner. At runtime, *INDRA* continuously monitors the in-vehicle network for deviations from learned normal system behavior to detect malicious messages. Moreover, *INDRA* aims to maximize the detection accuracy with minimal false alarms and overhead on the ECUs.

Our novel contributions in this work are as follows:

1. We introduced a Gated Recurrent Unit (GRU) based recurrent autoencoder network to learn the normal system behavior during the offline phase;
2. We proposed an intrusion score (IS) metric to measure deviation from the normal operating system behavior;
3. We presented a comprehensive analysis of the selection of thresholds for the intrusion score metric;
4. Lastly, we compared our proposed *INDRA* framework with the best-known prior works in the area to demonstrate its effectiveness.

2 Related Work

Several techniques have been proposed in the literature to design IDS for protecting time-critical automotive CPS. These works try to detect various attacks by monitoring the in-vehicle network traffic. In this section, we first discuss the key rule-based IDSs and then discuss machine learning based IDSs.

Rule-based IDS detects known attacks by using the information from previously observed attack signatures. In [18], a language theory-based model was introduced to derive attack signatures. However, this technique fails to detect attacks when it misses the packets transmitted during the early stages of the attack interval. A transition matrix-based attack detection scheme for CAN bus systems was proposed in [19], but this approach only works for simple attacks and fails to detect advanced replay attacks. In [20], the authors identified key attack signatures such as increased message frequency and missing messages to detect cyber-attacks. In [21], the authors proposed a specification-based approach to detect cyber-attacks, which analyzes the system behavior and compares it with the predefined attack patterns to detect anomalies. However, their approach fails to detect unknown attacks. The authors in [22] propose an ADS technique using the Myers algorithm [23] under the map-reduce framework. A time-frequency analysis of CAN messages is used to detect multiple anomalies in [24]. In [25], the authors analyzed message frequency at design time to derive a regular operating mode region, which is used as a baseline during runtime to detect cyber-attacks. In [26], the sender ECU's clock skew, and the messages are fingerprinted at design time and used at runtime to detect attacks by observing for variations. The authors in [27] presented a formal analysis of clock-skew-based IDS and evaluated it on a real vehicle. In [28], a memory heat map is used to characterize the memory behavior of the operating system to detect anomalies. An entropy-based IDS that observes the change in system entropy to detect anomalies was proposed in [29]. Nonetheless, the technique fails to detect complex attacks for which the entropy change is minimal. In conclusion, rule-based IDSs offer a fast solution to the intrusion detection problem with lower false positive rates but fail to detect more complex and novel attacks. Moreover, obtaining signatures of every possible attack pattern is not practical.

On the other hand, machine learning-based IDSs aim to learn the normal system behavior in an offline phase and observe for any deviation from the learned normal behavior to detect anomalies at runtime. In [30], the authors proposed a sensor-based IDS that utilizes attack detection sensors in the vehicle to monitor various system events and observe for deviations from normal behavior. However, this approach is expensive and suffers from poor detection rates. In [31], a One-Class Support Vector Machine (OCSVM) based IDS was introduced, but it suffers from poor detection latency. In [32], an ensemble of different nearest neighbor classifiers was used to distinguish between normal and an attack-induced CAN messages. A decision-tree-based detection model to monitor the physical features of the vehicle was proposed in [33] to detect cyber-attacks. However, this model is impractical and suffers from high anomaly detection latencies. In [34], a Hidden Markov Model (HMM) based technique was proposed to monitor the temporal relationships between messages to detect cyber-attacks. In [35], a deep neural network-based approach was proposed to scan the messages in the in-vehicle network to detect attacks. This approach is finetuned for a low-priority tire pressure monitoring system (TPMS), which makes it hard to adapt to high-priority powertrain applications. In [36], a Long Short-Term Memory (LSTM) based IDS for multi-message ID detection was proposed. However, due to the high complexity of model architecture, this approach has a high computational overhead on the ECUs. In [37], an LSTM-based IDS was proposed to detect insertion and dropping attacks (explained in Sect. 4.3). In [38], an LSTM-based predictor model is proposed to predict the next time step message value at a bit level and observe for large variations to detect anomalous messages. A recurrent neural network (RNN) based IDS to learn the normal CAN message pattern in the in-vehicle network is proposed in [39]. A hybrid IDS was proposed in [40], which utilizes a specification-based system in the first stage and an RNN-based model in the second stage to detect anomalies in time-series data. Several other machine learning models, such as the stacked LSTMs and temporal convolutional neural networks (TCNs) based techniques, were proposed in [7, 8], respectively. However, none of these techniques provides a complete system-level solution that is scalable, reliable, and lightweight to detect various attacks for in-vehicle networks.

In this chapter, we introduce a lightweight recurrent autoencoder-based IDS using gated recurrent units (GRUs) to monitor the in-vehicle network messages at a signal level to detect various attacks with higher efficiency than various state-of-the art works in this area. A summary of some of the state-of-the-art works' performance under different metrics and our proposed *INDRA* framework is shown in Table 1.

Table 1 Performance metrics comparison between our proposed *INDRA* framework and state-of-the-art machine learning-based intrusion detection works

Technique	Performance metrics			
	Lightweight model	Low false positive rate	High detection accuracy	Fast inference time
PLSTM [25]	X	✓	X	X
RepNet [26]	✓	X	X	✓
CANet [23]	X	✓	✓	X
INDRA	✓	✓	✓	✓

3 Background on Sequence Learning

The availability of increased compute power from GPUs, and custom hardware accelerators enabled the training of deep neural networks with many hidden layers, which led to the creation of powerful models for solving complex problems in many domains. One such problem is detecting cyber-attacks in automotive CPS. In an automotive CPS, the communication between ECUs occurs in a time-dependent manner. Therefore, the temporal relationship between the messages in the system can be exploited in order to detect cyber-attacks. However, this cannot be achieved using traditional feedforward neural networks as the output of any input at any instance is independent of the other inputs. This makes sequence models appropriate for such problems, as they inherently handle sequences and time-series data.

3.1 Sequence Models

A sequence model is a function that ensures that the outcome is reliant on both current and prior inputs. The recurrent neural network (RNN), which was introduced in [41], is an example of such a sequence model. Other sequence models, such as gated recurrent unit (GRU) and long short-term memory (LSTM), have also become popular in recent years.

3.1.1 Recurrent Neural Networks (RNN)

An RNN is a form of artificial neural network that takes the sequential data as input and tries to learn the relationships between the input samples in the sequence. The RNNs use a hidden state to allow learned information from previous time steps to persist over time. A single RNN unit with feedback is shown in Fig. 1a, and an RNN unit unrolled in time is shown in Fig. 1b.

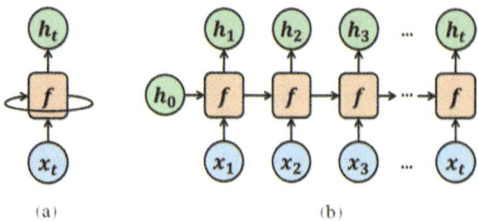

(a)   (b)

Fig. 1 (**a**) A single RNN unit and (**b**) RNN unit unrolled in time; f is the RNN unit, x is the input, and h represents hidden states

The output (h_t) of an RNN unit is a function of both the input (x_t) and the previous output (h_{t-1}):

$$h_t = f(Wx_t + Uh_{t-1} + b) \tag{1}$$

where f is a nonlinear activation function (e.g., sigmoid or tanh), U and W are weight matrices, and b is the bias term. One of the major limitations of RNNs is that they are very hard to train. As RNNs and other sequence models handle time-series data or sequences as inputs, backpropagation happens through various time steps (commonly known as backpropagation through time (BPTT)). During the BPTT step, the feedback loop in RNNs causes the errors to expand or shrink rapidly thereby creating exploding or vanishing gradients respectively. This destroys the information in backpropagation and makes the training process obsolete. Moreover, the vanishing gradient problem prohibits RNNs from learning *long-term dependencies*. To solve this problem, additional states and gates were introduced in the RNN unit in [42] to remember long-term dependencies, which led to the development of LSTM Networks.

3.1.2 Long Short-Term Memory (LSTM) Networks

LSTMs use cell state, hidden state information, and multiple gates to capture long-term dependencies between messages. The cell state can be visualized as a freeway that carries relevant information throughout the processing of an input sequence. The cell state stores information from previous time steps and passes it to the subsequent time steps to reduce the effects of short-term memory. Moreover, the information in the cell state is modified by the gates in the LSTM unit, which helps the model in determining which information should be retained and which should be ignored.

An LSTM unit contains 3 gates: (*i*) forget gate (f_t) (*ii*) output gate (o_t), and (*iii*) input gate (i_t) as shown in Fig. 2a. The forget gate is a binary gate that determines which information to retain from the previous cell state (c_{t-1}). The output gate uses information from the previous two gates to produce an output. Lastly, the input gate

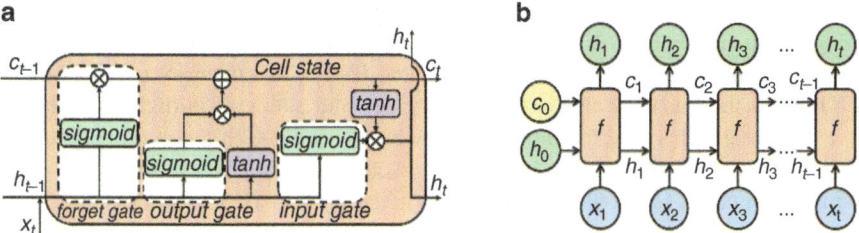

Fig. 2 (**a**) A single LSTM unit with different gates and activations, and (**b**) LSTM unit unrolled in time; f is an LSTM unit, x is input, c is cell state, and h is the hidden state

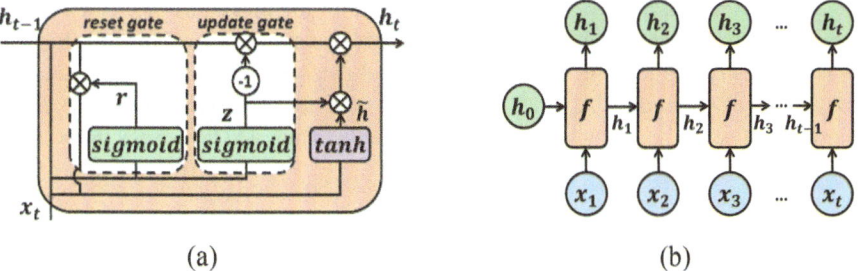

Fig. 3 (**a**) A single GRU unit with different gates and activations, and (**b**) GRU unit unrolled in time; f is a GRU unit, x is input, and h is the hidden state.

adds relevant information to the cell state (c_t). An illustration of an LSTM unit unrolled in time is shown in Fig. 2b.

LSTMs learn long-term dependencies in a sequence by using a combination of different gates and hidden states. However, they are computationally expensive due to the complex sequence path from having multiple gates (compared to RNNs), and require more runtime memory. Moreover, training LSTMs have a high computation overhead even when advanced training methods such as truncated backpropagation are employed. To overcome the above-mentioned limitations, a simpler sequence model called gated recurrent unit (GRU) was introduced in [43]. GRUs can be trained faster than LSTMs and also capture dependencies in long sequences with minimal overhead (in both memory and runtime) while solving the vanishing gradient problem.

3.1.3 Gated Recurrent Unit (GRU)

Unlike LSTMs, a GRU unit takes a different route for gating information. The input and forget gate in the LSTM unit are combined into a solitary *update* gate. Moreover, hidden and cell states are combined into one state, as shown in Fig. 3a, b.

A GRU unit consists of two gates *(i)* reset gate and *(ii)* update gate. The reset gate combines new input with previous memory, while the update layer determines how much relevant information should be stored. Thus, a GRU unit controls the data stream similar to an LSTM by uncovering its hidden layer contents. Moreover, GRUs are computationally more efficient and have a low memory overhead than LSTMs as they achieve this using fewer gates and states. It is highly crucial to use lightweight machine learning models when working with automotive systems, as real-time automotive ECUs are highly resource-constrained embedded systems with strict energy and power budgets. This makes GRU-based networks an ideal fit for inference in resource-constrained automotive systems. Thus, *INDRA* utilizes a lightweight GRU-based model to implement the IDS (explained in detail in Sect. 5).

One of the significant advantages of sequence models is that they can be trained in both supervised and unsupervised learning fashion. Due to the large volume of CAN message data in a vehicle, high variability in the messages between vehicle models from the same manufacturer, and the proprietary nature of this information make it highly challenging and tedious to label messages correctly. However, due to the ease of obtaining CAN message data via onboard diagnostics (OBD-II), large amounts of unlabeled data can be collected easily. Thus, *INDRA* uses GRUs in an unsupervised learning setting.

3.2 Autoencoders

Autoencoders are a special class of neural networks that try to reconstruct the input by learning the latent input features in an unsupervised fashion. They achieve this by encoding the input data (x) to a hidden layer which produces the embedding, and finally decoding the embedding to produce a reconstruction $\tilde{x}$ (as shown in Fig. 4). The layers used to create this embedding are called the encoder, and the

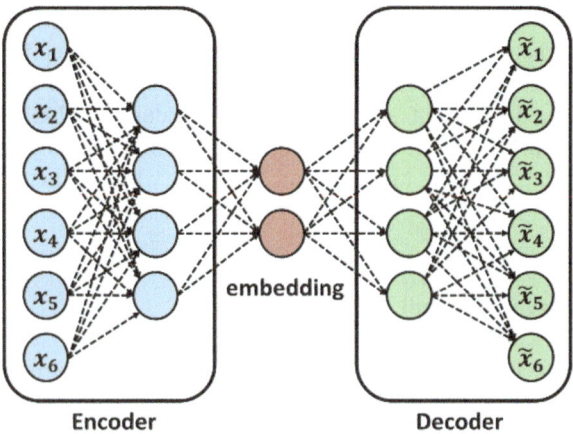

Fig. 4 An autoencoder network with encoder, decoder, and embedding layers

layers used in reconstructing the embedding into the original input (decoding) are called the decoder. During the training process, the encoder tries to learn a nonlinear mapping of the inputs, while the decoder tries to learn the nonlinear mapping of the embedding to the inputs. The encoder and decoder achieve this using various non-linear activation functions such as tanh and rectified linear unit (ReLU). Moreover, the autoencoder aims to recreate the input as closely as possible by extracting the key features from the inputs with the goal of minimizing reconstruction loss. The most commonly used loss functions in autoencoders include mean squared error (MSE) and Kullback-Leibler (KL) divergence.

As autoencoders aim to reconstruct the input by learning the underlying distri-bution of the input data, they are an excellent choice for efficiently learning and re-constructing highly correlated time-series data by learning the temporal relations between messages. *Hence, our proposed INDRA framework uses lightweight GRUs in an autoencoder to learn latent representations of CAN message data in an unsupervised learning setting.*

4 Problem Definition

4.1 System Model

In this chapter, we consider a generic automotive system consisting of multiple ECUs connected using an in-vehicle network, as shown in Fig. 5. Each ECU in the system runs a specific set of automotive applications that are hard-real time in nature (i.e., they have strict timing and deadline constraints). Moreover, we assume that each ECU also runs intrusion detection applications (IDS) that are responsible for monitoring and detecting cyber-attacks in the in-vehicle network.

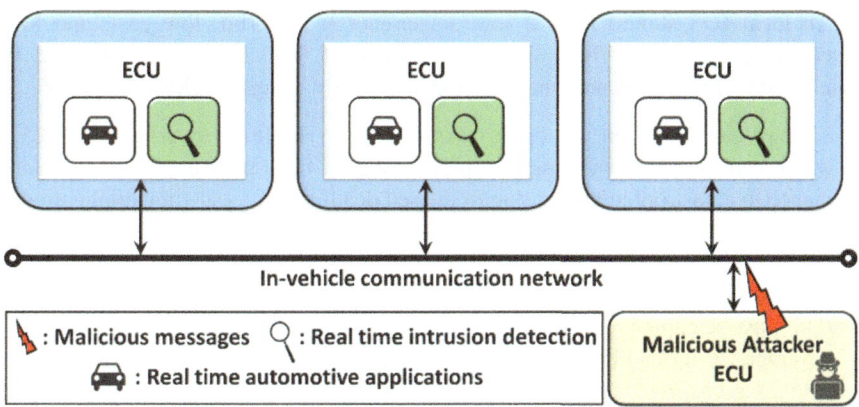

Fig. 5 Overview of the system model considered in *INDRA*

In the *INDRA* framework, we consider a distributed IDS approach (where intrusion detection applications are collocated with automotive applications) as opposed to a centralized IDS approach in which one central ECU handles all intrusion detection tasks due to the following reasons:

- A centralized IDS approach is prone to single-point failures, which can completely expose the system to the attacker.
- In extreme scenarios such as during a flooding attack (explained in Sect. 4.3), the in-vehicle network would get highly congested, and the centralized system might not be able to communicate with the victim ECUs.
- If an attacker successfully tricks the centralized IDS ECU, the attacks can go undetected by the other ECUs, compromising the entire system; however, in the case of a distributed IDS, it requires fooling multiple ECUs (which is more difficult) to compromise the system. Moreover, the decentralized intelligence in a distributed IDS scenario can still detect the attacks, even if one of the ECU is compromised.
- In a distributed IDS, ECUs can stop accepting messages as soon as an intrusion is detected. This results in significantly faster reaction times as there is no need for a notification from a centralized system.
- Lastly, in a distributed IDS, the computation load of IDS is divided among the ECUs, and monitoring can be limited to only required messages. As a result, multiple ECUs can independently monitor a subset of messages with lesser overhead.

Distributed IDS approach has been adopted in many state-of-the-art works, such as [18, 25], for the above-mentioned reasons. Moreover, with the increasing computation power of automotive ECUs, the collocation of IDS applications with real-time automotive applications in a distributed manner should not be a problem if the IDS has minimal overhead. *INDRA* framework is not only lightweight but also highly scalable and achieves superior intrusion detection performance (discussed in detail in Sect. 6).

An ideal IDS should have a low power/energy footprint, low cost, and low susceptibility to noise. The following are some of the key characteristics of an efficient IDS, that were taken into consideration when designing the *INDRA* IDS:

- *Lightweight*: Intrusion detection tasks can incur additional overhead on ECU, which can have a broad range of impact ranging from poor application performance to catastrophic events due to missed deadlines for real-time applications. Therefore, *INDRA* aims to have a lightweight IDS that incurs minimal overhead on the ECU.
- *Coverage*: This is defined as the range of attacks that an IDS can detect. A good IDS must be capable of detecting more than one type of attack. Moreover, high coverage for IDS will make the system resilient to multiple attack surfaces.
- *Few false positives*: This is a highly desired quality in any IDS (even outside of the automotive domain), as dealing with false positives can quickly become costly. Thus, a good IDS is expected to have few false positives or false alarms.

- *Scalability*: As the number of ECUs in emerging vehicles is growing along with software and network complexity, this is an essential requirement. A good IDS should be highly scalable and capable of supporting multiple system sizes.

4.2 Communication Model

In this subsection, we discuss the communication model that was considered for the *INDRA* framework. *INDRA* primarily focuses on detecting anomalies in Controller Area Network (CAN) bus-based automotive CPS, as CAN is the most commonly used in-vehicle network protocol. CAN offers a low-cost, lightweight, event-triggered communication where messages are transmitted in the form of frames. A standard CAN frame structure with the length of each field (in bits) on the top is shown in Fig. 6. The standard CAN frame consists of *(i)* header, *(ii)* payload, and *(iii)* trailer segments. The header contains information about the message identifier (ID) and the length of the message, whereas the payload segment contains the actual data that needs to be transmitted. The trailer section is mainly used for error checking at the receiver. More recently, a new variation of the CAN protocol, called CAN-extended or CAN 2.0B, is also being deployed increasingly in modern vehicles. The key difference is that CAN-extended has a 29-bit identifier which allows for a greater number of message IDs.

The *INDRA* IDS focuses on monitoring the payload of the CAN frame and observes for anomalies within the payload segment to detect cyberattacks. This is because most modern-day attacks involve an attacker modifying the payload to accomplish malicious activities. On the other hand, if an attacker targets the header or trailer segments, the message would get rejected at the receiver. The typical payload segment of a CAN message comprises of multiple data entities called signals. Figure 7 illustrates a real-world example CAN message with the list of signals within the message. Each signal has a particular data type, fixed size (in bits), and a start bit which specifies the signal's location in the 64-bit payload segment of the CAN frame.

INDRA focuses on monitoring individual signals within CAN payload to observe for anomalies and detect attacks. During training, *INDRA* learns the temporal relationships between the messages at a signal level and observes for deviations at runtime to detect attacks. This ability to detect attacks at a signal level enables

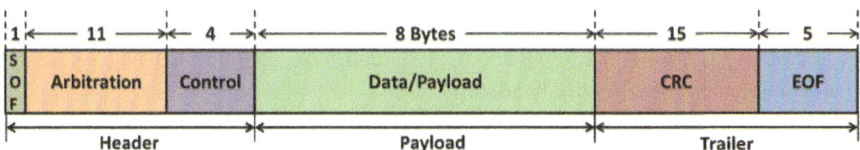

Fig. 6 Standard frame format of a Controller Area Network (CAN) message

Signal Name	Message	Start bit	Length	Byte Order	Value Type
Battery_Current	Status	0	16	Intel	Signed
Battery_Voltage	Status	16	16	Intel	Unsigned
Motor_Current	Status	32	16	Intel	Signed
Motor_Speed	Status	48	8	Intel	Signed
Motor_Direction	Status	56	8	Intel	Unsigned

Fig. 7 A real-world example CAN message with signal information [44]

INDRA to not only detect the presence of an attacker but also help in identifying the signal under attack. This can provide valuable information related to the attack and help in understanding the intentions of the attacker, which can be used to initiate appropriate countermeasures. The signal level monitoring technique employed in *INDRA* IDS is discussed in detail in Sect. 5.2.

Note: Even though the *INDRA* framework focuses on detecting attacks by monitoring CAN messages, our approach is protocol-agnostic and can be used with other in-vehicle network protocols (such as FlexRay and LIN) with minimal changes.

4.3 Attack Model

Our proposed *INDRA* IDS aims to protect the vehicle from various types of state-of-the-art attacks that are most commonly seen and difficult to detect in automotive CPS. Moreover, these attacks have been widely studied in literature to evaluate IDSs.

1. *Plateau attack*: In this attack, an attacker overwrites a signal value with a constant value for the entirety of the attack interval. The severity of this attack is determined by the magnitude of change in signal value and the duration for which the signal magnitude is changed. Larger changes in signal values are easier to detect compared to shorter changes.
2. *Flooding attack*: This is the most common and simple to launch attack, as it requires no knowledge of the system. In this attack, the attacker continuously floods the in-vehicle network with random or specific messages with the goal of preventing other ECUs from accessing the bus and rendering the bus unusable. These attacks are typically detected by the gateways and network bridges in the vehicle and often do not reach the last line of defense (the IDS). However, it is crucial to consider these attacks as they can have serious security and safety consequences when poorly handled.
3. *Playback attack*: In this attack, the attacker attempts to trick the IDS by replaying a valid series of message transmissions from the past. This attack is hard to detect if the IDS lacks the ability to capture the temporal relationships between messages and detect when they are violated.

4. *Continuous attack*: In this attack, an attacker gradually overwrites the signal value to some target value while avoiding the activation of an IDS. These attacks are difficult to detect and can be sensitive to the IDS parameters (discussed in Sect. 5.2).

5. *Suppress attack*: In this attack, the attacker suppresses the signal value(s) by either disabling the target ECU's communication controller or shutting down the ECU. These attacks are easy to detect when they disrupt message transmission for long durations but are harder to detect for shorter durations.

Moreover, in this work, we assume that the attacker can gain access to the in-vehicle network using the most common attack vectors, such as connecting to the OBD-II port, connecting to V2X systems that communicate with the outside world (for e.g., infotainment and connected ADAS systems), probe-based snooping on the in-vehicle bus, and by replacing an existing ECU with a malicious ECU. We also assume that the attacker has access to the network parameters (such as parity, flow control, and BAUD rate) that can further assist in gaining access to the in-vehicle network.

Objective The goal of our proposed *INDRA* framework is to implement a lightweight IDS that can detect a variety of attacks (discussed above) in a CAN bus-based automotive CPS, with a high detection accuracy and low false positive rate while having a large attack coverage.

5   *INDRA* Framework Overview

INDRA framework utilizes a machine learning-based signal level IDS for monitoring real-time CAN messages in automotive CPS. An overview of the *INDRA* framework is depicted in Fig. 8. The *INDRA* framework consists of design-time and runtime steps. During design time, *INDRA* uses CAN message data from a trusted vehicle to train a recurrent autoencoder-based model to learn the normal system behavior. At runtime, the trained recurrent autoencoder model observes for the deviation from the learned normal system behavior using the proposed intrusion score metric to detect cyberattacks. These steps are described in detail in the subsequent subsections.

5.1   *Recurrent Autoencoder*

Recurrent autoencoders are powerful neural networks that are similar to an encoder-decoder structure but can handle time-series or sequence data inputs. They typically consist of units such as RNNs, LSTMs, or GRUs (discussed in Sect. 3). Similar to regular autoencoders, recurrent autoencoders have an encoder and a decoder stage. The encoder generates a latent representation of the input data in an n-dimensional

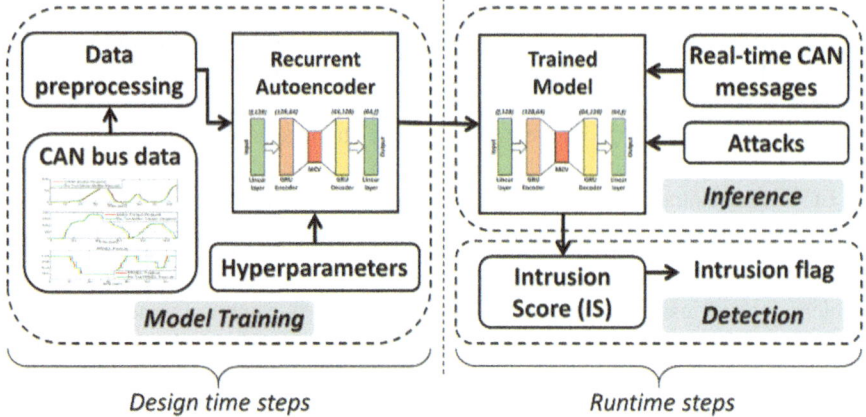

Fig. 8 Overview of the *INDRA* IDS framework

space, and the decoder uses this latent representation from the encoder output and tries to reconstruct the input data with minimal reconstruction loss. In *INDRA*, we propose a novel lightweight recurrent autoencoder model that is tailored for the design of IDS to detect cyberattacks in the in-vehicle network. The details of the proposed neural network architecture and the various steps involved in its training and evaluation are discussed in the subsequent sections.

5.1.1 Model Architecture

Our proposed recurrent autoencoder model architecture is illustrated in Fig. 9, with each layer's input and output dimensions on the top. The model comprises of a linear layer at the input, a GRU-based encoder, a GRU-based decoder, and a linear layer before the final output. The input time-series CAN message data with signal level information consisting of f features (where f is the number of signals in the message) is given as input to the first linear layer. The output from the first linear layer is then passed to the GRU-based encoder, which generates the latent representation of the time-series signal inputs, which is referred to as a message context vector (MCV) in this chapter. The MCV captures the context of various signals in the input message as a vector. Each MCV can be viewed as a point in an n-dimensional space containing the context of the series of signal values provided as input. The MCV is fed into a GRU-based decoder, which is then followed by a linear layer to generate the reconstruction of the input CAN message data with individual signal values. The loss between the input and the reconstructed input is calculated using mean square error (MSE), and the weights are updated using backpropagation through time. *INDRA* designs a recurrent autoencoder model for each message ID.

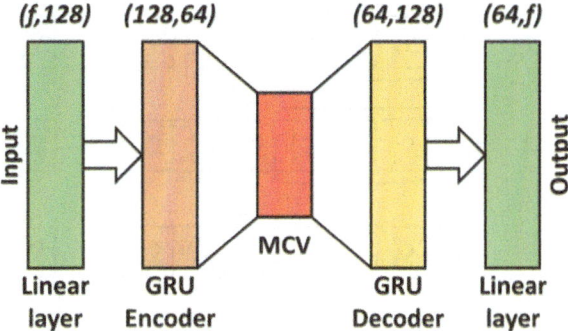

Fig. 9 Proposed recurrent autoencoder model used in *INDRA* (f is the number of features, i.e., number of signals in the input CAN message, MCV is message context vector)

5.1.2 Training Process

The training procedure starts with pre-processing the CAN message data collected from a trusted vehicle. Each sample in the CAN message dataset consists of a message ID and the corresponding signal values contained within that message ID. As signals represent a wide variety of information in the vehicle, the range of signal values can also be very large. This can make the training process extremely slow or unstable. To prevent this, we scale the signal values between 0 to 1 for each signal type. Moreover, scaling signal values also helps to avoid the problem of exploding gradients (as discussed in Sect. 3).

The pre-processed CAN dataset is divided into training data (85%) and validation data (15%), which is then prepared for training. We use a rolling window-based approach, which involves choosing a fixed-size window and rolling it to the right by one sample every time step. An example rolling window approach with a window size of three samples and its movement for the three consecutive time steps is illustrated in Fig. 10. The term S_i^j represents the *ith* signal value at *jth* sample. The elements in the rolling window are referred to as a subsequence, and the size of the rolling window is defined as the subsequence length. As each subsequence consists of a set of signal values over time, our proposed recurrent autoencoder model attempts to learn the temporal relationships between the series of signal values. These signal-level temporal relationships aid in detecting more complex attacks such as continuous and playback (as discussed in Sect. 4.3). The process of training using subsequences is done iteratively until the end of the training data.

Each training iteration consists of a forward pass and a backward pass (using backpropagation through time to update the weights and biases of the neurons based on the error value (as discussed in Sect. 3)). At the end of the training, the model's performance is evaluated (forward pass only) using the validation data, which was not seen by the model during the training. The end of the validation step marks the completion of one epoch during which the model has seen the complete dataset once. The model is trained for a set number of epochs until the model reaches

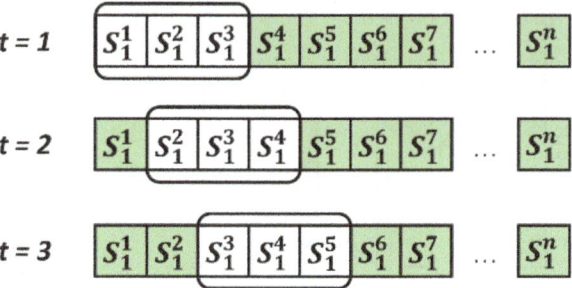

Fig. 10 An example of a rolling window approach and its movement for three consecutive time steps

convergence. Moreover, the process of training and validation using subsequences is sped up by training the input subsequences data in groups known as mini-batches. Each mini-batch is made up of several consecutive subsequences that are given as the input to the model in parallel. The size of each mini-batch is referred to as a batch size. Lastly, to control the rate of update of the model parameters during the backpropagation phase, a learning rate is defined. These hyperparameters, such as subsequence size, batch size, learning rate, etc., are covered in detail in Sect. 6.1.

5.2 Inference and Detection

At runtime, the trained model is set to evaluation mode, where only forward passes are performed, and the weights are not updated. During this phase, various attack conditions are simulated in the CAN message dataset, and the trained model is tested under multiple attack scenarios (mentioned in Sect. 4.3).

During inferencing, each data sample that passes through the model is reconstructed, and the reconstruction loss is computed. This reconstruction loss is sent to the detection module to compute the proposed *intrusion score* (IS) metric, which helps in determining whether a signal is malicious or normal. The IS is calculated at a signal level to predict which signal is under attack. The IS is calculated as a squared error during each iteration of the inference to estimate the prediction deviation from the input signal value, as shown in (2).

$$IS_i = \left(S_i^j - \hat{S}_i^j\right)^2 \quad \forall i \in [1, m] \tag{2}$$

where, S_i^j denotes the *ith* signal value at *jth* sample, $\hat{S}_i^j$ represents its reconstruction, and m is the number of signals in the message. We observe a large deviation for predicted value from the input signal value (i.e., large IS value) when the current signal pattern is not seen by the model during the training phase and a smaller IS value otherwise. This serves as the foundation for our detection phase.

Since the dataset lacks a signal-level attack label information, *INDRA* combines the signal level IS information into a message-level IS by calculating the maximum IS of the signals in that message, as shown in (3).

$$MIS = \max(IS_1, IS_2 \dots, IS_m) \tag{3}$$

To achieve adequate detection accuracy, it is critical to choose the intrusion threshold (IT) for flagging messages. *INDRA* investigates multiple choices for IT, using the best model (model with the lowest running validation loss) from the training phase. From this model, multiple metrics such as maximum, mean, median, 99.99%, 99.9%, 99%, and 90% validation loss are recorded across all iterations as the potential choices for the IT. The analysis for the selection of the IT metric is presented in detail in Sect. 6.2.

A working snapshot of *INDRA* IDS is illustrated in Fig. 11a, b, with a plateau attack on a message with three signals between time 0 and 50. Figure 11a compares the input (true) vs. IDS predicted signal value for three signals, and the attack interval is highlighted in red. It can be observed that the reconstruction is close for almost all signals except during the attack interval for the majority of the time. Signal 3 is subjected to a plateau attack in which the attacker maintains a constant value until the end of the attack interval. This is illustrated in the third subplot of Fig. 11a (note the larger difference between the predicted and actual input signal values in that subplot, compared to signals 1 and 2). Figure 11b depicts the signal intrusion scores for all three signals, and the dotted black line represents the intrusion threshold (IT). As stated previously, the maximum of signal intrusion scores is chosen as message intrusion score (MIS), which in this case is the IS of signal 3. As seen in Fig. 11b, the intrusion score of signal 3 is above the IT for the entire duration of the attack interval, which clearly highlights *INDRA's* ability to detect such attacks. The value of IT (equal to 0.002) in Fig. 11b is calculated using the method discussed in Sect. 6.2. However, it is important to note that this value is specific to the example case shown in Fig. 11 and is not the IT value used for the remaining experiments. The details of IT selection is discussed in detail in Sect. 6.2.

6 Experiments

6.1 *Experimental Setup*

A series of experiments have been conducted to evaluate the real-time performance of our proposed *INDRA* IDS. We begin by presenting an analysis for the selection of intrusion threshold (IT). The derived IT is used to contrast against two variants of the same framework known as *INDRA*-LED and *INDRA*-LD. The *INDRA-LED*

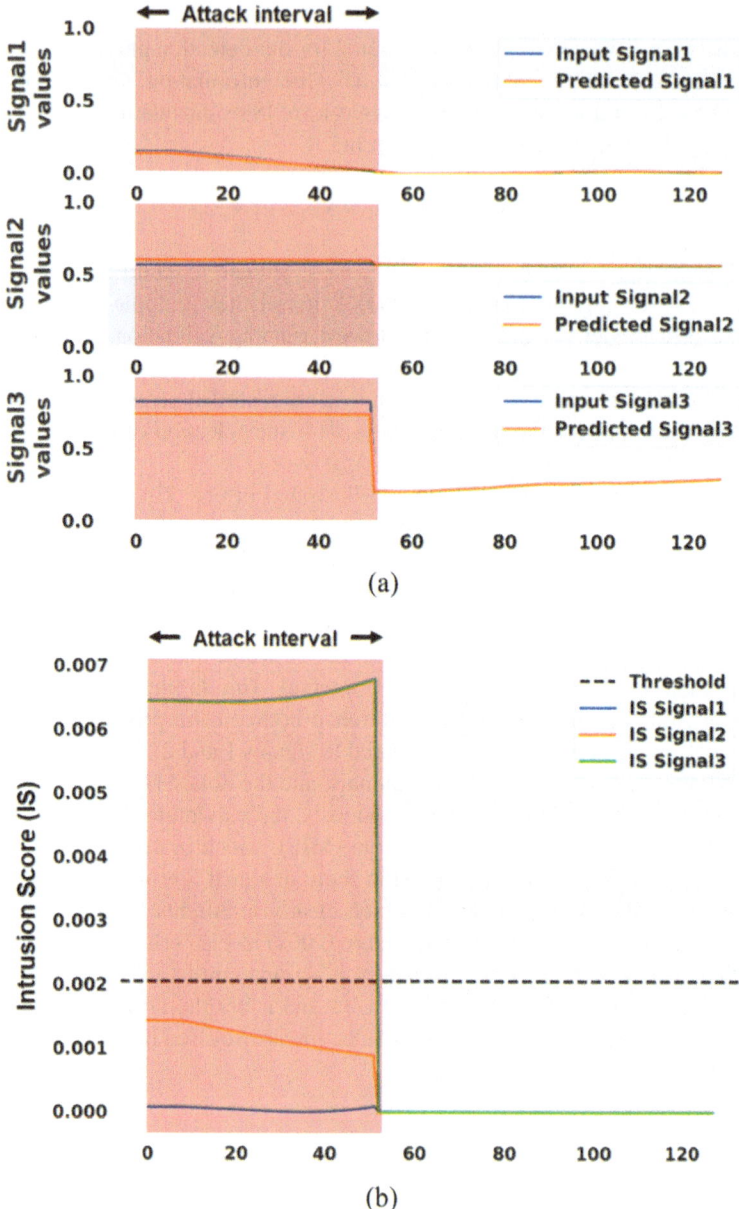

Fig. 11 Internal working of *INDRA* IDS checking a message with three signals under a plateau attack, where (**a**) shows the signal comparisons and (**b**) shows IS of three signals and the IT

removes the linear layer before the output, essentially leaving the task of decoding the message context vector (i.e., reconstructing the input) to GRU based decoder. The abbreviation LED stands for (L)linear layer, (E) encoder GRU, and (D) decoder GRU. The *INDRA-LD* variant replaces the GRU and the linear layer at the decoder with a series of linear layers (LD stands for linear decoder). These experiments were carried out to assess the importance of different layers in the network. However, the encoder part of the network is not changed because it is required to generate an encoding (MCV) of the input time-series data. *INDRA* investigates other variants as well, but they were not included in the discussion as their performance was subpar compared to that of *INDRA-LED* and *INDRA-LD* variants.

Subsequently, the best *INDRA* variant is compared with three state-of-the-art prior works that use different machine learning-based techniques to detect intrusions: *(i)* Predictor LSTM (PLSTM [38]), *(ii)* Replicator Neural Network (RepNet [39]), and *(iii)* CANet [36]. The first comparison work (PLSTM) employs an LSTM-based network that has been trained to predict the signal values in the following message transmission. PLSTM accomplishes this by taking the 64-bit CAN message payload as the input and learning to predict the signal at a bit-level granularity by minimizing prediction loss. The bit level deviations between the actual and the predicted next signal values are computed using a log loss or binary cross-entropy loss function. Additionally, PLSTM uses the prediction loss values at runtime to decide whether a particular message is malicious or not. The second comparison work (RepNet) employs a series of RNN layers to increase the dimensionality of the input data and reconstruct the input signal values by decreasing back to the original dimensionality. RepNet accomplishes this by reducing the mean squared error (MSE) between the input and the reconstructed signal values. At runtime, RepNet uses large deviations between the input received signal and the reconstructed signal values to detect cyberattacks. Lastly, CANet uses a quadratic loss function to minimize the signal reconstruction error by combining multiple LSTMs and linear layers in an autoencoder architecture. All experiments conducted with *INDRA* and its variants and prior works are discussed in detail in subsequent subsections.

In this work, we use the SynCAN dataset developed by ETAS and Robert Bosch GmbH [36] to evaluate the effectiveness of the *INDRA* framework with its variants and against the above-mentioned prior works. The SynCAN dataset consists of CAN message data for ten different IDs that have been modeled after real-world CAN message data. Furthermore, the dataset consists of both training and test data with multiple attacks (discussed in Sect. 4.3). Each row in the dataset consists of a timestamp, message ID, and individual signal values. Additionally, the test data contains a label column with either 0 or 1 values indicating normal or malicious messages. However, the label information is only available on per message basis and does not specify which signal within the message is under attack. It is important to note that the label information in the training data is not used to train the *INDRA* model, as the *INDRA* model learns the patterns in the input data in an unsupervised manner. This label information in the dataset is only used to evaluate the performance of the proposed IDS using several metrics such as detection accuracy and false positive rate. Moreover, to simulate a more realistic attack scenario in the in-vehicle networks, the test data also contains normal CAN traffic between the attack injections.

All the machine learning-based frameworks, including the *INDRA* framework and its variants as well as comparison works, are implemented using Pytorch 1.4 with CUDA support. We conducted several experiments to select the best-performing model hyperparameters (number of layers, hidden unit sizes, and activation functions). The final model discussed in Sect. 5.1 was trained using the SynCAN data set, with 85% of train data used for training and the remaining for validation. The validation data is primarily used to assess the model performance at the end of each epoch. The model is trained for 500 epochs, using a rolling window approach (as discussed in Sect. 5.1.2) with a subsequence size of 20 messages and a batch size of 128. Moreover, an early stopping mechanism is employed to monitor the validation loss across epochs and stop the training process if there is no improvement after 10 (patience) epochs. The initial learning rate is chosen as 0.0001, and *tanh* activations are applied after each linear and GRU layer. Furthermore, the ADAM optimizer is used with the mean squared error (MSE) as the loss criterion to compute the reconstruction loss. The trained model is used during testing and subjected to multiple simulated attack scenarios using the test dataset. The intrusion score metric (as stated in Sect. 5.2) was used to calculate the intrusion threshold to flag the message as malicious or normal. Lastly, to evaluate the performance of the IDS, several performance metrics such as detection accuracy and false positive rate were considered. All the simulations were executed on an AMD Ryzen 9 3900X server with an Nvidia GeForce RTX 2080Ti GPU.

Additionally, we present the following definitions in the context of IDS before discussing the experimental results:

- True Positive (TP)- when the IDS detects an actual malicious message as malicious;
- False Negative (FN)- when the IDS detects an actual malicious message as normal;
- False Positive (FP)- when the IDS detects a normal message as malicious (aka false alarm);
- True Negative (TN)- when the IDS detects an actual normal message as normal.

INDRA framework primarily focuses on two key performance metrics: *(i) Detection accuracy*- a measure of IDS's ability to detect malicious messages correctly, and *(ii) False positive rate*: also known as false alarm rate. These metrics are computed using (4) and (5), respectively.

$$Detection\ Accuracy = \frac{TP + TN}{TP + FN + FP + TN} \tag{4}$$

$$False\ Positive\ Rate = \frac{FP}{FP + TN} \tag{5}$$

6.2   *Intrusion Threshold Selection*

In this subsection, we present a detailed analysis on the selection of intrusion threshold (IT) by investigating various options such as maximum (max), median, mean, and different quantile bins of validation loss of the final model. As the model is trained only on attack-free data, the reconstruction error for the malicious message will be much larger than the error for normal messages. Hence, *INDRA* explores several candidate options for IT to achieve this goal that would work across multiple attack and no-attack scenarios. A high threshold value can make it harder for the model to detect the attacks that change the input pattern minimally (e.g., continuous attack). On the other hand, having a small threshold value can cause multiple false positives, which is highly undesirable. Thus, it is crucial to select an appropriate intrusion threshold value to achieve optimal model performance.

The detection accuracy and false positive rate for various candidate options used to calculate IT is shown in Fig. 12a, b, respectively, under different attack scenarios. The results from the Fig. 12 indicates that selecting a higher validation loss as the IT can lead to high accuracy and a low false alarm rate. However, selecting a very high value (such as 'max' or '99.99 percentile') may result in missing small variations in the input patterns that are found in more sophisticated attacks. We empirically conclude that the maximum and 99.99 percentile values are very close. Moreover, to capture attacks that produce small deviations, a slightly smaller threshold value is selected that would still perform similar to the max and 99.99 percentile thresholds under all attack scenarios. Thus, the 99.9th percentile value of the validation loss is chosen as the intrusion threshold (IT) value, and the same IT value is used for the remainder of the experiments (discussed in the following subsections).

6.3   *Comparison of INDRA Variants*

After selecting the appropriate intrusion threshold from the previous subsection, we use that same criterion for evaluating against two other variants: *INDRA*-LED and *INDRA*-LD. The main intuition behind evaluating different variants of *INDRA* is to study the impact of different layers in the model on the performance metrics discussed in Sect. 6.1.

Figure 13a illustrates the detection accuracy for the *INDRA* framework and its variants on the y-axis with multiple types of attacks and a no-attack scenario (normal) on the x-axis. It can be clearly seen that *INDRA* outperforms the other two variants and has high accuracy in most attack scenarios.

The false positive rate or false alarm rate of *INDRA* and other variants under different attack scenarios is illustrated in Fig. 13b. When compared to other variants, *INDRA* has the lowest false positive rate and highest detection accuracy. Moreover, *INDRA*-LED, which is just short of a linear layer on the decoder side, is the second-best performing model after *INDRA*. The ability of *INDRA*-LED to use

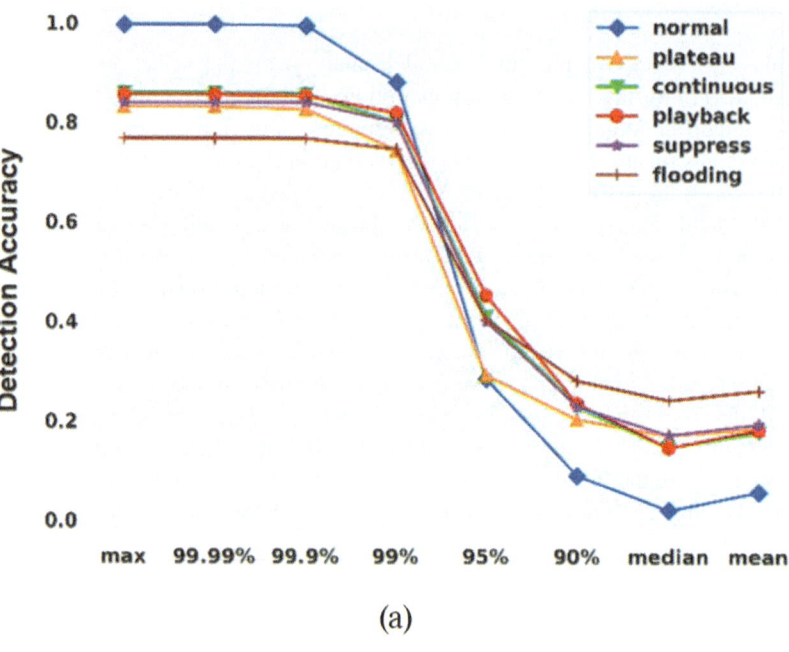

(a)

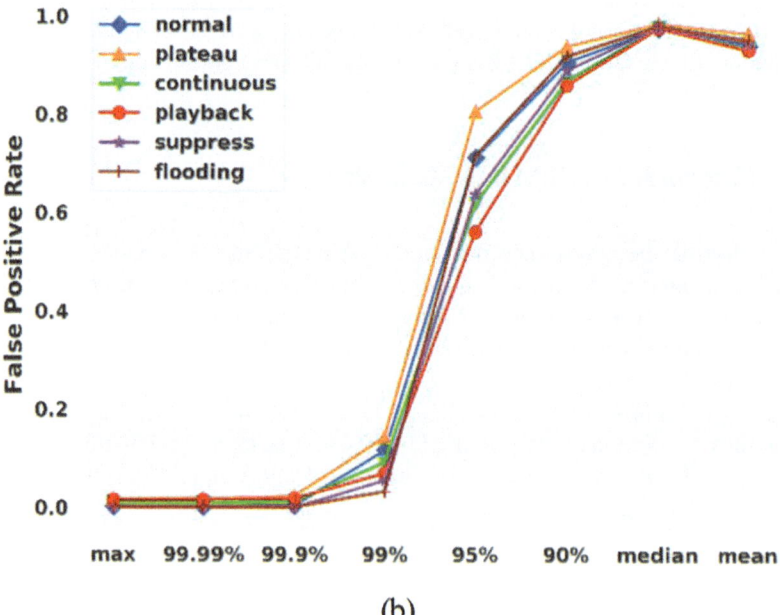

(b)

Fig. 12 Comparison of (**a**) detection accuracy and (**b**) false positive rate for various choices of intrusion threshold (IT) as a function of validation loss under different attack scenarios. (% refers to percentile, not percentage)

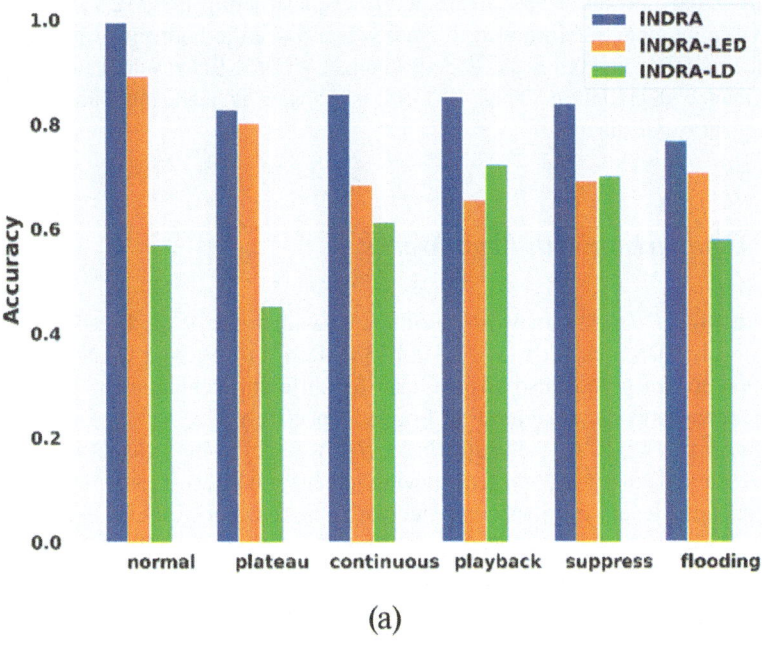

(a)

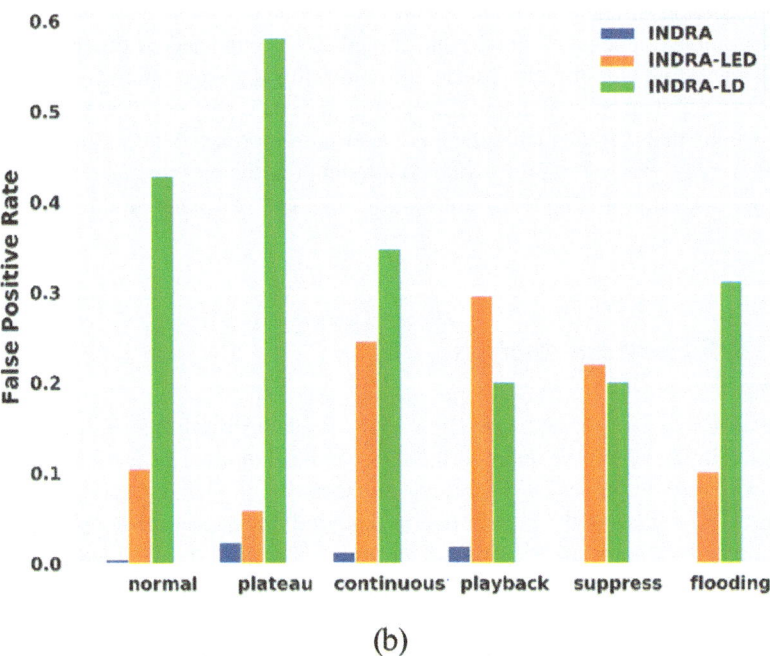

(b)

Fig. 13 Comparison of (**a**) detection accuracy and (**b**) false positive rate under different attack scenarios for *INDRA* and its variants (*INDRA*-LED and *INDRA*-LD)

a GRU-based decoder helps in efficiently reconstructing the MCV back to the original input signals. Moreover, it can be clearly seen in both Fig. 13a, b that the absence of GRU layers on the decoder end of *INDRA*-LD resulted in significant performance degradation. Thus, *INDRA* is chosen as the candidate model for subsequent experiments.

6.4 Comparison with Prior Works

Our proposed *INDRA* framework is compared with some of the best-known prior works in the IDS area, such as PLSTM [38], RepNet [39], and CANet [36]. The detection accuracy and false positive rate for different techniques under different attack scenarios is illustrated in Fig. 14a, b, respectively.

From Fig. 14a, b, it is evident that *INDRA* achieves high detection accuracy under each attack scenario while achieving lower false positive rates. The ability to monitor signal level variations combined with a more cautious selection of intrusion threshold gives *INDRA* an advantage over comparison works. PLSTM and RepNet use the maximum validation loss in the final model as the threshold, whereas CANet uses interval-based monitoring to detect malicious messages. Choosing a higher threshold helped PLSTM to achieve slightly lower false positive rates for some scenarios, but it hurt the ability of both PLSTM and RepNet to detect attacks with minor variations in the input data. This is because the deviations produced by some of the complex attacks are small, and the attacks go undetected due to the large thresholds. Moreover, the interval-based monitoring approach employed in CANet struggles to find an optimal threshold resulting in subpar performance. It is essential to highlight that *INDRA* achieves this superior performance by monitoring at a signal level as opposed to prior works that monitor at the message level. Lastly, the false positive rates of *INDRA* remain significantly low, with a maximum of 2.5% for plateau attacks.

6.5 IDS Overhead Analysis

In this section, we present a detailed analysis of the overhead incurred by our proposed *INDRA* IDS. We quantify the IDS overhead in terms of memory footprint and time taken to process an incoming message, i.e., inference time. The former metric is important as the automotive ECUs are highly resource-constrained and have limited memory and compute capacities. Therefore, it is critical to have a low memory overhead to avoid interference with real-time automotive applications. The inference time metric provides important information about the time it takes to detect the attacks and can also be used to compute the utilization overhead on the ECU. Hence, the above-mentioned two metrics are analyzed to study the overhead and quantify the lightweight nature of *INDRA* IDS.

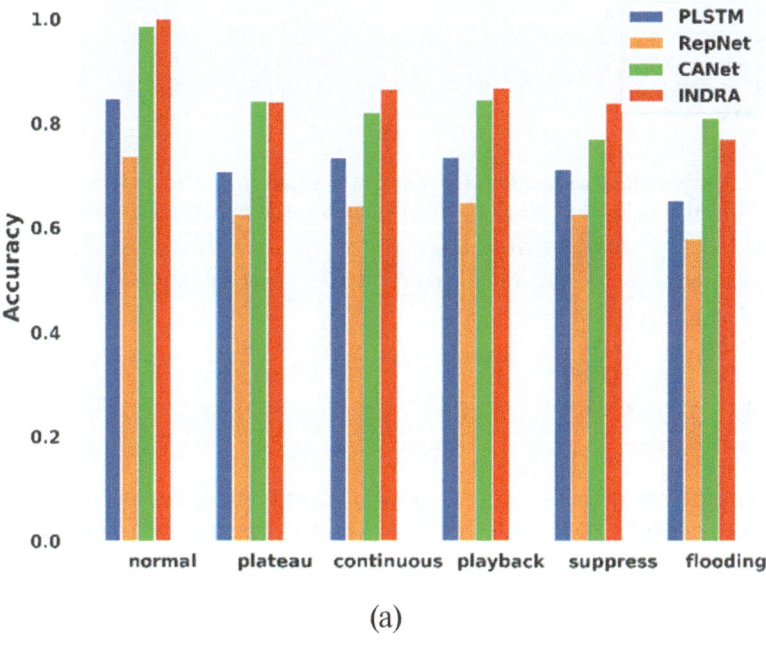

(a)

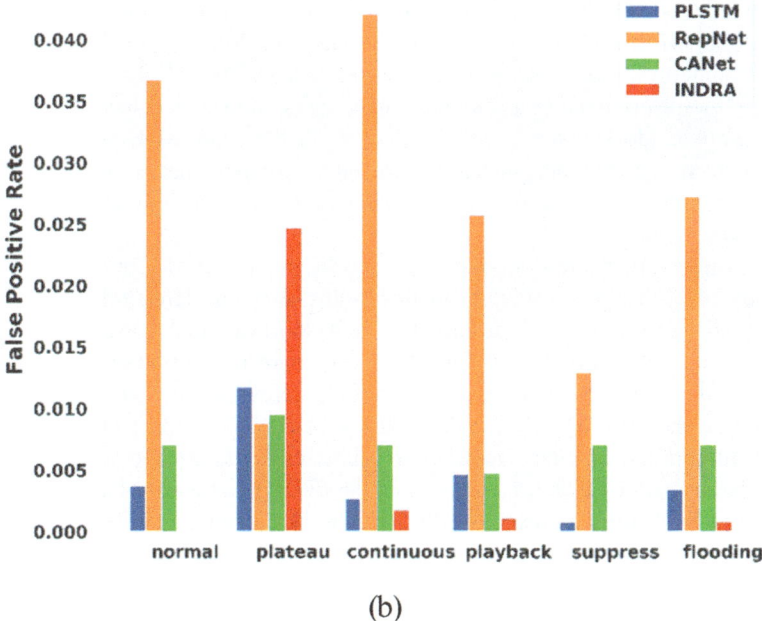

(b)

Fig. 14 Comparison of (**a**) detection accuracy and (**b**) false positive rate of *INDRA* and the prior works PLSTM [38], RepNet [39], and CANet [36]

Table 2 Memory footprint comparison between our proposed *INDRA* framework and the prior works PLSTM [38], RepNet [39], and CANet [36]

IDS framework	Memory footprint (KB)
PLSTM [38]	13,417
RepNet [39]	55
CANet [36]	8718
INDRA	443

Table 3 Inference time comparisons between our proposed *INDRA* framework and the prior works PLSTM [38], RepNet [39], and CANet [36] using single and dual-core configurations

IDS framework	Average inference time (μs)	
	Single core ARM Cortex A57 CPU	Dual core ARM Cortex A57 CPU
PLSTM [38]	681.18	644.76
RepNet [39]	19.46	21.46
CANet [36]	395.63	378.72
INDRA	80.35	72.91

To quantify the overhead of our proposed *INDRA* framework and the prior works, we implemented the IDSs on the NVIDIA Jetson Tx2 board, consisting of an ARM Cortex- A57 CPU, which has similar specifications to the state-of-the-art multi-core ECUs. The memory footprint of the *INDRA* framework and the prior works mentioned in the previous subsections is shown in Table 2. It is evident that the *INDRA* framework has a low memory footprint compared to the prior works, except for the RepNet [39]. However, it is important to observe that even though the *INDRA* framework has a slightly higher memory footprint compared to the RepNet [39], *INDRA* outperforms all the prior works, including RepNet [39], in all performance metrics under various attack scenarios, as shown in Fig. 14. The heavier (high memory footprint) models can capture a wide range of system behaviors; however, they are not an ideal choice for resource-constrained automotive CPS. On the other hand, a much lighter model (such as RepNet) fails to capture necessary details about the system behavior due to its limited model parameters, which in turn suffers from performance issues.

We benchmarked different IDS frameworks on an ARM Cortex- A57 CPU to study the inference overhead. In this study, we considered different system configurations to explore a wide variety of ECU hardware that is available in state-of-the-art vehicles. Based on the available hardware resources, single-core (uses only one CPU core) and dual-core (uses two CPU cores) system configurations were selected on the Jetson TX2. The IDS frameworks are executed 10 times for each CPU configuration, and the average inference times (in μs) are recorded in Table 3. From the results in Table 3, it is clear that the *INDRA* framework has significantly faster inference times compared to the prior works (excluding RepNet) under all system configurations. From the results in Fig. 14, it can be seen that RepNet has the worst performance of any comparison framework, despite having a lower inference time. The large inference times for the better-performing frameworks can have a significant impact on the real-time performance of the vehicle and can be catastrophic in the event of deadline misses. We also believe that using a dedicated deep learning accelerator (DLA) further enhances the performance of the IDS models.

Thus, from Fig. 14, Table 2 and 3, it is clear that *INDRA* achieves a clear balance of having superior intrusion detection performance while maintaining a low memory footprint and fast inference times, making it a powerful and lightweight IDS solution.

6.6 Scalability Results

In this subsection, we present a detailed analysis on the scalability of the *INDRA* framework by studying the system performance using the ECU utilization metric as a function of increasing system complexity (i.e., number of ECUs and messages). Each ECU in the system has a real-time utilization (U_{RT}) and an IDS utilization (U_{IDS}) from running real-time and IDS applications, respectively. We focus on analyzing the IDS overhead (U_{IDS}), as it is a direct measure of the compute efficiency of the IDS. Moreover, as the safety-critical messages monitored by the IDS are periodic, the IDS can be modeled as a periodic application with a period that is the same as the message period [5]. As a result, monitoring an *ith* message (m_i) results in an induced IDS utilization ($U_{IDS,\ mi}$) at an ECU, which can be calculated using (6).

$$U_{IDS,m_i} = \left(\frac{T_{IDS}}{P_{m_i}} \right) \tag{6}$$

where, T_{IDS} and P_{mi} represent the time taken by the IDS to process one message (inference time) and the period of the monitored message, respectively. Moreover, the sum of all IDS utilizations because of monitoring different messages is the overall IDS utilization at that ECU (U_{IDS}) and is computed using (7).

$$U_{IDS} = \sum_{i=1}^{n} U_{IDS,m_i} \tag{7}$$

To evaluate the scalability of the *INDRA* IDS, six different system sizes were considered. Moreover, a set of commonly used message periods {1, 5, 10, 15, 20, 25, 30, 45, 50, 100} (all periods in ms) in automotive CPS is considered to sample uniformly, when assigning periods to the messages in the system. These messages are distributed evenly among different ECUs, and the IDS utilization is calculated using (6) and (7). *INDRA* assumes a pessimistic scenario where all the ECUs in the system have only a single core, which would allow us to analyze the worst-case overhead of the IDS.

The average ECU utilization for different system sizes denoted by {p, q}, where p is the number of ECUs and q is the number of messages in the system, is illustrated in Fig. 15. In this study, a very pessimistic estimate of 50% real-time ECU utilization

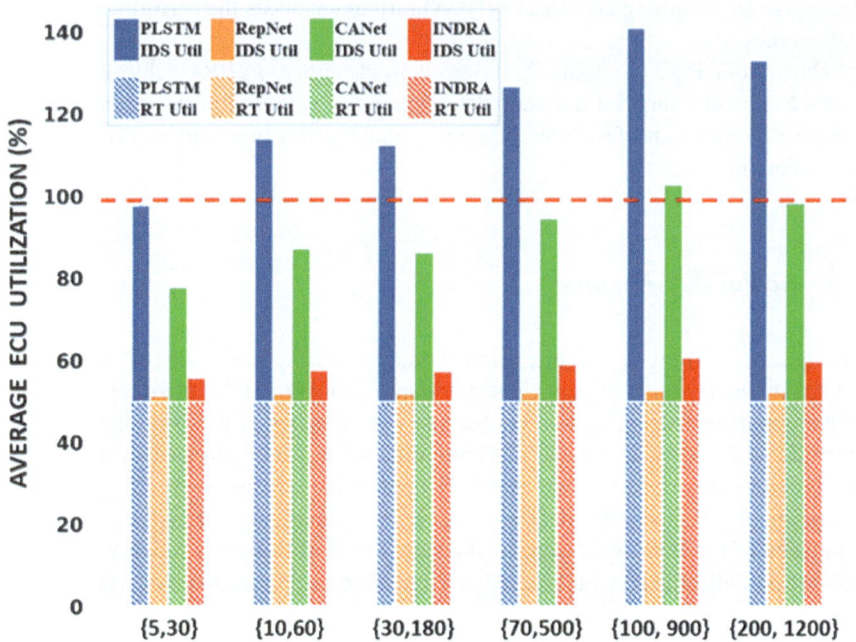

Fig. 15 Scalability analysis of our proposed *INDRA* IDS for different system sizes and the prior works PLSTM [38], RepNet [39], and CANet [36]

for real-time automotive applications ("RT Util", as shown in the dotted bars) is assumed. The solid bars on top of the dotted bars represent the IDS overhead on the ECUs, and the horizontal dotted line in red represents the 100% ECU utilization mark. It is critical to ensure that the ECU utilization does not exceed 100% under any scenario, as it could introduce undesired latencies resulting in missing deadlines for time-critical automotive applications, which can be catastrophic. It is clear from the results that the prior works, such as PLSTM and CANet, incur heavy overhead on the ECUs, while RepNet and our proposed *INDRA* framework have a very minimal overhead that is favorable to increasing system sizes. Thus, from the results in this section (Figs. 14 and 15; Tables 2 and 3), it is apparent that *INDRA* not only achieves better performance in terms of detection accuracy and false positive rate for intrusion detection than state-of-the-art prior works, but it is also lightweight and highly scalable.

7 Conclusion

In this chapter, we presented a novel recurrent autoencoder-based lightweight real-time intrusion detection system called *INDRA* for automotive CPS. *INDRA* framework uses the intrusion score (IS) metric to measure the deviation from the learned system behavior to detect intrusions. Moreover, we presented a thorough analysis on the intrusion threshold selection process and compared the *INDRA* IDS with the best-known prior works in this area. The promising results indicate a compelling potential for utilizing our proposed *INDRA* IDS in emerging automotive platforms.

References

1. Kukkala, V.K., Bradley, T., Pasricha, S.: Priority-based multi-level monitoring of signal integrity in a distributed powertrain control system. In: Proceedings of IFAC Workshop on Engine and Powertrain Control, Simulation and Modeling (2015)
2. Kukkala, V.K., Bradley, T., Pasricha, S.: Uncertainty analysis and propagation for an auxiliary power module. In: Proceedings of IEEE Transportation Electrification Conference (TEC) (2017)
3. Kukkala, V.K., Pasricha, S., Bradley, T.: JAMS: Jitter-aware message scheduling for flexray automotive networks. In: Proceedings of IEEE/ACM International Symposium on Network-on-Chip (NOCS) (2017)
4. Kukkala, V.K., Pasricha, S., Bradley, T.: JAMS-SG: A framework for jitter-aware message scheduling for time-triggered automotive networks. ACM Trans. Design Autom. Electron. Syst. (TODAES). **24**(6) (2019)
5. Kukkala, V., Pasricha, S., Bradley, T.: SEDAN: Security-aware design of time-critical automotive networks. IEEE Trans. Vehic. Technol. (TVT). **69**(8) (2020)
6. Kukkala, V.K., Thiruloga, S.V., Pasricha, S.: INDRA: Intrusion detection using recurrent autoencoders in automotive embedded systems. IEEE Trans. Comput.-Aided Des. Integr. Circuits Syst. (TCAD). **39**(11) (2020)
7. Kukkala, V.K., Thiruloga, S.V., Pasricha, S.: LATTE: LSTM self-attention based anomaly detection in embedded automotive platforms. ACM Trans. Embed. Comput. Syst. (TECS). **20**(5s), Article 67 (2021)
8. Thiruloga, S.V., Kukkala, V.K., Pasricha, S.: TENET: Temporal CNN with attention for anomaly detection in automotive cyber-physical systems. In: Proceedings of IEEE/ACM Asia & South Pacific Design Automation Conference (ASPDAC) (2022)
9. Kukkala, V.K., Thiruloga, S.V., Pasricha, S.: Roadmap for cybersecurity in autonomous vehicles. In: IEEE Consum. Electron. Magaz. (CEM) (2022)
10. Tunnell, J., Asher, Z., Pasricha, S., Bradley, T.H.: Towards improving vehicle fuel economy with ADAS. SAE Int. J. Connect. Autom. Veh. **1**(2) (2018)
11. Tunnell, J., Asher, Z., Pasricha, S., Bradley, T.H.: Towards improving vehicle fuel economy with ADAS. In: Proceedings of SAE World Congress Experience (WCX) (2018)
12. Asher, Z., Tunnell, J., Baker, D.A., Fitzgerald, R.J., Banaei-Kashani, F., Pasricha, S., Bradley, T.H.: Enabling prediction for optimal fuel economy vehicle control. In: Proceedings of SAE World Congress Experience (WCX) (2018)
13. Dey, J., Taylor, W., Pasricha, S.: VESPA: A framework for optimizing heterogeneous sensor placement and orientation for autonomous vehicles. IEEE Consum. Electron. Magaz. (CEM). **10**(2) (2021)
14. Kukkala, V.K., Pasricha, S., Bradley, T.: Advanced driver-assistance systems: A path toward autonomous vehicles. IEEE Consum. Electron. Magaz. **7**(5) (2018)
15. Koscher, K., Czeskis, A., Roesner, F., Patel, S., Kohno, T., Checkoway, S., McCoy, D., Kantor, B., Anderson, D., Shacham, H., Savage, S.: Experimental security analysis of a modern automobile. In: Proceedings of IEEE Symposium on Security and Privacy (SP) (2010)
16. Miller, C., Valasek, C.: Remote exploitation of an unaltered passenger vehicle. Black Hat USA (2015)

17. Izosimov, V., Asvestopoulos, A., Blomkvist, O., Törngren, M.: Security-aware development of cyber-physical systems illustrated with automotive case study. In: Proceedings of IEEE/ACM Design, Automation & Test in Europe & Exhibition (DATE) (2016)
18. Studnia, I., Alata, E., Nicomette, V., Kaâniche, M., Laarouchi, Y.: A language-based intrusion detection approach for automotive embedded networks. Int. J. Embed. Syst. (IJES). **10**(8) (2018)
19. Marchetti, M., Stabili, D.: Anomaly detection of CAN bus messages through analysis of ID sequences. In: Proceedings of IEEE Intelligent Vehicle Symposium (IV) (2017)
20. Hoppe, T., Kiltz, S., Dittmann, J.: Security threats to automotive CAN networks- practical examples and selected short-term countermeasures. Reliab. Eng. Syst. Saf. **96**(1) (2011)
21. Larson, U.E., Nilsson, D.K., Jonsson, E.: An approach to specification-based attack detection for in-vehicle networks. In: Proceedings of IEEE Intelligent Vehicles Symposium (IV) (2008)
22. Aldwairi, M., Abu-Dalo, A.M., Jarrah, M.: Pattern matching of signature-based IDS using Myers algorithm under MapReduce framework. EURASIP J. Inf. Secur. **1** (2017)
23. Myers, E.W.: An O(ND) difference algorithm and its variations. Algorithmica (1986)
24. Hoppe, T., Kiltz, S., Dittmann, J.: Applying intrusion detection to automotive IT-early insights and remaining challenges. J. Inf. Assur. Secur. (JIAS). **4**(6) (2009)
25. Waszecki, P., Mundhenk, P., Steinhorst, S., Lukasiewycz, M., Karri, R., Chakraborty, S.: Automotive electrical and electronic architecture security via distributed in-vehicle traffic monitoring. IEEE Trans. Comput.-Aided Des. Integr. Circuits Syst. (TCAD). **36**(11) (2017)
26. Cho, K.T., Shin, K.G.: Fingerprinting electronic control units for vehicle intrusion detection. In: Proceedings of USENIX (2016)
27. Ying, X., Sagong, S.U., Clark, A., Bushnell, L., Poovendran, R.: Shape of the Cloak: Formal analysis of clock skew-based intrusion detection system in controller area networks. IEEE Trans. Inf. Forensics Secur. (TIFS). **14**(9) (2019)
28. Yoon, M.K., Mohan, S., Choi, J., Sha, L.: Memory heat map: Anomaly detection in real-time embedded systems using memory behavior. In: Proceedings of IEEE/ACM/EDAC Design Automation Conference (DAC) (2015)
29. Müter, M., Asaj, N.: Entropy-based anomaly detection for in-vehicle networks. In: Proceedings of IEEE Intelligent Vehicles Symposium (IV) (2011)
30. Müter, M., Groll, A., Freiling, F.C.: A structured approach to anomaly detection for in-vehicle networks. In: Proceedings of IEEE International Conference on Intelligent and Advanced System (ICIAS) (2010)
31. Taylor, A., Japkowicz, N., Leblanc, S.: Frequency-based anomaly detection for the automotive CAN bus. In: Proceedings of World Congress on Industrial Control Systems Security (WCI-CSS) (2015)
32. Martinelli, F., Mercaldo, F., Nardone, V., Santone, A.: Car hacking identification through fuzzy logic algorithms. In: Proceedings of IEEE International Conference on Fuzzy Systems (FUZZ-IEEE) (2017)
33. Vuong, T.P., Loukas, G., Gan, D.: Performance evaluation of cyber-physical intrusion detection on a robotic vehicle. In: Proc. of IEEE International Conference on Computer and Information Technology; Ubiquitous Computing and Communications; Dependable, Autonomic and Secure Computing; Pervasive Intelligence and Computing (CIT/IUCC/DASC/PICOM) (2015)
34. Levi, M., Allouche, Y., Kontorovich, A.: Advanced analytics for connected car cybersecurity. In: Proceedings of IEEE Vehicular Technology Conference (VTC) (2018)
35. Kang, M.J., Kang, J.W.: A novel intrusion detection method using deep neural network for in-vehicle network security. In: IEEE Proceedings of Vehicular Technology Conference (VTC) (2016)
36. Hanselmann, M., Strauss, T., Dormann, K., Ulmer, H.: CANet: An unsupervised intrusion detection system for high dimensional CAN bus data. IEEE Access. (2020)
37. Loukas, G., Vuong, T., Heartfield, R., Sakellari, G., Yoon, Y., Gan, D.: Cloud-based cyber-physical intrusion detection for vehicles using deep learning. IEEE Access. (2018)

38. Taylor, A., Leblanc, S., Japkowicz, N.: Anomaly detection in automobile control network data with long short-term memory networks. In: Proceedings of IEEE International Conference on Data Science and Advanced Analytics (DSAA) (2016)

39. Weber, M., Wolf, G., Sax, E., Zimmer, B.: Online detection of anomalies in vehicle signals using replicator neural networks. In: Proceedings of ESCAR USA (2018)

40. Weber, M., Klug, S., Sax, E., Zimmer, B.: Embedded hybrid anomaly detection for automotive can communication. In: Embedded Real Time Software and Systems (ERTS) (2018)

41. Schmidhuber, J.: Habilitation Thesis: System Modeling and Optimization (1993)

42. Hochreiter, S., Bengio, Y., Frasconi, P., Schmidhuber, J.: Gradient Flow in Recurrent Nets: The Difficulty of Learning Long-Term Dependencies. IEEE Press (2001)

43. Cho, K., Van Merriënboer, B., Gulcehre, C., Bahdanau, D., Bougares, F., Schwenk, H., Bengio, Y.: Learning phrase representations using RNN encoder-decoder for statistical machine translation. arXiv Preprint, arXiv:1406.1078, 2014

44. DiDomenico, G.C, Bair, J., Kukkala, V.K, Tunnell, J., Peyfuss, M., Kraus, M., Ax, J., Lazarri, J., Munin, M., Cooke, C., Christensen, E.: Colorado State University EcoCAR 3 final technical report. In: SAE World Congress Experience (WCX) (2019)

Stacked LSTM Based Anomaly Detection in Time-Critical Automotive Networks

Vipin Kumar Kukkala, Sooryaa Vignesh Thiruloga, and Sudeep Pasricha

1 Introduction

The increased interest in enabling full self-driving cars and the growing demand for the integration of advanced safety features in today's vehicles have rapidly increased the complexity of embedded systems being integrated into various vehicular subsystems. The aggressive competition between automakers to reach autonomy goals is further driving the complexity of Electronic Control Units (ECUs) and the in-vehicle network that connects them [1]. Moreover, recent solutions for Advanced Driver Assistance Systems (ADAS) require interactions with various external systems using a variety of advanced communication standards such as Wi-Fi, 5G, and Vehicle-to-X (V2X) protocols [2]. The V2X communication enables a spectrum of connections such as vehicle-to-vehicle (V2V), vehicle-to-infrastructure (V2I), vehicle-to-pedestrian (V2P), and vehicle-to-cloud (V2C) [3]. These new solutions are transforming modern vehicles by making them highly connected to the external environment. Moreover, to support the increasingly sophisticated ADAS functions and connectivity to the outside world, ECUs in today's vehicles run highly complex software to handle various highly safety-critical and time-sensitive automotive applications, e.g., pedestrian and traffic sign detection, lane changing, automatic

V. K. Kukkala (✉)
NVIDIA, Santa Clara, CA, USA
e-mail: vipin.kukkala@colostate.edu

S. V. Thiruloga
Department of Electrical and Computer Engineering, Colorado State University, Fort Collins, CO, USA
e-mail: sooryaa@colostate.edu

S. Pasricha
Colorado State University, Fort Collins, CO, USA
e-mail: sudeep.pasricha@colostate.edu

parking, and path planning. This increased software and hardware complexity of the automotive electrical/electronic (E/E) architecture and increased connectivity with external environment resulted in various challenged related to reliability [49–52], security [14, 32, 45–47], and real-time performance [53–56]. In this work, we mainly focus on improving security in automotive networks. The above-mentioned advances have one crucial security implication: they provide a large attack surface and thus give rise to more opportunities for attackers to gain unauthorized access to the in-vehicle network and execute cyber-attacks. Additionally, the high complexity in emerging vehicles resulted in poor attack visibility over the in-vehicle network, making it hard to detect attacks that can be easily hidden within normal operational activities. Many cyber-attacks on vehicles can induce various anomalies in the network, resulting in a change in the normal behavior of the network as well as the ECU behavior. Due to the highly safety-critical and time-sensitive nature of automotive applications, any minor instability in the system due to these induced anomalies could lead to a major catastrophe, e.g., preventing an airbag from deploying in the case of a collision, delaying the perception of a pedestrian, or erroneously changing lanes into oncoming traffic, due to maliciously corrupted sensor readings.

An attack via an externally linked component or compromised ECU can manifest in several forms over the in-vehicle network. One of the most commonly observed and easy-to-launch attacks is flooding the in-vehicle network with random or specific messages. This increases the overall network traffic and results in halting any useful activity over the network. An advanced attack on an ECU could involve remotely sending a kill command to the engine during normal driving. More sophisticated attacks could involve installing malware (e.g., trojan) on the ECU and using it to achieve malicious goals. State-of-the-art cyber-attacks on vehicles have used a variety of attack vectors, such as infotainment systems to launch buffer overflow and denial of service attacks [5] and reverse engineering keyless entry systems to wirelessly lock pick the vehicle immobilizer [6]. Researchers in [4] foresees a much more severe attack involving potentially targeting the U. S. electric power grid by using public electric vehicle charging stations as an attack vector to infect vehicles that use these stations with malware. Several other cyber-attacks on different real-world vehicles are presented in [7–10]. The common aspect of these attacks is that they involve gaining unauthorized access to the in-vehicle network and modifying certain fields in the message frames, thereby tricking the receiving ECU into thinking that the malicious message is legitimate. All these attacks can have catastrophic consequences and must be detected before they are executed. This problem will get exacerbated with the onset of connected and autonomous vehicles. Thus, it is crucial to restrict the attackers via early detection to achieve a secure automotive system.

Traditional computer networks utilize different protective mechanisms such as firewalls (software) and isolation units such as gateways and switches (hardware) to protect from external cyber-attacks [11]. However, advanced persistent attackers have been coming up with various novel attacks that leverage the increased compute and communication capabilities in modern ECUs, causing the traditional protection systems to become obsolete. Thus, it is essential to have an advanced

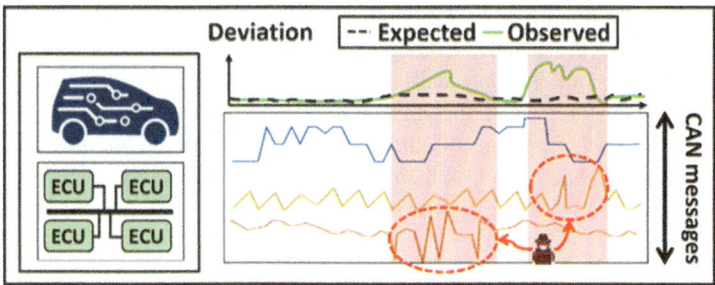

Fig. 1 Illustration of an example anomaly detection framework that monitors the in-vehicle network traffic and detects deviations from expected normal system behavior during the attack intervals shown in red

solution that can continuously monitor the vehicle network, to detect cyber-attacks. One promising solution is to deploy an anomaly detection system (ADS), which continuously monitors the network for unusual activities and raises the alarm when suspicious activity is detected. The ADS frameworks learn the normal system behavior at design time and monitor the network for anomalies at runtime. This approach can be extended to detect and classify various cyber-attacks on the in-vehicle network. A traditional approach for anomaly detection uses rule-based techniques such as, monitoring message frequency [12] and memory heat map [13] to detect known attack signatures. However, due to the increased complexity of today's cyber-attacks, such traditional rule-based systems fail to recognize new and complex attack patterns, rendering these approaches ineffective. Fortunately, recent advances in deep learning and the availability of in-vehicle network data have paved the path to using sophisticated deep learning models for anomaly detection.

In this chapter, we present a novel anomaly detection framework called *LATTE*, which was first introduced in [45] to detect cyber-attacks in time-critical automotive networks. Our proposed *LATTE* framework uses advanced deep learning models (discussed in Sect. 5.2) in an unsupervised setting to learn the normal behavior of the system. *LATTE* leverages that information at runtime to detect anomalies by observing for any deviations from the learned normal behavior. This process is illustrated in Fig. 1. The plot on the top right in Fig. 1 shows the expected deviation (computed using the model that was trained at design time) vs. the observed deviation. The deviation in signal values during the attack intervals (shown in the red area) can be used to detect cyber-attacks as anomalies. Our proposed *LATTE* framework aims to maximize the anomaly detection accuracy, precision, and recall, while minimizing the false-positive rate. Our novel contributions in *LATTE* are:

- We introduce a stacked Long-Short Term Memory (LSTM) based predictor model that integrates a novel self-attention mechanism to learn the normal system behavior at design time;
- We devise a one class support vector machine (OCSVM) based detector model that works in tandem with the predictor model to detect various cyber-attacks at runtime;

- We present modifications to the communication controller in the ECU that can help in realizing the proposed anomaly detection system;
- We perform a comprehensive analysis on the selection of deviation measures that quantify the deviation from the normal system behavior;
- We explored several variants of our proposed *LATTE* framework and selected the best-performing one, which is then compared with the best-known prior works in the area to show *LATTE*'s effectiveness.

2 Related Work

Researchers studied various cyber-attacks on vehicles to discover vulnerabilities in automotive systems. Recent attacks such as [15] exploit the vulnerability in security access algorithms to deploy airbags without any actual impact. In [16], the attackers reverse-engineered a telematics control unit to exploit a memory vulnerability in the firmware to circumvent the existing firewall and remotely send diagnostic messages to control an ECU. Other attacks that compromised the ADAS camera sensor were studied in [17]. All these attacks introduce anomalous behavior during vehicle operation, which a good anomaly detection framework must detect.

Anomaly detection has been a popular research topic in the domain of computer networks, and several solutions have been proposed to detect cyber-attacks in large-scale computer networks [18]. However, these solutions require high compute power, which makes them hard to adapt to resource-constrained automotive cyber-physical systems for detecting cyber-attacks in in-vehicle networks. Several solutions were developed in the past decade to tackle the problem of anomaly detection in automotive systems [19–34]. These works can be broadly categorized into two types *(i)* heuristic-based and *(ii)* machine learning based. Heuristic-based anomaly detection approaches observe for traces of known attack signatures. In contrast, a machine-learning-based approach learns the normal system behavior during an offline phase and observes for any deviation from the learned normal behavior at runtime to detect anomalies. The heuristic-based techniques can be simple and have fast detection times compared to machine learning-based techniques. However, machine learning based techniques can detect both known and unknown attacks, which is not possible with heuristic-based techniques. Some of the key prior works in these categories are discussed in the subsequent subsections.

2.1 Heuristi-Based Anomaly Detection

A language theory-based model is used to obtain signatures of known attacks from the vehicle's CAN bus in [19]. However, this approach fails to detect anomalous messages when the model misses the packets transmitted during the early stages of an attack. The transition matrices-based anomaly detection scheme was introduced in [20] to detect anomalous sequences in CAN bus-based systems. Although this approach was able to achieve low false-positive rates for simple attacks, it failed to detect realistic replay attacks. In [21], the authors proposed a Hamming-distance based model which monitors the CAN network to detect cyber-attacks. However, the model had very limited attack coverage. A specification-based approach was presented in [22], which compared the messages with predefined attack signatures to detect anomalies. A time-frequency analysis model is used to continuously monitor CAN message frequency to detect anomalies in [23]. In [24], a heuristic-based approach is used to build a normal operating region by analyzing the messages at design time and a message-frequency-based in-vehicle network monitoring system to detect anomalies at runtime. In [25], clock-skew based fingerprints are recorded at design time, and at runtime, the variations in clock-skew at sender ECUs are used to detect anomalies. An anomaly detection system introduced in [26], monitors the entire system for changes in entropy to detect anomalies. However, their approach fails to detect smaller anomalous sequences that result in minimal change in the entropy. *In summary, heuristic-based anomaly detection systems provide low-cost and high-speed detection techniques but fail to detect complex and novel attacks. Additionally, modeling every possible attack signature is impractical, and hence heuristic-based anomaly detection approaches have a limited scope.*

2.2   Machine Learning Based Anomaly Detection

Many recent works leverage advances in machine learning to build highly efficient anomaly detection systems. In [27], a deep neural network (DNN) based approach that continuously monitors the network is used to observe for changes in communication patterns to detect anomalies. However, this approach is only designed and tested for a low-priority system (a tire pressure monitoring system), which limits us from directly adapting this technique to safety-critical systems. A recurrent neural network (RNN) based intrusion detection system that attempts to learn the normal behavior of CAN messages in the in-vehicle network was introduced in [28]. In [29], a hybrid approach, which utilizes both specification and RNN-based systems in two stages is used to detect anomalies. In [30], an LSTM-based predictor model that predicts the next time step message value at a bit level and detects intrusions by observing for large deviations in prediction errors. A long short-term memory (LSTM) based multi message-id detection model was proposed in [31]. However, the model is highly complex and has a high implementation overhead when deployed on an ECU. A GRU-based lightweight recurrent autoencoder and a static threshold-based detection scheme to detect cyber-attacks was introduced in [32]. However, using static thresholds limits the system to detecting only simple attacks. In [33], a deep convolutional neural network (CNN) model was proposed to detect anomalies in the vehicle's CAN bus. However, the model does not consider the temporal relationships between messages, which limits the model from predicting certain attacks. In [34], an LSTM framework with a hierarchical attention mechanism and a non-parametric kernel density estimator, along with a k-nearest

Table 1 Comparison between our proposed *LATTE* framework and the state-of-the-art machine learning-based anomaly detection works

Technique	ADS task	Requires labeled data?	Network architecture	Attention type	Detection model
BWMP [30]	Bit level prediction	Yes	LSTM network	–	Static threshold
RepNet [28]	Input recreation	No	Replicator network	–	Static threshold
HAbAD [34]	Input recreation	Yes	Autoencoder	Hierarchical	KDE and KNN
LATTE [45]	Next message value prediction	No	Encoder-decoder	Self-attention	OCSVM

neighbors classifier, is used to detect anomalies. *Although most of these techniques attempt to increase detection accuracy and attack coverage, none of them offers the ability to process very long sequences with relatively low runtime and memory overhead and still achieve reasonably high performance.*

In this chapter, we introduce a robust deep learning model that integrates a stacked LSTM-based encoder-decoder model with a self-attention mechanism to learn normal system behavior by learning to predict the next message instance. At runtime, we continuously monitor in-vehicle network messages and provide a reliable detection mechanism using a non-linear classifier. A summary of the state-of-the-art anomaly detection works and their key features, and the unique characteristics of our proposed *LATTE* framework is presented in Table 1. The details of the proposed model and the overall framework are presented in Sects. 4 and 5. In Sect. 6, we present different experimental results that demonstrate the efficiency of our *LATTE* framework compared to various state-of-the-art anomaly detection works in identifying a variety of attack scenarios.

3 Background

Solving complex problems using deep learning was made possible due to advances in computing hardware and the availability of large high-quality datasets. Anomaly detection is one such problem that can leverage the power of deep learning. In an automotive system, ECUs exchange safety-critical messages periodically over the in-vehicle network. These periodic messages have temporal relationships between them, which can be exploited to detect anomalies. However, this requires a special type of neural network called Recurrent Neural Network (RNN). Unlike traditional feed-forward neural networks, where the output at any point is independent of any previous inputs, RNNs use information from previous sequences when computing the output, which makes them an ideal choice for handling time-series data.

3.1   Recurrent Neural Network (RNN)

An RNN [35] is the most basic sequence model that takes sequential or time-series data as the input and learns the underlying temporal relationships between data samples. An RNN block consists of an input, an output, and a hidden state that allows it to remember the learned temporal information. The input, output, and hidden state all correspond to a particular time step in the sequence. The hidden-state information can be visualized as a latent space data point containing important temporal information about the inputs from previous time steps. RNNs compute the current stage output by taking the previous hidden-state information along with the current input. Moreover, the backpropagation in RNNs occurs through time, resulting in the error value shrinking or growing rapidly, leading to vanishing or exploding gradients. This severely hampers RNN's ability to learn patterns in the input data that have long-term dependencies [36]. To overcome this problem, long short-term memory (LSTM) networks [37] were introduced.

3.2   Long Short-Term Memory (LSTM) Network

LSTMs are enhanced RNNs that use a combination of cell state, hidden state, and multiple gates to learn long-term dependencies in the sequences. The cell state carries the relevant long-term dependencies throughout the processing of an input sequence, whereas the hidden state contains relevant information from the recent time steps accommodating short-term dependencies. The gates in LSTM are used to regulate the flow of information from the hidden state to the cell state. These combinations of gates and states give LSTM an edge over the simple RNN in remembering long-term dependencies in sequences. LSTMs have therefore replaced simple RNNs in the areas of time-series forecasting, natural language processing, and machine translation [36].

In general, LSTMs overcome many of the limitations of RNNs and provide a more than acceptable solution for the vanishing and exploding gradient problems. However, they suffer a significant performance drop when handling very long input sequences (e.g., with 100 or more time steps). This is mainly because the predictions of an LSTM unit at the current time step t, are heavily influenced by the hidden and cell states at the previous time step $t-1$ compared to the past time steps. Hence, for a very long input sequence, the representation of the input at the first time step tends to diminish as the LSTM processes inputs at the future time steps. To overcome this limitation, we need a mechanism that can look back and identify the information that can influence future sequences. One such look-back mechanism is neural attention.

3.3 *Attention*

Attention or neural attention is a mechanism in neural networks that mimics the visual attention mechanism in humans [38]. The human eye can focus on certain objects or regions with higher resolution compared to their surroundings. Similarly, the attention mechanism in neural networks allow focusing on the relevant parts of the input sequence and selectively output only the most relevant information. Although LSTMs take the previous hidden state information and the input at the current time step to compute the current output, they suffer a significant drop in performance when processing very long input sequences. This is because the information from the first time step is less representative in the hidden states than the information from the recent time steps. To overcome this problem, we can incorporate attention mechanisms with LSTMs, which allows them to capture the crucial information from any past time steps of the input sequence.

Attention mechanisms are frequently used in encoder-decoder architectures [36]. An encoder-decoder architecture mainly consists of three major components *(i)* encoder, *(ii)* latent vector, and *(iii)* decoder. The encoder takes the input sequence and converts it to a fixed-size latent representation called a latent vector. The latent vector consists of all the information representing the input sequence in a latent space. The decoder takes the latent vector as input and converts it to the desired output. However, as the latent vector uses a fixed length to represent the input sequence, the fixed-length vector fails to encapsulate all the information from a very long input sequence, resulting in poor performance. To address this problem, the authors in [39] introduced an attention mechanism in sequence models that enabled encoders to build a context vector by creating customized shortcuts to parts of the inputs. This ensures that the context vector represents the crucial parts and learns the very long-term dependencies in the input sequence leading to improved decoder outputs. A self-attention mechanism for an LSTM encoder-decoder model was presented in [40], which consumes all the encoder hidden states to compute the attention weights.

The input to the decoder in an LSTM-based encoder-decoder model (rolled out in time for 4 time steps) using no attention and self-attention is illustrated in Fig. 2. The input to the LSTM encoder at each time step is represented as x_t, and the initial hidden vector is represented as h_0. The colored rectangle next to each LSTM unit for every time step represents the hidden state information, and the height of each color in the rectangle signifies the amount of information from each time step. At each time step, a square filled with a different color inside the LSTM cell is used to represent the hidden state information of that time step. Moreover, for this example, we consider a scenario where the output at the last time step ($t = 4$) has a high dependency on the input at the second time step (x_2). We can see that in the case of <u>no attention</u>, shown in Fig. 2a, the LSTM hidden state at $t = 4$ largely comprises of information from the third (blue) and fourth (orange) time steps. This results in providing the decoder with an inaccurate representation of current time step dependency, which leads to poor results at the output of the decoder. On

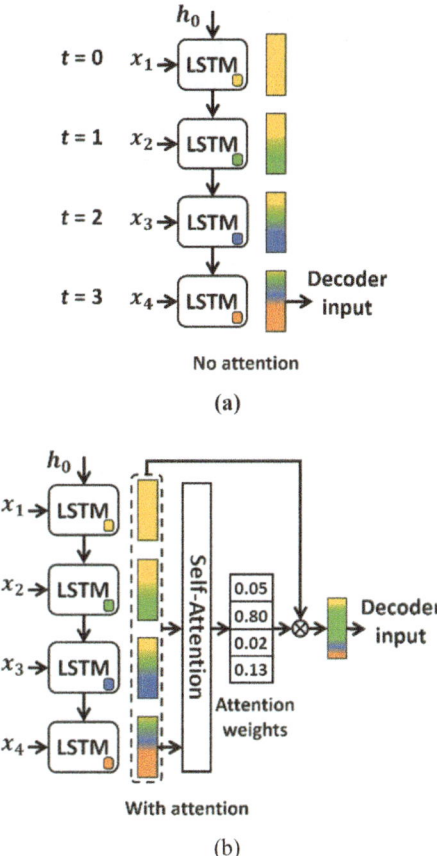

Fig. 2 Decoder input in case of (**a**) no attention, (**b**) with attention in sequence models using LSTMs

the contrary, in the case <u>with attention</u>, shown in Fig. 2b, the self-attention block consumes all hidden state representations at each time step as well as the current time step ($t = 4$) and generates the context vector (decoder input). It can be observed that the self-attention mechanism clearly captures the high dependency of output at $t = 4$ on the output at $t = 2$ (shown in the hidden state information at the output of self-attention). This can also be seen in the attention weights computed by the attention mechanism, where the information from the second (green) time step is given high weightage compared to others. Therefore, by accurately representing the key portions of the input sequence in the decoder input, the self-attention mechanism is able to facilitate better decoder outputs. Also, unlike other attention mechanisms such as [41], the attention vector in self-attention aligns encoder outputs to encoder hidden states, thereby eliminating the need for any feedback from previous decoder predictions. This lack of feedback loop enables the self-attention mechanism to quickly learn the temporal dependencies in the long input sequences. Hence, *for the first time, LATTE adapts the self-attention mechanism to a stacked LSTM-based encoder-decoder network to learn the temporal relationships between messages in the* automotive network.

4 Problem Formulation

4.1 System Overview

In this chapter, we consider an automotive system that consists of multiple ECUs connected using a CAN-based in-vehicle network, as shown in Fig. 3. Each ECU consists of a *processor, communication controller*, and *transceiver*. A processor can have single or multiple cores that are used to run various real-time automotive applications. Most of these automotive applications are hard real-time in nature, i.e., they have strict timing and deadline constraints that must be satisfied. Each application can be modeled as a set of data dependent and independent tasks mapped to different ECUs. The dependent tasks communicate by exchanging messages over the CAN network. A communication controller bridges the compute and communication fabric, facilitating data movement from the processor to the network and vice versa. Some of the key functions of a communication controller include packing of data from the processor into CAN frames, managing the transmission and reception of CAN frames, and filtering CAN messages based on the pre-programmed CAN filters (setup by the original equipment manufacturer (OEM) when programming the communication controller). Lastly, a transceiver acts as an interface between the physical CAN network and the ECU. It facilitates the actual transmission and reception of CAN frames to and from the network. Moreover, in this work, we do not consider monitoring the execution within the CAN hardware IPs as it would require access to proprietary information that is only available to OEMs.

To realize anomaly detection system in today's ECUs, we propose a few modifications to existing CAN communication controllers, as shown in Fig. 3. A traditional CAN communication controller consists of message filters that are used to filter out unwanted CAN messages and message buffers to temporarily store the messages before they are sent to the processor. This can be observed in the right region of Fig. 3. We introduce message counters inside the communication controller, which keeps track of message frequencies. This bookkeeping helps in the observation of any abnormal message rates that may occur during a distributed denial of service (DDoS) attack (see Sect. 4.3). After confirming the message rate, the message is sent to the deployed anomaly detection system, which uses a two-step process to determine whether the message is anomalous or not. We chose the communication controller instead of the processor to avoid jitter in real-time application execution.

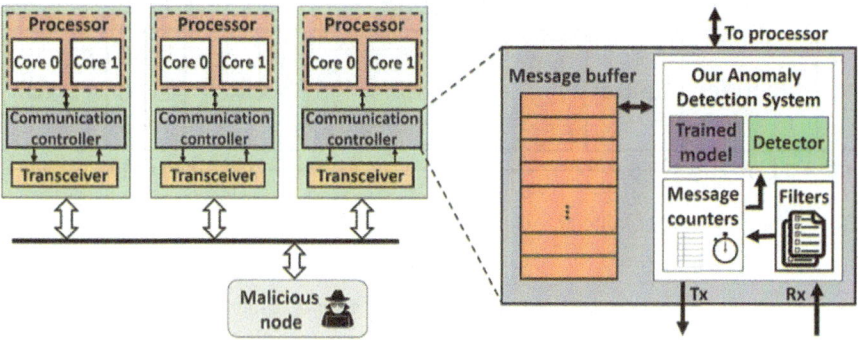

Fig. 3 Overview of the system model and proposed modifications to the communication controller in *LATTE* [45]

In the first step, our trained LSTM-based attention model is used to predict the next message instance, which is then used to compute the deviation from the true message. This deviation measure is given as the input to a detector model that uses a non-linear classifier to determine if a given deviation measure represents a normal or an anomalous message. The details related to the models and the deviation metrics used in our framework are discussed in Sects. 5.2 and 5.3, respectively. Messages are temporarily stored in the message buffer while they are being evaluated. If the anomaly detection system determines a particular message to be anomalous, it is discarded from the message buffer and will not be sent to the processor, thereby avoiding the execution of attacker messages.

Our anomaly detection system is implemented in the communication controller instead of a centralized ECU because of the following reasons:

(i). Avoid single-point failures;
(ii). Prevent scenarios where the in-vehicle network load increases significantly due to high message injection (e.g., during a DDoS attack, explained in Sect. 4.3), where the centralized ECU will not be able to communicate with a target ECU;
(iii). Enable independent and immediate detection without additional delay compared to relying on a message from a centralized ECU.

4.2   Communication Overview

In this study, we consider Controller Area Network (CAN) as the in-vehicle network protocol used to exchange time-critical messages between ECUs. CAN is a low-cost, lightweight, event-triggered in-vehicle network protocol and is the defacto industry standard. Several variants of CAN have been proposed over time, but the CAN standard 2.0B is the most widely used in-vehicle network protocol [48].

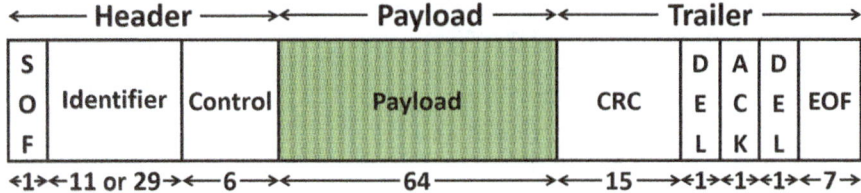

Fig. 4 Frame format of Controller Area Network (CAN) 2.0B

A CAN message consists of one or multiple signal values. Each signal contains independent information corresponding to a sensor value, actuator control, or computation output of a task on an ECU. Signals are grouped with additional information to form CAN frames, which are exchanged between ECUs. The frame format of a CAN 2.0B is illustrated in Fig. 4. Each CAN frame consists of a header, payload, and trailer segments. The header segment consists of an 11-bit (CAN standard) or 29-bit (CAN extended) unique message identifier and a 6-bit control field. This is followed by a 64-bit payload segment and a 15-bit cyclic redundancy check (CRC) field in the trailer segment. The payload segment (in green) consists of multiple signals that are arranged in a predetermined order as per the definitions in the CAN database (.dbc) files. In addition to the above-mentioned fields, the CAN frame also has a 1-bit start of the frame (SOF) field at the beginning of the header, two 1-bit delimiters separating the 1-bit acknowledgment (ACK) field and a 7-bit end of frame (EOF) field in the trailer segment.

In this study, our proposed *LATTE* framework operates on the payload segment of the CAN frame, i.e., signals within each message. The primary motivation for monitoring the payload field is that the attacker needs to modify the bits in the payload to achieve malicious goals. A modification in the header or trailer segments would simply result in the frame getting invalidated at the receiving ECU. Since we mainly focus on monitoring the payload segment, our technique is agnostic to the in-vehicle network protocol. It can be extended to other in-vehicle network protocols, such as CAN-FD and FlexRay, with minimal changes.

4.3   Threat Model

In this subsection, we present details of the threat model considered in *LATTE*. We assume that the attacker can gain access to the in-vehicle network using the most common attack vectors, such as connecting to the vehicle OBD-II port, probing into the in-vehicle network, and via advanced attack vectors such as connected V2X ADAS systems, insecure infotainment systems, or by replacing a trusted ECU with a malicious ECU. We also assume that the attacker has access or can easily gain access to the in-vehicle network parameters such as BAUD rate, flow control, channel information, and parity. This information can be obtained by using a simple

CAN data logger and can help in the transmission of malicious CAN messages. We further assume a pessimistic situation where the attacker can access the in-vehicle network at any instance and try to send malicious messages.

Considering the above assumptions, in this work, we try to protect the in-vehicle network from various cyber-attacks listed below. These attacks are modeled based on the most common and hard-to-detect attacks in the automotive domain.

1. *Constant attack*: In this attack, the attacker overwrites the signal value to a constant value for the entire duration of the attack interval. A small change in the magnitude of the signal value is more challenging to detect than larger changes.
2. *Continuous attack*: In this attack, the attacker tries to trick the anomaly detection system by continuously overwriting the signal value in small increments until a target value is achieved. The complexity of detecting this attack depends on the rate at which the signal value is overwritten. Larger change rates are easier to detect than smaller rates.
3. *Replay attack*: In this attack, the attacker plays back a valid message transmission from the past, tricking the anomaly detection system into believing it to be a valid message. The complexity of detecting this attack depends mainly on the frequency and sometimes the duration of the playbacks. High-frequency replays are easier to detect compared to low-frequency replays.
4. *Dropping attack*: In this attack, the attacker disables the transmission of a message or group of messages resulting in the dropping of communication frames. Longer durations of this attack are easier to detect due to missing messages for a prolonged time compared to shorter durations.
5. *Distributed Denial of Service (DDoS) attack*: This is the most common and easy-to-launch attack as it requires no information about the nature of the message. In this attack, the attacker floods the in-vehicle network with an arbitrary or specific message with the goal of increasing the overall network traffic and rendering the network unusable for other ECUs. These attacks are fairly simple to detect, even using a rule-based approach, as the message frequencies are fixed and known at design time for automotive systems. Any deviation in this message rate can be used as an indicator for detecting this attack.

Thus, the main objective of our work is to develop a real-time anomaly detection framework that can detect various cyber-attacks in CAN-based automotive networks that has *(i)* high detection accuracy, *(ii)* low false-positive rate, *(iii)* high precision and recall, *(iv)* large attack coverage, and *(v)* minimal implementation overhead (low memory footprint and fast runtime) for practical anomaly detection in resource-constrained ECUs.

5 Proposed Framework

The overview of our proposed *LATTE* framework is illustrated in Fig. 5. *LATTE* begins by collecting trusted in-vehicle network data under a controlled environment

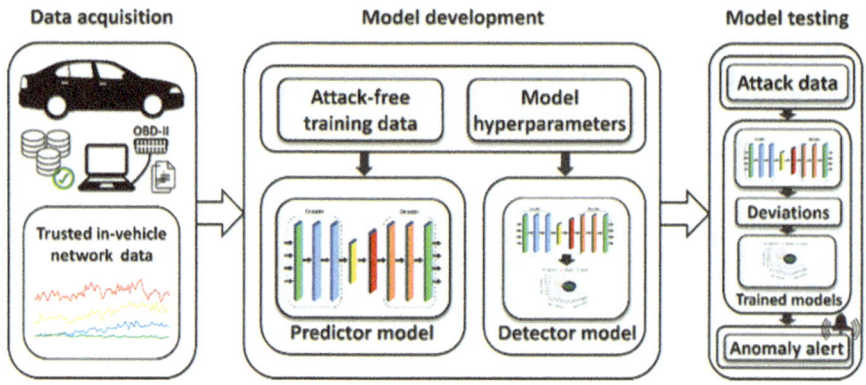

Fig. 5 Overview of *LATTE* framework

in the data acquisition step and processes the data for training. The processed data is used to train a novel stacked LSTM predictor model with an integrated self-attention mechanism in an unsupervised setting to learn the normal operating behavior of the system. We also developed a one class support vector machine (OCSVM) based detector model that utilizes the predictions from the earlier trained LSTM predictor model to detect various cyber-attacks as anomalies at runtime. After training, the framework is tested by being subjected to multiple attacks. The details of this framework are presented in the subsequent subsections.

5.1 Data Acquisition

This is the first step of the *LATTE* framework and involves collecting the in-vehicle network data from a trusted vehicle under a controlled environment. It is essential to ensure that the in-vehicle network and the ECUs in the vehicle are free from attackers. This is mainly to avoid logging corrupt in-vehicle network data that falsely represents the normal operating conditions, leading to learning an inaccurate representation of the normal system behavior with our proposed models. Moreover, to ensure high confidence in the collected data, it is crucial to cover a wide range of normal operating conditions and have the data collected over multiple intervals. Since the performance of the anomaly detection system is highly dependent on the quality of the collected data, this step is a crucial part of the *LATTE* framework. Additionally, the type of data collected depends on the functionalities or ECUs that are subjected to monitoring by the anomaly detection system. The OBD-II port is the most common access point to collect the in-vehicle network data, as it gives access to the diagnostic and most commonly used messages. However, we recommend logging the messages by probing into the CAN network, as it provides unrestricted access to the in-vehicle network, unlike the OBD-II port.

The collected in-vehicle network data is prepared for pre-processing to make it easier for the training models to learn the temporal relationships between messages. The full dataset is split into groups based on the unique CAN message identifier, and each group is processed independently. The data entries in the dataset are arranged as rows and columns, with each row representing a single data sample corresponding to a particular timestamp and each column representing a unique feature of the message. The dataset consists of the following features (i.e., columns): *(i)* timestamp at which the message was logged, *(ii)* message identifier, *(iii)* number of signals in the message, *(iv)* individual signal values (one per column), and *(v)* a binary field representing the label of the message. The label column is 0 for non-anomalous samples and 1 for anomalous samples. In the training and validation datasets, the label column is set to 0 for all samples, as all the data samples are non-anomalous and collected from a trusted vehicle. In the test dataset, the label column will have a value of 1 for the samples during the attack interval and 0 for the other cases. Since we train our models in an unsupervised setting, it is important to highlight that we do not use this label information while training our predictor and detector models. Moreover, the signal values are scaled between 0 to 1, as there can be high variance in the signal magnitudes. Such high variance in the input data can result in very slow or unstable training. Additionally, in this study, we do not consider time stamps as a unique feature. We use the concept of time in a relative manner when training (to learn patterns in sequences) and during deployment. We use the dataset presented in [31] to train and evaluate our proposed *LATTE* framework. The dataset consists of both normal and attack data. Details related to the models and the training procedure are discussed in the following subsections, while the dataset is discussed in Sect. 6.1.

5.2 Predictor Model

In this work, we designed predictor and detector models that work in tandem to detect cyber-attacks as anomalies in the in-vehicle network. At design time, the predictor model learns the normal system behavior in an unsupervised learning approach to predict the next message instance with high accuracy using the normal (non-anomalous) data. During the training process, our predictor model learns the underlying distribution of the normal data and relates it to the normal system behavior. This knowledge of the learned distribution is used to make accurate predictions of the next message instances at runtime for normal messages. In the event of a cyber-attack, the message values no longer represent the learned distribution or maintain the same temporal relationships between messages, leading to large deviations between the predictions and the true (observed) messages. In this study, we develop a detector model using a non-linear classifier to learn the deviation patterns that correspond to the normal messages, which is then used to detect anomalies (i.e., attacks that cause anomalous deviations) at runtime. The details related to the detector model are discussed in detail in Sect. 5.3.

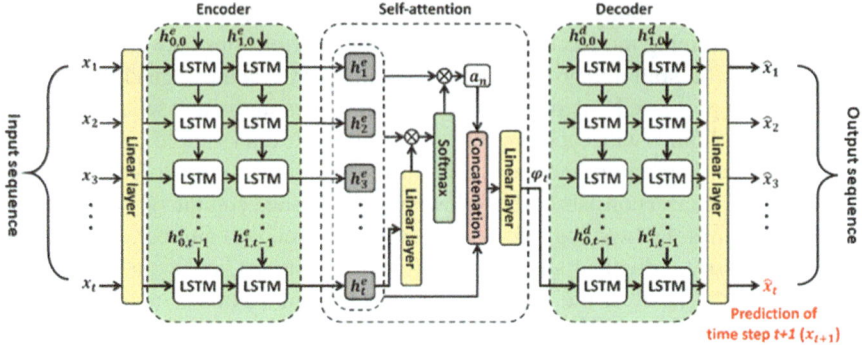

Fig. 6 Our proposed predictor model for the *LATTE* anomaly detection framework showing the stacked LSTM-based encoder-decoder model rolled out in time for t time steps along with the self-attention mechanism generating context vector (φ_t) for time step t. The output at time step t ($\hat{x}_t$) is the prediction of the input at time step $t + 1$ (x_{t+1})

Our proposed predictor model consists of a stacked LSTM-based encoder-decoder architecture with the self-attention mechanism, as shown in Fig. 6. The first linear layer in the predictor model takes the time series CAN message data and generates a 128-dimension embedding for each input. Each input sample consists of k features, where each feature represents a particular signal value within that message. The output embedding from the linear layer is given as input to the stacked two-layer LSTM encoder to produce a 64-dimension encoder output ($h_1^e, h_2^e \dots h_t^e$). The encoder output is the latent representation of the input time-series signal values that encompass the temporal relationships between messages. The encoder outputs are passed to the self-attention block to generate the context vector (φ_t). The self-attention mechanism begins by applying a linear transformation to the encoder's current hidden state (h_t^e) and multiplies the result with the encoder output. The output from the multiplication is given to a softmax activation to compute the attention weights. The attention weights represent the importance of each hidden state information from the earlier time steps at the current time step. The attention weights are scalar multiplied with the encoder outputs to compute the attention applied vector (a_n), which is then combined with the encoder output to generate the context vector (φ_t). The context vector, along with the previous decoder's hidden state (h_{t-1}^d) is given as input to the stacked two-layer decoder. The decoder block produces a 64-dimension output that is passed through the last linear layer to obtain a k-dimensional output. This k-dimension output represents the signal values of the next message instance. Thus, given an input sequence $X = \{x_1, x_2, \dots x_t\}$, our predictor model predicts the sequence $\hat{X} = \{\hat{x}_1, \hat{x}_2, \dots, \hat{x}_t\}$, where the output at time step t ($\hat{x}_t$) is the prediction of the input at time step $t + 1$ (x_{t+1}). Moreover, the last prediction ($\hat{x}_t$) is generated by consuming the complete input sequence (X).

The predictor model is trained using non-anomalous (normal) data in an unsupervised manner (i.e., without any labels). We employ a rolling window approach, with a window of fixed size length (known as subsequence length) consisting of signal values over time. And the sequence of signal values within the window is called a subsequence. Our predictor model learns the temporal dependencies that exist between the signal values within the subsequence and uses them to predict the signal values in the next subsequence (i.e., window shifted to the right by one-time step). The signal values corresponding to the last time step in the output subsequence represent the final prediction, as the model consumes the entire input subsequence to generate them. The prediction error is computed by comparing this last time step in the output subsequence with the actual signal values using the mean square error (MSE) loss function. This process is repeated until the end of the training dataset. The subsequence length is a hyperparameter related to the LSTM network, that needs to be selected before training the model and is independent of the vehicle and message data. We conducted multiple experiments with different model parameters and selected the hyperparameters that gave us the best performance results. The predictor model is trained by splitting the dataset into training (80%) and validation (20%) data without shuffling, as shuffling would destroy the existing temporal relationships between messages. During the training process, the model tries to minimize the prediction error (i.e., MSE loss) in each iteration (a forward and backward pass) by adjusting the weights of the neurons in each layer using backpropagation through time. At the end of each training epoch, the model is validated (forward pass only) using the validation dataset to evaluate the model performance. We employ mini-batches to speed up the training process and use an early stopping mechanism to avoid overfitting. The details related to the non-anomalous dataset and the hyperparameters selected for the model are presented in Sect. 6.1.

5.3 Detector Model

In this subsection, we present the details of the detector model used in *LATTE* to detect cyber-attacks. After training the predictor model, we train a separate non-linear classifier (detector model) that utilizes the information from the predictor to detect cyber-attacks. In this study, we treat the anomaly detection problem as a binary classification problem since we are mainly interested in distinguishing between normal and anomalous messages. Due to the large volumes of data being exchanged on the in-vehicle network, the network data recordings can grow very rapidly in size, which makes labeling this data very expensive. Additionally, due to the low frequency of attack scenarios, the number of attack samples would be significantly smaller compared to normal samples even when the dataset is labeled. This results in having a highly imbalanced dataset that would result in poor performance when trained using a traditional binary classifier in a supervised learning setting. However, a popular non-linear classifier known as a support vector machine (SVM) can be altered to make it work with unbalanced datasets where there is only one class. Hence, in this work, we employ a one class support vector machine (OCSVM) to classify the messages as anomalous or normal. At design

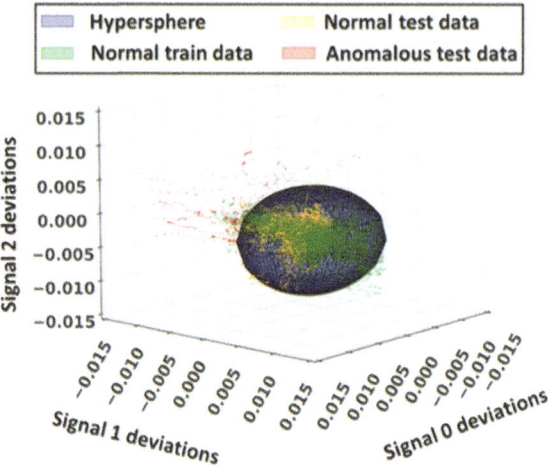

Fig. 7 OCSVM decision boundary shown in the blue sphere with the green dots showing the normal samples from training data and yellow and red dots showing the normal and anomalous samples, respectively, from test data

time, the OCSVM learns the distribution of the training dataset by constructing the smallest hypersphere that contains the training data. We train an OCSVM by using the output from the previously trained predictor model. At design time, we begin by giving the previously used normal training dataset as the input to the predictor model to generate the predictions. We then compute the deviations (prediction errors) for all the training data and pass them as input to the OCSVM. The OCSVM tries to generate the smallest hypersphere that can fit most of the deviation points and uses it at runtime to detect anomalies. Any sample deviation that falls outside the hypersphere is treated as an anomaly at runtime. An example hypersphere generated by training an OCSVM for a message with three signals is illustrated in Fig. 7. Each axis in the figure represents the relevant signal deviation, and the dark blue sphere represents the decision boundary. It can be observed that almost the entirety of training data (shown as green dots) is confined to within the decision boundary (shown as a blue sphere). The yellow and red dots in Fig. 7 represent the OCSVM classified normal and anomalous samples in the test dataset.

In this work, the deviation of a message is represented as a vector where each element of the vector corresponds to the difference between the true and predicted signal value. Therefore, for a message m with k_m number of signals, the deviation vector ($\Delta_{m,t}$) at time step t can be computed using (1).

$$\Delta_{m,t} = \left(\hat{S}_{i,t} - S_{i,t+1} \right) \in \mathbb{R}^2, \forall i \in [1, k_m] \tag{1}$$

where $\hat{S}_{i,t}$ is the prediction of the next true i^{th} signal value ($S_{i,t+1}$) made at time step t. We also experimented with other deviation measures that are modeled using

(2), (3), and (4).

$$\Delta_{m,t}^{sum} = \sum_{i=1}^{k_m} \left| \Delta_{m,t} \right|, \forall i \in [1, k_m] \tag{2}$$

$$\Delta_{m,t}^{avg} = \frac{1}{k_m} \sum_{i=1}^{k_m} \left| \Delta_{m,t} \right|, \forall i \in [1, k_m] \tag{3}$$

$$\Delta_{m,t}^{max} = \max \left(\left| \Delta_{m,t} \right| \right), \forall i \in [1, k_m] \tag{4}$$

Moreover, there can be scenarios where some signal deviations in a message can be positive while others are negative. This could potentially result in making the sum or mean of signal deviations zero or near zero, falsely indicating no deviation or very small deviation. To avoid these situations, we use absolute signal deviations to compute the deviations for the variants. It is important to note that, unlike Eq. (1) which uses a vector of k-dimensions to represent the message deviation, Eqs. (2), (3), and (4) use different reduction operations to reduce the vector to a scalar value. We explored all four deviation measures to determine the best one (discussed in Sect. 6.2.)

In summary, our predictor model predicts normal samples with very small deviations and anomalous samples with high deviations. The OCSVM considers this unique property of the predictor when constructing the hypersphere. It can be observed that when the test data with anomalies is given as input to the OCSVM, it generally correctly classifies the normal samples (shown as yellow dots in Fig. 7) within the hypersphere and anomalous samples (shown as red dots in Fig. 7) outside the hypersphere. Thus, both predictor and detector models work collectively to detect cyber-attacks as anomalies.

5.4 Model Testing

In this step, we present a test dataset consisting of anomalous samples representing multiple attacks (discussed in Sect. 4.3) along with the normal samples to the *LATTE* framework. The normal messages have a label value of 0, and the attack or anomalous messages have a label value of 1. During this step, each sample (signal values in a message) is first sent to the predictor model to predict the signal values of the next message instance, and the deviation is computed based on the true message data. This deviation vector is then passed to the OCSVM detector model to compute the position of the deviation vector in the k-dimensional space, where k represents the number of signals in the message. The message is marked as non-anomalous when the point corresponding to the deviation vector falls completely inside the learned hypersphere. Otherwise, the message is marked as anomalous, and an anomaly alert is raised. This alert can be used to invoke an appropriate remedial action to suppress further actions from the attacker. However, the design of remedial measures and response mechanisms falls outside the scope of this chapter. The performance evaluation of our proposed *LATTE* framework under various attack scenarios is presented in detail in Sects. 6.2 and 6.3.

5.5 Anomaly Detection System Deployment

Our proposed anomaly detection system can be deployed in a real-world vehicle in two different approaches. The first is a centralized approach, where a powerful centralized ECU monitors all messages on the in-vehicle network to detect cyber-attacks. The second is a distributed or decentralized approach, where the anomaly detection task is distributed across ECUs and is limited to monitoring only the messages that are relevant to that particular ECU. Both choices have pros and cons, but we believe that distributed monitoring has multiple advantages over the centralized approach because of the following reasons:

- A centralized anomaly detection approach is prone to single-point failures, which can completely expose the system to the attacker;
- If an attacker succeeds in fooling the centralized ECU, attacks can go undetected by the other ECUs, resulting in compromising the entire system. On the other hand, with a distributed detection scheme, fooling multiple ECUs is required, which is much more challenging. Even if an ECU is compromised, this can still be detected by the decentralized intelligence in a distributed detection approach;
- In extreme scenarios, such as during a DDoS attack (explained in Sect. 4.3), the in-vehicle network can get highly congested, and the centralized ECU might not be able to communicate with the victim ECUs;
- A distributed detection scheme has faster response times compared to centralized detection, as ECUs can stop accepting messages as soon as an anomaly is detected without waiting for a centralized system to notify them;
- Moreover, the overall computation load of detection is split among the ECUs with a distributed approach, and the monitoring can be limited to only the required messages. This enables multiple ECUs to monitor a subset of messages independently, with relatively low overhead;

For these reasons, many prior works, such as [19, 24], considered a distributed detection approach. Moreover, with automotive ECUs becoming increasingly powerful, the collocation of anomaly detection tasks with real-time automotive applications in a distributed manner should not be a problem, provided the overhead from the detection is minimal. The lightweight nature and anomaly detection performance of our proposed *LATTE* framework are discussed in detail in Sect. 6. Since the detector model looks at the payload segment individually, it needs to keep track of the previous messages to detect anomalous patterns. We can cache the previous normal samples and predictions (in the case of anomalies) and use them to preserve the dependencies within the data. This can be later used in determining whether the next sample is normal or anomalous. Moreover, to minimize the storage overhead, we can employ a circular buffer of size equal to the subsequence length (configured at design time). This approach would enable us to look into the message dependencies in the past.

6 Experiments

6.1 Experimental Setup

We conducted a series of experiments to evaluate the effectiveness of our proposed *LATTE* framework. We first explored five variants of the same framework with different deviation criteria: *LATTE*-ST, *LATTE*-Diff, *LATTE*-Sum, *LATTE*-Avg, and *LATTE*-Max. *LATTE*-ST uses our proposed predictor model with a static threshold (ST) value to determine whether a given message is normal or anomalous based on the deviation. The other four variants use the same predictor model but different criteria for computing the deviations, which are then given as the input to the OCSVM-based detector model. *LATTE*-Diff uses the difference in signal values (Eq. (1)), while *LATTE*-Sum and *LATTE*-Avg use a sum and mean of absolute signal deviations, respectively (Eqs. (2) and (3)). *LATTE*-Max uses the maximum absolute signal deviation (Eq. (4)).

Subsequently, we compare the best variant of our framework with four state-of-the-art works: Bitwise Message Predictor (BWMP [30]), Hierarchical Attention-based Anomaly Detection (HAbAD [34]), a variant of [34] called Stacked HAbAD (S-HAbAD [34]), and RepNet [28]. The first comparison work, BWMP [30], trains an LSTM-based network that aims to predict the next 64 bits of the payload of a CAN message by minimizing the bitwise prediction error using a binary cross-entropy loss. At runtime, BWMP uses prediction loss as a measure to detect anomalies. The second comparison work, HAbAD [34], uses an LSTM-based autoencoder model with hierarchical attention. The HAbAD model attempts to recreate the input message sequences at the output with the goal of minimizing reconstruction loss. Additionally, HAbAD uses supervised learning in the second step to model a detector using the combination of a non-parametric kernel density estimator (KDE) and k-nearest neighbors (KNN) algorithm to detect cyber-attacks at runtime. The S-HAbAD is a variant of HAbAD that uses stacked LSTMs as autoencoders and uses the same detection logic used by the HAbAD. The S-HAbAD variant is compared against to show the effectiveness of using stacked LSTM layers. Lastly, RepNet [28] uses vanilla RNNs to increase the dimensionality of input signal values and tries to reconstruct the signal values at the output by minimizing the reconstruction error using mean squared error. At runtime, RepNet monitors for large reconstruction errors to detect anomalies. The results of all experiments are discussed in detail in Sects. 6.2, 6.3 and 6.4.

In this work, all experiments are conducted using an open-source CAN message dataset called SynCAN [31], that was developed by ETAS and Robert Bosch GmbH. The SynCAN dataset consists of CAN message data for different message IDs with various fields such as timestamps, message ID, and individual signal values. Additionally, the dataset consists of a training dataset with only normal data and a labeled test dataset with multiple attacks (as discussed in Sect. 4.3). The attacks in the dataset are modeled from the real-world attacks that are commonly seen in automotive systems. It is important to note that we do not use any labeled data during the training or validation of our models and learn the normal system behavior in an unsupervised manner. The labeled data is used only during the testing phase, mainly to study the performance of our proposed *LATTE* framework. Moreover, high-frequency messages in the in-vehicle network pose a significant challenge to the anomaly detection system, as they incur high computational overhead. In this work, we considered the highest frequency message with a period of 15 ms for all of our experiments.

We implemented all of the machine learning models, including *LATTE* and its variants, and the models from the comparison works using PyTorch 1.5 with CUDA support. The 80% of the available normal data is used to train our proposed predictor model, and the remaining 20% is used for validating the model. We conducted multiple experiments with different model parameters and selected the hyperparameters that gave us the best performance results. The model training is repeated for 500 epochs with an early stopping mechanism that monitors the validation loss after the end of each epoch and stops if there is no improvement after 10 (patience) epochs. We used the ADAM optimizer with mean squared error (MSE) as the loss function. Additionally, we employed a rolling window approach (discussed in Sect. 5.2) with a subsequence length of 32 time steps, a batch size of 256, and a starting learning rate of 0.0001. We implemented the OCSVM-based detector model (Sect. 5.3) using the scikit-learn package. The OCSVM uses a radial basis function (RBF) kernel with a kernel coefficient (*gamma*) equal to the reciprocal of the number of features (i.e., the number of signals in the message). Moreover, to speedup OCSVM training, we set the kernel cache size to 400 MB and enabled the shrinking technique to avoid solving redundant optimizations. All the simulations are run on an AMD Ryzen 9 3900X server with an Nvidia GeForce RTX 2080Ti GPU.

The following definitions are used in the context of anomaly detection to compute different performance metrics. A true positive is when an actual attack is detected as an anomaly by the anomaly detection system, and a true negative is a situation where an actual normal message is detected as normal. Additionally, a false positive would be a false alarm where a normal message is incorrectly classified as an anomaly, and a false negative would occur when an anomalous message is incorrectly classified as normal. Based on these definitions, we evaluate our *LATTE* framework using four different metrics: *(i) Detection accuracy*: a measure of the anomaly detection system's ability to detect anomalies correctly, *(ii) False positive rate*: i.e., false alarm rate, *(iii) F1 score*: a harmonic mean of precision and recall; and *(iv) receiver operating characteristic (ROC) curve with area under the curve (AUC)*: a popular measure of classifier performance. We use the F1-score instead of individual precision and recall values as it captures the combined effect of both precision and recall metrics. In summary, an efficient anomaly detection system will have high detection accuracy, F1 score, and ROC-AUC while having a very low false-positive rate.

6.2   Comparison of LATTE Variants

In this subsection, we present the comparison results of the five variants of *LATTE*, namely, *LATTE*-ST, *LATTE*-Sum, *LATTE*-Avg, *LATTE*-Max, and *LATTE*-Diff. All the variants of *LATTE* use the trained predictor model (discussed in Sect. 5.2) to make the predictions and use OCSVM as a detector, except in the case of *LATTE*-ST, which uses a fixed threshold scheme introduced in [32] to predict the given message as normal or anomalous. The main purpose of this experiment is to study the effect of different deviation criteria on the OCSVM detection performance. The deviations for any given message in *LATTE*-Diff ($\Delta_{m,\,t}$), *LATTE*-Sum $\left(\Delta_{m,t}^{sum}\right)$, *LATTE*-Avg $\left(\Delta_{m,t}^{avg}\right)$ and *LATTE*-Max $\left(\Delta_{m,t}^{max}\right)$ are computed using the Eqs. (1), (2), (3), and (4), respectively. Additionally, we aim to analyze the impact of using a non-linear classifier such as OCSVM on the model performance instead of a simple static threshold scheme (*LATTE*-ST).

Figure 8a–c shows the detection accuracy, false-positive rate, and F1 score for the five different variants of *LATTE* under five different attack scenarios discussed in Sect. 4.3. Under the 'No attack' case, the model is tested with new normal (non-anomalous) data that the model has not seen before. Firstly, from Fig. 8a–c, it is evident that the OCSVM-based detection models clearly outperform the static threshold models (*LATTE*-ST). This is mainly because of their ability to process complex attack patterns and generate non-linear decision boundaries that can distinguish better between normal and anomalous data. Moreover, it can be seen that *LATTE*-Diff outperforms all the OCSVM-based models in detection accuracy, false-positive rate, and F1 score. The ROC curves and the corresponding AUC values in the brackets next to each legend is illustrated in Fig. 8d. Among various attacks considered in this study, we show ROC-AUC results for continuous attacks, as it is the most challenging attack to detect. This is because, during this attack, the attacker constantly tries to fool the anomaly detection system into thinking that the signal values in the messages are legitimate. This requires careful monitoring and the ability to learn complex patterns to differentiate between normal and anomalous samples. On average, across all attacks, LATTE-Diff was able to achieve an average of 13.36% improvement in accuracy, 11.34% improvement in F1 score, 17.86% improvement in AUC, and 47.9% reduction in false positive rate, and up to 42% improvement in accuracy, 32.6% improvement in F1 score, 29.4% improvement in AUC and 95% decrement in false positive rate, compared to the other variants. Therefore, we selected *LATTE*-Diff as our candidate model for subsequent experiments where we present comparisons with state-of-the-art anomaly detection systems. Henceforth, we refer to *LATTE*-Diff as *LATTE*.

6.3 Comparison with Prior Works

We compared our *LATTE* framework with four comparison works, namely, BWMP [30], HAbAD [34], S-HAbAD [34] (a variant of HAbAD), and RepNet [28]. The detection accuracy, false-positive rate, and F1 score of our proposed *LATTE* framework and the comparison works under different attack scenarios are illustrated in Fig. 9a–c. It can be observed that *LATTE* outperforms all the prior works in terms of detection accuracy, false-positive rate, and F1 score. This is mainly because of three factors. Firstly, the stacked LSTM encoder-decoder structure provides adequate depth to the model to learn complex time-series patterns. This can be seen when comparing HAbAD with S-HAbAD, as the latter differs only in terms of stacked LSTM layers in comparison to the former. Secondly, the integrated self-attention mechanism in *LATTE* helps in learning very long-term dependencies in message sequences. Lastly, the use of powerful OCSVMs as non-linear classifiers helps in designing a highly efficient classifier. These factors together resulted in the superior performance of *LATTE* compared to all the comparison works. On average, across all attacks, *LATTE* achieved an average of 18.94% improvement in accuracy, 19.5% improvement in F1 score, 37% improvement in AUC, and a 79% reduction in false positive rate. Moreover, *LATTE* achieved up to 47.8% improvement in accuracy, 37.5% improvement in F1 score, 76% improvement in AUC, and a 95% reduction in false positive rate.

To highlight the effectiveness of our proposed *LATTE* framework, we further compared *LATTE* with statistical and proximity-based anomaly detection techniques. We selected Bollinger bands, which is a popular statistical technique used in the finance domain, as the candidate for a statistical technique to detect anomalies in time series data. Bollinger bands generate envelopes with two standard deviation levels above and below the moving average and flag the samples that fall outside the bands as anomalous. In this work, we considered two different moving average-based variants of this approach: *(i)* simple moving average (SMA) and *(ii)* exponential weighted moving average (EWMA), similar to [42]. We also compared *LATTE* against a popular proximity-based anomaly detection technique called local outlier factor (LOF) [43]. The LOF algorithm measures the local deviation of each point in the dataset with respect to the neighbors (given by KNN) to detect anomalies. Figure 10 illustrates the F1 score results for SMA-based Bollinger bands (SMA-BB), EWMA-based Bollinger bands (EWMA-BB), LOF, and *LATTE* under different attack scenarios. From Fig. 10, it is clear that *LATTE* outperforms both statistical and proximity-based anomaly detection techniques under all attack scenarios. This is mainly because the complex patterns in CAN message data are hard to capture using statistical and proximity-based techniques. On the other hand, our proposed *LATTE* framework uses an LSTM-based predictor model to

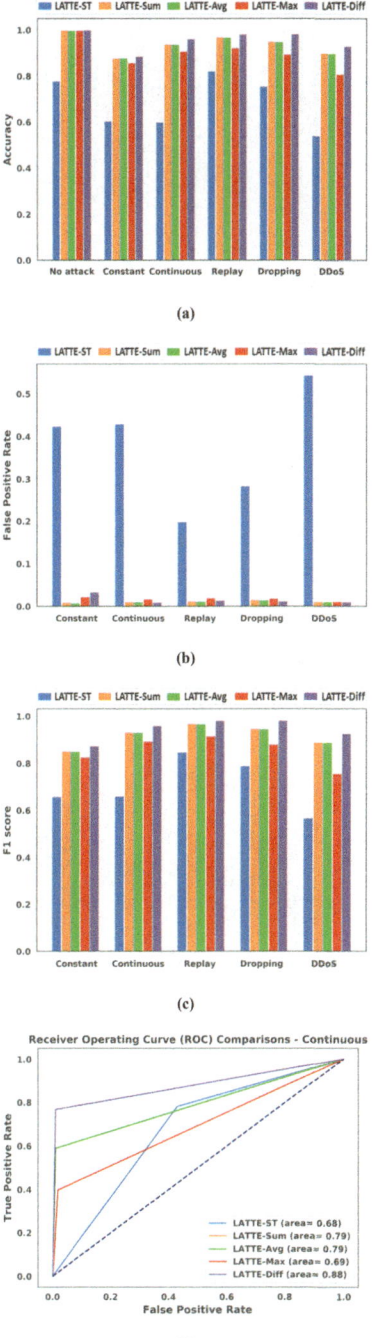

(a)

(b)

(c)

(d)

Fig. 8 Comparison of (**a**) detection accuracy, (**b**) false-positive rates, (**c**) F1 score of *LATTE* variants under different attack scenarios, and (**d**) ROC curve with AUC for continuous attack

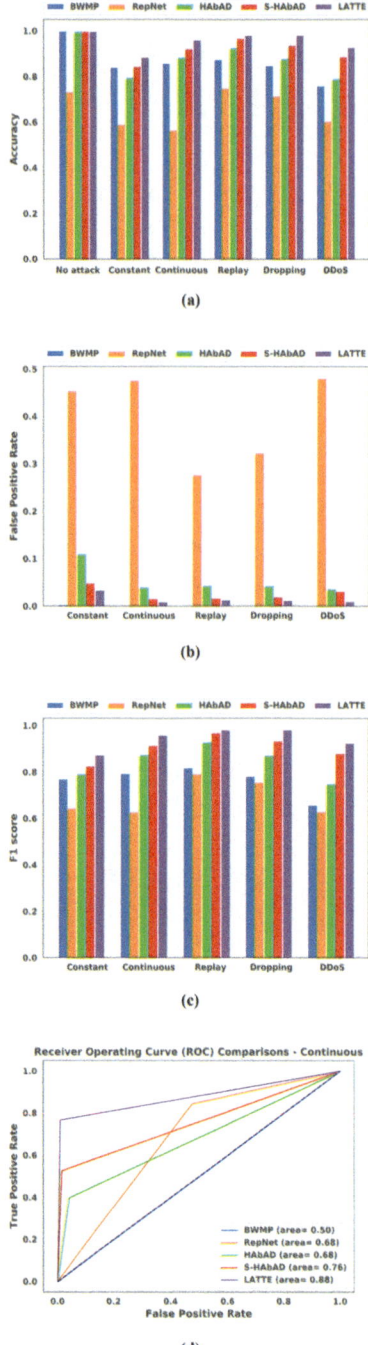

Fig. 9 Comparison of (**a**) accuracy, (**b**) false-positive rates, (**c**) F1 score of *LATTE* and the comparison works under different attack scenarios, and (**c**) ROC curve with AUC for continuous attack

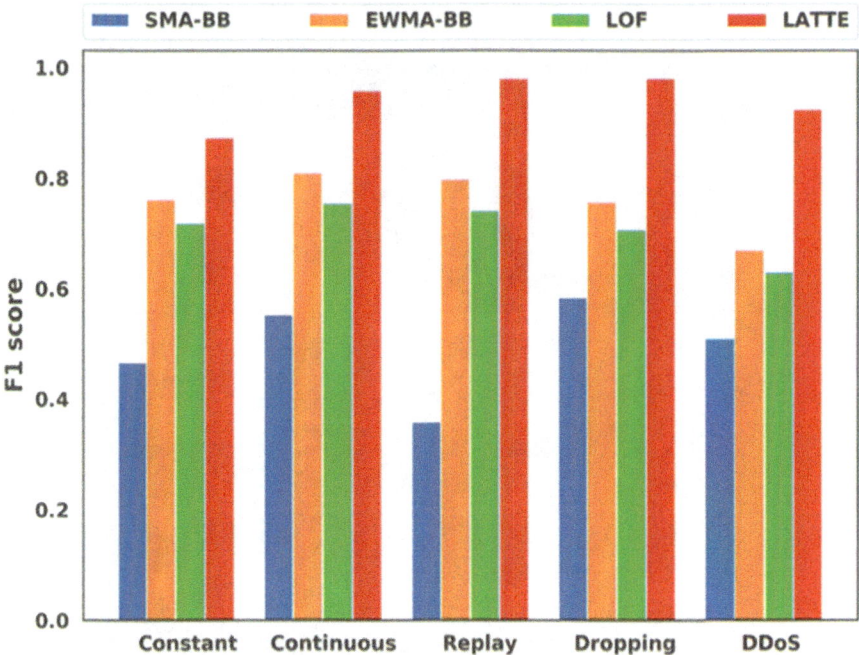

Fig. 10 Comparison of F1 score for SMA-BB [42], EWMA-BB [42], LOF [43], and *LATTE* under different attack scenarios

efficiently learn these complex patterns and is thus able to detect various attacks more efficiently.

6.4   Overhead Analysis

In this subsection, we quantify the overhead of our proposed *LATTE* framework, and the comparison works using memory footprint, the number of model parameters, and the inference time metrics. We implemented each framework on a dual-core ARM Cortex- A57 CPU on an NVidia Jetson TX2 board (shown in Fig. 11), which has similar specifications to that of a real-world ECU. The inference experiment is repeated 10 times to compute the average inference time. Moreover, in this study, we consider a total message buffer size of 2.25 KB. This comprises of storage space for 32 CAN message payloads (0.25 KB assuming a worst-case max payload of 8 Bytes) that represent the subsequence length number of past messages, and storage space for 16 CAN message frames (2 KB assuming the CAN extended protocol and a worst-case max payload of 8 Bytes) that is used by the transceiver. It is important to note that we only introduce the 0.25 KB storage as the 2 KB transceiver buffer space is already available in the traditional CAN communication controllers.

Fig. 11 Nvidia Jetson TX2 board

The 2 KB transceiver buffer space is based on the most commonly seen sizes in many real-world automotive ECUs, such as Woodward SECM 112 and dSpace MicroAutoBox. We also computed the area overhead of the 0.25 KB buffer using the CACTI tool [44] by modeling the buffer as a scratchpad cache using a 32 nm technology node. Our additional 0.25 KB buffer resulted in a minimal area overhead of around 581.25 μm^2. From Table 2, it is evident that our *LATTE* framework has minimal overhead compared to both attention-based prior works (HAbAD and S-HAbAD) and non-attention-based work (BWMP except RepNet). The high runtime and memory overhead in HAbAD and S-HAbAD is associated with using KNNs. Since KNNs scan through each training data sample to make a prediction, they are slower and consume high memory overhead (due to the requirement of having training data available at runtime). Moreover, it needs to be noted that, even though RepNet has the lowest memory and runtime overhead, it fails to capture the complex attack patterns due to the smaller model size and the lack of ability of vanilla RNNs to learn long-term dependencies, leading to poor performance (as shown in Fig. 9).

In this work, we factor this additional latency into our real-time constraints for messages (i.e., modeled a constant time overhead). But since the latency overhead (shown in Table 2) is very minimal, we envision that our proposed *LATTE* framework will have a minimal change in the timing constraints when compared to the prior works. Moreover, the inference overhead of our *LATTE* framework

Table 2 Compute and memory overhead of *LATTE*, BWMP [30], HAbAD [34], S-HAbAD [34], RepNet [28]

ADS framework	Memory footprint (KB)	#Model parameters ($\times 10^3$)	Average inference time (μs)
BWMP [30]	13,147	3435	644.76
HAbAD [34]	4558	64	685.05
S-HAbAD [34]	5600	325	976.65
LATTE	1439	331	193.90
RepNet [28]	5	0.8	68.75

(~194 μs) is much smaller compared to the deadline constraints of some of the fastest (i.e., most stringent) safety-critical applications, which are around 10 ms. Thus, the additional latency due to anomaly detection should not violate any safety-critical deadlines. In summary, from Fig. 9 and Table 2, we can observe that *LATTE* achieves superior performance compared to all comparison works across diverse attack scenarios while maintaining relatively low memory and runtime overhead.

7 Conclusion

In this chapter, we proposed a novel anomaly detection framework called *LATTE* that uses stacked LSTMs with a self-attention mechanism to learn the normal system behavior by learning to predict the next message instance under normal operating conditions. We presented a one class support vector (OCSVM) based detector model to detect cyber-attacks by monitoring the message deviations from the normal operating behavior. We presented a comprehensive analysis by evaluating our proposed model against multiple variants of our model and the best-known prior works in this area. Our *LATTE* framework surpasses all the variants and the best-known prior works under various attack scenarios while having a relatively low memory and runtime overhead. This makes *LATTE* a promising anomaly detection framework for next-generation connected and autonomous vehicles.

References

1. Renub Research: Self driving car market global forecast by levels, hardware, software, and companies. In: Research and Markets – Market Research Reports (2018)
2. Kukkala, V.K., Tunnell, J., Pasricha, S., Bradley, T.: Advanced Driver-Assistance Systems: A Path Toward Autonomous Vehicles, Proc. IEEE CEM (2018)
3. Hasan, M., Mohan, S., Shimizu, T., Lu, H.: Securing Vehicle-to-Everything (V2X) communication platforms. In: IEEE Transactions on Intelligent Vehicles (2020)
4. Acharya, S., Dvorkin, Y., Karri, R.: Public plug-in electric vehicles + Grid data: is a new cyberattack vector viable? IEEE Trans. Smart Grid. **11**(6), 5099–5113 (2020)
5. Braeckel, P.: Feeling bluetooth: from a security perspective. In: Advances in Computers (2011)

6. Verdult, R., Garcia, F.D., Ege, B.: Dismantling megamos crypto: wirelessly lockpicking a vehicle immobilizer. In: Proceeding of USENIX (2013)
7. Koscher, K., Czeskis, A., Roesner, F., Patel, S., Kohno, T., Checkoway, S., McCoy, D., Kantor, B., Anderson, D., Shacham, H., Savage, S.: Experimental security analysis of a modern automobile. In: Proceeding of IEEE SP (2010)
8. Valasek, C., Miller, C., Remote exploitation of an unaltered passenger vehicle. In Black Hat USA (2015)
9. Izosimov, V., Asvestopoulos, A., Blomkvist, O., Törngren, M.: Security-aware development of cyber-physical systems illustrated with automotive case study. In: Proceeding of IEEE/ACM DATE (2016)
10. Francillon, A., Danev, B., Capkun, S.: Relay attacks on passive keyless entry and start systems in modern cars. In: Proceeding of NDSS (2011)
11. Dastres, R., Soori, M.: A review in recent development of network threats and security measures. In: International Journal of Information Sciences and Computer Engineering (2021)
12. Taylor, A., Japkowicz, N., Leblanc, S.: Frequency-based anomaly detection for the automotive CAN bus. In: Proceeding of WCICSS (2015)
13. Yoon, M., Mohan, S., Choi, J., Sha, L.: Memory heat map: anomaly detection in real-time embedded systems using memory behavior. In: IEEE/ACM/EDAC DAC (2015)
14. Kukkala, V.K., Pasricha, S., Bradley, T.: SEDAN: Security-Aware Design of Time-Critical Automotive Networks. In: IEEE TVT (2020)
15. Dürrwang, J., Braun, J., Rumez, M., Kriesten, R., Pretschner, A.: Enhancement of automotive penetration testing with threat analyses results. In: SAE (2018)
16. Keen Lab: Experimental Security Assessment of BMW Cars: A Summary Report. [Online] (2017). Available: https://keenlab.tencent.com/en/whitepapers/Experimental_Security_Assessment_of_BMW_Cars_by_KeenLab.pdf
17. Petit, J., Stottelaar, B., Feiri, M.: Remote attacks on automated vehicles sensors: experiments on camera and LiDAR. In: Black Hat Europe (2015)
18. Raiyn, J.: A survey of cyber attack detection strategies. In: International Journal of Security and Its Applications (2014)
19. Studnia, I., Alata, E., Nicomette. V., Kaâniche, M., Laarouchi, Y.: A language-based intrusion detection approach for automotive embedded network. In: Proceeding of IEEE PRISDC (2015)
20. Marchetti, M., Stabili, D.: Anomaly detection of CAN bus messages through analysis of ID sequences. In: Proceeding of IEEE IV (2017)
21. Stabili, D., Marchetti, M., Colajanni, M.: Detecting attacks to internal vehicle networks through hamming distance. In: Proceeding of AEIT (2017)
22. Larson, U.E., Nilsson, D.K., Jonsson, E.: An approach to specification-based attack detection for in-vehicle networks. In: Proceeding of IEEE IV (2008)
23. Hoppe, T., Kiltz, S., Dittmann, J.: Applying intrusion detection to automotive IT-early insights and remaining challenges. In JIAS (2009)
24. Waszecki, P., Mundhenk, P., Steinhorst, S., Lukasiewycz, M., Karri, R., Chakraborty, S.: Automotive Electrical and Electronic Architecture Security via Distributed In-Vehicle Traffic Monitoring. In: IEEE TCAD (2017)
25. Cho, K.T, Shin, K.G.: Fingerprinting electronic control units for vehicle intrusion detection. In: Proceeding of USENIX (2016)
26. Müter, M. Asaj, N.: Entropy-based anomaly detection for in-vehicle networks. In: Proceeding of IEEE IV (2011)
27. Kang, M., Kang, J.: A novel intrusion detection method using deep neural network for in-vehicle network security. In: Proceeding of IEEE VTC Spring (2016)
28. Weber, M., Wolf, G., Zimmer, B., Sax, E.: Online Detection of Anomalies in Vehicle Signals using Replicator Neural Networks. In: Proceeding of ESCAR (2018)
29. Weber, M., Klug, S., Sax, E., Zimmer, B.: Embedded hybrid anomaly detection for automotive can communication. In: Proceeding of ERTS (2018)

30. Taylor, A., Leblanc, S., Japkowicz, N.: Anomaly detection in automobile control network data with long short-term memory networks. In: IEEE DSAA (2016)
31. Hanselmann, M., Strauss, T., Dormann, K., Ulmer, H.: CANet: an Unsupervised Intrusion Detection System for High Dimensional CAN Bus Data. In: Proceeding of IEEE Access (2020)
32. Kukkala, V.K., Thiruloga, S.V., Pasricha, S.: INDRA: Intrusion Detection using Recurrent Autoencoders in Automotive Embedded Systems. In: IEEE TCAD (2020)
33. Song, H.M., Woo, J., Kim, H.K.: In-vehicle network intrusion detection using deep convolutional neural network. In: Proceeding of Vehicular Communications (2020)
34. Ezeme, M.O., Mahmoud, Q.H., Azim, A.: Hierarchical Attention-Based Anomaly Detection Model for Embedded Operating Systems. In: Proceeding of IEEE RTCSA (2018)
35. Bengio, Y., Simard, P., Frasconi, P.: Learning long-term dependencies with gradient descent is difficult. In: Proceeding of IEEE Transactions on Neural Networks (1994)
36. Elsworth, S., Güttel, S.: Time series forecasting using LSTM networks: a symbolic approach. [Online]. Available: https://arxiv.org/abs/2003.05672 (2020)
37. Hochreiter, S., Schmidhuber, J.: Long short-term memory. In: Proceeding of Neural Computation (1997)
38. Sood, E., Tannert, S., Frassinelli, D., Bulling, A., Vu, N.T.: Interpreting attention models with human visual attention in machine reading comprehension [online]. Available: https://arxiv.org/abs/2010.06396 (2020)
39. Bahdanau, D., Cho, K., Bengio, Y.: Neural Machine Translation by Jointly Learning to Align and Translate [Online]. Available: https://arxiv.org/abs/1409.0473 (2016)
40. Jing, R.: A Self-attention based LSTM network for text classification. In: Proceeding IEEE CCEAI (2019)
41. Luong, M.T., Pham, H., Manning, C.D.: Effective approaches to attention-based neural machine translation [Online]. Available: https://arxiv.org/abs/1508.04025 (2015)
42. Vergura, S.: Bollinger bands based on exponential moving average for statistical monitoring of multi-array photovoltaic systems. In: Energies (2020)
43. Breunig, M.M., Kriegel, H.P., Ng, R.T., Sander, J.: LOF: identifying density-based local outliers. In: Proceeding of ACM MOD (2000)
44. Muralimanohar, N., Balasubramonian, R., Jouppi, N.P.: CACTI 6.0: a tool to model large caches. In: HP laboratories, [Online]. Available: https://github.com/HewlettPackard/cacti (2014)
45. Kukkala, V.K., Thiruloga, S.V., Pasricha, S.: LATTE: LSTM Self-Attention based anomaly detection in embedded automotive platforms. In: ACM Transactions on Embedded Computing Systems (TECS), Vol. 20, No. 5s, Article 67. (2021)
46. Thiruloga, S.V., Kukkala, V.K., Pasricha, S.: TENET: temporal CNN with attention for anomaly detection in automotive cyber-physical systems. In Proceeding of IEEE/ACM Asia & South Pacific Design Automation Conference (ASPDAC) (2022)
47. Kukkala, V.K., Thiruloga, S.V., Pasricha, S.: Roadmap for cybersecurity in autonomous vehicles. In IEEE Consumer Electronics Magazine (CEM) (2022)
48. DiDomenico, G.C., Bair, J., Kukkala, V.K., Tunnell, J., Peyfuss, M., Kraus, M., Ax, J., Lazarri, J., Munin, M., Cooke, C., Christensen, E.: Colorado state university EcoCAR 3 final technical report. In: SAE World Congress Experience (WCX) (2019)
49. Kukkala, V.K., Bradley, T., Pasricha, S.: Priority-based Multi-level monitoring of signal integrity in a distributed powertrain control system. In: Proceeding of IFAC Workshop on Engine and Powertrain Control, Simulation and Modeling (2015)
50. Kukkala, V.K., Bradley, T. Pasricha, S.: Uncertainty analysis and propagation for an auxiliary power module. In Proceeding of IEEE Transportation Electrification Conference (TEC) (2017)
51. Kukkala, V.K., Pasricha, S., Bradley, T.: JAMS: Jitter-Aware Message Scheduling for FlexRay Automotive Networks. In Proceeding of IEEE/ACM International Symposium on Network-on-Chip (NOCS) (2017
52. Kukkala, V.K., Pasricha, S., Bradley, T.: JAMS-SG: a framework for jitter-aware message scheduling for time-triggered automotive networks. In ACM Transactions on Design Automa-

tion of Electronic Systems (TODAES), Vol. 24, No. 6, (2019)

53. Tunnell, J., Asher, Z., Pasricha, S., Bradley, T.H., Towards improving vehicle fuel economy with ADAS. In SAE Inter-national Journal of Connected and Automated Vehicles, Vol. 1, No. 2, (2018)

54. Tunnell, J., Asher, Z., Pasricha, S., Bradley, T.H.: Towards improving vehicle fuel economy with ADAS. In Proceeding of SAE World Congress Experience (WCX) (2018)

55. Asher, Z., Tunnell, J., Baker, D.A., Fitzgerald, R.J., Banaei-Kashani, F., Pasricha, S., Bradley, T.H.: Enabling prediction for optimal fuel economy vehicle control. In Proceeding of SAE World Congress Experience (WCX), 2018

56. Dey, J., Taylor, W., Pasricha, S.: VESPA: a framework for optimizing heterogeneous sensor placement and orientation for autonomous vehicles. In: IEEE Consumer Electronics Magazine (CEM), Vol. 10, No. 2 (2021)

Deep AI for Anomaly Detection in Automotive Cyber-Physical Systems

Sooryaa Vignesh Thiruloga, Vipin Kumar Kukkala, and Sudeep Pasricha

1 Introduction

Vehicles are becoming increasingly autonomous and highly connected to achieve improved vehicle safety and fuel efficiency goals. New technologies such as vehicle-to-vehicle (*V2V*), advanced driver assistance systems (*ADAS*), 5G vehicle-to-infrastructure (*5G V2I*), and others have emerged to support this evolution [1]. Due to these advancements, the complexity of electronic control units (*ECUs*) and the in-vehicle network that connects them has significantly increased. Thus, today's vehicles represent a massively complex time-critical cyber-physical system (CPS). This introduced several new challenges related to reliability [30–33], security [2, 21, 24, 29, 34], and real-time performance [35–39] of the vehicles. Additionally, the increased connectivity of vehicles to various external electronic systems has made modern vehicles highly vulnerable to various cyber-attacks [2].

Attackers can gain unauthorized access to the in-vehicle network by exploiting various access points (known as an attack surface) in a vehicle, such as Bluetooth and USB ports, telematics systems, and OBD-II ports. After gaining access to the in-vehicle network, an attacker can inject malicious messages in an attempt to gain control of the vehicle. Recent automotive attacks have included tampering with

S. V. Thiruloga
Department of Electrical and Computer Engineering, Colorado State University, Fort Collins, CO, USA
e-mail: sooryaa@colostate.edu

V. K. Kukkala (✉)
NVIDIA, Santa Clara, CA, USA
e-mail: vipin.kukkala@colostate.edu

S. Pasricha
Colorado State University, Fort Collins, CO, USA
e-mail: sudeep.pasricha@colostate.edu

© The Author(s), under exclusive license to Springer Nature Switzerland AG 2023
V. K. Kukkala, S. Pasricha (eds.), *Machine Learning and Optimization Techniques for Automotive Cyber-Physical Systems*, https://doi.org/10.1007/978-3-031-28016-0_12

speedometer and indicator signals [3], unlocking doors [4], tampering with the fuel level indicator [4], etc. These attacks confuse the driver and can be dangerous but are not fatal. More sophisticated machine learning-based attacks can result in incorrect traffic sign recognition by targeting the vehicle's camera-connected ECU [5]. Other researchers analyzed vulnerabilities in airbag systems and were able to remotely deploy airbags in a vehicle [6]. These types of attacks can have catastrophic consequences and can be highly fatal.

Traditional security mechanism such as firewalls can only detect simple attacks but fails to detect more complex attacks such as those described in [5, 6]. With increasing vehicle complexity, the attack surface is only going to increase, paving the way for more complex and novel attacks in the future. As a result, there exists an urgent need for a solution that can actively monitor the in-vehicle network and detect complex cyber-attacks. Deploying an anomaly detection system is one of the many approaches to achieving this goal. An ADS is a hardware or software-based system that continuously monitors the in-vehicle network for attacks without human intervention. Many state-of-the-art ADS employ machine learning techniques to detect cyber-attacks. The ability to collect large amounts of in-vehicle network data and ECUs with high computation complexity facilitated the deployment of machine learning models in automotive systems. At a very high level, the machine learning model in an ADS attempts to learn the normal operating behavior of the in-vehicle network during the design and testing. This learned knowledge of the normal system behavior is then used at runtime to continuously monitor for any anomalous/malicious behavior, allowing the detection of both known and unknown attacks. Due to its high attack coverage and ability to detect complex attack patterns, we focus on (and make new contributions to) machine learning-based ADS for detecting cyberattacks in vehicles.

In this chapter, we provide an overview of a novel ADS framework called *TENET* [29] that actively monitors the in-vehicle network and observes for any deviation from the normal behavior to detect cyber-attacks. *TENET* attempts to increase the detection accuracy, receiver operating characteristic (ROC) curve with area under the curve (AUC), Mathews correlation coefficient (MCC) metrics, and minimize false negative rate (FNR) with minimal overhead. The key contributions of the *TENET* framework can be summarized as follows:

- We propose a temporal convolutional neural attention (TCNA) architecture to learn very-long term temporal dependencies between messages in the in-vehicle network;
- We introduce a metric called divergence score (DS) to quantify the deviation from expected behavior;
- We present a decision tree-based classifier to detect various cyberattacks at runtime using the proposed DS metric;
- We present a comprehensive analysis of the *TENET* framework [7] with various state-of-the-art ADS frameworks to demonstrate its effectiveness.

2   Related Work

Several solutions have been proposed to detect in-vehicle network attacks. Most of these solutions can be mainly divided into either signature-based or anomaly-based.

Signature-based IDS reckon on detecting known and pre-modeled attack signatures. In [7], the authors proposed a language-theory-based model to derive attack signatures. However, their proposed approach fails to detect attack packets at the starting stages of the attack interval. The authors in [8, 9] proposed a message frequency-based technique to detect cyberattacks. In [10], a transition matrix-based ADS was proposed to detect attacks on the controller area network (CAN) bus in the vehicles. However, their approach fails to detect complex attacks (such as replay attacks (discussed in section III-C.1)). The authors in [11, 12] presented an entropy-based ADS to detect in-vehicle network attacks. However, entropy-based techniques fail to detect small variations in the entropy and minor modifications in the CAN message data. The authors in [13] introduced a novel approach that monitors the hamming distance between messages to detect cyberattacks. However, this approach has a high computational overhead on the ECU. In [14], ECUs were fingerprinted using their voltage measurements during message transmission and reception. However, this method cannot detect attacks at the application layer as it is only applied at the physical layer. In conclusion, signature-based ADS approaches can detect vehicle network attacks with high accuracy and a low false-positive rate. However, obtaining all possible attack signatures and consistently updating them is highly impractical.

On the other hand, anomaly-based solutions aim to learn the normal system behavior and observe any abnormal behavior in the system to detect both known and unknown attacks. The authors in [15] used deep neural networks (DNNs) to capture low dimensional features of transmitted packets to differentiate between normal and attack-injected messages. In [16], the authors used a recurrent neural network (RNN) to learn the normal behavior of the network and leveraged that information to detect attacks during runtime. The authors in [17–20] have employed long short-term memory (LSTM) based ADS to learn the relationship between messages traversing the in-vehicle networks. However, these LSTM-based models are complex and incur very high ECU overheads. Moreover, the effectiveness of these ADSs was not tested on complex attack patterns. In [21], a gated recurrent unit (GRU) based autoencoder ADS was proposed to learn the normal system behavior. An LSTM-based encoder-decoder ADS with an integrated attention mechanism was proposed in [22]. Furthermore, the approach uses k-nearest neighbors (KNN) with a kernel density estimator (KDE) to detect anomalies. However, this approach incurs a high memory overhead on the ECU. The authors in [23] proposed an approach that combined an LSTM with a convolutional neural network (CNN) to learn the dependencies between messages in a CAN network. The model was, however, trained on a labeled dataset in a supervised manner. In addition, due to the large volume of in-vehicle CAN message data, labeling the data is highly impractical. In [24], the authors proposed a novel deep neural network architecture comprising stacked LSTMs and a self-attention mechanism to learn the normal system behavior during the training phase and employed a one-class support vector machine (OCSVM) based classifier to identify anomalous messages.

All of these recent machine learning-inspired works suggest that sequence models such as LSTMs and GRUs are popular for detecting vehicle attacks. However, since the functional complexity of vehicles has grown significantly, there exist very long dependencies between messages exchanged among ECUs. The traditional sequence models such as GRUs and LSTMs fail to capture them effectively. This is mainly because the current time step output of both LSTMs and GRUs is heavily influenced by the recent time steps compared to time steps in the distant past, making it highly challenging to capture very-long term dependencies. Moreover, processing long sequences also exacerbate the computational and memory overhead of LSTMs and GRUs.

In summary, none of the existing ADS provide a comprehensive approach that can efficiently learn very-long term dependencies between in-vehicle network messages with a low memory overhead and accurately detect a multitude of simple and complex attacks on the in-vehicle network. This chapter discusses an efficient ADS framework called *TENET* that was first introduced in [29]. *TENET* employs a novel TCNA network architecture to overcome the shortcomings of state-of-the-art ADS. The subsequent section describes the *TENET* framework in detail, and the detailed performance analysis results are presented in section IV.

3   *TENET* Framework: Overview

The *TENET* [29] framework consists of three key phases: *(i)* data collection and pre-processing, *(ii)* learning, and *(iii)* evaluation. In the first phase, in-vehicle network data is collected from a trusted vehicle and pre-processed. In the learning phase (*offline*), the pre-processed data is utilized for training the Temporal Convolutional Neural Attention (TCNA) network in an unsupervised manner to learn the normal operating system behavior. The trained TCNA network is then used in the final evaluation phase (*online*) to calculate a divergence score (DS). Lastly, a decision tree-based classifier uses this computed DS to detect various cyberattacks. The overview of the proposed *TENET* framework is shown in Fig.1.

3.1   *Data Collection and Preprocessing*

TENET [29] framework's first phase involves collecting in-vehicle network data from a trusted vehicle. It is critical to ensure that the data is collected from a trusted vehicle and is malware-free, as any anomalies in the collected data can make the design of the ADS obsolete. Moreover, data needs to be collected under various vehicle operating states to get more comprehensive coverage of network traffic.

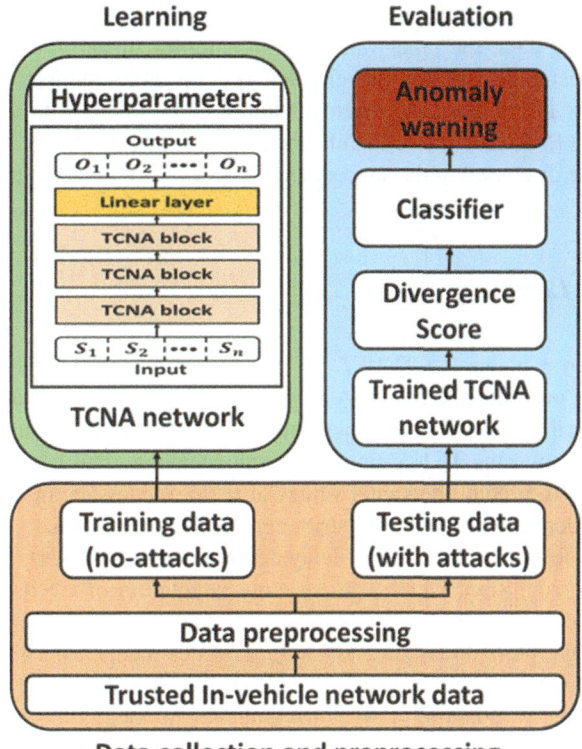

Fig. 1 Overview of the *TENET* framework

Failing to do so could result in the model learning an improper representation of the normal operating condition. *TENET* recommends splicing into the in-vehicle network and directly logging the messages using a standard logger such as Vector GL 1000 [25]. This facilitates the recording of any message traversing the network.

After collecting trusted data, it is prepared for pre-processing to facilitate easy and efficient training of the machine learning models. A typical vehicle network protocol such as CAN, or Flexray, has a unique identifier (ID), and each message in the dataset is grouped by this unique ID and processed independently. The processed records are arranged in a 2-dimensional table where each row represents a single data sample, and each column represents various unique attributes of the message. Each message has the following attributes (columns): *(i)* message ID, *(ii)* a unique timestamp corresponding to the time at which the message was logged, *(iii)* individual signal values in the message (which together comprise the message payload), *(iv)* number of signals in the message, and *(v)* label of the message ('0' for no-attack and '1' for attack). Moreover, all signal values are scaled between 0 and 1 before beginning the model training to minimize the impact of high variance in message signal values. The learning phase (shown in the green colored box in Fig. 1) and evaluation phases (shown in the blue colored box in Fig. 1) of the *TENET* framework use training and testing data, respectively. Since the model is trained using attack-free data, the label values of all samples in the training dataset are set to 0. The test data contains a label value of 1 for attack samples and 0 for no-attack samples. In addition, the original training data is split into training (85%) and validation (15%) data. More details about the training procedure and the model architecture are discussed in the subsequent subsections of this chapter.

3.2 Model Learning

This subsection explains the *TENET* [29] framework's novel TCNA architecture and its training procedure. The TCNA network within the *TENET* framework helps to learn the normal system behavior of the in-vehicle network in an *unsupervised* manner. The input to the TCNA model is a sequence of signal values in a message and uses CNNs to predict the signal values of the next message instance by trying to learn the underlying probability distribution of the normal data.

In [26], a convolution-based time-delay neural network (TDNN) was introduced for phoneme recognition, which was an early adaptation of CNNs for sequence modeling tasks. Moreover, traditional CNNs require a very deep network of CNN layers with large filters to capture very-long term dependencies. This significantly increases the number of convolutional operations, resulting in a high computational cost. Thus, directly adapting CNNs to sequence modeling tasks in resource-constrained automotive systems is not a viable solution. However, recent advances have enabled the use of CNNs to capture very-long term dependencies using dilated causal convolution (DCC) layers [27]. The *dilation factor* of each DCC layer determines the number of input samples skipped by that layer. The *receptive field* is the total number of samples influencing the output at a particular time step. A larger dilation factor allows the output to represent a broader range of inputs, which aids in the learning of very-long term dependencies. Unlike traditional sequence models such as RNNs, LSTMs, and GRUs, CNNs do not have to wait for the previous time step output before processing input at the current time step. This enables CNNs to process the input sequences in parallel, making them more computationally efficient during both training and testing. Due to these promising properties, the TCNA network in the *TENET* framework adapts dilated CNNs for learning dependencies between messages in the vehicle.

The TCNA network consists of three novel *TCNA blocks*. Each TCNA block consists of an attention block and a TCN residual block (TRB), as shown in Fig. 2a. The first TCNA block's input is a time series of message data with n signal values as features. A partial sequence from the entire time-series dataset (called a subsequence) is fed to the model during each training step. The TRB is inspired by [27] and employs two DCC layers, two ReLU layers stacked together, and two weight normalization layers, as shown in Fig. 2b. This residual architecture facilitates efficient backpropagation of gradients and encourages the reuse of learned features. *TENET* enhanced the TRB from [27] by: *(i)* adding an attention layer (discussed later); *(ii)* not employing dropout layers to avoid thinning the network

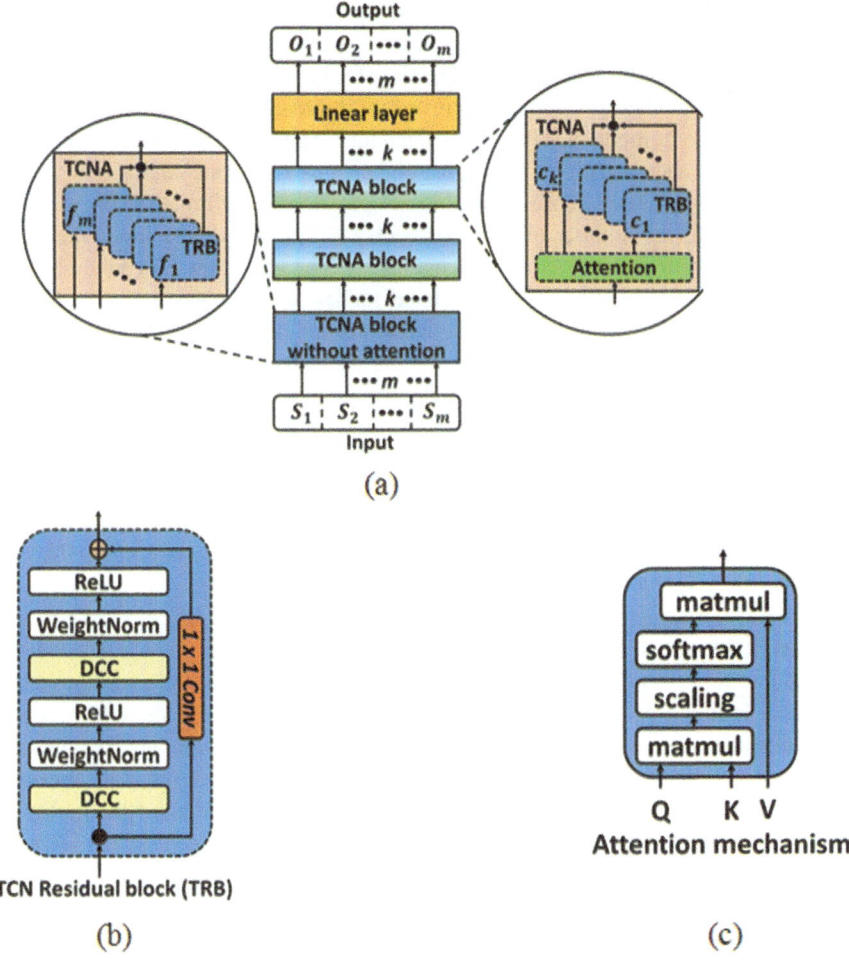

Fig. 2 (**a**) Overview of the *TENET's* TCNA network architecture and the internal structure of the TCNA block, (**b**) architecture of the TCN residual block (TRB) and different layers inside it, and (**c**) the attention mechanism

and provide the attention blocks with non-sparse inputs; and *(iii)* avoiding zero-padding the input time-series by computing the length of the subsequence using (1).

$$R = (k - 1) * 2^l \tag{1}$$

where R is the subsequence length, k is the kernel size, and l is the number of DCC layers in the networks. *TENET* performs this operation to mitigate sequence distortion of zero padding in in-vehicle time series data.

The first TCNA block in the network does not contain an attention block, and the inputs are given directly to the TRB. This is illustrated in Fig. 2a, where $\{f_1, f_2, \ldots, f_m\}$ represent multiple channels of the first TRB block, m is equal to the number of input features, $\{c_1, c_2, \ldots, c_k\}$ represent multiple channels of the TRB, and k representing the number of TRB input channels. The first DCC layer within the TRB processes each feature of the input sequence as a separate *channel*, as shown in Fig. 2a. A 1-D dilated causal convolution operation is performed using a kernel of size two, and the number of filters in each DCC layer is three times the input features (m). The input and output dimensions are the same other than the first TRB. The DCC layer output is weight normalized for fast convergence and to avoid the explosion of weight values. A ReLU activation function is applied to the weighted outputs, and this process is repeated once more inside the TRB. A convolution layer with a filter size of 1×1 is added to make the dimensions of the outputs from the last ReLU activation and the TRB input consistent. Each DCC layer in the TRB learns temporal relationships between messages by applying filters to its inputs and updating filter weight values.

The TCNA block also contains an attention block that helps the DCC layers focus on the input sequences' important aspects when producing outputs [28]. *TENET* computes a scaled dot product attention mechanism and models the attention as a mapping of three vectors called key (K), query (Q), and value (V). A weight vector is generated by comparing the similarities between the Q and K vectors, and the output attention weights are generated by a dot product between the weight vector and the V vector. This type of attention mechanism is referred to as the self-attention mechanism because it does not use the previous output information when generating the attention weights. In the context of the proposed TCNA network, self-attention mechanisms help in identifying important feature maps and enhance the quality of intermediate inputs received by the DCC layers. This also aids in efficiently learning very-long term dependencies between messages in an in-vehicle network.

The TRB's output feature maps are fed into the attention block, shown in Fig. 2c. To obtain the Q, K, and V vectors, the attention block repeats its inputs. A scalar-dot product is performed between Q and transpose of key (K^T) to calculate the similarities between each Q and K vector. The resulting dot product is scaled by a factor of $1/\sqrt{d_k}$ and then passed through a *softmax* layer to calculate attention weights as shown below:

$$Attention\,(Q, K, V) = softmax\left(QK^T \Big/ \sqrt{d_k} \right).V \qquad (2)$$

where the term d_k represents the dimension of the K vector. The computed attention weights represent the importance of each feature map of the previous DCC layer. The attention weights are then scalar multiplied by V to produce the attention block's output. Therefore, the attention block employs a self-attention mechanism to improve the quality of feature maps received by the subsequent TRBs. Similarly, the input sequence flows through the TCNA network and is fed to a final linear layer which generates an n-dimensional output. The n-dimensional output represents the predicted signal values of each dimension.

The TCNA network model is trained using a rolling window approach where each window contains signal values from the current subsequence. The proposed TCNA network learns the temporal dependencies between messages within a subsequence and attempts to predict the signal values of the subsequence that are shifted to the right by one time step. *TENET* uses a mean squared error (MSE) loss function to compute the prediction error between the signal values of the last time step in the predicted subsequence and the last time step in the input subsequence. The error is backpropagated to update the filter weights. This process is repeated for each subsequence until the training data is exhausted, which equals one epoch. To speed up the training, *TENET* employs a mini-batch training approach and trains the model for multiple epochs. At the end of each epoch, the model is evaluated using the unseen validation data. Additionally, *TENET* employs an early stopping mechanism to prevent the model from overfitting. The details related to the model hyperparameters are discussed later in section IV-A.

3.3 Model Testing

3.3.1 Attack Model

This subsection details various attack scenarios considered by *TENET* [29] framework. They assume that the attacker has access to the in-vehicle network and can alter signal values and network parameters at any instance of the vehicle operation. Based on these assumptions *TENET* framework attempts to detect the following complex and commonly observed attack scenarios in the in-vehicle network:

1. *Plateau attack:* This is an attack scenario in which the attacker sets a constant value for one or more signals over the attack interval. This attack is difficult to detect, especially when the set constant value is close to the true signal value.
2. *Playback attack:* In this attack, the attacker uses previously observed sequences of signal values and attempts to replay them again at a later time to trick the ADS. If the ADS is not trained to understand patterns in the transmitted message sequence, it will fail to detect these types of advanced attacks.
3. *Continuous attack:* This is a scenario in which the attacker gradually overwrites the true signal value. The attacker then eventually will reach the target value without being detected by most ADS frameworks. These attacks are difficult to detect and require a robust ADS.
4. *Suppress attack:* This type of attack involves the attacker attempting to suppress a signal value by either disabling the ECU or deactivating the communication controller, effectively resulting in no message being transmitted. It is challenging to detect short bursts of these attacks as they could be confused for a missing or delayed message.

3.3.2 Evaluation Phase

In the evaluation phase, the trained TCNA network along with a detection classifier is employed to efficiently detect in-vehicle network attacks at runtime. Due to the high frequency of messages in the in-vehicle network, there is a need for a lightweight detection classifier that can quickly classify messages while maintaining high detection accuracy and low overhead. *TENET* uses a categorical variable decision tree-based classifier to detect between normal and attack samples (binary classification) due to their speed, precision, and simpler nature. A decision tree begins with a single node (*root* node), which then branches into various outcomes. Each of these outcomes results in the formation of additional nodes called *branch nodes*. Each branch node branches off into other possibilities before ending in a *leaf node*, giving the structure a treelike appearance. During training, the decision tree algorithm creates the tree structure by determining the set of rules in each branch node based on its input. During testing, the decision tree takes the input data and traverses the tree structure until it reaches a leaf node. The evaluation phase starts by dividing the test data (attacks samples included) into two parts: *(i)* calibration data, and *(ii)* evaluation data. Only the calibration data is fed to the trained TCNA network in the first stage, to generate the predicted sequences. Then for each signal in every message, a divergence score (DS) is calculated using (3):

$$DS_i^m(t) = \left(\hat{S}_i^m(t) - S_i^m(t+1) \right) \forall \; i \in [\, 1, N_m \,], m \in [\, 1, M] \tag{3}$$

where m denotes the *mth* message sample and M denotes the total number of message samples, i denotes the *ith* signal of the *mth* message sample and N_m denotes the total number of signals in the *mth* message, t represents the current time step, $\hat{S}_i^m(t)$ denotes the *ith* predicted signal value of the *mth* message at time step t, and $S_i^m(t+1)$ denotes the true *ith* signal value of the *mth* message sample at time step $t+1$.

The DS is higher during an attack because the TCNA model is trained on the no-attack data and fails to accurately predict the signal values in the event of an attack. The sensitiveness of DS to attacks makes it a good candidate for the input to our detection classifier. Furthermore, the DS vector is created by staking the signal level DS for each message sample. Therefore, the decision tree classifier is trained using the DS vector as input to learn the distribution of both no-attack and attack samples. To assess the performance of the *TENET* framework, the unseen evaluation data with both attack and no-attack samples is used.

4 Experimental Setup

In this section, we discuss different experiments that were conducted to evaluate the effectiveness of the *TENET* [29] framework. Moreover, *TENET* is compared against three state-of-the-art prior works on automotive ADS: RN [16], INDRA [21], and HAbAD [22], which collectively cover a wide range of sequence modeling architectures for anomaly detection. RN [16] employs vanilla RNNs to increase the dimensionality of input signal values and tries to reconstruct the input signal at the output by minimizing MSE. The trained RN model scans continuously for large reconstruction errors at runtime to detect anomalies over in-vehicle networks. HAbAD [22] detects anomalies in real-time embedded systems using an LSTM-based autoencoder model with attention. This model tries to reduce the MSE reconstruction loss by replicating the input message signal at the output. HAbAD combines a kernel density estimator (KDE) and k-nearest neighbors (KNN) algorithm and detects anomalies in a supervised fashion. Finally, INDRA [21] employs a GRU-based autoencoder that minimizes MSE loss by reconstructing input sequences at the output. INDRA uses a pre-computed static threshold to flag anomalous messages at runtime. The comparisons of *TENET* with the above-mentioned ADS are presented in subsections IV-B and IV-C.

TENET adopts an open-source CAN message dataset developed by ETAS and Robert Bosch GmbH [17] to train their proposed TCNA network, and the comparison works. The dataset contains a variety of CAN messages with multiple signals that were modeled after real-world vehicular network data. Furthermore, the dataset includes a separate training dataset that has attack-free CAN messages and a labeled testing dataset with different types of attacks. For both training and validation of the model, *TENET* uses the training dataset from [17] without any attack scenarios in an unsupervised manner. The proposed *TENET* framework and all comparison works are tested by modeling various real-world attacks (discussed in section III-C.1) using the test dataset in [17]. It is essential to highlight that *TENET* can be easily adapted to other in-vehicle network protocols such as Flexray and Ethernet, as it relies only on the message payload information.

PyTorch 1.8 was used to model and train various machine learning models, including *TENET* and the comparison works. *TENET* framework uses 85% of data for training and the remaining 15% for validation. The *TENET* framework was trained for 200 epochs with an early stopping mechanism that constantly monitors the validation loss after each epoch. The training is terminated if no improvement in validation loss is observed in the past 10 (patience) epochs. *TENET* uses MSE to compute the prediction error and the ADAM optimizer with a learning rate of 1e-4. *TENET* employs a rolling window-based approach (discussed in section III-B) with a batch size of 256 and a subsequence length of 64. The authors used a scikit-learn library-based decision tree classifier with the *gini* criterion, and *best* splitter to detect anomalies based on the divergence score.

The proposed *TENET* framework classifies a message as a true positive *(TP)* only if the model detects an actual attack as an anomalous message. A true negative *(TN)* is when a normal message is detected as a no-anomalous message. A false positive *(FP)* occurs when the model misinterprets a normal message as an anomalous message. False negative (FN) occurs when the model fails to detect an actual anomalous message as an anomaly. Using these definitions, the *TENET* framework is evaluated based on <u>four</u> different performance metrics:

(i) *Detection Accuracy*: this metric measures ADS ability to detect an anomaly correctly, as defined below:

$$Detection\ accuracy = \frac{TP + TN}{TP + FP + TN + FN} \tag{4}$$

(ii) *Receiver Operating Characteristic (ROC) curve with area under the curve (AUC)*: this metric quantifies the classifier's performance as the area under the curve in a plot between the true positive rate (TPR) and false positive rate (FPR):

$$TPR = \frac{TP}{TP + FN} \quad FPR = \frac{FP}{FP + TN} \tag{5}$$

(iii) *False Negative Rate (FNR)*: this metric measures the probability that a *TP* will be missed by the model (a lower value is better):

$$FNR = \frac{FN}{FN + TP} \tag{6}$$

and *(iv) Mathews Correlation Coefficient (MCC)*: this metric determines an accurate evaluation of the model performance while working with imbalanced datasets, as defined below:

$$MCC = \frac{(TP * TN) - (FP * FN)}{\sqrt{(TP + FP)(TP + FN)(TN + FP)(TN + FN)}} \tag{7}$$

Another widely used metric in the literature is the F-1 score, which is the harmonic mean of precision and recall. However, the F-1 score fails to represent the model's true performance because both precision and recall do not include the true negatives in their computation. Unlike the F-1 score metric, the MCC metric considers all the cells of the confusion matrix, thus providing a much more accurate evaluation of the frameworks. Hence, *TENET* replaces the F-1 score with MCC as a performance evaluation metric.

A. *Receptive Field Length Sensitivity Analysis*

In the first experiment, the performance of the proposed TCNA architecture is compared with four different receptive field lengths while retaining other hyperparameters. This analysis is conducted to evaluate whether very long receptive lengths can help better understand normal system behavior. All the model variants were evaluated based on their performance on two training metrics, *(i)* average training loss and *(ii)* average validation loss. Lastly, the best model is selected and used for further comparisons. The average training loss value is the difference between the average loss of predicted and observed behavior of each iteration in the training data.

Table 1 TCNA variants with different receptive field lengths

	Receptive field lengths			
	16	32	64	128
Average training loss	4.1e-4	3e-4	**2.5e-4**	6.8e-4
Average validation loss	5.5e-4	4.3e-4	**2.9e-4**	9.3e-4

The average validation loss, on the other hand, represents the average loss between the predicted behavior and the observed behavior of each iteration in the validation data.

Table 1 represents the average training and validation loss of the three variants of TCNA architecture of the *TENET* framework. From the table, it can be observed that a receptive length of 64 has the lowest average training and validation loss. Therefore, a receptive field length of 64 was selected for the proposed TCNA architecture, which is twice the maximum receptive field length presented in the comparison works (sequence length of 32 in [22]). This long receptive field length enables the TCNA architecture to effectively learn very long-term dependencies in the input time series data and allows it to better understand the normal vehicle operating behavior.

B. *Prior Work Comparison*

In this subsection, a comparison of *TENET* [29] framework with the state-of-the-art ADS works RN [16], INDRA [21], and HAbAD [22] is presented. The comparison results on the metrics discussed in subsection C.1 are as shown in Fig. 3.

TENET [29] outperforms all comparison works in all four metrics under various attack scenarios, as shown in Fig. 3a–d. Table 2 summarizes the average relative percentage improvement of *TENET* over the comparison works for all attack scenarios. In comparison to the best performing prior work (INDRA), *TENET* improves detection accuracy by 3.32%, ROC-AUC by 17.25% for playback attacks (as it is the most difficult attack to detect), MMC by 19.14%, and FNR by 32.70%.

In summary, the proposed *TENET* framework with a customized TCNA network outperforms all previous recurrent architectures inclusive and not inclusive of attention mechanisms due to its ability to capture very-long term dependencies in time-series data. Furthermore, the attention mechanism within the TCNA helps to improve the quality of the outputs of the TRB thereby enabling efficient learning of very-long term dependencies. Thus, the proposed TCNA network with a decision tree classifier presents a promising anomaly detection framework.

C. *Memory Overhead and Latency Analysis*

Lastly, to understand the inference and latency overhead, different IDS frameworks are analyzed in this subsection. Table 3 shows the memory footprint, model parameters, and inference timings of *TENET* and other ADS comparison works. Since automotive ECUs have limited resources, the ADS models must not interfere with the normal operation of safety-critical applications. All the results were obtained by deploying the ADS models on an NVIDIA Jetson TX2 with dual-core ARM Cortex-A57 CPUs, which have specifications similar to real-world ECUs.

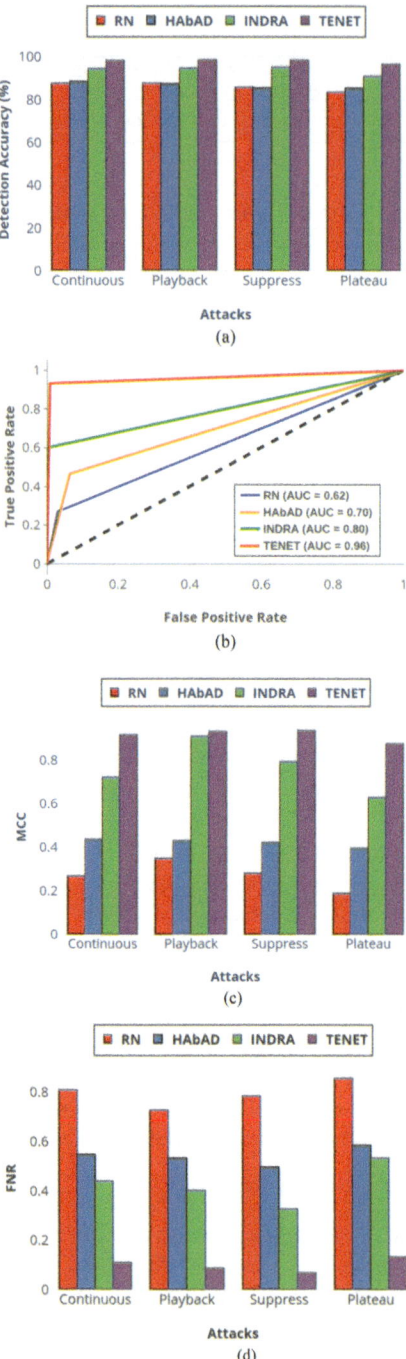

Fig. 3 Comparison of (**a**) detection accuracy, (**b**) ROC-AUC for playback attack, (**c**) MCC, and (**d**) FNR for *TENET* [29] and other comparison ADS works

Table 2 Relative % improvement of *TENET* vs. other ADS

Prior ADS Works	Detection accuracy	ROC-AUC	MCC	FNR
INDRA [23]	3.32	17.25	19.14	32.70
HABAD [24]	9.07	26.50	49.26	44.05
RN [17]	9.48	37.25	64.3	69.47

Table 3 Memory and latency analysis

ADS framework	Memory footprint (KB)	Model parameters	Inference time (ms)
TENET [29]	59.62	6064	**250.24**
RN [16]	**7.2**	**1300**	412.50
INDRA [21]	453.8	112900	482.10
HAbAD [22]	261.63	64484	1370.10

Except for RN [16], *TENET* [29] has the second lowest number of model parameters and memory footprint of all the comparison works. Even though RN has the fewest model parameters and the smallest memory footprint, it fails to effectively capture the temporal dependencies between messages, resulting in poor performance, as shown in Fig. 3a–d. *TENET* reduces the memory footprint by 86.86% and 77.21% respectively, and the number of trainable model parameters by 94.62% and 90.59% compared to INDRA and HAbAD. *TENET* is able to attain this higher performance with significantly fewer model parameters due to the fewer filters used by each DCC layer in the TCNA network. The attention block in TCNA improves the quality of the outputs of each TRB, thus obliviating the need for additional filters. Furthermore, compared to all comparison works, *TENET* has the lowest inference time with an average reduction of 56.43%. *TENET* achieves faster inferencing rates, unlike recurrent architectures, because *TENET* employs CNNs to process multiple subsequences in parallel, thereby significantly reducing the inference time. As a result, *TENET* can achieve superior performance in automotive platforms across a variety of attack scenarios with minimal memory and computational overhead.

5   Conclusion

In this chapter, a novel anomaly detection framework called *TENET* [29] for automotive cyber-physical systems based on Temporal Convolutional Neural Attention (TCNA) networks was presented. The *TENET* framework introduces a metric called the divergence score (DS), which measures the deviation of the predicted signal value from the true signal value. Moreover, the *TENET* framework is then compared with best-known prior works that employ a variety of sequence model architectures for anomaly detection. These promising results indicate a compelling potential for using *TENET* for anomaly detection in emerging automotive platforms.

References

1. Kukkala, V.K., Tunnell, J., Pasricha, S., Bradley, T.: Advanced driver-assistance systems: a path toward autonomous vehicles. In: Proceeding of IEEE CEM (2018)
2. Kukkala, V.K., Pasricha, S., Bradley, T.: SEDAN: security-aware design of time-critical automotive networks. In IEEE TVT (2020)
3. Valasek, C., Miller, C.: Remote exploitation of an unaltered passenger vehicle. In BlackHat USA (2015)
4. Adventures in Automotive Networks and Control Units, [online]: https://ioactive.com/pdfs/IOActive_Adventures_in_Automotive_Networks_and_Control_Units.pdf (2013)
5. Sitawarin, C., Bhagoji, A.N., Mosenia, A., Chiang, M., Mittal, P.: DARTS: deceiving autonomous cars with toxic signs. In: ArXiv (2018)
6. Dürrwang, J., Braun, J., Rumez, M., Kriesten, R., Pretschner, A.: Enhancement of Automotive Penetration Testing with Threat Analyses Results. In: SAE (2018)
7. Song, H.M., Kim, H.R., Kim, H.K.: Intrusion detection system based on the analysis of time intervals of can messages for in-vehicle network. In: ICOIN (2016)
8. Gmiden, M., Gmiden, M.H., Trabelsi, H.: An intrusion detection method for securing in-vehicle CAN bus. In: IEEE Proceeding of STA (2016)
9. Lee, H., Jeong, S.H., Kim, H.K.: OTIDS: A novel intrusion detection system for in-vehicle network by using remote frame. In: PST (2017)
10. Marchetti, M., Stabili, D.: Anomaly detection of CAN bus messages through analysis of ID equences. In: IEEE IV (2017)
11. Müter, M., Asaj, N.: Entropy-based anomaly detection for in-vehicle networks. In: IV (2011)
12. Wu, W., Huang, Y., Kurachi, R., Zeng, G., Xie, G., Li, R., Li, K.: Sliding window optimized information entropy analysis method for intrusion detection on in-vehicle networks. In: IEEE Access (2018)
13. Stabili, D., Marchetti, M., Colajanni, M., Detecting attacks to internal vehicle networks through hamming distance. In: AEIT (2017)
14. Shin, K.G., Shin, K.G.: Viden: attacker identification on in-vehicle networks. In: ACM SIGSAC (2017)
15. Kang, M., Kang, J.: A novel intrusion detection method using deep neural network for in-vehicle network security. In: VTC (2016)
16. Weber, M., Wolf, G., Zimmer, B., Sax, E.: Online detection of anomalies in vehicle signals using replicator neural networks. In: ESCAR (2018)
17. Hanselmann, M., Strauss, T., Dormann, K., Ulmer, H.: CANet: an unsupervised intrusion detection system for high dimensional can bus data. In: IEEE Access (2020)
18. Loukas, G., Vuong, T., Heartfield, R., Sakellari, G., Yoon, Y., Gan, D.: Cloud-based cyber-physical intrusion detection for vehicles using deep learning. In: IEEE Access (2018)
19. Hossain, M.A., Inoue, H., Ochiai, H., Fall, D., Kadobayashi, Y. LSTM-based intrusion detection system for in-vehicle can bus communications. In: IEEE Access (2020)
20. Taylor, A., Leblanc, S., Japkowicz, N.: Anomaly detection in automobile control network data with long short-term memory networks. In: DSAA (2016)
21. Kukkala, V.K., Thiruloga, S.V., Pasricha, S.: INDRA: intrusion detection using recurrent autoencoders in automotive embedded systems. In: TCAD (2020)
22. Ezeme, M., Mahmoud, Q.H., Azim, A.: Hierarchical attention-based anomaly detection model for embedded operating systems. In: RTCSA (2018)
23. Tariq, S., Lee, S., Woo, S.S.: CANTransfer: Transfer learning based intrusion detection on a controller area network using convolutional lstm network. In: ACM SAC (2020)
24. Kukkala, V.K., Thiruloga, S.V., Pasricha, S.: LATTE: LSTM self-attention based anomaly detection in embedded automotive platforms. In: ACM TECS (2021)
25. [online]   https://assets.vector.com/cms/content/products/gl_logger/Docs/GL1000_Manual_EN.pdf

26. Waibel, T., Hanazawa, G., Hinton, K.S., Lang, K.J.: Phoneme recognition using timedelay neural networks. In: IEEE Transactions on Acoustics, Speech, and Signal Processing (1989)
27. Bai, S., Kolter, J.Z., Koltun, V.: An empirical evaluation of generic convolutional and recurrent networks for sequence modeling. In: ArXiv (2018)
28. Vaswani, A., Shazeer, N., Parmar, N., Uszkoreit, J., Jones, L., Gomez, A.N., Kaiser, L., Polosukhin, I.: Attention is all you need. In: ArXiv (2017)
29. Thiruloga, S.V., Kukkala, V.K., Pasricha, S.: TENET: temporal cnn with attention for anomaly detection in automotive cyber-physical systems. In: IEEE/ACM ASPDAC (2022)
30. Kukkala, V.K., Bradley, T., Pasricha, S.: Priority-based multi-level monitoring of signal integrity in a distributed powertrain control system. In: Proceeding of IFAC Workshop on Engine and Powertrain Control, Simulation and Modeling (2015)
31. Kukkala, V.K., Bradley, T., Pasricha, S.: Uncertainty Analysis and Propagation for an Auxiliary Power Module. In: Proceeding of IEEE Transportation Electrification Conference (TEC) (2017)
32. Kukkala, V.K., Pasricha, S., Bradley, T.: JAMS: Jitter-Aware Message Scheduling for FlexRay Automotive Networks. In: Proceeding of IEEE/ACM International Symposium on Network-on-Chip (NOCS) (2017)
33. Kukkala, V.K., Pasricha, S., Bradley, T.: JAMS-SG: a framework for jitter-aware message scheduling for time-triggered automotive networks. In: ACM Transactions on Design Automation of Electronic Systems (TODAES), Vol. 24, No. 6 (2019)
34. Kukkala, V.K., Thiruloga, S.V., Pasricha, S.: Roadmap for cybersecurity in autonomous vehicles. In: IEEE Consumer Electronics Magazine (CEM) (2022)
35. Tunnell, J., Asher, Z., Pasricha, S., Bradley, T.H.: Towards improving vehicle fuel economy with ADAS. In: SAE Inter-national Journal of Connected and Automated Vehicles, Vol. 1, No. 2 (2018)
36. Tunnell, J., Asher, Z., Pasricha, S., Bradley, T.H.: Towards Improving Vehicle Fuel Economy with ADAS. In Proceeding of SAE World Congress Experience (WCX) (2018)
37. Asher, Z., Tunnell, J., Baker, D.A., Fitzgerald, R.J., Banaei-Kashani, F., Pasricha, S., Bradley, T.H.: Enabling prediction for optimal fuel economy vehicle control. In Proceeding of SAE World Congress Experience (WCX) (2018)
38. Dey, J., Taylor, W., Pasricha, S.: VESPA: a framework for optimizing heterogeneous sensor placement and orientation for autonomous vehicles. In IEEE Consumer Electronics Magazine (CEM), Vol. 10, No. 2 (2021)
39. DiDomenico, G.C., Bair, J., Kukkala, V.K., Tunnell, J., Peyfuss, M., Kraus, M., Ax, J., Lazarri, J., Munin, M., Cooke, C., Christensen, E.: Colorado state university EcoCAR 3 final technical report. In: SAE World Congress Experience (WCX) (2019)

Physical Layer Intrusion Detection and Localization on CAN Bus

Pal-Stefan Murvay, Adriana Berdich, and Bogdan Groza

1 Introduction

While concerns on vehicle cybersecurity were raised as early as 2004 [32], more recent demonstrations regarding security issues and their consequences coming from comprehensive security analyses of modern vehicles [2, 17, 22], led to an increased research interest in this area. Many of the identified issues come from the use of in-vehicle communication protocols which lack security mechanisms. One such protocol is the Controller Area Network (CAN) [10] which is still the most widely employed protocol that links Electronic Control Units (ECUs) even after more than three decades since its introduction. To address these issues researchers have focused on two main lines of work. A consistent body of works look at securing CAN communication by introducing cryptographic authentication or related mechanisms [8], while, more recently, many works are focusing on designing intrusion detection systems (IDS) for CAN. Reactions from the automotive industry sector and international organizations are also visible through their efforts in standardising various aspects related to vehicle cybersecurity [1, 12, 31].

As stated, the development of intrusion detection systems for CAN is a research topic that attracted considerable interest in the recent years. While many of the existing proposals adopt statistical tests and machine learning mechanisms, the various lines of work that focus on this topic generally adopt one of two approaches when it comes to sourcing data employed in the detection process. On one hand we have systems which use CAN traffic-related data (e.g., frame content, periodicity or arrival timing) that can be obtained at the application layer, from the CAN controller. Since in-vehicle CAN communication is often based on proprietary protocols which

P.-S. Murvay · A. Berdich · B. Groza (✉)
Faculty of Automatics and Computers, Politehnica University of Timisoara, Timisoara, Romania
e-mail: pal-stefan.murvay@aut.upt.ro; adriana.berdich@aut.upt.ro; bogdan.groza@aut.upt.ro

© The Author(s), under exclusive license to Springer Nature Switzerland AG 2023
V. K. Kukkala, S. Pasricha (eds.), *Machine Learning and Optimization Techniques for Automotive Cyber-Physical Systems*, https://doi.org/10.1007/978-3-031-28016-0_13

are not made public, intrusion detection systems that fall into this category generally attempt to extract meaningful behavioral data from captured CAN traces and use it to detect potential misuse [19]. Some works even go further and attempt to reverse engineer CAN frames in an attempt to extract information on signals encoded in the payload [20]. On the other hand, there are mechanisms that employ physical layer characteristics (e.g., voltage levels, propagation delays, signal rise/fall times) related to CAN communication. They rely on the well known fact that minute, uncontrollable, differences in the production process of electronic circuits introduce unique characteristics in their behavior. Therefore, this uniqueness in the signalling behavior could be used to identify transmitters.

In this chapter we focus on the former approach and discuss two approaches for intruder detection in CAN networks. The first is based on the use of timing characteristics of the CAN bus which influences signal propagation. Since detection is only the first step in thwarting potential attacks, we also cover the use of signal propagation delays for intruder localization in CAN-based networks. The second approach discussed here is based on voltage characteristics of CAN signals and makes use of machine learning algorithms to improve node identification accuracy.

The rest of this chapter is organized as follows. In Sect. 2 we provide some background on CAN, voltage based intrusion detection and voltage propagation delays. Then in Sect. 3 we discuss localization methods that use signal propagation time to localize ECUs on the bus. Section 4 contains experimental results regarding ECU identification from physical layer data with the help of machine learning algorithms. Finally, Sect. 5 contains the conclusion of this chapter.

2 Background

In this section we provide some background on the CAN bus and its physical layer signalling. We also discuss some related works that use voltage to detect intrusions on the bus.

2.1 The CAN Protocol

The CAN protocol was designed by Bosch as a solution for reliable communication for in-vehicle networks. Version 2.0 of the CAN specification [27], released in 1999, was later standardised as ISO 11898 and describes the data-link [10] and physical [11] layers which make up the CAN protocol. The data-link layer is implemented by the CAN controller, that can be used as a stand-alone chip or as a module integrated in a microcontroller (as suggested in Fig. 1), and is responsible for medium access, framing and error handling. The standard CAN frame, depicted in Fig. 2, can accommodate a maximum payload of 8 bytes. Larger payloads, of up to 64 bytes, can be transmitted using CAN-FD (CAN with Flexible Data-rate)

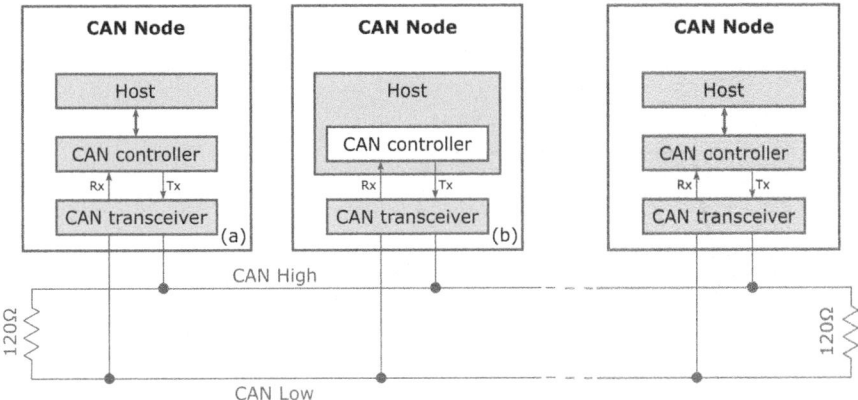

Fig. 1 CAN bus implemented with nodes using stand-alone controller chips (**a**) and controllers integrated in the host microcontroller (**b**)

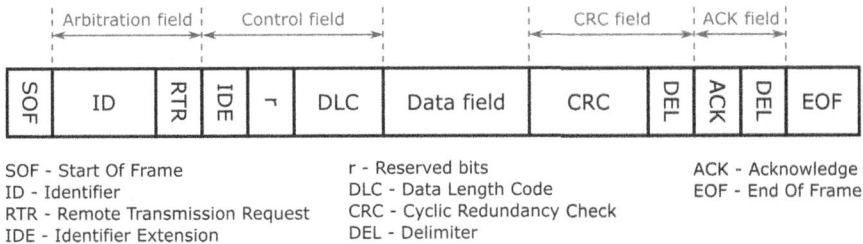

SOF - Start Of Frame
ID - Identifier
RTR - Remote Transmission Request
IDE - Identifier Extension

r - Reserved bits
DLC - Data Length Code
CRC - Cyclic Redundancy Check
DEL - Delimiter

ACK - Acknowledge
EOF - End Of Frame

Fig. 2 Standard CAN frame format

a more recent extension of the original CAN protocol [10]. Each frame includes an identifier (ID) field which is usually an indicative of the frame content type or sender. While specific ID values that can be transmitted by each network node are defined at design time, the CAN protocol offers no mechanism for preventing ID misuse.

The standard high-speed CAN physical layer supports bit rates of up to 1 Mbit/s (500 kbit/s is usually used for in-vehicle communication) while its newer embodiment CAN-FD is able to deliver bit rates of up to 8 Mbit/s (the higher bit rate is only used for payload transmission). The CAN physical layer is implemented by the CAN transceiver which connects to the CAN High and CAN Low lines that form a two wire differential bus, as illustrated in Fig. 1. The bus is terminated at the ends with 120 Ω resistors (matching the characteristic impedance of the bus) to suppress signal reflections. The CAN physical layer specification [11] defines ranges for the two differential voltage ($V_{diff} = V_{CAN\_H} - V_{CAN\_L}$) levels used to encode logical information as shown in Fig. 3. The two logical bus states, called *dominant* and *recessive*, are used to implement a wired-AND signalling behavior. That is, a dominant state is set when at least one transceiver is actively driving the bus, while the recessive state is obtained when none of the network nodes is driving

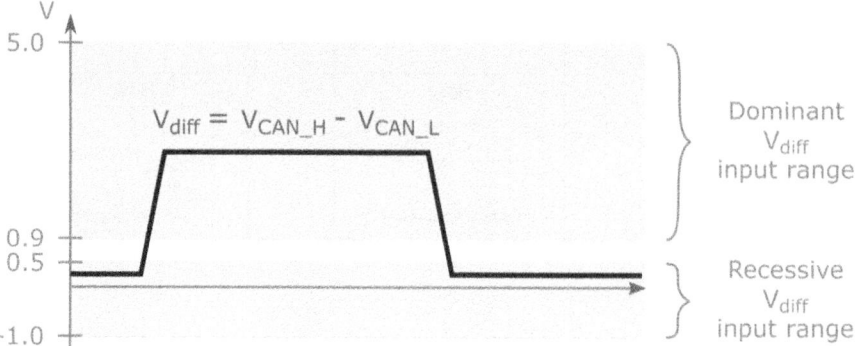

Fig. 3 Differential voltage ranges defined for the dominant and recessive states, according to ISO 11898-2

the bus. As a result, a logical "0" bit is encoded as a dominant state, while a logical "1" represents a recessive state.

While a CAN frame represents a successful transmission from a single CAN node, other nodes are allowed to actively drive the bus during the *arbitration* and *acknowledgment* fields (indicated in Fig. 2). The arbitration field is dedicated to the arbitration mechanism implemented to resolve contention (i.e. the case when two or more nodes try to transmit a CAN frame at the same time). During the arbitration field, nodes competing over bus access make simultaneous bit by bit transmissions and monitor resulting values on the bus. A node stops when it transmits a recessive bit and reads back a dominant value. Consequently, transmission priorities can be set based on the ID field, with lower values indicating higher priorities. As its name suggests, the acknowledgment (ACK) bit is used by receivers to acknowledge the successful reception of a CAN frame. Transmitters send a recessive value during this bit while all receivers are expected to transmit a dominant value if they were successful in correctly decoding the received frame.

2.2 Voltage-Based Intrusion Detection

CAN intrusion detection mechanisms based on physical layer voltages rely on features that can be extracted from the characteristics of physical signals generated by CAN nodes. The signalling behavior of CAN nodes display unique characteristics determined by minute, uncontrollable, differences in the production process of electronic components involved (e.g., transceivers and power supply circuitry). Table 1 lists works that use various physical layer features for detecting intrusions on CAN. While some use simple threshold comparison or for matching new transmission to existing fingerprints, other works use various machine learning algorithm to achieve classification. A more recently emerging body of works, which

Table 1 Comparison of existing proposals for CAN intrusion detection mechanisms based on the physical layer

Paper	Year	Sampling rate	CAN bit rate (max)	Methodology
Murvay et al. [23]	2014	2 GS/s	125 kbps	Statistical distributions
Cho et al. [4]	2017	50 kS/s	500 kbps	Machine learning
Choi et al. [5]	2018	2.5 GS/s	500 kbps	Machine learning
Choi et al. [6]	2018	2.5 GS/s	500 kbps	Machine learning
Kneib et al. [13]	2018	20 MS/s	500 kbps	Machine learning
Foruhandeh et al. [7]	2019	50 MS/s	500 kbps[a]	Statistical distributions
Rumez et al. [28]	2019	$\geq$2GS/s[b]	Any	Statistical distributions
Kneib et al. [15]	2020	2 MS/s	500 kbps	Machine learning
Murvay et al. [24]	2020	250 MS/s	Any	Threshold comparison
Groza et al. [9]	2021	250 MS/s	Any	Signal slope

[a] Extracted from the associated dataset [7]
[b] Estimated based on paper details

is also discussed in the next section, i.e., [24] and [9], uses physical signal to locate ECUs on the bus.

The dominant voltage level was the first among the characteristics used for uniquely identifying CAN transmitters and is still the most commonly employed. The idea was introduced in [23], which applies basic signal processing tools (i.e., mean squared error, convolution and mean-value) to extract unique sender characteristics from samples captured at the start of the arbitration field of CAN frames. The detection accuracy of this approach is later improved by Choi et al. [5] which apply classification algorithms on a set of 17 features extracted from samples obtained during the ID field of extended CAN frames (i.e., CAN data frames that use a 29 bit ID field instead of the 11 bit found in standard CAN frames). Another line of work by Choi et al. [6] brings further improvements by considering not only the dominant level voltage for feature extraction but also the rising and falling edges generated by transitions between the recessive and dominant state as these can contain transients with a potential to reveal additional unique transmitter features.

The first works on CAN physical layer intrusion detection did not consider which of the frame fields are more appropriate for sampling. As explained in the previous section, CAN signals generated during the arbitration field can be the result of more than one node actively driving the bus which affects the resulting dominant voltage levels. Figure 4a illustrates three dominant bits generated by the simultaneous transmissions of up to three transceiver circuits all from MC33742 system basis chips. While the resulting dominant bus value increases with each additional transceiver driving the bus (as indicated in Fig. 4b), the value is always correctly decoded by receiving transceivers (i.e., CAN Rx pin value) as long as the bus voltage levels stay within the specified ranges (Fig. 3). Therefore, even under normal CAN bus usage conditions, samples acquired during the arbitration field might not be representative for the characteristics of a single node while the dominant bit in the ACK field is generated by receiver nodes. Such aspects are

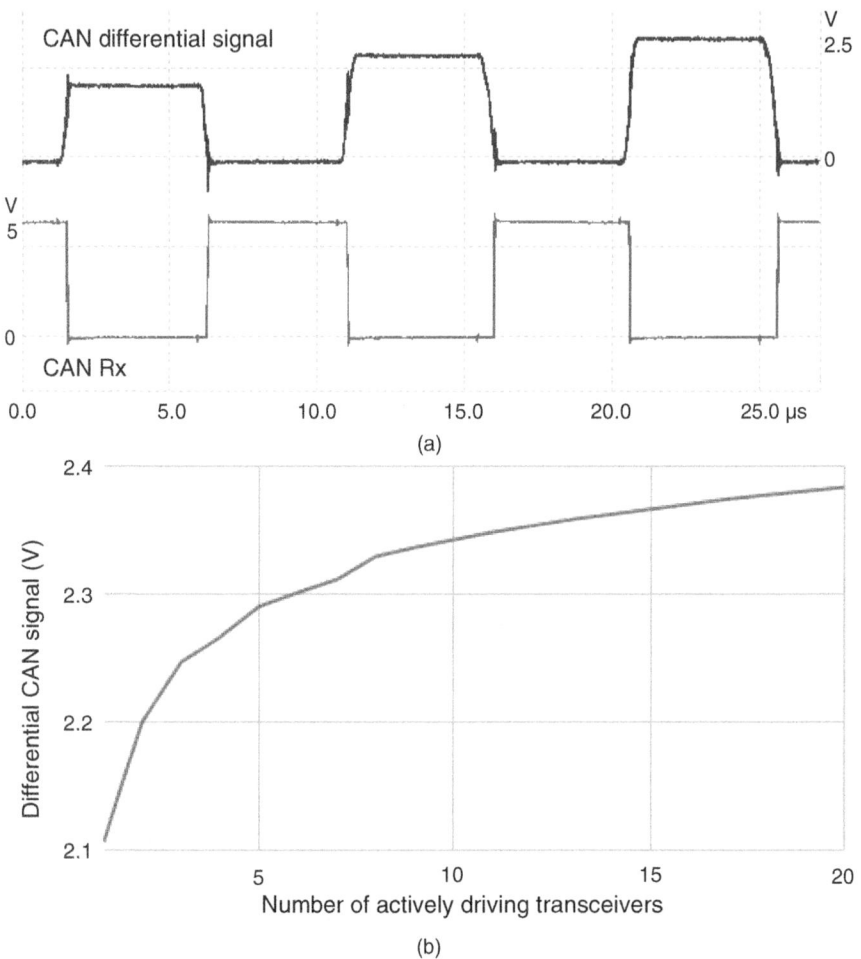

Fig. 4 Influence of simultaneous transmissions on bus voltage levels: (**a**) Example of bus voltage and received bit values for 1–3 simultaneous CAN transmissions of dominant bits, (**b**) Effect of the number of transceivers actively driving the bus line at the same time on dominant voltage level

first considered in Viden [4], a first practical implementation of a physical layer IDS using a sampling rate of 50 kS/s which is very low in contrast with other solutions that require sampling rates in the order of Ms/s or GS/s. In Viden the ACK field thresholds are isolated when generating ID-based dominant voltage profiles, however, the arbitration field can still be used as a sampling area.

Scission [13] is the first line of work to carefully consider the frame fields to be used for reliable transmitter-related feature extraction. By using samples around the rising and falling edges of dominant bits and simple machine learning support, Scission is able to achieve better detection accuracy than previous works. Kneib et al. improve on their initial approach and propose EASI [15] reducing the sample rate requirements to 2 MS/s which should be feasible for analog-to-digital converters in common automotive-grade microcontrollers with the use of random interleaved sampling.

A completely different approach and a first step towards location based intrusion detection is presented in the work of Rumez et al. [28]. The authors use time domain reflectometry (TDR) for measuring the network response to a pulse sent by the IDS. The pulse response is an indicative of the network structure (i.e., network nodes and their location on the bus) and is compared to prerecorded reference responses to determine potential changes. This approach is able to detect when nodes are added or removed from the bus and can correlate the response signal with network node locations. On the downside, using TDR will not be effective in detecting existing network nodes that were compromised.

2.3   Signal Propagation Delays

As they propagate along the bus, signals generated by CAN nodes travel through a non-ideal medium which introduces propagation delays. Sources for such delays can be found in the characteristics of the physical transmission medium as well as in local alterations of the transmission medium characteristic behavior caused by the nodes connected to the bus.

A transmission line is characterised by a specific propagation speed which is the main responsible for propagation delays. A common way of approximating the line delay is by using the distributed model of the transmission line. As illustrated in Fig. 5, transmission lines can be modeled as an infinite number of elementary line components connected in series. This is the model of a lossy transmission line in which each elementary component represents a line segment of infinitely small length with its behavior characterised by a series resistance R, a series inductance L, a parallel capacitance C and a conductance G, caused by imperfect insulation between line conductors. The values of these parameters are defined per line unit length and can be used to calculate the complex characteristic line propagation constant $\gamma(\omega) = \sqrt{(R + j\omega L)(G + j\omega C)} = \alpha(\omega) + j\beta(\omega)$. Here $\alpha(\omega)$ represents the line attenuation factor and $\beta(\omega)$ represents the propagation coefficient of the transmission line and are both dependent on frequency ($\omega = 2\pi f$). In practice, the lossless line model is used more often and it is obtained by considering that the conductance G and line resistance R are negligibly small. These assumptions can be safely made in the case of CAN lines based on the

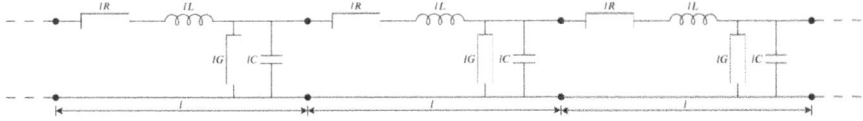

Fig. 5 Generic model of a lossy transmission line

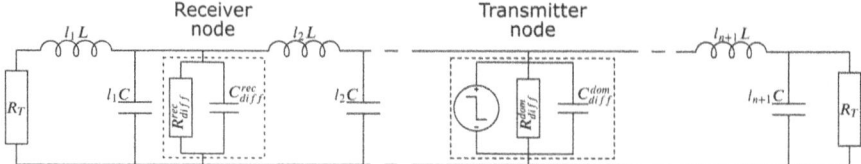

Fig. 6 Equivalent model of a CAN bus with receiver and transmitter nodes

fact that G is very small in comparison to the ωC component, while R is in the order of tens of $m\Omega/m$ ($70m\Omega/m$ according to ISO 11898-2 [11] or $25m\Omega/m$ according to SAE J1939 [29]). This simplifies the propagation constant, making it purely imaginary $\gamma(\omega) = j\omega\sqrt{LC}$ which, in this form, only represents the propagation coefficient. As a result, the characteristic propagation delay of the line can be calculated as $t_{pd} = \sqrt{LC}$ (s/m). According to ISO11898-2 [11] the nominal value for the propagation delay along a high speed CAN bus transmission medium is 5 ns/m (considering a homogeneous transmission medium). This offers a good approximation of propagation delay for a section of CAN bus not considering the presence of loads. However, for a better approximation of delays the loads along the bus must also be considered.

In most cases, in transmission line models, loads (i.e., bus nodes participating in communication) are considered to be uniformly distributed along the transmission line. The resistive and capacitive loads are factored into calculations as additional distributed components per line unit length. However, when looking at in-vehicle networks, and the CAN bus in particular, there is considerable variability in the function and manufacturer of nodes sharing the same network which translates into the variability of bus interface circuitry. Moreover, bus nodes are not uniformly distributed along the line since their physical location is usually restricted to specific areas inside the vehicle.

Each CAN node connected to the bus behaves like a load connected in parallel to the bus lines. In addition, sender nodes act as voltage sources during transmission of dominant bits. The load represented by each CAN node has a resistive and a capacitive component. The resistive component mainly consists of the transceiver differential input resistance R_{diff} which is expected to be in the 10–100 kΩ interval range according to the CAN specification. The capacitive load mainly consists of the transceiver internal differential capacitance C_{diff} which is expected to have a nominal value of 10pF [11] while the node is in the recessive state but should not exceed 50pF (measured with the node disconnected from the bus) [29]. The stub and connector used to link the node to the bus can also add to the capacitive load but this component is usually negligible. With this in mind, we can define the equivalent model of a loaded CAN bus as illustrated in Fig. 6, where R_T are the bus termination resistors, each bus line segment connecting nodes is represented as a lossless transmission line component and nodes are represented as a parallel RC load, with an additional voltage source added as a component of a transmitting node.

To estimate the propagation delay based on this model, the work presented in [24] considers the loads caused by CAN network nodes to be mainly capacitive due to the reduced effect of the resistive load on propagation delays. Based on this assumption, the propagation delay is estimated as the sum of delays for each bus segment, where a bus segment is considered to span the distance between two nodes or between a node and the bus end. The differential capacitance of the node included in each segment is considered to be distributed along that line segment and factored in the calculations along with the characteristic line capacitance. Therefore, the estimated propagation delay on a segment of CAN bus can be calculated as

$$t_{pd} = \sum_{i=1}^{n} l_i \sqrt{L(C + C_{diff_i}/l_i)}, \tag{1}$$

where L and C are the characteristic line inductance and capacitance, C_{diff_i} is the differential capacitance of the CAN node included in segment i and l_i is the length of the ith segment.

3 Localization Methods Based on Physical Layer Signals

In this section we discuss two intrusion detection mechanisms based on the differential propagation delays of the signals recorded at the physical layer which can be used to estimate the location of the transmitter node.

3.1 Transmitter Identification by Propagation Delays

Based on the loaded CAN bus model discussed in the previous section it is evident that the propagation delays of CAN signals, as viewed from a fixed observation point, are directly influenced by the transmitter location on the bus. This suggests that the propagation delay of CAN signals could be used to identify transmitters and estimate their location on the bus. However, the problem comes down to how can these propagation delays be recorded. Using a single, fixed, observation point on the bus for measuring propagation delays would require additional information regarding the actual transmission time of the message which can only be recorded at the transmitter node location. To obtain this information from sender nodes, the receiver node, in charge with delay measurements, needs to be synchronized with the transmitters and trust the timing information they provide. Moreover, there should also be a way to determine if the transmission came from the left or right-hand side of the bus relative to the receiver location.

To alleviate these problems, the authors of [24] and [9] proposed a novel intrusion detection mechanism based on signal propagation delays that does not require knowledge about the message transmission time which also eliminates the need for time synchronization between nodes. They achieve this by measuring the differential

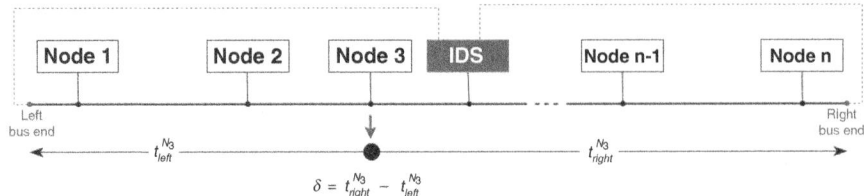

Fig. 7 Concept for recording differential propagation delays on a CAN bus

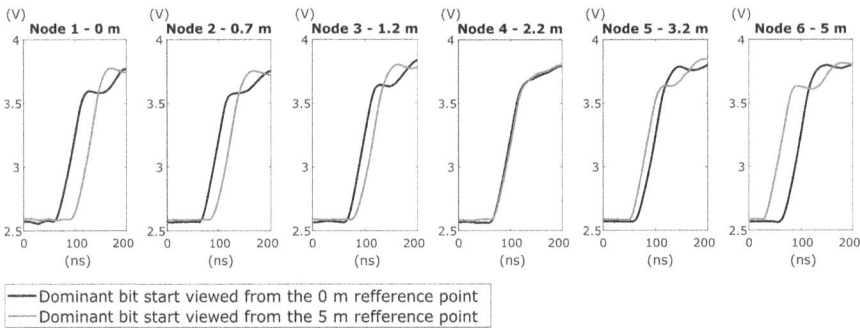

Fig. 8 Difference in dominant bit arrival time at CAN bus ends depending on transmitter location

propagation time, that is, the difference between the time required for a signal to reach one end of the bus and the time required for it to reach the other end. This requires monitoring CAN signals at the two ends of the CAN bus (considering the network is based on a bus topology) and recording signal arrival time at each end, as suggested in Fig. 7. The differential propagation time can then be computed as $\delta = t_{right}^{N_i} - t_{left}^{N_i}, i = 1, n$, where $t_{right}^{N_i}$ and $t_{left}^{N_i}$ are the arrival times of the signal generated by node N_i at the right and left end of the bus respectively. This is equivalent to measuring the propagation time on the two signal propagation paths relative to the transmission point and calculating $\delta = t_{pd_{right}} - t_{pd_{left}}$.

Like propagation delays, the differential propagation time is directly influenced by the location of the transmitter node on the bus. The absolute value of δ increases as the transmission node location is farther away from the point which represents the bus center of mass with respect to propagation delay. The sign of the differential propagation time indicates the bus end closer to the node location (i.e., a negative value suggests a transmitter closer to the right bus end while a positive value indicates a transmitter closer to the left bus end).

The relation between differential propagation time and transmitters' location on the bus is illustrated by Fig. 8 which shows dominant bit arrival times recorded on the CAN-High line at the two bus ends of a 5 m long transmission line having 10 nodes distributed unevenly along the CAN bus. As expected the difference in signal arrival time is close to zero for transmitter located toward the center of the bus and it increases as senders are located closer to either end of the bus.

3.2 Signal Acquisition

For intrusion detection based on voltage-related signal characteristics, samples must be recorded via a direct connection to the CAN bus physical layer. Since CAN uses differential signalling for increased noise immunity, it would be preferable to sample both the CAN-High and CAN-Low lines and use the resulting differential signal. Existing physical layer CAN IDS proposals use either the differential CAN signal for sampling or only one of the two CAN bus lines.

As discussed in the previous section an intrusion detection mechanism based on differential propagation delays requires the ability to sample physical layer signals as they are seen at the bus ends. This involves connecting the IDS sampling circuitry to the physical bus lines at both of the bus ends. Sampling the differential CAN signal would require a two wire connection to each bus end which increases wiring complexity. TIDAL-CAN [24] and CAN-SQUARE [9] proved to be efficient in extracting the differential propagation delay using a single CAN wire connection at each bus end (either CAN-High or CAN-Low). An alternative would be to use extra circuitry that connects to the two bus lines and outputs the differential signal. For example, PLI-TDC [25], a time-based physical layer intrusion detection mechanism, uses samples from the Rx pin of a transceiver connected to the bus for time measurements instead of directly sampling the bus lines. This approach would only be appropriate for timing based intrusion detection mechanisms since physical layer voltage characteristics are not transmitted through the transceiver.

Many of the more recent voltage-based IDSs use rising and falling edges as a main target area for sampling due to the presence of transients that may reveal more unique features. The use of rising or falling edges is also required for measuring differences in signal arrival times at the bus ends since they provide clear indication for the start of a bit. Therefore, the signal areas targeted by sampling should be recessive-to-dominant and dominant-to-recessive transitions. Moreover, sampled signals must be the result of a single target node actively driving the bus. As discussed in previous sections, it is expected to have multiple nodes transmitting during the *arbitration* and *acknowledge* fields of a CAN frame. Therefore, the CAN frame areas targeted for sampling should be the *control*, *data* and *CRC* fields. In addition to the physical layer signal, the IDS also needs to capture the actual frame content. This is required for extracting information needed to identify the expected frame sender which is compared against the actual transmitter as inferred by the IDS.

The datasets used in the intrusion detection mechanisms discussed in this section were obtained from experimental models of a CAN bus. A PicoScope device from the 5000 series, along with the associated PC application, were used for sample acquisition.

3.3   The TIDAL-CAN Methodology

TIDAL-CAN [24] introduces the concept of using differential propagation delays
for intrusion detection and transmitter location estimation. By using the TIDAL-
CAN mechanism it is possible to detect and distinguish between various attack
strategies, i.e., compromised nodes, replaced nodes and node insertion. Sender
location estimation is possible with an accuracy of several tens of centimeters,
depending on the attack strategy employed. The CAN-TIDAL methodology is
evaluated on an experimental setup comprising a 5 m CAN bus with 10 nodes as
illustrated in Fig. 9 represented by up to 5 different device types.

Differential delays are measured from the rising/falling edges of the signals
captured at the ends of the bus. Since the shapes of the rising/falling edges are not
ideal, a threshold is used to specify the voltage level at which the differential delay δ
should be measured. A common threshold is established for all transmitters so that
it assures the best separation accuracy of differential delays from known network
nodes. The differential delays of known network nodes are prerecorded in a training
phase and associated with frame IDs which are usually uniquely assigned to specific
senders. Data recorded in the training phase is then used during normal run-time to
identify transmitter nodes. Failure to correlate a newly recorded differential delay
with prerecorded values expected for a specific frame triggers an intrusion alarm.
This will happen when a compromised node attempts to transmit a frame associated
with a different node as well as when the network structure is altered by removing
or inserting a node since this will alter the characteristic propagation delay behavior
of the network which is reflected in the propagation delays of other networks nodes.

Sender location estimation is made based on differential delays using simple
linear interpolation considering two nodes with known locations. While identifying
transmitter location is straightforward for attacks that do not lead to the alteration
of the network, attacks involving node replacement or node insertion pose more
challenges. This is caused by the fact that altering the network structure results in
changes in the characteristic propagation behavior of signals sent along the bus.
As a result, differential delays, including those produced by legit nodes, will be
affected. Therefore, locating transmitters in this case cannot rely on prerecorded
fingerprints. Node legitimacy must be reassessed at the bus level in order to allow
sender location estimation and this requires processing multiple transmissions from
all network nodes.

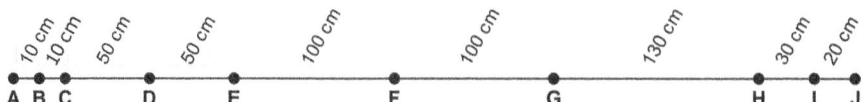

Fig. 9 Node positioning along the experimental bus model employed for evaluating TIDAL-CAN
and CAN-SQUARE

3.4 The CAN-SQUARE Methodology

CAN-SQUARE [9] proposes a simple algorithm that improves on the threshold based separation of differential delays that is presented in TIDAL-CAN [24]. The separation accuracy is in the range of 10 cm which proves good localization accuracy while requiring only elementary arithmetic operations, i.e., additions and multiplications of the sampled amplitudes. What is notable about the methodology is that it proves resilience against replacement and insertion attacks as well as against temperature variations in the range of 0–60 °C. Previous works on physical layer identification of in-vehicle ECUs had a hard time with environmental variations under which the fingerprint of each ECU drastically changes. Apparently, the propagation time of the signal is far less influenced by environmental changes and localization is still sufficiently accurate even when the geometry of the cable changes as well when the temperature changes. The authors from [9] present experiments with the cable heated in enclosed box or cooled down in a fridge using four temperature check points 0, 24, 50 and 60 °C.

How the methodology works is quite easy to explain. Two sampling points v_i, v_{i+w} are selected, separated by a window of size w, and using the sampling period δ multiplied by the size of the window w, the slope of the line that leads through the two sampling points is extracted, i.e.,

$$s[i] = \frac{v[i+w] - v[i]}{w\delta} \tag{2}$$

When the slope exceeds a fixed threshold τ the point is marked as the start time of the bit. The start time of the bit is computed both to the left and right side of the bus allowing to extract the exact location of the ECU from which the bit originates as:

$$\pi = \frac{(\lambda_l - \lambda_r)\delta}{5 \times 10^{-9}} \tag{3}$$

Here λ_l and λ_r are the recorded indexes of the sample when the angle s reaches threshold τ at the left and right sides of the bus respectively, δ is the sampling time while 5×10^{-9} is a constant representing the default propagation time of the signal (representing the nominal propagation speed on a CAN bus which is 5 ns/m).

The paper presents two algorithms, the forward-square and backward-square algorithms, which parse the signal from the left-to-right or right-to-left respectively [9]. The backward algorithm gives slightly better results than the forward algorithm. As a suggestive depiction for the accuracy, we graphically show localization results in Fig. 10 for the case of ECU replacements in the 10 ECU network configuration that is also used in CAN-TIDAL [24]. When replacements are done with identical ECUs (dashed-dotted line marked with triangles in Fig. 10) the determined distances by the BCQ-SQUARE method generally fluctuate under 1 dm precision. Due to impedance changes, when replacements are done with distinct ECUs (dotted line

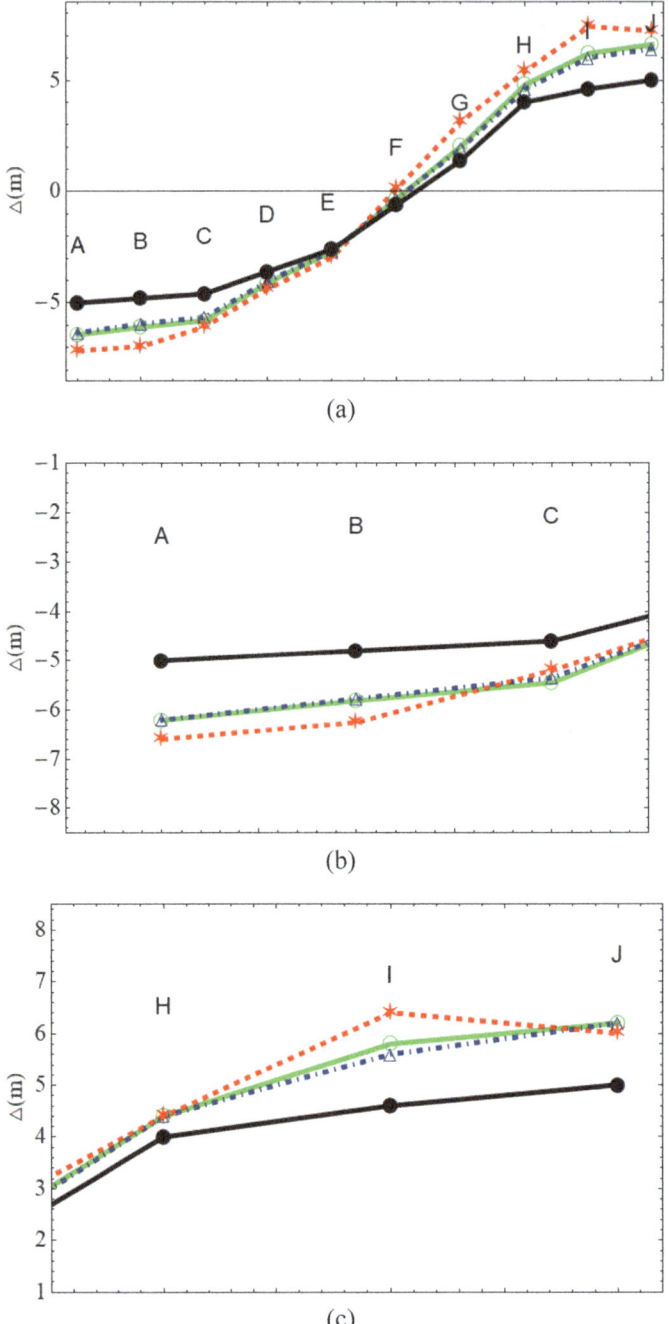

Fig. 10 Localization with BCW-SQUARE of the 10 EUCs in case of replacements with identical ECUs (blue line) and distinct ECUs (red line): (**a**) complete view, (**b**) left side detail, (**c**) right side detail

marked with stars in Fig. 10) the determined distances by the BCQ-SQUARE method may deviate by 2–3 dm but the nodes are still easy to locate. Figure 10b and c give a detailed view of the distances at left and right sides of the bus. Notably, in case of replacements with distinct ECUs, only a single node out of the 10 nodes, i.e., ECU I, appears to be further than its real position, the rest of the 9 ECUs come close to their original location. The black line marked with filled circles denotes the actual physical position of the ECUs. We also note that in case of replacements with distinct ECUs, 6 out of the 10 ECUs were replaced with distinct ECUs, which is quite an extreme case and not an usual situation for a real in-vehicle network. For more results, in case of ECU insertions as well as temperature changes, we refer the reader to the original work [9].

4   Machine Learning on Physical Layer Signals

In this section we present concrete experimental results on ECU identification by using physical layer data. The results show that using single voltage features, like the maximum or minimum voltages, may help for smaller pools of ECUs but are insufficient as the pool becomes larger. In this case traditional machine learning algorithms give better results but only neural networks seem to separate between samples with excellent accuracy.

4.1   The ECUPrint Dataset

A recently published work has released in the public domain a comprehensive physical layer dataset, ECUPrint [26] that contains data collected from 10 vehicles, i.e., nine passenger vehicles and one heavy-duty vehicle compliant to the J1939 standard. It is also relevant to note that the ECUPrint paper [26] advocates the use of physical fingerprints for forensics purposes, i.e., identification of vehicles that may be subject to theft or VIN cloning or the illegal replacement/modification of in-vehicle ECUs, an application of physical fingerprints which has not been previously considered. The authors propose the use of four features: (1) mean voltage, (2) maximum voltage, (3) bit time and (4) plateau time, each of them being extracted from isolated bits, i.e., a dominant bit between two recessive bits. The paper acknowledges that the use of a single feature out of the four leads to great overlaps between ECUs and multiple features should be combined. The use of all four features results in a sufficiently good identification of the ECUs with only slight overlaps. As we will show later, the use of machine learning algorithms allow an identification with a very high accuracy, above 99.9%.

Each car from the dataset has between 3 and 9 ECUs and each ECU uses distinct IDs. For each ID several bits are extracted leading to between 20 and 20,187 sample files for each ECU in the dataset. One measurement represents 2000 sampling points for the nine passenger cars and 2700 sampling points in case of the measurements from the heavy-duty vehicle. This is due to the distinct data-rate of the bus from the heavy-duty vehicle. In this work we will use only measurements from the nine passenger vehicles which contain a total of 51 ECUs. The number of measurements is not equal for each ECU, but for each measurement there are exactly 2000 sampling points which allow us to use the same architecture for the classification algorithms.

4.2 Results with Traditional Classifiers and Neural Networks

We evaluate the detection performance based on the ECUPrint dataset for 5 machine learning algorithms, i.e., Decision Trees (Tree), Linear Discriminant (LD), K-Nearest Neighbors (KNN), Support Vector Machines (SVM) and a simple neural network (NN) available in the Matlab toolset [21]. A few words on these algorithms may be in order. Here is a short description based on the Matlab documentation [21]:

- *Decision Trees (Tree).* Decision tree is a supervised classification algorithm which organizes data as a tree to provide fast and easy to visualize classification results.
- *Linear Discriminant (LD).* Discriminant analysis is a classification algorithm based on the Gaussian distribution. The linear discriminant creates linear boundaries between classes.
- *K-Nearest Neighbors (KNN).* KNN is a commonly used classifier based on distances (in our case Euclidean distances) between the training samples and the test samples.
- *Support Vector Machines (SVM).* Support Vector Machines is used to train binary or multiclass models. SVM is a supervised machine learning algorithm commonly used to solve distinct classification problems. Matlab offers support for 6 types of SVM classifiers: Linear SVM, Cubic SVM, Quadratic SVM, Fine Gaussian SVM, Coarse Gaussian SVM and Medium Gaussian SVM. In this work we use the Fine Gaussian SVM classifier, which uses a Gaussian kernel function.
- *Neural Network (NN).* The Neural Network (NN) that we use is a simple wide neural network available in Matlab. It contains an input layer, one fully connected layer with 100 neurons, a rectified linear unit (ReLU) a final fully connected layer with 51 outputs (corresponding to the 51 ECUs used in our evaluation) and a Softmax function.

To evaluate the performance of the employed classifiers we used the following metrics, which derive from the true positives TP, false negatives FN, true negatives TN and the false positives FP rates:

1. *Accuracy* which is computed using the *kfoldLoss* classification error using k-fold cross validation (for our dataset we used fivefold cross validation)

$$Accuracy = 1 - kfoldLoss,$$

2. *False Acceptance Rate (FAR)* and *False Rejection Rate (FRR)* which give a better understanding of the success and failure rate in identifying a node and are computed as:

$$FAR = \frac{FP}{TN + FP}, \qquad FRR = \frac{FN}{TP + FN},$$

3. The traditionally used *Precision, Recall* and *F1-score* computed as

$$Precision = \frac{TP}{TP + FP}, \qquad Recall = TPR = \frac{TP}{TP + FN},$$

$$F1 - score = \frac{2 \times precision \times recall}{precision + recall}.$$

We evaluated the performance of the 5 machine learning algorithms mentioned above using various percentages of the dataset for training and testing. We employ the implementation of these classifiers provided in Matlab 2021a. The tests were performed on a laptop equipped with an Intel Core i7-9850H processor and 32Gb RAM.

To begin with, we show that the use of two voltage features, i.e., two sampling points from a single bit, is insufficient when using machine learning classifiers. The authors in [26] already argued that four features (the mean voltage, maximum voltage, bit time and plateau time) would be required for such separation. By using 2 voltage features, the maximum and minimum voltage, the KNN classifiers failed at least half of the times when identifying a node, leading to an accuracy of only 48.21%, mean values for FAR, FRR, precision and recall were 1.06%, 57.29%, 42.71% and 45.75% respectively, when using 80% of the data for training. In Fig. 11 we depict the confusion matrix for KNN when 80% of the data is used for training and 20% used for testing in the case of using 2 features, i.e., maximum and minimum voltage. The true and the predicted class axis represent the ECU classes, with the letter indicating a distinct vehicle and the number denoting a particular ECU inside a vehicle, e.g., A1 is ECU1 from car A, B2 is ECU2 for car B, etc. The correctly identified ECUs which are marked on the main diagonal and misidentifications are highlighted outside the main diagonal. Locally, inside a single car, the ECUs may be correctly identified but there are clear overlaps between ECUs from different vehicles from the pool of 51 ECUs. The KNN classifier gave better results when compared to the rest of the classifiers on these two features alone. Still, two voltage features are insufficient for separation.

We now extend the classifiers over all the 2000 sampling points for each bit extracted from the ECUPrint dataset [26]. Firstly we split the dataset from each ECU randomly in 20% training data and 80% as testing data and then increase the training percentages up to 80% in steps of 20%. The bar charts shown in Fig. 12 depict the

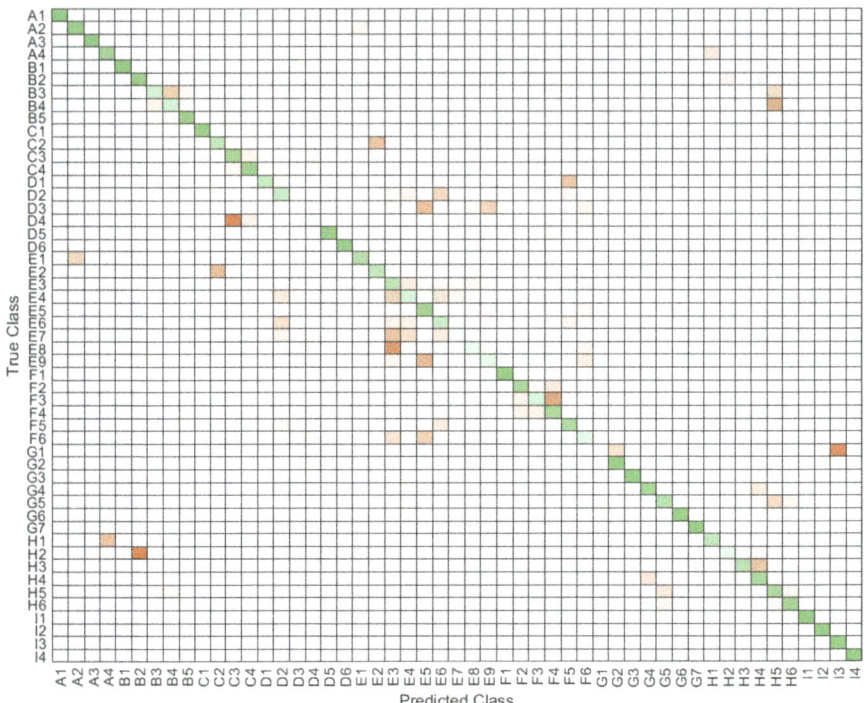

Fig. 11 Confusion matrix for KNN at 80% training on 2 features (max and min value) for the 51 ECUs

FAR obtained when using 20% (a) and 80% (b) as training data for each of the 51 ECUs while applying all 5 machine learning algorithms selected for evaluation. When using 20% of the dataset as training data, the Tree and LD classifiers show poor performance, with the FAR reaching 1.3% in case of Tree and 1.4% in case of LD. The KNN, SVM and NN exhibit better results. For KNN, the FAR is up to 0.046% while in case of SVM, the FAR goes up to 0.05%. In the case of NN the results are slightly better in terms of FAR with the values falling in the 0–0.01% range. When increasing the training data percentage to 80% the FAR values increase for two ECUs in the case of the Tree classifier reaching 0.28% for one of them. Improvements can be observed for the rest of the classifiers with the FAR going up to 0.006% in the case of NN.

In Fig. 13 we depict the FRR for 20% (a) and 80% (b) of the dataset used as training data. The FRR for the Tree and LD classifiers, when using 20% of the dataset for training, is far from acceptable with the FRR reaching 100% for several ECUs. Results obtained for KNN, SVM and NN look more promising with a FRR of up to 39% in the case of KNN, below 20% in the case of SVM (with the exception of one node for which we get a FRR of 62%) and below 10% for NN (except for one ECU with a FRR of 37.5%). In the case of 80% of the data used for training,

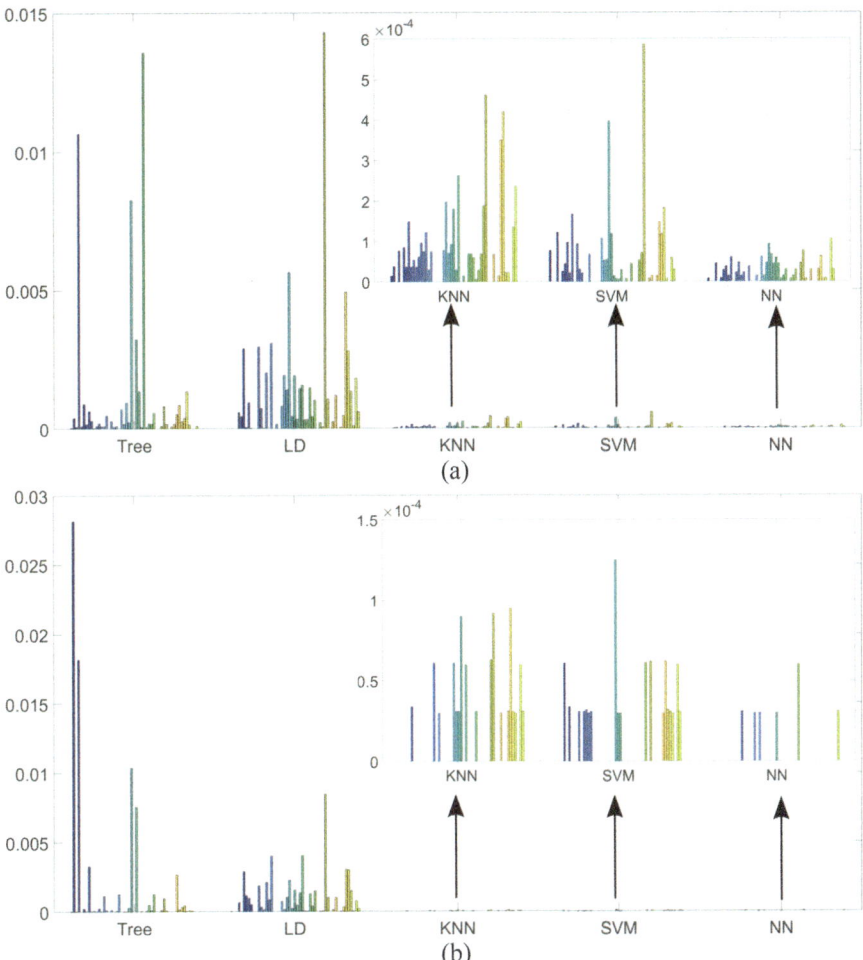

Fig. 12 FAR for 20% and 80% training for 51 ECUs. (**a**) FAR for 20% training and 80% testing. (**b**) FAR for 80% training and 20% testing

the FRR for the Tree classifier reaches 100% for even more ECUs, while in case of LD the results show slight improvements. The KNN and SVM algorithms show now considerable improvements with FRR values below 12% with the exception of a node that exhibits a FRR of 50%. The use of NN proves to be the most reliable with FRRs below 1% at 80% training.

In Fig. 14a we illustrate the confusion matrix obtained when using NN with 20% of the data employed for training. Occasionally, the ECUs may be misidentified, at this lower training rate, but this is a rare event in general. To complete the image, in Fig. 14b we illustrate the confusion matrix for NN when of 80% of the data is used for training and the remaining 20% for testing. In this case misidentifications are

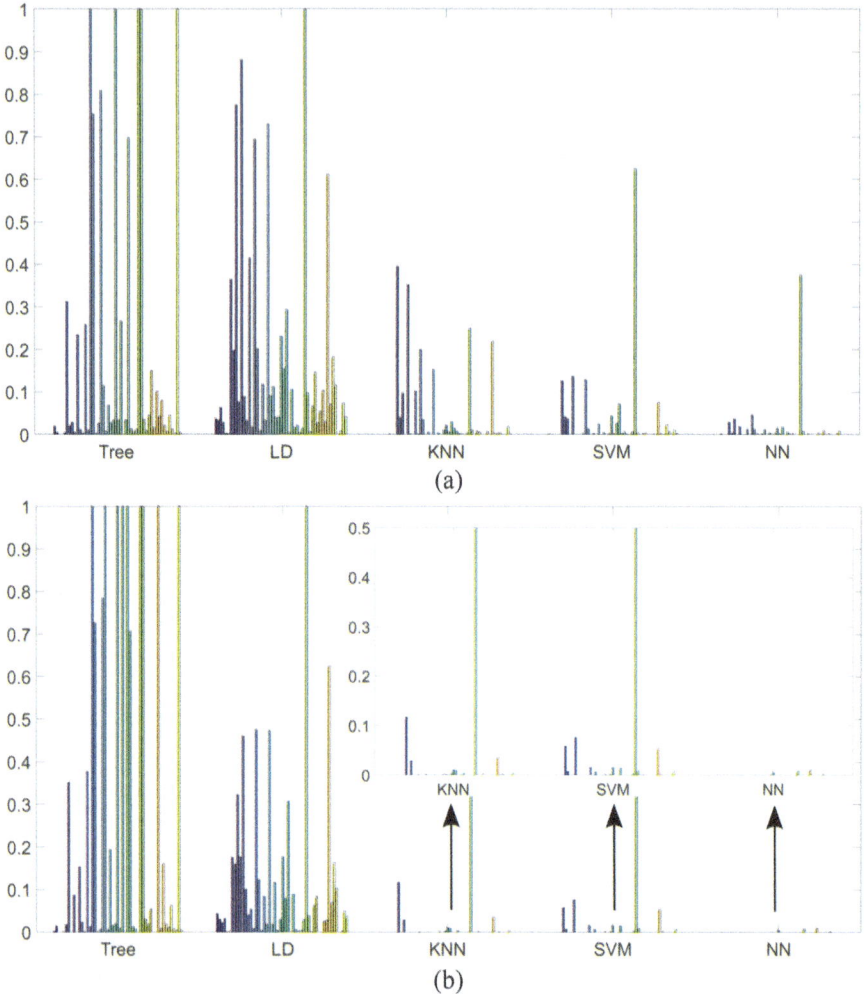

Fig. 13 FRR for 20% and 80% training for 51 ECUs. (**a**) FRR for 20% training and 80% testing. (**b**) FRR for 80% training and 20% testing

very rare, for example C2 is rarely misidentified as D1 and D1 is rarely misidentified as B2. Since this identification is based on samples from a single bit and multiple bits are available in each frame, the misidentification rate will essentially drop to 0 when multiple bits are used.

In Table 2 we summarize as numerical values the results obtained for all metrics, i.e., minimum, mean and maximum value of FAR, FRR, precision, recall and F1-score, for the 5 classifiers when using 20%, 40%, 60% and 80% of the dataset for training. The value *NaN* means division by zero, i.e., for some ECUs the sum between true positive and false negative is zero. It can be easily seen that the results

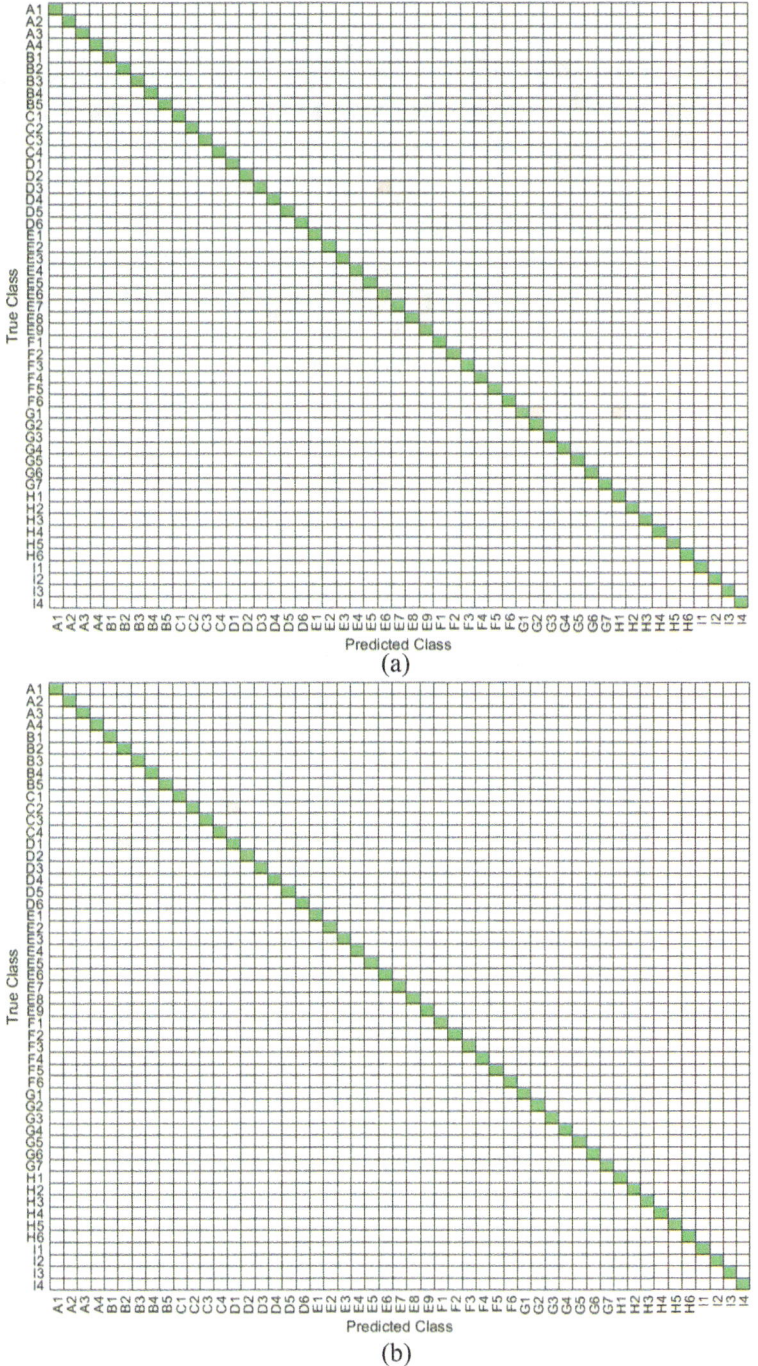

Fig. 14 Confusion matrix for NN with (**a**) 20% and (**b**) 80% training for the 51 ECUs

do improve when increasing the training percentage, but not as significantly as one would expect, which suggests that a small pool of data should be sufficient. Also the KNN, SVM and NN are the classifiers which give the best results with NN clearly outperforming the rest with an accuracy over 99.9%.

5   Discussion and Conclusion

Physical layer based intrusion detection mechanisms show promising results and, as suggested by the results above, neural networks seem to perform much better compared to traditional machine learning algorithms at such tasks. But there are still some challenges that remain to be addressed before such approaches can be included in real-life in-vehicle networks. We now discuss some limitations. The first challenge comes from the high sampling rate required by most of the proposed approaches and the costs involved by integrating the needed HW in a vehicle. The ECUPrint dataset [26] contains data collected at a 500 Ms/s sampling rate which is available on high-end oscilloscopes but not on regular microcontrollers. While some works investigate the use of lower sampling rates [4] or cost-effective HW solutions [15] such approaches have yet to be validated as real-life implementation inside vehicles.

Another challenge comes from the effect of environmental factors on physical layer characteristics. Temperature is one of the factors known to influence the characteristic signalling behavior of electronic circuits. A follow-up paper investigating the robustness of the Scission IDS [14] against environmental factors illustrates the importance of considering these elements in the design of physical layer based CAN IDSs. The authors consider the effect of temperature variations and improve their initial proposal to improve detection accuracy in such circumstances. The only methodology that proved surprisingly good resilience against environmental changes is the localization methodology from CAN-SQUARE [9]. Another IDs proposal entitled SIMPLE [7] accounts for both temperature and voltage variations and their effects. To compensate for these effects the authors of SIMPLE implement a secure update procedure for node fingerprints. And this leads us to the third challenge which is updating fingerprints to account for legit variations in physical layer signalling behavior while reliably protecting the IDS from malicious attempts at compromising the procedure to evade IDS detection.

It is no doubt that voltage information can be used to separate between ECUs, but having in mind the previous challenges, there is still room for further investigations. It is also worth noting that other physical features can be used to fingerprint ECUs. For example, clock skews which were commonly used to fingerprint computers [16] have been also recently used to fingerprint [3] or map ECUs inside a car [18]. Unfortunately, clock skews are very easy to clone [30] by adjusting the local clock of the controller, a reason for which clock skews do not seem to be as secure as voltage fingerprints.

Table 2 Results for the five classifiers at 20–80% training

Trn.	Alg.	Acc.	FAR			FRR			Precision			Recall			F1-score		
			Min	Avg	Max	Min	Avg	Max	Min	Avg	Max	Min	Avg	Max	Min	Avg	Max
20%	Tree	0.949	0	$<10^{-3}$	0.013	0	0.184	1.0	0	0.815	1.0	0.418	NaN	1.0	0.319	NaN	0.999
	LD	0.940	0	0.001	0.0143	0	0.168	1.0	0	0.831	1.0	0.161	NaN	1.0	0.201	NaN	0.995
	KNN	0.994	0	$<10^{-4}$	$<10^{-3}$	0	0.039	0.395	0.604	0.960	1.0	0.868	0.985	1.0	0.736	0.971	1.0
	SVM	0.995	0	$<10^{-4}$	$<10^{-3}$	0	0.0280	0.625	0.375	0.971	1.0	0.951	0.994	1.0	0.545	0.980	1.0
	NN	0.998	0	$<10^{-4}$	$<10^{-4}$	0	0.012	0.375	0.625	0.987	1.0	0.833	0.992	1.0	0.714	0.989	1.0
40%	Tree	0.951	0	$<10^{-3}$	0.013	0	0.194	1.0	0	0.805	1.0	0.279	NaN	1.0	0.229	NaN	1.0
	LD	0.948	0	0.001	0.0128	0	0.132	1.0	0	0.867	1.0	0.394	NaN	1.0	0.434	NaN	1.0
	KNN	0.997	0	$<10^{-4}$	$<10^{-3}$	0	0.023	0.250	0.750	0.976	1.0	0.939	0.992	1.0	0.841	0.983	1.0
	SVM	0.997	0	$<10^{-4}$	$<10^{-3}$	0	0.018	0.416	0.583	0.981	1.0	0.921	0.994	1.0	0.736	0.987	1.0
	NN	0.999	0	$<10^{-4}$	$<10^{-4}$	0	0.001	0.031	0.968	0.998	1.0	0.857	0.996	1.0	0.923	0.997	1.0
60%	Tree	0.941	0	0.001	0.018	0	0.236	1.0	0	0.763	1.0	0.269	NaN	1.0	0.205	NaN	0.999
	LD	0.953	0	$<10^{-3}$	0.008	0	0.127	1.0	0	0.872	1.0	0.127	NaN	1.0	0.142	NaN	1.0
	KNN	0.998	0	$<10^{-4}$	$<10^{-3}$	0	0.0169	0.192	0.807	0.983	1.0	0.898	0.992	1.0	0.875	0.987	1.0
	SVM	0.998	0	$<10^{-4}$	$<10^{-3}$	0	0.011	0.375	0.625	0.988	1.0	0.928	0.996	1.0	0.769	0.991	1.0
	NN	0.999	0	$<10^{-4}$	$<10^{-4}$	0	0.002	0.027	0.972	0.997	1.0	0.973	0.998	1.0	0.923	0.997	1.0
80%	Tree	0.928	0	0.001	0.028	0	0.253	1.0	0	0.746	1.0	0.343	NaN	1.0	0.265	NaN	1.0
	LD	0.951	0	$<10^{-3}$	0.008	0	0.117	1.0	0	0.882	1.0	0.275	NaN	1.0	0.360	NaN	1.0
	KNN	0.998	0	$<10^{-4}$	$<10^{-3}$	0	0.0141	0.500	0.500	0.985	1.0	0.984	0.998	1.0	0.666	0.990	1.0
	SVM	0.999	0	$<10^{-4}$	$<10^{-3}$	0	0.015	0.500	0.500	0.984	1.0	0.973	0.998	1.0	0.666	0.989	1.0
	NN	0.999	0	$<10^{-4}$	$<10^{-4}$	0	$<10^{-3}$	0.010	0.989	0.999	1.0	0.666	0.993	1.0	0.800	0.995	1.0

References

1. AUTOSAR. Specification of secure onboard communication, 4.3.1 edition (2017)
2. Checkoway, S., McCoy, D., Kantor, B., Anderson, D., Shacham, H., Savage, S., Koscher, K., Czeskis, A., Roesner, F., Kohno, T., et al. Comprehensive experimental analyses of automotive attack surfaces. In USENIX security symposium. San Francisco (2011)
3. Cho, K.-T., Shin, K.G.: Fingerprinting electronic control units for vehicle intrusion detection. In 25th {USENIX} security symposium ({USENIX} Security 16), pp. 911–927 (2016)
4. Cho, K.-T., Shin, K.G.: Viden: attacker identification on in-vehicle networks. In Proceedings of the 2017 ACM SIGSAC conference on computer and communications security, CCS '17, pp. 1109–1123. ACM, New York (2017)
5. Choi, W., Jo, H.J., Woo, S., Chun, J.Y., Park, J., Lee, D.H.: Identifying ecus using inimitable characteristics of signals in controller area networks. IEEE Trans. Veh. Technol. 67(6), 4757–4770 (2018)
6. Choi, W., Joo, K., Jo, H.J., Park, M.C., Lee, D.H.: Voltageids: low-level communication characteristics for automotive intrusion detection system. IEEE Trans. Inf. Forens. Secur. 13(8), 2114–2129 (2018)
7. Foruhandeh, M., Man, Y., Gerdes, R., Li, M., Chantem, T.: Simple: single-frame based physical layer identification for intrusion detection and prevention on in-vehicle networks. In Proceedings of the 35th annual computer security applications conference, ACSAC '19, pp. 229–244. Association for Computing Machinery, New York (2019)
8. Groza, B., Murvay, P.: Security solutions for the controller area network: Bringing authentication to in-vehicle networks. IEEE Veh. Technol. Mag. 13(1), 40–47 (2018)
9. Groza, B., Murvay, P.-S., Popa, L., Jichici, C.: Can-square-decimeter level localization of electronic control units on can buses. In European symposium on research in computer security, pp. 668–690. Springer (2021)
10. ISO. 11898-1, Road vehicles - Controller area network (CAN)–Part 1: Data link layer and physical signalling, International Organization for Standardization, Geneva, Switzerland (2015)
11. ISO. 11898-2, Road vehicles - Controller area network (CAN) Part 2: High-speed medium access unit, International Organization for Standardization, Geneva, Switzerland (2016)
12. ISO/SAE. 21434, Road vehicles - Cybersecurity engineering, International Organization for Standardization, Geneva, Switzerland (2021)
13. Kneib, M., Huth, C.: Scission: signal characteristic-based sender identification and intrusion detection in automotive networks. In Proceedings of the 2018 ACM SIGSAC conference on computer and communications security, CCS '18, pp. 787–800. ACM, New York (2018)
14. Kneib, M., Schell, O., Huth, C.: On the robustness of signal characteristic-based sender identification. Preprint. arXiv:1911.09881 (2019)
15. Kneib, M., Schell, O., Huth, C.: EASI: edge-based sender identification on resource-constrained platforms for automotive networks. In Proceedings of the 2020 network and distributed system security symposium, San Diego, CA (2020)
16. Kohno, T., Broido, A., Claffy, K.C.: Remote physical device fingerprinting. IEEE Trans. Depend. Secure Comput. 2(2), 93–108 (2005)
17. Koscher, K., Czeskis, A., Roesner, F., Patel, S., Kohno, T., Checkoway, S., McCoy, D., Kantor, B., Anderson, D., Shacham, H., et al.: Experimental security analysis of a modern automobile. In: 2010 IEEE symposium on security and privacy, pp. 447–462. IEEE (2010)
18. Kulandaivel, S., Goyal, T., Agrawal, A.K., Sekar, V.: Canvas: fast and inexpensive automotive network mapping. In 28th {USENIX} security symposium ({USENIX} Security 19), pp. 389–405 (2019)
19. Limbasiya, T., Teng, K.Z., Chattopadhyay, S., Zhou, J.: A systematic survey of attack detection and prevention in connected and autonomous vehicles. Preprint. arXiv:2203.14965 (2022)
20. Marchetti, M., Stabili, D.: Read: reverse engineering of automotive data frames. IEEE Trans. Inf. Forens. Secur. 14(4), 1083–1097 (2019)

21. Mathworks. Choose classifier options. https://www.mathworks.com/help/stats/choose-a-classifier.html [Online]; Accessed 1 Apr 2022
22. Miller, C., Valasek, C.: Remote exploitation of an unaltered passenger vehicle. Black Hat U S A **2015**, 91 (2015)
23. Murvay, P., Groza, B.: Source identification using signal characteristics in controller area networks. IEEE Signal Process. Lett. **21**(4), 395–399 (2014)
24. Murvay, P.-S., Groza, B.: Tidal-can: differential timing based intrusion detection and localization for controller area network. IEEE Access **8**, 68895–68912 (2020)
25. Ohira, S., Desta, A.K., Arai, I., Fujikawa, K.: PLI-TDC: super fine delay-time based physical-layer identification with time-to-digital converter for in-vehicle networks. In Proceedings of the 2021 ACM Asia conference on computer and communications security, pp. 176–186 (2021)
26. Popa, L., Groza, B., Jichici, C., Murvay, P.-S.: ECUprint—physical fingerprinting electronic control units on CAN buses inside cars and SAE J1939 compliant vehicles. IEEE Trans. Inf. Forens. Secur. **17**, 1185–1200 (2022)
27. Robert Bosch GmbH. CAN Specification, Version 2.0, Robert Bosch GmbH.Postfach 50, D-7000 Stuttgart. **1**, (1991)
28. Rumez, M., Dürrwang, J., Brecht, T., Steinshorn, T., Neugebauer, P., Kriesten, R., Sax, E.. CAN radar: sensing physical devices in CAN networks based on time domain reflectometry. In 2019 IEEE vehicular networking conference (VNC), pp. 1–8. IEEE (2019)
29. SAE International. J1939-11 – Physical layer, 250K bits/s, twisted shielded pair, Sept. (2006)
30. Sagong, S.U., Ying, X., Clark, A., Bushnell, L., Poovendran, R.: Cloaking the clock: emulating clock skew in controller area networks. In 2018 ACM/IEEE 9th international conference on cyber-physical systems (ICCPS), pp. 32–42. IEEE (2018)
31. UNECE. WP.29 Addendum 154 – UN Regulation No. 155, Uniform provisions concerning the approval of vehicles with regards to cyber security and cyber security management system, March (2021)
32. Wolf, M., Weimerskirch, A., Paar, C.: Security in automotive bus systems. In Workshop on embedded security in cars, pp. 1–13. Citeseer (2004)

Spatiotemporal Information Based Intrusion Detection Systems for In-Vehicle Networks

Xiangxue Li, Yue Bao, and Xintian Hou

1 Introduction

Modern vehicles become more networked and intelligent, which not only brings passengers better driving experiences, but also introduces more attack surfaces. Folklore attack surfaces include universal serial bus (USB), Bluetooth, Wi-Fi, cellular network, and other communication interfaces, as well as software vulnerabilities of in-vehicle operating systems [1–6]. For these attack surfaces, there exist complete penetration testing methods to making practical exploit [7]. For in-vehicle electronic control units (ECU)—essential components on CAN (controller area network) bus [8] however, there does not exist mature penetration testing methods.

Crucial information, such as diagnostic, informative, and controlling data, is transmitted over CAN bus to implement various vehicle services, such as autonomous driving and assisted driving [9–11]. Since communication security is not a primary concern at the beginning of CAN design, it is not a surprise that in-vehicle networks are exposed to numerous security threats [12–14]. For example, an arbitrary ECU can hear from any other ECU on the CAN bus without the capability of identifying real sender's identity. This may easily lead to attacks (e.g., packet injection and data manipulation) and risk passenger safety.

X. Li (✉)
East China Normal University and Shanghai Key Laboratory of Trustworthy Computing, Shanghai, China

Shanghai Key Laboratory of Privacy-Preserving Computation, MatrixElements Technologies, Shanghai, China
e-mail: xxli@cs.ecnu.edu.cn

Y. Bao · X. Hou
CATARC Software Testing (Tianjin) Co. Ltd, Tianjin, China
e-mail: baoyue@catarc.ac.cn; houxintian@catarc.ac.cn

© The Author(s), under exclusive license to Springer Nature Switzerland AG 2023
V. K. Kukkala, S. Pasricha (eds.), *Machine Learning and Optimization Techniques for Automotive Cyber-Physical Systems*, https://doi.org/10.1007/978-3-031-28016-0_14

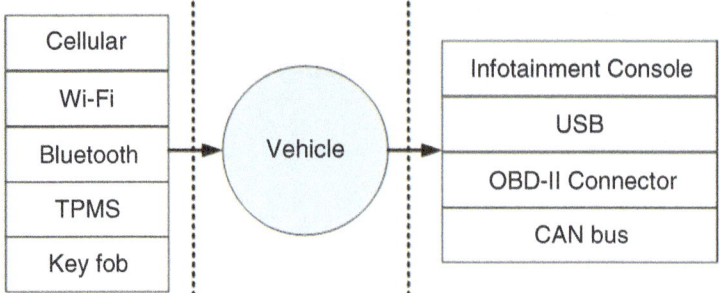

Fig. 1 Attack surface of modern vehicles

The community has made many attempts to secure in-vehicle networks. One research line is to use cryptographic primitives such as Message Authentication Code (MAC), and another is on in-vehicle intrusion detection systems (IDS) [15–22]. Cryptographic methods provide strong security, yet sacrifice system performance (e.g., occupying data fields in CAN packets). In contrast, in-vehicle IDSs [12] do not need to change in-vehicle system structure and can be installed simply (e.g., at in-vehicle gateway) to detect malicious behavior in real time. Given computational power of in-vehicle ECUs, lightweight IDS is preferred.

Attackers usually determine attack surfaces first when they want to crack a car [2, 23, 24]. Modern vehicles have many possible components for attackers to conduct their attack. These components can be simply divided into two types of external and internal as showed in Fig. 1 and might be exploited by the attackers to affect vehicle behaviour. For example, by accessing subscriber identity module or eavesdropping cellular communication channels, the attackers can track vehicle movement [25]. Wi-Fi connection can be built to gain access to in-vehicle network far away (e.g., up to 300 yards) or to break Wi-Fi password and install malicious code on infotainment unit [26]. Key fob can also be exploited by the attackers [27]. They may first crack key fob algorithm with brute force trick and then knockoff the key fob to unlock the door. With malformed key fob, they can send requests so that vehicle immobilizer would be degenerated into a precarious state. There exist many other attack surfaces (e.g., CAN bus, infotainment system, USB port, Bluetooth access, and even tire pressure monitor sensor) for the attackers to perform their attacks successfully [2].

2   In-Vehicle CAN Network

Controller area network (CAN) is a de facto standard for in-vehicle communication [28]. Important vehicle driving information such as diagnostics, sensors, and control data is transmitted via CAN bus for a variety of vehicle driving functions. Two types of CAN packets are defined: standard and extended.

2.1   CAN Packets

In general, each packet has four key fields: Arbitration ID, Identifier extension (IDE), Data length code (DLC), and Data. CAN bus packets also contain two segments with insignificant sense: Cyclic Redundancy Check (CRC) and End of Frame. Figure 2 shows CAN frame format. In CAN communications, there are no addresses and every node (i.e., ECU) connected to CAN bus receives all messages sent over the bus. Messages are identified via their identification fields. Arbitration ID characterizes device ID intended to communicate, and any device can communicate with multiple devices by appending multiple arbitration IDs. If two packets are sent out over the bus simultaneously, the one with lower arbitration ID would win. For standard CAN bus packets, IDE field is set to zero. DLC field specifies data size ranging from 0 to 8 bytes, meaning that maximum data size is 8 bytes at most (if less, some system might pad the packet). Extended CAN bus packets are similar to standard packages, except that they can be concatenated together so that longer IDs will be supported.

From functionality perspective, CAN packets can be divided into two main types: normal and diagnostic. Normal packets carry broadcast message, either command or information. Attackers might reveal these messages with specific semantics and then inject them into CAN bus with specific frequency to affect targeted automobile behavior. Diagnostic packets can put ECUs into diagnostic mode and prevent the devices from communicating on CAN bus (Table 1). This is usually feasible when the vehicle is not moving (or with low speed). ISO-14229 (Table 2) defines a group of diagnostic services used to communicate with vehicles via CAN bus [4]. For example, some security access service would be invoked when an ECU needs to authenticate some sender (e.g., diagnostic tool) sending sensitive diagnostic actions [29].

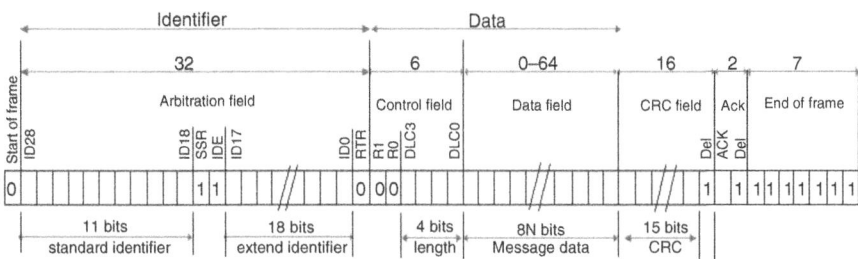

Fig. 2 CAN frame

Table 1 Diagnostic packet example

IDH	IDL	Len	Data							
07	60	08	03	14	FF	00	00	00	00	00
07	68	08	03	7F	14	00	00	00	00	00
07	68	08	03	54	FF	00	00	00	00	00

Table 2 Services defined in ISO-14229

ID	Service name	ID	Service name
10	Diagnostic session control	30	I/O control by local ID
11	ECU reset	31	Routine control
14	Clear diagnostic information	34	Request download
19	Read DTC information	35	Request upload
22	Read data by ID	36	Transfer data
23	Read memory by address	37	Request transfer exit
24	Read scaling data by ID	3d	Write memory by address
27	Security access	3e	Tester present
28	Communication control	83	Access timing parameter
2a	Read data by periodic ID	84	Secured data transmission
2c	Dynamically define data ID	85	Control DTC setting
2e	Write data by ID	86	Response on event
2f	I/O control by ID	87	Link control

Fig. 3 Threat model

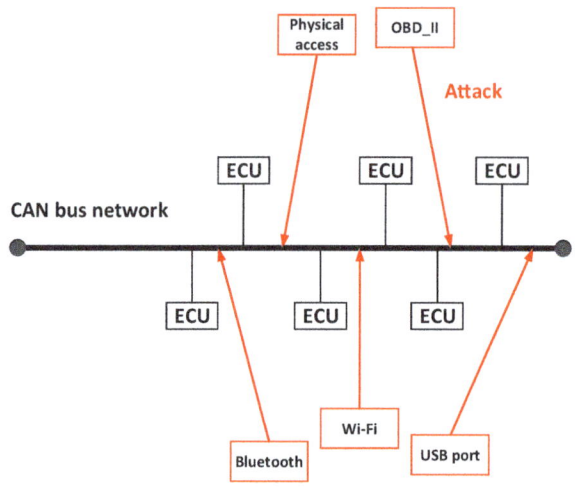

2.2   *Threat Model*

We assume that an attacker can access CAN bus (Fig. 3) and other device interfaces connected to it such as Bluetooth, OBD-II,Wi-Fi, physical access, and USB ports.

There exist several types of attacks. The first one is flooding attack by injecting massive forged messages in a short time period to block in-vehicle network communication. In Fig. 4a, an attacker can send messages withe the highest priority ID 0000 in a short cycle. Then normal messages with ID 02b0 and ID 0316 are delayed.

In fuzzy attack, an attacker randomly sends instructions to cause the vehicle to perform unexpected behavior. In order to implement a fuzzy attack, the attacker

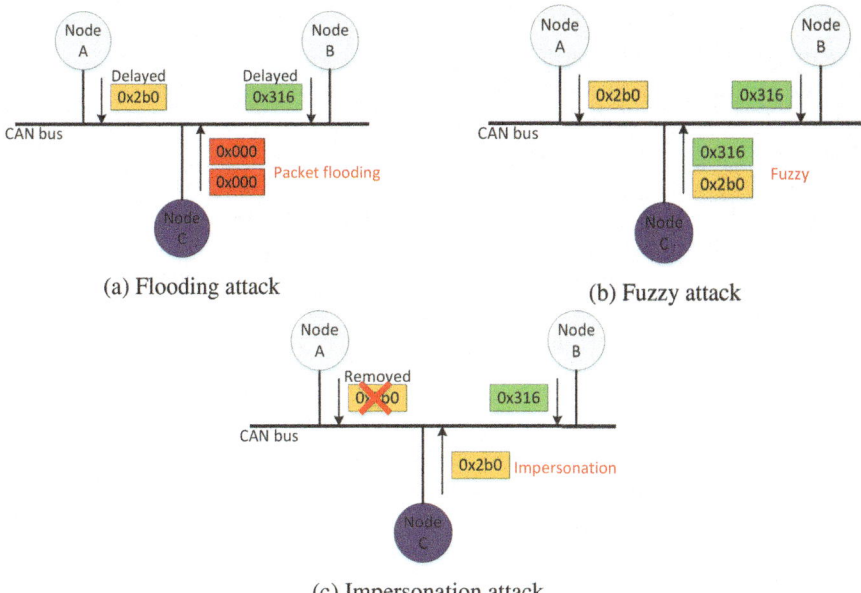

Fig. 4 Attack types. (**a**) Flooding attack. (**b**) Fuzzy attack. (**c**) Impersonation attack

needs to find specific information about the vehicle and ID that can produce unexpected behavior. In Fig. 4b, an attacker injects messages of spoofed random CAN ID and DATA value. Unlike the flooding attack, it paralyzes functions of a vehicle rather than delaying normal messages via bus occupancy.

In impersonation attack, an attacker ceases the function of some ECU and then inserts an ECU for specific purpose. In Fig. 4c, the inserted ECU is disguised as normal one (that stopped working) and can periodically reply to remote frames [13].

3   Related Work

Many attempts have been made in the industry and academia communities to provide security solutions on CAN bus. In-vehicle IDS can detect anomaly behavior on CAN network without the need of changing in-vehicle network structure.

Lan et al. [30] propose an anomaly detecting algorithm based on survival analysis. Song et al. [31] suggest statistical CAN information time interval in anomaly detecting. Although these methods have low computational overhead, they can only detect specific attacks and find unfavorable performance for many attack modes. Machine-learning-based anomaly detecting can solve this problem.

Seo et al. [32] propose double discriminators via generative adversarial networks. It uses two different data sets to train two discriminators, greatly improving

Table 3 Some IDS solutions for in-vehicle CAN network

Key references	Detection strategy	Method
Müter et al. [12]	Anomaly-based	Statistical-based (entropy-based)
Lan et al. [30]	Anomaly-based	Survival
Song et al. [31]	Signature-based	Frequency-based
Seo et al. [32]	Anomaly-based	Generative adversarial networks
Tariq et al. [33]	Anomaly-based	Heuristic algorithm and RNN
Larson et al. [34]	Security rules	Object dictionary of the CANopen protocol
Wang et al. [35]	Anomaly-based	Time series prediction model
Hu et al. [36]	Anomaly-based	Support vector machine
Li et al. [37]	Anomaly-based	Clock drift
Xiao et al. [38]	Anomaly-based	Time series prediction model

detection accuracy. Tariq et al. [33] propose an IDS combining heuristic algorithm and recurrent neural network (RNN). In this method, some features of CAN data are first counted to make a preliminary judgment, then RNN is used for the final judgment. Larson [34] proposes a CAN attack detecting method by security rules. The method is based on the object dictionary of the CANopen protocol; it uses protocol-level security rules to detect illegal ECU behavior, and Larson provides a set of example security rules. Wang et al. [35] propose a time series prediction model that trains different types of instructions in the CAN protocol separately before combining them in the final system. Müter et al. [12] detect attacks by calculating the entropy of normal traffic and abnormal traffic on the CAN bus. Hu et al. [36] use SVM model to detect abnormal state of a vehicle. In [37], Li et al. use the clock drift of an ECU to detect abnormal conditions in abnormal vehicles. In [38], Xiao et al. propose an early warning using convLSTM to predict time series deviations.

Table 3 compares some existing IDS solutions. In [35], the instruction ID of CAN frame is separated, resulting in a loss of part of the information. In [12], the entropy-based method can only perform preliminary statistics and detect only some attack methods (missing many others). In [36], directly detecting abnormal state of a vehicle with SVM requires excessive computational resources, and it is difficult to ensure real-time monitoring. For the method in [37], the vehicles produced by different manufacturers have different ECU clock drift features and it would be indispensable to reanalyze these features for each vehicle brand.

4   Weighted State Graph from CAN Frames

Modeling in-vehicle data plays a fundamental role for detecting vehicle anomalies. The model should reflect the relationships of different types of data and their patterns. Different CAN systems have different bit transfer rates, and the bit transfer rate is unique and fixed for a given system. This section introduces a model (built offline based on historical data) to identify (online) newly arrived data stream.

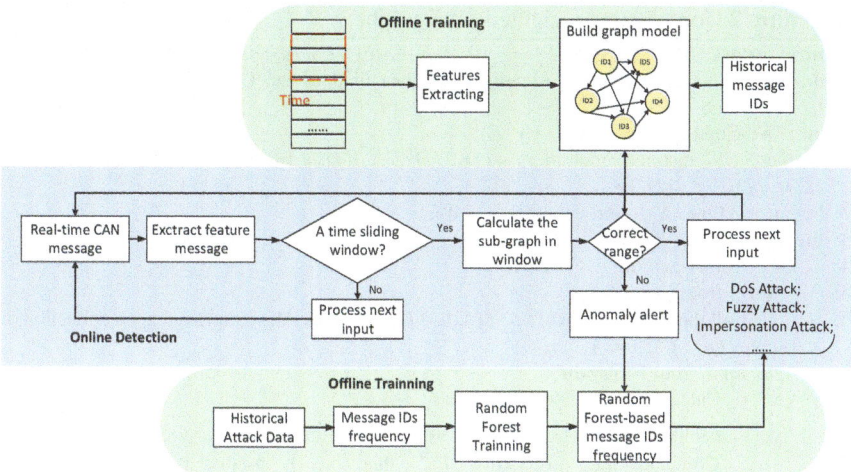

Fig. 5 Model architecture

Figure 5 shows the model architecture de-coupled into two stages, offline training and online detection. The former further includes weighted state graph construction and random forest model training, and the latter can be performed in a lightweight manner.

4.1 Constructing Weighted State Graph: Offline

The state relations of streamed in-vehicle data can be modeled as a weighted state graph $G = (V, \mathcal{E})$. Herein, the state represents the value of message ID and all states constitute the vertex set V. Specifically, we adopt discrete states to quantify status (ID values) and employ weighted edges (connections between states) to measure the relationship between states. The graph is constructed based on statistics of historical data and can be regarded logically as an accurate model that delineates the underlying vehicle. Next we narrate three steps in constructing G.

(1) States extracting: extract all different message IDs from the vehicle historical data as distinct states. Thereby we get the vertex set V.

(2) Feature extracting: extract features from historical data, i.e. timestamps, message IDs, data segments for each packet. We denote $y = [t, id, data]$, where t represents the timestamp, id the ID value, and $data$ the data field.

(3) Weight computing: fix a time sliding window (e.g., 100 ms) for data extraction and calculation. Within 100 ms, count the time offset between two message IDs, the number of IDs, and the probability of occurrence of bit 1 in all data segments between the two messages. A message ID may appear multiple times within the sliding window. For this, we reckon up distinct or identical message IDs that appear

---

Algorithm 2 Constructing weighted state graph

Require: Attack free sample $\mathcal{M} = \{(t_1, id_1, data_1), (t_2, id_2, data_2), \ldots\}$ where id_ℓ and $data_\ell, \ell = 1, 2, \ldots$ are the identifier and data field of the CAN frame at time t_ℓ, and $t_1 < t_2 < \ldots$;

Ensure: Weighted state graph $\mathcal{G} = (\mathcal{V}, \mathcal{E})$;

1: Extract all distinct identifiers $ID_1, ID_2, \ldots, ID_n$ from $\mathcal{M}$ and set $\mathcal{V} := \{ID_1, ID_2, \ldots, ID_n\}$;

2: Segment $\mathcal{M}$ into L sliding windows $W_1, W_2, \ldots, W_L$;

3: **for** ℓ from 1 to L **do**

4:     **for** i from 1 to n **do**

5:         **for** j from i to n **do**

6:             Call the function STATTRANS(ID_i, ID_j, W_ℓ) (see Algorithm) and get a collection of m 2-tuples $(p_\lambda, q_\lambda), \lambda = 1, \ldots, m$;

7:             **for** λ from 1 to m **do**

8:                 Calculate time offset $TO_{W_\ell}^\lambda := t_{q_\lambda} - t_{p_\lambda}$;

9:                 Count the number of transited states $N_{W_\ell}^\lambda := q_\lambda - p_\lambda$;

10:                Compute the probability $P_{W_\ell}^\lambda$ that bit 1 occurs in the data fields $data_{p_\lambda}, data_{p_\lambda+1}, \ldots, data_{q_\lambda}$;

11:                Set $(ID_i \to ID_j)_{W_\ell}^\lambda := [TO_{W_\ell}^\lambda, N_{W_\ell}^\lambda, P_{W_\ell}^\lambda]$;

12:             **end for**

13:             Set $(ID_i \to ID_j)_{W_\ell} := \frac{[TO_{W_\ell}^{(i \to j)}, N_{W_\ell}^{(i \to j)}, P_{W_\ell}^{(i \to j)}]}{m}$ where $TO_{W_\ell}^{(i \to j)} = \sum_{\lambda=1}^{m} TO_{W_\ell}^\lambda$, $N_{W_\ell}^{(i \to j)} = \sum_{\lambda=1}^{m} N_{W_\ell}^\lambda$, $P_{W_\ell}^{(i \to j)} = \sum_{\lambda=1}^{m} P_{W_\ell}^\lambda$;

14:         **end for**

15:     **end for**

16: **end for**

17: Set the vectorized weight $w_{ij} := \frac{[TO^{(i \to j)}, N^{(i \to j)}, P^{(i \to j)}]}{L}$ where $TO^{(i \to j)} = \sum_{\ell=1}^{L} TO_{W_\ell}^{(i \to j)}$, $N^{(i \to j)} = \sum_{\ell=1}^{L} N_{W_\ell}^{(i \to j)}$, $P^{(i \to j)} = \sum_{\ell=1}^{L} P_{W_\ell}^{(i \to j)}$, $i = 1, \ldots, n, j = i, \ldots, n$;

18: $\mathcal{E} = \{e_{ij} | e_{ij} := ID_i \to ID_j \wedge |e_{ij}| := w_{ij}, i = 1, \ldots, n, j = i, \ldots, n\}$;% $|e_{ij}|$ denotes the vectorized weight of the directed edge e_{ij};

19: **return** $\mathcal{G}$.

---

next to each other, and then average their features. Take Fig. 6 as an example, there are four ID states, and the message sequence is $[ID_1, ID_2, ID_3, ID_1, ID_2, ID_3, ID_4, \ldots]$. So we can construct a weighted directed graph that has the edges $ID_1 \to ID_1$, $ID_1 \to ID_2/ID_3/ID_4$, $ID_2 \to ID_2/ID_3/ID_4$, $ID_3 \to ID_3/ID_4$, $ID_4 \to ID_4$. As formulated in Eq. (1), the (vectorized) weight is determined by three features, time offset (TO), the number of IDs (N), and the probability of occurrence of data bit 1 (P). Then we get $\mathcal{G} = (\mathcal{V}, \mathcal{E})$: $\mathcal{V}$ has four states and ten weighted edges constitute $\mathcal{E}$.

$$(ID_i \to ID_j)_W^k = [TO_W^k, N_W^k, P_W^k] \qquad (1)$$

where W represents a sliding window, k represents the interval of the k-th ID_i and ID_j in a window. If the order (that two distinct IDs emerge in the message sequence) is irregular (e.g., Fig. 7), we mainly extract the two ID intervals in the dashed block. That is, the ID_B closest to ID_A is always selected in chronological order. We address that there does not exist the state explosion problem according

---

Algorithm 3 Function STATTRANS(ID_i, ID_j, W_ℓ)

Require: Two identifiers ID_i, ID_j, and a sliding window W_ℓ = $\{(t_{\ell_1}, id_{\ell_1}, data_{\ell_1}), (t_{\ell_2}, id_{\ell_2}, data_{\ell_2}), \ldots, (t_{\ell_\omega}, id_{\ell_\omega}, data_{\ell_\omega})\}, t_{\ell_1} < t_{\ell_2} < \ldots < t_{\ell_\omega}$;

Ensure: A collection Φ of 2-tuples $(p_\lambda, q_\lambda), \lambda = 1, \ldots, m$;

1: $\Phi = \emptyset$;
2: **for** κ from 1 to ω **do**
3:    **if** $id_{\ell_\kappa} = ID_i$ **then**
4:        Find the first ID_j (say $id_{q_{\lambda_1}}$) from $\{id_{\ell_{\kappa+1}}, \ldots, id_{\ell_\omega}\}$;
5:        Find the last ID_i (say $id_{p_{\lambda_1}}$) from $\{id_{\ell_K}, id_{\ell_{\kappa+1}}, \ldots, id_{q_{\lambda_1}}\}$;
6:        $\Phi = \Phi \cup \{(p_{\lambda_1}, q_{\lambda_1})\}$
7:    **end if**
8: **end for**

Fig. 6 weighted state graph

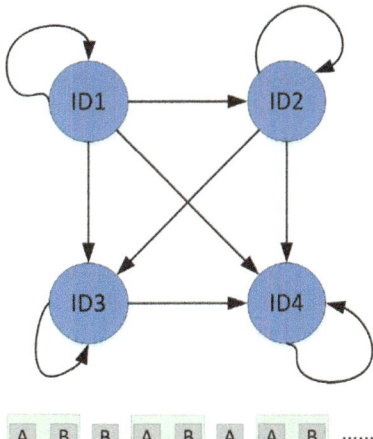

Fig. 7 two IDs' sequence

to the rationale of our trick (which can manipulate any patters of CAN traffic effectively).

Herein, the weighted state graph does not need the directed edge from ID_B to ID_A (once we obtain the directed edge from ID_A to ID_B, meaning that the vectorized weight is already produced). In fact, the frequency at which each ECU sends a message is fixed, given a specific CAN system. So as time goes on, the connection from ID_A to ID_B suffices to imply the connection from ID_B to ID_A. In order to save time and space complexity, we only require one-way connections in the graph.

Algorithm 2 illustrates the concrete steps in constructing $\mathcal{G}$. Its time complexity is related to the number of identifiers n, the number of sliding windows L, and the number of edges m, and bounded as $O(n^2 mL)$.

4.2 Segmenting and Learning: Offline

Random forest (or random decision forest) is an ensemble learning method for classification tasks by constructing a multitude of decision trees at training phase and outputting the class mode of individual trees. For unbalanced data set, it balances the error well, and during the training process, it is possible to detect the interaction between features and achieve a simpler implementation [39].

Given original dataset, i.e., historical CAN traffic, we segment the traffic into a slice of sliding windows. For each sliding window, we squeeze out n features which are then bundled together to form a row vector. Suppose that there are m sliding windows and we get an $m \times n$ matrix (called it as dataset D as a slight abuse of term). We feed D into random forest algorithm and thus accomplish the learning process.

(1) Data preprocessing: given a sliding window, we vectorize the message IDs (Fig. 8). We have three types of intrusion samples, namely flooding attack, fuzzy attack, and impersonation attack, which are labeled as 0, 1, and 2 respectively. In our experiment, we have 46 different message IDs from historical data. Therefore, there are 46 features totally (i.e., $n = 46$).

(2) Model training: we use ID3 (Iterative Dichotomiser 3) algorithm (by iterating through every unused attribute of the set D and calculating the information gain of that attribute) to train random forests. For the features $\{x_1, x_2, \ldots, x_n\}$, the information gain of D is: $G(D, x_i) = H(D) - H(D|x_i)$, where

$$H(D) = -\sum_{k} \frac{|c_k|}{D} log_2 \frac{|c_k|}{D} \tag{2}$$

$$H(D|x_i) = \sum_{j=1}^{l} \frac{D_j}{D} H(D_j) \tag{3}$$

Herein, D_j is sample subset of D where the feature x_i takes the j-th value, $H(D_j)$ is the entropy of D_j, $H(D|x_i)$ is conditional entropy of D for x_i, c_k is a subset of samples belonging to the k-th class in D, l is the number of the feature x_i.We use bootstrap sampling. For a given dataset containing m attack samples, we first randomly take a sample into the sampling set, and then put the sample back into

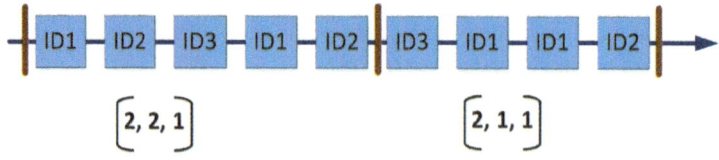

Fig. 8 Data preprocessing of random forest model

Algorithm 4 Online detection

1: Read the CAN network message online and store in the cache. Once the traffic reaches the size of the sliding window, a weighted state subgraph S is constructed.
2: Compare S with $\mathcal{G}$ (produced in Sect. 4.1) to obtain the intrusion detection result. The coming section will present more details.
3: Update system intrusion detection logs and trigger an alert if an attack is detected.
4: Pass the alert data into the trained random forest model to further determine which attack type.

---

the initial dataset, that is, simple random sampling with a return. In such a way, the specified number of samples are repeated to obtain a set of samples that meet the requirements. The training parameters of the classifier are detailed below: Max depth = 8, Max features = 6, Min samples leaf = 100, and entropy is used as the criterion.

4.3   Segmenting and Detecting: Online

We aim to perform online intrusion detection on streaming data. Not all message IDs appear in the data generated over a period of time, and the resulting weighted state graph is a subgraph (of $\mathcal{G}$ generated in Sect. 4.1). Once the stream data is generated, we construct a subgraph S from the data recursively. The concrete construction method is similar to that in constructing $\mathcal{G}$. The weighted edges of S are connected in exactly the same way as $\mathcal{G}$, but the weights need to be recalculated.

Given real-time CAN traffic, we recursively performs Algorithm 4 for online detection. The detecting is lightweight as only one sliding window is involved.

4.4   Scoring the Subgraphs for Sliding Windows

We evaluate driving process for each short period of time. For S, we compute an anomaly score $f(S)$.

$$f(S) = w_1 \frac{\sum_{i=1}^{m}(TO_1^i - TO_0^i)}{m} + w_2 \frac{\sum_{i=1}^{m}(N_1^i - N_0^i)}{m} + w_3 \frac{\sum_{i=1}^{m}(P_1^i - P_0^i)}{m}$$

$$(4)$$

Herein, m is the number of edges in the subgraph S. TO_1^i, N_1^i, P_1^i denote the connection value of the stream data ($i = 1, \ldots, m$), and TO_0^j, N_0^j, P_0^j represent the edge connection value in the graph $\mathcal{G}$ ($j = 1, 2, \ldots, |\mathcal{E}|$). w_1, w_2, and w_3 represent the proportion of TO, N, and P, respectively. The score is compared with a predefined threshold δ. If $f(S) > \delta$, then the subgraph S is marked as an anomaly. In practice, it is suggested to choose the threshold δ according to the distribution of score $f(S)$.

We use the stochastic gradient decent (SGD) method to optimize w_1, w_2, w_3. S is compared with $\mathcal{G}$, and the difference is recorded (i.e., $[TO_1 - TO_0, N_1 - N_0, P_1 - $

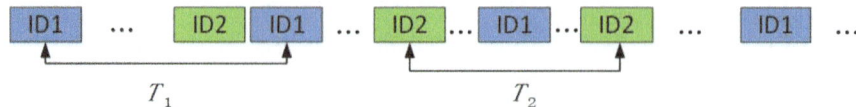

Fig. 9 The distance between two different IDs over time

P_0]). The plurality of sets of stream data form a matrix $\mathbf{X}$, the normal data is marked as 1, and abnormal as -1. We can use the SGD method to obtain the fractions by support vector machine (SVM): $min \; \frac{1}{2} \|\mathbf{w}\|^2 \; s.t. \; Y_i(\mathbf{w}^T\mathbf{X}_i+b) \geq 1, i = 1, \ldots, N$. Herein, $\mathbf{w}=[w, w_2, w_3]$, and b represents the intercept. $\mathbf{X}_i$ represents difference vector (i.e., i-th row vector of $\mathbf{X}$ and of the form $[TO_1 - TO_0, N_1 - N_0, P_1 - P_0]$). N indicates the number of samples. Y_i denotes the classification, the normal data is 1, and the abnormal data is -1.

If no such message ID exists in $\mathcal{G}$, the graph would be updated automatically. There is one more point. For a wide variety of commercial CAN bus systems in practice, one can not rule out the possibility that captured at the choke points, the CAN traffic might be parameterized differently and thus show polymorphic forms in data-driven continuous experimentation. As an example, the beginning of the messages sent by the ECUs may not coincide at the choke points of keying in the ignition and starting the engine. It is thus expected that the weighted state graph can be evolved over time. The feature is known as concept drift in anomaly detection for streaming data [40] and exemplified in Fig. 9. Assume that the order of the two IDs appears as shown above, and T_1 and T_2 are the fixed intervals of ID_1 and ID_2, respectively. However, the alternation between ID_1 and ID_2 is not constant and may vary considerably (even regularly or in cycle). In this scenario, for example, given two states, ID_i and ID_j, when measured in different times, the vectorized weight of the edge $(ID_i \rightarrow ID_j)$ could be different. As a result, the vectorized weight of $(ID_i \rightarrow ID_j)$ of the newly constructed subgraph S may not be much different from that in $\mathcal{G}$ when an attack occurs. And this may catalyze the unexpected consequence that the system cannot function properly (as such change could lower the performance of the detector by causing a multitude of false alarms). We address this problem by recourse to a unit outsourced to the cloud to maintain and update the model. The cloud maintains an array that records the connection number between states. It updates the array when each data arrives and then periodically calculates the connection weights of the model according to the array. The fractions w_1, w_2, w_3 also need to be reformed periodically.

4.5 Evaluation

Our data is retrieved from public source [41]. It can be categorized as four types (and each contains 2,369,868, 656,579, 591,990, 2,350,827 samples respectively). The corresponding three kinds of attack scenarios (flooding attack, fuzzy attack, impersonation attack) are illustrated in Fig. 4. Attack free samples (generated by the vehicle under normal driving and no attack behavior) are used for training, and the remaining for evaluation. For the last three types, the attacks are injected from 250 s.

4.5.1 Evaluation Metrics

For anomaly detection problems, evaluation indicators are much involved. Main reason is that the dataset of general anomaly detection is unbalanced, i.e., the number of normal data might be much larger than that of abnormal data. Misclassification of normal samples into attack samples is beyond the function the system should have and might cause huge losses. Further, IDS mainly gears toward anomaly detection. In our experiments, we use the detection precision, recall rate and F_1-score to judge the effectiveness for the random forest model:

$$precision = \frac{TP}{TP + FP} \qquad recall = \frac{TP}{TP + FN}$$

$$F_1 = 2 \times precision \times recall/(precision + recall)$$

(5)

where TP is true positive, FP is false positive, and FN is false negative.

4.5.2 Experiment Analysis

We hope that IDS detection accuracy would be as high as possible and false positive rate (FPR) as low as possible. An attack block of a time sliding window is detected each time. Here, we particularize the sliding window as 50, 100, 150 and 200 ms respectively. As exhibited in Table 4, the model shows promising results. For attack block = 50 ms and the threshold = 93, the detection precision is 0.988 for each detection, FPR is 0.02, but the recall is relatively low (0.68). For attack block = 100 ms and the threshold = 99, the detection precision is 0.993, FPR is 0.02, at the same time the recall is 0.975. For time sliding windows of 150 and 200 ms, the detection precision and recall reach 100%, FPR is 0 at $\delta = 114, 109$ respectively.

To better grasp model applicability, we manually scrutinize the detection results. Figure 10 illustrates the FPR evaluation of the model under different score threshold δ and different time sliding windows. A long tail effect can be perceived from the curve as the FPR shows a trend of sharp decreasing with the increase of δ. When setting a smaller threshold, many false alarms would be raised. When δ is relatively small, the FPRs of time sliding windows of 50 and 100 ms are lower than those of 150 and 200 ms. The discrimination between normal and abnormal of 50 and 100 ms time sliding windows is higher than that of 150 and 200 ms.

The trend of the FPR curve also shows the distribution characteristic of the anomaly score. In practice, a rational suggestion is to set δ based on statistical principles. E.g., let $\mu - k*\sigma < \delta < \mu + k*\sigma$, where μ is the mean value of

Table 4 Performance w.r.t. sliding window sizes

W	Accuracy	FPR	recall	δ
50 ms	0.988	0.02	0.68	93
100 ms	0.993	0.02	0.975	99
150 ms	1	0.0	1	114
200 ms	1	0.0	1	109

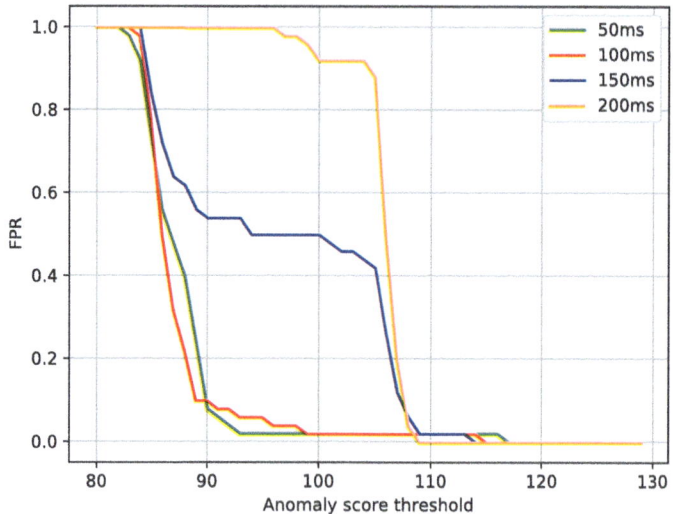

Fig. 10 FPR w.r.t. threshold values

Table 5 Random forest model w.r.t. different sliding window sizes

Attack type	Precision				Recall			
	50 ms	100 ms	150 ms	200 ms	50 ms	100 ms	150 ms	200 ms
Flooding	0.90	0.93	1	1	0.86	0.98	0.99	0.99
Fuzzy	0.93	0.98	0.99	1	0.95	0.99	0.99	1
Impersonation	0.94	0.95	1	0.99	0.95	0.96	0.98	1

anomaly score and σ is the standard deviation of the score. Usually, one may take $k \in [0.1, 2]$.

For random forest model, we also define the time sliding window as 50, 100, 150, and 200 ms to classify the abnormal data blocks. Due to the difference in information gain characteristics of different attack flows, random forest classification has significant effect. As demonstrated in Table 5, 50 ms time sliding window has lower classification precision (≤ 0.95); classification effect is improved when time sliding window is 100 ms, recall of three attack types are above 0.95; and for 150 and 200 ms of time sliding windows, almost all attacks are correctly identified.

Information entropy primarily reflects the changes within the data. We further compare the model with an entropy-based detector proposed in [42], which establishes a fixed time (or number) as a sliding window. Consider fixed time sliding

Table 6 Comparisons between the model and entropy-based detector [42]

Attack type	The model			Entropy-based model		
	Precision	Recall	F_1	Precision	Recall	F_1
Flooding	0.98	1	0.99	1	1	1
Fuzzy	0.98	0.98	0.98	0.89	0.74	0.81
Impersonation	0.98	0.94	0.96	0.93	0.97	0.95

window. We activate 100 ms as the size of the sliding window. As manifested in Table 6, the information entropy method has a detection rate of 100% for flooding attack, yet the detection rate is slightly lower for both fuzzy attack and impersonation attack. For fuzzy attack, our model exhibits advantageous metrics of precision, recall and F_1 value over the entropy-based method. One can also see that for impersonation attack, the precision and F_1 value of our model are more absorbing than those of the entropy-based method (yet the recall metric has the reverse effect).

5 Spatiotemporal Information Based IDS for In-Vehicle Networks

Next we see an IDS using the Convolutional LSTM Network-ConvLSTM [43]. The IDS utilizes comprehensively the information of multiple dimensions (e.g., time and space information) in regards to CAN protocol so that intrusion detection can be implemented effectively. It can not only practically detect traditional multiple attack modes (e.g., injection attack, data manipulation attack, etc.), but also support efficient detection of unknown attacks. The detection model is fully data-driven and does not require any domain knowledge about CAN protocol. It only needs a small number of iterations in the training process. More precisely, it provides strong robustness and can perform fast online learning of packet features (even in the vehicle's running state) to implement new protocols for intrusion detection (enabling to capture fresh vehicle states) even if the underlying communication protocol of the vehicle was updated. The model is based on real attack free data and attack data is only used as a criterion for model evaluation. Recall that most existing methods need to make judgments based on attack data during the detection process, but the used attack data (in experiments) is unlikely to be in correspondence with real attack data.

Figure 11 shows anomaly detection process for in-vehicle network. First, Con-vLSTM model is trained with attack free dataset (i.e., no attack occurs), then CAN data predicted by the ConvLSTM model is used to calculate correlation coefficients with real generated data, and finally abnormal attack is detected according to the difference range of the correlation coefficients.

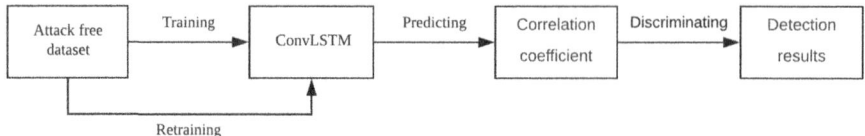

Fig. 11 Anomaly detection process

5.1 Unknown Attack and Self-evolving Model

As a vehicle travels under distinct road conditions, CAN data generated is subject to real-time changes due to factors such as weather and traffic conditions. Namely, CAN bus data exhibits distinct characteristics under different road conditions. For example, CAN data of sunny days is different from that of snowy days. If some model just learns CAN data that runs normally on sunny days, it is much likely that the data (being normal on rainy days) will be regarded as intrusion. Surely we can't collect driving data of a vehicle under all road conditions, so our ConvLSTM model should have the ability to be self-evolving in real time as the vehicle travels. On the other hand, it is impossible to exhaust and collect CAN data under all attack modes, and smart attackers keep inventing different attack modes. To ensure the ability of detecting unknown attack, it is necessary to use the past running state of in-vehicle network for warning purpose. We thus introduce retraining step to our model.

Our IDS can adjust the features of attack free data in real time as a vehicle travels. This adjustment allows better definition of attack free data to identify attack data. We use ConvLSTM model to generate predicted CAN data and real CAN data in calculating Pierce correlation coefficient, and identify intrusion data by coefficient threshold. At the same time, new ConvLSTM model is trained to adapt to the change of the threshold under different road conditions. Once an in-vehicle IDS model is fixed, it cannot meet safety requirements of vehicles in a distinct or more complex environment. We propose a model by calculating the correlation coefficient of CAN data, which can not only meet driving requirements of vehicles in myriad road conditions, but also detect unknown attack data.

Given CAN dataset, our concern is on the arbitration field, control field, data field and timestamp. A timestamp is the time elapsed from starting vehicle ignition to the generation of the data. All data are normalized after a conversion from binary to decimal according to $y_{normali} = \frac{x - \min}{\max - \min}$, where x represents data being normalized, min (max) the minimum (maximum, resp.) value of an attribute in the dataset.

5.2 Spatiotemporal Information from CAN Frames

Long short-term memory (LSTM) units are units of a recurrent neural network (RNN) [44]. An RNN composed of LSTM units is often called an LSTM network (that could solve long-term memory dependence problems). A common LSTM

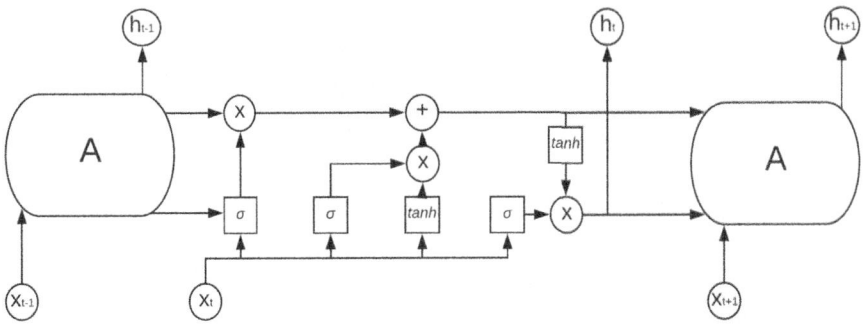

Fig. 12 LSTM model

unit is composed of a cell, an input gate, an output gate and a forget gate. The cell remembers values over arbitrary time intervals and the three gates regulate information flow into and out of the cell. LSTM networks are well-suited to classifying, processing and making predictions based on time series data, since there can be lags of unknown duration between important events in a time series. LSTM deals with the exploding and vanishing gradient problems in training traditional RNNs. Relative insensitivity to gap length is an advantage of LSTM over RNNs, hidden Markov models and other sequence learning methods in a sea of applications. Multiple LSTMs can be stacked and temporally concatenated to form more complex structures.

The LSTM model is shown in Fig. 12. One of its main innovation lies in that memory cells c_t are added to collect state information. Once new input data arrives, the information will be accumulated in the cell if the inputs gate i_t is activated. At the same time, if the forgotten gate f_t is open, the past information c_t will be forgotten. The output gate O_T determines whether the final output c_t is transmitted to h_t. Equations (6)–(10) defines detailed functions.

$$i_t = \sigma(W_{xi}x_t + W_{hi}h_{t-1} + W_{ci}c_{t-1} + b_i) \tag{6}$$

$$f_t = \sigma(W_{xf}x_t + W_{hf}h_{t-1} + W_{cf}c_{t-1} + b_f) \tag{7}$$

$$C_t = f_t \circ c_{t-1} + i_t \circ tanh(W_{xc}x_t + W_{hc}h_{t-1} + b_c) \tag{8}$$

$$o_t = \sigma(W_{xo}x_t + W_{ho}h_{t-1} + W_{co} \circ c_t + b_o) \tag{9}$$

$$H_t = o_t \circ tanh(c_t) \tag{10}$$

5.2.1 Convolutional LSTM Network

Traditional IDS for in-vehicle network based on time series prediction can only perform training according to the order in which each ECU generates data packets [35] (not according to the sequence of data packets sent by all ECUs to CAN bus).

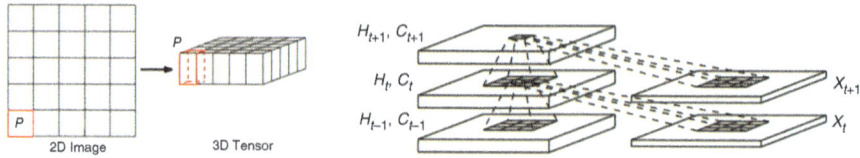

Fig. 13 ConvLSTM model

Thus each ECU has to build a model for its own, and finally all models are combined to implement intrusion detection. However, different ID data packets in the CAN protocol need to work together, so the training method based on go-it-alone IDs might lose too much information. Convolutional LSTM Network (ConvLSTM) [43] can take into account of both timing and spatial correlations.

Traditional LSTM appears incompetent in handling spatiotemporal data (i.e., full connections in input-to-state and state-to-state transitions). The Convolutional LSTM network (ConvLSTM) is expected to remove the obstacle (Fig. 13) [43]. Main difference between the two methods is that all the inputs $\chi_1, \ldots, \chi_t$, cell outputs $C_1, \ldots, C_t$, $H_1, \ldots, H_t$, and gates i_t, f_t, o_t are 3D tensors (Eqs. (11)–(15)). ConvLSTM can predict future data by drawing on current input data and past input data. Convolution operations can obtain combined information between multiple data packets.

$$i_t = \sigma(W_{xi} * \chi_t + W_{hi} * H_{t-1} + W_{ci} \circ C_{t-1} + b_i) \qquad (11)$$

$$f_t = \sigma(W_{xf} * \chi_t + W_{hf} * H_{t-1} + W_{cf} \circ C_{t-1} + b_f) \qquad (12)$$

$$C_t = f_t \circ C_{t-1} + i_t \circ tanh(W_{xc} * \chi_t + W_{hc} * H_{t-1} + b_c) \qquad (13)$$

$$o_t = \sigma(W_{xo} * \chi_t + W_{ho} * H_{t-1} + W_{co} \circ C_t + b_o) \qquad (14)$$

$$H_t = o_t \circ tanh(C_t) \qquad (15)$$

where '$*$' denotes the convolution operator, and '$\circ$' the Hadamard product.

ConvLSTM can combine data information from multiple packets and further perform intrusion detection by convolution operations. The larger the convolution kernel is, the more data packets are combined. The coming experiments define the convolution kernel size as 3×3. The number of cells at first/second/third layer is set as 128/64/1 respectively, and the corresponding dropout (the probability that cells in this layer will be activated) is 0.5, 0.5, and 1.

5.2.2   Robust and Self-evolving IDS for In-Vehicle Network

We use arbitration field, control field, data field and timestamp of CAN frames. The data field is of 8 dimensions, and each of the other fields is 1-dimension. Thus, a data packet is 11-dimension. We organize training sample array into a form of (10, 10, 11): the first array represents 10 time steps, the second represents a combination of 10 data packets, 11 represents a packet with 11 dimensions. One thousand samples

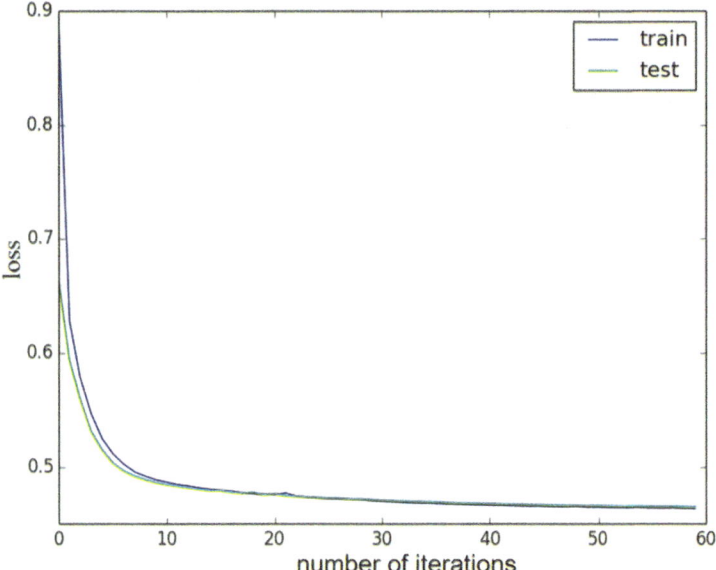

Fig. 14 ConvLSTM model training

are used as training set to produce ConvLSTM model optimized by using cross entropy as loss function: $L_H(x, \bar{x}) = -\sum_{i=1}^{n} x_i log \overline{x_i} + (1 - x_i) log (1 - \overline{x_i})$, where x ($\bar{x}$) denotes the real (predicted, resp.) data of next time step generated by current input sample (ConvLSTM, resp.). To achieve online learning, the model only iterates 60 times (Fig. 14) during training, and need not a complete convergence point.

Data prediction is performed using a trained ConvLSTM model. We calculate Pearson correlation between the predicted data and the real data:

$$r = \frac{\sum_{i=1}^{n}(X_i - \bar{X})(Y_i - \bar{Y})}{\sqrt{\sum_{i=1}^{n}(X_i - \bar{X})^2}\sqrt{\sum_{i=1}^{n}(Y_i - \bar{Y})^2}} \tag{16}$$

where X (and Y) denotes real (respectively, predicted) data sequence, and $\bar{X}$ (and $\bar{Y}$) denotes the average of the real (resp., predicted) data sequence.

We calculate the correlation coefficients for the first 100 consecutive samples in each dataset (Fig. 15). It can be seen that Dos attack dataset periodically appears the same data packet, resulting in a simple dataset structure and easy prediction so that the correlation coefficient value is the highest. Fuzzy attack has a lower correlation coefficient than attack free due to random modification of the values in the data packet, which makes the data too confusing and difficult to predict. The correlation coefficient values of impersonation attack and attack free dataset are very close, because impersonation attack will mimic normal ECU sending packets to the target

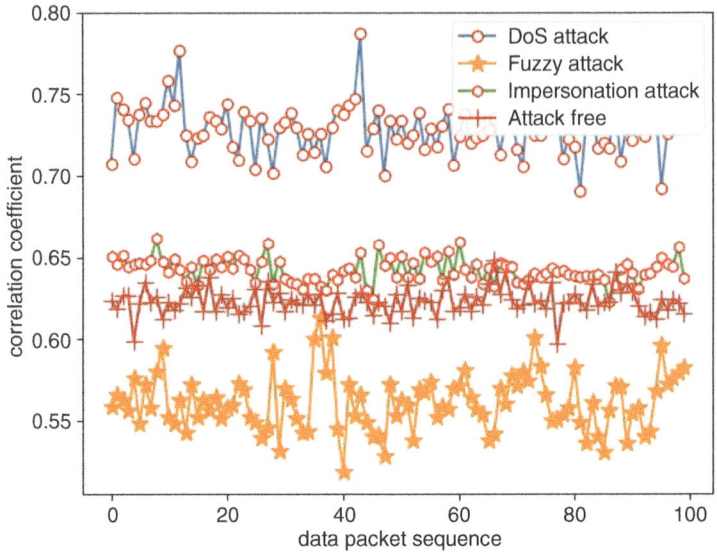

Fig. 15 Distribution of correlation coefficients for the first 100 samples

of deceiving IDS when launching an attack. The correlation coefficient between impersonation attack and attack free dataset is too close, but it can be seen that the correlation coefficient of impersonation attack is larger.

Time series prediction based model will gradually decrease its prediction ability over time. Figure 16 shows the correlation coefficient after 2500 sample sequences. Comparing Fig. 15 with Fig. 16, one may find that all correlation coefficients of four datasets are reduced, but attack free data is slower as ConvLSTM model learns potential features of attack free dataset to get the best predictability of the dataset. In Fig. 17, attack free correlation coefficient is already the highest.

Correlation coefficients of Figs. 15, 16, and 17 are used to illustrate when the model completes the training and enters into detection stage. Figure 15 shows the correlation coefficients of various datasets generated after the new model has just been trained. It can be seen that the correlation coefficient of attack free data is relatively small. However, after the 2500 CAN data sample sequence (Fig. 16), the correlation coefficient of attack free data is relatively stable, and the correlation coefficients of other attack data are reduced (in varying degrees). After the 3750 CAN data sample sequence (Fig. 17), the correlation coefficient of attack free data is the largest. At this time, the recognition threshold can be determined, previous model is discarded, and the new model can be deployed to undertake the task of intrusion detection. This is a key step in our IDS. Once abnormal CAN data passes through several data packets, the correlation coefficient will drop rapidly to ensure the maximum correlation coefficient of attack free. In this way, abnormal data can be identified. If higher detection accuracy is required, the new model will be postponed for more time and then enter into detection stage. Nonetheless, this will

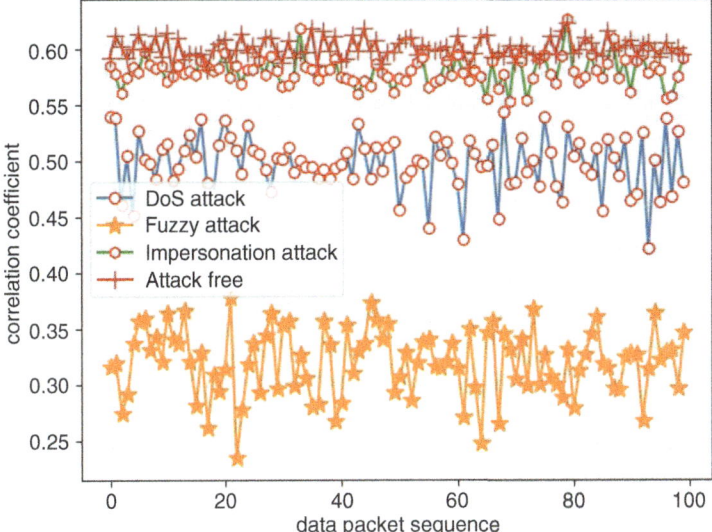

Fig. 16 Distribution of correlation coefficients for 2500th to 2600th samples

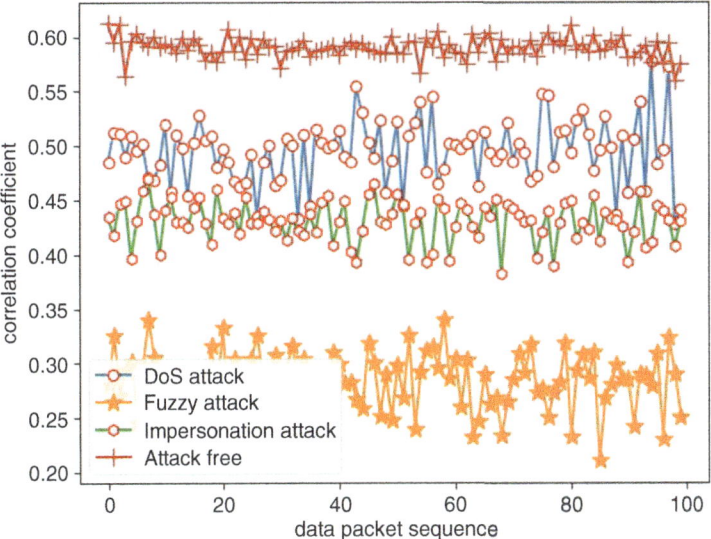

Fig. 17 Distribution of correlation coefficients for 3750th to 3850th samples

increase computational cost, and users can make trade-offs based on their security requirements.

To detect intrusions, an appropriate threshold of correlation coefficient must be found to distinguish attack data with attack free data. However, in a large dataset, there exist special points where the correlation coefficient might vary dramatically. For this, we calculate an average for every 25 consecutive correlation coefficients, and each average is used as a detection point for intrusion detection. This can not only avoid the influence of extreme data values on detection accuracy, but also reduce the frequency of detection and the amount of calculation.

5.3 Evaluation

5.3.1 Threshold Selection for Classification

To ensure that attack free has the highest correlation coefficient, the packet samples in Fig. 15 are already qualified for intrusion detection. First, we calculate the average correlation coefficient for each dataset using the first 100 consecutive packets. Then, we use an average of two values with the highest average correlation coefficient in the four datasets to calculate an average value that is used as the classification threshold.

The model based on time series prediction will gradually decrease the correlation coefficient with time, so we will re-determine the threshold every 50 detection points. To ensure detection accuracy, the ConvLSTM model needs to be retrained after every 500 detection points. Since the model only needs to be iterated 60 times, it is practically feasible to train the model during intrusion detection process.

5.3.2 Experiment

We take 210,000 packets from the four datasets for IDS performance evaluation. The evaluation data of the model is shown in Fig. 18. The precision represents the ratio in correctly classifying the dataset, which reaches 0.97 in our model. And the recall rate is 0.96, representing the rate at which the attack is detected. F1-score reaches 0.96 in our model.

Next we compare our method with SVM and decision tree model. The training set contains 375,000 packets generated before the test set used by ConvLSTM. Attack data is constructed from attack free data and the number of the former is three times that of the latter. Through dataset analysis, SVM performance is dreadful because the number of attack free data is too small, which makes the model learn attack data too much and thus lack the capability of identifying attack free data. Increasing the amount of attack free data in the training set can improve the performance of the SVM. In particular, if we have 1.5 million attack free packets in the training, then the detection precision of SVM could be 0.91. It is unlikely to further improve the detection precision only by increasing the number of attack free packets (and attack packets) but overlooking spatiotemporal information. The decision tree can extract excellent attack free data features, but the attack data is generally detectable, not to mention the detection of unknown attacks.

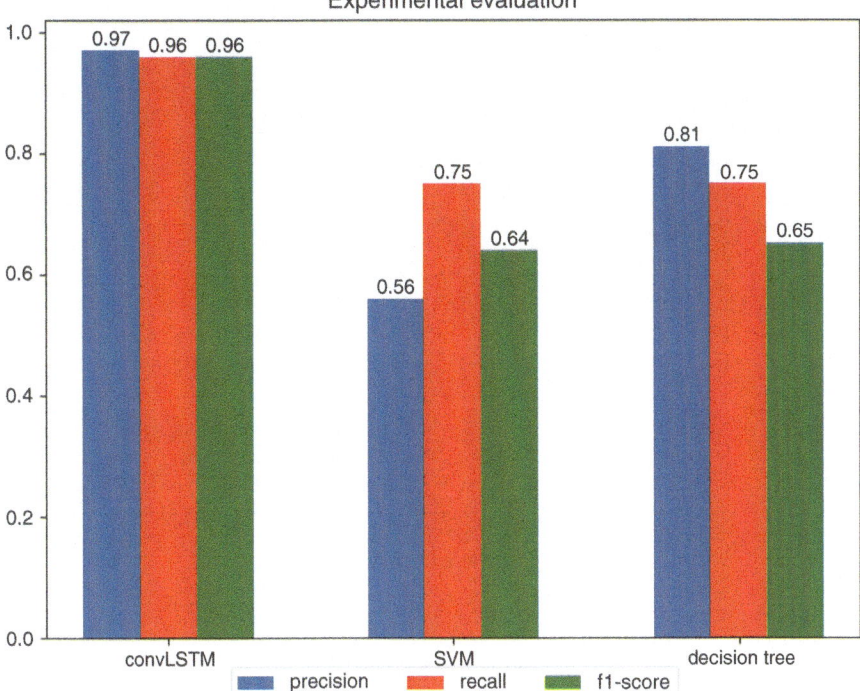

Fig. 18 Model evaluation

The receiver operating characteristic (ROC) curve can also be used to evaluate the performance of our model (Fig. 19). True Positive Rate (TPR) is the ratio of finding attack state correctly. False Positive Rate (FPR) indicates the rate at which the model misjudges attack state (of attack free data). The area under the ROC curve represents the Area Under Curve (AUC) value, and the greater the AUC is, the better the performance of the model. The ROC curve serves as an intuitive method for comprehensive evaluation models. It can be seen that the models trained by the SVM and the decision tree on the small-scale training set can detect attacks little or nothing. On the contrary, the ConvLSTM can effectively detect various attack modes only by periodically training attack free data for a period of time.

5.3.3 Comparing the ConvLSTM Model with the LSTM Model

As the ConvLSTM model improves the LSTM model [45], we will use LSTM as a comparative model to evaluate our model performance.

As shown in Fig. 20, we calculate the correlation coefficients of 500,000 data packets for the four datasets using LSTM model. We first train LSTM with 1000 packets, then use the model to calculate the correlation coefficient of the subsequent

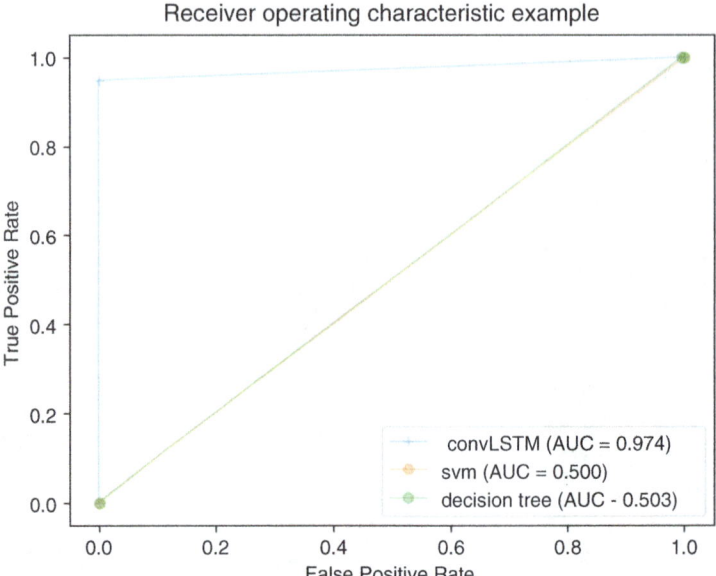

Fig. 19 ROC curve

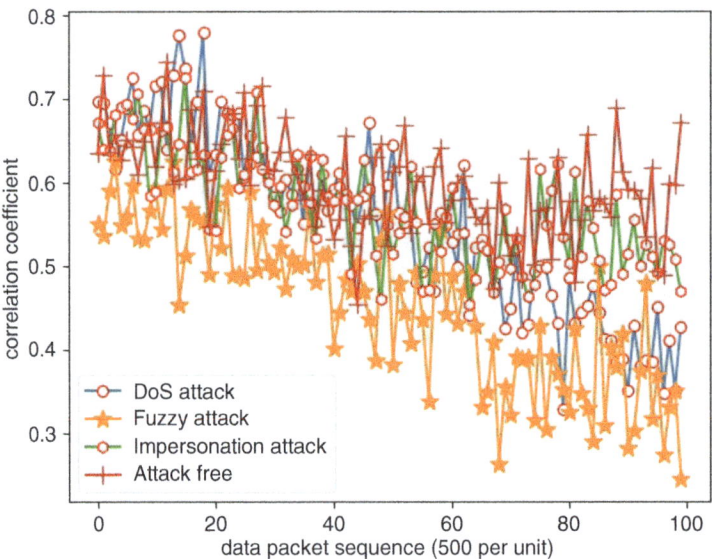

Fig. 20 Overall situation of the correlation coefficients in LSTM prediction

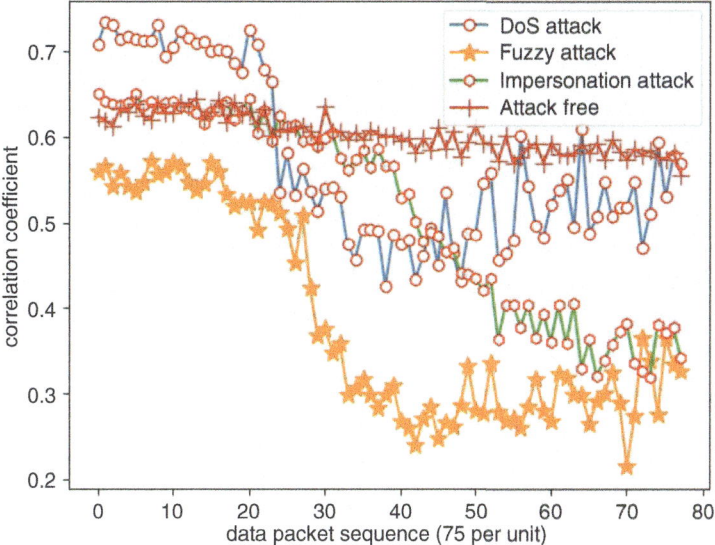

Fig. 21 Overall situation of the correlation coefficients in ConvLSTM prediction

packets, and print a correlation coefficient for every 500 packets. Although it can be seen that the correlation coefficient tends to be declining asymptotically, but the correlation coefficient between attack packet and attack free packet is not well distinguished, and the correlation coefficient of attack free is not very stable.

Figure 21 shows the overall correlation coefficient of the four datasets of the ConvLSTM model. A correlation coefficient point is printed for every 75 packets. For the trained ConvLSTM model, we see favorable detection results after approximately 2250 data packets, and the correlation coefficient of attack free is relatively stable.

6   Conclusion

The chapter introduces the applications of machine learning algorithms to in-vehicle IDS mechanism. In designing these ML-based IDS systems, one can rely on the spatiotemporal information in the CAN frames. Different exploit of these spatiotemporal information leads to different IDS systems characterized by different features. Besides the IDS accuracy, one should pay special attention to the IDS computation and storage consumption due to the limited power of in-vehicle ECU.

Acknowledgments The author is supported by the National Natural Science Foundation of China (61971192), Shanghai Municipal Education Commission (2021-01-07-00-08-E00101), and Shanghai Trusted Industry Internet Software Collaborative Innovation Center.

References

1. Nie, S., Liu, L., Du, Y.: Free-fall: Hacking tesla from wireless to can bus. https://www.blackhat.com/docs/us-17/thursday/us-17-Nie-Free-Fall-Hacking-Tesla-From-Wireless-To-CAN-Bus-wp.pdf
2. Checkoway, S., McCoy, D., Kantor, B., Anderson, D., Shacham, H., Savage, S., Koscher, K., Czeskis, A., Roesner, F., Kohno, T.: Comprehensive experimental analyses of automotive attack surfaces. In: USENIX Security Symposium, pp. 447–462 (2011)
3. Miller, C., Chris, C.: A survey of remote automotive attack surfaces. Black Hat USA (2014)
4. Miller, C., Chris, C.: Remote exploitation of an unaltered passenger vehicle. Black Hat U S A **2015**, 91 (2015)
5. Hunt, T.: Controlling vehicle features of Nissan LEAFs across the globe via vulnerable APIs. Blog Post. February (2016). https://www.troyhunt.com/controlling-vehicle-features-of-nissan/
6. Currie, R.: Developments in car hacking. SANS Institute. https://www.sans.org/white-papers/36607/
7. KEEN Security Lab. Experimental security assessment of BMW cars: a summary report. https://keenlab.tencent.com/en/whitepapers/Experimental_Security_Assessment_of_BMWCars_by_KeenLab.pdf
8. CAN specification Version 2.0, Robert BOSCH GmbH, Stuttgart, Germany (1991)
9. Liu, J., Zhang, S., Sun, W., Shi, Y.: In-vehicle network attacks and countermeasures: challenges and future directions. IEEE Netw. **31**, 50–58 (2017)
10. Koscher, K., Czeskis, A., Roesner, F., Patel, S., Kohno, T., Checkoway, S., McCoy, D., Kantor, B., Anderson, D., Shacham, H., et al.: Experimental security analysis of a modern automobile. In: 2010 IEEE symposium on security and privacy, Oakland, pp. 447–462 (2010)
11. Hoppe, T., Kiltz, S., Dittmann, J.: Security threats to automotive CAN networks–practical examples and selected short-term countermeasures. In: International conference on computer safety, reliability, and security, Newcastle Upon Tyne, UK (2008)
12. Müter, M., Asaj, N.: Entropy-based anomaly detection for in-vehicle networks. In: IEEE Intelligent Vehicles Symposium (IV), pp. 1110–1115 (2011)
13. Cho, K.T., Shin, K.G.: Fingerprinting electronic control units for vehicle intrusion detection. In: 25th USENIX security symposium, pp. 911–927 (2016)
14. Marchetti, M., Stabili, D.: Anomaly detection of CAN bus messages through analysis of ID sequences. In: IEEE intelligent vehicles symposium, pp. 1577–1583 (2017)
15. Schweppe, H., Roudier, Y., Weyl, B., Apvrille, L.: Car2X communication: securing the last meter - a cost-effective approach for ensuring trust in Car2X applications using in-vehicle symmetric cryptography. In: 2011 IEEE VTC Fall, pp. 1–5 (2011)
16. Schweppe, H., Gendrullis, T., et al.: Securing Car2X applications with effective hardware-software co-design for vehicular on-board networks. In: 27th Joint VDI/VW automotive security conference, Berlin, Germany, October (2011)
17. Groza, B., Murvay, P.S.: Efficient protocols for secure broadcast in controller area networks. IEEE Trans. Ind. Inf. **9**(4), 2034–2042 (2013)
18. Groza, B., Murvay, P.S.: Secure broadcast with one-time signatures in controller area networks. In: 6th International conference on availability, reliability and security (2011)
19. Nilsson, D.K., Larson, U.E., et al.: Efficient in-vehicle delayed data authentication based on compound message authentication codes. In: IEEE 68th vehicular technology conference (2008)
20. Woo, S., Jo, H.J., Lee, D.H.: A practical wireless attack on the connected car and security protocol for in-vehicle CAN. IEEE Trans. Intell. Transp. Syst. **16**, 993 (2014)
21. Kurachi, R., Matsubara, Y., Takada, H., et al.: CaCAN - Centralized authentication system in CAN. In: Embedded security in cars (ESCAR) Europe conference, Hamburg (2014)
22. Herrewege, A.V., Singelee, D., Verbauwhede, I., et al.: CANAuth - a simple, backward compatible broadcast authentication protocol for CAN bus. In: ECRYPT workshop on lightweight cryptography (2011)

23. Smith, C.: The car hacker's handbook: a guide for the penetration tester. No Starch Press, 401 China Basin Street Suite 108 San Francisco, CA United States, ISBN: 978-1-59327-703-1 (2016)
24. Saldivar-Sali, A., Einstein, H.: A landslide risk rating system for Baguio, Philippines. Eng. Geol. **91**, 85 (2007)
25. Golde, N., Redon, K., et al.: Weaponizing femtocells: the effect of rogue devices on mobile telecommunications. In: NDSS (2012)
26. Tsugawa, S.: Inter-vehicle communications and their applications to intelligent vehicles: an overview. In: IEEE intelligent vehicle symposium (2002)
27. Shukla, S.: Embedded security for vehicles: ECU hacking. Uppsala University (2016)
28. Farsi, M., Ratcli, K., Barbosa, M.: An overview of controller area network. Comput. Control Eng. J. **10**, 113–120 (1999)
29. Miller, C., Valasek, C.: Adventures in automotive networks and control units. In: DEF CON 21 hacking conference (2013)
30. Lan, H.M., Kwak, B.I., Kim, H.K.: Anomaly intrusion detection method for vehicular networks based on survival analysis. Veh. Commun. **14**, 52–63 (2018)
31. Song, H.M., Kim, H.K.: Intrusion detection system based on the analysis of time intervals of CAN messages for in-vehicle network. In Proceedings of the international conference on information networking (ICOIN), Kota Kinabalu, Malaysia, 13–15 January (2016)
32. Seo, E., Song, H.M., Kim, H.K. GIDS: GAN based intrusion detection system for in-vehicle network. In Proceedings of the 16th annual conference on privacy, security and trust (PST), Belfast, UK, 28–30 August (2018)
33. Tariq, S., Lee, S., Kim, H.K., Woo, S.S.: Detecting In-vehicle CAN message attacks using heuristics and RNNs. In Proceedings of the international workshop on information and operational technology security systems, Heraklion, Greece, 13 September (2018)
34. Larson, U.E., Nilsson, D.K., Jonsson, E.: An approach to specification-based attack detection for in-vehicle networks. In Proceedings of the IEEE intelligent vehicles symposium, Eindhoven, The Netherlands, 4–6 June, pp. 220–225 (2008)
35. Wang, C., Zhao, Z., Gong, L., Zhu, L., Cheng, X.: A distributed anomaly detection system for in-vehicle network using HTM. IEEE Access **6**, 9091–9098 (2018)
36. Hu, W., Liao, Y., Vemuri, V.R.: Robust anomaly detection using support vector machines. In Proceedings of the international conference on machine learning, Washington, DC, USA, 21–24 August, pp. 282–289 (2003)
37. Li, H., Wang, Y., Qin, H., Xinkai, W.: Investigating the effects of attack detection for in-vehicle networks based on clock drift of ECUs. IEEE Access **6**, 49375–49384 (2018)
38. Xiao, J., Wu, H., Li, X.: Robust and self-evolving IDS for in-vehicle network by enabling spatiotemporal information. In Proceedings of the IEEE 21st international conference on high performance computing and communications, Zhangjiajie, China, 10–12 August (2019)
39. Ho, T.K.: The random subspace method for constructing decision forests. IEEE Trans. Pattern Anal. Mach. Intell. **20**, 832–844 (1998)
40. Chandola, V., Banerjee, A., Kumar, V.: Anomaly detection: a survey. ACM Comput. Surv. **41**(3), 1–58 (2009)
41. Lee, H., Jeong, S.H., Kim, H.K.: A novel intrusion detection system for in-vehicle network by using remote frame. In: 15th Annual conference on privacy, security and trust, pp. 57–66 (2017)
42. Wu, W., Huang, Y., et al.: Sliding window optimized information entropy analysis method for intrusion detection on in-vehicle networks. IEEE Access **6**, 45233–45245 (2018)
43. Shi, X., Chen, Z., Wang, H., et al.: Convolutional LSTM network: a machine learning approach for precipitation nowcasting. In: Proceedings of the 28th International Conference on Neural Information Processing Systems **1**, 802–810 (2015)
44. Sutskever, I., Vinyals, O., Le, Q.V.: Sequence to sequence learning with neural networks. In: NIPS, pp. 3104–3112 (2014)
45. Hochreiter, S., Schmidhuber, J.: Long short-term memory. Neural Comput. **9**(8), 1735–1780 (1997)

In-Vehicle ECU Identification and Intrusion Detection from Electrical Signaling

Xiangxue Li, Yue Bao, and Xintian Hou

1 Introduction

Controller area network (CAN) protocol has strong anti-interference ability and can effectively suppress electromagnetic interference [1]. It relies on differential signals to transmit messages. Differential signals with dominant state (logical 0) and recessive state (logical 1) are transmitted through high (CAN-H) and low (CAN-L) lines. When the signal represents dominant state, CAN-H voltage is approximately 3.5 V, and CAN-L voltage is approximately 1.5 V, which results in a dominant differential voltage of approximately 2.0 V on CAN bus. For recessive state, both CAN-H and CAN-L voltages are approximately 2.5 V, yielding the differential voltage ≈ 0 V [2].

There are growing instances of hacking vehicles due to loose security protection of CAN protocol [3–7]. We have seen various IDSs of in-vehicle CAN networks for decades [8–13]. These suggestions cannot determine which ECU launches the particular attacks. Moreover, a smart attacker might mimic certain characteristics of the target ECU to launch an attack [9, 10, 12, 13]. Fortunately, some seminal work [1, 14–17] can not only detect malicious frames but identify their sender ECUs. The strategy counts on CAN signal unique characteristics, e.g., the hardware and

X. Li (✉)
East China Normal University and Shanghai Key Laboratory of Trustworthy Computing, Shanghai, China

Shanghai Key Laboratory of Privacy-Preserving Computation, MatrixElements Technologies, Shanghai, China
e-mail: xxli@cs.ecnu.edu.cn

Y. Bao · X. Hou
CATARC Software Testing (Tianjin) Co. Ltd, Tianjin, China
e-mail: baoyue@catarc.ac.cn; houxintian@catarc.ac.cn

© The Author(s), under exclusive license to Springer Nature Switzerland AG 2023
V. K. Kukkala, S. Pasricha (eds.), *Machine Learning and Optimization Techniques for Automotive Cyber-Physical Systems*, https://doi.org/10.1007/978-3-031-28016-0_15

topology information (delineated by the signal's characteristics so that even if two ECUs send identical message, corresponding signals are divergent).

Signal characteristics are not only affected by vehicle power supply but also related to the hardware characteristics of the sending device itself. It is difficult for an attacker to imitate some particular device's signal characteristics. Thus CAN signals show special functionality in detecting attack messages and identifying sender ECUs. Murvay and Groza [18] pioneered the methodology of studying the differences in CAN signals (sent by ECUs), which are significant for ECU identification. However, they only used the signals corresponding to the CAN frame's identifier field and did not account for the blended signals caused by the collisions between ECUs' simultaneous messages. The limitation was tackled in [15] where 18-bit identifier extension was used as the ECU's fingerprint.

One more interesting work-Sample was proposed in [17] with low time complexity and the advantage of robustness and recognition rates. Kneib and Huth [1] proposed Scission with in-depth analysis of CAN signals. Scission uses the rising and falling edges of CAN signal to design IDS. However, their method could be affected by CAN topology easily. Once the number of ECUs or the length of stub lines change, the characteristics of rising and falling edges would become different.

2   System Model and Ringing Effect

We can further look into the ringing generation mechanism and recognize the fuzzy discrepancy between transitions *from dominant to recessive state* and those *from recessive to dominant state*. Ringing intensity is related to the number of ECUs and the stub line length of CAN topology [19–21]. When we fix the number of electronic control units, longer stub line results in more intense ringing. The fluctuation of ringing intensity would further tweak falling edges' voltage, which might set off false alarms of IDS (i.e., not triggered by real attacks). We will investigate the factors that enlarge ringing effect and demonstrate the discrepancy between rising edges and falling edges. Our attempt is to design ECU identification scheme and IDS only from the characteristics of dominant states and rising edges (D.R for short).

Figure 1 shows CAN bus topology deployed widely in automotive applications. In particular, Fig. 1a presents linear topology and Fig. 1b depicts start-like topology. A twisted wire is commonly used for CAN bus, and the twisted wire's characteristic

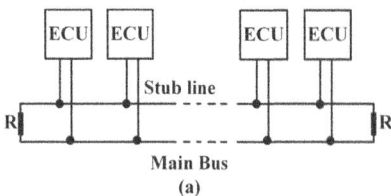
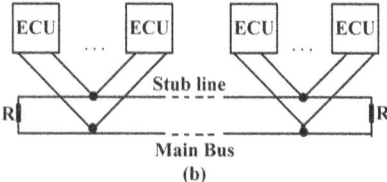

Fig. 1 CAN topology

impedance is marked as **R**. For linear topology (as recommended in the standards [22, 23]), the longest bus is called main bus. Two terminating resistors are arranged at left end and right end, and the resistor resistance is set to **R** to match the bus's characteristic impedance. In the automotive field, terminating resistors are commonly installed in the two farthest CAN nodes (called the terminal CAN nodes) to improve productivity. Other nodes are referred to as non-terminal nodes. ECUs are connected to the main bus using a twisted wire (i.e., stub lines). Stub line is attached to the main bus through a connector (indicated by the black circle in Fig. 1) called junction [21].

2.1 Threat Models

We consider two types of in-vehicle attacks: known-ECUs attack manipulates existing ECUs, and unknown-ECUs attack inserts extra devices to CAN bus.

Automotive manufacturers install ECUs during vehicle production. Attackers rely on additional interfaces to compromise a known ECU to transmit malicious CAN frames. These interfaces include WiFi, Bluetooth, and cellular communication modules. Telematics ECU [6, 24] is a prevalent example, installed widely in modern vehicles to enable supplementary functions. This kind of ECUs are connected to an external network (e.g., a cellular network), providing a target for the attackers.

Instead of exploiting existing ECU's vulnerability, an attacker can connect an unknown ECU to attack CAN network directly. Alternatively, he may plugin a special device to the network via the vehicle's On-Board Diagnostics (OBD)-II port.

2.2 Difference Between ECUs Voltage Outputs

The differences in voltage stabilizing ability of the regulator inside an ECU results in different outputs (V_{OUT}) [14], even for the same power supply (V_{IN}). ECU output voltage variations may stem from the differences in ground voltage and capacitors (denoted C_1,C_2 and C_3 in Fig. 2). Further, industrial typical 5% error tolerance is employed in CAN transceiver resistors, which leads to voltage changes.

2.3 Ringing Effect

The impedance mismatch occurs at two points over the CAN bus (Fig. 3) [20, 21], one at the junction and another at the front of non-terminal ECUs. Non-terminal ECU causes positive reflection as its impedance can be up to several tens of kΩ, significantly larger than the stub line characteristic impedance which is further larger than the junction's impedance, resulting in negative reflection.

Fig. 2 CAN application
schematic

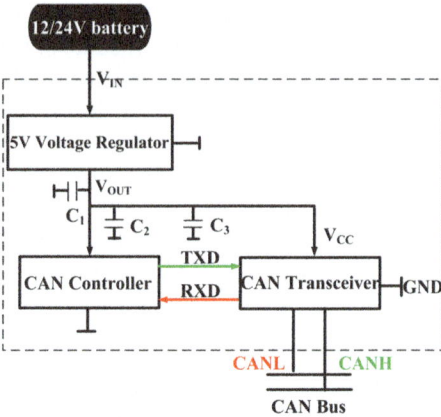

Fig. 3 Reflection

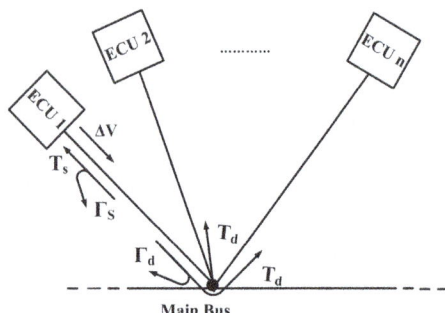

2.3.1 From Dominant to Recessive States

Let n denote the number of ECUs connected to the junction through stub lines and
ECU1 a transmitter whose signal voltage would be reduced by ΔV to transfer from
dominant state to recessive state. Since the dominant state's value is approximately
2 V, ΔV has a negative polarity. In Fig. 3, a total of $(n + 2)$ lines are connected to the
junction (i.e., the overall number of connected stub lines and the two main bus lines).
The signal transmitted from ECU1 to the junction follows $(n + 1)$ lines in parallel.
Thus, the stub lines have the same impedance $\frac{Z_R}{n+1}$, where the Z_R's nominal value is
120 Ω. The reflectance (Γ_d) and transmittance (T_d) at the junction are calculated as:

$$\Gamma_d = \frac{\frac{Z_R}{n+1} - Z_R}{\frac{Z_R}{n+1} + Z_R} = -\frac{n}{n+2}, \quad T_d = 1 + \Gamma_d = \frac{2}{n+2} \tag{1}$$

Since Γ_d has a negative polarity, a larger portion of the incident signal is reflected
as n increases, and its small part is delivered into other ECUs.

Fig. 4 Ringing for signals from dominant to recessive states

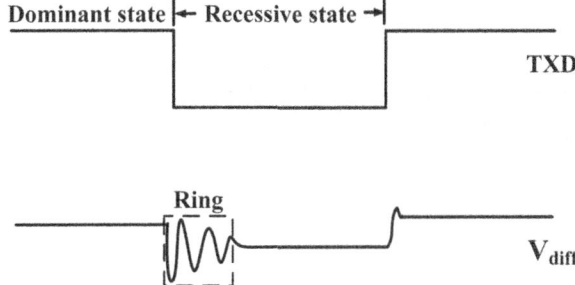

Denote Z_{diff} as ECU1's differential input impedance. Now, we have ECU1's front reflectance and transmittance (i.e., Γ_s and T_s):

$$\Gamma_s = \frac{Z_{diff} - Z_R}{Z_{diff} + Z_R}, \quad T_s = 1 + \Gamma_s = \frac{2Z_{diff}}{Z_{diff} + Z_R} \tag{2}$$

When the signal is at the recessive state, Z_{diff} is much larger than Z_R. Consequently, Γ_s has a positive polarity, and equals approximately one. Thus, ECU1 front end reflection direction is the same as the incident signal direction, and the incident signal and reflected signal superposition is about twice the original incident signal.

For a dominant-to-recessive transition, the negative transition signal ΔV is transmitted from ECU1 to the junction, undergoing partial transmission and reflection. The signals are transmitted to other ECUs through the junction and are partially reflected on the other ECUs' front end without changing the direction. At the ECU1's front, the signal returned from the connection is partially transmitted to ECU1. These reflections and transmissions are repeated, resulting in ringing (Fig. 4).

2.3.2 From Recessive to Dominant States

In the transitions from recessive state to dominant state, ECU1's output impedance is very low. In the recessive state, the electrical energy is released on the network. However, when the signal transfers from recessive to dominant states, ECU1's differential output impedance becomes lower and starts charging the network. ECU1 generates the signal of 2 V, whose polarity is inverted at the junction and reflected onto ECU1. Unlike the dominant-to-recessive transition, the reflection signal is partly received at ECU1 due to the low impedance of ECU1. Since there are no reflections' repetitions, we have small ringing at the recessive-to-dominant state transition.

3 Dominant States and Rising Edges for Source Identification

3.1 Signal Measurement and Preprocessing

In order to measure the differential signal on the CAN bus, we connect two channels of an oscilloscope CAN-H and CAN-L, respectively. Each CAN frame's differential signal would be obtained based on the oscilloscope's differential function.

Several preprocessing steps are applied to each CAN signal captured by the oscilloscope. First, all dominant states are extracted from the signals. We set a voltage threshold value as 0.9 V: voltage greater than the threshold marks the start of the dominant state. The dominant states are then classified into five sets (denoted as L_1, L_2, L_3, L_4, and L_5) based on the number of contained bits. Let L_i represent all dominant states containing exactly i bits (see Fig. 5). Note that CAN standard specifies that a recessive bit is automatically inserted whenever five consecutive dominant bits appear in a CAN signal. Thus, no dominant state can contain more than five consecutive dominant bits. By dividing a CAN frame into 5 sets, we have the following gains: (a) redundant features can be eliminated (the dominant states with the same number of dominant bits in a CAN frame have similar characteristics, and these dominant states with similar characteristics are in the same set); and (b) the influence of outliers might be eliminated to make the classification more accurate.

3.2 Feature Extraction

The sets obtained above are subjected to feature extraction, where the measured voltages are discrete values. Feature extraction is essential in ECU identification and needs to be time-efficient. Domain transformations should be avoided if possible. To reflect the characteristics of these discrete values, Table 1 qualifies the features that reflect the characteristics of a group of discrete values from the time domain feature quantities (x is time domain representation of data and N its dimension). Some work also discussed various features for ECU identification [15].

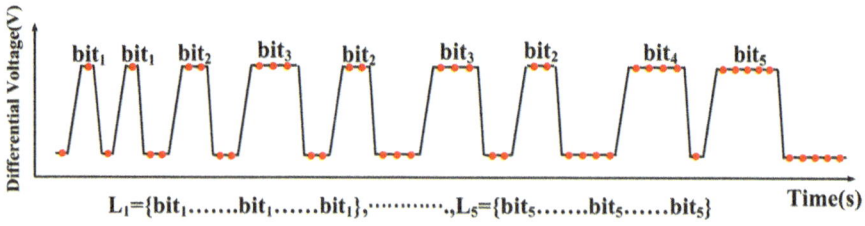

Fig. 5 A CAN frame is divided into 5 sets

Table 1 Features

Feature	Description
Maximum	$Max = Max(x(i))_{i=1....N}$
Minimum	$Min = Min(x(i))_{i=1....N}$
Mean	$\mu = \frac{1}{N}\sum_{i=1}^{N} x(i)$
Range	$R = Max - Min$
Average deviation	$adv = \frac{1}{N}\sum_{i=1}^{N} |x(i) - \mu|$
Variance	$\sigma^2 = \frac{1}{N}\sum_{i=1}^{N}(x(i) - \mu)^2$
Standard deviation	$\sigma = \sqrt{\frac{1}{N}\sum_{i=1}^{N}(x(i) - \mu)^2}$
Root mean square	$rms = \sqrt{\frac{1}{N}\sum_{i=1}^{N} x(i)^2}$

Table 2 Selected features ordered by ranks

Order	Feature	Order	Feature
1	$rms(L_5^{40})$	11	$max(L_1^1)$
2	$adv(L_2^{13})$	12	$min(L_4^{26})$
3	$\sigma^2(L_4^{30})$	13	$R(L_3^{20})$
4	$rms(L_3^{21})$	14	$rms(L_4^{32})$
5	$mean(L_1^3)$	15	$max(L_4^{25})$
6	$\sigma(L_4^{31})$	16	$adv(L_2^5)$
7	$\sigma^2(L_3^{22})$	17	$mean(L_2^{11})$
8	$\sigma(L_2^{15})$	18	$rms(L_2^{16})$
9	$R(L_4^{28})$	19	$max(L_3^{17})$
10	$min(L_3^{18})$	20	$\sigma(L_5^{39})$

Dominant states with the same number of dominant bits indicate analogous characteristics. Thus, CAN frames are divided into five sets (Sect. 3.1). For the unlikely case of empty sets (all characteristics obtain null values), one may replace the missing values with statistical properties, e.g., mean or median. For each set, eight features (Table 1) are extracted, yielding 40 CAN signal features in total.

Relief-F [25] can evaluate the features by calculating a score for each one and selecting the most important ones. We finally opt for 20 feature subsets for each CAN frame (Table 2). The order column represents the sequence number that Relief-F sorts in descending order according to the scores of the features, and these sequence numbers correspond to the dimension of each feature in the input feature set.

3.3 Training and Testing

We view ECU identification from a received CAN frame as a classification problem and use supervised learning to identify the signals' sender. The training set comprises 200 CAN frames for each ECU. After the training phase, a classifier is created that can be used to identify the sender of a CAN frame.

Algorithm 1 ECU identification and IDS

```
 1: function TRAINING(S: original CAN signal)
 2:     for i=1 to len(S) do
 3:         /*Divide the signal Sᵢ into ECUₗ*/
 4:         ECUₗ ← DECODE(Sᵢ)
 5:         /*dominant state and rising edge*/
 6:         [L₁, L₂, ..., L₅] ← PREPROCESSING (Sᵢ ∈ S)
 7:         Fᵢ ← EXTRACTION(L₁, L₂, ..., L₅)
 8:         TrainingSet(i)←[Fᵢ: ECUₗ]
 9:     end for
10:     Classifier←GET_TRAINING_
        ALGORITHM(TrainingSet)
11:     return Classifier
12: end function
13:
14: function TESTING(S: a new CAN signal)
15:     /*Divide the signal S into ECUₗ*/
16:     ECUₗ ← DECODE(S)
17:     /*dominant state and rising edge*/
18:     [L₁, L₂, ..., L₅] ← PREPROCESSING (S)
19:     F ← EXTRACTION(L₁, L₂, ..., L₅)
20:     [Result, Probability]← IDENTIFICATION
        (F, classifier)
21:     if Probability < threshold then
22:         return Unknown ECU Adversary
23:     else if Result ≠ ECUₗ then
24:         return Known ECU Adversary
25:     else
26:         return Normal
27:     end if
28: end function
```

Table 3 Comparison among voltage-based approaches

	Choi et al. [15]	Scission [1]	Simple [17]	Our system
Sampling rate	2.5 GS/s	20 MS/s	50 MS/s	50 MS/s
Identification rate	96.48%	99.85%	99.10%	99.15%
False positive	3.52%	0%	0.899%	0.85%
Signal type	Differential	Differential	Differential	Differential
Domain transformations	Yes	Yes	No	No
Unknown ECU	No	Yes	Yes	Yes

The training phase results in a classifier, which is then used to predict new frames in testing phase. The testing phase includes two tasks. ECU identification tests whether the system correctly identifies frames' source and examines the impact of stub lines' length on the execution ability. Intrusion detection assesses the system's capability of detecting attacks (Sect. 2.1). The system performance on identification and intrusion detection will be discussed in Sect. 4. Algorithm 1 describes the training and testing processes. Table 3 compares some voltage-based proposals.

4   Evaluation

Four CAN bus prototypes (each containing 3/6/9/13 ECUs) are equipped to simulate different CAN networks. All prototypes have the same configuration (except the number of ECUs). Take the prototype with 3 ECUs and 1-m stub lines as example. We first assemble 3 ECUs, each containing an Arduino UNO board and a CAN shield. The shield comprises MCP2515 CAN controller [26] and MCP2551 CAN transceiver [27]. Each ECU is connected through a stub line to the main bus of length 3 m. Two 120Ω resistors are connected to the two ends of the main bus. Use an oscilloscope, one of its probes being connected to the CAN-H line of the main bus and another to CAN-L. Adjust the sampling rate of the oscilloscope to 20 MS/s.

The system can also be evaluated on real vehicles, e.g., Nissan Sentra 2016 and Subaru Outback 2011 [17].

4.1   ECU Identification

CAN signals are acquired using the digital storage in the oscilloscope PicoScope 5244D MSO with a sampling rate of 1 GS/s (the oscilloscope captures 1G data points from the signal waveform in one second) and a flexible resolution. Set the sampling rate as 20 MS/s (higher sampling rate increases data volume and hardware costs).

4.1.1   Classification Algorithms

To evaluate the influence of classification algorithms on system performance, two algorithms are employed: Linear Regression (LR) and Support Vector Machine (SVM). For each ECU in the prototype, approximately 200 frames are collected. Table 4 demonstrates that the model accuracy on a simple topology (i.e., only 3 ECUs in the entire network) averages above 99.99% irrespective of the classification algorithms. When the topology becomes complicated (e.g., 13 ECUs, Table 5), the average SVM and LR accuracies are above 98.25%. When the stub line length equals 3 m, the SVM average accuracy is 98.01%, and LR's is 98.11%. In other words, the proposed model accuracy remains high for complex network structures.

4.1.2   CAN Topology

We also explore the effect of changes in stub lines' length on the developed system performance. The system identification rate is tested for the stub line's length = 1, 2, or 3 m. Figure 6 just shows the topology for 1-m stub line.

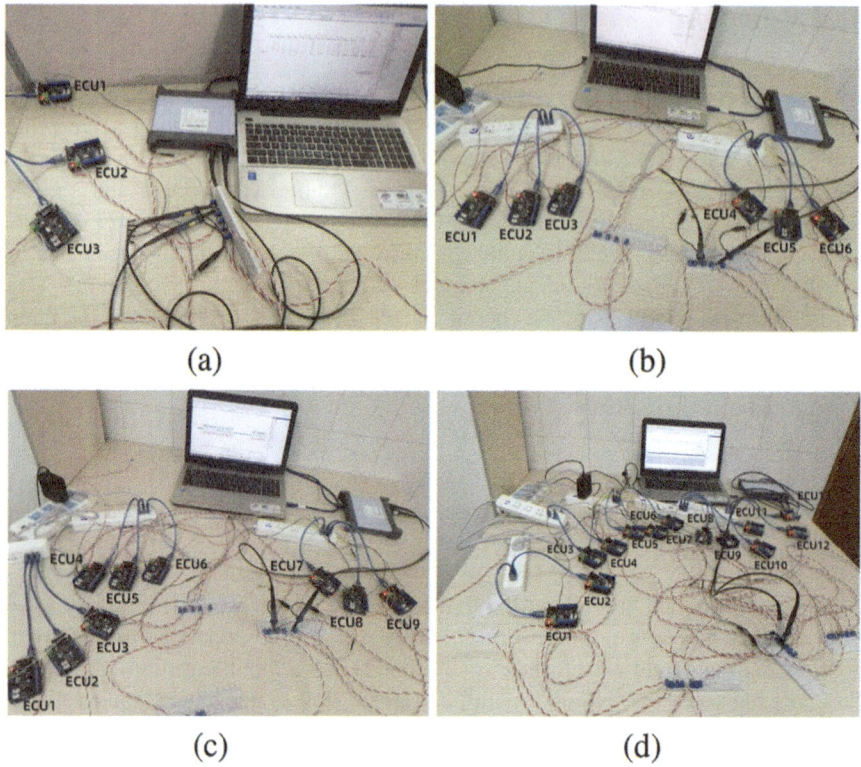

(a)   (b)

(c)   (d)

Fig. 6 Topologies of 1-m stub line and 3/6/9/13 ECUs

Same feature extraction and classification algorithms are utilized for each topology (i.e., 3/6/9/13 ECUs). Tables 4 and 5 show that if the number of ECUs is fixed, increasing stub lines' length has no impact on the system recognition accuracy. When using (falling edges and recessive states, F.R for short) and ((dominant states and rising edges) and (falling edges and recessive states), D.R.F.R for short) respectively, recognition rates decrease with the increasing of stub lines' length.

4.1.3 CAN Signal States

A series of comparative experiments are conducted to inspect the influence of CAN signal states on the system using F.R only, or D.R.F.R. As discussed above, the ringing mainly occurs in dominant-to-recessive states transitions, and the more complex the CAN bus topology, the more intense the ringing effect. Table 5 shows the results when the system uses D.R, and the average minimum recognition rates are 98.25% (for 13 ECUs and one-meter stub-line), 98.21% (for 13 ECUs and two-meter stub line), and 98.01% (for 13 ECUs and 3-meter stub line). When the system

Table 4 Recognition rate and recognition rate for 3/6 ECUs

Stub line	Algorithm	Signal state (3 ECUs)			Signal state (6 ECUs)		
		D.R	F.R	D.R.F.R	D.R	F.R	D.R.F.R
1-m	SVM Min	100	71.05	96	99.31	69.75	95.15
	SVM Avg	100	74.42	97.6	99.97	71.23	95.89
	LR Min	99.99	79.61	96	99.41	70.11	94.63
	LR Avg	99.99	80.66	98.4	99.99	72.36	95.99
2-m	SVM Min	99.35	70.09	96.21	98.89	68.11	94.21
	SVM Avg	99.98	72.32	97.32	99.85	69.87	94.32
	LR Min	99.98	79.61	96.12	99.01	69.11	94.21
	LR Avg	99.99	81.32	97.21	99.49	70.55	95.1
3-m	SVM Min	99.98	72.27	95.41	98.89	67.99	92.01
	SVM Avg	99.99	73.43	96.67	99.25	69.43	92.85
	LR Min	99.99	77.1	96.21	99.31	68.01	92.21
	LR Avg	99.99	78.29	97.32	99.55	70.53	93.32

Table 5 Recognition rate and recognition rate for 9/13 ECUs

Stub line	Algorithm	Signal state (9 ECUs)			Signal state (13 ECUs)		
		D.R	F.R	D.R.F.R	D.R	F.R	D.R.F.R
1-m	SVM Min	99.21	58.75	88.51	97.99	59.57	83.98
	SVM Avg	99.51	59.23	89.91	98.45	61.89	84.21
	LR Min	98.89	59.51	89.01	97.76	58.01	83.9
	LR Avg	99.25	60.35	90.11	98.25	62.58	84.11
2-m	SVM Min	98.8	57.35	87.11	97.51	54.25	80.99
	SVM Avg	99.01	58.25	88.26	98.21	56.75	81.21
	LR Min	98.38	57.11	88.11	97.55	54.91	79.21
	LR Avg	98.89	58.55	88.39	98.25	55.35	81.68
3-m	SVM Min	98.59	53.99	86.21	97.35	47.99	75.21
	SVM Avg	98.99	54.43	87.77	98.01	48.43	78.77
	LR Min	98.19	53.01	85.81	97.45	46.01	73.99
	LR Avg	98.71	55.53	86.34	98.11	47.53	77.34

uses F.R, the average minimum recognition rates are 61.89%, 55.35%, and 47.53%, respectively. If the system uses D.R.F.R, the average minimum recognition rates are 84.11%, 81.21%, and 77.34%. This demonstrates that using D.R is not affected by ringing and enables a higher recognition rate than other states.

4.1.4 On Real Vehicles

We can also check the method on real vehicles, Nissan Sentra 2016 and Subaru Outback 2011 and [17]. There will be 11 rounds of CAN signal, collected from these two vehicles. Table 6 shows that, for Nissan Sentra, using F.R yields the lowest

Table 6 Minimum/average recognition rate in Nissan Sentra and Subaru Outback

Vehicle	Algorithm	Signal state		
		D.R	F.R	D.R.F.R
Nissan Sentra	SVM Min	98.85	44.4	96.01
	SVM Avg	99.15	46.42	96.87
	LR Min	98.34	58.61	94.01
	LR Avg	99.08	60.20	95.41
Subaru Outback	SVM Min	98.37	62.05	91.80
	SVM Avg	99.10	63.88	93.05
	LR Min	98.05	77.23	88.52
	LR Avg	99.09	80.13	91.47

Table 7 IDS for known ECUs (support vector machine and logistic regression)

Vehicle	True	Predicted (SVM)		Predicted (LR)	
		No attack	Yes	No attack	Yes
Prototype	No attack	98.11	1.89	97.98	2.02
	Yes	2.15	97.85	2.59	97.41
Nissan Sentra	No attack	99.12	0.88	99.16	0.84
	Yes	1.89	98.11	1.79	98.21
Subaru Outback	No attack	98.99	1.01	99.01	0.99
	Yes	1.69	98.31	1.75	98.25

average accuracy of 46.42%. When D.R.F.R is used, the lowest average accuracy is 95.41%. For the Subaru outback, the lowest average accuracy equals 99.09% for D.R, 63.88% when using F.R, and 91.47% when using D.R.F.R.

4.2 Intrusion Detection

4.2.1 Known ECUs

We assume that the system has the knowledge: which identifiers are used, which ECUs are allowed to use them. If the ECU selected by the model as source is not allowed to send frames with the identifier of the received frame, an attack will be assumed. It is not allowed that multiple ECUs use same identifier.

We consider the most complex topology (i.e., 13 ECUs and 3-m stub lines) as a prototypical setup. 11 out of the 13 ECUs are seen as legitimate, and the remaining two as attackers. More than 1400 frames are collected, 500 of which are valid, and more than 900 counterfeit. Table 7 show the detection rate 97.85%.

Similar tests are conducted on real data of Nissan Sentra and Subaru Outback. The compromised ECUs are simulated on the real vehicles by adding two additional ECUs. These ECUs consist of an Arduino board and a CAN shield. Adding these ECUs differs from the situation of unknown ECU attack. Namely, unknown ECUs' electrical signals are not trained by the algorithm. In contrast, the two

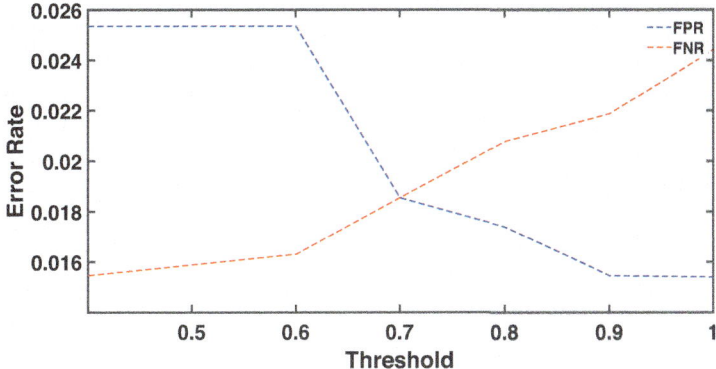

Fig. 7 FN and FP rates at varying thresholds (Subaru outback)

additional ECUs are based on the five original ECUs, and the CAN electrical signals' characteristics of the seven ECUs can be extracted and retrained. In the Nissan Sentra case, the additional ECUs imitate ECU A and ECU B, and 400 counterfeit frames are collected for each one. For Subaru Outback, the ECUs are used to fake ECU I and ECU J, and generate two sets of 400 counterfeit frames. The detection results for Nissan Sentra and Subaru Outback are shown in Table 7.

4.2.2   Unknown ECUs

Unknown ECUs' identification is related to novelty detection, i.e., the identification of new or unknown data not used in a machine learning algorithm's training [28]. We use the threshold trick (Fig. 7) and the instances with probabilities lower than the threshold are classified into unknown class.

We set the number of ECUs as 13, and the stub line's length as 3 m. To evaluate whether the system is capable of detecting unknown ECUs, the network is configured using 12 ECUs, and the 13th ECU is removed. Then monitor the resulting network and collect approximately 500 frames from each ECU for feature extraction. Now a new model can be trained (without the knowledge of the 13th ECU in the signals). Once the model training completes, ECU #13 is re-inserted to the network. Then, 3290 frames are acquired from the network with all 13 ECUs. The appropriate threshold is obtained by calculating the false positive (FP) and false negative (FN) rates (Fig. 8). The threshold 0.83 yields approximately equal values of FP and FN. And the system's identification rate is 97.89%.

For Nissan Sentra and Subaru Outback, 400 normal frames from each vehicle are selected. For Nissan Sentra, ECU M is added with the message ID {1201}, and 200 corresponding frames are collected. Overall, 600 frames are obtained from Nissan Sentra. Again, FN and FP are used to calculate the appropriate threshold (resulting in threshold value = 0.8). Figure 9 shows the results. The system achieves 98.54%

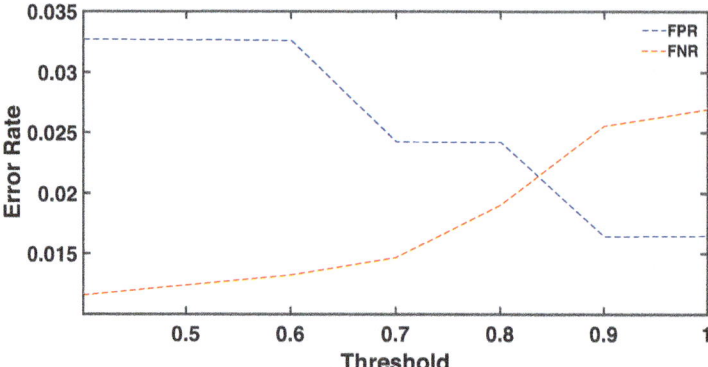

Fig. 8 FN and FP rates at varying thresholds (Prototype)

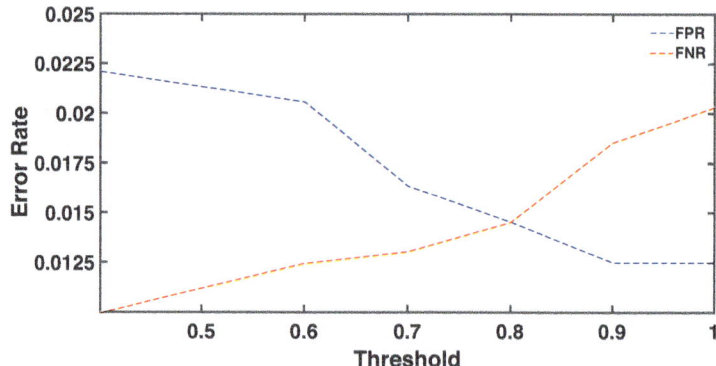

Fig. 9 FN and FP rates at varying thresholds (Nissan Sentra)

identification rate. Similarly, for Subaru Outback, ECU N sending e messages with IDs{537, 538} is inserted. Six hundred frames are collected and utilized to calculate the appropriate threshold (i.e., 0.7, see Fig. 7), and the system's accuracy is 98.15%.

4.3 Discussions

4.3.1 Environmental Factors

The voltage signal is really sensitive to environmental factors, such as temperature change. To pursue robustness against environmental factors, we adopt a method of threshold-based online model update. When the recognition rate of the model is lower than a threshold, the IDS composes an update batch with already classified fingerprints from all ECUs and thus does not require additional computing capacity.

We use the data Nissan sentra (on 02/01/2019 and 02/18/2019 [17]) to evaluate the system: select the frames sent by ECU A and ECU B from these two data sets, and then perform preprocessing and feature extraction.

We first see whether robust sender identification can be kept up over the entire training data without performing update operations. We extract approximately 2500 normal frames from the 02/01/2019 data set, the first 200 frames per ECU of the set are used for the initial training and the remaining 1000 frames of the data set for test, and this leads to the average recognition accuracy 99.31%. Then we select 1200 frames from the 02/18/2019 data set, and all the frames are classified using the already trained classifiers. The classification accuracy is 95.23%.

Next, we introduce automatic update mechanism to improve the recognition rate. The following metrics are used: recognition rate, false positive rate, false negative rate, and F-Score. Recognition rate represents the source of how many frames the model can correctly identify. False positive rate refers to the case that an unknown ECU is incorrectly classified as valid. False negative rate refers to the case that an valid ECU is classified as unknown. F-Score represents the comprehensive classification ability used to evaluate the model. We update the model online according to F-Score. When the F-Score is lower than the threshold (0.9, as demonstrated in experiments), the model will be automatically updated: the 02/01/2019 data is used to train the model and the 02/18/2019 data is used to verify the average recognition rate of the updated model. Now we manage the average recognition rate 99.12%.

4.3.2   Sample Rate

We duplicate the experiments at various sample rates to inspect system effectiveness, especially in a complex network environment (i.e., 13 ECUs and 3-m stub lines). Note that at different sample rate one will be at different position of sample sizes (which might convey tight relationship with system performance). The approach manifests robustness as expected (due to the contribution of rising edges and dominant states). Table 8 shows the average identification and false positive rates at the sample rates 2~20 MS/s. The experiments allow each ECU to use 1000 frames.

Table 8 Performance at various sample rates for Linear Regression

Sample rate (MS/s)	2	5	10	15	20
Identification rate	97.11	97.85	98.11	98.15	98.21
False positive rate	2.89	2.15	1.89	1.85	1.79

4.3.3 Limitation and Battery/ECU Aging

The method can detect compromised ECUs by monitoring CAN bus. An attack will be detected once a known ECU professes some message identifier affiliated with another normal ECUs. However, if a known ECU abuses its own identifier (that is permitted under normal circumstances) to launch some attack, our system cannot recognize the attack. We mention that this is an open problem in signaling-based ECU identification schemes [1, 14–16] and our focus of the work is on the connection between signal ringing and ECU identification.

Generally, the service life of car battery is of $3 \sim 5$ years and its real usage duration is also related to the driver's driving habits. Therefore, the aging of the car's battery might affect the characteristics of the electrical signal sent by each ECU, and one would see different impact level for different position of the ECU in the CAN network [29]. On the other hand, ECU has a relatively long service life and the aging process is really slow. One may thus not consider the impact of aging on electrical signals.

5 Source Identification on In-Vehicle CAN-FD Networks

Controller area network with flexible data rate (CAN-FD) is supposed to be the next generation of in-vehicle network to dispose of CAN limitations of data payload size and bandwidth. The section discusses ECU identification on CAN-FD network from bus signaling. If a model shows robustness to source identification, then we get convincing evidence on its applicability to forthcoming real vehicles set up by CAN-FD network. ECU identification can be easily extended to intrusion detection against attacks not only initiated by external devices but also internal devices.

5.1 CAN-FD

Robert Bosch GmBH recommends CAN-FD [30] to dispose of CAN limitations of data payload size and bandwidth. Besides its compatibility with CAN, CAN-FD has the advantages: the maximum length of the data field is 64 bytes; it supports variable rates (namely, a frame can use different transmission rates in different stages) and the maximum rate can reach 5Mbit/s (the maximum rate of CAN is 1Mbit/s).

CAN-FD itself does not convey security protection either (similar to CAN) and existing attacks on CAN might also be feasible on CAN-FD. Take masquerade attack on CAN network [13] as an example. Initiating a masquerade attack and not being detected by the system, an adversary needs to stop the transmission of targeted ECU and imitate it to inject attack messages. The attack also works on in-vehicle CAN-FD network. We should explore ECU identification on in-vehicle CAN-FD network.

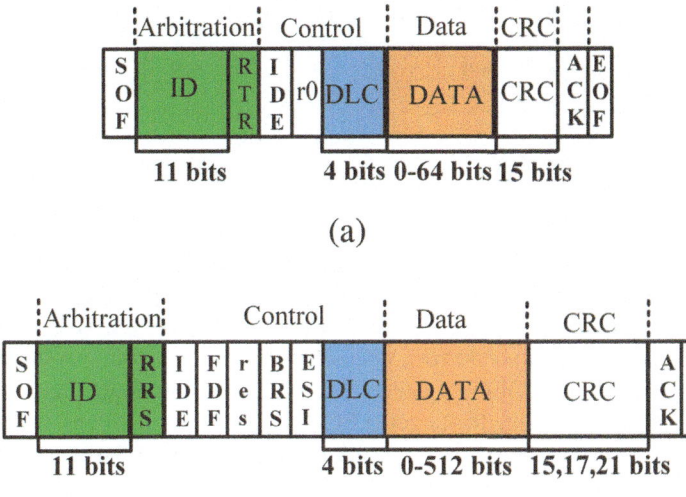

Fig. 10 CAN/CAN-FD frames with 11-bit identifier. (**a**) CAN data frame format. (**b**) CAN-FD data frame format

Comparing CAN-FD with CAN CAN-FD is defined to be compatible with CAN at the physical layer. All CAN-FD controllers can handle a mix of CAN frames and CAN-FD frames. One might use CAN-FD controllers in conjunction with CAN controllers on in-vehicle network. Thus one might see pure CAN frames or both CAN and CAN-FD frames on the bus.

CAN-FD and CAN differ in the format and the length of the data frame (Fig. 10). Compared with CAN frame, CAN-FD adds FDF (Flexible Data Rate Format), BRS (Bit Rate Switch) and ESI (Error State Indicator) fields (see Fig. 10b) [30]. Therein, FDF indicates whether the sent frame is a CAN frame or a CAN-FD frame and BRS stands for bit rate conversion. When the bit is a recessive bit (1), the rate is variable, and when the bit is a dominant bit (0), it is transmitted at a constant rate. ESI is an error status indicator: when ESI is a recessive bit (1), it means that the sending node is in a passive error (otherwise active error) state. A CAN-FD frame is divided into different fields (Fig. 10b). For example, we can set the rate of 2Mbit/s for the data field and 1Mbit/s for the arbitration field, control field and CRC field. The length of the CAN-FD data field is up to 64 bytes, increasing available load.

The maximum rate of CAN arbitration field and data field is no more than 1Mbit/s [3]. However, CAN-FD supports variable rates, and the bit rate of its arbitration field and data field might be different. The arbitration and the ACK stages continue to use CAN2.0 specification (i.e., the highest rate does not exceed 1Mbit/s), and the data field can reach 5Mbit/s through hardware setting, or even higher.

CAN-FD Security For CAN-FD, security experts can pursue stronger security tricks via its higher transmission rates and larger loads. In [31], an IDS was proposed

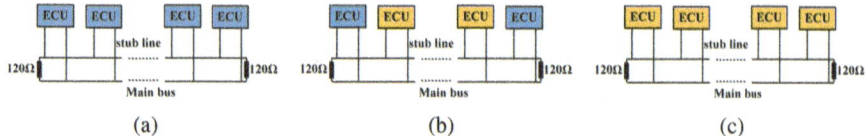

Fig. 11 Network topology. (a) CAN-FD network. (b) CAN/CAN-FD hybrid. (c) CAN network

for in-vehicle CAN-FD network based on topology verification. It uses variations of network topology to identify intrusions by external intruding devices (XIDs), but it cannot detect attacks via the vulnerabilities of existing ECUs. Woo et al. [32] proposed a security architecture for in-vehicle CAN-FD according to ISO 26262. This method may cause GECU (gate ECU) to generate excessive load as it has to encrypt data packets using the targeted ECU's unique key. To relieve pressure on GECU, Agrawal et al. [33] proposed a group-based approach for the communication among different ECUs. However, it should manage a large number of keys which requires a large amount of computing resources of the ECUs, making it beyond instant communication.

Ringing on CAN-FD Bus For CAN-FD, internal components of an ECU mainly include CAN-FD controller, CAN-FD transceiver, and voltage regulator and we have the same rationale of the dominant voltages of (CAN-FD)-H and (CAN-FD)-L on the bus. As in Sect. 2.3, ringing might exist on CAN-FD bus [19, 34, 35].

5.2 System Model

CAN-FD is designed to transmit large amounts of data at a faster rate and to replace CAN in future design. For possible transition mechanism from CAN to CAN-FD, we allow a hybrid topology of CAN and CAN-FD, namely, there exist on the network ECUs sending purely CAN frames, ECUs sending purely CAN-FD frames, and ECUs sending both CAN and CAN-FD frames. In Fig. 11a, the ECUs can send both CAN-FD and CAN frames. In Fig. 11b, blue nodes represent the ECUs that can send both CAN-FD frames and CAN frames, and yellow nodes only send CAN frames. In Fig. 11c, the ECUs only send CAN frames.

Signal Acquisition and Preprocessing To obtain differential signals from CAN-FD/CAN bus prototypes, we first link two probes of an oscilloscope to (CAN-FD)-H/CAN-H and (CAN-FD)-L/CAN-L lines respectively. Then we use the *difference* function in the software of the oscilloscope to calculate the differential signal. As in Sect. 3.1 (and Fig. 5), the trick of five sets L_1, L_2, L_3, L_4, and L_5 is used as well (Fig. 12).

Feature Extraction Statistical features could be extracted from the preprocessed electrical CAN-FD/CAN signals. We use the features in Table 1 as well for each set

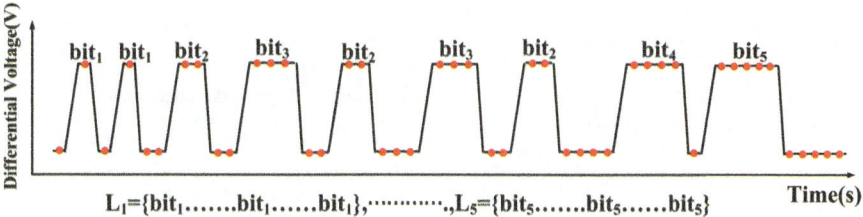

Fig. 12 A CAN-FD/CAN frame is divided into 5 sets

and a total of 40 features for each electrical CAN-FD/CAN signal. Relief-F [25] is also used to weight these features and the feature set in Table 2 can thus be obtained.

Identifying ECUs We use supervised learning, logistic regression (LR) and SVM, to identify the source of CAN-FD/CAN signal. The training phase generates fingerprints from multiple CAN-FD/CAN frames of each ECU. The resulting fingerprints are then used together to train the classifiers. For the testing phase, we have two types of tests. The first is to evaluate the trained model (i.e., whether or not it can determine the source of newly received frames), and the second is on intrusion detection.

5.3 Source Identification and Intrusion Detection

5.3.1 Experiment Setup

The system adapts to different bus prototypes (Fig. 13). Type A (Fig. 13a) contains five CAN-FD nodes that can send both CAN-FD and CAN frames. Type B (Fig. 13b) contains five CAN-FD nodes (the same as in Type A) and four extra CAN nodes that send purely CAN frames. Type C (Fig. 13c) contains five CAN nodes. Although the total number of ECUs in real cars might be up to 70 or even larger, in-vehicle networks are physically divided into several subnets, e.g., power-related or comfort-related. As ringing mainly exists between ECUs and junctions, the rationale of fingerprinting ECUs in real cars is the same as that in our experiments. CAN protocol defines low-speed CAN and high-speed CAN. High-speed CAN connects the ECUs related to the important functions of the vehicles. For example, the ECU that controls the brakes and the ECU that controls acceleration are both on high-speed CAN, and the data transmission speed of high-speed CAN is 500kbit/s. Our CAN bus prototype takes high-speed CAN network topology.

Each CAN node consists of an Arduino UNO board and a CAN shield from Seed Studio. Each CAN shield consists of an MCP2515 controller [26] and an MCP2551 transceiver [27], and the bit rate is 500kbit/s. For CAN-FD nodes, each one consists of a STM32F105 shield and a MCP2517FD controller [36]. MCP2517FD is known as compact, cost-effective and efficient CAN-FD controller and uses SPI interface

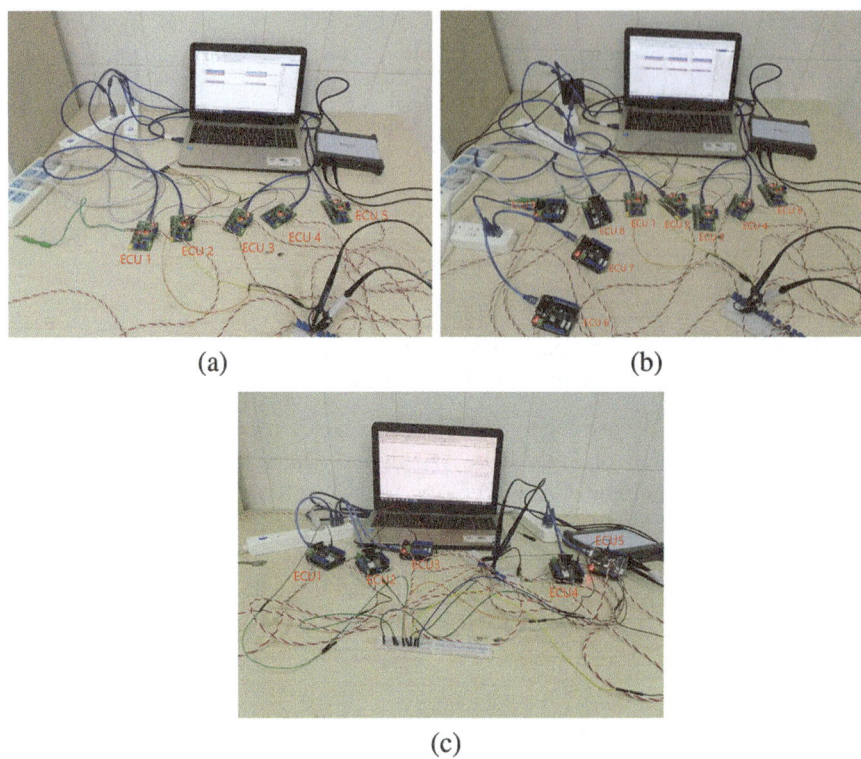

(a)   (b)

(c)

Fig. 13 Three prototypes. (**a**) Type A: CAN-FD nodes, (**b**) Type B: CAN-FD nodes and CAN nodes, (**c**) : CAN nodes

and MCU (Microcontroller Unit) communication. In the experiments, we set the bit rate of MCP2517FD as 1Mbit/s in the arbitration phase, control phase and CRC phase, and 2Mbit/s in the data transmission phase. We mention that using signal characteristics sampled at high bit rate to identify devices is more difficult than at low bit rate. If our method shows effectiveness on the high-speed CAN-FD (and CAN), it would also function well on the low-speed CAN-FD (and CAN, respectively). To maintain the consistency of experimental environments, we require that all the stub lines, oscilloscope, and other components used in the experiments are the same in all three prototypes (except the nodes of different functions).

All ECUs are powered by a battery which supplies electric power to each ECU via USB ports. Main bus (twisted pair as well) should be longer than any other stub line on the network (our configuration sets the length of main bus as the sum of those of stub lines). There is a 120 ohm resistor at each of the two ends of main bus. CAN-FD/CAN signals are measured by the oscilloscope PicoScope 5244D MSO with a sampling rate of 25 MS/s and a resolution of 8 bits. Two probes of the oscilloscope are connected to (CAN-FD)-H/CAN-H and (CAN-FD)-L/CAN-L respectively. For each ECU (CAN-FD or CAN node), we use 200 frames as training set (its size could be adjusted according to the performance of the model).

5.3.2 Sender Identification

5.3.2.1 Sender Identification on Pure CAN

For Type C (Fig. 13c), we consider ringing effect. We execute SVM and LR by using D.R, F.R, and D.R.F.R. The results are shown in Tables 9, 10, and 11. Each diagonal cell represents the accuracy of the two classification algorithms. As expected, D.R suffice to fingerprint ECUs.

5.3.2.2 Using Dominant States and Rising Edges (D.R)

We then evaluate whether the system can correctly classify ECUs for Type A and Type B. Table 12 lists the confusion matrix for 5 ECUs that send CAN-FD frames (Type A). The recognition rate of the system is sufficient to correctly recognize ECUs, and the error rate is very low. Table 13 lists the confusion matrix of 9 ECUs

Table 9 SVM/LR for Type C and D.R

	ECU 1	ECU 2	ECU 3	ECU 4	ECU 5
ECU 1	99.89/99.77	0/0	0/0	0.11/0.23	0/0
ECU 2	0/0	99.59/99.79	0/0	0.41/0.21	0/0
ECU 3	0.14/0.46	0/0	99.76/99.54	0/0	0/0
ECU 4	0/0	0/0	0.2/0.02	99.8/99.98	0/0
ECU 5	0.2/0.08	0/0	0/0	0/0	99.8/99.92

Table 10 SVM/LR for Type C and F.R

	ECU 1	ECU 2	ECU 3	ECU 4	ECU 5
ECU 1	86.52/84.66	0/0	5.23/6.01	8.25/9.33	0/0
ECU 2	0/0	88.21/87.11	6.47/7.56	0/0	5.32/5.33
ECU 3	14.34/11.46	0/0	85.66/88.54	0/0	0/0
ECU 4	0/0	0/0	15.12/14.62	84.88/85.38	0/0
ECU 5	4.32/5.01	0/0	4.66/3.84	5.17/6.23	85.85/84.92

Table 11 SVM/LR for Type C, D.R.F.R

	ECU 1	ECU 2	ECU 3	ECU 4	ECU 5
ECU 1	96.12/95.34	1.81/2.56	0/0	2.07/2.1	0/0
ECU 2	4.79/5.03	95.21/94.97	0/0	0/0	0/0
ECU 3	5.44/4.16	0/0	94.56/95.84	0/0	0/0
ECU 4	0/0	0/0	4.12/5.02	95.88/94.98	0/0
ECU 5	2.81/2.9	0/0	2.34/2.18	0/0	94.85/94.92

Table 12 SVM/LR for Type A and D.R

	ECU 1	ECU 2	ECU 3	ECU 4	ECU 5
ECU 1	99.12/99.34	0/0	0/0	0.88/0.66	0/0
ECU 2	0/0	99.21/99	0/0	0/0	0.79/1
ECU 3	0.24/0.46	0/0	99.76/99.54	0/0	0/0
ECU 4	0/0	0/0	0.12/0.02	99.88/99.98	0/0
ECU 5	0.15/0.08	0/0	0/0	0/0	99.85/99.92

(Type B), of which 5 ECUs send CAN-FD frames, and the remaining 4 ECUs send CAN frames. One may see the system can still correctly classify and recognize ECUs in hybrid network.

5.3.2.3   Using Falling Edges and Recessive States (F.R)

We also consider the recognition rate if F.R are used. As ringing intensity of falling edges of signals is higher than that of rising edges, recognition rate would be affected when falling edges are used. Table 14 shows the results for Type B and Table 15 shows the recognition rates 81.54~86.21% for Type A. We can see really low recognition rates.

5.3.2.4   Using (Dominant States and Rising Edges) and (Falling Edges and Recessive States) (D.R.F.R)

We also compare the execution rates when the system uses D.R.F.R. Tables 16 and 17 show the results of Type A and Type B respectively, both lower than that using D.R.

5.3.3   Detecting Known ECUs

Now we evaluate whether our system can recognize malicious frames sent by an attacker using known ECUs. For Type C (Fig. 13c), we assume that ECU 1 is normal and an attacker can use other ECUs to send messages with the same identifier as ECU 1. We collect a total of 500 frames, of which 300 are used as attack frames and the rest as normal. Table 18 shows a detection rate 99.01%. For Type A (Fig. 13a), we use the same assumptions and operations as for Type C and achieve a detection rate of 98.5% (Table 18). For Type B (Fig. 13b), we regard ECU 7, ECU 8 and ECU 9 as attackers (capable of sending both CAN and CAN-FD frames). We collect 1000 frames, of which 600 are used as attack frames and the rest are normal. Table 18 shows the results with comparable performance to Type A and Type C.

Table 13 Confusion matrix using SVM/LR respectively for Type B and D.R

	ECU 1	ECU 2	ECU 3	ECU 4	ECU 5	ECU 6	ECU 7	ECU 8	ECU 9
ECU 1	98.89/99.15	0/0	0/0	0/0	0.91/0.7	0.01/0.03	0/0	0/0	0.19/0.12
ECU 2	0/0	98.01/99.21	0/0	1.2/0.78	0/0	0/0	0.79/0.01	0/0	0/0
ECU 3	0/0	0/0	98.99/99.01	0.92/0.89	0/0	0/0	0/0	0/0	0.09/0.1
ECU 4	0/0	0/0	0/0	99.29/99.11	0/0	0/0	0.7/0.89	0.01/0	0/0
ECU 5	0/0	0/0	0/0	0/0	98.99/99.31	0/0	0/0	0/0	1.01/0.69
ECU 6	1.01/0.9	0/0	0/0	0.01/0.1	0/0	98.98/99	0/0	0/0	0/0
ECU 7	1.32/0.98	0/0	0/0	0/0	0.01/0.01	0/0	98.67/99.01	0/0	0/0
ECU 8	0/0	0/0	0.9/0.96	0.01/0.03	0/0	0/0	0/0	99.09/99.01	0/0
ECU 9	1.11/1.8	0/0	0/0	0/0.03	0/0	0/0	0/0	0/0	98.89/98.17

Table 14 Confusion matrix using SVM/LR respectively for Type B and F.R

	ECU 1	ECU 2	ECU 3	ECU 4	ECU 5	ECU 6	ECU 7	ECU 8	ECU 9
ECU 1	79.89/78.15	15.98/16.51	3.12/4.32	0/0	0/0	0/0	0/0	0/0	1.01/1.02
ECU 2	0/0	80.01/79.21	0/0	0/0	16.01/17.99	0/0	0/0	3.78/2.8	0/0
ECU 3	0/0	0/0	78.01/79.1	0/0	18.53/17.01	0/0	0/0	3.73/3.89	0/0
ECU 4	16.01/15.99	0/0	0.01/0.19	80.29/80.11	3.6/3.71	0/0	0/0	0/0	0/0
ECU 5	0/0	0/0	16.48/15.91	0/0	78.99/79.31	0/0	1.32/1.01	0/0	3.21/3.77
ECU 6	15.01/14.98	0/0	0/0	3.1/3.25	0.91/0.76	80.98/81.01	0/0	0/0	0/0
ECU 7	15.32/15.91	0/0	0/0	0/0	1.01/1.1	0/0	83.67/82.99	0/0	0/0
ECU 8	0/0	0/0	15.91/14.99	2.01/2.18	5.9/6.86	0/0	0/0	80.09/81.01	1.99/1.82
ECU 9	14.11/15.01	0/0	0/0	1.01/1.99	0/0	0/0	0/0	0.99/0.83	83.89/82.17

Table 15 SVM/LR for Type A and F.R

	ECU 1	ECU 2	ECU 3	ECU 4	ECU 5
ECU 1	84.12/85.34	12/13.14	0/0	3.88/1.52	0/0
ECU 2	0/0	86.21/85	11.79/12.78	2/2.22	0/0
ECU 3	5.14/6.46	4.12/4.36	82.76/81.54	3.51/3.96	4.47/3.68
ECU 4	0/0	15.82/16.62	0/0	84.18/83.38	0/0
ECU 5	0/0	12.32/12.01	2.93/3.17	0/0	84.75/84.82

Table 16 SVM/LR for Type A, D.R.F.R

	ECU 1	ECU 2	ECU 3	ECU 4	ECU 5
ECU 1	94.32/95.24	3.36/3.14	0/0	0/0	2.32/1.62
ECU 2	0/0	93.21/94.21	5.78/5.01	0/0	1.01/0.78
ECU 3	5.14/1.46	0/0	93.76/94.54	1.1/0.45	0/0
ECU 4	0/0	5.2/6.33	0/0.09	94.8/93.58	0/0
ECU 5	5.05/5.15	0.2/0.23	0/0	0/0	94.75/94.62

5.3.4 Detecting Unknown ECUs

We adopt a threshold-based method. For Type A, we first remove ECU 5 and obtain about 500 frames from the remaining ECUs to train a model. Then we plug ECU 5 back to the network and sample a total of 600 frames now. The obtained model is used to classify newly collected data and Fig. 14 shows False Positive (FP) and False Negative (FN) rates. The recognition rate can be up to 99.36% at threshold = 0.8. For Type B, we remove ECU 8, use the remaining ECUs to train a new model, and then plug ECU 8 back to the network. We collect now a total of 1000 data which will be classified by the obtained model. Figure 15 shows recognition rate 99% at 0.7. Type C uses similar method and Fig. 16 shows 99.1% recognition rate at 0.83.

5.4 Discussions

Sample Rate The experiments could be reproduced at various sample rates, especially for Type B. At different sample rate one will be at different position of sample sizes (which might be closely related to system performance). Table 19 shows the average identification and false positive rates at the sample rates 10 ~25 MS/s (1000 frames for each ECU).

Comparable Performance Between Type A and Type C For same topology, one may note considerable performance for Type A (CAN-FD) and Type C (CAN) by using any signal characteristics (rising edges, dominant states, falling edges, and recessive states). In fact, Type C could obtain generally a tiny little better recognition rate than Type A. First, CAN-FD supports data size up to 512 bits, drastically larger than 64 bits in CAN specification, thus the cumulative effect of ringing for Type A might be more powerful than for Type C. Second, CAN-FD provides variable

Table 17 SVM/LR for Type B and D.R.F.R

	ECU 1	ECU 2	ECU 3	ECU 4	ECU 5	ECU 6	ECU 7	ECU 8	ECU 9
ECU 1	93.89/94.15	5.98/5.51	0.13/0.34	0/0	0/0	0/0	0/0	0/0	0/0
ECU 2	0/0	92.01/93.21	0/0	0/0	6.01/5.89	0/0	0/0	1.98/0.9	0/0
ECU 3	0/0	0/0	94.01/93.1	0/0	5.53/6.01	0/0	0/0	0.46/0.89	0/0
ECU 4	3.9/4.01	0/0	0/0	95.29/95.11	0.81/0.88	0/0	0/0	0/0	0/0
ECU 5	0/0	0/0	5.8/6.91	0/0	93.99/92.31	0/0	0/0	0/0	0.21/0.78
ECU 6	6.01/6.4	0/0	0/0	2.1/1.5	0/0.01	91.98/92.09	0/0	0/0	0/0
ECU 7	5.32/4.91	0/0	0/0	0/0	1.08/1.01	0/0	93.67/94.01	0/0	0/0
ECU 8	0/0	0/0	0.9/0.2	0.01/0.03	5.9/6.86	0/0	0/0	93.09/92.01	1.01/1.82
ECU 9	1.11/1.8	0/0	0/0	1.01/0.03	0/0	0/0	0/0	5.1/6.01	93.89/92.17

Table 18 IDS using Support Vector Machines/Logistic Regression

Prototype	True	Predicted (SVM)		Predicted (LR)	
		No attack	Yes	No attack	Yes
CAN-FD	No attack	99.38	0.62	99.85	0.42
	Yes	1.5	98.5	1.88	98.12
CAN-FD&CAN	No attack	99.01	0.99	99.11	0.89
	Yes	1.18	98.82	1.89	98.11
CAN	No attack	99.58	0.52	99.44	0.56
	Yes	0.99	99.01	0.89	99.11

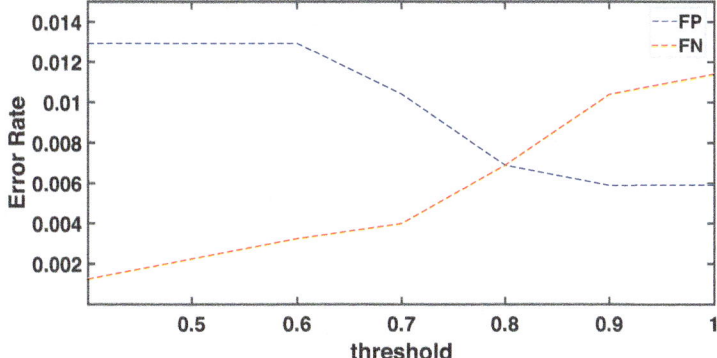

Fig. 14 Error rates at varying thresholds (Type A)

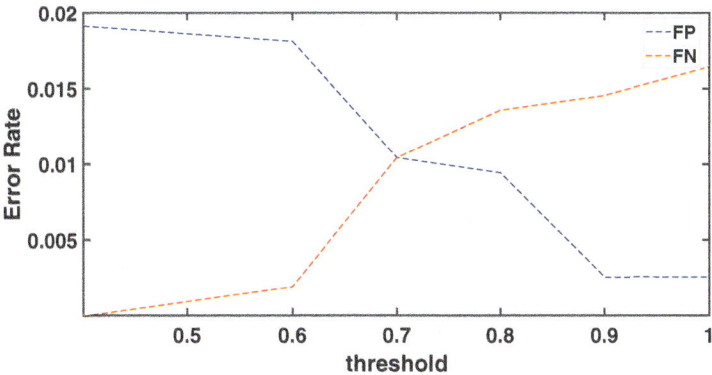

Fig. 15 Error rates at varying thresholds (Type B)

transmission rate and the experiments specify 2Mbit/s for data field of CAN-FD frames and 1Mbit/s for other fields (e.g., arbitration, control and CRC), whereas Type C regulates 500kbit/s. Namely, we have the bit width 2000 ns in a CAN frame, and 1000 ns in non-data field of and 500 ns in data field of a CAN-FD frame. Now, it is more likely for Type A (than Type C) that ringing of recessive states functions

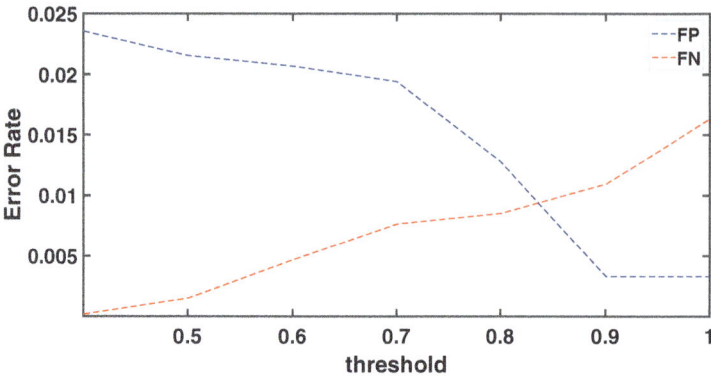

Fig. 16 Error rates at varying thresholds (Type C)

Table 19 LR Performance at various sample rates

Sample rate (MS/s)	10	15	20	25
Identification rate	97.11	98.95	99.01	99.15
False positive rate	2.89	1.05	0.99	0.85

unceasing (even though the bit itself was already completed on the network)[1] and thus involves the coming dominant states before it attenuates to be unnoticeable.

Applicability to CAN-FD Network in Real Vehicles The controllers used herein conform to ISO11898-1:2015 and support CAN-FD [36]. Possible transition mechanism from CAN to CAN-FD (i.e., Type A and Type B) is also considered. The results show expressive evidence on the applicability to forthcoming real vehicles set up by CAN-FD network. These results could be used as a step forward and a guidance on securing the commercialization and batch production of in-vehicle CAN-FD network in the near future.

Environmental Factors In real vehicles, the changes of internal temperature will affect the characteristics of electrical signals. A typical example is that the voltage output may deviate from 0.012 to 0.026 V [1] when we start the vehicle from a cooled turn-off engine to warmed-up. This may also exist for CAN-FD network. Howbeit, CAN-FD frames are longer than 512 bits, and the number of dominant states contained would be much likely greater than that in CAN frame. We might thus expect an acceptable impact of temperature changes on signal characteristics (and further on the system).

---

[1] It is reported [34, 37] that for CAN-FD, high-speed data phase and low-speed arbitration phase challenge the same ringing surrounds (as ringing does not depend on transmission rate), and ring of some recessive bit might not converge until criterion and interfere with the next dominant bit.

Battery/ECU Aging Battery aging might affect the characteristics of the electrical signals. For now, however, we can not track the impact of battery aging on the system by simulating CAN-FD nodes and car battery as there is no CAN-FD vehicle for real driving. This interesting topic might be explored in the coming future. On the other hand, ECU has a relatively long service life and the aging process is really slow. It is thus rational not to consider the impact of ECU aging on electrical signals.

6 Conclusion

The chapter introduces in-vehicle ECU identification by using CAN electrical Signaling. This can be viewed as side-channel information exploit on CAN networks. In designing the identification algorithms, signal characteristics of different phases in the signals has different impacts on the algorithm accuracy. The problem of source identification is also important on in-vehicle CAN-FD networks. ECU identification algorithms can be trivially extended to in-vehicle IDS systems.

Acknowledgments The author is supported by the National Natural Science Foundation of China (61971192), Shanghai Municipal Education Commission (2021-01-07-00-08-E00101), and Shanghai Trusted Industry Internet Software Collaborative Innovation Center.

References

1. Kneib, M., Huth, C.: Scission: signal characteristic-based sender identification and intrusion detection in automotive networks. In: Proceedings of the 2018 ACM SIGSAC conference on computer and communications security, pp. 787–800 (2018)
2. ISO 11898-2. Road vehicles - Controller area network (CAN) - Part2: High-speed medium access unit. ISO Standard-11898, International Standards Organisation (ISO) (Dec. 2016).
3. Robert Bosch GmbH. CAN specification version 2.0, Robert Bosch GmbH, Stuttgart, Germany, 1991. Available: http://www.bosch.com (1991)
4. Koscher, K., Czeskis, A., Roesner, F., Patel, S., Kohno, T., Checkoway, S., McCoy, D., Kantor, B., Anderson, D., Shacham, H., Savage, S.: Experimental security analysis of a modern automobile. In: IEEE symposium on security and privacy (2010)
5. Miller, C., Valasek, C.: Adventures in automotive networks and control units. Def Con **21**, 15 (2013)
6. Miller, C., Valasek, C.: Remote exploitation of an unaltered passenger vehicle. Black Hat U S A **2015**, 91 (2015)
7. Tencent Keen Security Lab. Experimental security assessment of Mercedes-Benz cars. https://keenlab.tencent.com/en/whitepapers/Mercedes_Benz_Security_Research_Report_Final.pdf
8. Kang, M., Kang, J.: A novel intrusion detection method using deep neural network for in-vehicle network security. In: IEEE 83rd vehicular technology conference (VTC Spring), pp. 1–5 (2016)
9. Muter, M., Asaj, N.: Entropy-based anomaly detection for in-vehicle networks. In: Intelligent vehicles symposium (IV). IEEE (2011)
10. Song, H.M., Kim, H.R., Kim, H.K.: Intrusion detection system based on the analysis of time intervals of CAN messages for in-vehicle network. In: 2016 International conference on information networking, pp. 63–68 (2016)

11. Taylor, A., Leblanc, S., Japkowicz, N.: Anomaly detection in automobile control network data with long short-term memory networks. In: DSAA 2016, pp. 130–139 (2016)
12. Guo, F., Wang, Z., Du, S., Li, H., Zhu, H., Pei, Q., Cao, Z., Zhao, J.: Detecting vehicle anomaly in the edge via sensor consistency and frequency characteristic. IEEE Trans. Veh. Technol. **68**(6), 5618–5628 (2019)
13. Cho, K.-T., Shin, K.G.: Fingerprinting electronic control units for vehicle intrusion detection. In: Proc. of the 25th USENIX security symposium, Aug. (2016)
14. Cho, K., Shin, K.G.: Viden: attacker identification on in-vehicle networks. In: Proceedings of 2017 ACM CCS, pp. 1109–1123 (2017)
15. Choi, W., Jo, H.J., Woo, S., Chun, J.Y., Park, J., Lee, D.H.: Identifying ECUs using inimitable characteristics of signals in controller area networks. IEEE Trans. Veh. Technol. **67**(6), 4757–4770 (2018)
16. Choi, W., Joo, K., Jo, H.J., Park, M.C., Lee, D.H.: VoltageIDS: low-level communication characteristics for automotive intrusion detection system. IEEE Trans. Inf. Forens. Secur. **13**, 2114 (2018)
17. Foruhandeh, M., Man, Y., Gerdes, R., Li, M., Chantem, T.: Simple: single-frame based physical layer identification for intrusion detection and prevention on in-vehicle networks. In: 35th Annual computer security applications conference, pp. 229–244 (2019)
18. Murvay, P.S., Groza, B.: Source identification using signal characteristics in controller area networks. IEEE Signal Process. Lett. **21**(4), 395–399 (2014)
19. Kim, G., Lim, H.: Ringing suppression in a controller area network with flexible data rate using impedance switching and a limiter. IEEE Trans. Veh. Technol. **68**(11), 10679–10686 (2019)
20. Lim, H., Kim, G., Kim, S., Kim, D.: Quantitative analysis of ringing in a controller area network with flexible data rate for reliable physical layer designs. IEEE Trans. Veh. Technol. **68**(9), 8906–8915 (2019)
21. Mori, H., Suzuki, Y., Maeda, N., Obata, H., Kishigami, T.: Novel ringing suppression circuit to increase the number of connectable ECUs in a linear passive star CAN. In: International symposium on electromagnetic compatibility - EMC EUROPE, Rome, pp. 1–6 (2012)
22. High-Speed CAN (HSC) for vehicle applications at 500 kbps, SAE J2284-3, SAE International, Warrendale, PA, USA (2002)
23. Studnia, I., Nicomette, V., Alata, E., Deswarte, Y., Kaniche, M., Laarouchi, Y.: Survey on security threats and protection mechanisms in embedded automotive networks. In: 2013 43rd Annual IEEE/IFIP conference on dependable systems and networks workshop, pp. 1–12 (2013)
24. Checkoway, S., McCoy, D., et al.: Comprehensive experimental analyses of automotive attack surfaces. In: 20th USENIX security symposium. USENIX Association (2011)
25. Kononenko, I.: Estimating attributes: analysis and extensions of RELIEF. In: Machine learning: ECML-94, pp. 171–182. Springer, Berlin Heidelberg (1994)
26. Microchip-Corporation: Stand-Alone CAN Controller with SPI Interface (2005). Microchip MCP2515. https://www.mouser.com/datasheet/2/268/MCP2515-Stand-Alone-CAN-Controller-with-SPI-200018-708845.pdf
27. Microchip-Corporation: MCP2551 High-Speed CAN Transceiver (2007). http://ww1.microchip.com/downloads/en/devicedoc/21667e.pdf
28. Muller, K.-R., Mika, S., Ratsch, G., Tsuda, K., Scholkopf, B.: An introduction to kernel-based learning algorithms. IEEE Trans. Neural Netw. **12**(2), 181–201 (2001)
29. Kneib, M., Schell, O., Huth, C.: On the robustness of signal characteristic-based sender identification. CoRR, vol. abs/1911.09881 (2019)
30. Robert Bosch GmbH. CAN with flexible data-rate (2012). https://www.can-cia.org/fileadmin/resources/documents/proceedings/2012_hartwich.pdf
31. Yu, T., Wang, X.: Topology verification enabled intrusion detection for in-vehicle CAN-FD networks. IEEE Commun. Lett. **24**(1), 227–230 (2019)
32. Woo, S., Jo, H.J., et al.: A practical security architecture for in-vehicle CAN-FD. IEEE Trans. Intell. Transp. Syst. **17**(8), 2248–2261 (2016)
33. Agrawal, M., Huang, T., et al.: CAN-FD-Sec: improving security of CAN-FD protocol. In: ESORICS 2018, Lecture notes in computer science 11552, pp. 77–93 (2018)

34. Mori, H., Suzuki, Y., et al.: Novel ringing suppression circuit to increase the number of connectable ECUs in a linear passive star CAN. In: International symposium on electromagnetic compatibility - EMC EUROPE, pp. 1–6 (2012)
35. Lim, H., Kim, G., et al.: Quantitative analysis of ringing in a controller area network with flexible data rate for reliable physical layer designs. IEEE Trans. Veh. Technol. **68**(9), 8906–8915 (2019)
36. Microchip-Corporation. External CAN FD Controller with SPI Interface MCP2517FD (2017). http://ww1.microchip.com/downloads/en/DeviceDoc/MCP2517FD-External-CAN-FD-Controller-with-SPI-Interface-20005688B.pdf
37. Islinger, T., Mori, Y.: Ringing suppression in CAN FD networks. CAN Newsl. Jan, pp. 12–16 (2016)

Machine Learning for Security Resiliency in Connected Vehicle Applications

Srivalli Boddupalli, Richard Owoputi, Chengwei Duan,
Tashfique Choudhury, and Sandip Ray

1 Resiliency Needs and Challenges in CAV Applications

Automotive systems have evolved over the last two decades from primarily mechanical and electro-mechanical systems into complex cyber-physical systems with a wide range of communication and sensory capabilities. In addition to a variety of sensors (e.g., Radar, Lidar, etc.), a modern vehicle is equipped with various interfaces for Internet connectivity, and vehicular communications (V2X) technology (e.g., Digital Short Range Radio) to interact with other vehicles (V2V), components of transportation infrastructure (V2I), or other electronic devices connected to the Internet (V2IoT). The combination of sophisticated sensors and communication enables *connected and autonomous vehicle* (CAV) applications, i.e., applications that exploit cooperative information sharing among vehicles and infrastructures for streamlining traffic movement, improving road safety, and efficient infrastructure utilization. CAV applications being developed today include platooning [5], cooperative dynamic route management [2, 3], intersection management [12], etc. With increasing proliferation of connectivity and autonomy of vehicles, the trend is towards increasing sophistication of such applications and the consequent potential to bring in transformative impact on road safety, passenger comfort, and environmental sustainability.

However, one critical challenge with CAV applications is their vulnerability to a spectrum of cyber-attacks. An adversary can easily compromise the sensory and communication inputs to disrupt traffic movement, cause catastrophic accidents, and bring down the transportation infrastructure. A key problem with these attacks is that an adversary no longer needs to actually hack into the hardware or software

S. Boddupalli · R. Owoputi · C. Duan · T. Choudhury · S. Ray (✉)
Department of ECE, University of Florida, Gainesville, FL, USA
e-mail: bodsrivalli12@ufl.edu; rowoputi@ufl.edu; duan.c@ufl.edu; choudhury.t@ufl.edu; sandip@ece.ufl.edu

© The Author(s), under exclusive license to Springer Nature Switzerland AG 2023
V. K. Kukkala, S. Pasricha (eds.), *Machine Learning and Optimization Techniques for Automotive Cyber-Physical Systems*, https://doi.org/10.1007/978-3-031-28016-0_16

of the vehicle that is the target of the cyber-attack: sending misleading or even malformed V2X messages or sensory data is often sufficient to disrupt the connected car ecosystem.

The focus of this chapter is the problem of *real-time resiliency* in CAV applications, against adversaries that target compromising V2X or sensory inputs. We refer to these inputs as "perception inputs" or "perception channels". The impact of a successful compromise can be a perturbation of some (subset of) perception channels involved in the application, such that the inputs received would be different from actual. For instance, in Cooperative Adaptive Cruise Control (CACC), a vehicle $\mathcal{E}$ receives the velocity, relative position, and acceleration of its preceding vehicle $\mathcal{P}$; during an attack, the values received by $\mathcal{E}$ would be perceived to be different from ground truth. The focus of *real-time resiliency* is to augment the application functionality so that $\mathcal{E}$ can perform safely and efficiently, even during attack.

1.1 Constraints

Designing real-time resiliency for practical CAV applications is a challenging proposition. A viable solution must address the following key issues (among others).

How Can We Identify a Suitable Threat Model for a Given Application? The security requirements vary from one cooperative driving application to the other based on the application objectives. Consequently, the relevant adversaries to defend against, and the impact of a given attack on the target vehicle also vary from one application to the other. For instance, an eavesdropping attack may be considered unimportant for cooperative collision detection application. However, for a routing service application it may be paramount to protect private navigation data of the target vehicles from an unauthorized entity. While it is important to consider a realistic threat model that can account for the relevant attack orchestrations, an all-powerful adversary with limitless capabilities cannot be defended against by any security solution. For instance, consider a CAV application where a vehicle computes driving decisions based on the sensory data specifying the states (position, velocity, acceleration, etc.) of all the vehicles in the vicinity. If an adversary collusively corrupts all the sensory data in a way that all kinematics equations remain valid but the values are different from ground truth, then it is impossible for the victim vehicle to determine if the values it receives are ground reality or corrupted. For instance, an multi-channel adversary could replace the velocity of the preceding vehicle with a different value while adjusting the position and acceleration accordingly so that the laws of kinematics are satisfied. Such an adversary is clearly "all powerful" in the sense that it is impossible to defend against. Therefore it becomes essential to strike the right balance between identifying a threat model that is practical while also accounting for the most relevant adversaries compromising the application objectives.

How Can We Make Sure that the Resiliency Is Viable? A viable solution should address the diverse spectrum of attacks on CAVs, while obeying the safety requirements and automotive platform constraints including limited computational resources, strict timing requirements, stringent cost and time to market constraints, etc. At the same time, it is required that the security solution should be capable of handling unknown attack scenarios.

How Can We Validate the Solution? Testing and validation are crucial in developing a security solution for CAV systems. However, real-world testing is not a feasible option due to road safety concerns. This may require the use of simulation environments for validation instead. Unfortunately, most automotive simulators available for the research community are not sophisticated enough to provide a flexible simulation environment to validate the solution in realistic attack scenarios. The diverse and evolving attack spectrum further complicates the validation process.

1.2 REDEM: *Vision for ML-Based Resiliency*

In recent work [6], we have put forward a generic approach, which we call REDEM (for "Real-time Detection and Mitigation"), to address real-time resiliency requirements in CAV applications to protect adversaries compromising perception inputs. REDEM is not a specific architecture: after all, note from above that a resiliency solution must be customized for different CAV applications and different target adversaries. Instead, REDEM represents a systematic methodology for architecting, tuning, and validating a resiliency solution. At the heart of REDEM is the idea that it is possible for a vehicle to detect adversarial actions through a machine learning model designed (and tuned) to predict normal behavior pattern when engaged in the targeted CAV application. The REDEM infrastructure includes a configurable, flexible "architectural skeleton" (described below) to realize this vision, together with recipes for (1) configuring the skeleton into an architectural solution for a given CAV application against a specific adversary model, and (2) providing a comprehensive validation of such architectures.

1.3 Overview of the Chapter

In this chapter, we provide the vision of REDEM and its realization in a fundamental but representative CAV application, Cooperative Adaptive Cruise Control (CACC). Our goal for this chapter is not necessarily to advocate REDEM as an instrument for designing CAV resiliency. Furthermore, we eschew rehashing technical results about the various REDEM incarnations, except as necessary for the completeness of the chapter or to explain the intuition behind a specific design choice; the readers interested in a more technical treatment of the REDEM architecture for specific CAV applications and their validation are referred to previous publications on the subject [7, 9–11]. Instead, in this chapter we endeavor to elucidate the thinking behind the many architectural decisions, challenges encountered, and the approaches taken to address them. We believe the lessons from REDEM could carry over to other applications of ML targeting real-time security resiliency in various critical infrastructures, particularly under computational resource limitations.

The remainder of the chapter is organized as follows. Section 2 introduces the high-level design of REDEM and explains the relevance (and requirements) of ML-based resiliency for CAV applications. Sections 3 and 4 present a variety of challenges involved in making such a solution work, and REDEM's approach to address these challenges. In Sect. 5 we demonstrate the efficacy of REDEM in an illustrative, foundational CAV application. We conclude in Sect. 6.

2 REDEM Basics

At the level of usage, REDEM can be envisioned as a vehicular service for connected vehicles. A vehicle can subscribe to the service as long as it includes a certain on-board architecture for ML-based anomaly detection described below. Figure 1 shows the overall setup of REDEM. We refer to the subscribing vehicle as the *ego vehicle*,"$\mathcal{E}$", and all of REDEM analysis is done from the point of view of this vehicle. Data from all subscribing vehicles is periodically uploaded to a trusted cloud server for progressively refining ML models used by the on-board hardware; $\mathcal{E}$ periodically updates the on-board system by downloading the latest ML models. The communication with cloud is performed when $\mathcal{E}$ is connected to Internet through a

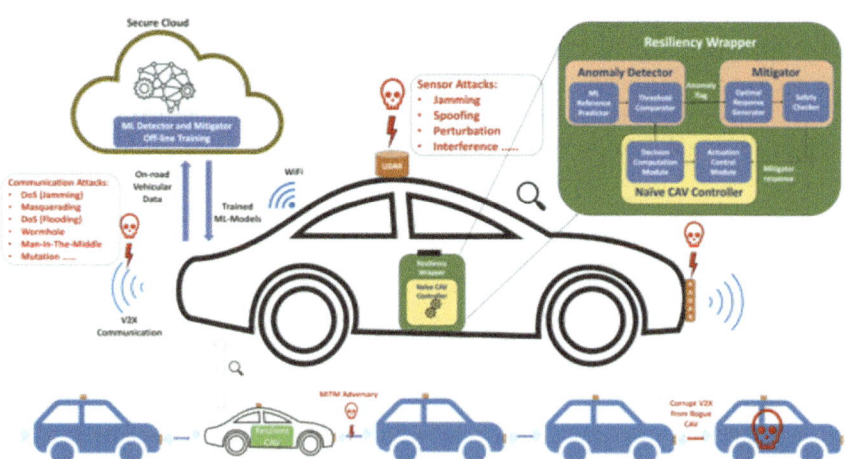

Fig. 1 REDEM-augmented CAV engaging in cooperative autonomous driving with neighbouring CAVs

trusted network, e.g., when stationary at the owner's residence; on-road connectivity with cloud is not necessary. During driving operations, the on-board hardware automatically detects anomalies using the trained ML model installed in $\mathcal{E}$, and performs mitigation.

2.1 Architecture

The key insight behind REDEM on-board design is that the architecture of most CAV applications follow a standard template with two major components: (1) *Decision Computation Module* and (2) *Actuation Control Module*. Given the sensory and V2X inputs pertaining to the application, Decision Computation Module computes the desired actions of the vehicle, and Actuation Control Module generates the control commands for the actuators. Correspondingly, REDEM augments this template with the following two resiliency components to defend against adversarial attacks.

1. *Anomaly Detector* is responsible for detecting suspicious communication or sensory inputs.
2. *Mitigator* is responsible for applying the appropriate alternate action to the vehicle in response to a detected anomaly.

The role of ML in REDEM is in the design of the Anomaly Detector and Mitigator components. More precisely, anomaly detection is implemented through deployment of an ML-based *predictor* model that is trained to learn the normal behavior of Decision Computation Module. The output of Predictor is compared against the (real) output Decision Computation Module. A deviation beyond a pre-defined threshold is classified as an anomaly. If no anomaly is detected, the output of Decision Computation Module is applied to the vehicle; otherwise, Mitigator is triggered.

2.2 Appropriateness of ML-Based Solution

CAV applications represent a domain where safety requirements are paramount. Given that the resiliency solution influences the driving behavior of a vehicle, safety requirements obviously extend to the resiliency solution as well. In particular, any driving decision generated from an automated source must not increase the risk of accident. This applies particularly to any system that performs real-time mitigation in response to detected anomalies: road safety should not be compromised by the mitigating action irrespective of whether the response is to an input classified as anomalous in the context of a real attack or simply due to the imprecision/inaccuracy in the detection algorithm. Given the criticality of safety requirement, it is natural to ask why one would consider ML-based solutions to address the resiliency question in CAV applications. After all, machine learning approaches are inherently probabilistic: even the best ML solution would incur errors in some cases. Would it not be more appropriate to consider a technology that would provide a more deterministic safety guarantee?

Unfortunately, the answer is "no". To understand the reason, note that a key requirement for resiliency solutions is that they must enable protection against a spectrum of attacks. In particular, it is infeasible to have a different solution for each individual attack. Aside of the fact that the number of potential attack mechanisms already available today is prohibitively large, we can anticipate several more to be discovered during the long life time of the vehicle. Since resiliency is a design solution, it will be difficult (and sometimes impossible) to patch the design in field in response to each new attack discovered after deployment. A corollary is that the resiliency solution must be equipped with mechanisms to address the so-called *zero-day attacks*, i.e., attacks not known at design time but subsequently discovered when the vehicle is in field. To our knowledge, machine learning is one of the only few known technologies that enable potential prediction and analysis of previously unknown scenarios, based on the similarity of the new scenario with those the model has been trained for. Furthermore, the need for zero-day attack resiliency undermines any argument of deterministic (or complete) protection against the spectrum of attacks: after all, if an attack is not known at design time one cannot directly guarantee that the resiliency solution protects against that attack.

Nevertheless, it is non-trivial to actually create a practically viable resiliency solution using the technology. In Sects. 3 and 4, we consider some of the challenges and considerations involved. For each of the challenges discussed, we briefly mention the REDEM approach to addressing the challenge. Note that the goal is not to specifically advocate the REDEM approach itself but to provide a sense of the kind of thinking that one has to carry on to make an ML-based resiliency solution viable for a safety-critical multi-agent cyber-physical application domain.

There has been significant research in developing resiliency solutions for CAVs against adversaries compromising the perception systems. In addition to machine learning approaches, there has been work on control-theoretic approaches for detecting attacks on CAV applications [1, 13, 17]. Control-theoretic solutions enable a more deterministic analysis than machine learning. However, these solutions indeed suffer from the problem of being point solutions to specific vulnerabilities alone. For instance, control-theoretic solutions proposed to defend Cooperative Adaptive Cruise Control against Denial-of-Service attacks on V2X communications or spoofing attacks on sensor systems are tightly coupled to specific attack mechanisms. Consequently, an adversary can easily evade these protections by tweaking the attack mechanisms to break the assumptions made in the solution design. Correspondingly, while ML-based solutions have been devised before to detect anomalies in cooperative connected vehicle applications [4, 14], they did not account for *real-time resiliency*. Rather, these techniques are used to detect a compromised execution off-line through post-analysis of the communications or sensory inputs provided to the vehicle vis-a-vis ground truth.

3   Architectural Considerations

Coming up with a resiliency architecture requires addressing a variety of challenges. While ML is the central component of a viable real-time resiliency, it is not the only thing. The resiliency solution must define a system that incorporates the ML prediction together with other components (e.g., the original application, mitigation, anomaly source identification, etc.). In this section, we consider the thought process behind coming up with the architecture of this overall system, the exploration challenges, and the REDEM approach for addressing them. Implementation, tuning, and validation of the ML components in particular will be discussed in Sect. 4.

3.1   Small Data Problem

The efficacy of any ML-based system depends upon the availability of high-quality data. So a critical question task is: how do we get copious high-quality data necessary to make the ML-based predictions viable? Note that in traditional applications of ML (e.g., recommendation systems) this problem is addressed simply by collecting data for a longer duration. Unfortunately, that does not work for a domain like cyber-security, since finding one (or a few) security vulnerabilities generally triggers a mitigation response (possibly through patching, point fixes, or sometimes design overhaul) resulting in the previous vulnerabilities being obsolete and possibly making ways for newer attacks and compromises. The lack of data represents a vexing problem in security and is known to be a bottleneck in the application of ML in cyber-security solutions.

To address the small data problem in REDEM, our key insight is that while the data on security attacks is indeed limited, normal behavior data is in fact plentiful. Furthermore, normal behavior data follows the typical characteristic of standard ML domains: more data can be obtained by simply collecting data for a longer duration. REDEM makes use of this observation by defining the resiliency problem in terms of *anomaly detection* (capturing deviations from normal behavior), rather than as *classification* (categorizing an input into normal or attack classes). The formulation as anomaly detection implies that the ML models need to be trained to predict only the *normal behavior*; attack or adversarial data is not necessary. Furthermore, data collector in REDEM enables progressive improvement of the ML model through continuous real-world data collection.

However, the small data problem does have repercussions on parameter tuning and validation, which must be done before the application is deployed in field. We discuss those challenges in Sect. 4.

3.2   *Resource Constraints*

Vehicular systems are resource-constrained in terms of performance and power. Although automotive systems can be considered relatively high-performance compared to many other Internet-of-Things devices, operations with high computational complexity are infeasible. The situation is exacerbated in the case of resiliency solutions because of the need for real-time response: if the response of a resiliency solution depends on the result of a computation, viability of the solution relies on the feasibility of carrying out the computation within limited computing resources under a tight upper bound on time. Indeed, resource constraints preclude traditional hardware security mitigations such as high-overhead cryptography-based approaches or authentication techniques. Resource limitations affect the choice of ML as a resiliency solution as well, given the computational needs of ML.

Addressing the resource limitation problem in REDEM requires a more careful dissection of the source of computational overhead in ML. Roughly, there are two sources of computational overhead in ML-based systems. First is the cost of training through a substantially large set of examples to ensure sufficient prediction accuracy. Second is the cost of inference (or prediction) of an input in field as normal or anomalous. The way REDEM architecture ameliorates the training cost is to separate the training from in-field inference. Training in REDEM is performed offline in the cloud, and no real-time communication is required with the trained model during prediction in-field: the trained model is downloaded periodically and deployed into the on-board architecture. Optimizing inference cost is more tricky. Inference has to be done in real time using in-vehicle electronics: reliance on a cloud-based infrastructure for this activity would result in a requirement of continuous connectivity which may not be viable for various terrains and geographical regions. Reduction of inference cost therefore requires reducing the complexity of the ML model itself: the more elaborate the model, the more likely that the inference entails increasingly sophisticated computation and consequently higher inference cost. On the other hand, prediction accuracy does require the ML model to be sufficiently elaborate both in terms of the sophistication of the underlying algorithm and in the number of features/parameters incorporated in the model. REDEM addresses this conundrum by making the trade-off between cost and accuracy explicit and providing the user the ability to tune their model to customize for the trade-off target for the application. The framework itself is agnostic to the specifics of the underlying ML model. Rather, the user chooses a target for prediction accuracy (in terms of metrics like precision, recall, and f1-score) and can select the simplest ML model that addresses that accuracy need.

3.3 Multi-Channel Adversary

In typical CAV applications, more than one perception channel can be compromised. For instance, in CACC application, there are three perception channels that correspond to the velocity, position, and acceleration of the preceding vehicle. Generally acceleration is communicated through V2X messages and velocity and position data computed by the follower vehicle through its on-board sensors. Different adversary models would be interesting for different implementations or even a specific instance of the application, e.g., it may be appropriate for some CACC implementation to consider an adversary to corrupt only acceleration information (V2X corruption), or velocity and position information (corruption of sensor data), or some combination thereof.[1] When an adversary can corrupt multiple channels, one crucial requirement for ML-based resiliency is *source identification*, i.e., determining which channels are "actually" corrupted. Considering the CACC example above, suppose the adversary actually corrupts the acceleration information. From the perspective of the following vehicle, however, all that can be perceived is that pattern of acceleration values and velocity/position values received from the preceding vehicle are mutually inconsistent based on standard kinematics equations. Without some contextual information about the environment (e.g., what acceleration values are feasible under a specific driving condition), it is not possible to derive which ones the acceleration or velocity/position channels correspond to ground truth and which ones are anomalous. Furthermore, if we want to enumerate all subsets of potentially compromised channels, we will quickly run into combinatorial explosion. For instance, consider a platooning scenario consisting of five non-lead vehicles following a leader to create a platoon string. Suppose each vehicle receives three inputs (e.g., position, velocity, acceleration) from the leader and the vehicle immediately preceding it in the platoon. Assuming that at most three of the six inputs each non-lead vehicle receives can be corrupted by an adversary, there are 15 possibly compromised channels in the platoon at any point. Consequently, the total number of possible subsets of corrupt channels will be 2^{15}. Clearly, a naive approach of systematically examining each subset of channels for possible anomalies would be computationally prohibitive.

REDEM addresses the problem of multi-channel adversaries through a process of source identification that exploits *selective sensitivity*. The key insight is that the same anomaly can affect behavior of different functions in different ways. For instance, an anomalous value of a preceding vehicle acceleration would not affect a machine learning model in the following vehicle that is trained to predict based on only the values of the velocity and position of the preceding vehicle. REDEM source identification creates a number of ML models with selective sensitivity to different

---

[1] When doing this, care has to be taken so that we are still considering an adversary against whom it is possible to have a viable defense, e.g., if the adversary can collusively corrupt all the perception channels of the ego vehicle it is easy to see that no resiliency solution is possible.

parameters which can then be used cumulatively to narrow down the number of adversarial channels.

3.4 Error Control and Recoverability

The error control and recovery challenges arise from the imperfections in ML-based systems. The ideal case for a CAV resiliency is that the resilient system, when provided with any input whether benign or malicious, **always** behaves the way that the application is targeted to behave when all inputs receive ground truth, i.e., with no adversarial action. However, since ML techniques can only provide accuracy with a certain probability, it is important for any ML-based resiliency solution to account for the situations when ML would perform misprediction. There are two different ways in which the misprediction can impact the application. One is the *direct* way, where the impact would be a risk to safety or efficiency of the application. Another, more subtle way is the *indirect* effect on the vehicle state after the attack is completed. Under the latter, consider a platooning application where a vehicle computes its acceleration at each instance based on the acceleration, velocity, and position of the preceding vehicle. Consider an attack in which the position and velocity channels are collusively corrupted, i.e., the values of these channels are changed such that the kinematics equations are satisfied. The upshot of this attack will be that the victim vehicle would receive values of velocity and position that are mutually consistent, but inconsistent with the acceleration values. By analysis of the inconsistency alone, the victim vehicle would have no reason to deduce a vel-pos attack instead of the acceleration attack. (Indeed, if a vehicle does in fact deduce this then it would likely mis-predict the complementary scenario where acceleration is the channel being corrupted and vel-pos channels provide the ground truth.) A good mitigation would likely be conservative and ensure safe operation irrespective of the channels corrupted. Nevertheless, the vehicle's "perception" of its environment would be different depending on whether it correctly identifies the source of corruption. Furthermore, if the source identification is erroneous, it is possible that a subsequent benign (ground-truth) input would then be deduced as malicious. In the platooning example, after having wrongly deduced that the acceleration value received is corrupted and the velocity-position values are ground truths, the resiliency system would have a perception of the preceding vehicle's state velocity, position, and acceleration which is different from reality. Consequently, when it receives benign (ground truth) values of these three parameters it might wrongly consider (any subset of) them anomalous.

From the discussion above, we see that a resiliency solution must have the property of *recoverability*, i.e., the ability to return to a state in which when it is provided benign inputs when its perception of the environment is not too far from ground truth. We also observe that in particular with multi-channel adversary, recoverability may be difficult to ensure.

REDEM introduces a variety of techniques to reduce inaccuracies for solutions abating the errors resulting from the imperfections in ML systems. This includes additional rule-based validation steps in each decision-making cycle, which can control errors from previous time steps to propagate and accumulate. In addition to checks, REDEM "exploits" adversary assumptions to address recoverability issues, e.g., the adversaries handled by REDEM are constrained in terms of the degree of bias they can introduce during a corruption at different time steps and the number of channels that can be corrupted at the same time. This permits REDEM solutions to correct mistakes, e.g., the scenario where velocity and position are continually corrupted can be precluded by the requirement that an adversary during one episode of continuous attack can only select one untrusted channel. We put the word "exploits" in quote, since in practice a resiliency solution clearly cannot get to choose the adversaries against which to defend; the quality and power of the adversary ought to be defined by the characteristics of the application and deployment. Nevertheless, as we argued in Sect. 1.1, it is impossible (in principle and practice) to develop resiliency against an adversary that is all powerful. Consequently, it is fair to constrain the set of adversaries that can be handled by a specific resiliency solution. Nevertheless, we must still ensure that the adversary is realistic, i.e., it is worth developing a resiliency solution to focus specifically on the adversary for a threat model. For each incarnation of REDEM for different applications, we define a threat model constraining adversary power and argue why it is a realistic adversary.

4 Design, Implementation, Tuning, and Validation of ML Component

The considerations discussed in the preceding section pertained to the design of the overall resiliency system. In this section we delve a bit more into the ML component of the system. Some representative questions we need to address here include: (1) *Which ML architecture should we choose?* (2) *How should we train it?* (3) *Where do we find valid data to train it?* (4) *How can we perform validation?* We discuss some of these issues here, and the methodologies developed in REDEM to address them.

4.1 Architecture Selection and Tuning

One of the key activities to enable ML application is to identify the appropriate ML model for the task and determine its parameters. In case of ML-based resiliency, selection and tuning of ML model incurs several interesting challenges. First, the complexity of the ML model itself is constrained by the available computation resource as discussed in Sect. 3.2. Second, accuracy of different ML models depends on the specifics of the scenario, e.g., a model may be more accurate under benign conditions or a specific type of corruption. Navigating this space of models to identify the accurate one for a target application is highly challenging. The small data problem discussed in Sect. 3.1 exacerbates the problem: given the very low amount of data available, it is difficult for any ML model to learn the features to make it generalizable for all target applications, with a real danger of over-fitting.

To address this problem, a key insight is that the goal of an ML-based resiliency is to ideally behave like the naive application (i.e., application with no resiliency introduced) when provided data corresponding to the ground reality. In other words, the efficiency (and accuracy) of the solution would be determined primarily by the prediction accuracy under benign scenario. With that in mind, the following steps provide a recipe for selecting the ML architecture.

1. Identify a set of candidate ML architectures that can be deployed under the resource constraints. The constraints preclude overtly complicated ML systems, and generally permit only a small set of simple candidate architectures. It is generally possible to effectively navigate the space of architectures left through quick sampling.
2. Train the candidate architectures with benign data, discarding ones with unacceptable prediction accuracy under benign conditions.
3. Among the architectures with acceptable prediction accuracy under benign scenarios, select the one with the highest prediction accuracy under malicious conditions.

Obviously, the above steps should be used as a guideline, not a procedure cast in stone. For instance, determining the accuracy entails tuning the right set of hyperparameters, which in turn requires trade-offs between time, cost, computation capability, and many others.

4.2 Data Preprocessing and Feature Selection

CAV application anomalies are contextual, i.e., determining whether a specific input is anomalous requires understanding of the driving environment. For instance, a speed of 70 Mph is normal in a rural highway during a clear summer evening but perhaps not in a snowy winter morning or during rush hours near a big city. To be able to accurately qualify some input as normal or anomalous, the ML system ideally must have access to a large quantity of fine-grained data (sampled at high frequency), together with sufficient context. Unfortunately, real-world datasets are generally incomplete and inadequate. For instance, HighD Dataset [15], which provides trajectory data corresponding to real vehicles driving in German highways, has individual vehicle trajectory data encompassing approximately 15 s. Data from real datasets also include noise and inconsistency, arising from errors in collection and measurement from the physical environment. One can augment this with synthetic data, collected from a variety of simulators. However, for synthetic data it is critical to ensure that data characteristics are consistent with what the application is expected to encounter in field. Note that the accuracy of ML can be harmed by having irrelevant features and lack of diversity in the data. Insufficient diversity in data may lead to overfitting. Poor feature selection may also affect generalizability. So, feature selection and feature engineering must be done, and it completely depends upon the purpose of the ML model used in the CAV application resiliency.

Data preprocessing is the idea of preparing the raw data to make it suitable for consumption by ML algorithms. This includes several components. Data *cleaning* entails filling in missing values, smoothing or removing noisy data and outliers, and resolving inconsistencies. Data *integration* involves integrating data from multiple sources such as databases, data cubes, files, etc. To solve the complexities arising due to feature selection, the input features should be selected such that they have actual impact on the learning behavior. In such cases redundant features and features having no importance for the prediction objective of the ML target in question should be ignored while training to boost up the performance. At the same time new features can be engineered from existing features in order to get better results.

REDEM addresses the data preprocessing issues by using "realistic synthetic data". In particular, REDEM uses a *physical* simulator platform, RDS1000® [16]. This platform can be used to acquire data as follows. The system permits configuration of various different driving environment, and an immersive environment for a human to have the experience of driving in the programmed environment. We can record the actions of humans as they perform driving, and use that as a proxy for what a vehicle does under similar situation. It also includes autonomous driving modules that can be used to study reaction of autonomous driving algorithms under similar situations. We curated an extensive dataset of vehicular behavior under 24 different driving conditions by first acquiring the data and then performing cleaning, reformatting and validation. This setup can produce fine-grained and real-time data. However, it leaves open the issue whether the environments programmed (and the vehicular behavior recorded) do in fact correspond to reality. To address this question, we show that the real dataset snippets in fact match in pattern with our curated dataset. The real dataset only includes short snippets of time as mentioned above. Nevertheless, if these snippets match the synthetic data for the corresponding environment, we can gain confidence that the synthetic data is indeed realistic.

Finally, note the apparent dichotomy in the discussion above and the discussion in Sect. 3.1. We argued that normal behavior data is in fact plentiful, and it is the anomaly data that is limited. The discussion here at cursory glance would appear in contradiction to that statement, *is that right?*

Actually, it is wrong: there is no contradiction. Normal behavior data is in fact plentiful once the application is deployed and can collect such data in field. However, when determining the parameter set and performing feature engineering, the application is not yet deployed in field, so we have to depend on synthetic data or available real-vehicle datasets.

4.3 Decision Threshold Selection

Given that ML in CAV resiliency is targeted specifically for anomaly detection, an important issue is to determine the anomaly threshold. A high threshold may result in reduced detection accuracy, whereas a low threshold may result in more false reports in detection. An ideal threshold is one that would provide both safety and efficiency under adversarial scenarios while incurring minimal performance overhead in benign conditions. A more subtle impact of the choice of the threshold is the robustness of the resiliency system to *subversion attacks*. The idea of subversion attacks is for the adversary to create anomalous data that is nevertheless accepted as normal by the detector, thereby bypassing any mitigation against the attack. A high anomaly threshold can make the CAV application vulnerable to subversion attacks, impacting the safety, efficiency, and recoverability of the resiliency solution.

To address this problem, REDEM includes systematic methodology for identifying and tuning anomaly threshold. REDEM accounts for the fact that the choice of threshold also depends on the operating environment and may require re-configuration as the driving conditions change during the application engagement. In REDEM methodology, the threshold is determined by analyzing the distribution of test-set error incurred by the ML prediction model under benign conditions as well as a finite set of representative attacks. This is achieved in two steps. First, the threshold is coarsely tuned to minimize false positives and false negatives under benign conditions determining a ball-park range. Subsequently, a series of special subversion attacks are orchestrated to fine-tune the threshold that can balance the trade-off between the conflicting design goals of minimizing inference cost, achieving required detection accuracy, and minimizing overhead due to false-positives/false-negatives. While achieving the ideal outcome for all the design goals simultaneously is impractical, REDEM identifies the tolerable imperfections that can still ensure overall resiliency (guaranteed safety and optimal efficiency at all times) for CAVs. The threshold selection then accounts for the re-defined practical design goals carefully allowing for a small amount of inference cost, false-positives, and false-negatives, while achieving the required detection accuracy.

4.4 Validation

Validation is crucial for a CAV resiliency system, since it targets highly safety-critical applications. However, the unique nature of ML-based resiliency makes it highly challenging to achieve effective validation. Obviously, validation is a broad topic with many different facets. Here we provide a very quick summary of the challenges involved, and our approach to address these challenges. The reader interested in a fuller discussion of validation in REDEM is referred to our companion publication [9] that provides an exclusive treatment of the subject.

Roughly, validation of ML-based prediction involves addressing three critical problems as described below.

1. **Inadequacy of data.** The challenge with data for validation is similar to that for parameter estimation and tuning discussed in Sect. 4.2: before the application is deployed, how can we obtain copious amount of realistic data to validate an ML system?

2. **Validation against zero-day attacks.** This issue arises from the fact that it is inadequate for a resiliency solution to only provide protection against a specific set of known attacks. New attacks not considered during resiliency design can become feasible during the life-time of the application after technology (and hence sophistication of attack) advances in ways not necessarily anticipated at deployment. This results in a conundrum for security validation: how can we ensure that the resiliency system is indeed effective, not only against known attacks but against a spectrum of attacks that are unknown at deployment time?

3. **Validation challenges for an inherently probabilistic system.** This challenge is the verification counterpart of the design challenge we discussed in Sect. 3.4. Since no ML system is accurate in 100% of cases, we must be able to verify that, either (1) no matter what attack is instigated, the victim vehicle's perception is always within tolerable limits of reality; or (2) if the perception of the victim vehicle deviates significantly from reality then its response still ensures safe and efficient operation, and after the attack is over it eventually returns to a state in which benign inputs are treated as benign.

REDEM addresses the first two problems discussed above through new validation techniques. The third problem (probabilistic system challenge) is relegated to design (and validation) of resiliency solutions with the property of recoverability as discussed in Sect. 3.4. We address the problem of data for validation in the same way we did for model training, e.g., by creation of realistic synthetic data from a physical simulator. The uniqueness of REDEM validation is in how it addresses the second problem, i.e., validating resiliency against unknown adversary. The key insight is that it is possible to develop a resiliency system that accounts for attacks based on its manifestation features, stealth, and impact rather than detailed attack mechanism. Furthermore, it is possible to comprehensively classify the spectrum of attacks in this manner simply from the threat model. For instance, consider a CACC application where a vehicle follows its preceding vehicle by maintaining a specific time headway. If the adversary is confined to V2V communications, the only choices for the adversary are to (1) mutate an existing message, (2) fabricate a new message, and (3) prevent the delivery of a message. Going through this argument enables us to create a taxonomy of V2X attacks. Note that if our validation covers attack space defined by the taxonomy then the above argument suggests that we indeed comprehensively cover the space of all attacks defined by the threat model, including unknown attacks. REDEM additionally includes an automated evaluation framework for systematically generating attacks from the adversary taxonomy [8].

5 REDEM Case Study: Resilient Cooperative Adaptive Cruise Control

We have instantiated REDEM to incorporate resiliency on platooning applications [6, 7, 10]. Here we quickly summarize the instantiation of REDEM architecture on Cooperative Adaptive Cruise Control (CACC). We also present a summary of results showing the overall efficacy of the REDEM resiliency for CACC.

5.1 CACC Overview

In CACC, the following vehicle autonomously adapts its velocity in accordance to the acceleration of the vehicle in front (received through V2V communication), as well as the relative velocity and inter-vehicle gap (obtained from the ranging sensor readings). CACC enables improved road safety and efficiency (e.g., a much smaller headway) compared to its non-cooperative counterpart, Adaptive Cruise Control (ACC). Figure 2a depicts vehicles engaged in CACC. Figure 2b demonstrates the high-level functionality of a CACC decision computation module that implements constant time headway policy. Following vehicle receives the preceding vehicle's instantaneous acceleration as a V2V message. It utilizes this information in addition to the on-board ranging sensor readings providing the relative position and velocity of the preceding vehicle. Consequently, the following vehicle efficiently adapts its velocity in accordance with the acceleration of the preceding vehicle achieving improved efficiency and safety. CACC forms the basis for several connected car applications such as multi-vehicle platooning, cooperative on-ramp merging, etc.

A vehicle engaging in CACC can be exploited by an adversary that is capable of manipulating the V2V communication or a malicious preceding vehicle that shares false information. For instance, a malicious preceding vehicle can report a fake acceleration value that is greater than its true value. This can mislead the vehicle to accelerate at a higher value than desired that can lead to an increased risk of a collision. Similarly, a Man-In-The-Middle (MITM) adversary can mutate the messages from the preceding vehicle by adding a negative bias. The vehicle receives false acceleration value and fails to maintain the optimal space gap. This leads to loss in efficiency or can cause string instability in the traffic.

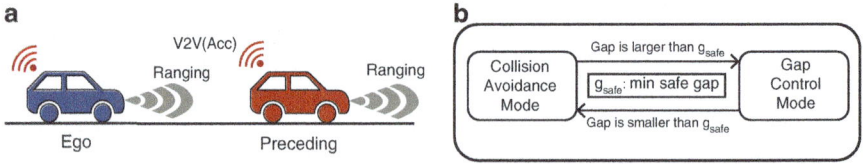

Fig. 2 (**a**) Two vehicles engaged in CACC; (**b**) Modes of operation of a conventional CACC decision computation module

5.2   *Evaluation of* REDEM *Resiliency on CACC*

We extended CACC with REDEM architecture to generate a resilient CACC solution which we call RACCON (for "Resilient Cooperative Adaptive Cruise Control). Our evaluation of RACCON also represents one of the most comprehensive resiliency evaluations performed on a connected vehicle application up to date. As pointed out in Sects. 3 and 4, this means consideration of a number of different factors. In summary, the evaluation of RACCON included the following components.

1. **Data Validation:** We validated that the vehicular driving patterns reflected in our simulation data conform to real-world patterns from a public dataset.
2. **Identification of Appropriate ML Model:** We developed a systematic evaluation methodology for identifying and tuning the optimal ML architecture.
3. **Attack Impact Analysis:** The viability of attack orchestration framework for RACCON evaluation depends on the quality of the orchestrated attacks themselves. We developed a methodology to analyze attacks, in terms of stealth and impact.
4. **Anomaly Detection Threshold:** A key factor in the effectiveness of RACCON is the identification of *anomaly threshold*, i.e., the extent of deviation from normal behavior pattern that would be classified as a potential threat. Selecting an appropriate threshold involves balancing the trade-off between maximizing attack detection accuracy and minimizing false alarms. We present a series of experiments to compute the optimal threshold, achieving the balance between maximizing attack detection accuracy and minimizing false alarms.
5. **V2V Attack Resiliency:** The central component of our evaluation shows the robustness of RACCON against various V2V attacks.
6. **Resiliency Against Detector Subversion:** We designed a set of experiments to address evaluating the robustness of RACCON against detector subversion, and tune anomaly threshold accounting for the trade-off between robustness to subversion and minimizing false alarms.

Here we show some representative plots from our experiments to give a flavor of the evaluation and the extent and quality of REDEM resiliency for CACC. Figure 3 shows a representative plot for the scenario (highway, windy, day), for discrete, cluster, and continuous attacks. The frequency of the malicious activity and the magnitude of deviation between the false V2V message received and the ground truth, determine the stealth of the attack. The detection system is capable of capturing attacks of varying stealth as can be seen from the figure. Figures 4, 5, and 6 show the conclusions from our experiments on Mitigator efficacy under collision-causing, efficiency-degrading, and delivery prevention attacks. Under each category, different types of attacks for discrete, cluster, and continuous adversaries are simulated. Mitigation guarantees safety while keeping the efficiency optimal. This is reflected in the time headway values achieved by the mitigation that closely resemble the ideal values.

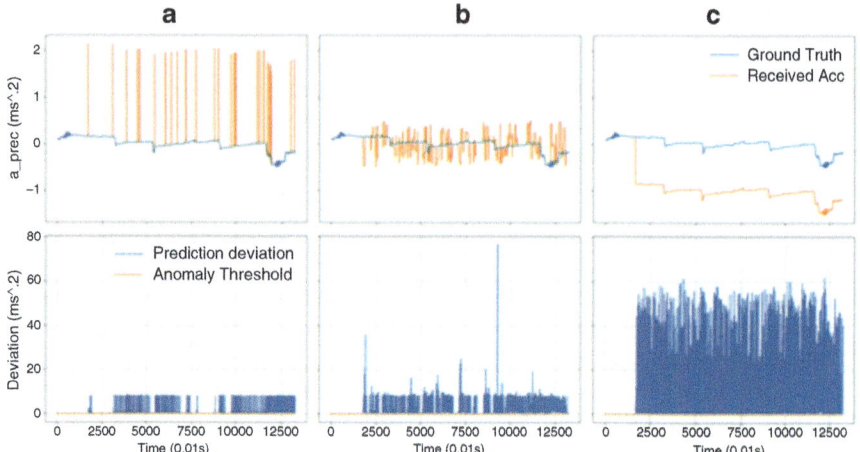

Fig. 3 Anomaly detection analysis: (**a**) Discrete attack (constant bias +2.0); (**b**) Cluster attack (bias [−0.5, +0.5]); (**c**) Continuous attack (constant bias −1.0)

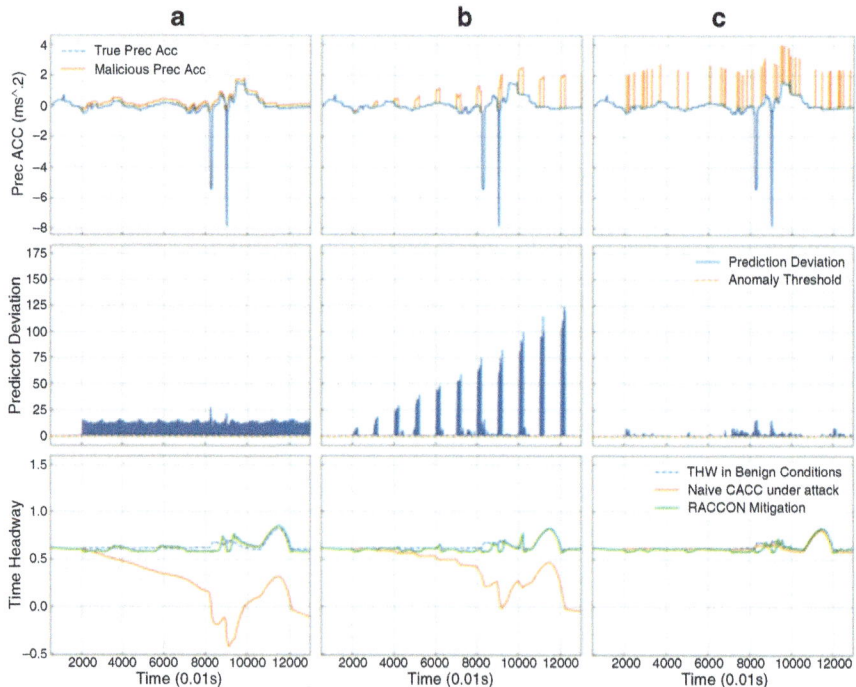

Fig. 4 Evaluation under collision attacks: Comparison of resultant time headway for resiliency augmented CACC and naive CACC with no resiliency; (**a**) Continuous attack (constant bias +0.25); (**b**) Cluster attack (linear bias +0.1t); (**c**) Discrete attack (constant bias +2.5)

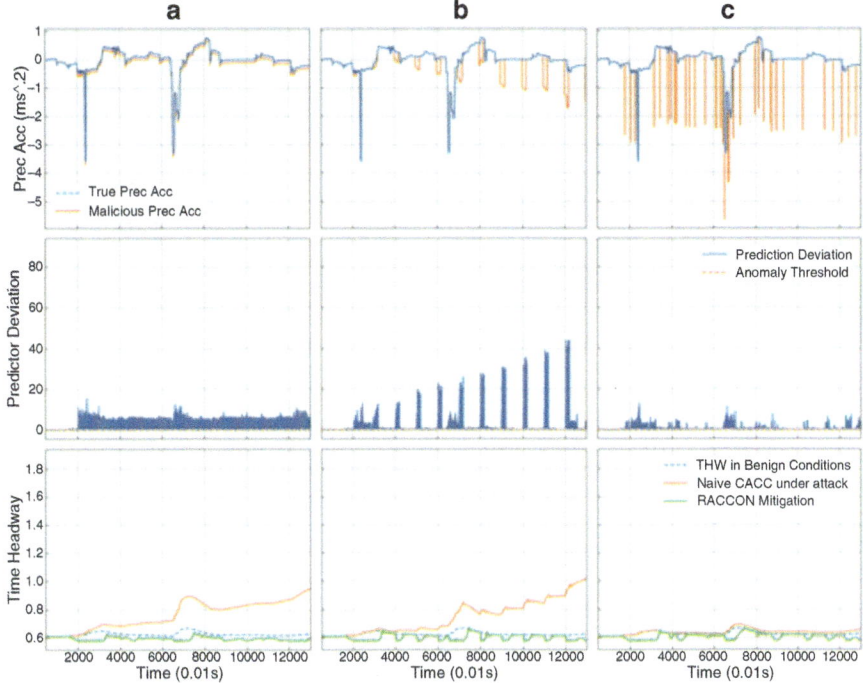

Fig. 5 Evaluation under efficiency degradation attacks: comparison of resultant time headway for resiliency augmented CACC and naive CACC with no resiliency; (**a**) Continuous attack (constant bias −0.1); (**b**) Cluster attack (linear bias −0.06t); (**c**) Discrete attack (constant bias −2.5)

6 Conclusion

We have considered the problem of introducing resiliency in CAV applications against attacks on perception inputs. With increasing proliferation of connectivity and autonomy in vehicles, perception inputs can create a large and highly vulnerable attack surface that can be easily exploited with catastrophic consequences. We discussed the promise and challenges in adopting ML-based solutions to achieve resiliency in this domain. We also discussed one effective framework, REDEM, to achieve this resiliency, and explained REDEM's approach to address the various challenges in system design, architecture, and validation. The efficacy of REDEM was demonstrated in Cooperative Adaptive Cruise Control application.

REDEM is very much a work in progress, and is under active development. What we presented is representative of our thinking at the time of this writing, but the thinking will inevitably evolve as we extend REDEM for newer applications. Indeed, it is important to try REDEM on applications of diverse flavors to identify weakness in the current line of thinking and determining how to expand the methodology to incorporate new challenges. Some critical applications that can provide such

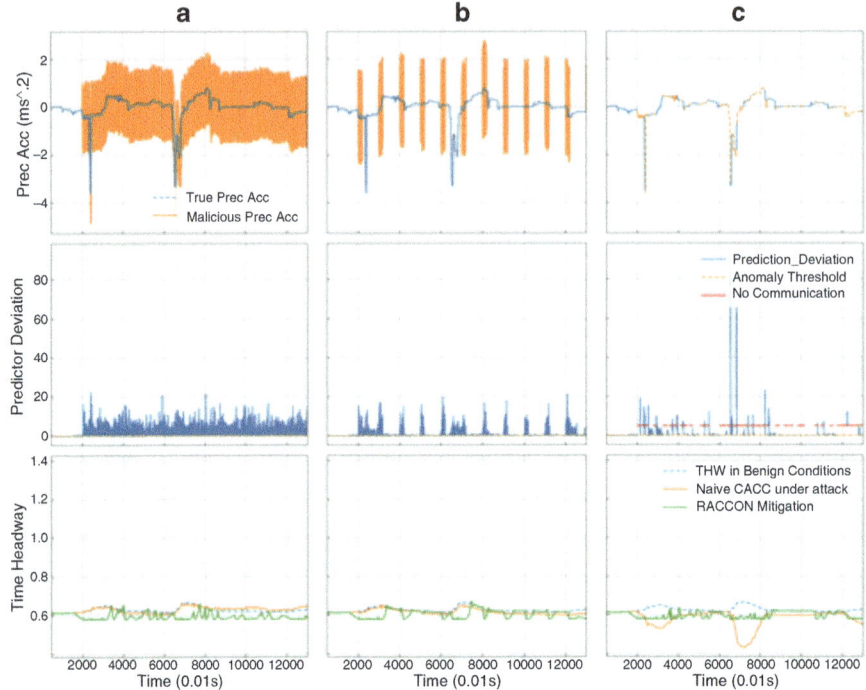

Fig. 6 Evaluation under random mutation and delivery prevention attacks: comparison of resultant time headway for resiliency augmented CACC and naive CACC with no resiliency; (**a**) Continuous attack (random bias −1.5, 1.5); (**b**) Cluster attack (random bias −2.0, 2.0); (**c**) Intermittent communication

challenges include Distributed Cooperative Collision Detection, Cooperative Route Management, etc. These applications are different from the current platooning applications in that they involve communication of perceived environment in addition to the state of the communicating vehicle. We will explore security challenges in perception of such scenarios and investigate the applicability of REDEM.

References

1. Abdollahi Biron, Z., Dey, S., Pisu, P.: Real-time detection and estimation of denial of service attack in connected vehicle systems. IEEE Trans. Intell. Transp. Syst. **19**(12), 3983–3902 (2018)
2. Adler, J.L., Blue, V.J.: A cooperative multi-agent transportation management and route guidance system. Transp. Res. C Emerg. Technol. **10**(5–6), 433–454 (2002)
3. Adler, J.L., Satapathy, G., Manikonda, V., Bowles, B., Blue, V.J.: A multi-agent approach to cooperative traffic management and route guidance. Transp. Res. B Methodol. **39**(4), 297–318 (2005)

4. Alotibi, F., Abdelhakim, M.: Anomaly detection for cooperative adaptive cruise control in autonomous vehicles using statistical learning and kinematic model. IEEE Trans. Intell. Transp. Syst. **22**(6), 1–11 (2020)

5. Bergenhem, C., Shladover, S., Coelingh, E., Englund, C., Tsugawa, S.: Overview of platooning systems. In: Proceedings of the 19th ITS world congress, Oct 22–26, Vienna (2012)

6. Boddupalli, S., Ray, S.: Redem: real-time detection and mitigation of communication attacks in connected autonomous vehicle applications. In IFIP international Internet of Things conference, pp. 105–122. Springer (2019)

7. Boddupalli, S., Hegde, A., Ray, S.: Replace: real-time security assurance in vehicular platoons against v2v attacks. In: 2021 IEEE international intelligent transportation systems conference (ITSC), pp. 1179–1185 (2021)

8. Boddupalli, S., Chamarthi, V.S.G., Lin, C.-W., Ray, S.: CAVELIER: Automated security evaluation for connected autonomous vehicle applications. In: 25th IEEE international conference on intelligent transportation (ITSC 2022) (2022)

9. Boddupalli, S., Owoputi, R., Duan, C., Choudhury, T., Ray, S.: Resiliency in connected vehicle applications: challenges and approaches for security validation. In: Proceedings of 22nd great lakes symposium on VLSI 2022 (GLSVLSI 2022) (2022)

10. Boddupalli, S., Rao, A.S., Ray, S.: Resilient cooperative adaptive cruise control for autonomous vehicles using machine learning. IEEE Trans. Intell. Transp. Syst., **23**(9), 15655–15672 (2022)

11. Casaca, A., Katkoori, S., Ray, S., Strous, L.: Internet of Things. In: A confluence of many disciplines: second IFIP international cross-domain conference, IFIPIoT 2019, Tampa, FL, USA, October 31–November 1, 2019, Revised Selected Papers, vol. 574. Springer Nature (2020)

12. Dresner, K., Stone, P.: A multiagent approach to autonomous intersection management. J. Artif. Intell. Res. **31**, 591–656 (2008)

13. Dutta, R.G., Yu, F., Zhang, T., Hu, Y., Jin, Y.: Security for safety: a path toward building trusted autonomous vehicles. In: ICCAD (2018)

14. Jagielski, M., Jones, N., Lin, C.-W., Nita-Rotaru, C., Shiraishi, S.: Threat detection for collaborative adaptive cruise control in connected cars. In: Proceedings of the 11th ACM conference on security & privacy in wireless and mobile networks, pp. 184–189. ACM (2018)

15. Krajewski, R., Bock, J., Kloeker, L., Eckstein, L.: The highd dataset: a drone dataset of naturalistic vehicle trajectories on German highways for validation of highly automated driving systems. In: 2018 21st International conference on intelligent transportation systems (ITSC), pp. 2118–2125 (2018)

16. Realtime-Technologies. Physical automotive simulator. See https://www.faac.com/realtime-technologies/products/rds-1000-single-seat-simulator

17. van Nunen, E., et al.: Robust model predictive cooperative adaptive cruise control subject to v2v impairments. In: IEEE 20th international conference on intelligent transportation systems (ITSC) (2017)

Part IV
Robust Perception

Object Detection in Autonomous Cyber-Physical Vehicle Platforms: Status and Open Challenges

Abhishek Balasubramaniam and Sudeep Pasricha

1 Introduction

Autonomous vehicles (AVs) have received immense attention in recent years, in large part due to their potential to improve driving comfort and reduce injuries from vehicle crashes. It has been reported that more than 36,000 people died in 2019 due to fatal accidents on U.S. roadways [1]. AVs can eliminate human error and distracted driving that is responsible for 94% of these accidents [2]. By using sensors such as cameras, lidars, and radars to perceive their surroundings, AVs can detect objects in their vicinity and make real-time decisions to avoid collisions and ensure safe driving behavior.

AVs are generally categorized into six levels by the SAE J3016 standard [3] based on their extent of supported automation (see Table 1). While level 0–2 vehicles provide increasingly sophisticated support for steering and acceleration, they heavily rely on the human driver to make decisions. Level 3 vehicles are equipped with Advanced Driver Assistance Systems (ADAS) to operate the vehicle in various conditions, but human intervention may be requested to safely steer, brake, or accelerate as needed. Level 4 vehicles are capable of full self-driving mode in specific conditions but will not operate if these conditions are not met. Level 5 vehicles can drive without human interaction under all conditions.

Automotive manufactures have been experimenting with AVs since the 1920s. The first modern AV was designed as part of CMU NavLab's autonomous land

A. Balasubramaniam (✉)
Department of Electrical and Computer Engineering, Colorado State University, Fort Collins, CO, USA
e-mail: abhishek.balasubramaniam@colostate.edu

S. Pasricha
Colorado State University, Fort Collins, CO, USA
e-mail: sudeep@colostate.edu

Table 1 SAE J3016 levels of automation

SAE level	Name	Driving environment monitor
0	No automation	Human driver
1	Driver assistance	
2	Partial driving automation	
3	Conditional driving automation	ADAS system
4	High driving automation	
5	Full driving automation	

vehicle project in 1984 with level 1 autonomy that was able to steer the vehicle while the acceleration was controlled by a human driver [4]. This was followed by an AV designed by Mercedes-Benz in 1987 with level 2 autonomy that was able to control steering and acceleration with limited human supervision [5]. Subsequently, most major auto manufacturers such as General Motors, Bosch, Nissan, and Audi started to work on AVs.

Tesla was the first company to commercialize AVs with their Autopilot system in 2014 that offered level 2 autonomy [6]. Tesla AVs were able to travel from New York to San Francisco in 2015 by covering 99% of the distance autonomously. In 2017, Volvo launched their Drive Me feature with level 2 autonomy, with their vehicles traveling autonomously around the city of Gothenburg in Sweden under specific weather conditions [7]. Waymo has been testing its AVs since 2009 and has completed 200 million miles of AV testing. They also launched their driverless taxi service with level 4 autonomy in 2018 in the metro Phoenix area in USA with 1000–2000 riders per week, among which 5–10% of the rides were fully autonomous without any drivers [8]. Cruise Automation started testing a fleet of 30 vehicles in San Francisco with level 4 autonomy in 2017, launched their self-driving Robotaxi service in 2021 [9]. Even though Waymo and Cruise support level 5 autonomy, their AVs are classified as level 4 because there is still no guarantee that they can operate safely in all weather and environmental conditions.

AVs rely heavily on sensors such as cameras, lidars, and radars for autonomous navigation and decision making. For example, Tesla AVs rely on camera data with six forward facing cameras and ultrasonic sensors. In contrast, Cruise AVs use a sensor cluster that consists of a radar in the front while camera and lidar sensors are mounted on the top of the AV to provide a 360-degree view of the vehicle surroundings [9]. One of the main tasks involved in achieving robust environmental perception in AVs is to detect objects in the AV vicinity using software-based object detection algorithms. Object detection is a computer vision task that is critical for recognizing and localizing objects such as pedestrians, traffic lights/signs, other vehicles, and barriers in the AV vicinity. It is the foundation for high-level tasks during AV operation, such as object tracking, event detection, motion control, and path planning.

The modern evolution of object detectors began 20 years ago with the Viola Jones detector [10] used for human face detection in real-time. A few years later, Histogram of Oriented Gradient (HOG) [11] detectors became popular for pedestrian detection. HOG detectors were then extended to Deformable Part-based

Models (DPMs), which were the first models to focus on multiple object detection [12]. With growing interest in deep neural networks around 2014, the Regions with Convolutional Neural Network (R-CNN) deep neural network model led to a breakthrough for multiple object detection, with a 95.84% improvement in Mean Average Precision (mAP) over the state-of-the-art. This development helped redefine the efficiency of object detectors and made them attractive for entirely new application domains, such as for AVs. Since 2014, the evolution in deep neural networks and advances in GPU technology have paved the way for faster and more efficient object detection on real-time images and videos [10]. AVs today rely heavily on these improved object detectors for perception, pathfinding, and other decision making.

This chapter discusses contemporary deep learning-based object detectors, their usage, optimization, and limitations for AVs. We also discuss open challenges and future directions.

2   Overview of Object Detectors

Object detection consists of two sub-tasks: localization, which involves determining the location of an object in an image (or video frame), and classification, which involves assigning a class (e.g., 'pedestrian', 'vehicle', 'traffic light') to that object. Figure 1 illustrates a taxonomy of state-of-the-art deep learning-based object detectors. We discuss the taxonomy of these object detectors in this section.

2.1   Two-Stage vs Single Stage Object Detectors

Two-stage deep learning based object detectors involve a two-stage process consisting of (1) region proposals and (2) object classification. In the region proposal stage, the object detector proposes several Regions of Interest (ROIs) in an input image that have a high likelihood of containing objects of interest. In the second stage, the most promising ROIs are selected (with other ROIs being discarded)

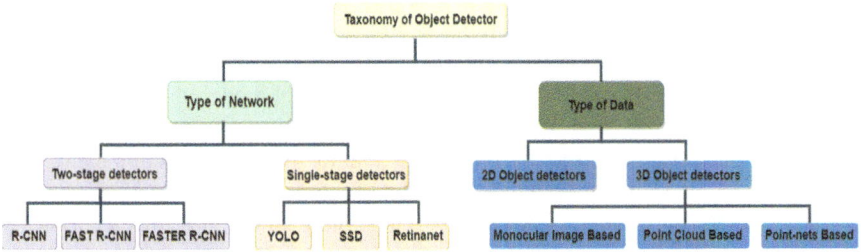

Fig. 1 Taxonomy of object detectors

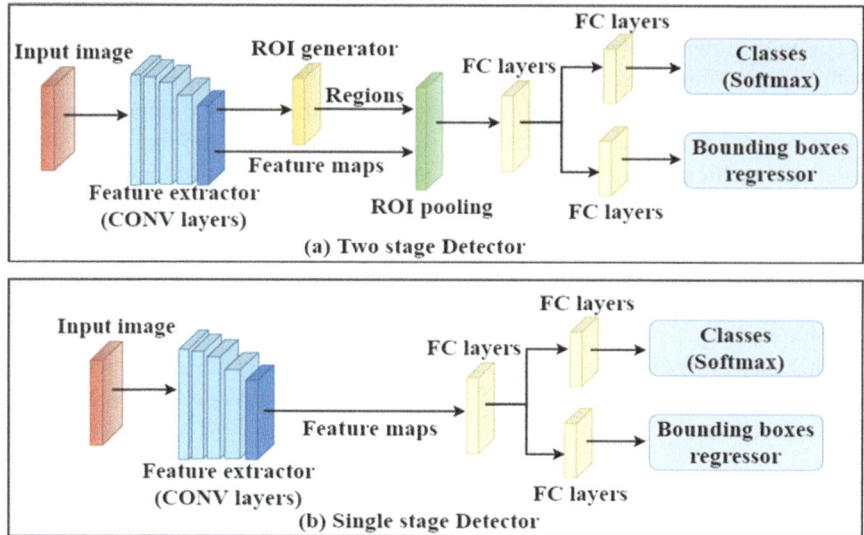

Fig. 2 Two-stage vs Single stage detector network diagram

and objects within them are classified [13]. Popular two-stage detectors include R-CNN, Fast R-CNN, and Faster R-CNN. In contrast, single-stage object detectors use a single feed-forward neural network that creates bounding boxes and classifies objects in the same stage. These detectors are faster than two-stage detectors but are also typically less accurate. Popular single-stage detectors include YOLO, SSD, EfficientNet, and RetinaNet.

Figure 2 illustrates the difference between the two types of object detectors. Both types of object detectors are typically evaluated using the mAP and Intersection over Union (IoU) accuracy metrics. mAP is the mean of the ratio of precision to recall for individual object classes, with a higher value indicating a more accurate object detector. IoU measures the overlap between the predicted bounding box and the ground truth bounding box. Formally, IoU is the ratio of the area of overlap between the (bounding and ground truth) boxes and the area of union between the boxes. Figure 3 illustrates the IoU of an object detector prediction and the ground truth. Figure 3a shows a highly accurate IoU and 3b shows a less accurate IoU.

R-CNN was one of the first deep learning-based object detectors and used an efficient selective search algorithm for ROI proposals as part of a two-stage detection [13]. Fast R-CNN solved some of the problems in the R-CNN model, such as low inference speed and accuracy. In the Fast R-CNN model, the input image is fed to a Convolutional Neural Network (CNN), generating a feature map and ROI projection. These ROIs are then mapped to the feature map for prediction using ROI pooling. Unlike R-CNN, instead of feeding the ROI as input to the CNN layers, Fast R-CNN uses the entire image directly to process the feature maps to detect objects [14]. Faster R-CNN used a similar approach to Fast R-CNN, but instead of using a

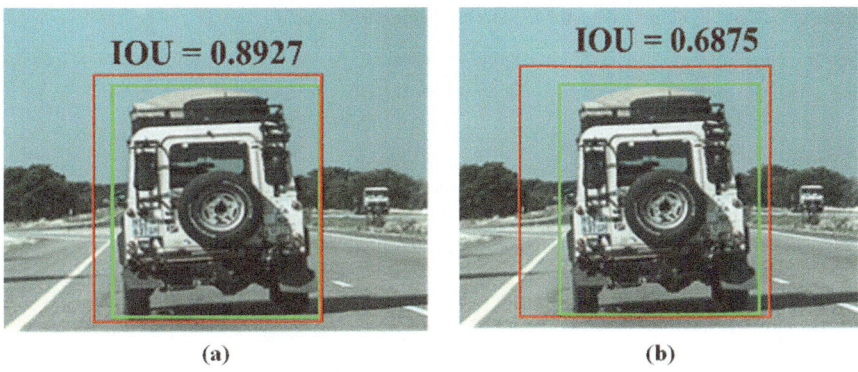

Fig. 3 Example of an IOU; green box: ground truth; red box: prediction

selective search algorithm for the ROI proposal, it employed a separate network that fed the ROI to the ROI pooling layer and the feature map, which were then reshaped and used for prediction [15].

Single-stage object detectors such as YOLO (You only look once) are faster than two-stage detectors as they can predict objects on an input with a single pass. The first YOLO variant, YOLOv1, learned generalizable representations of objects to detect them faster [16]. In 2016, YOLOv2 improved upon YOLOv1 by adding batch normalization, a high-resolution classifier, and use of anchor boxes to create bounding boxes instead of using a fully connected layer like YOLOv1 [17]. In 2018, YOLOv3 was proposed with a 53 layered backbone-based network that used an independent logistic classifier and binary cross-entropy loss to predict overlapping bounding boxes and smaller objects [18]. Single-Shot Detector (SSD) models were proposed as a better option to run inference on videos and real-time applications as they share features between the classification and localization task on the whole image, unlike YOLO models that generate feature maps by creating grids within an image. While the YOLO models are faster than SSD, they trail behind SSD models in accuracy [19]. Even though YOLO and SSD models provide good inference speed, they have a class imbalance problem when detecting small objects. This issue was addressed in the RetinaNet detector that used a focal loss function during training and a separate network for classification and bounding box regression [20].

In 2020, YOLOv4 introduced two important techniques: 'bag of freebies' which involves improved methods for data augmentation and regularization during training and 'bag of specials' which is a post processing module that allows for better mAP and faster inference [21]. YOLOv5, which was also introduced in 2020, proposed further data augmentation and loss calculation improvements. It also used auto-learning bounding box anchors to adapt to a given dataset [22]. Another variant called YOLOR (You Only Learn One Representation) was proposed in 2021 and used a unified network that encoded implicit and explicit knowledge to predict the output. YOLOR can perform multitask learning such as object detection, muti-label image classification, and feature embedding using a single model [23]. The YOLOX

Table 2 2D and 3D object detector models and their performance

Name	Year	Type	Dataset	mAP	Inference rate (fps)
R-CNN [13]	2014	2D	Pascal VOC	66%	0.02
Fast R-CNN [14]	2015		Pascal VOC	68.8%	0.5
Faster R-CNN [15]	2016		COCO	78.9%	7
YOLOv1 [16]	2016		Pascal VOC	63.4%	45
YOLOv2 [17]	2016		Pascal VOC	78.6%	67
SSD [19]	2016		Pascal VOC	74.3%	59
RetinaNet [20]	2018		COCO	61.1%	90
YOLOv3 [18]	2018		COCO	44.3%	95.2
YOLOv4 [21]	2020		COCO	65.7%	62
YOLOv5 [22]	2021		COCO	56.4%	140
YOLOR [23]	2021		COCO	74.3%	30
YOLOX [24]	2021		COCO	51.2%	57.8
Complex-YOLO [27]	2018	3D	KITTI	64.00%	50.4
Complexer-YOLO [28]	2019		KITTI	49.44%	100
Wen et al. [29]	2021		KITTI	73.76%	17.8
RAANet [30]	2021		NuScenes	62.0%	–

model, also proposed in 2021, uses an anchor-free, decoupled head technique that allows the network to process classification and regression using separate networks. Unlike the YOLOv4 and YOLOv5 models, YOLOX has reduced number of parameters and increased inference speed [24]. The performance of each model in terms of mAP and inference speed is summarized in Table 2.

2.2 2D vs 3D Object Detectors

2D object detectors typically use 2D image data for detection, but recent work has also proposed a sensor-fusion based 2D object detection approach that combines data from a camera and radar [25]. 2D object detectors provide bounding boxes with four Degrees of Freedom (DOF). Figure 4 shows the most common approach for encoding bounding boxes 4a: [x, y, height, width] and 4b: [xmin, ymin, xmax, ymax] [26]. Unfortunately, 2D object detection can only provide the position of the object on a 2D plane but does not provide information about the depth of the object. Depth of the object is important to predict the shape, size, and position of the object to enable improved performance in various self-driving tasks such as path planning, collision avoidance, etc.

Figure 5 shows the difference between a 2D and 3D object detector output on real images. 3D object detectors use data from a camera, lidar, or radar to detect objects and generator 3D bounding boxes. These detectors provide bounding boxes with (x, y, z) and (height, width, length) along with yaw information [26]. These object detectors use several approaches, such as point clouds and frustum pointnets,

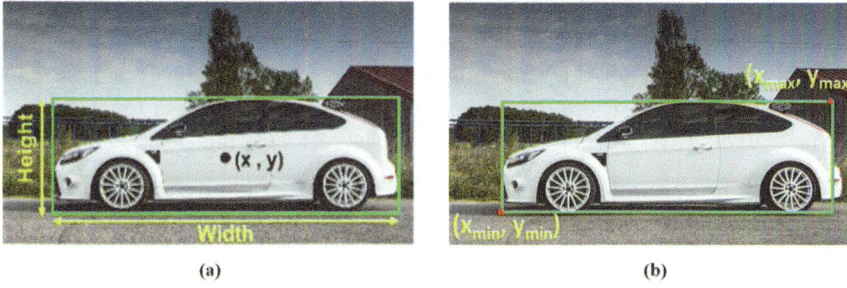

Fig. 4 Commonly used bounding box encoding methods

Fig. 5 Object detection modalities: (**a**) 2D vs. (**b**) 3D

for predicting objects in real-time. Point cloud networks can directly use 3D data, but the complexity and cost of computing are very high, so some networks use 2D to 3D lifting while compensating for the loss of information. Pointnets are used along with RGB images, where 2D bounding boxes are obtained using RGB images. Then these boxes are used as ROIs for 3D object detection which reduces the search effort [26]. Monocular image-based methods have also been proposed that use an RGB image to predict objects on the 2D plane and then perform 2D to 3D lifting to create 3D object detection results.

Recent years have seen growing interest in 3D object detection with deep learning. Complex-YOLO, an extension of YOLOv2, used a Euler Region Proposal Network (E-RPN), based on an RGB Birds-Eye-View (BEV) map from point cloud data to get 3D proposals. The network exploits the YOLOv2 network followed by E-RPN to get the 3D proposal [27]. Later in 2019, Complexer-YOLO achieved semantic segmentation and 3D object detection using Random Finite Set (RFS) [28]. The more recent work on 3D object detection by Wen et al. [29] in 2021 proposed a lightweight 3D object detection model that consists of three submodules: (1) point transform module, which extracts point features from the RGB image based on the raw point cloud, (2) voxelization, which divides the features into equally spaced voxel grids and then generates a many-to-one mapping between the voxel grids and the 3D point clouds, and (3) point-wise fusion module, which fuses the features using two fully connected layers. The output of the point-wise fusion module is encoded and used as input for the model. Another 3D detector proposed in 2021 called RAANet used only lidar data to achieve 3D object detection [30]. It used the BEV lidar data as input for a region proposal network which was then used to create shared features. These shared features were used as the input for an anchor free network to detect 3D objects. The performance of these models is summarized in Table 2.

3 Deploying Object Detectors in AVs

Deploying deep learning-based object detector models in AVs has its own challenges, mainly due to the resource-constrained nature of the onboard embedded computers used in vehicles. These computing platforms have limited memory availability and reduced processing capabilities due to stringent power caps, and high susceptibility to faults due to thermal hotspots and gradients, especially during operation in the extreme conditions found in vehicles. As the complexity of the object detector model increases, the memory and computational requirements, and energy overheads also increase. In this section, we discuss techniques to improve object detector model deployment efficiency. The performance of some of the latest works on this topic is summarized in Table 3.

3.1 Pruning

Pruning a neural network model is a widely used method for reducing the model's memory footprint and computational complexity. Pruning was first used in the 1990s to reduce neural network sizes for deploying them on embedded platforms [31]. Pruning involves removing redundant weights and creating sparsity in the model by training the model with various regularization techniques (L1, L2, unstructured, and structured regularization). Sparse models are easier to compress, and the zero weights created during pruning can be skipped during inference, reducing inference time, and increasing efficiency. While most pruning approaches target deep learning models for the simpler image classification problem, relatively fewer works have attempted to prune the more complex object detector models. Wang et al. [32] proposed using a channel pruning strategy on SSD models in which they start by creating a sparse normalization and then prune the channels with a small scaling factor followed by fine-tuning the network. Zhao et al. [33] propose a compiler aware neural pruning search on YOLOv4 which uses an automatic neural pruning search algorithm that uses a controller and evaluator. The controller is used to select the search space, pre-layer pruning schemes, and prune the model whereas the evaluator evaluates the model accuracy after every pruning step.

Table 3 Different object detector model optimization techniques and their performance

Name	Technique used	Model compression achieved	Object detector	Latency improvement	Hardware used
Wang et al. [32]	Pruning	32.40%	SSD	33.61%	GTX 1080Ti
Zhao et al. [33]	Pruning	93.27%	YOLOv4	80.70%	Qualcomm Adreno 64
Fan et al. [34]	Quantization	75%	SSDLite-MobileNetV2	85.67%	Zynq ZC706
LCDet [35]	Quantization	77.79%	YOLOv2	13.66%	Snapdragon 835
Kang et al. [36]	Knowledge distillation	37.11%	RetinaNet	32.98%	–
Chen et al. [37]	Knowledge distillation	83.82%	RCNN	7.93%	–

3.2 Quantization

Quantization is the process of approximating a continuous signal by a set of discrete symbols or integer values. The discrete set is selected as per the type of quantization such as integer, floating-point, and fixed-point quantization. Quantizing deep learning-based object detector models involves converting the baseline 32-bit parameters (weights, activations, biases) to fewer (e.g., 16 or 8) bits, to achieve lower memory footprint, without significantly reducing model accuracy. Fan et al. [34] proposed an 8-bit integer quantization of all the bias, batch normalization, and activation parameters on SSDLite-MobileNetV2. LCDet [35] proposed a fully quantized 8-bit model in which parameters of each layer of a YOLOv2 object detector were quantized to 8-bit fixed point values. To achieve this, they first stored the minimum and maximum value at each layer and then used relative valued to linearly distribute the closest integer value to all the reduced bitwidth weights.

3.3 Knowledge Distillation

Knowledge Distillation involves transferring learned knowledge from a larger model to a smaller, more compact model. A teacher model is first trained for object detection, followed by a smaller student model being trained to emulate the prediction behavior of the teacher model. The goal is to make the student model learn important features to arrive at the predictions that are very close to that of the original model. The resulting student model reduces the computational power and memory footprint compared to the original teacher model. Kang et al. [36] proposed an instance-conditional knowledge decoding module to retrieve knowledge from the teacher network (RetinaNet with a ResNet-101 classifier model as backbone) via query-based attention. They also used a subtask that optimized the decoding module and feature maps to update the student network (RetinaNet with a simpler ResNet-50 model as backbone). Chen et al. [37] proposed a three-step knowledge distillation process on R-CNN with a Resnet-50 model as backbone. The first step used a feature pyramid distillation process to extract the output features that can mimic the teacher network features. They then used these features to remove the output proposal to perform Regional Distillation (RD), enabling the student (RCNN with a much simpler ResNet-18 model as backbone) to focus on the positive regions. Lastly, Logit Distillation (LD) on the output was used to mimic the final output of the teacher network.

4   Open Challenges and Opportunities

While there has been significant work on effective object detection for AVs, there are significant outstanding challenges that remain to be solved. Here we discuss some of the key challenges and opportunities for future research in the field.

Neural Architecture Search (NAS)   In recent years, NAS based efforts have gained much attention to automatically determine the best backbone architecture for a given object detection task. Recent works such as NAS-FCOS [38], MobileDets [39], and AutoDets [40] have shown promising results on image classification tasks. Using automated NAS methods can help identify better anchors boxes and backbone networks to improve object detector performance. The one drawback of these efforts is that they take significantly longer to discover the final architecture. More research is needed to devise efficient NAS approaches targeting object detectors.

Real-Time Processing   Object detectors deployed in AV's use video inputs from AV cameras, but the object detectors are typically trained to detect objects on image datasets. Detecting an object on every frame in a video can increase latency of the detection task. Correlations between consecutive frames can help identify the frames that can be used for detecting new objects (while discarding others) and reduce the latency of the model. Creating models that can correlate spatial and temporal relationships between consecutive frames is an open problem. Recent work on real-time object detection [41, 42], has begun to address this problem, but much more work is needed.

Sensor Fusion   Sensor fusion is one of the most widely used methods for increasing accuracy of 2D and 3D object detection. Many efforts fuse lidar and RGB images to perform object detection for autonomous driving. But there are very few works that consider fusion data from ultrasonic sensors, radar, or V2X communication. The fusion of data from more diverse sensors is vital to increasing the reliability of the perception system in AVs. Fusing additional sensor data can also increase stability and ensure that the perception system does not fail when one of the sensors fails due to environmental conditions. Recent efforts [43, 44] are beginning to design object detectors that work with data from various sensors, which is a step in the right direction for reliable perception in AVs.

Time Series Information   Most conventional object detection models rely on a CNN-based network for object detection that does not consider time series information. Only a few works, such as [45, 46], consider multi-frame perception that uses data from the previous and current time instances. Correlating time series information about vehicle dynamics can increase the reliability of the model. Some works such as Sauer et al. [47] and Chen et al. [48] have used time-series data such as steering angle, vehicle velocity, etc. with object detector output to create a closed loop autonomous driving system. Research on combining these efforts with time-series object detector outputs can enable us to make direct driving decisions from these multi-modal models for safer and more reliable driving.

Semi-supervised Object Detection Supervised machine learning methods which are used in all object detectors today require an annotated dataset to train the detector models. The major challenge in supervised object detection is to annotate data for different scenarios such as, but not limited to, weather conditions, terrain, variable traffic, and location, which is a time-consuming task to ensure improved safety and adaptability of these models in real-world AV driving scenarios. Due to the evolving changes in driving environments, the use of semi-supervised learning for object detection can reduce training time of these models. Some recent efforts, e.g., [49–51] advocate for performing object detection using semi-supervised transformer models. Due to the high accuracy of transformer-based models, they can yield better performance when detecting object for autonomous driving tasks. Even though transformer-based models yield higher accuracy, deploying them on embedded onboard computers is still a challenge due to their large memory footprint, which requires further investigation.

Open Datasets Object detector model performance can vary due to changing lighting, weather, and other environmental conditions. Data from different weather conditions during training can help fit all the environmental needs to address this problem. Adding new data to accommodate these weather conditions changes when training and testing these models can help overcome this issue. The Waymo open dataset [52] has a wide variety of data that focus on different lighting and weather conditions to overcome this issue. More such open datasets are needed to train reliable object detectors for AVs to ensure robust performance in a variety of environmental conditions.

Resource Constraints Most object detectors have high computational and power overheads when deployed on real hardware platforms. To address this challenge, prior efforts have adapted pruning, quantization, and knowledge distillation techniques (see Sect. 3) to reduce model footprint and decrease the model's computational needs. New approaches for hardware-friendly pruning and quantization, such as recent efforts [53, 54], can be very useful. Techniques to reduce matrix multiplication operations, such as [55–57] can also speed up object detector execution time. Hardware and software co-design, by combining pruning, quantization, knowledge distillation etc. along with hardware optimization such as parallel factors adjustment, resource allocation etc. also represents an approach to improve object detector efficiency. Results from recent work [58–60] have been promising, but much more research is needed on these topics.

5 Conclusion

In this chapter, we discussed the landscape of various object detectors being considered and deployed in emerging AVs, the challenges involved in using these object detectors in AVs, and how the object detectors can be optimized for lower computational complexity and faster inference during real-time perception. We also presented a multitude of open challenges and opportunities to advance the state-of-the-art with object detection for AVs. As AVs are clearly the transportation industry's future, research to overcome these challenges will be crucial to creating a safe and reliable transportation model.

References

1. Automated Vehicles for Safety, NHTSA Report (2021)
2. Kukkala, V.K., Tunnell, J., Pasricha, S., Bradley, T.: Advanced driver-assistance systems: a path toward autonomous vehicles. IEEE Consum. Electron. Mag. **7**(5), 18–25 (2018)
3. J3016B: Taxonomy and definitions for terms related to driving automation systems for on-road motor vehicles – SAE international., https://www.sae.org/standards/content/j3016_201806
4. Staff, R.D.: Navlab: The self-driving car of the '80s. Rediscover the '80s (2016)
5. Dickmanns, E.D.: Dynamic Vision for perception and control of Motion (2010)
6. Lawler, R.: Riding shotgun in Tesla's fastest car ever (2014)
7. Drive Me, the World's most ambitious and advanced public autonomous driving experiment, starts today. Volvo Cars Global Media Newsroom, 2016
8. Safety report and Whitepapers, Waymo., https://waymo.com/safety/
9. McEachern, S.: Cruise Founder Takes Company's First Driverless Ride on SF Streets: Video. GM Authority (2021)
10. Jiao, L., Zhang, F., Liu, F., Yang, S., Li, L., Feng, Z., Qu, R.: A survey of deep learning-based object detection. IEEE Access. **7**, 128837–128868 (2019)
11. Dalal, N., Triggs, B.: Histograms of oriented gradients for human detection. In: 2005 IEEE computer society conference on computer vision and pattern recognition (CVPR'05) (Vol. 1, pp. 886–893). IEEE (2005)
12. Felzenszwalb, P.F., Girshick, R.B., McAllester, D., Ramanan, D.: Object detection with discriminatively trained part-based models. IEEE Trans. Pattern Anal. Mach. Intell. **32**(9), 1627–1645 (2009)
13. Girshick, R., Donahue, J., Darrell, T., Malik, J.: Rich feature hierarchies for accurate object detection and semantic segmentation. In: Proceedings of the IEEE conference on computer vision and pattern recognition, pp. 580–587 (2014)
14. Girshick, R.: Fast R-CNN. In: Proceedings of the IEEE International Conference on Computer Vision (ICCV), pp. 1440–1448 (2015)
15. Ren, S., He, K., Girshick, R., Sun, J.: Faster r-cnn: towards real-time object detection with region proposal networks. Adv. Neural Inf. Proces. Syst. **28** (2015)
16. Redmon, J., Divvala, S., Girshick, R., Farhadi, A.: You only look once: Unified, real-time object detection. In: Proceedings of the IEEE conference on computer vision and pattern recognition, pp. 779–788 (2016)
17. Redmon, J., Farhadi, A.: YOLO9000: better, faster, stronger. In Proceedings of the IEEE conference on computer vision and pattern recognition, pp. 7263–7271 (2017)
18. Redmon, J., Farhadi, A.: Yolov3: An incremental improvement. arXiv preprint, arXiv:1804.02767 (2018)
19. Liu, W., Anguelov, D., Erhan, D., Szegedy, C., Reed, S., Fu, C.Y., Berg, A.C.: SSD: Single shot multibox detector. In European conference on computer vision, pp. 21–37. Springer, Cham (2016)
20. Lin, T.Y., Goyal, P., Girshick, R., He, K., Dollár, P.: Focal loss for dense object detection. In: Proceedings of the IEEE international conference on computer vision, pp. 2980–2988 (2017)
21. Bochkovskiy, A., Wang, C.Y., Liao, H.Y.M.: Yolov4: optimal speed and accuracy of object detection. arXiv preprint, arXiv:2004.10934 (2020)

22. Ultralytics; Ultralytics/yolov5: Yolov5 in PyTorch & ONNX & CoreML & TFLite, GitHub. https://github.com/ultralytics/yolov5
23. Wang, C.Y., Yeh, I.H., Liao, H.Y.M.: You only learn one representation: unified network for multiple tasks. arXiv preprint, arXiv:2105.04206 (2021)
24. Ge, Z., Liu, S., Wang, F., Li, Z., Sun, J.: Yolox: Exceeding yolo series in 2021. arXiv preprint, arXiv:2107.08430 (2021)
25. Nobis, F., Geisslinger, M., Weber, M., Betz, J., Lienkamp, M.: A deep learning-based radar and camera sensor fusion architecture for object detection. In: 2019 Sensor Data Fusion: Trends, Solutions, Applications (SDF), pp. 1–7, IEEE (2019)
26. Fang, J., Zhou, L., Liu, G.: 3d bounding box estimation for autonomous vehicles by cascaded geometric constraints and depurated 2d detections using 3d results. arXiv preprint, arXiv:1909.01867 (2019)
27. Simony, M., Milzy, S., Amendey, K., Gross, H.M.: Complex-yolo: an euler-region-proposal for real-time 3d object detection on point clouds. In: Proceedings of the European Conference on Computer Vision (ECCV) Workshops pp. 0–0 (2018)
28. Simon, M., Amende, K., Kraus, A., Honer, J., Samann, T., Kaulbersch, H., Michael Gross, H.: Complexer-yolo: Real-time 3d object detection and tracking on semantic point clouds. In: Proceedings of the IEEE/CVF Conference on Computer Vision and Pattern Recognition Workshops, pp. 0–0 (2019)
29. Wen, L.H., Jo, K.H.: Fast and accurate 3D object detection for lidar-camera-based autonomous vehicles using one shared voxel-based backbone. IEEE Access. **9**, 22080–22089 (2021)
30. Lu, Y., Hao, X., Sun, S., Chai, W., Tong, M., Velipasalar, S.: RAANet: range-aware attention network for LiDAR-based 3D object detection with auxiliary density level estimation. arXiv preprint, arXiv:2111.09515 (2021)
31. Liang, T., Glossner, J., Wang, L., Shi, S., Zhang, X.: Pruning and quantization for deep neural network acceleration: a survey. Neurocomputing. **461**, 370–403 (2021)
32. Wang, Q., Zhang, H., Hong, X., Zhou, Q.: Small object detection based on modified FSSD and model compression. arXiv preprint, arXiv:2108.10503 (2021)
33. Zhao, P., Yuan, G., Cai, Y., Niu, W., Liu, Q., Wen, W., Ren, B., Wang, Y., Lin, X.: Neural pruning search for real-time object detection of autonomous vehicles. ACM/IEEE DAC. (2021)
34. Fan, H., Liu, S., Ferianc, M., Ng, H.C., Que, Z., Liu, S., Niu, X., Luk, W.: A real-time object detection accelerator with compressed SSDLite on FPGA. FPT. (2018)
35. Tripathi, S., Dane, G., Kang, B., Bhaskaran, V., Nguyen, T.: LCDet: low-complexity fully-convolutional neural networks for object detection in embedded systems. IEEE CVPRW. (2017)
36. Kang, Z., Zhang, P., Zhang, X., Sun, J., Zheng, N.: Instance-conditional knowledge distillation for object detection. Adv. Neural Info. Proces. Syst. (2021)
37. Chen, R., Ai, H., Shang, C., Chen, L., Zhuang, Z.: Learning lightweight pedestrian detector with hierarchical knowledge distillation. IEEE ICIP. (2019)
38. Wang, N., Gao, Y., Chen, H., Wang, P., Tian, Z., Shen, C., Zhang, Y.: NAS-FCOS: "efficient search for object detection architectures". Int. J. Comput. Vis. **129**, 3299–3312 (2021)
39. Xiong, Y., Liu, H., Gupta, S., Akin, B., Bender, G., Wang, Y., Kindermans, P.J., Tan, M., Singh, V., Chen, B.: MobileDets: searching for object detection architectures for mobile accelerators. IEEE/CVF CVPR. (2021)
40. Li, Z., Xi, T., Zhang, G., Liu, J., He, R.: AutoDet: pyramid network architecture search for object detection. Int. J. Comput. Vis. **129**, 1087–1105 (2021)
41. Zhu, H., Wei, H., Li, B., Yuan, X., Kehtarnavaz, N.: Real-time moving object detection in high-resolution video sensing, vol. 20. Sensors (2020)
42. Dai, X., Yuan, X., Wei, X.: TIRNet: object detection in thermal infrared images for autonomous driving. Appl. Intel. (2021)
43. Chavez-Garcia, R.O., Aycard, O.: Multiple sensor fusion and classification for moving object detection and tracking. IEEE TITS. **17**(2) (2016, Feb)

44. Cho, H., Seo, Y.W., Kumar, B.V., Rajkumar, R.R.: A multi-sensor fusion system for moving object detection and tracking in urban driving environments. IEEE ICRA. (2014)
45. Casas, S., Luo, W., Urtasun, R.: IntentNet: learning to predict intention from raw sensor data. Proc. 2nd Annu. Conf. Robot Learn. (2018)
46. Luo, W., Yang, B., Urtasun, R.: Fast and furious: real time endto-end 3D detection, tracking and motion forecasting with a single convolutional net. Proc. IEEE/CVF CVPR. (2018, Jun)
47. Sauer, A., Savinov, N., Geiger, A.: Conditional affordance learning for driving in urban environments. Proc. 2nd Annu. Conf. Robot Learn. (2018)
48. Chen, C., Seff, A., Kornhauser, A., Xiao, J.: DeepDriving: learning affordance for direct perception in autonomous driving. Proc. ICCV. (2015, Dec)
49. Xie, E., Ding, J., Wang, W., Zhan, X., Xu, H., Sun, P., Li, Z., Luo, P.: DetCo: unsupervised contrastive learning for object detection. Proc. IEEE/CVF Int. Conf. Comp. Vision. (2021)
50. Dai, Z., Cai, B., Lin, Y., Chen, J.: UP-DETR: unsupervised pre-training for object detection with transformers. IEEE/CVF CVPR. (2021)
51. Bar, A., Wang, X., Kantorov, V., Reed, C.J., Herzig, R., Chechik, G., Rohrbach, A., Darrell, T., Globerson, A.: DETReg: unsupervised pre-training with region priors for object detection. arXiv preprint, arXiv:2106.04550 (2021)
52. Sun, P., Kretzschmar, H., Dotiwalla, X., Chouard, A., Patnaik, V., Tsui, P., Guo, J., Zhou, Y., Chai, Y., Caine, B., Vasudevan, V.: Scalability in perception for autonomous driving: Waymo open dataset. IEEE/CVF CVPR. (2020)
53. Chen, P., Liu, J., Zhuang, B., Tan, M., Shen, C.: AQD: towards accurate quantized object detection. IEEE/CVF CVPR. (2021)
54. Kim, S., Kim, H.: Zero-centered fixed-point quantization with iterative retraining for deep convolutional neural network-based object detectors. IEEE Access. **9**, 20828–20839 (2021)
55. Pilipović, R., Risojević, V., Božič, J., Bulić, P., Lotrič, U.: An approximate GEMM unit for energy-efficient object detection. Sensors. **21**, 4195 (2021)
56. Winograd, S.: Arithmetic complexity of computations, vol. 33. SIAM (1980)
57. Kala, S., Mathew, J., Jose, B.R., Nalesh, S.: UniWiG: unified Winograd-GEMM architecture for accelerating CNN on FPGAs. IEEE VLSID, 209–214 (2019)
58. Zhang, X., Lu, H., Hao, C., Li, J., Cheng, B., Li, Y., Rupnow, K., Xiong, J., Huang, T., Shi, H., Hwu, W.M.: SkyNet: a hardware-efficient method for object detection and tracking on embedded systems. Proc. Mach. Learn. Syst. **2**, 216–229 (2019)
59. Zhu, Y., Liu, Y., Zhang, D., Li, S., Zhang, P., Hadley, T.: Acceleration of pedestrian detection algorithm on novel C2RTL HW/SW co-design platform. IEEE ICGCS. (2010)
60. Ma, Y., Zheng, T., Cao, Y., Vrudhula, S., Seo, J.S.: Algorithm-hardware co-design of single shot detector for fast object detection on FPGAs. IEEE/ACM ICCAD. (2018)

Scene-Graph Embedding for Robust Autonomous Vehicle Perception

Shih-Yuan Yu, Arnav Vaibhav Malawade, and Mohammad Abdullah Al Faruque

1 Introduction

Automotive CPS, also called *Autonomous Vehicles* (AV), aims to revolutionize personal mobility, logistics, and road safety [17]. However, accidents involving perception errors in modern self-driving cars are still a regular occurrence, highlighting that the development of safe and robust AVs remains a difficult challenge [26–28]. What is worse is that these perception errors often seem completely irrational from our perspective. In one crash, a self-driving car failed to perceive a semi-truck that was utterly obstructing the highway [28]. Another crash involved a vehicle steering directly into the freeway divider in broad daylight [27]. These events cast serious doubt on the ability of current AV perception systems to understand the state of the road. According to a statistic [24], perception and prediction errors were the primary factors in over 40% of driver-related crashes between conventional vehicles. In complex urban environments, navigation is particularly challenging because the scenarios are highly variable and involve pedestrians and bicyclists, heavy traffic, blind driveways, blocked roadways, etc. [23, 30, 46]. Within this context, the effectiveness of understanding the driving scenes becomes particularly crucial, leading researchers and industry leaders to race to address these problems via more advanced AV perception systems.

One might ask, how are humans able to perceive the state of the road effectively without succumbing to the common mistakes of AV perception systems? Recent research suggests that humans rely on cognitive mechanisms to identify the structure of a scene and reason about inter-object relations when performing complex tasks such as identifying risk during driving [5]. However, existing AV perception

S.-Y. Yu · A. V. Malawade (✉) · M. A. Al Faruque
The University of California, Irvine, Irvine, CA, USA
e-mail: shihyuay@uci.edu; malawada@uci.edu; alfaruqu@uci.edu

© The Author(s), under exclusive license to Springer Nature Switzerland AG 2023   525
V. K. Kukkala, S. Pasricha (eds.), *Machine Learning and Optimization Techniques for Automotive Cyber-Physical Systems*, https://doi.org/10.1007/978-3-031-28016-0_18

architectures use road geometry information and vehicle trajectory models for estimating the state of the road scene (model-based methods) [29, 37]. More recently, architectures using deep learning techniques that leverage *Convolutional Neural Networks* (CNNs), *Long-Short Term Memory Networks* (LSTMs), or *Multi-Layer Perceptrons* (MLPs) [6, 19, 20, 38, 41, 47] have proven effective at capturing features essential for modeling subjective risk in both spatial and temporal domains [47]. However, these approaches cannot obtain a high-level, human-like understanding of complex road scenarios due to their inability to explicitly capture inter-object relationships or the overall structure of the road scene. Failing to capture these relationships can result in poor perception performance in complex scenarios. Overall, designing a robust perception system for automotive CPSs using data-driven approaches poses the following challenges:

1. Designing a reliable method that can handle a wide range of complex and unpredictable traffic scenarios,
2. Building a model that is transferable from the simulation setting to the real-world setting because the real-world datasets for supervised training are limited,
3. Building a model that can provide explainable decisions.

Take risk assessment tasks as an example. To overcome the first challenge, deep learning-based methods must be trained on large datasets covering a wide range of "corner cases" (especially risky driving scenarios), which are expensive and time-consuming to generate [9]. In this case, many researchers resort to using synthesized datasets containing many examples of these corner cases to address this issue. However, as mentioned in the second challenge, for these to be valuable, a model must be able to transfer the knowledge gained from simulated training data to real-world situations. A standard method for measuring a model's ability to generalize is *transferability*, where a model's accuracy on a dataset different from the training dataset is evaluated. Suppose a model can effectively transfer the knowledge gained from a simulated training set to a real-world testing set. In that case, it will likely perform better in unseen real-world scenarios. Even if these existing methods can transfer knowledge well, the predictions of such methods lack *explainability*, which is crucial for establishing trust between ADSs and human drivers [1, 2, 4]. In the third research challenge, *Explainability* refers to the ability of a model to effectively communicate the factors that influenced its decision-making process for a given input, particularly those that might lead the model to make incorrect decisions [1, 13]. Suppose a model can give attention to the aspects or entities in a traffic scene that make the scenario risky or non-risky. In that case, it can improve its decision, and its decisions become more explainable [39].

2   Scene-Graph Representation of Road Scenes

2.1   ADS Design Philosophies and Intermediate Representation

Many design philosophies for ADS have been proposed over the years, such as the *modular* design and the *end-to-end* design. Most *modular* design approaches comprise a pipeline of separate components from the sensory inputs to the actuator outputs. In contrast, *end-to-end* approaches generate output directly from their sensory inputs [6, 31]. One advantage of a *modular* design approach is the division of a task into an easier-to-solve set of sub-tasks that have been addressed in other fields such as robotics [14], computer vision [12, 19, 20] and vehicle dynamics [32]. As a result, prior knowledge from these fields can be leveraged when designing the components corresponding to the sub-tasks. However, one disadvantage of such an approach is the complexity of the whole pipeline [46]. *End-to-end* design approaches can achieve good performance with a smaller network size because they perform feature extraction from sensor inputs implicitly through the network's hidden layers [6, 18]. However, the authors in [8] point out that the needed level of supervision is too weak for the end-to-end model to learn critical controlling information (e.g., from image to steering angle), so it can fail to handle complicated driving maneuvers.

Recently, few methodologies have leveraged the benefits of an *intermediate representation* (IR). *DeepDriving* [8], called the *direct perception*, was one of the first approaches to use an IR methodology. In their methodology, a set of *affordance* indicators, such as the distance to lane markings and cars in the current and adjacent lanes, are extracted from an image and serve as an IR for generating the final control output. The authors of [8] prove that the use of this IR is effective for simple driving tasks such as lane following and for generalizing the learned knowledge from simulation to real-world environments, thus improving *transferability*. Authors in [3] use a collection of filtered images, each representing a piece of distinct information, as the IR. They state that the IR used in their methodology allows the training to be conducted on real or simulated data, facilitating testing and validation in simulations before testing on a real car. Moreover, they show that it is easier to synthesize perturbations to the driving trajectory at the mid-level representations than at the level of raw sensors, enabling them to produce non-expert behaviors such as off-road driving and collisions. As such, the capability to capture and identify the complex relationships between road objects is critical in designing an effective human-like perception system for automotive CPS.

2.2   Graph-Based Driving Scene Understanding

In literature, several groups have adopted a variant of *Knowledge Graphs* known as *scene-graphs* to model the road state and the relationships between objects [16,

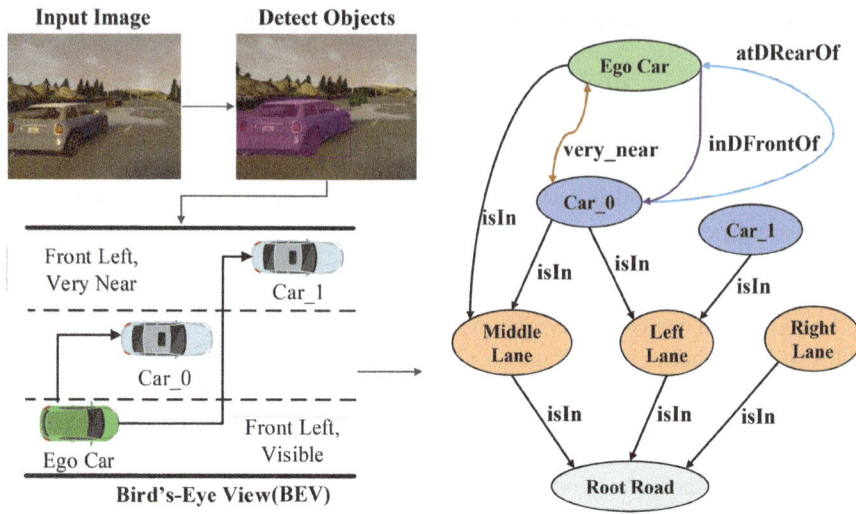

Fig. 1 How camera data can be used to construct a road *scene-graph* representation

21, 22, 25, 45]. A *scene-graph* representation encodes rich semantic information of an image or observed scene, essentially bringing an abstraction of objects and their complex relationships as illustrated in Fig. 1. While each of these related works proposes a different form of *scene-graph* representation, all demonstrate significant performance improvements over conventional perception methods. In [16], the authors propose a 3D-aware egocentric spatio-temporal interaction framework that uses both an *Ego-Thing* graph and an *Ego-Stuff* graph, which together encode how the ego vehicle interacts with both moving and stationary objects in a scene, respectively. In [25], the authors propose a pipeline using a multi-relational graph convolutional network (MR-GCN) for classifying the driving behaviors of traffic participants. The MR-GCN combines spatial and temporal information, including relational information between moving objects and landmark objects. Our prior work has demonstrated that the use of spatio-temporal *scene-graph* embeddings improves performance at subjective risk assessment and collision prediction versus state-of-the-art methods [21, 22, 45]. In addition, our method can better transfer knowledge and is more explainable.

2.3 Scene-Graph Extraction from Driving Scenes

In literature, several approaches have been proposed for extracting *scene-graphs* from images by detecting the objects in a scene and then identifying their visual relationships [42, 44]. However, these works focus on extracting *scene-graphs* for single general images for tasks like automated image captioning instead of modeling these graphs to maximize performance over a temporally-correlated sequence of images as are typically used for autonomous driving. Thus, we adopted a partially rule-based process to extract objects and their attributes from images. Object attributes and bounding boxes are extracted directly from images using state-of-the-art image processing techniques. As Fig. 1 shows, we first convert each image I_t into a collection of objects O_t using Faster RCNN [34], a state of the art object detection algorithm in the *Detectron2* [40] computer vision library. Next, we use OpenCV's perspective transformation library to generate a top-down perspective of the image, commonly known as a "birds-eye view" projection [7]. This projection lets us approximate each object's location relative to the road markings and the ego vehicle. Next, for each detected object in O_t, we use its estimated location and class type (cars, motorcycles, pedestrians, lanes, etc.) to compute the attributes required in building the *scene-graph*.

After collecting the list of objects in each image and their attributes, we can begin constructing the corresponding *scene-graphs*. For each image I_t, we denote the corresponding *scene-graph* by $G_t = \{O_t, A_t\}$ and model it as a directed multi-graph where multiple types of edges connect nodes. The nodes of a *scene-graph*, denoted as O_t, represent the objects in a scene such as lanes, roads, traffic signs, vehicles, pedestrians, etc. The edges of G_t are represented by the adjacency matrix A_t, where each value in A_t represents the type of the corresponding edge in G_t. The edges between two nodes represent the different kinds of relations between them (e.g., near, Front_Left, isIn, etc.). For assessing the risk of driving behaviors, we consider both distance and directional relations between traffic participants useful. We assume that one object's local proximity and positional information will influence the other's motion only if they are within a certain distance. Therefore, in this work, we extract only the location information for each object and adopt a simple rule to determine the relations between the objects using their attributes (e.g., relative location to the ego car), as shown in Fig. 1. For distance relations, we assume two objects are related by one of the relations $r \in \{Near\_Collision$ (4 ft.), *Super_Near* (7 ft.), *Very_Near* (10 ft.), *Near* (16 ft.), *Visible* (25 ft.)$\}$ if the objects are physically separated by a distance that is within that relation's threshold. In the case of the directional relations, we assume two objects are related by the relation $r \in \{Front\_Left, Left\_Front, Left\_Rear, Rear\_Left, Rear\_Right,$ *Right_Rear*, *Right_Front*, *Front_Right*$\}$ based on their relative positions if they are within the *Near* threshold distance from one another.

In addition to directional and distance relations, we also implement the *isIn* relation that connects vehicles with their respective lanes. Specifically, we use each vehicle's horizontal displacement relative to the ego vehicle to assign cars to either the *Left Lane, Middle Lane,* or *Right Lane* based on known lane width. Our abstraction only includes these three-lane areas, and, as such, we map vehicles in all left lanes to the same *Left Lane* node and all vehicles in right lanes to the *Right Lane* node. If a vehicle overlaps two lanes (i.e., during a lane change), we assign it an *isIn* relation to both lanes. Figure 1 illustrates an example of resultant *scene-graph*.

3   Spatio-Temporal Scene-Graph Embedding Approach for Robust Automotive CPS Perception

To tackle the research challenges, we propose a *scene-graph* augmented data-driven approach for assessing the subjective risk of driving maneuvers, where the *scene-graphs* serve as intermediate representations (IR) as shown in Fig. 1. The key advantage of using *scene-graph* as IR is that they allow us to model the relationships between the participants in a traffic scene, thus potentially improving the model's understanding of a scene. Our proposed architecture consists of three major components: (1) a pipeline to convert the images of a driving clip to a sequence of *scene-graphs*, (2) a Multi-Relational Graph Convolution Network (MR-GCN) to convert each of the *scene-graphs* to an embedding (a vectorized representation), and (3) an LSTM for temporally modeling the sequence of embeddings of the respective *scene-graphs*. Our model also contains multiple attention layers: (1) a node attention layer before the embedding of a *scene-graph* is computed, and (2) an attention layer on top of the LSTM, both of which can further improve its performance and explainability.

3.1   Problem Formulation

For training the model, we formulate the problem of subjective risk assessment as a supervised *scene-graph* sequence classification problem. Our approach makes the same assumption used in [47] that the set of driving sequences can be partitioned into two jointly exhaustive and mutually exclusive subsets: risky and safe. We denote the sequence of images of length T by $\mathbf{I} = \{I_1, I_2, I_3, \ldots, I_T\}$. We assume the existence of a spatio-temporal function f that outputs whether a sequence of driving actions x is safe or risky via a risk label y, as given in Eq. (1).

$$y = f(\mathbf{I}) = f(\{I_1, I_2, I_3, \ldots, I_{T-1}, I_T\}), \tag{1}$$

where

$$y = \begin{cases} (1, 0), & \text{if the driving sequence is safe} \\ (0, 1), & \text{if the driving sequence is risky.} \end{cases} \tag{2}$$

The goal of our approach is to propose a suitable model for approximating the function f. Here, the first step is the extraction of the *scene-graph* G_t from each image I_t of the video clip $\mathbf{I}$. This step is achieved by a series of processes that we collectively call the *Scene-Graph Extraction Pipeline* (described in Sect. 2.3). In the second step, these *scene-graphs* are passed through graph convolution layers and an attention-based graph pooling layer. The graph-level embeddings of each *scene-graph*, $\mathbf{h}_{G_t}$, are then calculated using a graph readout operation. Next, these

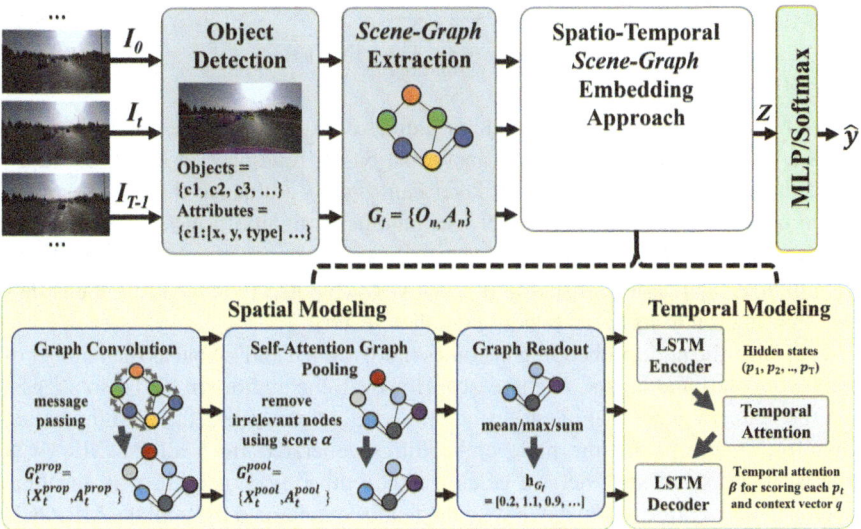

Fig. 2 An illustration of spatio-temporal *scene-graph* embedding approach

scene-graph embeddings are passed sequentially to LSTM cells to acquire the *spatio-temporal* representation, denoted as **Z**, of each *scene-graph* sequence. Lastly, we use a Multi-Layer Perceptron (MLP) layer with a *Softmax* activation function to acquire the final inference, denoted as $\hat{y}$, of the risk for each driving sequence **I**.

To sum up, the model of our approach consists of three major components: a spatial model, a temporal model, and a risk inference component. The spatial model outputs the embedding h_{G_t} for each scene-graph G_t. The temporal model processes the sequence of *scene-graph* embeddings $\mathbf{h}_I = \{h_{G_1}, h_{G_2}, \ldots, h_{G_T}\}$ and produces the spatio-temporal embedding **Z**. The risk inference component outputs each driving clip's final risk assessment, denoted as $\hat{y}$, by processing the Spatio-temporal embedding **Z**. The overall network architecture is shown in Fig. 2. We discuss each of these components in detail below.

3.2   Spatial Modeling

The spatial model we propose uses MR-GCN layers to compute the embedding for a *scene-graph*. The use of MR-GCN allows us to capture multiple types of relations on each *scene-graph* $G_t = \{O_t, A_t\}$. In the *Message Propagation* phase, a collection of node embeddings and their adjacency information serve as the inputs to the MR-GCN layer. Specifically, the l-th MR-GCN layer updates the node embedding, denoted as $\mathbf{h}_v^{(l)}$, for each node v as follows:

$$\mathbf{h}_v^{(l)} = \mathbf{\Phi}_0 \cdot \mathbf{h}_v^{(l-1)} + \sum_{r \in A_t} \sum_{u \in N_r(v)} \frac{1}{|\mathbf{N}_r(v)|} \mathbf{\Phi}_r \cdot \mathbf{h}_u^{(l-1)}, \tag{3}$$

where $N_r(v)$ denotes the set of neighbor indices of node v with the relation $r \in A_t$. $\mathbf{\Phi}_r$ is a trainable relation-specific transformation for relation r in MR-GCN layer. Since the information in $(l-1)$-th layer can directly influence the representation of the node at l-th layer, MR-GCN uses another trainable transformation $\mathbf{\Phi}_0$ to account for the self-connection of each node using a special relation [35]. Here, we initialize each node embedding $\mathbf{h}_v^{(0)}$, $\forall v \in O_t$, by directly converting the node's type information to its corresponding one-hot vector.

Typically, the node embedding becomes more refined and global as the number of graph convolutional layers, L, increases. However, the authors in [43] also suggest that the features generated in earlier iterations might generalize the learning better. Therefore, we consider the node embeddings generated from all the MR-GCN layers. To be more specific, we calculate the embedding of node v at the final layer, denoted as $\mathbf{H}_v^L$, by concatenating the features generated from all the MR-GCN layers, as follows,

$$\mathbf{H}_v^L = \mathbf{CONCAT}(\{\mathbf{h}_v^{(l)}\}|l = 0, 1, \ldots, L). \tag{4}$$

We denote the collection of node embeddings of *scene-graph* G_t after passing through L layers of MR-GCN as $\mathbf{X}_t^{prop}$ (L can be 1, 2 or 3).

The node embedding $\mathbf{X}_t^{prop}$ is further processed with an attention-based graph pooling layer. As stated in [13], such an attention-based pooling layer can improve the explainability of predictions and is typically considered a part of a unified computational block of a graph neural network (GNN) pipeline. In this layer, nodes are pooled according to the scores predicted from either a trainable simple linear projection [10] or a separate trainable GNN layer [15]. We denote the graph pooling layer that uses the **SCORE** function in [10] as *TopkPool* and the one that uses the **SCORE** function in [15] as *SAGPool*. The calculation of the overall process is presented as follows:

$$\alpha = \mathbf{SCORE}(\mathbf{X}_t^{prop}, \mathbf{A_t}), \tag{5}$$

$$\mathbf{P} = \text{top}_k(\alpha), \tag{6}$$

where α stands for the coefficients predicted by the graph pooling layer for nodes in G_t and $\mathbf{P}$ represents the indices of the pooled nodes, which are selected from the top k of the nodes ranked according to α. The number k of the nodes to be pooled is calculated by a pre-defined pooling ratio, pr, and using $k = pr \times |O_t|$, where we consider only a constant fraction pr of the embeddings of the nodes of a scene-graph to be relevant (i.e., 0.25, 0.5, 0.75). We denote the node embeddings and edge adjacency information after pooling by $\mathbf{X}_t^{pool}$ and $\mathbf{A_t^{pool}}$ and are calculated as follows:

$$\mathbf{X}_t^{pool} = (\mathbf{X}_t^{prop} \odot \tanh(\alpha))_{\mathbf{P}}, \tag{7}$$

$$\mathbf{A}_t^{pool} = \mathbf{A}_t^{prop}{}_{(\mathbf{P},\mathbf{P})}. \tag{8}$$

where $\odot$ represents an element-wise multiplication, $()_{\mathbf{P}}$ refers to the operation that extracts a subset of nodes based on P, and $()_{(\mathbf{P},\mathbf{P})}$ refers to the formation of the adjacency matrix between the nodes in this subset.

Finally, our model aggregates the node embeddings of the graph pooling layer, $\mathbf{X}_t^{pool}$, using a graph **READOUT** operation, to produce the final graph-level embedding $\mathbf{h}_{G_t}$ for each *scene-graph* G_t as given by

$$\mathbf{h}_{G_t} = \mathbf{READOUT}(\mathbf{X}_t^{pool}), \tag{9}$$

where the **READOUT** operation can be either summation, averaging, or selecting the maximum of each feature dimension, over all the node embeddings, known as *sum-pooling*, *mean-pooling*, or *max-pooling*, respectively. The process until this point is repeated across all images in $\mathbf{I}$ to produce the sequence of embedding, $\mathbf{h}_I$.

3.3 Temporal Modeling

The temporal model we propose uses an LSTM for converting the sequence of scene-graph embeddings $\mathbf{h}_I$ to the combined spatio-temporal embedding $\mathbf{Z}$. For each timestamp t, the LSTM updates the hidden state p_t and cell state c_t as follows,

$$p_t, c_t = \mathbf{LSTM}(\mathbf{h}_{G_t}, c_{t-1}), \tag{10}$$

where $\mathbf{h}_{G_t}$ is the final *scene-graph* embedding from timestamp t. After the LSTM processes all the scene-graph embeddings, a temporal readout operation is applied to the resultant output sequence to compute the final Spatio-temporal embedding Z given by

$$Z = \mathbf{TEMPORAL\_READOUT}(p_1, p_2, \ldots, p_T) \tag{11}$$

where the **TEMPORAL_READOUT** operation could be extracting only the last hidden state p_T (LSTM-last), or be a temporal attention layer (LSTM-attn).

In [2], adding an attention layer b to the encoder-decoder based LSTM architecture is shown to achieve better performance in Neural Machine Translation (NMT) tasks. For the same reason, we include *LSTM-attn* in our architecture. *LSTM-attn* calculates a context vector q using the hidden state sequence $\{p_1, p_2, \ldots, p_T\}$ returned from the LSTM encoder layer as given by

$$q = \sum_{t=1}^{T} \beta_t p_t \tag{12}$$

where the probability β_t reflects the importance of p_t in generating q. The probability β_t is computed by a *Softmax* output of an energy function vector e, whose component e_t is the energy corresponding to p_t. Thus, the probability β_t is formally given by

$$\beta_t = \frac{\exp(e_t)}{\sum_{k=1}^{T} \exp(e_k)}, \tag{13}$$

where the energy e_t associated with p_t is given by $e_t = b(s_0, p_t)$. The temporal attention layer b scores the importance of the hidden state p_t to the final output, which in our case is the risk assessment. The variable s_0 in the temporal attention layer b is computed from the last hidden representation p_T. The final Spatio-temporal embedding for a video clip, Z, is computed by feeding the context vector q to another LSTM decoder layer.

3.4 Risk Inference

The last piece of our model is the risk inference component that computes the risk assessment prediction $\hat{Y}$ using the spatio-temporal embedding $\mathbf{Z}$. This component is composed of a MLP layer followed by a *Softmax* activation function. Thus, the prediction $\hat{Y}$ is given by

$$\hat{Y} = Softmax(\mathbf{MLP}(Z)) \tag{14}$$

During training, the loss for the prediction is calculated as follows,

$$\mathbf{CrossEntropyLoss}(Y, \hat{Y}) \tag{15}$$

For training our model, we use a mini-batch gradient descent algorithm that updates its parameters by training on a batch of *scene-graph* sequences. To account for label imbalance, we apply class weighting when calculating loss. Besides, several dropout layers are inserted into the network to reduce overfitting.

4 Experimental Results

To illustrate the benefits of our *scene-graph* augmented approach, we present experimental results for assessing the risk of several common driving tasks, including lane changes, turns, and merges into (merging) and out of (branching) the traffic flow. We also evaluate a state-of-the-art SMT+CNN+LSTM based risk assessment model [47] on these tasks to serve as the *baseline*. We evaluate several different aspects of performance, including risk assessment accuracy, capability to transfer knowledge from synthetic data to real-world data, and explainability. Next, let us discuss the experimental setup.

4.1 Experimental Setup

We prepare two types of datasets for the experiments (1) synthesized datasets and (2) real-world driving datasets. To create the synthesized datasets, we collected data from various driving conditions simulated in the CARLA driving simulator.[1] We generated the real-world dataset by extracting various driving actions from the Honda Driving Dataset (HDD) [33]. We generated a wide range of simulated lane changes using the various presets in CARLA that allowed us to specify the number of cars, pedestrians, weather and lighting conditions, driver behavior, etc. The lane changes that resulted in collisions, near collisions, or otherwise dangerous conditions are considered our *risky* samples, while the safe lane changes are labeled as *safe*. Common factors that can affect the risk of a driving action include the distance to other cars and the side curbs, the speed relative to other vehicles, the sizes of adjacent vehicles, the presence of bikers or pedestrians, and the traffic light status.

We generated two synthesized datasets: a *271-syn* dataset and a *1043-syn* dataset, containing 271 and 1043 lane-changing clips, respectively. In addition, we sub-sampled the *271-syn* and *1043-syn* datasets further to create two balanced datasets that have a 1:1 distribution of risky to safe lane changes: *96-syn* and *306-syn*. Our synthesized driving datasets are available online in both raw image and *scene-graph* format [11]. For real driving datasets, we processed the HDD dataset to create a dataset called *1361-honda* composed of 571 lane changing, 350 turning, 297 branching, and 149 merging video clips. For evaluating the capability of the model to transfer knowledge after training on the synthesized lane change datasets, we subsampled *1361-honda* to create a lane-changing dataset that contains 571 real-world lane changing clips, called *571-honda*. The final score of a model on a dataset is computed by averaging over the testing set scores for ten different train-test splits, where 30% of the dataset is reserved as the testing set.

In our experiments, we trained each model for 500 epochs. From our experimentation, we found that the best configuration of our model consisted of two MR-GCN layers with 64 hidden units, a SAGPool pooling layer with a ratio of 0.5, *sum-pooling* for graph readout operation, and *LSTM-attn* for temporal modeling.

---

[1] https://github.com/carla-simulator/carla.

4.2   Experiments on Risk Assessment

We evaluate each model's performance on each dataset by measuring its classification accuracy and the Area Under the Curve (AUC) of the Receiver Operating Characteristic (ROC). The classification accuracy is the ratio of the number of correct predictions on the test set of a dataset to the total number of samples in the testing set. AUC, sometimes referred to as a *balanced* accuracy measure [36], quantifies the likelihood that a binary classifier ranks a positive sample more highly than a random negative sample. This metric is especially useful for imbalanced datasets (i.e. *271-syn*, *1043-syn*, *571-honda*).

Figure 3 shows the comparison between our model's performance and the baseline [47] for the synthetic datasets. The results show that our approach performs best across all datasets.

The results also show that the performance difference between our approach and the baseline increased when the training datasets were smaller. This result indicates that our approach can learn an accurate model even from a smaller dataset, likely resulting from its use of a *scene-graph* based IR. We also found that our approach performs better than the baseline on balanced datasets, meaning that our approach is better at discriminating between the two classes in general. For context, the datasets *271-syn* and *306-syn* contain roughly the same number of clips but differ in the distribution of safe to risky lane changes (2.30:1 for *271-syn* vs. 1:1 for *306-syn*).

Although these results are impressive, we must ask, how much does each component in the model contribute to the overall performance? One easy way to answer this question is with an *ablation study*, where we measure the performance of our model after adding each modeling component one at a time, as is shown in Table 1. From Table 1 we find that the simplest of the models, with no MR-GCN layer (replaced with an MLP layer) and a simple average of the embeddings in $\mathbf{h}_l$ for the temporal model (denoted as *mean* in Table 1), achieves a relatively low classification accuracy of 75%. Starting from this base model, we find that replacing *mean* with an LSTM layer for temporal modeling yields a 10.5% increase in performance. Next, we try adding a single MR-GCN layer with 64 hidden units and *sum-pooling* to the base model, resulting in a 14.8% performance gain. The performance gain achieved by including the MR-GCN layer alone demonstrates the effectiveness of explicitly modeling the relations between objects. Now, we try the single MR-GCN layer with *sum-pooling* and the LSTM model together, which yields the maximum performance gain of 18.1% over the simplest model. This result clearly illustrates that our model's spatial and temporal components are both crucial for maximizing performance.

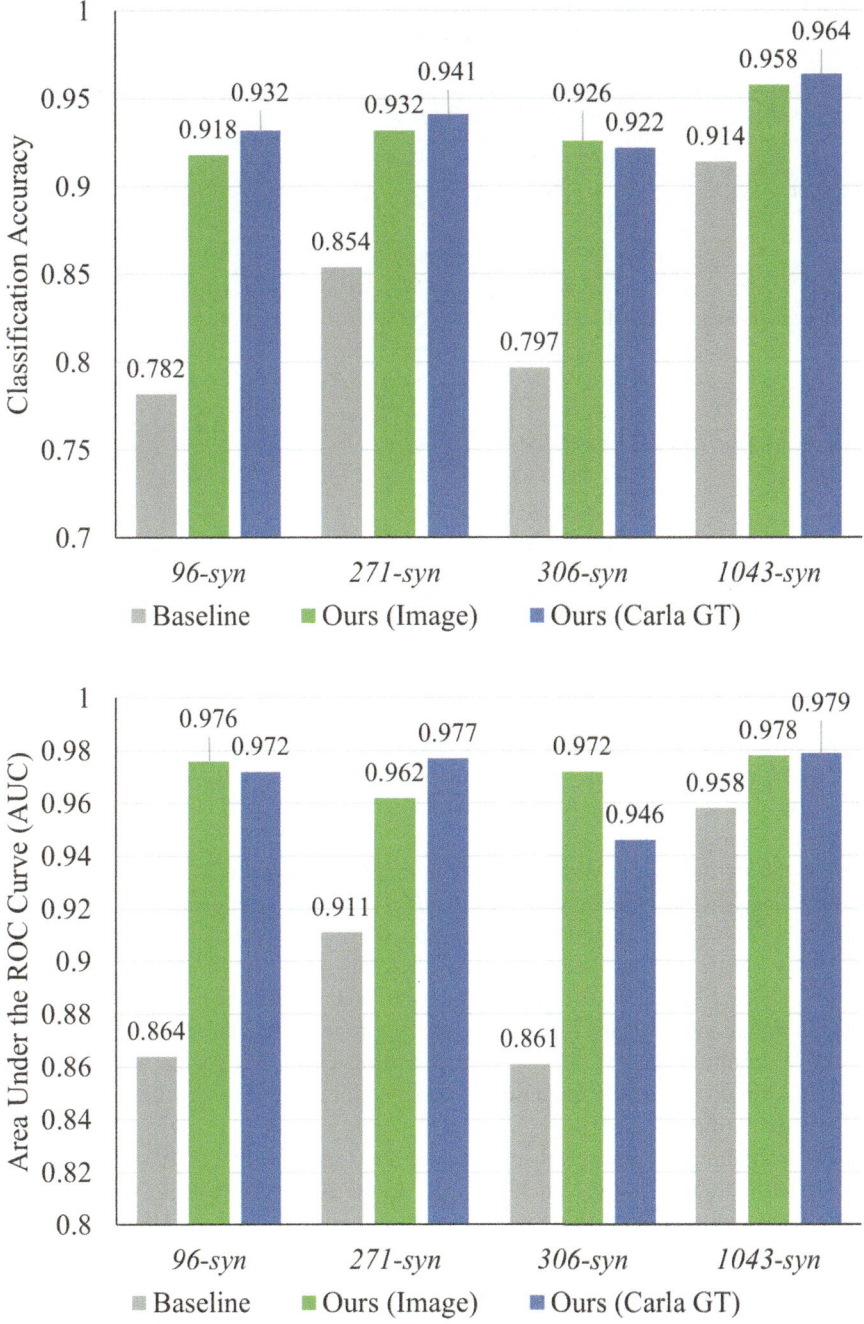

Fig. 3 Accuracy and AUC comparison between our approaches (Real Image and CARLA GT) and [47] on different datasets. Our approach outperforms the baseline across datasets

Table 1 The results of the CARLA GT approach on *1043-syn* dataset with various spatial and temporal modeling settings. In these experiments, we used MR-GCN layers with 64 hidden units and *sum-pooling* as the graph readout operation. The bolded numbers indicate the highest performing configuration in terms of Average Accuracy (Avr. Acc.) and Average AUC Score (Avr. AUC) for the grouping indicated by the leftmost column

	Spatial modeling	Temporal modeling	Avr. Acc.	Avr. AUC
Ablation study	No MR-GCN	*mean*	0.762	0.823
	No MR-GCN	*LSTM-last*	0.867	0.929
	1 MR-GCN	*mean*	0.910	0.960
	1 MR-GCN	*LSTM-last*	**0.943**	**0.977**
Temporal attention	No MR-GCN	*LSTM-last*	0.867	0.929
	No MR-GCN	*LSTM-attn*	0.868	0.928
	1 MR-GCN	*LSTM-last*	0.943	0.977
	1 MR-GCN	*LSTM-attn*	**0.950**	**0.977**
Spatial attention	1 MR-GCN	*mean*	0.910	0.960
	1 MR-GCN, *TopkPool*	*mean*	0.886	0.930
	1 MR-GCN, *SAGPool*	*mean*	**0.937**	**0.968**

4.3 Evaluation of Attention Mechanisms on Risk Assessment

Next, we evaluate the various attention components of our proposed model. To evaluate the benefit of attention over the spatial domain, we tested our model with three different graph attention methods: no attention, *SAGPool*, and *TopkPool*. To evaluate the impact of attention on the temporal domain, we tested our model with the following temporal models: *mean*, *LSTM-last*, and *LSTM-attn*. The results of this analysis are also shown in Table 1.

For evaluating the benefits of graph attention, we start with an attention-free model: one MR-GCN layer with *sum-pooling + mean*. In comparison, the model that uses *SAGPool* for attention on the graph shows a 2.7% performance gain over the attention-free model because using attention over both nodes and relations allows *SAGPool* to better filter out irrelevant nodes from each *scene-graph*. We found that the model using *TopkPool* as the graph-attention layer became relatively unstable, resulting in a 2.4% performance drop compared to the attention-free model. This drop is likely because *TopkPool* ignores the relations between nodes when calculating α.

For evaluating the impact of attention on the temporal model, we assessed the effects of adding a temporal attention layer to the following two models: (1) with no MR-GCN layers and no temporal attention and (2) with one MR-GCN layer and no temporal attention. Our model with no MR-GCN and no temporal attention performed nearly the same as our model with no MR-GCN and *LSTM-attn*. We also find that adding *LSTM-attn* to the model with one MR-GCN layer increases its performance by 0.7% over the same model with no temporal attention. These results demonstrate that the inclusion of temporal attention improves performance, though only marginally compared to the benefits of spatial attention. This might be because

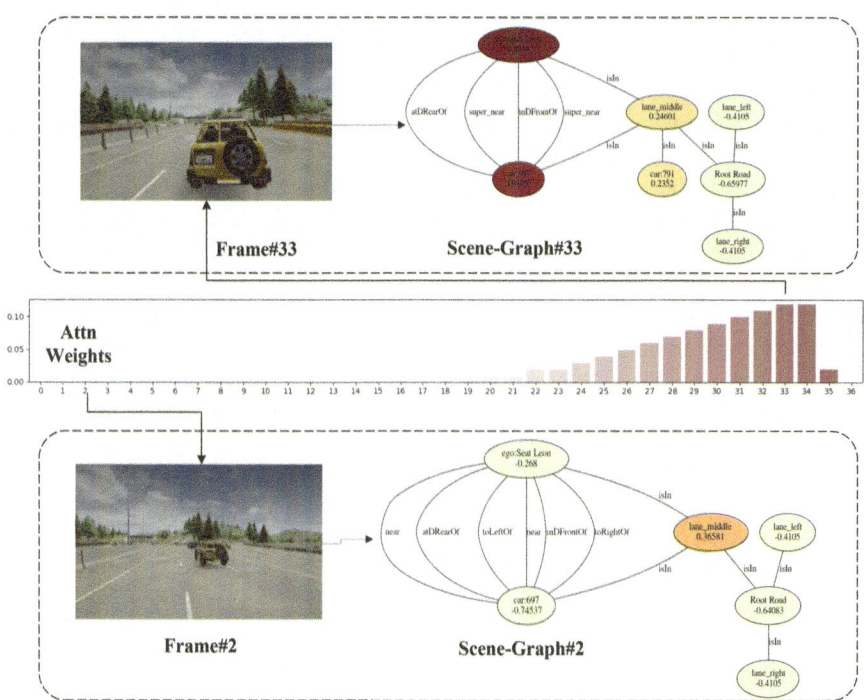

Fig. 4 The visualization of attention weights in both spatial (α) and temporal (β) domains using a risky lane changing clip as an example. We used a gradient color from light yellow to red to visualize each node's projection score indicating its relative importance. The white to red bar chart visualizes the temporal attention scores of each frame

LSTM-*last* learns a good enough temporal model that LSTM-*attn* can only slightly improve on it.

Figure 4 demonstrates how we can use the attention weights of our model to pinpoint the critical factors related to driving risk in both temporal and spatial domains, thus enabling it to *explain* its decisions. As described previously in Eq. (7), the node attention weights α are used by our graph pooling layer to filter out the objects in a *scene-graph* that are less relevant to the overall risk of the scene. Meanwhile, the temporal attention weights, β, allow the LSTM encoder to score each intermediate hidden state (p_t) and retain only the most useful information in Z for the final risk assessment. We demonstrate our model's capability to explain its decisions better using the visualization of both spatial and temporal attention shown in Fig. 4. The figure shows a clearly increasing trend of temporal attention scores $\beta_1, \beta_2, \ldots \beta_T$ as the lane-changing scenario becomes riskier over time. Intuitively, the frames with higher attention scores are weighted more heavily when calculating Z and thus contribute more to the final risk assessment decision. In this risky lane changing example, the temporal attention scores progressively increase between frames 19 and 32 during the lane change; and the highest frame attention weights appear in frames 33 and 34, which are the frames immediately before the collision occurs. Figure 4 also shows the projection scores for the node attention layer, where a higher score for a node indicates that it contributes more to the final decision of risk assessment. As shown in this example, as the ego car approaches the yellow vehicle, the node attention weights for the ego car and the yellow vehicle are increased proportionally to the scene's overall risk. In the first few frames, the risk of collision is low; thus, the node attention weights are low; however, in the last few frames, a collision between these two vehicles is imminent; thus, the attention weights for the two cars are much higher than for any other nodes in the graph. This example clearly shows how graph representations and models, when used with attention, can effectively explain their decision-making process. This capability can be valuable for debugging edge cases at design time, thus reducing the chances of ADS making unexpected, erroneous decisions in real-world scenarios and improving human trust in the system.

4.4 Transferability from Virtual To Real Driving

This section demonstrates our approach's capability to effectively transfer the knowledge learned from a simulated dataset to a real-world dataset. As mentioned previously, this capability is vital since little real-world data exists for rare scenarios. Models must primarily rely on simulation data to improve driving safety in the real world. To demonstrate this capability, we use the model weights and parameters learned from training on the *271-syn* dataset or the *1043-syn* dataset directly for testing on the real-world driving dataset: *571-honda*. We also compare the transferability of our model with that of the baseline method [47]. The results are shown in Fig. 5.

As expected, the performance of both our approach and the baseline degrades when tested on *571-honda* dataset. However, as Fig. 5 shows, the accuracy of our approach only drops by 6.7% and 3.5% when the model is trained on *271-syn* and *1043-syn*, respectively, while the baseline's performance drops drastically by a much higher 21.3% and 14.9%, respectively. The results show that our proposed model can transfer knowledge more effectively than the baseline.

4.5 Risk Assessment By Action Type

This section shows results from evaluating our model's performance on other kinds of driving scenarios available in the HDD besides lane changes: turning, branching, merging, etc. The results for training and evaluating our model on the *1361-honda* dataset are shown in Table 2. From Table 2, we can see that our graph-based approach significantly outperforms [47] in both overall accuracy (0.86 v.s. 0.58) and overall AUC (0.91 v.s. 0.61), indicating that our approach can better

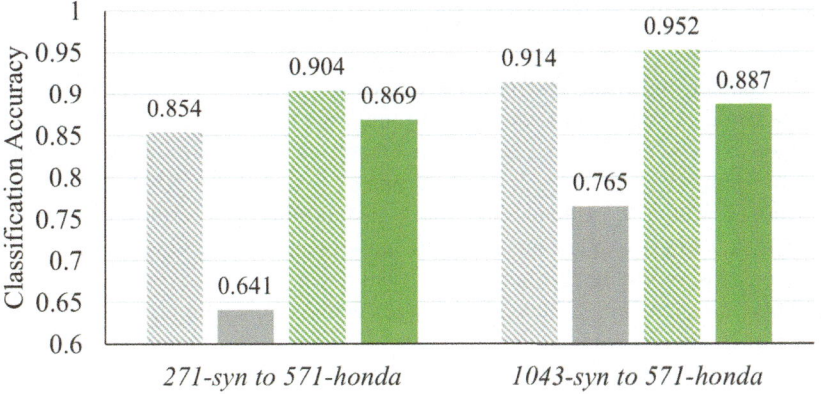

§ Baseline (Original) ■ Baseline (Transfer) § Ours (Original) ■ Ours (Transfer)

Fig. 5 Transferability comparison between our real image model and the baseline [47]. In this experiment, we trained each model on both *271-syn* dataset and *1043-syn* dataset. Then we tested the accuracy of each model on both original dataset and *571-honda* dataset

Table 2 Breakdown of risk assessment performance by driver action types (Lane Changing, Merging, Branching, and Turning) evaluated on *1361-Honda* dataset. The bolded numbers indicate the highest score between *Ours* and the baseline [47] on each of the categories of driver actions (rows)

Metric	Action type	**Ours**	[47]
Accuracy	Overall	**0.8655**	0.5844
	Lane changing	**0.8710**	0.5714
	Merging	**0.8462**	0.5854
	Branching	**0.9101**	0.5556
	Turning	**0.8211**	0.6218
AUC	Overall	**0.9124**	0.6078
	Lane changing	**0.9105**	0.5877
	Merging	**0.9395**	0.6526
	Branching	**0.9462**	0.5807
	Turning	**0.8645**	0.6400

assess risk across diverse driving scenarios and driving action types. In Table 2 we also show the performance for each action type. The results show that our approach also outperforms [47] on each class of driving action. Our approach slightly under-performs on turning scenarios compared to its performance on other action types. This discrepancy is likely because turning scenarios are intrinsically more complicated than straight-road driving scenarios (lane change, branch, merge). Another reason could be that the heading of vehicles is a more significant factor in complicated scenarios, while the *scene-graph* used in our work contains only distance and directional relations.

5 Conclusion

In this chapter, we discovered how the expressive power of graph representations of data could be leveraged to significantly improve the perception performance of automotive CPS. There were clear improvements across experiments and datasets, with our graph-based approach outperforming conventional CNN-based methods in terms of accuracy, explainability, and transferability. All of these benefits can be attributed to the explicit modeling of inter-object relationships via the graph's topology, thus improving the model's ability to semantically *understand* each scene. Although the approach presented here was effective at modeling risk, several other problems in the AV domain remain unsolved, including motion prediction, object detection, and control. When adapted to fit these problems, graph-based methods could potentially provide the same benefits over existing methods.

References

1. Adadi, A., Berrada, M.: Peeking inside the black-box: a survey on explainable artificial intelligence (xai). IEEE Access **6**, 52138–52160 (2018)
2. Bahdanau, D., Cho, K., Bengio, Y.: Neural machine translation by jointly learning to align and translate. Preprint (2014). arXiv:14090473
3. Bansal, M., Krizhevsky, A., Ogale, A.: Chauffeurnet: learning to drive by imitating the best and synthesizing the worst. Preprint (2018). arXiv:181203079
4. Bao, N., Yang, D., Carballo, A., Özgüner, Ü., Takeda, K.: Personalized safety-focused control by minimizing subjective risk. In: 2019 IEEE Intelligent Transportation Systems Conference (ITSC), pp. 3853–3858. IEEE (2019)
5. Battaglia, P.W., Hamrick, J.B., Bapst, V., Sanchez-Gonzalez, A., Zambaldi, V., Malinowski, M., Tacchetti, A., Raposo, D., Santoro, A., Faulkner, R., et al.: Relational inductive biases, deep learning, and graph networks. Preprint (2018). arXiv:180601261
6. Bojarski, M., Del Testa, D., Dworakowski, D., Firner, B., Flepp, B., Goyal, P., Jackel, L.D., Monfort, M., Muller, U., Zhang, J., et al.: End to end learning for self-driving cars. Preprint (2016). arXiv:160407316
7. Bradski, G.: The OpenCV Library. Dr Dobb's Journal of Software Tools (2000)
8. Chen, C., Seff, A., Kornhauser, A., Xiao, J.: Deepdriving: learning affordance for direct perception in autonomous driving. In: Proceedings of the IEEE International Conference on Computer Vision, pp 2722–2730 (2015)
9. Dosovitskiy, A., Ros, G., Codevilla, F., Lopez, A., Koltun, V.: Carla: an open urban driving simulator. Preprint (2017). arXiv:171103938
10. Gao, H., Ji, S.: Graph u-nets. Preprint (2019). arXiv:190505178
11. Hsu, B., Yu, S.Y., Malawade, A.V., Muthirayan, D., Khargonekar, P., Al Faruque, M.A.: Scene-graph-risk-assessment dataset (2021). https://doi.org/10.21227/c0z9-1p30
12. Jain, R., Kasturi, R., Schunck, B.G.: Machine Vision, vol. 5. McGraw-Hill, New York (1995)
13. Knyazev, B., Taylor, G.W., Amer, M.: Understanding attention and generalization in graph neural networks. In: Advances in Neural Information Processing Systems, pp 4202–4212. Curran Associates, Inc. (2019)
14. Laumond, J.P., et al.: Robot Motion Planning and Control, vol 229. Springer, Berlin (1998)
15. Lee, J., Lee, I., Kang, J.: Self-attention graph pooling. Preprint (2019). arXiv:190408082
16. Li, C., Meng, Y., Chan, S.H., Chen, Y.T.: Learning 3d-aware egocentric spatial-temporal interaction via graph convolutional networks. Preprint (2019). arXiv:190909272

17. Litman, T.: Autonomous Vehicle Implementation Predictions. Victoria Transport Policy Institute, Victoria (2017)
18. Malawade, A.V., Odema, M., Lajeunesse-DeGroot, S., Al Faruque, M.A.: Sage: a split-architecture methodology for efficient end-to-end autonomous vehicle control. ACM Trans. Embedd. Comput. Syst. **20**(5s), 1–22 (2021)
19. Malawade, A.V., Mortlock, T., Faruque, M.A.A.: Ecofusion: energy-aware adaptive sensor fusion for efficient autonomous vehicle perception. In: Design Automation Conference (DAC). ACM (2022)
20. Malawade, A.V., Mortlock, T., Faruque, M.A.A.: Hydrafusion: context-aware selective sensor fusion for robust and efficient autonomous vehicle perception. In: International Conference on Cyber-Physical Systems (ICCPS). IEEE (2022)
21. Malawade, A.V., Yu, S.Y., Hsu, B., Kaeley, H., Karra, A., Al Faruque, M.A.: roadscene2vec: a tool for extracting and embedding road scene-graphs. Knowl.-Based Syst. **242**, 108245 (2022)
22. Malawade, A.V., Yu, S.Y., Hsu, B., Muthirayan, D., Khargonekar, P.P., Al Faruque, M.A.: Spatio-temporal scene-graph embedding for autonomous vehicle collision prediction. IEEE Internet Things J. **9**(12), 9379–9388 (2022) https://doi.org/10.1109/JIOT.2022.3141044
23. Montgomery, W.D., Mudge, R., Groshen, E.L., Helper, S., MacDuffie, J.P., Carson, C.: Securing America's future energy. 52 p. Available: https://avworkforce.secureenergy.org/
24. Mueller, A.S., Cicchino, J.B., Zuby, D.S.: What humanlike errors do autonomous vehicles need to avoid to maximize safety? J. Saf. Res. **75**, 310–318 (2020) ISSN 0022-4375, https://doi.org/10.1016/j.jsr.2020.10.005
25. Mylavarapu, S., Sandhu, M., Vijayan, P., Krishna, K.M., Ravindran, B., Namboodiri, A.: Towards accurate vehicle behaviour classification with multi-relational graph convolutional networks. Preprint (2020). arXiv:200200786
26. National Transportation Safety Board: Collision between vehicle controlled by developmental automated driving system and pedestrian. Technical Report, NTSB/HAR-19/03, National Transportation Safety Board, 2019
27. National Transportation Safety Board: Collision between a sport utility vehicle operating with partial driving automation and a crash attenuator. Technical Report, NTSB/HAR-20/01, National Transportation Safety Board, 2020
28. National Transportation Safety Board: Collision between car operating with partial driving automation and truck-tractor semitrailer. Technical Report, NTSB/HAB-20/01, National Transportation Safety Board, 2020
29. Nistér, D., Lee, H.L., Ng, J., Wang, Y.: The safety force field. NVIDIA White Paper (2019)
30. Paden, B., Čáp, M., Yong, S.Z., Yershov, D., Frazzoli, E.: A survey of motion planning and control techniques for self-driving urban vehicles. IEEE Trans. Intell. Veh. **1**(1), 33–55 (2016)
31. Pomerleau, D.A.: Alvinn: an autonomous land vehicle in a neural network. In: Advances in Neural Information Processing Systems, pp 305–313. Morgan Kaufmann Publishers (1989)
32. Rajamani, R.: Vehicle Dynamics and Control. Springer Science & Business Media, New York (2011)
33. Ramanishka, V., Chen, Y.T., Misu, T., Saenko, K.: Toward driving scene understanding: A dataset for learning driver behavior and causal reasoning. In: Conference on Computer Vision and Pattern Recognition (2018)
34. Ren, S., He, K., Girshick, R., Sun, J.: Faster r-cnn: Towards real-time object detection with region proposal networks. In: Advances in Neural Information Processing Systems, pp 91–99. Curran Associates, Inc. (2015)
35. Schlichtkrull, M., Kipf, T.N., Bloem, P., Van Den Berg, R., Titov, I., Welling, M.: Modeling relational data with graph convolutional networks. In: European Semantic Web Conference, pp 593–607. Springer (2018)
36. Sokolova, M., Lapalme, G.: A systematic analysis of performance measures for classification tasks. Inf. Process. Manag. **45**(4), 427–437 (2009)
37. Sontges, S., Koschi, M., Althoff, M.: Worst-case analysis of the time-to-react using reachable sets. In: 2018 IEEE Intelligent Vehicles Symposium (IV), pp 1891–1897. IEEE (2018)

38. Tao, C., He, H., Xu, F., Cao, J.: Stereo priori rcnn based car detection on point level for autonomous driving. Knowl.-Based Syst. **229**, 107346 (2021)
39. Vaswani, A., Shazeer, N., Parmar, N., Uszkoreit, J., Jones, L., Gomez, A.N., Kaiser, Ł., Polosukhin, I.: Attention is all you need. In: Advances in Neural Information Processing Systems, pp 5998–6008. Curran Associates, Inc. (2017)
40. Wu, Y., Kirillov, A., Massa, F., Lo, W.-Y., Girshick, V.: Facebook AI research (FAIR). Available: https://github.com/facebookresearch/detectron2. (2019)
41. Xiao, D., Yang, X., Li, J., Islam, M.: Attention deep neural network for lane marking detection. Knowl. Based Syst. **194**, 105584 (2020)
42. Xu, D., Zhu, Y., Choy, C.B., Fei-Fei, L.: Scene graph generation by iterative message passing. In: Proceedings of the IEEE Conference on Computer Vision and Pattern Recognition, pp 5410–5419 (2017)
43. Xu, K., Hu, W., Leskovec, J., Jegelka, S.: How powerful are graph neural networks? Preprint (2018). arXiv:181000826
44. Yang, J., Lu, J., Lee, S., Batra, D., Parikh, D.: Graph r-cnn for scene graph generation. In: Proceedings of the European Conference on Computer Vision (ECCV), pp 670–685 (2018)
45. Yu, S.Y., Malawade, A.V., Muthirayan, D., Khargonekar, P.P., Al Faruque, M.A.: Scene-graph augmented data-driven risk assessment of autonomous vehicle decisions. IEEE Trans. Intell. Trans. Syst. **23**(7), 7941–7951 (2021) https://doi.org/10.1109/TITS.2021.3074854
46. Yurtsever, E., Lambert, J., Carballo, A., Takeda, K.: A survey of autonomous driving: common practices and emerging technologies. Preprint (2019). arXiv:190605113
47. Yurtsever, E., Liu, Y., Lambert, J., Miyajima, C., Takeuchi, E., Takeda, K., Hansen, J.H.: Risky action recognition in lane change video clips using deep spatiotemporal networks with segmentation mask transfer. In: 2019 IEEE Intelligent Transportation Systems Conference (ITSC), pp 3100–3107. IEEE (2019)

Sensing Optimization in Automotive Platforms

Joydeep Dey and Sudeep Pasricha

1 Introduction

The increasing maturity of Advanced Driver Assistance Systems (ADAS) [30] is enabling the introduction of vehicles with greater levels of autonomy. The degree to which ADAS can effectively reduce human intervention during driving is classified by SAE according to the J3016 standard [1], into five levels of autonomy.

Level 0 characterizes vehicles that have no assistive features. Level 1 autonomy encompasses vehicles that have the ability to share control between the driver and the vehicle. Adaptive cruise control and park assist are examples of features that can assist the driver in this level. Level 2 autonomy vehicles have the capability to perform all acceleration, steering, and braking tasks that require longitudinal and lateral control. Examples of features supported in this level include forward collision warning and blind spot warning, in addition to features from level 1. Level 3 autonomy vehicles can assess the risk of a situation and additionally perform path planning. At Level 4 autonomy, no driver intervention is required in most cases, unless requested, in contrast to level 3. Level 5 autonomy requires no human intervention or safety driver in the vehicle, unlike in level 4. *Most vehicles today are beginning to support level 2 autonomy.*

The higher autonomy levels require support for increasingly sophisticated ADAS features such as Lane Keep Assist (LKA) and Forward Collision Warning (FCW), which in turn defines requirements for sensing capabilities and perception

J. Dey (✉)
Department of Electrical and Computer Engineering, Colorado State University, Fort Collins, CO, USA
e-mail: Joydeep.Dey@colostate.edu

S. Pasricha
Colorado State University, Fort Collins, CO, USA
e-mail: sudeep@colostate.edu

© The Author(s), under exclusive license to Springer Nature Switzerland AG 2023
V. K. Kukkala, S. Pasricha (eds.), *Machine Learning and Optimization Techniques for Automotive Cyber-Physical Systems*, https://doi.org/10.1007/978-3-031-28016-0_19

Table 1 ADAS sensor trade offs

Characteristics	Camera	LiDAR	Radar
Perception reliability	Medium	High	Medium
Spatial resolution	High	High	Low
Noise susceptibility	High	Low	Low
Velocity detection	Low	Low	High
Weather durability	Low	Low	High

performance of the vehicle. This increased demand for vehicle autonomy resulted in various challenges related to reliability [31–34], security [35–39], and real-time perception [40–44] of the vehicle. In this chapter, we focus on real-time perception, specifically, challenges associated with sensor configuration for achieving vehicle autonomy goals. Table 3 summarizes the trade-offs between popular sensors used to support ADAS features and their relative performance. Using a camera as a vision sensor is a widely used approach to perform the classification and detection of objects on the road. However, cameras have high susceptibility to noise and are not reliable in extreme weather or lighting conditions [2]. A radar sensor is also capable of object detection and is particularly suited for accurate velocity detection of neighboring vehicles even under harsh weather and poor visibility conditions. Long-range radars (typically at 77GHz) used to support ADAS features such as adaptive cruise control (ACC) and automatic emergency braking (AEB) have a shorter azimuth than mid or short-range radars (typically at 24GHz), to prioritize monitoring vehicle velocity and approaching distance. However, long range radars can also detect more number of objects than short or mid-range radars. A drawback of the radar is their high false positive rate when detecting objects, and an upper bound on the number of objects that can be detected at the same time, e.g., the Bosch midrange radar with a maximum range of 160 meters can only detect up to 32 objects simultaneously [3]. A LiDAR sensor uses invisible laser light to measure the distance to objects in a similar way to radars. It can create an incredibly detailed 3D view (point cloud) of the environment around the vehicle. However, LiDAR data processing is computationally very expensive and relies on moving parts which can make it more vulnerable to damage. Ultrasonic sensors listed in Table 3 use the principle of 'time of flight' to measure distance from targets by computing the travel time of the ultrasonic echo from a neighboring vehicle or obstacle [4]. Usage of ultrasonic sensors for ADAS feature implementation are not uncommon, however they require accurate modelling for their use case, since their performance is highly dependent on the physical properties (shape, surface material) of the target being tracked [5] (Table 1).

Most level 2 and higher autonomy vehicles today rely on a combination of sensors, to overcome their individual drawbacks (see Table 3). For example, Waymo (a subsidiary of Alphabet Inc., originally started as a project by Google in 2009) combines 3 different types of LiDAR sensors, 5 radar sensors, and 8 cameras. Tesla's vehicles avoid LiDARs due to their high costs and instead their Autopilot uses 8 surround cameras, 12 ultrasonic sensors (primarily for short-range self-parking support), and 1 forward-facing radar. Each of the cameras has a maximum visibility range of up to 250 meters, so this configuration ensures a 360-degree coverage up to 250 meters around the vehicle.

An important challenge facing emerging vehicles is to determine a sensor configuration that can be responsible for environment perception as per the SAE autonomy level supported by the vehicle. An optimal sensor configuration should consist of carefully selected location and orientation of each sensor in a heterogeneous suite of sensors, to maximize coverage from the combined field of view obtained from the sensors, and also maintain a high object detection rate. Today there are no generalized rules for the synthesis of sensor configurations, as the location and orientation of sensors depends heavily on the target features and use cases to be supported in the vehicle.

In this chapter, we propose a novel framework called *VESPA* (VEhicle Sensor Placement and orientation for Autonomy) (first introduced in [43]), to optimize heterogeneous sensor synthesis. More precisely, for a given set of heterogeneous sensors and ADAS features to be supported, *VESPA* performs intelligent algorithmic design space exploration to determine the optimal placement and orientation for each sensor on the vehicle, to support the required ADAS features for SAE level 2 autonomy systems. The *VESPA* framework can be easily utilized to generate optimal sensor configurations across different vehicle types. Our experimental results indicate that the proposed framework is able to optimize perception performance across multiple ADAS features for the 2019 Chevrolet Blazer and 2016 Chevrolet Camaro vehicles.

2   Related Work

State-of-the-art SAE level 2 autonomy systems require the selection and placement of sensors based on the assistive target features required to be supported, e.g., forward collision warning (FCW) and lane keep assist (LKA). While several prior works evaluate the performance of a specific sensor configuration and its deployment, very few works have explored the problem of generating optimal sensor configurations for vehicles.

An optimal sensor placement approach was proposed in [6] for a blind spot detection and warning system. The work recognizes the inability of the camera to perform in non-ideal lighting conditions and selects an ultrasonic sensor to measure distance of vehicles trailing in the vehicle's blind spot. The time response of the system with the position of the sensor above the rear tire is analyzed for two scenarios: when the vehicle is at rest and when it is moving at a constant velocity. The sensor selection identifies price as a constraint and optimizes the price of the total sensor setup through usage of an ultrasonic sensor instead of a more expensive camera sensor. The work in [7] focuses on generating a LiDAR configuration from a set of LiDARs with the goal of reducing occurrences of dead zones and improving point cloud resolution. A LiDAR occupancy grid is constructed for a homogenous set of LiDARs and the configuration is generated using a genetic algorithm. An approach for optimal positioning and calibration of a three LiDAR system is proposed in [8] that uses a neural network to qualify the effectiveness of different sensor location and orientations. Unlike these prior works that focus on generating configurations for a homogenous set of sensors, our work in this chapter presents a novel sensor placement and orientation optimization framework for a heterogeneous set of sensors. Moreover, our framework is also shown to be capable of easily adapting to different vehicle types.

3 Background

3.1 ADAS Features for Level 2 Autonomy

We target four ADAS features in this chapter that need to be supported by a deployed sensor configuration on a vehicle (henceforth referred to as an ego vehicle). A sensor configuration consists of the location and orientation of each sensor within a heterogeneous set of sensors. Our *VESPA* framework optimizes the sensor configuration to support four features: adaptive cruise control (ACC), lane keep assist (LKA) forward collision warning (FCW), and blind spot warning (BW). Each of the features discussed above, require varying degrees of sensing and control along longitudinal (i.e., within the same lane as the ego vehicle) and lateral (i.e., along neighboring lanes) regions.

SAE J3016 defines ACC and LKA individually as level 1 features, as they only perform the dynamic driving task in either the latitudinal or longitudinal direction of the vehicle. FCW and BW are defined in SAE J3016 as level 0 active safety systems, as they only enhance the performance of the driver without performing any portion of the dynamic driving task. However, when all four features are combined, the system can be described as a level 2 autonomy system. Many new vehicles being released today support level 2 autonomy. For instance, Volvo announced that its upcoming Level 2+ vehicles will use surround sensors for 360-degree perception, as well as deep neural networks running in parallel for robust object detection [9]. It is not only relevant, but also important to optimize sensor placement for ADAS systems as more and more vehicles with these features become available. Figure 1 shows an overview of the four features we focus on for level 2 autonomy, which are discussed next.

Adaptive cruise control (ACC) was first introduced in the Mercedes-Benz S-Class sedan in 1999, with the goal of increased driver comfort. ACC causes the ego vehicle to follow a lead vehicle at a specified distance (Fig. 1) without exceeding the speed limit specified by the operator upon activation of the feature [10]. If the lead vehicle slows down, then it is the responsibility of ACC to slow down the ego vehicle to maintain the specified distance. Although implementations differ, all ACC systems take over longitudinal control from the driver (Fig. 1). The challenge in ACC is to maintain an accurate track of the lead vehicle with a forward facing sensor and using longitudinal control to maintain the specified distance while maintaining driver comfort (e.g., avoiding sudden velocity changes).

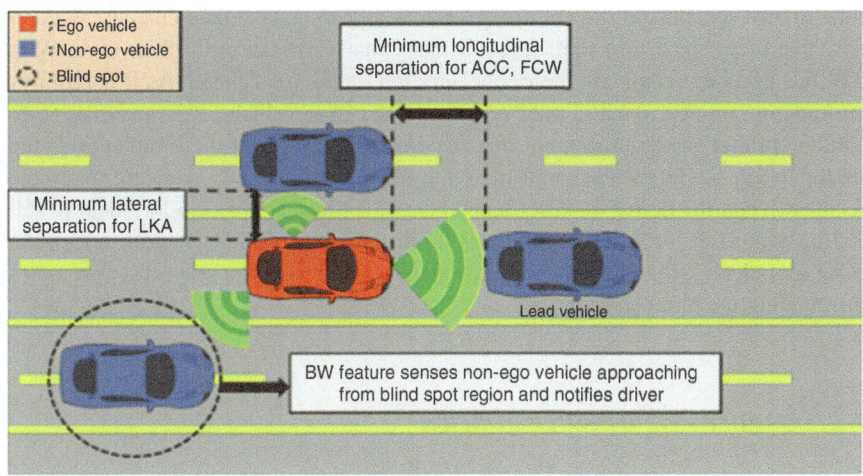

Fig. 1 Visualization of common scenarios in ACC, FCW, LKA, and BW

Lane keep assist (LKA) is an evolution of lane departure warning systems. It involves a forward-facing sensor (often a camera) to identify where the lane lines exist in front of the ego vehicle. Once the lane lines have been detected (e.g., using Canny edge detection and Hough transforms on forward-facing images), LKA can then determine if the ego vehicle lies between those lines (Fig. 1). If the ego vehicle appears to be drifting toward a position where it will cross lane line boundaries, LKA engages steering torque to steer the vehicle in the opposite direction of the lane line until it no longer has the trajectory to cross that lane. LKA systems have been known to over-compensate, creating a "ping-pong" effect where the vehicle oscillates back and forth between the lane lines [11]. The main challenges in LKA are to reduce this ping-pong effect and the accurate detection of lane lines on obscured (e.g., dirt covered) roads.

Forward collision warning (FCW) uses information gathered via various forward facing sensors to determine whether the ego vehicle is going to collide with an object in front of it (Fig. 1). As objects approach the boundary where the vehicle can no longer come to a stop, an audio-visual warning notifies drivers instructing them to apply the brakes. As this is a safety-critical system, it is important that FCW avoids false positives as well as false negatives to improve driver comfort, safety, and reduce rear end accidents [12]. For this to be achieved, it is a necessary prerequisite that the sensors used by the FCW system be placed where they have an accurate view of the vehicle in front of them. The United States National Transportation Safety Board has recommended that FCW be included in all new vehicles [13].

Lastly, blind spot warning (BW) uses sensors mounted on the sides of the ego vehicle to determine whether there is a vehicle towards the rear on either side of the ego vehicle in a location the driver cannot see with their side mirrors [14] (Fig. 1). This area is typically referred to as the "blind spot" and must be verified as clear of any vehicles before the driver can attempt to make a lane change. Without BW, the driver must turn their head to make that verification on their own. With BW, the driver can maintain their concentration on the road ahead. As BW requires information about a specific area near the rear of the vehicle, it is a challenge to find an optimal sensor placement that maximizes the view of the blind spot. If the sensor is too far forward, it will miss the blind spots entirely, causing a vehicle accident when the driver makes a lane change. If the sensor is too far back, it will end up capturing information for areas around the ego vehicle that are not in the blind spot, decreasing the sensor's effectiveness at viewing the presence of vehicles surrounding the blind spot.

3.2 Feature Performance Metrics

To quantify the performance of a sensor configuration on a vehicle being evaluated over drive cycle test cases (i.e., across various driving scenarios; see Section V), we define eight metrics (m1–m8) that are characteristic of the configuration's ability to track and detect non-ego vehicles across various road geometries and traffic scenarios. The eight metrics are defined as follows:

$$\text{Longitudinal Position Error (m1)} = \frac{\sum (y - ygroundtruth)}{Number\ of\ non\ ego\ vehicle} \tag{1}$$

$$\text{Lateral Position Error (m2)} = \frac{\sum (x - xgroundtruth)}{Number\ of\ non\ ego\ vehicle} \tag{2}$$

$$\text{Object Occlusion Rate (m3)} = \frac{\text{Number of non ego vehicle undetected}}{Total\ number\ of\ passing\ non\ ego\ vehicles} \tag{3}$$

$$\text{Velocity Uncertainty (m4)} = \frac{\text{Number of invalid detected non ego vehicle velocities}}{Total\ number\ of\ non\ ego\ velocities} \tag{4}$$

$$\text{Rate of late detection (m5)} = \frac{\text{Number of late non ego vehicle detection}}{Total\ number\ of\ non\ ego\ vehicles} \tag{5}$$

$$\text{False positive lane detecion rate (m6)} = \frac{\text{Number of false positive lane detections}}{Total\ number\ of\ lane\ detections} \tag{6}$$

$$\text{False negative lane detecion rate (m7)} = \frac{\text{Number of false negative lane detections}}{Total\ number\ of\ lane\ detections}$$
(7)

$$\text{False positive object detecion rate (m8)} = \frac{\text{Number of false positive non ego vehicle detections}}{\text{Total number of non ego vehicle detections}}$$
(8)

The longitudinal position error *(m1)* and lateral position error *(m2)* are computed as the deviation of the positional data detected by the sensor configuration from the ground truth of non-ego vehicle positions along the y and x axes respectively. The lateral position error is relevant for LKA, while longitudinal position error is most relevant for ACC and FCW. The object occlusion rate *(m3)* measures the percentage of passing non-ego vehicles that go undetected in the vicinity of the ego vehicle. The minimization of this metric optimizes BW capabilities of a sensor configuration. The velocity uncertainty *(m4)* is the fraction of times that the velocity of a non-ego vehicle is measured incorrectly, which matters for ACC and FCW. The rate of late detection metric *(m5)* is computed as a fraction of the number of 'late' non ego vehicle detections made by the total number of non-ego vehicles, which matters for BW. A detection is classified as late if it is made after the non-ego vehicle crosses the minimum safe longitudinal or lateral distance defined by Intel RSS (Responsibility Sensitive Safety) models on NHTSA for pre-crash scenarios [15]. When a lane marker is detected but there exists no ground truth lane in simulation it is classified as a false positive lane detection, conversely, if a ground truth lane exists in simulation but is not detected, it is classified as a false negative lane detection [16]. Metrics 6 and 7 *(m6 and m7)* characterize the perception system's ability to make a correct case for lane keep assist by taking into account the false positive and false negative lane detection rate. False positive object detection rate *(m8)* measures the fraction of total vehicle detections which were classified as non-ego vehicle detections but did not actually exist in ground truth in the test cases.

4 *VESPA* Framework

The following section describes the proposed *VESPA* framework in detail.

4.1 Overview

Figure 2 shows an overview of our proposed *VESPA* framework. The physical dimensions of the vehicle model and the number and type of sensors to be considered are inputs to the framework. A design space exploration algorithm is

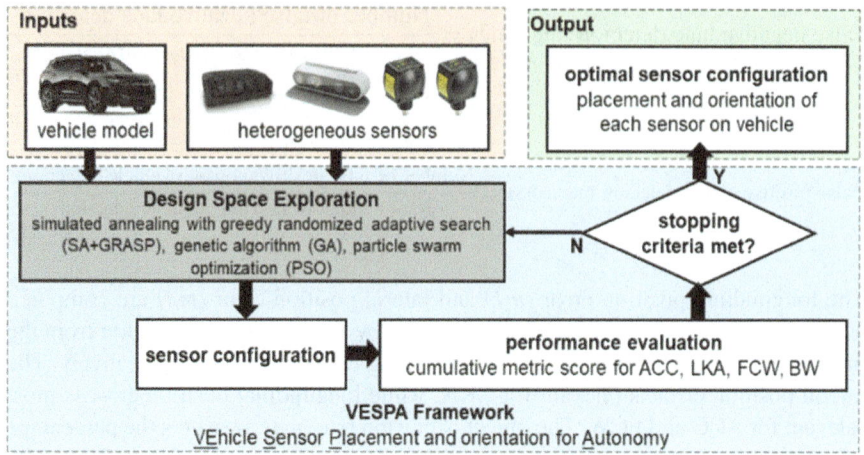

Fig. 2 Overview of *VESPA* framework

used to generate a sensor configuration which is subsequently evaluated based on
a cumulative score from the performance metrics presented in the previous section.
We evaluate three design space exploration algorithms: simulated annealing with
greedy randomized adaptive search (SA + GRASP), genetic algorithm (GA), and
particle swarm optimization (PSO). The process of sensor configuration generation
and evaluation continues until an algorithm-specific stopping criteria is met, at
which point the best configuration is output. The following subsections describe
our framework in more detail.

4.2 Inputs

Each of the design space exploration algorithms generates sensor configurations that
consider feature to field of view (FOV) zone correlations around the ego vehicle.
Figure 3a shows the FOV zones around the ego-vehicle. These zones of interest are
defined as the most important perception areas in the environment for a particular
feature. Figure 3b shows the regions on the vehicle on which sensors can be mounted
(in blue). Regions F and G (in yellow) are exempt from sensor placement due to the
mechanical instability of placing sensors on the door of a vehicle.

The correlation between features, zones, regions, and performance metrics shown
in Fig. 3 is summarized in Table 4. For example, in Fig. 3a, for ACC, the zones of
interests are 6, and 7, and the corresponding regions for possible sensor placement
are A and C. For exploration of possible locations within a region, a fixed step
size of 5 cm in two dimensions across the surface of the vehicle is considered,
which generates a 2D grid of possible positions in each zone shown in Fig. 3b,
c. The orientation exploration of each sensor involves rotation at a fixed step size

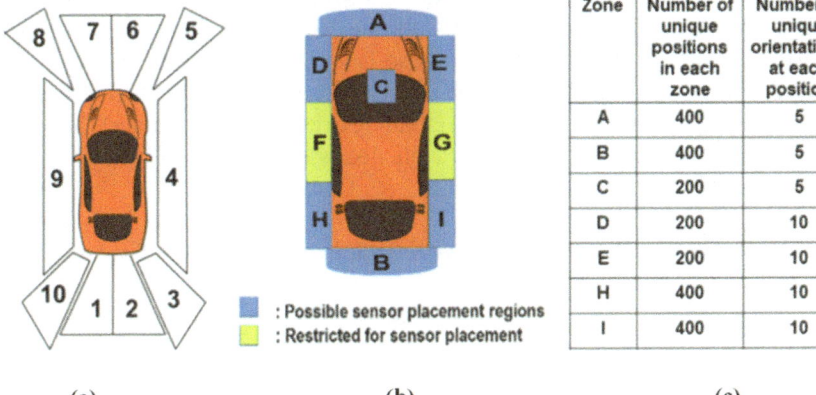

Zone	Number of unique positions in each zone	Number of unique orientations at each position
A	400	5
B	400	5
C	200	5
D	200	10
E	200	10
H	400	10
I	400	10

: Possible sensor placement regions
: Restricted for sensor placement

(a)   (b)   (c)

Fig. 3 (a) Field of view (FOV) zones; (b) sensor placement regions; (c) design space breakdown

Table 2 Feature, region, zone and performance metric relationship

Feature	Region	Zone	Associated metrics
BW	B.H.I	1, 2,3,10	(m3, m5, m8)
LKA	E, I	3,4,5	(m2, m3, m6, m7)
	D, H	8, 9, 10	
ACC, FCW	A, B,C	6, 7, 11	(m1, m4, m8)

of 1 degree between an upper and lower bounding limit for roll, pitch and yaw respectively, at each of these possible positions within the 2D grid.

The orientation exploration limits were chosen with caution to the caveat that long range radars with extreme orientations increase the number of recorded false positives. The combined position and orientation exploration generates an intractably large design space as discussed next (Table 2).

4.3 Design Space Exploration

All of the metrics (m1 – m8) defined in 2.3.2 represent good performance at lower values. We create a cost function that combines these metrics and frame our sensor placement and optimization problem as a minimization problem. The most important metrics are identified and grouped for each feature, as shown in Table 4, and are used to model the cost function as a weighted sum of these five metrics, where the weights are chosen on the basis of their total cardinality across all feature. By searching through the design space of sensor configurations for a minimum cost function value, a sensor configuration can thus be generated where the metrics are cumulatively minimized.

The design space considered in this chapter uses 4 radars and 4 cameras that can be placed in any zone. With a fixed step size of 5 cm in each dimensions and 1 degree rotation in orientation, the number of ways 8 sensors can be placed in all unique locations and orientations is 2.56e+23C8 for the 2019 Blazer and 6.4e+22C8 for the 2016 Camaro. As this design space is so large that it cannot be exhaustively traversed in a practical amount of time, we explore the use of intelligent design space search algorithms that support hill climbing to escape local minima. The three algorithms implemented as part of *VESPA* are discussed next.

4.3.1 SA + Greedy Random Adaptive Search Procedure (SA + Grasp)

Simulated annealing (SA) is a search algorithm that is useful in finding the global optima when the design space has multiple local optima [17]. The process is analogous to the way metals cool and anneal [18]. Typically, SA picks the best solution at each iteration, but can also pick the worst solution based on a temperature-dependent probability, which can allow it to climb out of local minima to arrive at global minima [19]. But SA suffers from the drawback of behaving like a greedy algorithm at lower temperatures as it tends to accepts only those solution configurations very close in cost function value to the previous solution, so it can get stuck in local minima in more complex design spaces [20]. The GRASP (Greedy Randomized Adaptive Search Procedure) algorithm is another search algorithm that is used in many exploration problems [21], but it does not always generate optimal solutions during the greedy construction phase and can get stuck in local optima easily. The SA + GRASP algorithm eliminates the inherent drawbacks of each algorithm. Specifically, the greedy randomized construction phase of the algorithm is used to create disturbances in the existing list of best sensor configurations in our problem, to generate better solutions. A new solution is generated in each iteration by selecting the better solution between the greedy solution from the greedy randomized construction phase and the configuration found from the local search. We decreased the SA temperature variable from Tmax = 10,000 to Tmin = 0 at the rate of 4 degrees per iteration. The search repeats by decreasing SA temperature till an optimal solution is found or a stopping criterion is achieved.

4.3.2 Genetic Algorithm (GA)

The GA is an evolutionary algorithm that can solve optimization problems by mimicking the process of natural selection [22]. It repeatedly selects a population of candidate solutions and then improves the solutions by modifying them. GA has the ability to optimize problems where the design space is discontinuous and also if the cost function is non differentiable [23]. The GA is adapted for our design space such that a chromosome is defined by the combined location and orientation of each sensor's configuration (consisting of six parameters: x, y, z, roll, pitch, and yaw). For a given set of N sensors, the number of parameters stored in each chromosomes is thus '6 N'. Next, in the selection stage, the cost function values are computed for 100 configurations at a time, and a roulette wheel selection method is used to select which set of chromosomes will be involved in the crossover step based on their cost function probability value, computed as a fraction of the cumulative cost function sum of all chromosomes considered in the selection. In the crossover stage, the crossover parameter is set to 0.5, which allows 50 out of the 100 chromosomes to produce offspring. The mutation parameter is set to 0.2 such that in the mutation stage, the mutation rate is set to 10, which is the number of new genes allowed for mutation in each iteration.

4.3.3 Particle Swarm Optimization (PSO)

PSO considers a group of particles where each particle has a position and velocity and is a solution to the optimization problem [24]. In our problem each sensor configuration in the design space is represented as a particle having a defined position and velocity. With a random start, the cost function in (5) evaluates the quality of the solution of a particle. The particle's velocity and position values are updated recursively using a linear update [24]. Each particle stores a trace of its best position within the group and globally as well. The history of the cost function values for this trace can explain the effectiveness of changing the position of a particular sensor from the set of heterogeneous sensors [25]. Unlike GA, PSO does not have any evolution operators like crossovers or mutation [26]. PSO also does not require any binary encoding of solution configurations like in GA [27]. The total number of particles considered were 50, and the importance of personal best and importance of neighborhood best parameters were both empirically selected to be 2.

5 Experiments

The following section describes the experimental setup and results involving the *VESPA* framework.

5.1 *Experimental Setup*

To evaluate our *VESPA* framework, we consider a scenario with a maximum of 8 sensors: 4 radars and 4 camera vision sensors. Many recent contributions such as the work presented in [28, 29] combine radar and camera modalities for ADAS applications. We did not include LiDARs in this heterogeneous set of sensors due to their relatively poor performance in adverse weather conditions as shown in Table 3. For the given set of test cases, it was observed that if less than 4 sensors were used, the ability of the perception system to make an accurate prediction was relatively poor. Conversely, on increasing the number of radars and cameras to more than

Table 3 *VESPA* generated solution vs baseline configuration

	VESPA Camaro	Baseline Camaro	*VESPA* Blazer	Baseline Blazer
Cost function	0.9971	2.1367	1.2841	2.4630
Longitudinal position error	0.0523	0.1427	0.0845	0.2419
Lateral position error	0.1810	0.2566	0.0958	0.2204
Object occlusion rate	0.1331	0.2351	0.2062	0.3158
Velocity uncertainty	0.0823	0.1851	0.0474	0.2056
Rate of late detection	0.1158	0.2123	0.1578	0.2315
False positive lane detection rate	0.0142	0.1335	0.0221	0.1571
False negative lane detection rate	0.0214	0.0236	0.0393	0.0412
False positive object detection rate	0.0431	0.1283	0.0976	0.0954

	Length	Width	Height	Front Overhang	Rear Overhang
Blazer	4.7498	1.9558	1.6764	1.016	0.9144
Camaro	4.7498	1.905	1.3716	0.9652	1.016

Fig. 4 2019 Chevrolet Blazer (Left) and 2016 Chevrolet Camaro (Right)

4 each, there was minimal improvement in cost function score. Hence to keep implementation cost low while still achieving good accuracy, we decided to use these 8 sensors. Please note that these modalities and number of sensors have been used to show a proof of concept for our *VESPA* framework, which can be extended to scenarios with different modalities and numbers of sensors. We considered two vehicles for evaluation: a 2019 Chevrolet Blazer and a 2016 Chevrolet Camaro. Figure 4 shows the dimensions for the vehicles. Figure 5 shows images of the sensor placements on both car models in our workspace.

Each configuration generated by the SA + GRASP, GA, and PSO algorithms was optimized on 40 test cases designed (10 test cases each for evaluating performance with ACC, FCW, LKA, and BW) using the Automated Driving Toolbox in Matlab. Half (20) of these test cases for each feature are used during the optimization phase and the remaining (20) test cases are used during the evaluation phase.

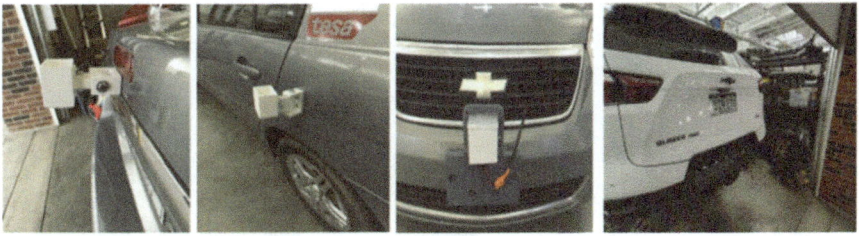

Fig. 5 Sensors mounted in workspace on both car models

Finally, the optimized configurations were evaluated on a different set of evaluation test cases. Each of the test cases was characterized by unique road geometry, variations in road elevation, curvature, banking, and different traffic densities. In some test cases, the number of lanes were varied to make the framework optimize the sensor configuration for challenging and realistic driving scenarios.

A Kalman filter sensor fusion algorithm was used to combine readings from sensors in a sensor configuration being evaluated, to make predictions. The longitudinal and lateral ground truth were defined for non-ego vehicles and the position error was calculated from the fused sensor measurements. The deviation of sensor measurements from ground-truth was used to calculate the values of metrics m1–m8, and hence the cost function over all test cases. Lastly, we set the stopping criterion for all three algorithms as the case when the cost function does not show a greater than 5% change over 200 iterations.

5.2 Experimental Results

In our first experiment we were interested in evaluating the efficacy of different optimization algorithms (SA + GRASP, GA, and PSO) in finding optimal sensor configurations as well as exploring the consistency of the quality of solution returned by each. The cost function values for the best solution found by each algorithm for the 2016 Camaro and 2019 Blazer are shown in Fig. 6 As shown in Fig. 6, GA returned the solution configuration with the lowest cost function score of 0.7648 for the Camaro and 0.9252 for the Blazer. GA was able to better traverse the complex design space for our problem to arrive at the global minima compared to the SA + GRASP and PSO algorithms.

Next, we compared the solution generated by *VESPA* (utilizing the GA algorithm which gives the best results) with a baseline sensor configuration selected manually, based on best practices by a vehicle design expert in our team. This baseline configuration involved coupling a radar and camera in zones A, B, E and H each such that every mutually perpendicular direction in the 2D plane of the ego vehicle was covered using a radar and camera combined. All 8 sensors were fixed in the orientation angle, which matched the orientation of surface normal vector of the

Fig. 6 Cost function values
for the best solution found by
the SA + GRASP, GA, and
PSO algorithms on the
Camaro and Blazer vehicles

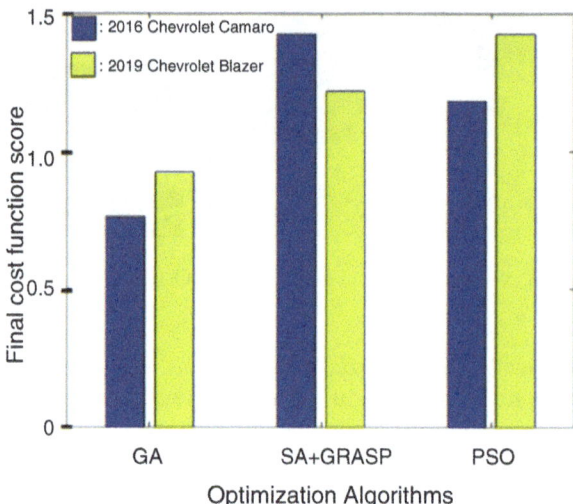

respective zone in which they were placed. The selected baseline configuration
maximizes coverage by considering feature to zone correlation. This is ensured by
placing at least one sensor in each region such that all zones dedicated to each of
the 4 selected features is covered in the field of view of that particular sensor.

Table 3 shows the results of the comparison between the *VESPA* generated
solution and the baseline configuration for the 2016 Camaro and 2019 Blazer. The
final cost function score was higher for the baseline approach, showing that *VESPA*
generated a significantly better (lower cost) solution for both vehicles.

Table 4 summarizes the specific locations and orientations of the eight sensors
on the two vehicles, generated by *VESPA*. The location and orientation information
of each sensor in Table 6 is measured with respect to a global co-ordinate frame for
the car model, whose origin is at the geometric center of the vehicle. An interesting
observation from the table is that the sensors in the Blazer's configuration favor
higher Z values than the Camaro, since the Blazer is 0.3 m taller than the Camaro.

Figure 7 visualizes sensor coverage in a bird's eye plot between the best
configuration generated by *VESPA* in Fig. 7a and the baseline configuration in
Fig. 7b for the Camaro (results for Blazer are omitted for brevity). The baseline
configuration was optimized with a conventional approach towards improving
sensor coverage, with a secondary focus on sensor reliability.

In contrast, the solution generated by *VESPA* took into account the unique
strengths and weaknesses of each sensor to obtain a configuration having significantly better performance for the features supported, despite having lower
overlap between field of view of different sensors than the baseline solution Fig.
7 and also uses lesser number of sensors. The superiority of the *VESPA* solution
configuration, despite using lesser number of sensors, can be accounted for by the
optimized placement of camera 1, radar 2 and radar 3 in zones A and C maximizing
performance of ACC and FCW. Further, in physical testing it was observed that

Table 4 Solution from *VESPA* for Camaro and Blazer (in meters, degrees)

	Radar 1		Radar 2		Radar 3		Radar 4	
	VESPA camaro	*VESPA* blazer	*VESPA* camaro	*VESPA* blazer	*VESPA* camaro	*VESPA* blazer	*VESPA* camaro	*VESPA* blazer
X	3.7	3.7	3.7	3.65	0	2.8	3.7	2.91
Y	−0.18	−0.4	0.45	−0.9	0.9	0.9	−0.9	0.9
Z	0.15	0.2	0.20	0.2	0.2	0.25	0.2	0.2
Roll	0	0	0	0	0	0	0	0
Pitch	0	0	0	0	0	0	0	0
Yaw	15	−20	−15	−40	50	45	−130	−130

	Camera 1		Camera 2		Camera 3		Camera 4	
	VESPA camaro	*VESPA* blazer	*VESPA* camaro	*VESPA* blazer	*VESPA* camaro	*VESPA* blazer	*VESPA* camaro	*VESPA* blazer
X	3.7	3.7	2.8	2.8	2.8	0	X	−1
Y	0	0	−0.9	−0.9	0.9	0.9	X	0.9
Z	1.1	1.1	1.1	1.15	1.1	1.25	X	1.1
Roll	0	0	0	0	1	0	X	0
Pitch	0	0	0	0	1	1	X	1
Yaw	0	0	−90	−100	120	60	X	170

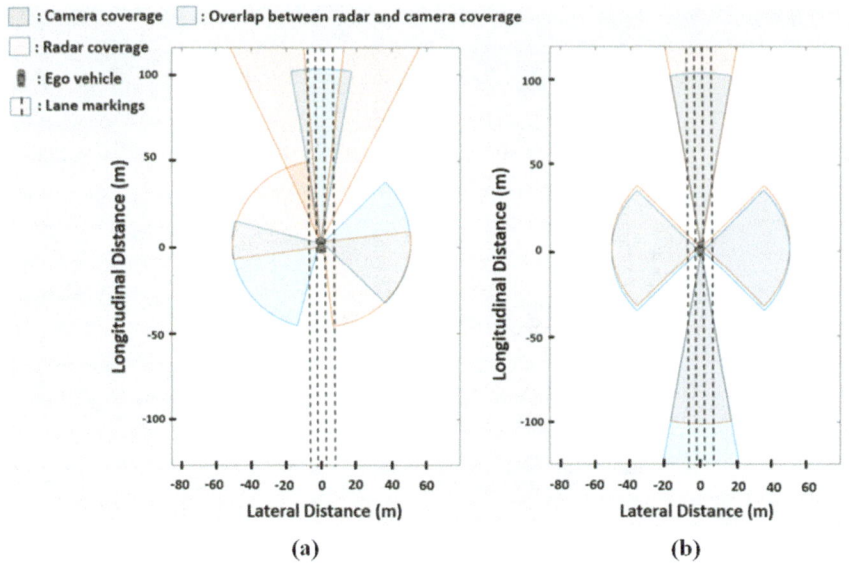

Fig. 7 Coverage for (**a**): *VESPA* Camaro solution (**b**) Baseline Camaro

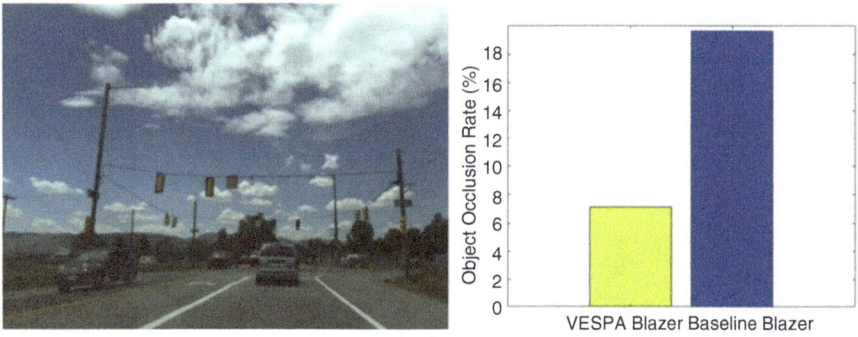

Fig. 8 Performance on real drive cycle in Colorado for best solution generated by *VESPA* and the baseline configuration for the 2019 Blazer

using a radar coupled with camera in zone B for LKA reduces the number of false positives during detections. In Fig. 7a, radars 3, 4 and cameras 2, 3 placed in zones D and E respectively were sufficient for improving performance of ACC and FCW by reducing the number of false positive object detections. The combined optimization of orientation and location with *VESPA* resulted in a sensor configuration that maximized performance for each feature.

Our last experiment involved testing the best sensor configuration from our *VESPA* framework and the baseline configuration for the 2019 Blazer on data from a real world drive cycle over 1 h in Colorado. We focus only on assessing performance for the ACC and FCW features. Figure 8 shows an image from the real drive cycle with data collected by the vehicle from the radar and camera sensors on it. The figure also shows a plot of the object occlusion rate (OOR). The OOR for the baseline configuration was 19.64% (it did not detect 11 out of 56 non ego vehicles), while the *VESPA* generated best solution had an OOR of 7.14% (it failed to detect only 4 out of 56 non ego vehicles). The results show the effectiveness of our proposed *VESPA* framework in generating higher quality sensor configurations.

6 Conclusions

In this chapter, we propose an automated framework called *VESPA* that is capable of generating sensor placement and orientation in modern semi-autonomous vehicles. *VESPA* has the ability to optimize locations and orientations for a set of heterogeneous sensors on a given target vehicle. The framework can be tuned to improve perception on a desired collection of test cases. *VESPA* is also scalable across different vehicle models as shown in our analysis on the Chevrolet Camaro and Blazer vehicles. Further, despite the sensor locations in the baseline configuration of Fig. 7b being the most intuitive, the best configuration is the one generated by *VESPA*, showing that even people skilled in the art of sensor placement may find it challenging to synthesize a significantly better placement than that generated by *VESPA*. We also validated *VESPA* with real drive cycle data to show its effectiveness for real-world scenarios.

References

1. SAE International Standard J3016: Taxonomy and definitions for terms related to driving automation systems for on-road motor vehicles, (2018)
2. Spinneker, R., Koch, C., Park, S.B, Yoon, J.J: Fast fog detection for camera based advanced driver assistance systems. In: IEEE International Conference on Intelligent Transportation Systems (ITSC) (2014)
3. Robert Bosch GmbH, Bosch MRR Data Sheet: [Online]. Available: https://www.bosch-mobility-solutions.com, (2019)
4. Xu, C., Zhang, P., Wang, H., Li, Y., Li, C.: Ultrasonic echo waveshape features extraction based on QPSO-matching pursuit for online wear debris discrimination. Mech. Syst. Signal Process. **60**, 301–315 (2015)
5. Li, S., Li, G., Yu, J., Liu, C., Cheng, B., Wang, J., Li, K.: Kalman filter based tracking of moving objects using linear ultrasonic sensor array for road vehicles. Mech. Syst. Signal Process. **60**, 301–315 (2018)
6. Jamaluddin, M., Shukor, A.Z., Miskon, M.F., Ibrahim, F.A., Redzuan, M.Q.A.: An analysis of sensor placement for vehicle's blind spot detection and warning system. J. Telecommun. Electron. Comput. Eng. **8**(7), 101–106 (2017)
7. Kim, T., Park, T.: Placement optimization of multiple Lidar sensors for autonomous vehicles. IEEE Trans. Intell. Transp. Syst. **21**(5), 2139–2145 (2019)
8. Meadows, W., Hudson, C., Goodin, C., Dabbiru, L., Powell, B., Doude, M., Carruth, D., Islam, M., Ball, J.E, Tang, B.: Multi-LIDAR placement, calibration, co-registration, and processing on a Subaru Forester for off-road autonomous vehicles operations. In: Autonomous Systems: Sensors, Processing and Security for Vehicles and Infrastructure (2019)

9. Shapiro, D.: Levelling up: what is level 2 automated driving? [Online]. Available: https://blogs.nvidia.com/blog/2019/02/06/what-is-level-2-automateddriving/, (2019)
10. Wenger, J.: Automotive radar – status and perspectives. In: IEEE Compound Semiconductor Integrated Circuit Symposium (2005)
11. Kirchner, C.: Lane keeping assist explained. In: Motor Review, [Online]. Available: https://motorreview.com/lane-keeping-assist-explained, (2014)
12. Consumer Reports: Guide to Forward Collision Warning, Consumer Reports, [Online]. Available: https://www.consumerreports.org/car-safety/forwardcollision-warning-guide/ (2019)
13. National Transportation Safety Board (NTSB): The use of forward collision avoidance systems to prevent and mitigate rear-end crashes (2015)
14. Consumer Reports: Guide to blind spot warning, consumer reports, [Online]. Available: https://www.consumerreports.org/car-safety/blind-spotwarning-guide/, (2019)
15. Mobileye, Intel Co.: Implementing RSS Model on NHTSA Pre-Crash scenarios, [Online]. Available: https://www.mobileye.com/responsibility-sensitivesafety/rss_on_nhtsa.pdf, (2019)
16. Kumar, A., Simon, P.: Review of lane detection and tracking algorithms in advanced driver assistance systems. Int. J. Comput. Sci. Inf. Technol. **7**(4), 65–78 (2015)
17. Aarts, E., Korst, J.: Simulated annealing and Boltzmann machines. In: A Stochastic Approach to Combinatorial Optimization and Neural Computing. Wiley, New York (1989)
18. Bohachevsky, O., Johnson, M.E., Stein, M.L.: Generalized simulated annealing for function optimization. Technometrics. **28**(3), 209–217 (1986)
19. Van Laarhoven, P.J.M., Aarts, E.H.L.: Simulated Annealing: Theory and Applications Mathematics and Its Applications. Springer, Dordrecht (1987)
20. Oliveira, E., Antunes, C.H., Gomes, Á.: A hybrid multi-objective GRASP+SA algorithm with incorporation of preferences. In: IEEE Symposium on Computational Intelligence in Multi-Criteria DecisionMaking (MCDM) (2014)
21. Sosnowska, D.: Optimization of a simplified fleet assignment problem with metaheuristics: simulated annealing and GRASP. In: Nonconvex Optimization and Its Applications, Springer (2000)
22. Reeves, C.: Genetic algorithms: handbook of metaheuristics, In: International Series in Operations Research & Management Science, Springer (2003)
23. Rattray, M., Shapiro, J.L.: The dynamics of a genetic algorithm for a simple learning problem. J. Phys. A Math. Gen. **29**(23), 7451 (1996)
24. Kennedy, J., Eberhart, R.: Particle swarm optimization. In: Proceedings of International Conference on Neural Networks (ICCN) (1995)
25. Clerc, M., Kennedy, J.: The particle swarm – explosion, stability, and convergence in a multidimensional complex space. In: IEEE Press (2002)
26. Huang, S.: A review of particle swarm optimization algorithm. In: Computer Engineering Design (1977)
27. Lei, X., Shi, Z.: Application and parameter analysis of particle swarm optimization algorithm in function optimization. In: Computer Engineering Applications (CEA) (2008)
28. Kumar, R., Jayashankar, S.: Radar and camera sensor fusion with ROS for autonomous driving. In: Fifth International Conference on Image Information Processing (ICIIP) (2019)
29. Zhong, Z., Liu, S., Matthew, M., Dubey, A.: Camera radar fusion for increased reliability in ADAS applications. In: Electronic Imaging, Autonomous Vehicles and Machines (2018)
30. Kukkala, V., Tunnell, J., Pasricha, S.: Advanced driver assistance systems: a path toward autonomous vehicles. In: IEEE Consumer Electronics, Vol. 7, No. 5 2018
31. Kukkala, V.K., Bradley, T., Pasricha, S.: Priority-based multi-level monitoring of signal integrity in a distributed powertrain control system. In Proceeding of IFAC Workshop on Engine and Powertrain Control, Simulation and Modeling (2015)
32. Kukkala, V.K., Bradley, T., Pasricha, S.: Uncertainty analysis and propagation for an auxiliary power module. In Proceeding of IEEE Transportation Electrification Conference (TEC) (2017)
33. Kukkala, V.K., Pasricha, S., Bradley, T.: JAMS: Jitter-aware message scheduling for FlexRay automotive networks. In: IEEE/ACM International Symposium on Networks-on-Chip (NOCS) (2017)

34. Kukkala, V.K., Pasricha, S., Bradley, T.: JAMS-SG: A Framework for Jitter-Aware Message Scheduling for Time-Triggered Automotive Networks. In: ACM Transactions on Design Automation of Electronic Systems (TODAES), Vol. 24, No. 6 (2019)
35. Kukkala, V.K., Pasricha, S., Bradley, T.: SEDAN: Security-aware design of time-critical automotive networks. In: IEEE Transaction on Vehicular Technology (TVT), Vol. 69, No. 8 (2020)
36. Kukkala, V.K., Thiruloga, S.V., Pasricha, S.: INDRA: intrusion detection using recurrent autoencoders in automotive embedded systems. In: IEEE Transactions on Computer-Aided Design of Integrated Circuits and Systems (TCAD), Vol. 39, No. 11 (2020)
37. Kukkala, V.K., Thiruloga, S.V., Pasricha, S.: LATTE: LSTM Self-Attention based Anomaly Detection in Embedded Automotive Platforms. In: ACM Transactions on Embedded Computing Systems (TECS), Vol. 20, No. 5s, Article 67 (2021)
38. Thiruloga, S.V., Kukkala, V.K., Pasricha, S.: TENET: temporal CNN with attention for anomaly detection in automotive cyber-physical systems. In: Proceeding of IEEE/ACM Asia & South Pacific Design Automation Conference (ASPDAC) (2022)
39. Kukkala, V.K., Thiruloga, S.V., Pasricha, S.: Roadmap for cybersecurity in autonomous vehicles. In: IEEE Consumer Electronics Magazine (CEM), (2022)
40. Tunnell, J., Asher, Z., Pasricha, S., Bradley, T.H.: Towards improving vehicle fuel economy with ADAS. In: SAE Inter-national Journal of Connected and Automated Vehicles, Vol. 1, No. 2 (2018)
41. Tunnell, J., Asher, Z., Pasricha, S., Bradley, T.H.: Towards improving vehicle fuel economy with ADAS. In: Proceeding of SAE World Congress Experience (WCX) (2018)
42. Asher, Z., Tunnell, J., Baker, D.A., Fitzgerald, R.J., Banaei-Kashani, F., Pasricha, S., Bradley, T.H.: Enabling prediction for optimal fuel economy vehicle control. In: Proceeding of SAE World Congress Experience (WCX) (2018)
43. Dey, J., Taylor, W., Pasricha, S.: VESPA: a framework for optimizing heterogeneous sensor placement and orientation for autonomous vehicles. In: IEEE Consumer Electronics Magazine (CEM), Vol. 10, No. 2 (2021)
44. DiDomenico, G.C., Bair, J., Kukkala, V.K., Tunnell, J., Peyfuss, M., Kraus, M., Ax, J., Lazarri, J., Munin, M., Cooke, C., Christensen, E.: Colorado state university EcoCAR 3 final technical report. In: SAE World Congress Experience (WCX) (2019)

Unsupervised Random Forest Learning for Traffic Scenario Categorization

Friedrich Kruber, Jonas Wurst, Michael Botsch, and Samarjit Chakraborty

1 Introduction

Looking at traffic scenarios microscopically, there is an infinite number of scenarios. A somewhat higher visual range shows that they nevertheless follow certain patterns and can be assigned to categories with a high degree of similarity. Testing representatives from each category ensures a broad scope, while minimizing the effort in the validation process. In order to perform a relevance evaluation, one has to memorize and structure traffic scenarios. Therefore, the vast amount of sensor data needs to be shrinked. This can be achieved by representing a traffic situation with a set of relevant features. These features can then be used in machine learning algorithms for analysis purposes.

A training dataset of recorded traffic scenarios is usually manually labeled in order to run supervised classification algorithms. In contrast to that, unsupervised learning yields to identify patterns in datasets, where the availability of labels for training machine learning models is absent. The main focus here is the introduction of an unsupervised learning procedure for the categorization of traffic scenarios, only given the input from arbitrary data sources.

The chapter is organized as follows. After the methodological introduction into Decision Trees and Random Forests in Sects. 2 and 3, the method is extended for its usage in the field of unsupervised learning in Sect. 4. Its application for the unsupervised clustering of real world traffic scenarios is discussed in Sect. 5. Finally, the versatility of Random Forests is demonstrated by another method, which

F. Kruber · J. Wurst · M. Botsch
Technische Hochschule Ingolstadt, Ingolstadt, Germany
e-mail: friedrich.kruber@thi.de; jonas.wurst@thi.de; michael.botsch@thi.de

S. Chakraborty (✉)
UNC Chapel Hill, Chapel Hill, NC, USA
e-mail: samarjit@cs.unc.edu

integrates a Random Forest into a Deep Learning architecture to tackle the problem of Open-Set recognition for traffic scenarios.

Before moving to the next section about Decision Trees, some notations will be introduced. A dataset $\mathcal{D}$ consists of M datapoints $x_m \in \mathbb{R}^N$, referred to as feature vectors. Supervised learning requires a training dataset containing a target vector y_m for each feature vector x_m. Assuming that the possible output is a scalar it yields

$$\mathcal{D}_s = \left\{ (x_1, y_1), \ldots, (x_{M_s}, y_{M_s}) \right\}. \tag{1}$$

Contrary to that, in unsupervised learning the dataset does not provide any information about the objective values. Therefore, those datasets are defined as

$$\mathcal{D}_u = \left\{ x_1, \ldots, x_{M_u} \right\}. \tag{2}$$

The feature vector x as well as the target y are not deterministic. Therefore, the feature vector x is a realization of the random variable $\mathbf{x}$ and y a realization of the random variable $\mathbf{y}$.

Supervised machine learning aims to find a function f based on $\mathcal{D}_s$, which performs the mapping from the input variable $\mathbf{x}$ to the target $\mathbf{y}$. Depending on the target value characteristics, supervised learning can be in the form of *classification* or *regression*. If $\mathbf{y}$ is of categorical type, the function is called classification. With $\mathbf{x} \in \mathbb{R}^N$, the classification is defined as

$$f : \mathbb{R}^N \rightarrow \{c_1, \ldots, c_K\}, \mathbf{x} \mapsto \hat{y}, \tag{3}$$

where $\hat{y} \in \{c_1, \ldots, c_K\}$ is the categorical predicted output. If the output is continuous, i.e. $\hat{y} \in \mathbb{R}$, the function is called regression:

$$f : \mathbb{R}^N \rightarrow \mathbb{R}, x \mapsto \hat{y}. \tag{4}$$

The aim of all supervised learning methods is to find a function f, which gains the highest performance. Therefore, the performance measurement called risk

$$R(f) = \underbrace{\mathrm{E}_{\mathbf{x},y} \left\{ \mathcal{L}(y, f(\mathbf{x})) \right\}}_{\text{expectation of } \mathcal{L}} = \int_{\mathbb{R}^N} \sum_{k=1}^{K} \mathcal{L}(c_k, f(x)) \underbrace{\mathrm{p}(\mathbf{x} = x, y = c_k)}_{\text{joint probability density function}} dx \tag{5}$$

$$R(f) = \mathrm{E}_{\mathbf{x},y} \mathcal{L}(y, f(\mathbf{x})) = \int_{\mathbb{R}^N} \int_{\mathbb{R}} \mathcal{L}(y, f(x)) \, \mathrm{p}(\mathbf{x} = x, y = y) \, dy dx, \tag{6}$$

for classification and regression is introduced. $\mathcal{L}$ denotes the loss and E the expectation. The goal is to find a function f_B, which gains the highest performance by minimizing $R(f)$

$$f_B = \arg\min_f \{R(f)\}, \tag{7}$$

where f_B is known as the Bayes classifier or Bayes regression function. The density functions are usually unknown. Instead, risk can be estimated through the empirical risk by employing the dataset $\mathcal{D}_s$,

$$R_{\text{emp}}(f, \mathcal{D}_s) = \frac{1}{M_s} \sum_{m=1}^{M_s} \mathcal{L}(y_m, f(\mathbf{x}_m)). \tag{8}$$

Various approaches try to minimize the empirical risk in order to find a good mapping f. The approaches differ in their architecture and hence the realized function. In the following, we focus on the proven-in-use ensemble method termed Random Forests, which is constructed from a set of Classification and Regression trees.

2   Classification and Regression Trees

Classification and Regression Trees (CART) have several benefical properties for real world applications. They can model relations between an input $\mathbf{x}$ and the output y independent of the number of features or dataset size. The input variables can be either categorial or ordered, and even both types may be apparent in the feature set. Probably the most important aspect to favor decision trees over many other methods, is the interpretability. All decisions and interactions among features can be interpreted by humans, and thus provide the white-box character of CARTs.

The CART algorithm, introduced in [8], is a specific form of binary decision trees. As depicted in Fig. 1 and explained in detail in what follows, a binary tree can be thought of a multi decision process, as well as a directed graph. In an algorithmic view, binary trees consist of many *if/else* queries. The intuition behind trees is simple, yet understanding the underlying principles is mandatory since they build the basis for the more sophisticated Random Forest algorithms.

Fig. 1 Example of a binary classification tree

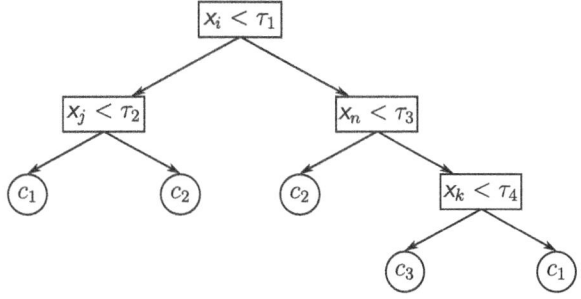

Before diving into the methodology, a set of definitions considering a binary tree is established first. A binary tree $\mathfrak{T}$ is constructed with *nodes* t, which are connected by *edges*. The top node is designated as the *root*, such that all edges are directed downwards from the root. Two nodes are connected by a single edge. The information flows from the parent node to the child node. In a binary tree, intermediate nodes have exactly two children, but only one parent node. If a node does not possess children, it is termed as *terminal node*, or *leaf*.

Starting from the input space $\mathfrak{X} \in \mathbb{R}^N$, at each *if/else* decision stage the input is split into two disjoint subspaces $\mathfrak{X}_i$ and $\mathfrak{X}_j$ of $\mathbb{R}^N$ for which $p(\mathbf{x} \in t) > 0$. These subspaces are represented by the two children nodes. When propagating through all stages of a tree, each datapoint $\mathbf{x}_m$ is assigned to a constant prediction $\hat{y}$ within that subspace. This subspace is corresponding to the terminal node in the tree.

2.1 Computing the Optimal Split

The aim of growing a tree is to minimize the risk according to Eq. (5) and Eq. (6) for classification or regression problems, respectively. This is achieved by finding the best *splits* and will be derived in the following.

We define $\tilde{\mathfrak{T}}$ as the subspace of $\mathfrak{X}$ representing all terminal nodes of a tree. The function, which assigns each input $\mathbf{x}$ to a terminal node $t \in \tilde{\mathfrak{T}}$ can be formalized as

$$\rho : \mathfrak{X} \to \tilde{\mathfrak{T}}, \mathbf{x} \mapsto t. \tag{9}$$

Given an assignment function $\upsilon(t) \in \{c_1, \dots, c_K\}$ for classification, or $\upsilon(t) \in \mathbb{R}$ for regression, a mapping f corresponds to $\tilde{\mathfrak{T}}$ so that $f(\mathbf{x}) = \upsilon(t)$ for all inputs $\mathbf{x} \in t$, then

$$f(\mathbf{x}) = \upsilon(\rho(\mathbf{x})). \tag{10}$$

Based on the previous equations, one obtains the risk of the mapping realized as

$$R(f) = \int_{\mathfrak{X}} E_{y|x} \{\mathcal{L}(y, f(x))\} p(\mathbf{x} = x) dx \tag{11}$$

$$= \sum_{t \in \tilde{T}} E_{y|x \in t} \{\mathcal{L}(y, \upsilon(t))\} p(\mathbf{x} \in t). \tag{12}$$

The global error in Eq. (11) is equal to sum of the local errors within all terminal nodes, Eq. (12). Hence, the risk can be minimized locally with

$$r_{\min}(t) = \min_{\upsilon} \{E_{y|x \in t} \{\mathcal{L}(y, \upsilon(t))\}\}, \tag{13}$$

leading to the overall minimum prediction risk for $\tilde{\mathfrak{T}}$ of $\mathbb{R}^N$

$$R_{\min}(\tilde{\mathfrak{T}}) = \sum_{\tilde{\mathfrak{T}}} r_{\min}(t) p(\mathbf{x} \in t). \tag{14}$$

The local minimum risk for regression in a node $t \in \tilde{\mathfrak{T}}$ is

$$r_{\min}(t) = \hat{\sigma}_{\tilde{\mathfrak{T}}}^2. \tag{15}$$

For classification, the minimum local risk is obtained with the class of highest posterior probability

$$r_{\min}(t) = \min_{c_l} \left\{ \sum_{k=1}^{K} (1 - \delta(c_k, c_l)) p(y = c_k | \mathbf{x} \in t) \right\} \tag{16}$$

$$= 1 - \max_{c_k} \{ p(y = c_k | \mathbf{x} \in t) \}, \tag{17}$$

where $\delta(\cdot, \cdot)$ is one if both arguments are equal, otherwise zero. Typically, classification is realized through a class probability estimation, which leads to the local risk in a node $t \in \tilde{\mathfrak{T}}$ as

$$r_{\min}(t) = \sum_{k=1}^{K} p(y = c_k | \mathbf{x} \in t)(1 - p(y = c_k | \mathbf{x} \in t)). \tag{18}$$

When growing a tree, the aim is to reduce the local risk as best as possible with an optimal split s_{opt} in order to divide a node $t \in \tilde{\mathfrak{T}}$ into a left t_L and right t_R childnode. We define this as a new fraction of $\tilde{\mathfrak{T}}'$ of $\mathbb{R}^N$. A split reduces the risk with

$$\Delta R(s, t) = R_{\min}(\tilde{\mathfrak{T}}) - R_{\min}(\tilde{\mathfrak{T}}') \tag{19}$$

$$= p(\mathbf{x} \in t)(r_{\min}(t) - p(\mathbf{x} \in t_L | \mathbf{x} \in t) r_{\min}(t_L) - p(\mathbf{x} \in t_R | \mathbf{x} \in t) r_{\min}(t_R)), \tag{20}$$

and the relative risk reduction is

$$\Delta r(s|t) = \frac{R_{\min}(\tilde{\mathfrak{T}}) - R_{\min}(\tilde{\mathfrak{T}}')}{p(\mathbf{x} \in t)} \tag{21}$$

$$= r_{\min}(t) - p(\mathbf{x} \in t_L | \mathbf{x} \in t) r_{\min}(t_L) - p(\mathbf{x} \in t_R | \mathbf{x} \in t) r_{\min}(t_R). \tag{22}$$

Now, the best split s_{opt} for a node t is achieved by maximizing $\Delta r(s|t)$

$$s_{opt}(t) = \underset{\tilde{s}}{\mathrm{argmax}} \left\{ \Delta r(\tilde{s}|t) \right\}. \tag{23}$$

In practice the required probability density functions are not known. Therefore, in the following we consider how to perform a split with a given dataset $\mathcal{D}_s$.

2.2 Growing a Tree

Since the probability function $p(\mathbf{x} \in t)$ is unknown, it has to be approximated with a training dataset $\mathcal{D}_s$ containing M datapoints. We define a set $\mathfrak{M}(t) = \{m \in \mathfrak{M} : x_m \in t\}$, which contains all datapoints of $\mathcal{D}_s$ in the node t. The empirical estimate $\hat{p}$ can then be computed with

$$\hat{p}(\mathbf{x} \in t) = \frac{|\mathfrak{M}(t)|}{M} = \frac{M(t)}{M}, \tag{24}$$

with the assumption that $\hat{p}(\mathbf{x} \in t) > 0$ for all $t \in \tilde{\mathfrak{T}}$. Now, we can reformulate Eq. (13) and Eq. (14)

$$\hat{r}_{min}(t) = \min_{\upsilon} \frac{\sum_{\mathfrak{M}(t)} \mathcal{L}(y_m, \upsilon(\rho(x_m)))}{M(t)}, \text{ and} \tag{25}$$

$$\hat{R}_{min}(f) = \sum_{\tilde{\mathfrak{T}}} \hat{r}_{min}(t)\hat{p}(\mathbf{x} \in t). \tag{26}$$

For regression we set $\upsilon = \hat{\mu}(t)$, so that $\hat{r}_{min}(t) = \hat{\sigma}_{\tilde{\mathfrak{T}}}^2(t)$, with μ and σ^2 denoting the expectation and variance.

For classification the estimate for $p(y = c_k|\mathbf{x} \in t), k = 1, \ldots, K$ has to be calculated to compute $\hat{r}_{min}(t)$. We define another set $\mathfrak{M}_k(t) = \{m \in \mathfrak{M} : x_m \in t \text{ and } y_m = c_k\}$ for all datapoints in the node t belong to a class c_k, so that the empirical probability estimation is

$$\hat{p}(y = c_k|\mathbf{x} \in t) = \frac{M_k(t)}{M(t)}. \tag{27}$$

Hence, Eq. (18) turns into

$$\hat{r}_{min}(t) = \sum_{k=1}^{K} \hat{p}(y = c_k|\mathbf{x} \in t)(1 - \hat{p}(y = c_k|\mathbf{x} \in t)) \tag{28}$$

$$= \sum_{k=1}^{K} \sum_{\substack{k'=1 \\ k' \neq k}}^{K} \hat{p}(y = c_k|\mathbf{x} \in t)\hat{p}(y = c_{k'}|\mathbf{x} \in t), \tag{29}$$

and Eq. (17) turns into

$$\hat{r}_{min}(t) = \min_{c_l} \left\{ \sum_k (1 - \delta(c_k, c_l)) \hat{p}(y = c_k | \mathbf{x} \in t) \right\} \qquad (30)$$

$$= 1 - \max_{c_k} \left\{ \hat{p}(y = c_k | \mathbf{x} \in t) \right\}. \qquad (31)$$

In order to compute the best split, we need to formulate the empirical relative risk reduction, which is

$$\Delta \hat{r}(s|t) = \hat{r}_{min}(t) - \hat{p}(\mathbf{x} \in t_L | \mathbf{x} \in t) \hat{r}_{min}(t_L) - \hat{p}(\mathbf{x} \in t_R | \mathbf{x} \in t) \hat{r}_{min}(t_R), \qquad (32)$$

where

$$\hat{p}(\mathbf{x} \in t_L | \mathbf{x} \in t) = \frac{\hat{p}(\mathbf{x} \in t_L)}{\hat{p}(\mathbf{x} \in t)} = \frac{M(t_L)}{M(t)}, \qquad (33)$$

and

$$\hat{p}(\mathbf{x} \in t_R | \mathbf{x} \in t) = \frac{\hat{p}(\mathbf{x} \in t_R)}{\hat{p}(\mathbf{x} \in t)} = \frac{M(t_R)}{M(t)}. \qquad (34)$$

Now, that all parts for $\Delta \hat{r}(s|t)$ are defined, the optimal empirical split at a node t can be computed with

$$\hat{s}_{opt}(t) = \underset{\tilde{s}}{\operatorname{argmax}} \left\{ \Delta \hat{r}(\tilde{s}|t) \right\}. \qquad (35)$$

Note, that the border of the partition of t is a hyperplane perpendicular to one of the axes of $\mathbf{x}_n$. The evaluation to set the threshold for the split along a feature n can be conducted with a brute-force approach. If all datapoints in t are distinct with respect to feature n, $M(t) - 1$ splits for the n-th feature have to be evaluated in order to find the optimal split.

Equation. (29) computes the relative risk reduction $\hat{r}_{min}(t)$ for classification, known as the *Gini impurity* $i(t)$. The purity gain due to a split can be formulated as

$$\Delta i(s, t) = i(t) - \frac{M(t_L)}{M(t)} i(t_L) - \frac{M(t_R)}{M(t)} i(t_R), \qquad (36)$$

such that the optimization task turns into

$$\hat{s}_{opt}(t) = \underset{\tilde{s}}{\operatorname{argmax}} \left\{ \Delta i(\tilde{s}, t) \right\}. \qquad (37)$$

Growing a tree until the impurity in terminal nodes becomes $i(t) = 0$, so that all datapoints belong to a single class, is likely to cause overfitting on the training dataset. On contrary, if the growing process is abrupted too early, the subspaces might not be defined well enough to separate the classes properly. In practice, a tree is first fully grown and then *pruned*. The goal is to prune those nodes, which only have a minor effect on the estimated risk. Several heuristics can be applied as pruning criterias. For example, one can define a minimum number of $M(t)$ samples per terminal node. Another criterion is to define a maximum tree depth to reduce complexity. Lastly, a threshold for the minimum purity gain, provided by a split, can be defined as pruning criterion. Tuning these parameters appropriately is task-specific and should be monitored with a validation dataset.

After pruning, assigning a class to a terminal node is achieved by choosing the class with the highest estimated probability $\hat{p}(y = c_k|\mathbf{x} \in t)$.

3   Ensemble Learning with Random Forests

A Random Forest is a randomized ensemble learning method, which uses a set of binary trees as base learners. Before addressing Random Forests, we will first take a brief look at the concepts of ensemble learning.

3.1   *Ensemble Learning*

Ensemble Learning methods use several base learning models and combine the predictions of each individual learner with the aim to improve the final prediction. In case of the Random Forest algorithm, each tree is grown independently of the other trees forming the ensemble. The final prediction of the ensemble method is computed as an average or majority vote of all independently constructed predictors.

The advantage of ensemble methods will be illustrated with a simple example based on the principle of *collective wisdom*. For our example, we assume two possible outputs, where one of the two is the correct answer. All voters predict independently of each other with the same error rate ϵ. Each voter is considered to be competent, i.e. the probability of a false prediction is $p(\epsilon) < 0.5$ [14]. Under these assumptions, the error limit is going towards zero for an infinite number of voters B

$$\lim_{B \to \inf} \epsilon(B) = 0. \tag{38}$$

Given these restrictions, a voter can be interpreted as a Bernoulli variable with $\mu = \epsilon$ and $\sigma = \epsilon(1 - \epsilon)$. The error rate for majority voting can then be calculated using the binominal distribution

$$\epsilon(B) = \sum_{b=\mathsf{b}}^{B} \binom{B}{b} \epsilon^{b}(1-\epsilon)^{B-b} \quad \text{with} \quad \mathsf{b} = \left\lfloor \frac{B}{2} \right\rfloor + 1. \tag{39}$$

In our example we assume $B = 50$ predictors and an individual error rate of $\epsilon = 0.35$ for all predictors. The ensemble error rate for this majority vote is then

$$\epsilon_{\text{ens}} = \sum_{b=26}^{50} \binom{50}{b} \epsilon^{b}(1-\epsilon)^{50-b} = 0.01 < \epsilon. \tag{40}$$

Although in practice the predictor models are not completely independent, alone due to the shared training dataset, this examples illustrates the benefits of ensemble methods. Generally, the predictive error is composed of the *bias* and *variance* components. The key behind the success of ensemble methods is related to the reduction of the variance component. Ensembles work effectively as long as the bias and correlation of the base learners is low. CARTs, for example, have a small bias but large variance. A set of CARTs reduce variance, and by inducing randomization techniques, the correlation between all base learners can be reduced as described in the next section. Interested readers are referred to [11] for a detailed explanation of the *bias-variance decomposition*.

Following [10], three fundamental reasons explain why ensembles perform well. The first reason is statistical. A learning algorithm tries to identify the best hypothesis in space. When the amount of training data available is too small, the algorithm can find many different hypotheses of similar accuracy. By averaging the hypotheses one can find a good approximation for f by avoid the risk of choosing a wrong hypothesis. The second reason is computational. Finding a split to grow decision trees is conducted in a brute-force manner. Running the local search from many different starting points often provides a better approximation to the true unknown function. The third reason is representational. Given a finite training dataset, none of the candidate models is able to find the true function. Due to the limited dataset, will explore only a finite set of hypotheses and stop searching when the hypothesis fits the training data. By combining several learners to an ensemble, it can be possible to expand the space of representable functions.

Figure 2 illustrates the potential performance gain of ensemble methods. The spiral training dataset $\mathcal{D}_{t,0}$ consists of $M_{t,0} = 20{,}000$ and the test dataset of $M_{v} = 500$ datapoints. Additional noise is added to the test dataset to increase the difficulty of the two-class classification task. The first two columns depict the training and test dataset, respectively. The two plots on the right-hand side depict the classification performance of a CART ($B = 1$) and a Random Forest constructed with $B = 10$ tree models. In the first row, the classification performance for both methods is comparable due to the relatively large dataset for the problem to be solved. The ensemble reduces the error rate ϵ by 1%. In the second and third row, the number of training datapoints is being reduced to $M_{t,1} = 2000$ and $M_{t,2} = 200$, where $\mathcal{D}_{t,2} \subset \mathcal{D}_{t,1} \subset \mathcal{D}_{t,0}$. With a decreasing training dataset size, the gap

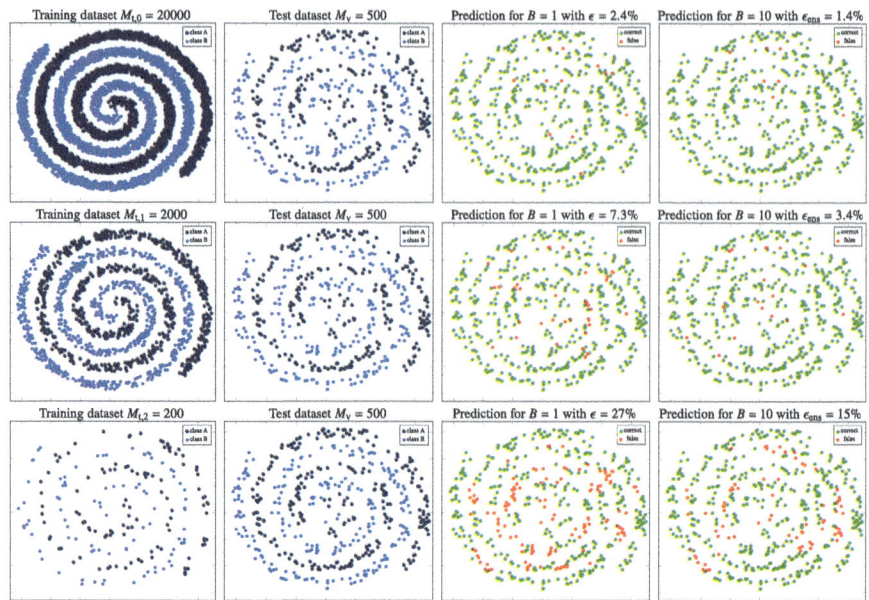

Fig. 2 Classification example on a spiral dataset. The error prediction gap between a single CART model with $B = 1$ and an ensemble method with $B = 10$ learners increases considerably, when the size of the training dataset is a limiting factor for the task to be solved

between a single CART and the ensemble method increases to 3.9% for $M_{t,1}$, and a considerable gain of 12% for $M_{t,2}$. The number of samples required for a good classification performance certainly depends on the difficulty of the problem to be solved. In practice, however, the dataset is often a limiting factor, making ensemble methods appealing.

3.2 Random Forests

The Random Forest algorithm [6] constructs a set of several individual CART as base learners, where each tree is grown independently. After growing the trees, the final ensemble prediction is made by taking the average over the predictions of all trees for regression problems, or by majority vote for classification. Since the Random Forest is composed of CARTs, it inherits the advantages of CARTs, such as the robustness to outliers and noise. The Random Forest is able to handle ordered and categorial variables. It is able to perform a prediction, even when some entries in the input data are missing. Interpretability, though, is not inherited due to the averaging approach for its prediction.

Let a Random Forest be the collection of B trees, where a tree is expressed by $\{\mathfrak{T}_b(\mathbf{x}, \boldsymbol{\theta}_b)\}$. $\boldsymbol{\theta}_b$ is an independent identically distributed random vector. In order to obtain a low variance, the trees should be as uncorrelated as possible. Therefore, the learning procedure is perturbed by two elements forming the vector $\boldsymbol{\theta}_b$, which largely define the characteristics of the Random Forest algorithm. The first element is *bagging* [5], a bootstrap aggregating technique. This means that B new bootstrapped datasets $\mathcal{D}_{s,b}$, $b = 1, \ldots, B$ are randomly drawn by sampling M_S datapoints with replacement from $\mathcal{D}_s$. On average, around 37% of the datapoints in $\mathcal{D}_s$ are omitted in each bootstrapped set $\mathcal{D}_{s,b}$

$$\lim_{M_S \to \inf} \left(1 - \frac{1}{M_S}\right)^{M_S} = \frac{1}{e} \approx 0.368. \tag{41}$$

Bagging constructs individual trees by learning with a different dataset $\mathcal{D}_{s,b}$. The second element defines a rule, how the process of splitting a node is conducted. Instead of searching the best split over all N features, only a subset of N_{RF} features is randomly chosen at each node, where $N_{RF} < N$ holds. A common choice is $N_{RF} = \left\lceil \sqrt{N} \right\rceil$. Limitating the list of candidates with accelerates the learning procedure as well.

Applying both strategies, bagging and the random subset of features for splitting a node, generate individual base learners and make the Random Forest algorithm likely to benefit from the averaging process. As shown in [6], a Random Forest does not overfit. Therefore, increasing the number of base learners decreases the generalization error, which converges to a limiting value. Another beneficial property of Random Forests and bagging in particular is, that the construction of the base learners is performed with bootstrap samples. The omitted samples of $\mathcal{D}_s$ enable an unbiased estimate of the generalization error during the building process. This is denoted as out-of-bag estimates (oob).

3.2.1   Out-of-Bag Estimates

Bagging allows an unbiased estimate of the generalization error while constructing the ensemble of trees. It comes for free and it can replace an additional validation dataset. As shown in Eq. (41), approximately 37% of all datapoints in $\mathcal{D}_s$ are not part of the bootstrap set $\mathcal{D}_{s,b}$. We denote the trees, which did not use a certain datapoint $\{\boldsymbol{x}_m, y_m\}$, as the set $\mathfrak{B}_m$. Furthermore, we define the out-of-bag class probability estimator as

$$\boldsymbol{f}_{\text{oob}}(\boldsymbol{x}_m, \boldsymbol{\theta}) = \frac{1}{|\mathfrak{B}_m|} \sum_{b \in \mathfrak{B}_m} \boldsymbol{f}_b(\boldsymbol{x}_m, \boldsymbol{\theta}_b), \tag{42}$$

with $\boldsymbol{f}_b(\boldsymbol{x}_m, \boldsymbol{\theta}_b) = \left[\hat{p}(y = c_1 | \mathbf{x} \in t_{\mathcal{I}_b}), \ldots, y = c_K | \mathbf{x} \in t_{\mathcal{I}_b})\right]^{\mathrm{T}}$ the vector of all K class probability estimates of the b-th tree. Hence, $\boldsymbol{f}_{\text{oob}}(\boldsymbol{x}_m, \boldsymbol{\theta})$ contains the average class probability over all trees. The majority voting is realized by selecting the class with the highest probability in $\boldsymbol{f}_{\text{oob}}(\boldsymbol{x}_m, \boldsymbol{\theta})$ as $f_{\text{oob}}(\boldsymbol{x}_m)$. Then, the oob estimate for the empirical prediction risk is the averaged sum of errors

$$\hat{R}_{\mathrm{oob}}(f) = \frac{1}{M} \sum_{m=1}^{M} \delta(f_{\mathrm{oob}}(\boldsymbol{x}_m), y_m). \tag{43}$$

The oob estimate has two attractive properties. First, it is equivalent to a test dataset [5]. Second, in contrast to the cross-validation technique, the oob estimate is unbiased, if the number of trees in the Random Forest is large enough.

3.2.2 Proximity Measure

The Random Forest algorithm allows to determine the similarity between two datapoints. Unlike other proximity measures such as the Euclidean, Manhattan or Mahalanobis distance, the Random Forest proximity follows a data-adaptive principle, since the trees are grown according to the training dataset. In order to evaluate the similarity between two datapoints $\boldsymbol{x}_i$ and $\boldsymbol{x}_j$, one observes if both datapoints end in the same terminal node of a tree. In this case, the similarity value is increased by one. This process is repeated over all trees and the final similarity measure is the average similarity over all trees

$$prox(\boldsymbol{x}_i, \boldsymbol{x}_j) = \frac{1}{B} \sum_{b=1}^{B} \delta(\mathsf{t}_{\mathfrak{I}_b}(\boldsymbol{x}_i), \mathsf{t}_{\mathfrak{I}_b}(\boldsymbol{x}_j)), \tag{44}$$

with $\mathsf{t}_{\mathfrak{I}_b}(\boldsymbol{x}_i)$ denoting the leaf of the b-th CART in which $\boldsymbol{x}_i$ terminates. Following the idea of proximity, the next Section demonstrates how the Random Forest algorithm can be adapted in order to establish an unsupervised learning method.

4 Random Forests for Unsupervised Learning

In the unsupervised learning method [15], which is described in what follows, the training data $\mathcal{D}_{\mathrm{u}}$ consists of a set of input vectors $\boldsymbol{x}_m$ without any corresponding target values. Furthermore, no assumptions are made about the number of clusters potentially present. The goal is to discover groups of similar examples within the data, which is called *clustering*. Intuitively, a cluster represents a group of datapoints whose distances are small compared with the distances to points outside of the cluster [4]. Hence, clustering aims to partition the dataset by finding K clusters $\{C_1, \ldots, C_K\}$ of unlabeled datapoints.

A general approach for the clustering process is depicted in Fig. 3. The dataset is often first subjected to pre-processing. An essential part of this pre-processing is the extraction of relevant features of $\boldsymbol{x}_m$. In the next step, all datapoints $\{\boldsymbol{x}_1, \ldots, \boldsymbol{x}_{M_{\mathrm{u}}}\}$

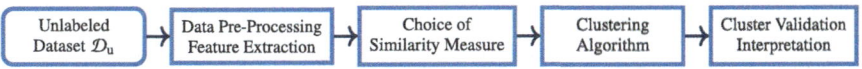

Fig. 3 A general approach for clustering an unlabeled dataset

of $\mathcal{D}_u$ are being compared with a similarity measure. This relation is described in terms of the proximity, which can be defined through the similarity or dissimilarity. The proximity measure strongly influences how the data is going to be clustered. Since the proximity between all datapoints of the dataset $\mathcal{D}_u$ has to be specified, the proximity is a $M_u \times M_u$ symmetric matrix $P = (P_{ij})$. The higher the value of P_{ij}, the more similar the datapoints x_i and x_j are. The diagonal elements P_{ii} always have to be one. After performing the similarity comparison between all datapoints, the matrix P is fed into a clustering algorithm, with the aim to partition the dataset into homogeneous groups and to identify patterns. Finally, the clustering results have to be validated and interpreted.

Given the general overview of a clustering procedure, in the following we introduce an unsupervised learning approach based on Random Forests and focus on defining a similarity measure and the choice of the clustering algorithm. A big advantage over many other clustering techniques is, that the proposed method does not require a pre-defined number of clusters.

4.1 Similarity Measure Based on Random Forests

In order to perform unsupervised learning with Random Forests, first a similarity measure is proposed. Therefore, the procedure of growing the trees has to be adapted. In a first step we define a classification task with two classes, A and S, where all datapoints x_m of $\mathcal{D}_u$ are labeled with A. The tricky part arises with the construction of the second class S. We define S as a synthetic dataset $S = \{z_1, \ldots, z_K\}$, $z \in \mathfrak{X}$ based on some distribution, such that the synthetic dataset can be considered as noise. Given A and S, the Random Forest is trained just as a normal classification task, described in the previous Sections. The aim here is to distinguish between the given dataset $\mathcal{D}_u$ and the generated noise dataset S. The underlying mechanism is that, if there is a structure in the data in $\mathcal{D}_u$ the Random Forest has to fit its leaves to it in order to achieve a low error. Figure 4 illustrates the mechanism of distinguishing between A and S, which leads to the separation of $\mathcal{D}_u$. Once the Random Forest is trained, two datapoints x_i and x_j at a time are run through all trees to determine the similarity P_{ij}, similarly as described in Sect. 3.2.2. But, instead of evaluating the proximity solely on leaves, the proposed similarity measure takes into account the full paths of the datapoints through the trees instead. Thereby the complete information provided by the Random Forest is captured, which makes the process robust. The principle behind this path proximity

Fig. 4 Assuming a separable
structure in the dataset $\mathcal{D}_{\mathrm{u}}$,
the separation between class
A and class S will implicitly
learn the underlying structure
of $\mathcal{D}_{\mathrm{u}}$

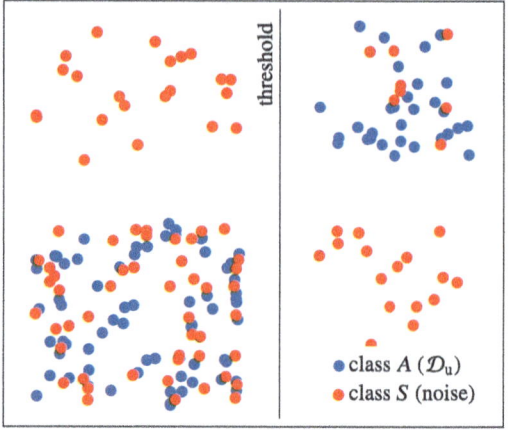

is detailed in Sect. 4.1.2. Before describing the path proximity, we will first examine
what needs to be considered when constructing the noise dataset S.

4.1.1   Constructing the Noise Dataset

In [7] the construction of S is realized by "independent sampling from the
one-dimensional marginal distributions of the original data" $\mathcal{D}_{\mathrm{u}}$, as depicted in
Fig. 4. Another construction principle is explained in [21], where S is build by
sampling randomly from an assumed uniform distribution within the Q-dimensional
hypercube defined by the minimum and maximum values of $\mathcal{D}_{\mathrm{u}}$. Independent of
the way how the synthetic data is generated, the appearing task is to distinguish
between noise and the actual data. One main drawback of the proposed noise
generation methods occur with high dimensional input spaces. In order to force
the Random Forest to fit properly to the inherent structure, the noise distribution
has to be very dense. If there is no more noise data left to perform the dividing
task, the resulting leaves become large. Due to the curse of dimensionality [4], for
high dimensional input spaces this leads to a huge amount of necessary synthetic
datapoints and accordingly to a highly unbalanced ratio between A and S. Even
though the sampling from marginal distributions addresses this issue by sampling
mainly in the regions of interest, the results in high dimensional spaces might not
be satisfactory.

A solution to this problem is to not explicitly generate the noise, but to estimate
the required number of noise datapoints in each split. The number of noise points
in a node of the tree is chosen to be equal to the number of datapoints from $\mathcal{D}_{\mathrm{u}}$ in
that node. When growing the trees, a randomly chosen distribution from a set of
predefined distributions is used in order to construct the noise dataset at each split.

We re-formulate the estimated Gini impurity for the node t in an arbitrary tree as

$$r_g(t) = \sum_{c=1}^{2} \frac{M_c(t)}{M(t)} \left(1 - \frac{M_c(t)}{M(t)}\right), \tag{45}$$

where $M_c(t)$ the number of datapoints in node t, which belong to class c. The gini gain resulting by splitting t into the child nodes t_L (left) and t_R (right) is then

$$\Delta i(t, t_L, t_R) = r_g(t) - \frac{M(t_L)}{M(t)} r_g(t_L) - \frac{M(t_R)}{M(t)} r_g(t_R). \tag{46}$$

The optimal split is given if $\Delta i(t, t_L, t_R)$ is maximal. Hence, the number of datapoints of each class in each node (t, t_L and t_R) is required. The number of original datapoints $M_{\mathcal{D}_{u,b}}(t_{j,b})$ of the bagged dataset $\mathcal{D}_{u,b}$ belonging to the b-th tree in the j-th node $t_{j,b}$ of this tree, as well as in the possible child nodes can simply be counted. The number of noise datapoints in the same node needs to be estimated for a given split value $\tau_{\tilde{n}}$. Let $z_{\tilde{n}}$ be the standardized value of $\tau_{\tilde{n}}$ (see Eq. (49)). Then, the number of noise datapoints for the left and right child nodes of a node $t_{j,b}$ is calculated as

$$M_{S,l_{j,b}}(z_{\tilde{n}}) = M_{\mathcal{D}_{u,b}}(t_{j,b}) P(z_{\tilde{n}} \le z_{\tilde{n}}) \text{ and} \tag{47}$$

$$M_{S,r_{j,b}}(z_{\tilde{n}}) = M_{\mathcal{D}_{u,b}}(t_{j,b}) - M_{S,l_{j,b}}(z_{\tilde{n}}), \tag{48}$$

where $\tilde{n}$ stands for the $\tilde{n}$-th dimension of the vector whose features are chosen randomly in each node when constructing the trees. The values $M_{S,l_{j,b}}$ and $M_{S,r_{j,b}}$ denote the number of corresponding noise points in the left and right child note of $t_{j,b}$, given that the split $\tau_{\tilde{n}}$ is chosen, since $z_{\tilde{n}}$ is the standardized value of $\tau_{\tilde{n}}$. $P(z_{\tilde{n}} \le z_{\tilde{n}})$ is the value of the cumulative density function (cdf) at the standardized threshold $z_{\tilde{n}}$. The standardized threshold $z_{\tilde{n}}$ in the $\tilde{n}$-th dimension is determined with

$$z_{\tilde{n}} = \frac{\tau_{\tilde{n}} - \mu_{\tilde{n}}}{\sigma_{\tilde{n}}}, \tag{49}$$

$$\mu_{\tilde{n}} = \frac{\max\left\{\mathfrak{X}_{t_{j,b}}\right\}_{\tilde{n}} + \min\left\{\mathfrak{X}_{t_{j,b}}\right\}_{\tilde{n}}}{2}, \tag{50}$$

$$\sigma_{\tilde{n}} = \frac{\max\left\{\mathfrak{X}_{t_{j,b}}\right\}_{\tilde{n}} - \min\left\{\mathfrak{X}_{t_{j,b}}\right\}_{\tilde{n}}}{6}, \tag{51}$$

where $\max\left\{\mathfrak{X}_{t_{j,b}}\right\}_{\tilde{n}}$ and $\min\left\{\mathfrak{X}_{t_{j,b}}\right\}_{\tilde{n}}$ yield the maximum or minimum value of the subspace $\mathfrak{X}_{t_{j,b}}$ in the dimension specified by $\tilde{n}$. The interval of a node covers $\pm 3\sigma_{\tilde{n}}$.

Next, we define a set of distributions. The first distribution used is the uniform distribution, where its cdf is given by

$$P_u(z_{\tilde{n}} \le z_{\tilde{n}}) = \frac{1}{6} z_{\tilde{n}} + \frac{1}{2}. \tag{52}$$

The standard normal distribution is the second used distribution and is approximated through [22]

$$P_n (z_{\tilde{n}} \leq z_{\tilde{n}}) = \frac{1}{1 + e^{-\sqrt{\pi}\left(\beta_1 z_{\tilde{n}}^5 + \beta_2 z_{\tilde{n}}^3 + \beta_3 z_{\tilde{n}}\right)}}, \tag{53}$$

where $\beta_1 = -0.0004406$, $\beta_2 = 0.04181198$ and $\beta_3 = 0.9$ holds. Third, a bimodal distribution is used, which is build as the sum of two shifted standard normal distributions P_n with

$$P_b (z_{\tilde{n}} \leq z_{\tilde{n}}) = P_n (z_{\tilde{n}} - 3 \leq z_{\tilde{n}} - 3) + P_n (z_{\tilde{n}} + 3 \leq z_{\tilde{n}} + 3). \tag{54}$$

The randomly selected noise distributions at each split relax the dependency of the proximity measure to one specific distribution. Obviously, the set of three proposed distributions can also be extended with or replaced by other distributions.

Up to this point, we know how to grow the forest and how to compute the noise data to solve the classification task. The last missing element to obtain the data adaptive similarity measure is to describe the proposed path proximity.

4.1.2 Path Proximity

The proposed proximity measure takes into account the full paths of the datapoints through the trees instead of just using the terminal nodes.

Let a Random Forest consist of B trees $\mathfrak{T}$, where the b-th tree $\mathfrak{T}_b$ is constructed based on the bagged dataset $\mathcal{D}_{u,b}$. Then a tree $\mathfrak{T}_b$ consists of N_b nodes $t_{n,b}$. A path of a datapoint through a tree can be defined by a set including all nodes the datapoint passed. This leads to the path formulation

$$\mathcal{T}_{i,b} = \left\{ t_{1,b}, t_{n_{i_2},b}, \dots, t_{N_i,b} \right\}, \tag{55}$$

where the index i represents the i-th datapoint x_i. The node $t_{1,b}$ is the root node of the b-th tree $\mathfrak{T}_b$ and hence the first node on the path of the i-th datapoint. The node $t_{n_{i_2},b}$ is the second node on the path, where $n_{i_2,b}$ represents the node number n the datapoint has passed. The last node on the path of the i-th datapoint in the b-th tree is $t_{N_i,b}$.

In order to compare the paths of two datapoints through the b-th tree, the corresponding sets $\mathcal{T}_{i,b}$ and $\mathcal{T}_{j,b}$ need to be compared. The Jaccard Index [13]

$$P_{ij} (b) = \frac{|\mathcal{T}_{i,b} \cap \mathcal{T}_{j,b}|}{|\mathcal{T}_{i,b} \cup \mathcal{T}_{j,b}|} \tag{56}$$

$$= \frac{|\mathcal{T}_{i,b} \cap \mathcal{T}_{j,b}|}{|\mathcal{T}_{i,b}| + |\mathcal{T}_{j,b}| - |\mathcal{T}_{i,b} \cap \mathcal{T}_{j,b}|} \tag{57}$$

Fig. 5 Path proximity
example

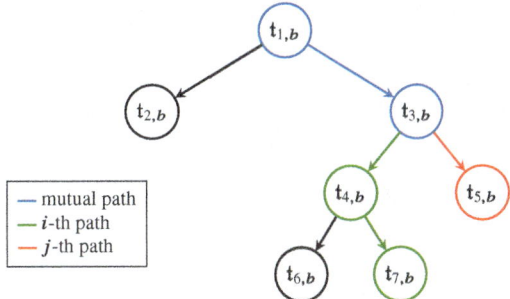

is used for this purpose. It holds that $P_{ij}(b) \in (0, 1]$, since at least the root node is present in both sets. This way, a similarity measure, given the two datapoints x_i and x_j, based on the b-th tree is defined. In Fig. 5 an example tree with two arbitrary paths is shown. The mutual path of both datapoints is colored in blue, and the single paths are depicted in green and red. Interpreting Eq. (57) based on Fig. 5 leads to $|\mathcal{T}_{i,b} \cap \mathcal{T}_{j,b}|$ being the length of the mutual path and $|\mathcal{T}_{i,b}|$ being the length of the i-th path (i-th datapoint) starting from the root node, $|\mathcal{T}_{j,b}|$ respectively. For the depicted example, the corresponding Jaccard index would be $2/5$. In other words, the i-th and j-th datapoints have a similarity of 0.4. A value of 1 indicates, that both datapoints are identical or very similar according to the given tree.

By averaging over all B values $P_{ij}(b)$ in a forest, we obtain the path proximity

$$P_{ij} = \frac{1}{B} \sum_{b=1}^{B} \frac{|\mathcal{T}_{i,b} \cap \mathcal{T}_{j,b}|}{|\mathcal{T}_{i,b}| + |\mathcal{T}_{j,b}| - |\mathcal{T}_{i,b} \cap \mathcal{T}_{j,b}|}. \tag{58}$$

If two datapoints have the same paths in all trees the proximity will be one. Contrary, if they only share the root nodes, the proximity will be very small tending towards zero. The path proximity enables one to cover more than just the leaf information of the forest within one scalar value. Extracting the Random Forest based path proximity for a given dataset we obtain a data adaptive similarity measure.

With the proximities between all datapoints structured in the similarty matrix P, we can advance to the next step according to Fig. 3 by applying a clustering algorithm, which has the task to group similar datapoints given P. Clustering algorithms can be categorized into several types and various algorithms, in the following we briefly discuss the *hierarchical clustering* method applied for the proposed unsupervised learning technique with Random Forests.

4.2 Hierarchical Clustering

Hierarchical clustering methods can be distinguished between *agglomerative* or *divisive*. Due to the computation complexity, divisive methods are not commonly used in practice [23]. At the beginning of the clustering process, agglomerative algorithms consider each datapoint as a single cluster, so that the number of datapoints equals the number of clusters. Then, in each iteration a pair of clusters is successively merged until one single cluster remains. This hierarchy can be visualized with a dendrogram. The methods differ in how the dissimilarity from a merged cluster to all the remaining is computed, the so-called *linkage* function. Among several other functions [17], two commonly used linkage functions are the single and average linkage.

For single linkage, the minimum dissimilarity between all the elements of both clusters is used as dissimilarity of the clusters

$$d_{kl} = \min_{\substack{i \in C_k \\ j \in C_l}} \{d_{ij}\}, \tag{59}$$

where d_{kl} denotes the dissimilarity between two arbitrary clusters C_k and C_l.

For average linkage, dissimilarity is determined through the average of all dissimilarities between the points of the two clusters

$$d_{kl} = \frac{1}{|C_k||C_l|} \sum_{i \in C_k} \sum_{j \in C_l} d_{ij}, \tag{60}$$

where $|C_k|$ denotes the number of objects in cluster C_k, and $|C_k|$ the number in C_k.

The agglomerative hierarchical clustering results in a hierarchy, the hierarchy can be visualized as dendrogram. When using the order of the leaves in the dendrogram, permutations on P can be performed, such that a reordered proximity matrix (P_o), is obtained. The matrix P_o represents the clusters in the data and provides a graphical interpretation of the data inherent structure.

4.3 Cluster Analysis and Visualization

Cluster analysis can be performed with a visualization as depicted in Fig. 6. On the right-hand side, the two-dimensional toy dataset, consisting of four clusters with different shapes and densities, is depicted. The proximity matrices are represented as squared images (a)–(d), where dark pixels represent zero entries ($P_{ij} = 0$) and bright pixels with higher similarity. The bright squares along the diagonal represent the four clusters. Squares, which are not aligned the diagonal represent the inter-cluster similarity. It should be noted, that these matrices represent the similarities before applying hierarchichal clustering. The main purpose of Fig. 6 is to show the beneficial effects of the path proximity and ensemble noise.

For example, the compact cluster no. 4 reveals a brighter square on the similarity matrix at the bottom right side along the diagonal axis compared to the widely spread cluster no. 2. That is, because the datapoints within cluster no. 4 share a

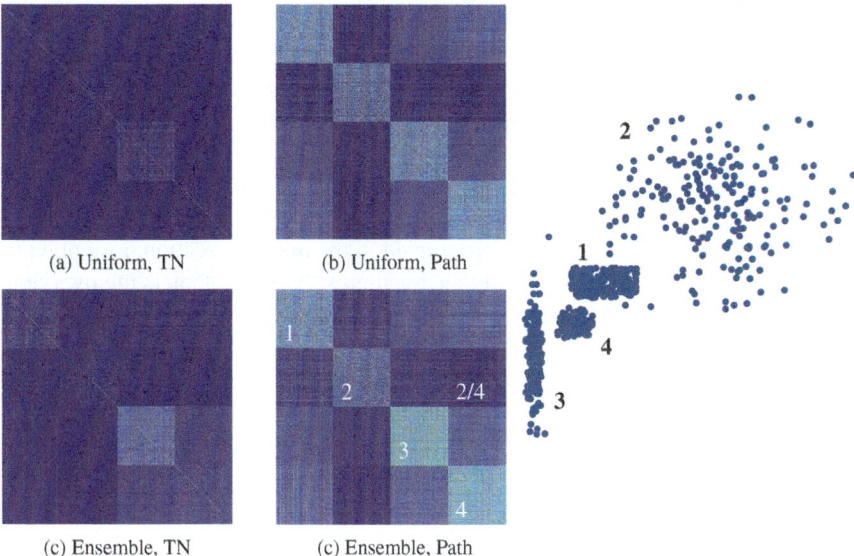

(a) Uniform, TN   (b) Uniform, Path

(c) Ensemble, TN   (c) Ensemble, Path

Fig. 6 Comparison of uniformly distributed noise versus ensemble noise, as well as path proximity versus terminal node (TN) for computing the similarity. The right hand-side depicts the two-dimensional toy dataset, consisting of four clusters with different shapes and densities. (**a**) Uniform, TN. (**b**) Uniform, Path. (**c**) Ensemple, TN. (**d**) Ensemple, Path

higher similarity in along both axes compared to cluster no. 2. Additionally, the inter-cluster similarity between both clusters, depicted with '2/4', is low, since the datapoints of the two clusters show almost no overlap along the axes.

In addition, the four subfigures (a)–(d) depict the effects of the noise ensemble compared to a single uniform distribution, as well as extracting the similarity out of the datapoints' paths through the trees instead of only taking the terminal nodes (TN) into account. The similarity of the four clusters can best be identified in subfigure (d), where the ensemble noise and especially the path proximity support an improved similarity measure.

5 Applications

Possible applications of Random Forests in the automotive domain are manifold. In the following, two methods for categorizing traffic scenarios are presented. The first case applies the unsupervised learning method from the previous section for the identification of similar traffic scenarios. The second case demonstrates how Random Forests can be integrated into Deep Learning architectures to tackle the problem of Open-Set recognition for traffic scenarios.

5.1 Traffic Scenario Clustering

Traffic scenario categorization is an important component for downstream tasks like trajectory planning, emergency braking and other functions for autonomous driving. Road traffic does not evolve completely random, since it's framed by the infrastructure, traffic rules, etc. [12], so that traffic scenarios do follow certain patterns, and can be categorized according to the set of features selected. In this section the unsupervised method, as presented in the previous section, is applied to real world traffic scenarios to identify such patterns. Figure 7 depicts the overview of the complete framework, starting from data generation and feature extraction, up to the cluster identification and validation.

First, a set of vehicle trajectories from a public roundabout is extracted with drone imagery by applying the method published in [16, 19]. A second trajectory dataset is recorded on a vehicle test track in order to create scenarios with critical driving maneuvers. The criticality is achieved via strong braking and cut-in maneuvers with small gaps between the vehicles, which are not commonly seen in public traffic. Since the two datasets are recorded at different places, a coordinate transformation is applied as well, so that both can be overlayed on a road map. All scenarios involve at least two vehicles and the timespan is set to 5 s. In total 110 critical scenarios are generated, the same quantity of scenarios of the public road is randomly selected. The extracted trajectories are then geo-referenced and coupled with a road map, which allows one to generate road-adaptive features. Especially for road sections with curvatures and crossings, one has to align the paths driven by vehicles in accordance to the road layout, in order to compute features such as time gaps. The aim of this demonstration is to identify several scenario categories and especially to distinguish between critical and non-critical scenarios.

Given the trajectories and the road information, a set of four features

$$x = [\bar{v}, v_{\mathrm{x}}, a_{\mathrm{x}}, \varrho] \tag{61}$$

is extracted from the two datasets. All features relate to the ego vehicle within that 5 s scenario length. $\bar{v}$ denotes the average speed, v_{x} and a_{x} denote the minimum

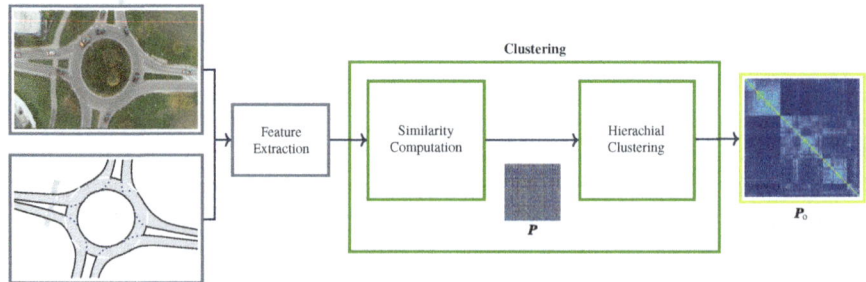

Fig. 7 Traffic scenario clustering: from data generation up to cluster visualization

Fig. 8 Proximity matrix: The scenarios are coarsely divided into five clusters A to E. Below, the normalized values for all variables are depicted, where bright coloring indicates high values

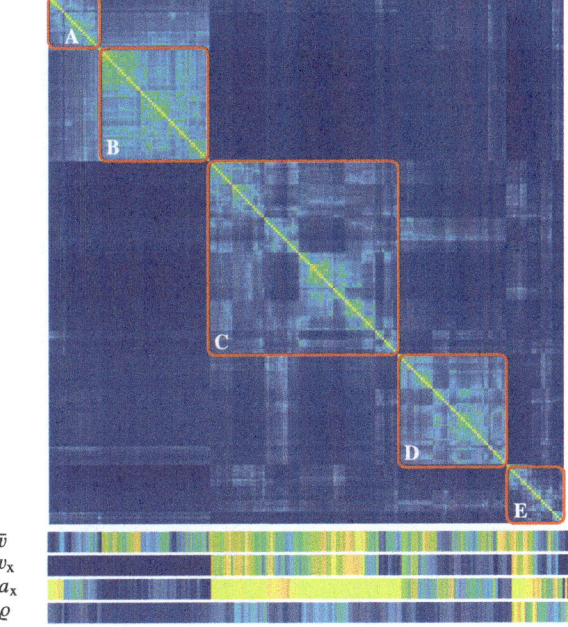

$\bar{v}$
v_x
a_x
ϱ

longitudinal velocity and acceleration. ϱ denotes the minimum time-headway between the ego and a target vehicle in a crossing scenario, assuming constant velocity, similarly as with the typical time-headway estimation for car-following scenarios. For demonstration purposes the number of features is limited to four, which should separate the critical from the non-critical ones. In general, the choice and number of features has to be aligned according to the application.

As depicted in Fig. 8, five clusters, A to E, are selected. One can recognize smaller clusters within these clusters and differentiate them more fine-grained accordingly. Instead of visually selecting the clusters, one could also use the *elbow* method instead. The clustering result can be physically validated by illustrating the feature values below the proximity matrix, as depicted in Fig. 8. For each cluster one typical scenario is depicted in Fig. 9. The left column in Fig. 9 shows the paths of the vehicles, the two right hand-side columns depict the start and the end of the scenario.

Cluster A and B represent critical crossing scenarios with two vehicles, one gray and one black vehicle. In both cases, the merging vehicle (gray) approaches the roundabout without considering the second vehicle (black). In cluster A, both vehicles are able to brake just before a potential collision. Whereas in cluster B, the merging vehicle continues its drive and violates the right-of-way, hence the black vehicle has to perform a braking. Similarly to B, in cluster D the merging vehicle violates the right-of-way, thus forcing a third vehicle (green) to react and brake.

Cluster C and E contain most of the casual driving scenarios, which were filmed on the public road. Since a few scenarios are randomly selected from a large dataset,

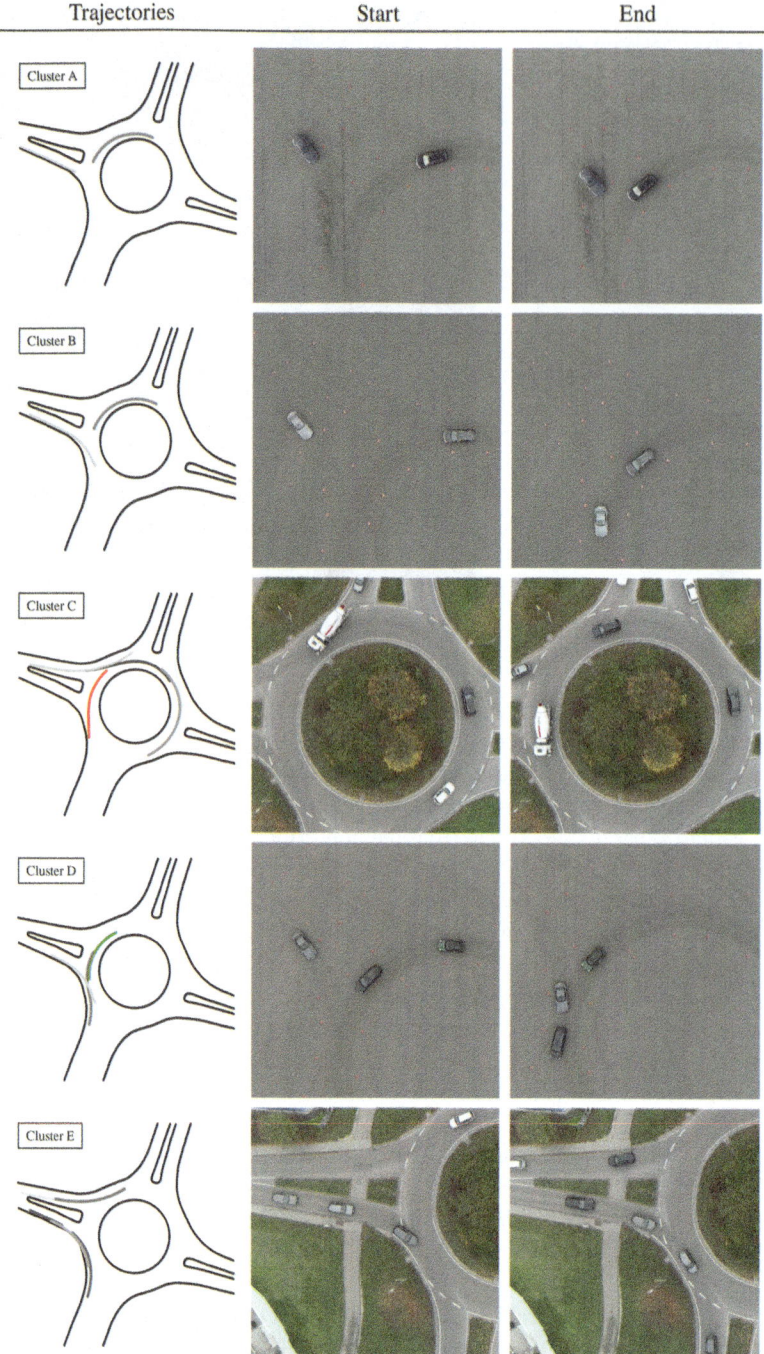

Fig. 9 Typical scenarios from each cluster: Cluster A,B and D represent critical scenarios, cluster C and E casual driving on a public road

these scenarios are correspondingly diverse. The selection and number of features only allows a rough separation. This can be confirmed with the proximity matrix in Fig. 8, where especially for cluster C several smaller clusters can be detected. Cluster C and E contain casual driving scenarios, such as car-following and leaving the roundabout, see the white car depicted in the example for cluster C. The example in cluster E depicts the gray colored ego vehicle approaching the roundabout behind another vehicle and entering the roundabout, while other cars are leaving the scene above.

After the cluster validation, these five groups can be defined as classes. If further data is collected during operation, the new scenarios can be assigned to already known classes. However, one must expect to find novel scenario types. The next section shows how the Random Forest, embedded in a deep learning architecture, can help to deal with new scenarios that cannot be assigned to any of the known classes.

5.2   Open Set Recognition for Traffic Scenarios

Typically, learning models proposed in the literatures [9, 18] work under the *closed-world* assumption, which means that the model will classify all the inputs only to one of the K classes used in the training. This is an issue in the real-world, as there are possibilities to encounter new scenario classes when the vehicle is driving on-road. The models trained with closed-world assumptions will fail in the cases where they encounter new classes as the models classify the inputs to only one of the K trained classes. This is a challenging and important problem to be addressed and leads to a new paradigm called *Open-Set Recognition* (OSR) [20].

An OSR model trained on K classes should be able to classify a given input to one of the K classes - or as an unknown. According to [2, 3], simply thresholding a closed-world model with a user defined threshold might not be satisfactory and the performance of such models deteriorates in an open-world case.

OSR models are either distance based, reconstruction based or extreme value based. In [1], a method based on a combination of *Convolutional Neural Networks* (CNN) and a Random Forest is proposed. An overview of the architecture is shown in Fig. 10. The scenarios are represented as a sequence of occupancy grids, with each occupancy grid representing the occupancy of objects and infrastructure in the scene at a time stamp. These grids are fed into the CNN to extract the features. Finally, the classification is done by the Random Forest algorithm combined with extreme value distributions. During the training phase, firstly a CNN is trained on a set of K labelled classes. As a second step, the fully connected layer of the CNN is removed and the flattened output is used as input for a Random Forest algorithm. The Random Forest is trained on a set of extracted features from the CNN for a given training set. In this second phase, the trained Random Forest and the CNN classify an input based on the majority voting scheme by the Random Forest algorithm. In the third step the class-specific vote patterns are modelled using Extreme Value Theory (EVT) based

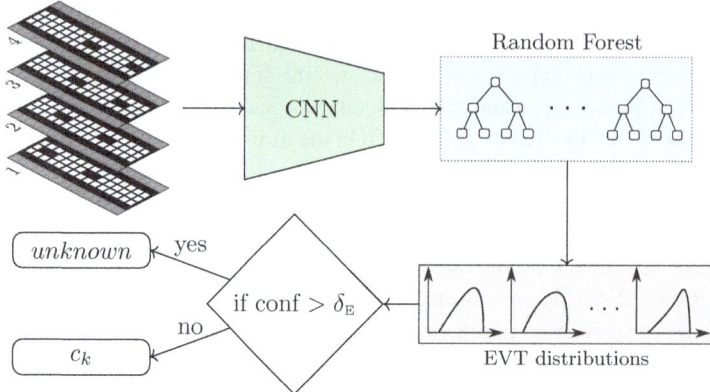

Fig. 10 Open-Set Recognition architecture: CNN for feature extraction, Random Forest and EVT based voting patterns for classification, Figure adapted from [1]

distributions. Class-specific vote patterns collected for the training dataset provide comprehensive information about the uncertainty of the classifier for each class. In the inference or the test phase, the class-specific EVT distributions are used to estimate the probability that a sample belongs to a class c_k or an unknown class based on the number of trees in the Random Forest voting for each class.

The method was tested on real world datasets. The ensemble nature of the Random Forest algorithm when combined with EVT distributions is shown to provide a much more robust OSR accuracy when compared to using other OSR methods and using standard scores like *Softmax* or majority voting.

6   Conclusion

This chapter proposes an unsupervised learning method in order to categorize traffic scenarios. The knowledge about traffic scenario categories is an important aspect for an efficient validation process for automated driving functions. Hence, the scenario categorization has the potential to accelerate the validation process by selecting representatives of each cluster and thereby reducing redundancies by avoiding to test very similar scenarios.

The proposed method is based on Random Forests and performs the pattern recognition only given the input data, i. e., where the availability of labels for training is absent. The goal is to discover groups of similar examples within the data. To achieve this, one has to memorize, compress and structure the data. This can be done by representing a traffic situation with a set of relevant features. These features can then be used to train the proposed method.

The core of the presented method lies in the data-adaptive similarity measure, so that data points can be compared in order to decide, whether they are similar and belong to the same cluster. The underlying mechanism is that, if a separable structure in the dataset is existent, the separation between the actual data and the second, synthetic data, will implicitly learn the underlying structure of the unlabeled dataset. Once the Random Forest is trained, two data points at a time are run through all trees to determine the similarity between both. The similarities between all traffic scenarios can be written in a similarity matrix. By applying hierarchical clustering techniques on that matrix, clusters of traffic scenarios with similar characteristics emerge, while being separated from those with low similarity. The chapter concludes with an exemplified application. It is shown how scenarios, represented by vehicle trajectories, can be categorized according to the vehicle dynamics, as well as the interaction between the traffic participants.

References

1. Balasubramanian, L., Kruber, F., Botsch, M., Deng, K.: Open-set recognition based on the combination of deep learning and ensemble method for detecting unknown traffic scenarios. In: 2021 IEEE Intelligent Vehicles Symposium (IV), pp. 674–681. IEEE (2021)
2. Bendale, A., Boult, T.E.: Towards open world recognition (2014). CoRR abs/1412.5687. https://doi.org/10.1109/cvpr.2015.7298799. http://arxiv.org/abs/1412.5687
3. Bendale, A., Boult, T.E.: Towards open set deep networks (2015). CoRR abs/1511.06233. https://doi.org/10.1109/cvpr.2016.173. http://arxiv.org/abs/1511.06233
4. Bishop, C.M.: Pattern Recognition and Machine Learning, vol. 4 (2006). https://doi.org/10.1117/1.2819119
5. Breiman, L.: Out-of-bag estimation, Technical report, Statistics Department, University of California Berkeley, Berkeley CA 94708 (1996)
6. Breiman, L.: Random forests. Mach. Learn. **45**(1) (2001)
7. Breiman, L.: Using random forests v3.0. Technical Report (2002)
8. Breiman, L., Friedman, J., Olshen, R., Stone, C.J.: Classification and Regression Trees. Chapman and Hall/CRC (1984)
9. Cara, I., Gelder, E.D.: Classification for safety-critical car-cyclist scenarios using machine learning. In: 2015 IEEE 18th International Conference on Intelligent Transportation Systems, pp. 1995–2000 (2015). https://doi.org/10.1109/ITSC.2015.323
10. Dietterich, T.G.: Ensemble methods in machine learning. In: Multiple Classifier Systems, LBCS-1857, pp. 1–15. Springer, Berlin (2000)
11. Domingos, P.: A unified bias-variance decomposition. In: Proceedings of 17th International Conference on Machine Learning, pp. 231–238. Morgan Kaufmann Stanford (2000)
12. Gindele, T., Brechtel, S., Dillmann, R.: Learning context sensitive behavior models from observations for predicting traffic situations. In: 16th International IEEE Conference on Intelligent Transportation Systems (ITSC 2013), pp. 1764–1771 (2013). https://doi.org/10.1109/ITSC.2013.6728484
13. Jaccard, P.: The distribution of the flora in the alpine zone.1. New Phytol. **11**(2) (1912)
14. Jain, B.: Condorcet's jury theorem for consensus clustering. In: Trollmann, F., Turhan, A.Y. (eds.) KI 2018: Advances in Artificial Intelligence. Springer International Publishing, New York (2018)
15. Kruber, F., Wurst, J., Morales, E.S., Chakraborty, S., Botsch, M.: Unsupervised and supervised learning with the random forest algorithm for traffic scenario clustering and classification. In: 2019 IEEE Intelligent Vehicles Symposium (IV), pp. 2463–2470 (2019). https://doi.org/10.1109/IVS.2019.8813994

16. Kruber, F., Sánchez Morales, E., Chakraborty, S., Botsch, M.: Vehicle position estimation with aerial imagery from unmanned aerial vehicles. In: IEEE Intelligent Vehicles Symposium (IV) (2020)
17. Murtagh, F., Contreras, P.: Algorithms for hierarchical clustering: an overview. WIREs Data Mining and Knowledge Discovery, pp. 86–97 (2012). https://doi.org/10.1002/widm.53
18. Reichel, M., Botsch, M., Rauschecker, R., Siedersberger, K.H., Maurer, M.: Situation aspect modelling and classification using the scenario based random forest algorithm for convoy merging situations. In: 13th International IEEE Conference on Intelligent Transportation Systems, pp. 360–366 (2010)
19. Sánchez Morales, E., Kruber, F., Botsch, M., Huber, B., García Higuera, A.: Accuracy characterization of the vehicle state estimation from aerial imagery. In: IEEE Intelligent Vehicles Symposium (IV) (2020)
20. Scheirer, W.J., Jain, L.P., Boult, T.E.: Probability models for open set recognition. IEEE Trans. Pattern Anal. Mach. Intell. **36**(11), 2317–2324 (2014). https://doi.org/10.1109/TPAMI.2014.2321392
21. Shi, T., Horvath, S.: Unsupervised learning with random forest predictors. J. Comput. Graph. Stat. **15**(1) (2006). https://doi.org/10.1198/106186006X94072
22. Waissi, G.R., Rossin, D.F.: A sigmoid approximation of the standard normal integral. Appl. Math. Comput. **77**(1) (1996)
23. Xu, R., WunschII, D.: Survey of clustering algorithms. IEEE Trans. Neural Netw. **16**(3) (2005)

Development of Computer Vision Models for Drivable Region Detection in Snow Occluded Lane Lines

Parth Kadav, Sachin Sharma, Farhang Motallebi Araghi, and Zachary D. Asher

1 Introduction

Advanced Driver Assistance Systems (ADAS) have the ability to prevent or reduce around 40% of all passenger vehicle incidents [1]. Some examples of ADAS include forward collision warning (FCW), automatic emergency braking (AEB), lane departure warning (LDW), lane-keeping assistance (LKA), and blind-spot warning assistance, among others. Since human error is the leading cause of road accidents [2], ADAS was designed to automate and improve aspects of the driving experience in order to increase road safety and safe driving habits. Lane-keeping systems detect reflective lane markers in front the vehicle and warn the driver via various audible, tactile, and/or visual cues if the vehicle deviates from its lane and no turn signals or steering movements are detected [3]. LDW/LKA systems can reduce head-on and single-vehicle collisions by 53% on highways with higher speed limits (45–75 mph) with visible lane markings, according to a study of 1853 driver injury crashes [4, 5]. 11%–23% of drift-out-of-lane events and 13%–22% of critically to fatally injured drivers could have been prevented if the technology had been implemented at lower operating speeds (5–20 mph), according to [6]. FCW and AEB alone significantly halve front-to-rear crashes [7]. By 2023, it is anticipated that the market for ADAS would be worth more than \$30 billion [8] and that ADAS will not be limited to safety but will also enable improvements in vehicle efficiency [9–14].

P. Kadav (✉) · S. Sharma · F. M. Araghi · Z. D. Asher
Department of Mechanical and Aerospace Engineering, Western Michigan University, Kalamazoo, MI, USA
e-mail: parth.kadav@wmich.edu; sachin.sharma@wmich.edu; farhang.motallebiaraghi@wmich.edu; zach.asher@wmich.edu

Despite the success of ADAS technology, there remains a glaring issue: adverse weather. In the United States, weather-related crashes accounted for 21% (1,235,145) of all recorded crashes, 16% (5376) of crash fatalities, and 19% (418,005) of crash injuries between 2007 and 2016 [15]. Fundamentally, adverse weather conditions can hinder situational awareness and vehicular maneuverability in a variety of ways, depending on the type of adverse weather [15]. It is critical to recognize how various weather conditions can affect the ground transportation infrastructure. A current research problem is to develop strategies for operating ADAS in bad weather. Because there are significant safety implications, the first research gap is to recognize and classify road lanes during inclement weather in order to aid in the location of both the ego vehicle and other vehicles [16]. The difficulty is that inclement weather, such as heavy rain, snow, or fog, reduces the maximum range and signal quality of ADAS sensors, such as cameras, as it obscures the lane markings [16]. This issue has been illustrated with cameras and lidars in particular [17]. According to [4], LDW/LKA could further reduce head-on and single-vehicle collisions on roads with operating speeds of 45–75 mph by 53% only if the roads had visible road markings and "the road surface was not coated by ice or snow." The performance of new sensor technologies is improving, but not enough to address the issue of reliable ADAS operation in inclement weather [9]. To address this research gap, this study concentrates on the snow covered roads to keep the research scope reasonable.

There are only a few significant studies that address the issue of reliable ADAS operation in snowy conditions. The first study created a customized snowy weather dataset and determined the driveable region using semantic segmentation [18]. When assessed on a non-snow dataset, the model's mean Intersection over Union (mIoU) was 80%; when trained on a snowy dataset, mIoU fell to 19%. When both models were combined, mIoU was 83.3%. The model must be improved and strengthened because it analyzes the entire road rather than just the Region of Interest (ROI), which can be computationally costly. The second study used a CNN model with a predefined architecture and sensor fusion between the camera, lidar, and radar [19]. A dataset test showed an increase in driveable region detection (81.35%) and non-driveable region detection (93.85%) after combining data from several sensors. This is an improvement, but it has downsides, the most notable of which is that the method necessitates the use of more sensors, raising the cost and computational power required. Additionally, like the first study, this technique examines the full driveable zone rather than just a ROI [19]. In a third study, "You Only Look Once" (YOLO) was combined with a CNN and Federated Learning (FL) architecture to increase detection in inclement weather [20]. The Canadian Adverse Driving Conditions (CADC) dataset was used to evaluate this method. The average test accuracy of the model used in their study was 82.4%–88.1% . This model is based on the FL technique, which utilizes an edge server. The edge server transmits the initial parameters to the AVs after training a global YOLO CNN model on a publically available dataset. Following that, the AVs utilize these parameters to locally train the model on their own data. Once the local models are trained on each vehicle, they are sent back to the edge server. The training time of the FL approach is influenced by the number of AVs collecting data, the connection between the edge server and each vehicle, and the processing power of each vehicle. In addition, the vehicle has been fitted with eight cameras, resulting in an increase in price [20]. All of the above mentioned research provides strategies for enhancing the identification of objects and regions in the full driveable environment, but not necessarily the lane information. These studies are computationally and monetarily expensive and rely on several sensors. None of these studies offer precise, implementable driveable region detection for snow-covered roads using a single camera sensor in ADAS systems. Furthermore, custom data acquisition and labeling methods on a custom dataset are not included in these studies. A study addressing these difficulties and discussing unique CNN architectures to improve drivable region prediction with limited data is required.

We devised a computationally efficient, cost-effective, and high-accuracy technique for extracting driveable region information from a single camera, a ubiquitous vehicle sensor, to address the adverse weather research gap for ADAS [17, 18]. Deep Learning (DL) approaches such as Convolutional Neural Network (CNN) have been established as the dominant paradigm in modern computer vision algorithms and applications, as well as in segmentation research. CNNs are a robust method of obtaining semantic segmentation, but are generally computationally intensive when compared to classical ML models. Classic ML models are faster at real-time compute speeds, but they require feature engineering and pre-processing, and they do not serve as an end-to-end solution for identifying the drivable region in snow-covered lane lines, which we know from previous work [21]. To solve this problem, we will investigate DL techniques that need little or no feature engineering. For semantic segmentation, both supervised classical ML models and custom CNN models were created. Then, these methods for detecting tire marks in snow were compared. To broaden the scope of the research, we will build five different CNN architectures for determining the drivable region in snow-occluded lane lines using a single camera sensor.

2 Methodology

In this section, we will first discuss the methods we used to collect and prepare the data. The data that has been processed is then used to develop the classical ML models and the Deep Neural Network models

2.1 Drive Cycles

Figure 1 shows the route we chose which consisted of two-lane arterial roads in Kalamazoo that met our criteria for road characteristics. This drive cycle included of roads that are rarely cleaned following winter and are maintained at a much lower

Fig. 1 Drive cycle for data collection in Kalamazoo, MI, USA which drives from the Western Michigan University's college of engineering and applied sciences to Kalamazoo Valley Community College which totals a distance of 5.56 miles along residential roads with speed limits of 35 mph

rate than freeways and other multi-lane routes. We gathered the data during the winter of 2020. The lanes were obscured by snow and featured distinct tire track patterns, with tire tracks visible to expose the tarmac beneath . The road portion was chosen for its low traffic volume, two-lane configuration, and clearly visible lane markings during non-snowy conditions.

2.2   Equipment and Instrumentation

2.2.1   Camera Sensor

The forward-facing ZED 2 RGB camera from Steroelabs was chosen for use in this study and is shown in Fig. 2a. The ZED 2 RGB camera was chosen firstly because it is a widely available commercial computer vision system. The ZED 2 also features a 120-degree wide-angle lens for collecting images and videos. These camera parameters are beneficial as we have a lot of information to work with, and the wide angle capability of this camera allows us to have a lot of spatial information. The camera was set to capture video at 29 frames per second at a resolution of 1280×720 pixels. This resolution was chosen because it was a fair compromise between image quality and image size. The ZED 2 was connected to the vehicle's onboard computer, and data was collected. The dataset was created by

(a)   (b)

Fig. 2 (**a**) The ZED 2 camera sensor [22] and (**b**) The instrumented WMU EEAV lab research vehicles platform

segmenting and extracting frames from the recorded videos of the drive cycle. The frames from the videos show the tire tracks and features on which the model must be trained.

2.2.2 Vehicle Type

The Energy Efficient and Autonomous Vehicles (EEAV) research vehicle platform, shown in Fig. 2b, was used to collect data. This platform is a 2019 Kia Niro and includes a forward-facing RGB camera, Polysync Drivekit, Neousys in-vehicle computer, vehicle Controller Area Network (CAN) bus interface and a Mobileye camera.

2.3 Data Pipeline

2.3.1 Data Preparation

Nearly 15,000 RGB images were acquired; however, when the images were resampled from 30 to 5 Hz, the quantity was reduced. Resampling is carried out to reduce the amount of frames for labeling, which is followed by more quality control assessments (i.e., eliminating over-exposed, occluded, or poor resolution images). This resulted in a final dataset of 1500 frames. The images were separated into three batches, each with 500 images. This was done to make the next step easier, as splitting the images into batches and obtaining labels for each batch will allow for easier error correction during the labeling process.

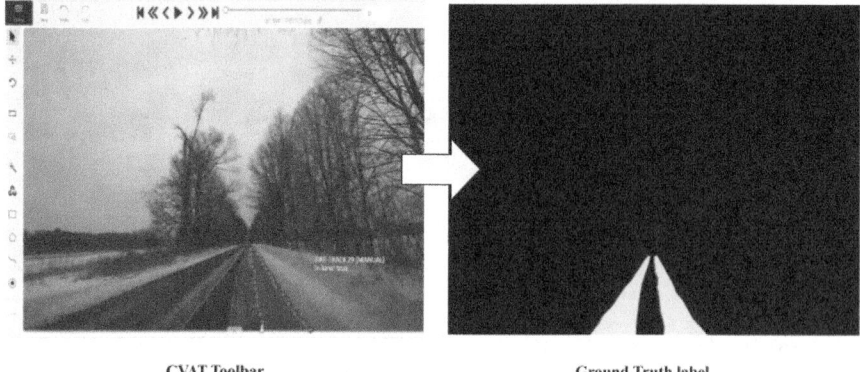

CVAT Toolbar   Ground Truth label

Fig. 3 A raw image annotated with CVAT's interface and the corresponding ground truth label, CVAT offers multiple options such as polygons, poly lines and points to create the labeled masks

2.3.2 Image Ground Truth Labeling

The frames were then labeled in batches. The tire tracks on each frame were manually annotated using the Computer Vision Annotation Tool (CVAT), an opensource web tool. Each batch's labeled dataset was exported with their matching raw images in the CVAT for images 1.1 format. The raw images and an Extensive Markup Language (XML) file including the attributes for the labels, such as the position of the tire-track with their corresponding pixel location on the image, image file name, and assigned tags, were included in each exported dataset (tire-track, road, road-edge boundary). The exported labels were then used for post-processing and inputs to model training. Figure 3 shows a camera image with a CVAT toolbar and its corresponding ground truth label after CVAT annotation.

2.3.3 Data Conditioning

To build ML models, we must first preprocess the data and then extract features. Feature extraction is the process of transforming raw data into numerical features that the model can process while retaining original data information. This is done because it generates better results than applying machine learning straight to the raw dataset [23, 24]. Deep Neural Networks can carry out some basic feature engineering on their own as it is hard-coded into their architecture so in some cases they do not require any feature engineering at all [25].

To improve feature detection and reduce computational load, images were masked with a Region of Interest (ROI) that only included the road surface. As described in [17, 18], this is a reasonable approach because there are many methods that can detect road surface regions with high precision. We built similar road

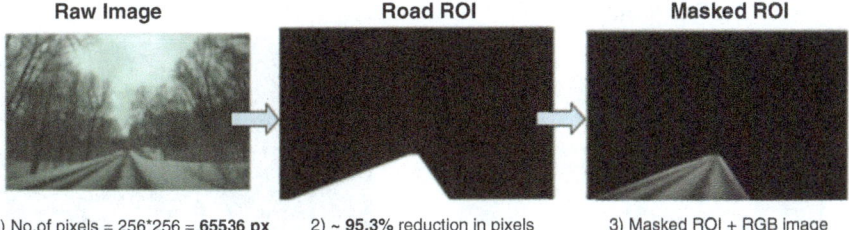

Raw Image	Road ROI	Masked ROI
1) No.of pixels = 256*256 = **65536 px**	2) ~ **95.3%** reduction in pixels	3) Masked ROI + RGB image

Fig. 4 Creating the static ROI and masking the ROI onto the raw image. The raw image includes $65,536$ pixels after resizing to 256×256, creating a static ROI which only focuses on the road removes 95.3% pixels. Finally, the raw image is masked with the road ROI to give the masked ROI

surface detection methods using a static ROI which works well for our chosen drive cycle. Figure 4 shows how to extract the ROI masked pictures.

The Road ROI is 3099 pixels in size, accounting for less than 5% of the total pixels in the raw image. Following that, the ROI mask was fused with the raw image to acquire all of the pixels contained within the ROI. This will serve as the model's input. Similar to our previous study, the different features recovered from the masked images include the red, green, blue, grayscale, and pixel X, Y values [21]. Figure 5 shows the overall process for data preparation for ML model training.

The feature vectors in Table 1 are organized into sets and selected as final inputs to the model. The results will indicate which features contribute the most to the model and perform the best. The dataset was split 55%–45% for training and testing. Input array $X = ((m \times p), n)$ was used to train the complete model where m is the number of images, p is the number of pixels in each image's ROI (3099 pixels for 256×256 images), and n is the number of feature vectors in the array.

2.4 Classical Machine Learning Models

2.4.1 Model Description

We used 6 different machine learning techniques to train the models. The first technique used is Decision Trees or **Dtrees**, which is a type of supervised machine learning technique that makes decisions and splits the dataset until all points/sets are isolated using a set of rules. The data is structured in a tree-like manner, with each dividing node representing a decision. When Dtrees is applied to our problem, it applies the rules and makes decisions based on these rules to classify pixels to be tire tracks or not tire tracks. The second technique used was Random Forest. Random forest is nothing but a number of decision trees on various subsets of the same dataset. It takes into consideration the average to improve the prediction accuracy of the dataset. The third technique used was the K-Nearest Neighbors (KNN). KNN, is based on the assumption that similar data points/classes occur

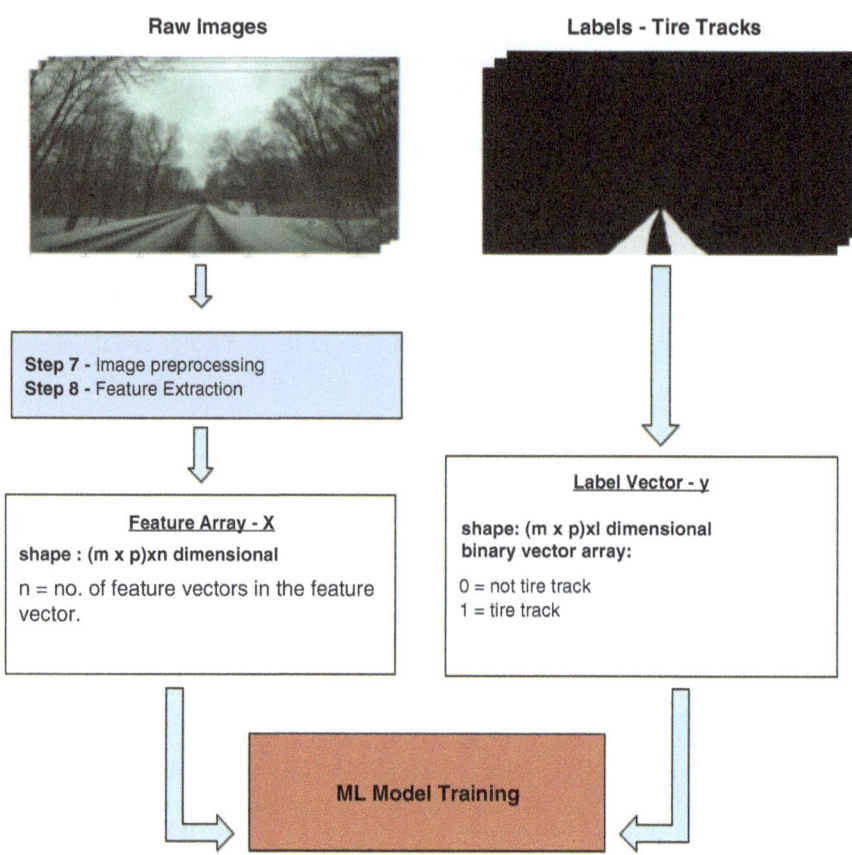

Fig. 5 This figure summarizes the image preprocessing and feature extraction from raw images. The feature array X, which contains the raw images as well as the number of feature vectors, and the label vector y, which contains a binary array with each value representing either a tire track or not a tire track, are the two inputs to the model training. This is known as the data preparation pipeline, and it will be used in the model training section

Table 1 Feature set properties

Feature set	Included feature vector	Train array shape (m = 1200)	Test array shape (m = 300)
0	Gray	(3,718,800, 1)	(929,700, 1)
1	Gray X loc, Y loc	(3,718,800, 1)	(929,700, 1)
2	Red, Green, Blue	(3,718,800, 3)	(929,700, 3)
3	Red, Green, Blue, X loc, Y loc	(3,718,800, 5)	(929,700, 5)

in close proximity. Classes with comparable properties are close to one another, which is the assumption by KNN. The user specifies the K value, where K is the desired number of nearest neighbors. We also used other techniques such as linear regression classifier, logistic regression classifier and naive bayes classifier. Both logistic regression and naive bayes are probabilistic classifiers, which means they calculate probabilities of each element in the dataset whereas linear regression predicts continuous values for the elements. These models were chosen for their characteristics and capabilities in commuting binary classification [26–28]. Other models such as support vector machines do not perform well with large datasets so they were not included.

2.4.2   Model Training

We trained a variety of machine learning models by using our input features which were defined in the data pipeline section and their associated labels. The image pre-processing and feature extraction block extracted the input feature array X and label vector y, which were then used as inputs to the machine learning model. Six distinct models, discussed in the classical ML model section were tested with each feature set (refer to Sect. 2.3.3) in order to discover the feature set/model combination that resulted in the best performance metrics.

In total we have 24 different classical ML models that can be tested. The models were trained on a desktop machine with 16 GB of RAM, an Intel i7 processor, and an Nvidia GeForce GTX 1060 graphics on Ubuntu 20.04 LTS as the operating system.

2.5   Deep Neural Network Models

A wide range of tasks, including image recognition, natural language processing, and speech recognition, have been proven to be significantly improved by deep learning approaches. When compared to classical machine learning methods, deep networks scale effectively with data, do not necessitate feature engineering, are adaptable and transferable, and perform better on larger datasets with unbalanced classes [29].

CNNs are a sort of deep neural network whose architecture is designed to do feature extraction automatically, obviating the need for this step [30]. CNNs produce feature maps by performing convolutions to the input layers, which are subsequently passed to the next layer. CNNs, unlike classical machine learning approaches, can extract relevant features from raw data, removing the need for manual image processing [31, 32]. As previously indicated, our ML models were not an end-to-end pipeline for tire track detection as they required feature engineering. In this study we look at using CNN's to simplify the process and enhance overall accuracy.

Figure 6 shows a basic convolutional neural network architecture with one convolutional layer and one max-pooling layer; we will discuss more about this

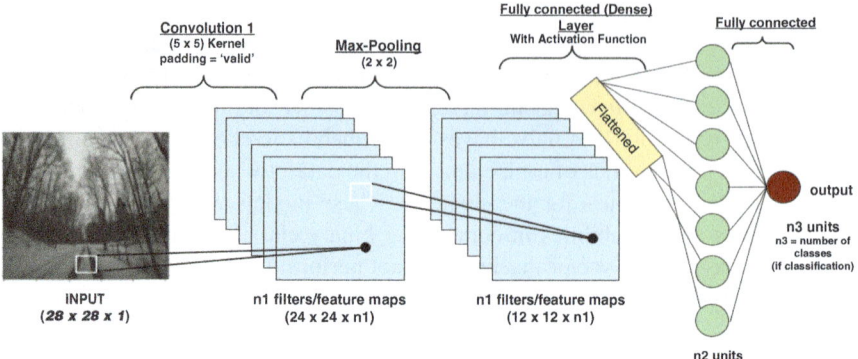

Fig. 6 An example of a simple Convolutional Neural Network. The input image goes through a convolutional layer which has a specified kernel, the convolutional operation makes a feature map which includes important feature information from the input image. The Max-Pooling operation reduces the dimensions (halves the dimensions in this case) of the feature map. The feature maps are then flattened and passed through a fully connected layer with the output neurons equalling the number of classes/desired outputs

in the coming sections. We only focus on CNNs in context of the images to keep the discussion simpler.

Before we examine the various CNN architectures, we should examine the various types of model blocks; to simplify things, we will examine model blocks that can be combined to form various models. The convolutional block consists of a convolutional layer and a pooling layer to perform feature extraction. The convolution operation with a given filter size or a kernel size slides over the input data to perform an element-wise multiplication which is essentially matrix multiplication over the 2-dimensional data, the results inside the kernel are summed up into a single output. The pooling layer down-samples the dimensions of the feature maps, which are the outputs from the convolutional layers. The fully connected block performs classification tasks based on input from previous operations [33]. Recurrent, residual, and attention operations, explained in the next section will be added to the convolutional block to make different model architectures.

2.6 Model Architectures

We have examined the fundamentals of a deep neural network in the context of images, which in our case is a convolutional neural network (CNN), as well as the numerous operations that a CNN is capable of performing. In the following subsections, a standard U-Net architecture, different convolutional model blocks such as Recurrent, Residual, and Attention, and the concept of Backbones will be discussed.

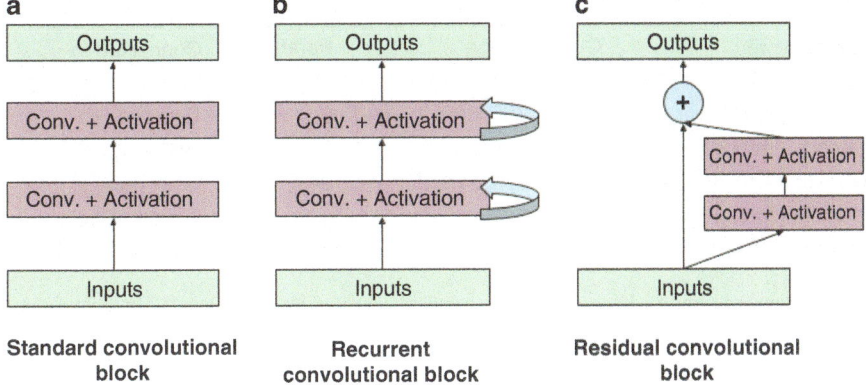

Fig. 7 (**a**) Standard convolutional block, (**b**) Recurrent convolutional block, and (**c**) Residual convolutional block

1. Standard U-Net
2. Recurrent U-Net (Rec U-Net)
3. Attention U-Net (Att U-Net)
4. Residual Attention U-Net (Res-Att U-Net)
5. Backbone U-Net

2.6.1 Standard U-Net

Figure 7a shows a standard convolutional block. The two red blocks are the convolutional layers with the respective activation function such as 'ReLu' or 'Sigmoid'. The inputs to these layers are tensors of shape $(w \times h \times c)$ where $w = width\ of\ the\ image, h = height\ of\ the\ image, c = number\ of\ channels$. The convolutional layers learn local patterns, which are patterns observed in the input windows. These windows are also known as kernels, and the patterns learned by these convolutions are transitionally invariant, which means that if the convolution learns one pattern somewhere, it may apply that knowledge in another place. This is why convolution layers outperform dense layers at recognizing image features. Figure 8 shows a sample code for a simple convolutional operation.

Now that we have introduced the concept of a standard convolutional block, we can look at the model architecture. The standard convolutional neural network provides an output based on the number of neurons in the output layer, if we want a binary output such as 0,1 or Cat and Dog, the output layer will only have one neuron which states that the output can only be either one of the classes. In our case, to have an end-to-end solution of obtaining tire tracks as the output image from the raw image input, we have to upsample/upscale the layers to have the same shape as the input layer and preserve the spatial information at the same time. To accomplish this, we look at a U-Net architecture. The U-Net architecture

```
>> model.summary()
```

Layer (type)	Output Shape	Param#	Connected to
input_1 (InputLayer)	(None, 256, 256, 3)	0	[]
lambda (Lambda)	(None, 256, 256, 3)	0	['input_1[0][0]']
conv2d (Conv2D)	(None, 256, 256, 32)	896	['lambda[0][0]']
dropout (Dropout)	(None, 256, 256, 32)	0	['conv2d[0][0]']
conv2d_1 (Conv2D)	(None, 256, 256, 32)	9248	['dropout[0][0]']

Fig. 8 Standard keras model summary for a standard convolutional block in a U-Net architecture

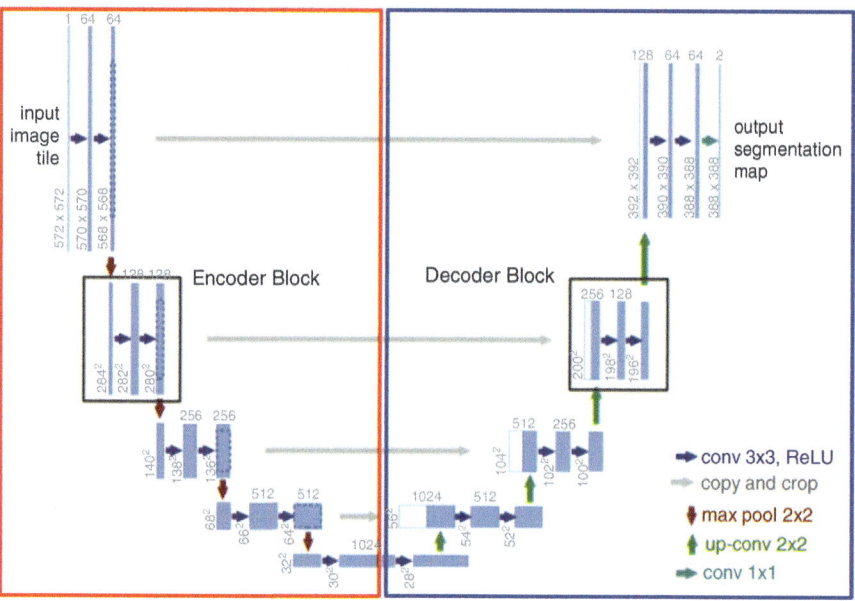

Fig. 9 Standard U-Net architecture from [35] modified to support the discussion

has been shown to perform exceptionally well in computer vision segmentation
[34]. CNN's fundamental assumption is to learn the feature mapping of an image
and then utilize that knowledge to construct more sophisticated feature maps. This
technique is effective for classification problems since it converts the image to a
vector, which is subsequently classified. However, image segmentation requires not
only the transformation of a feature map to a vector but also the reconstruction of
an image from the vector. Figure 9 shows the standard U-Net architecture. The red
box shows the encoder path and the blue box shows the decoder path.

A standard convolutional block can either serve as an encoder or a decoder. The
encoder path makes the input array smaller (also known as downsampling) with

every max-pooling operation and doubles the feature maps. Conversely, the decoder path scales the input back to its original shape with every up-convolution operation.

While converting an image to a vector, the U-Net architecture learns the image's feature maps, which are then utilized to convert it back to an image. The contracting path or the encoder path is on the left side of the U-Net architecture, while the expansive path or the decoder path is on the right. After each downsampling block, the number of feature channels/filters doubles in order to learn more intricate structures from the previous layer's output, while the image size reduces. This path is filled with numerous contraction blocks. Each block accepts the input and applies it to a 3×3 convolutional layer (where $n \times n$ is also known as the kernel, n can be any number, usually it is common to see $n = 3$ or 5) and with an activation function and padding (usually rectified linear unit or 'ReLU'). A 2×2 max-pooling layer is used for downsampling. We begin with 32 feature channels and increase them by a factor of two with each contraction block until we reach 512 feature channels, at which point we reach the expansive path. Each block in the expansive path (shown on the right) is composed of two 3×3 convolution layers and one 2×2 upsampling or up-convolution layer with an activation function and padding. The input is concatenated by appending the feature maps of the matching encoder block to the corresponding decoder block as represented by the gray arrow connecting the two layers. Each block in the expansive path reduces the number of feature channels by half. In the final layer, a 1×1 convolution layer is applied, with the number of feature maps corresponding to the number of needed classes/segments. Additionally, we add a dropout layer between each convolution layer in the encoder and decoder blocks to combat overfitting. Note the number of feature channels and input size shown in the figure are not the same for every model. Depending on the requirements such as the input shape, the kernel size, feature channels, the parameters can be modified in the architecture.

These general concepts of how a convolutional layer works and how it's used in a neural network architecture like a U-Net to achieve image segmentation are important for development of the Recurrent and Residual Deep Neural Networks discussed next. We will now discuss the various convolutional blocks and operations that will result in different model architectures.

2.6.2 Recurrent U-Net

Figure 7b. shows an example of a recurrent convolutional block; the recurrent network can store information over time by using the feedback connection represented by the arrows on the convolution layer. Even though the input is constant, the network in a recurrent convolutional layer can evolve over time. We can specify the number of iterations that the recurrent block must undergo. We simply substitute the standard convolution blocks with recurrent convolutional blocks in the encoder and the decoder path.

Figure 10 shows a sample code for a recurrent convolutional operation. If we combine the recurrent convolutional block with a standard U-Net we get a recurrent

\>> model.summary()

Layer (type)	Output Shape	Param #	Connected to
input_1 (InputLayer)	[(None, 256, 256, 3)]	0	[]
conv2d (Conv2D)	(None, 256, 256, 32)	128	['input_1[0][0]']
conv2d_1 (Conv2D)	(None, 256, 256, 32)	9248	['conv2d[0][0]']
dropout (Dropout)	(None, 256, 256, 32)	0	['conv2d_1[0][0]']
add (Add)	(None, 256, 256, 32)	0	['dropout[0][0]', 'conv2d[0][0]']
conv2d_2 (Conv2D)	(None, 256, 256, 32)	9248	['add[0][0]']
add_1 (Add)	(None, 256, 256, 32)	0	['conv2d_2[0][0]', 'conv2d[0][0]', 'conv2d_3[0][0]', 'conv2d[0][0]']
conv2d_3 (Conv2D)	(None, 256, 256, 32)	9248	['add_1[0][0]']
max_pooling2d (MaxPooling2D)	(None, 128, 128, 32)	0	['add_1[1][0]']

Fig. 10 Model summary of a recurrent convolutional operation

convolutional U-Net (RCU-Net) which is shown in Fig. 11. In Fig. 11 we can see that the recurrent convolutional layers replace the standard convolutional layers to make the RCU-Net.

The recurrent convolutional layers will look at the same features throughout the provided recurrency number, in our instance the layers will look at the same characteristics of pixels having a tire track multiple times, which will help the model reinforce when its learning process is taking place.

2.6.3 Attention U-Net (Att U-Net)

In image segmentation training, attention is used to highlight only relevant activations. This saves processing resources and improves the network's generalization power. Basically, the network may "focus" on selected areas of the image. We use Soft attention. Soft attention weighs different parts of the image. High relevance areas are given to areas of higher weight, whereas low relevance areas are given a lower weight. As the model learns, higher weighted regions get more attention [36, 37].

Figure 12 shows the overall layout of an attention gate along with the gating signal (g) and skip connection (x) Two inputs are required for the attention gate: x and g, g is the gating signal that originates at the network's sub-layer. Since g originates from a deeper layer of the network, it contains a more complete

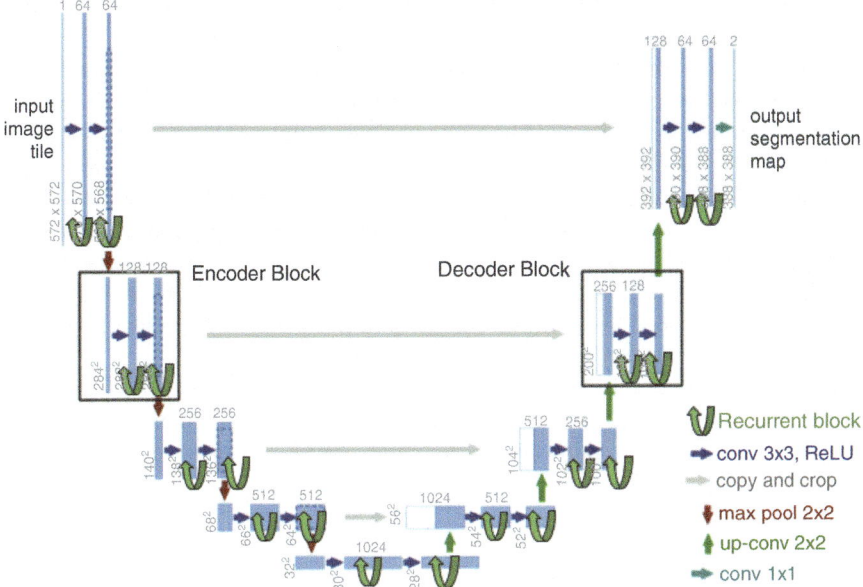

Fig. 11 Recurrent U-Net architecture obtained from modifying the standard U-Net by replacing the standard convolutional blocks with recurrent convolutional blocks, original figure modified from [35]

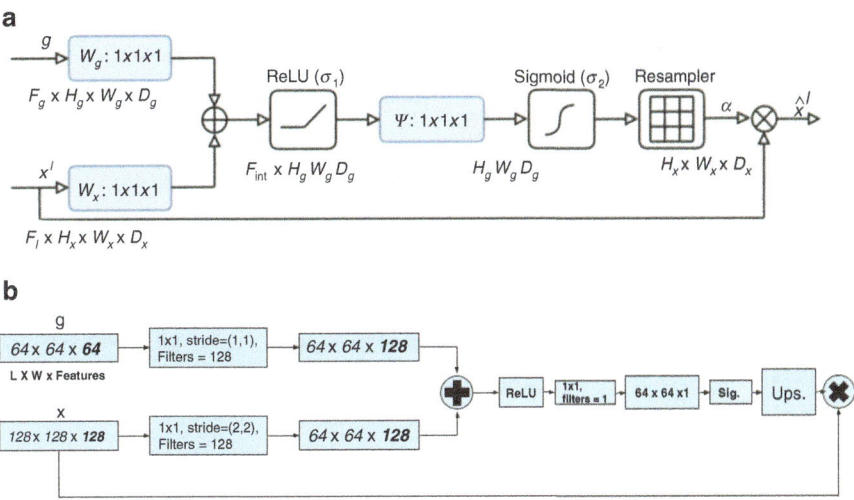

Fig. 12 (**a**) Attention gate, obtained from [37] and (**b**) Attention gate with two inputs x and g having different input dimensions

representation of features.While x originates in the early levels (concatenation of encoder blocks), and so contains more spatial information. Consider the first attention gate, which is at the topmost part of the decoder block (output layer). Input x is the encoder block's output, which is $64 \times 64 \times 64$ ($height \times width \times filters$). The output from the preceding layers (decoder block) is input g, which has dimensions of $128 \times 128 \times 128$ ($height \times width \times filters$). To make x have the same dimensions and feature numbers as g, we pass it through a convolutional layer with a stride of $(2, 2)$ and a filter count of 128, halving the dimensionalities while maintaining the same filter count for both x and g. We can perform the operations on both inputs because they have the same dimensions. The addition operation adds aligned weights and makes them larger. Upsampling is used to restore the dimensions to their original values (128×128 in this case). Finally, the output of the upsample is multiplied by the input x to perform the attention operation. Figure 12 summarizes the operation performed by the attention gate.

If we combine the attention operation with a standard U-Net we get an Attention U-Net which is shown in Fig. 13. Since we are using soft attention, the key activations would be the contrasting regions between tire tracks and the snowy road surface.

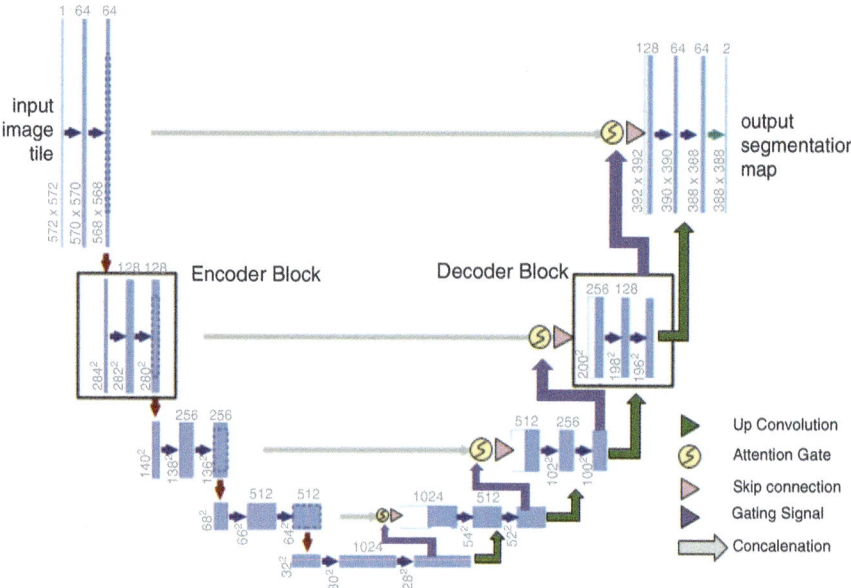

Fig. 13 Attention U-Net architecture obtained from modifying the standard U-Net by adding attention gates and skip connections to each convolutional block in the decoder path, original figure modified from [35]

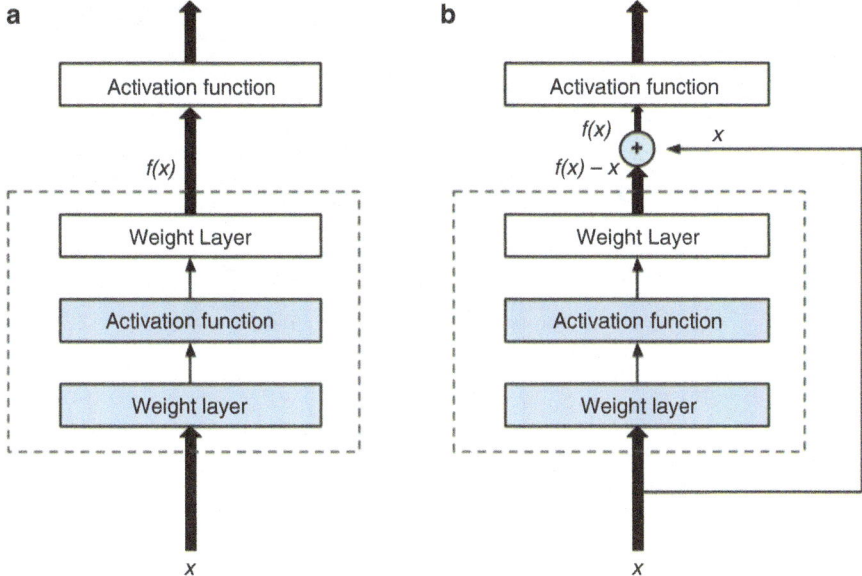

Fig. 14 (**a**) Traditional network with a single input which goes through the weight layers and specified activation functions such as 'ReLu' and (**b**) Network with residual function which uses the idea of skip connections to learn from inputs provided by previous layers

2.6.4 Residual Operation

Having more convolutional layers and making the model deeper hurts the generalization ability of the network which causes overfitting. To address this issue we use the residual operation which is shown in Fig. 7c. The residual network addresses this issue by introducing the concept of skip connections [38]. The skip connections address the vanishing gradient problem. One group of researchers [39] discusses this problem and how Residual-Net reduces the risk of overfitting and smoothens the loss surfaces [39]. Figure 14a shows the traditional feedforward network, where the block is trying to learn $f(x)$, so learning true output $f(x)$, whereas the residual block in Fig. 14b is trying to learn the residual $R(x) = f(x) - x$. The x which is being added to the residual from the input is also known as the identity. So essentially, in networks with residual blocks, each layer feeds into the next layer and directly into the layers about 2–3 hops away. Inputs can forward propagate faster through residual (shortcuts) across layers.

2.6.5 Residual + Attention U-Net (Res-Att U-Net)

Additionally, it is possible to combine two distinct blocks, such as a residual convolutional block with an attention operation. This generates a Residual Attention

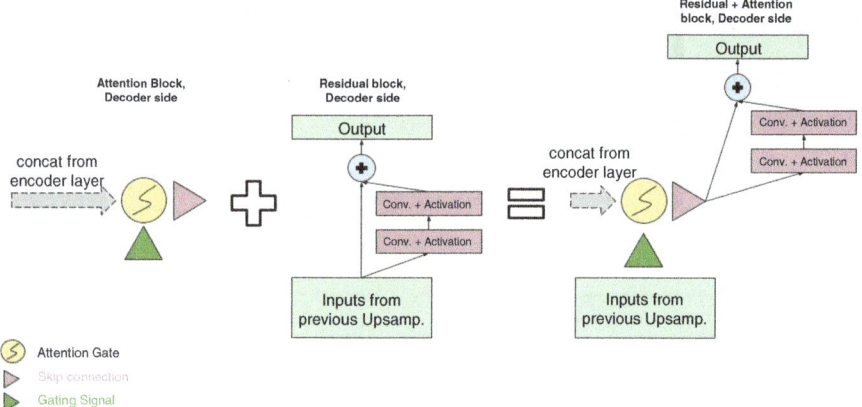

Fig. 15 We can combine two operations such as attention and residual. This results in the Residual Attention block or Res-Att. block. This block can only be used on the decoder path as it needs the spatial information from the previously concatenated layers by the use of skip connections

Convolutional Neural Network, or ResAtt-U-Net. Figure 15 illustrates the combination of the attention block and the residual convolutional block. The residual convolutional blocks can be substituted for the standard convolutional blocks on both the encoder and decoder ends of the model, whereas the attention operation can only be applied to the decoder path/blocks. And hence, the encoder path contains the residual convolutional blocks and the decoder path contains the Residual + Attention convolutional blocks.

Figure 16 shows the architecture for the ResAtt U-Net. Combining attention gates with residual convolutional blocks could increase the model's ability to detect features and reduce overfitting. This should improve the model's ability to generalize image feature recognition, in our instance tire track detection, with little overfitting.

2.6.6 Backbone U-Net

Another way of making model architectures is by using backbones. Backbones are pre-made architectures that can be used to replace the encoder path of our U-Net. A few of them are VGG, ResNet, and Inception [40]. These backbones are trained on datasets for example ImageNet [41] and we can benefit from transfer learning by using the pre-trained weights.

We used the segmentation models library that contains various Python libraries with Neural Networks for Image segmentation tasks[40]. This library consists of 4 model architectures for binary and multi-class image segmentation. Each architecture has 25 available backbones. All backbones have pre-trained weights for faster and better convergence. We used the resnet34 as our model architecture

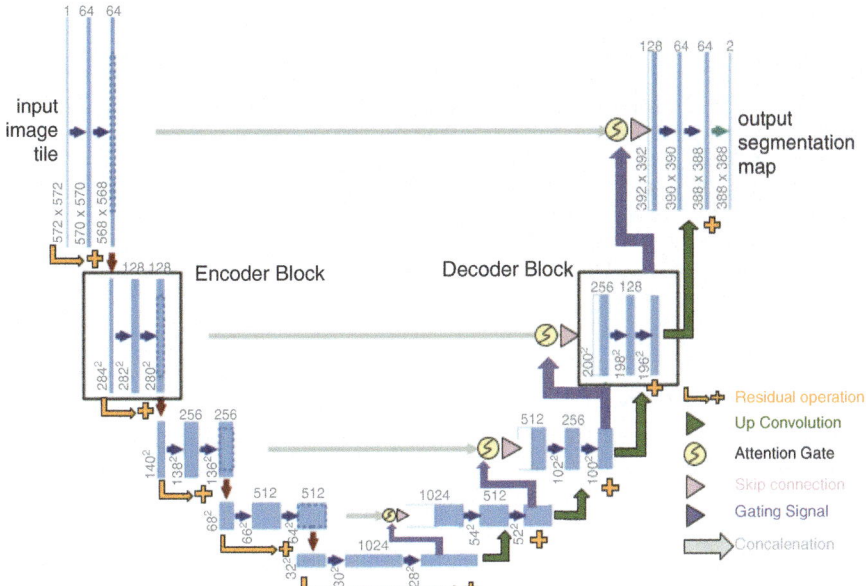

Fig. 16 Residual attention U-Net or Res-Att U-Net architecture obtained from adding residual blocks to the encoder path and Res-Att. blocks to the decoder path mentioned in Fig. 15, original figure modified from [35]

```
>> BACKBONE = 'resnet34'
>> model = sm.Unet(BACKBONE,
    classes=1,
    activation='sigmoid',
    encoder_weights ='imagenet')
```

Listing 1 Model backbone and encoder weights used from segmentation models library

and ImageNet as encoder weight. ResNet34 is a 34-layer residual network [38, 41]. ImageNet is a large dataset containing over 1000 classes, 1.28 million training images, and 50 thousand validation images. The encoder weights which are set to ImageNet are the pre-trained weights from the same network, which will make training faster. Listing 1 shows the model backbone and encoder weights used from the segmentation models library.

2.7 Model Training

The inputs to the model are an image with (width x height x channels). As we are using the raw RGB image (feature set 2, refer to data pipeline section) which has been resized to the desired size for training. In our case, the inputs are of shape (256

```
>> import tensorflow as tf
   from tensorflow import keras

>> model = tf.keras.Model(inputs=[inputs], outputs=[outputs])
>> model.compile(optimizer = 'adam',
loss = 'binary_crossentropy',
metrics = [IoU, tf.keras.metrics.Accuracy(),
       tf.keras.metrics.Recall(),
       tf.keras.metrics.Precision()])
```
Listing 2 Lines of code used for compilation of CNN models

$\times$ 256 $\times$ 3). Unlike the classical machine learning models, no feature engineering is used to train the CNN models, we can directly feed in the raw RGB image as the input to the model. We resize the images to make the training process faster and is a standard practice while training CNNs. We split the dataset into 1200 images for training and 300 images for testing. We compiled the 5 CNN models with the same optimizer, loss function and metrics. We set the optimizer to 'adam' and the loss function as 'binary cross entropy', both have been applied successfully to similar semantic segmentation tasks [42–44]. Listing 2 shows the line to compile the CNN models.

We can evaluate both the classical ML models and the different CNN models using different metrics. These metrics should serve as good evaluations to test the output of the predicted model y_{pred} with the ground truth. Intersection over union (IoU), pixel prediction accuracy, precision, recall, F1 score, and frame per second (FPS) were the evaluation metrics. These measures were evaluated based on the ability to make conclusive inferences from the performance of the model [26]. Below are the equations explaining these metrics and the four corners of a confusion matrix, which determine the true positives, true negatives, false positives, and false negatives, respectively. We only predict tire tracks, hence it's a binary classification task, hence classes $= 1$

1. True Positive (TP): no. of pixels which were a tire track and correctly identified as a tire track
2. False Positive (FP): no. of pixels which were not a tire track but identified as a tire track
3. True Negative (TN): no. of pixels which were not a tire tracks and identified as not a tire track
4. False Negative (FN): no. of pixels which were a tire track but identified as not a tire track

$$\text{Accuracy} = \frac{\text{total correct predictions}}{\text{all predictions}} = \frac{TP + TN}{TP + TN + FP + FN} \tag{1}$$

$$IoU\,(Intersection\ over\ Union) = \frac{|A \cap B|}{|A \cup B|} = \frac{|A \cap B|}{|A| + |B| - |A \cap B|} \tag{2}$$

$$mIoU = \frac{1}{n} \times \sum_{n=1}^{n} \frac{intersection}{union} = \frac{1}{n} \times \sum_{n=1}^{n} \times \frac{TP_i}{TP_i + FP_i + FN_i} \qquad (3)$$

where n is the number of classes

$$Precision = \frac{TP}{TP + FP} \qquad (4)$$

$$Recall = \frac{TP}{TP + FN} \qquad (5)$$

$$F/F1\ Score = 2 \times \frac{precision \times recall}{precision + recall} \qquad (6)$$

The accuracy Eq. 1 is the proportion of total accurate predictions made by our model over all the predictions. But accuracy alone does not tell the whole story when working with a dataset with an imbalance class distribution [45]. Accuracy is calculated over all classes. In our sample, there is a significant imbalance between the tire tracks and not tire tracks (background), therefore accuracy is not an appropriate evaluation metric. In terms of pixel-wise accuracy, this implies that the inaccuracy of minority classes is dominated by the accuracy of majority classes. IoU, also known as the Jaccard Index or the Jaccard coefficient, is significantly more indicative of success for segmentation tasks, particularly when input data is sparse and there is a high class imbalance. When training labels consist of 80 to 90% background and a small number of positive labels, a basic metric such as accuracy can acquire a high score by being dominated by the larger class. This naive problem will never arise with IoU, since IoU is unconcerned about true negatives, even with incredibly limited data. IoU calculates the overlapping region for the true and predicted labels by comparing the similarity of finite sample sets A, B as the IoU [46]. According to Eq. 7, T represents the true label image and P represents the output prediction. This is used as a measure, giving us a more precise means of quantifying IoU in the segmentation region of our model. The mIoU or mean intersection over union is nothing but the IoU computed over each class. We would only be looking at IoU because we only have one class.

$$Jaccard\ Index\ (IoU) = \frac{|T \cap P|\ (Area\ of\ Overlap)}{|T \cup P|\ (Area\ of\ Union)} \qquad (7)$$

Listing 3 shows the implementation of IoU as a metric in the model and then used to compile the model.

2.8 Results

In this section, we will set forth the results, beginning with the metrics for the different ML models and their feature sets, and then moving on to the metrics for

```
>> from tensorflow import keras
>> def IoU(y_true,y_pred):
        y_true_f = keras.backend.flatten(y_true)
        Y_pred_f = keras.backend.flatten(y_pred)
        inter = keras.backend.sum(y_true_f * y_pred_f)
        return (inter + 1.0)/(keras.backend.sum(y_true_f) +
     keras.backend.sum(y_pred_f) - inter + 1.0)
```
Listing 3 Jaccard coefficient/ Intersection over Union (IoU) as a metric

the CNN models. As described in the previous section, IoU is the relevant metric since, unlike accuracy, it provides better and complete information about the model.

2.8.1   Classical Machine Learning Models

We obtained the metrics for the 24 different model combinations, which included the 6 different ML models with 4 feature sets each. We are mainly interested in IoU scores for each model. We used the standard scaling method to plot the IoU of each model and feature set as shown in Fig. 17, where *Standard scale value* $=$ $(IoU_x - IoU_{mean})/IoU_{std.dev}$. The random forest model performed the best using feature set 1 containing grayscale pixel values and pixel X,Y locations as the feature set input. All models that use pixel locations outperform those that do not. In

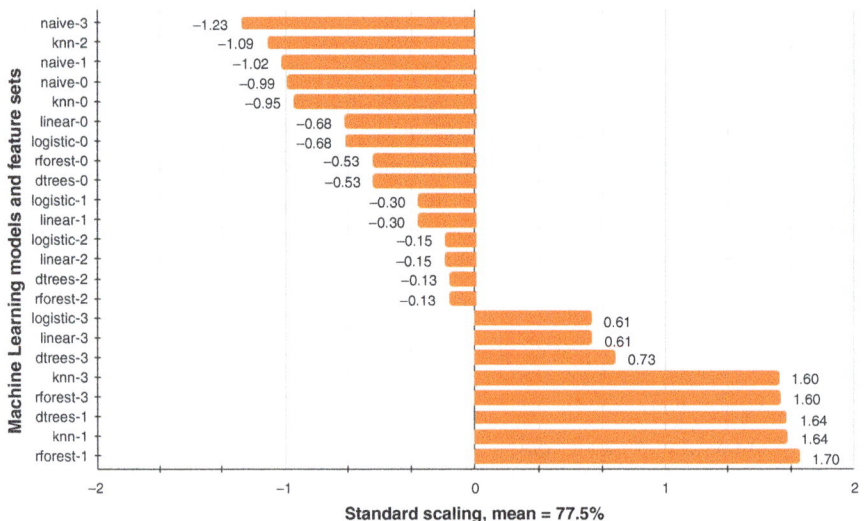

Fig. 17 Standard scaled IoU for all the classical ML models, standard scaling centers all the values around the mean with a unit standard deviation. The model/feature set combinations with positive values are good performing models, where Random forest with feature set 1 obtains the highest IoU score. This technique allows us to rule out models that perform poorly

addition, the image demonstrates that grayscale pixels provide a higher IoU than RGB pixels, as the three highest-performing models are all grayscale. Random forest seems to be the most effective method for every feature set. This is another indicator that feature engineering improves the performance of our machine learning models.

2.8.1.1 Performance Comparison Between Classical ML Models

Figure 18 shows the metrics for the best performing classical ML models. KNN with feature set 1 obtained an IoU score of 83.2%, Accuracy of 90% and an F1 score of 91.0%. Naive Bayes with feature set 0 obtained an IoU score of 74.1%, Accuracy of 82% and an F1 score of 85.1%. Random Forest with feature set 1 attained the highest IoU score at 83.4% with an Accuracy of 90% and F1 score of 91%. From an initial analysis this might indicate that Random forest with feature set 1 is the best performing model/feature-set combination. Decision trees with feature set 1 follows Random forest with an IoU score of 83.2%. Regression based classifiers such as linear regression classifier and logistic regression classifier achieved the same scores and performed well on feature set 3. Both of these models needed more feature information than the other models.

Random Forest with feature set 1 performed best in terms of key metrics like IoU, Accuracy, and F1 score, followed by Decision trees with feature set 1. As described in section 2.8, the IoU score provides a more comprehensive assessment

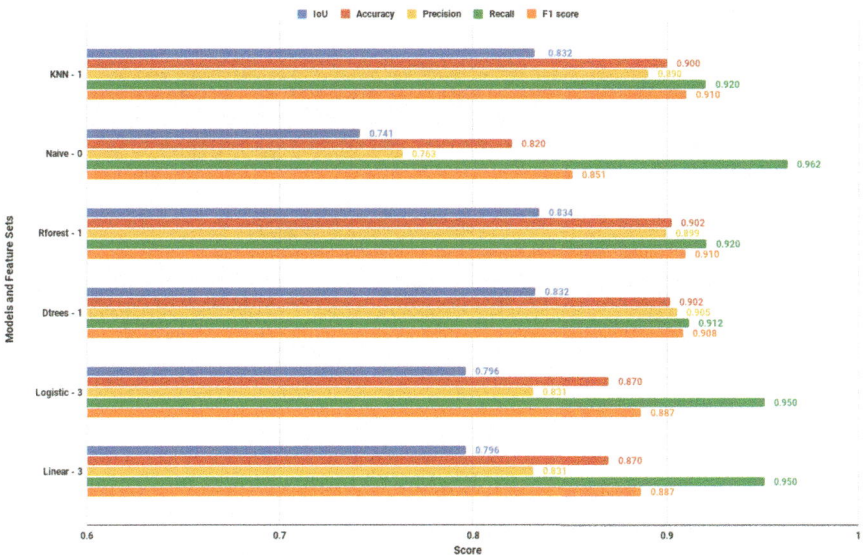

Fig. 18 Classical ML model metrics for the best performing models, where models with high IoU score, Accuracy and F1 score are of interest

of a model's performance. A high training IoU score indicates that there is a greater overlap between the predicted and ground-truth tire track pixels. Since accuracy is calculated across all classes, it does not account for the imbalance between classes and is not the metric of interest. By computing their harmonic mean, the F1 score accounts for both precision and recall. When other metrics are taken into account, random forest, decision trees, and KNN achieve a high F1 score.

2.8.1.2   Real Time Compute Speed Comparison

We may state that models like Random forest, Decision trees, and KNN, along with their provided feature sets, are suitable for our application based on the previous metrics, however real-time computation is important as well since the inability to provide outputs in time removes the approach from realistic implementation. In our case, we can use the relationship between compute speeds and feature sets to determine the best model/feature-set combination. The model with the greatest IoU score performed poorly in real-time computation at 11.3 FPS, whereas Decision Trees, which achieved an IoU of 83.2%, just 0.2% below the best model, performed at 1084 FPS. KNN, which performed well on key metrics, struggled in real-time compute performance. Based on the metrics and real-time compute speed, we can say that Decision trees with feature set 1 is a good fit for our application. The real time compute speeds for all the models is shown in Fig. 19.

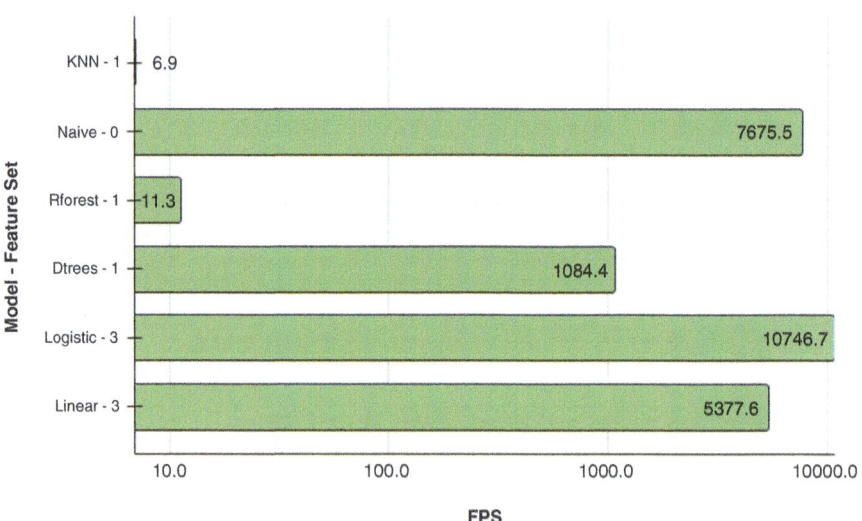

Fig. 19 Real time compute speeds in FPS for the best model/feature set combination. Low computational cost algorithms have a high FPS and high computational cost algorithms have a low FPS. More efficient models might yield faster a FPS score

Fig. 20 Qualitative prediction from our classical ML model. This was produced by overlaying the predictions from the Decision trees with feature set 1 onto the raw image

In addition to quantitative analysis, we must also consider qualitative analysis for the models. Performing both of these procedures will ensure a thorough review of the models and aid in selecting the most appropriate model for our application. Figure 20 displays a qualitative model output. The anticipated array of tire track pixels within the ROI was then overlaid on the raw image. This was derived from Decision trees with feature set 1, our most effective ML model.

2.8.2   Convolutional Neural Network Models

The CNN model's output is shown in Fig. 19; all of the models will produce an image that reflects the segmentation mask for the predicted tire track. Unlike the classical ML models, where the output is a flattened array of points which include the prediction values for each pixel in the ROI, the CNN models output a segmentation mask of the predicted tire track. Semantic segmentation means that each pixel is assigned a label based on the prediction. The output from the CNN models gives out a segmentation mask which is of the same image as the input to the model which tells us where the tire tracks lie given a new image. These prediction masks can be used to obtain pixel values in terms of labels for the image. By changing the input dimensions of the image, we can obtain a predicted segmentation mask with the same input dimensions. Figure 21a shows the raw image which is the input to the model obtained from the test set, this image was resized to the shape of 256×256 to make the prediction faster. Figure 21b shows the ground truth label

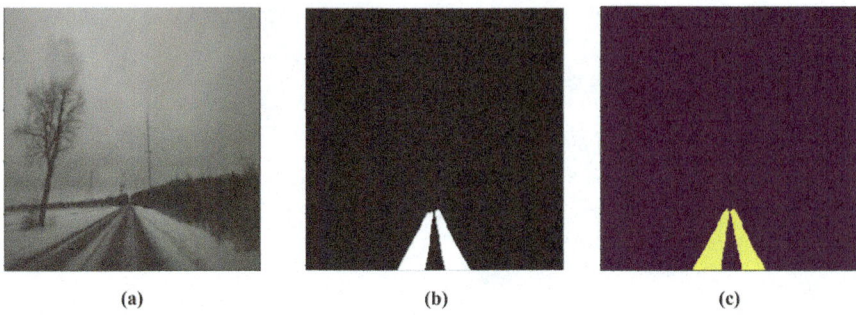

Fig. 21 (**a**) Raw image, (**b**) Ground truth label and (**c**) Predicted tire track. The CNN prediction is the image of the segmentation mask with the same size as that of the input image

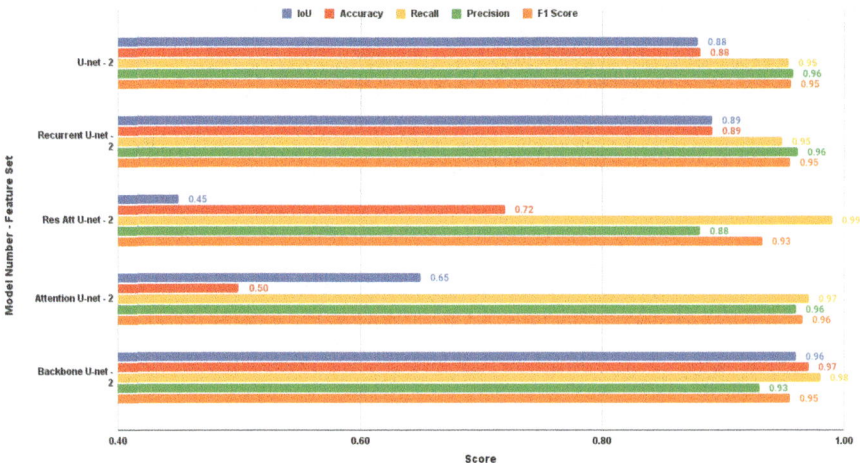

Fig. 22 CNN model metrics for the best performing models, where models with high IoU score, Accuracy and F1 score are of interest

that was annotated using CVAT and Figure 21c shows the output from the standard CNN U-Net model. The prediction resembles the ground truth label.

2.8.2.1 Performance Comparison Between CNN Models

The metrics for each CNN model are displayed in Fig. 22. All of the CNN models use feature set, as mentioned previously, CNN models do not require feature engineering, the input to the models is the raw image, which is feature set 2. The Standard U-Net model obtains an IoU score of 88%, Accuracy of 89%, and F1 score of 95%. The Recurrent U-Net model achieved an IoU score of 89%, Accuracy of 89% and F1 score of 95%. The Residual Attention U-Net and the Attention U-Net both performed poorly in terms of IoU and Accuracy. The Backbone U-Net attained the highest IoU, Accuracy and F1 score among all the other models. This might indicate that the Backbone U-Net model is the best performing CNN model. However, real time compute speeds also need to be considered as part of a qualitative analysis.

2.8.2.2   Real Time Compute Speed Comparison

Figure 23 shows the real time compute speed of the five different CNN models. The Recurrent U-Net model achieved the fastest real-time compute speeds, followed by the U-Net. Backbone U-Net, which had the best IoU score, had the slowest compute speed of 25 FPS. A qualitative investigation is required to determine which model produces good results.

The outputs from all of the CNN models on new images are shown in Fig. 24, along with the IoU score earned on each of the models during training. On the training set, all of the models perform well, but when tested on new images, the results in Fig. 24 demonstrate which model produces good results. Model 1 and 2 perform well and output diverse tire tracks as their predictions complement their IoU scores. Models 3 and 4 have poor performance. Model 5, which has the highest IoU, performs well, but it has a tendency to overfit the tire tracks by merging the space between them and does not distinguish between the left and right tracks like models 1 and 2. This could also explain why Model 5 has the highest IoU score and shows evidence of overfitting. Looking at the real time compute speeds, both Model 1 and 2 perform better then model 5. Based on the metrics and real time compute

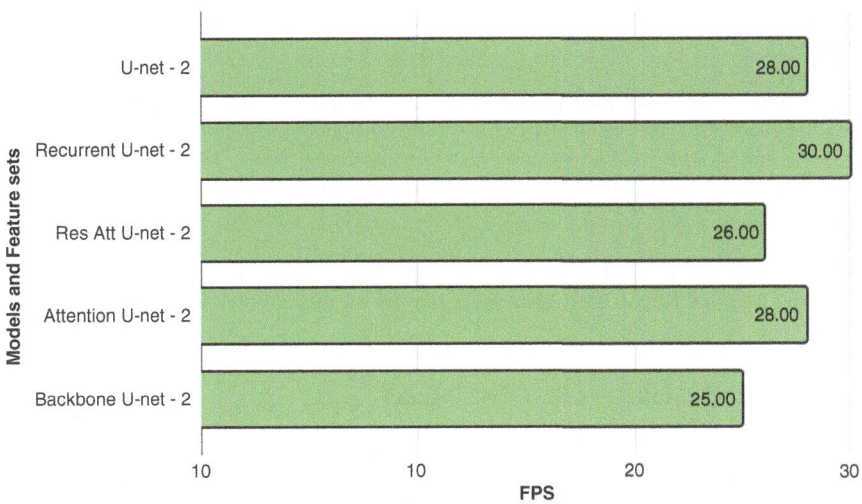

Fig. 23 Real time compute speeds in FPS for the best CNN model. Low computational cost CNN models have a high FPS and high computational cost algorithms have a low FPS. More efficient architectures might yield faster a FPS score

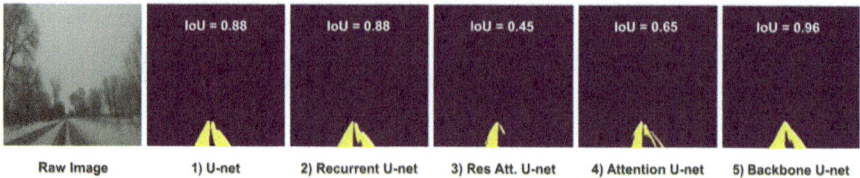

Fig. 24 Qualitative analysis of the outputs from the 5 CNN models. A high IoU score means the model performs better, which is true in case of models 1 and 2, their outputs show distinct tire tracks. The highest IoU which is attained by model 5 shows signs of over fitting as the left and the right tracks have merged into one solid body. Models 3 and 4 with low IoU scores show poor tire tracks

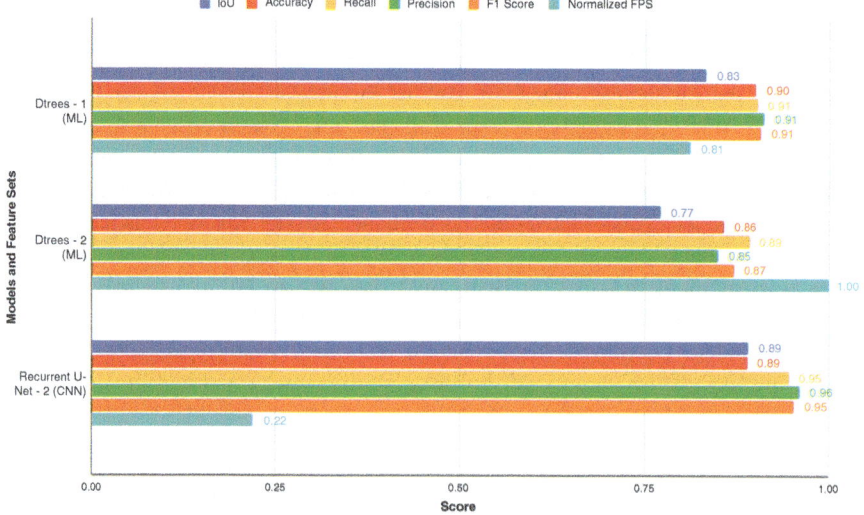

Fig. 25 Best classical ML model and best CNN model metrics comparison. The FPS values have been normalized between 0 and 1. The CNN model performs much better in terms of IoU, Accuracy and F1 score without using any kind of feature engineering. The classical ML models outperform the CNN model in real time compute speeds (FPS)

speeds shown in Fig. 22 and Fig. 23, and a qualitative analysis shown in Fig. 24, we can conclude that Recurrent U-Net is a good fit for our application.

2.8.3 Best ML Models vs Best CNN Model

When comparing the best model from the classical ML model section, Decision Trees with feature set 1, to CNN models that use feature set 2, we should also compare Decision Trees with feature set 2, which is the raw RGB image as input. We compare these to the Recurrent U-Net, which is the best performing CNN model. We look at all the key metrics and normalized real time compute values.

| Raw Image | Predicted Tire Track | Drivable Region Extraction | Lane Line Extraction |

Fig. 26 The overall process of using this system to obtain the drivable region. By implementing a few CV transformations, we can extract the drivable region from tire tracks, this can be further expanded to get lane line information

Figure 25 shows that the CNN performs much better at metrics such as IoU, Accuracy, Recall, Precision, and F1 score. To perform a fairer comparison, Decision Trees with feature set 2 and the Recurrent U-Net with feature set 2 should be compared, as both have the same feature sets. Recurrent U-Net outperforms the Decision Trees in all of the key metrics except for real time compute speeds.

2.9 Drivable Region Extraction from Tire Tracks

Once the tire tracks are identified, the drivable region can be extracted using standard computer vision transformations. Figure 26 illustrates an example of overlaying the predicted tire tracks on the raw image to generate the drivable region. Likewise, we can extract the lane lines. Our results show that using tire tracks, we have an alternate method in obtaining the drivable region unlike the predictions from the leading CV provider.

Figure 27 depicts the three cases: (a) Detections from the leading CV provider without lane line occlusion. (b) Detections from the leading CV provider with snow occlusion on lane lines and (c) Detections from our algorithm to extract the drivable lane (Fig. 26). In Fig. 27a, the leading CV provider is able to detect the lane lines, which are indicated by the two green lines that show the left lane line and right lane line while the third red line indicates the road boundary. In Fig. 27b both the left and right lane lines appear red, indicating that the system lacks confidence in detecting the lane lines. Figure 27c shows the drivable lane detection from our model.

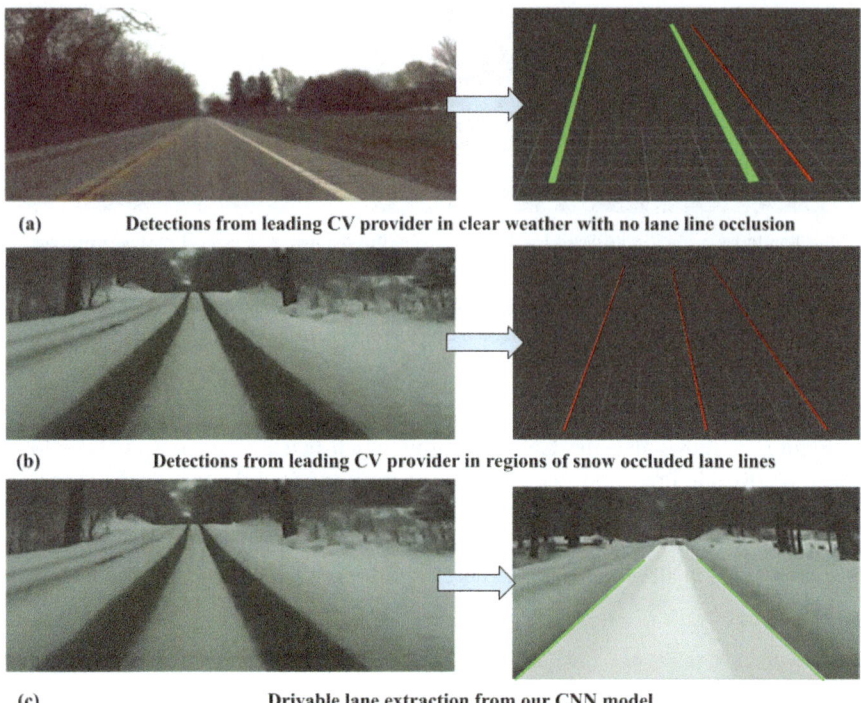

(a)   Detections from leading CV provider in clear weather with no lane line occlusion

(b)   Detections from leading CV provider in regions of snow occluded lane lines

(c)   Drivable lane extraction from our CNN model

Fig. 27 We looked at a road section from our drive cycle, where we collected camera data and detections from a leading CV provider in two conditions (1) clear weather with no lane line occlusion and (2) snowy weather with lane line occlusion. In (**a**) we can see that the leading CV provider system is able to detect lane lines with full confidence. In (**b**) the system is misidentifying lane lines and has poor confidence in detecting the drivable region whereas in (**c**) our algorithm is able to detect the drivable region using the predictions and transformations

3   Conclusion

This study investigates the research gap in driveable region detection for snow-covered roads with a single camera sensor that can be incorporated in current ADAS systems. We proposed a new method for identifying the drivable region in snowy road conditions when lane lines are occluded by focusing on tire tracks and extracting the drivable region with that information. Data was first acquired using our instrumented vehicle, and then processed by extracting frames from videos, segmenting them into batches, and labeling them with CVAT. That data was then utilized to build a CV model. We explored both classical ML approaches and Deep Neural Networks, specifically CNN, for detecting the driveable region based on tire tracks. We developed 5 different neural network architectures and compared their performance to that of classical machine learning methods. We evaluated the U-Net based CNN models for IoU, Accuracy, Recall, F1 score, and FPS using only the raw image with no image pre-processing or feature extraction. The Recurrent U-Net model had an IoU score of 89%, followed by the U-Net model which achieved 88%. The best performing ML model was Random Forest with feature set 1 with an IoU of 83.4%, however when we looked at the FPS, we chose Decision Trees with feature set 1 that had an IoU of 83.2%. We also examined F1 score, Accuracy, Recall, and Precision. The classical ML models performed much better in terms of real-time computational speeds (FPS) but at the expense of considerable pre- and post-processing processing effort as well as extensive feature engineering. The CNN models provide an end-to-end solution for detecting drivable regions in snowy road surfaces by feeding in the raw image and predicting tire tracks without any feature engineering at the cost of slower real time compute speeds. The classical ML models do not handle variation and noise as well as the CNN models do. The CNN models offer a more mature solution to identify tire tracks in regions of snow-occluded lane lines. This study demonstrates that it is possible to detect drivable regions for specific scenarios of lane line occlusion due to snow using a single camera and existing technology. By enhancing image processing and tuning the CNN hyper-parameters, the results can be further improved. Additionally, having more data would significantly improve the CNN models and offer a more flexible model. Running the CNN models on a powerful computing machine would also result in faster compute speeds and allow data scalability. Future work to expand this study includes addressing other circumstances such as traffic lights, intersections, road curvature, turns, lane changes, active snowfall, and various lighting conditions. Overall, the problem of automated driving in adverse weather needs to be addressed in order to reduce the fatalities and economic costs that occur annually.

References

1. Benson, A.J., Tefft, B.C., Svancara, A.M., Horrey, W.J.: Potential Reductions in Crashes, Injuries, and Deaths from Large-Scale Deployment of Advanced Driver Assistance Systems, pp. 1–8. Res. Brief (2018) [Online]. Available: https://trid.trb.org/view/1566022
2. Wenwen, S., Fuchuan, J., Qiang, Z., Jingjing, C.: Analysis and control of human error. Proc. Eng. **26**, 2126–2132 (2011)
3. Varghese, J.Z., Boone, R.G., et al.: Overview of autonomous vehicle sensors and systems. In: International Conference on Operations Excellence and Service Engineering, pp. 178–191 (2015)
4. Sternlund, S., Strandroth, J., Rizzi, M., Lie, A., Tingvall, C.: The effectiveness of lane departure warning systems-a reduction in real-world passenger car injury crashes. Traffic Inj. Prev. **18**, 225–229 (2017)
5. Kusano, K., Gabler, H., Gorman, T.: Fleetwide safety benefits of production forward collision and lane departure warning systems, SAE Int. J. Passeng. Cars - Mech. Syst. **7**(2), 514–527 (2014). https://doi.org/10.4271/2014-01-0166
6. Kusano, K.D., Gabler, H.C.: Comparison of expected crash and injury reduction from production forward collision and lane departure warning systems. Traffic Inj. Prev. **16**(Suppl 2), S109–14 (2015)
7. IIHS-real-world-CA-benefits.pdf, [Online]. Available: https://www.iihs.org/media/259e5bbd-f859-42a7-bd54-3888f7a2d3ef/e9boUQ/Topics/ADVANCED%20DRIVER%20ASSISTANCE/IIHS-real-world-CA-benefits.pdf

8. Advanced driver assistance systems: global revenue growth 2020-2023, Statista. https://www. statista.com/statistics/442726/global-revenue-growth-trend-of-advanced-driver-assistance-systems/ (accessed Apr. 24, 2023)
9. Jiménez, F., Naranjo, J.E., Anaya, J.J., García, F., Ponz, A., Armingol, J.M.: Advanced driver assistance system for road environments to improve safety and efficiency. Trans. Res. Proc. **14**, 2245–2254 (2016)
10. Asher, Z.D., Tunnell, J.A., Baker, D.A., Fitzgerald, R.J., Banaei-Kashani, F., Pasricha, S., Bradley, T.H.: Enabling prediction for optimal fuel economy vehicle control. Technical Report, SAE Technical Paper, 2018
11. Motallebiaraghi, F., Yao, K., Rabinowitz, A., Hoehne, C., Garikapati, V., Holden, J., Wood, E., Chen, S., Asher, Z., Bradley, T.: Mobility energy productivity evaluation of prediction-based vehicle powertrain control combined with optimal traffic management. Technical Report, 2022-01-0141, SAE Technical Paper, 2022
12. Kadav, P., Asher, Z.D.: Improving the range of electric vehicles. In: 2019 Electric Vehicles International Conference (EV), pp. 1–5 (2019)
13. Rabinowitz, A., Araghi, F.M., Gaikwad, T., Asher, Z.D., Bradley, T.H.: Development and evaluation of velocity predictive optimal energy management strategies in intelligent and connected hybrid electric vehicles. Energies **14**, 5713 (2021)
14. Mahmoud, Y.H., Brown, N.E., Motallebiaraghi, F., Koelling, M., Meyer, R., Asher, Z.D., Dontchev, A., Kolmanovsky, I.: Autonomous Eco-Driving with traffic light and lead vehicle constraints: An application of best constrained interpolation. IFAC-PapersOnLine **54**, 45–50 (2021)
15. How Do Weather Events Impact Roads? https://ops.fhwa.dot.gov/weather/q1_roadimpact.htm. Accessed 08 Oct. 2022
16. Gern, A., Moebus, R., Franke, U.: Vision-based lane recognition under adverse weather conditions using optical flow. In: Intelligent Vehicle Symposium, 2002, vol. 2, pp. 652–657. IEEE (2002)
17. Brandon, S.: Sensor fusion: a comparison of capabilities of human highly automated, [Online]. Available: http://websites.umich.edu/~umtriswt/PDF/SWT-2017-12.pdf
18. Lei, Y., Emaru, T., Ravankar, A.A., Kobayashi, Y., Wang, S.: Semantic image segmentation on snow driving scenarios. In: 2020 IEEE International Conference on Mechatronics and Automation (ICMA) (2020). https://doi.org/10.1109/icma49215.2020.9233538
19. Rawashdeh, N.A., Bos, J.P., Abu-Alrub, N.J.: Drivable path detection using CNN sensor fusion for autonomous driving in the snow. In: Autonomous Systems: Sensors, Processing, and Security for Vehicles and Infrastructure 2021, vol. 11748, pp. 36–45. SPIE (2021)
20. Rjoub, G., Wahab, O.A., Bentahar, J., Bataineh, A.S.: Improving autonomous vehicles safety in snow weather using federated YOLO CNN learning. In: Mobile Web and Intelligent Information Systems, pp. 121–134. Springer International Publishing, New York (2021)
21. Goberville, N.A., Kadav, P., Asher, Z.D.: Tire track identification: a method for drivable region detection in conditions of Snow-Occluded lane lines. Technical Report, SAE Technical Paper, 2022
22. ZED 2 - AI stereo camera: https://www.stereolabs.com/zed-2/. Accessed 19 May 2022
23. Uddin, M.F., Lee, J., Rizvi, S., Hamada, S.: Proposing enhanced feature engineering and a selection model for machine learning processes. NATO Adv. Sci. Inst. Ser. E Appl. Sci. **8**, 646 (2018)
24. Duboue, P.: The Art of Feature Engineering: Essentials for Machine Learning. Cambridge University Press, Cambridge (2020)
25. Puget, J.-F.: Feature engineering for deep learning (2017). https://medium.com/inside-machine-learning/feature-engineering-for-deep-learning-2b1fc7605ace, Accessed 19 May 2022
26. Shetty, S.H., Shetty, S., Singh, C., Rao, A.: supervised machine learning: algorithms and applications, fundamentals and methods of machine and deep learning. Wiley, pp. 1–16 (2022). https://doi.org/10.1002/9781119821908.ch1

27. Chourasiya, S., Jain, S.: A study review on supervised machine learning algorithms. International Journal of Computer Science and Engineering. **6**(8), 16–20 (2019). https://doi.org/10. 14445/23488387/ijcse-v6i8p104
28. Osisanwo, F.Y., Akinsola, J.E.T., Awodele, O., Hinmikaiye, J.O., Olakanmi, O., Akinjobi, J.: Supervised machine learning algorithms: classification and comparison. International Journal of Computer Trends and Technology (IJCTT). **48**(3), 128–138 (2017)
29. Seif, G.: Deep learning vs classical machine learning (2018). https://towardsdatascience.com/ deep-learning-vs-classical-machine-learning-9a42c6d48aa, Accessed 19 May 2022
30. Long, J., Shelhamer, E., Darrell, T.: Fully convolutional networks for semantic segmentation. In: 2015 IEEE Conference on Computer Vision and Pattern Recognition (CVPR). IEEE (2015)
31. Albawi, S., Mohammed, T.A., Al-Zawi, S.: Understanding of a convolutional neural network. In: 2017 International Conference on Engineering and Technology (ICET), pp. 1–6 (2017)
32. Why convolutional neural networks are the go-to models in deep learning, Analytics India Magazine, (2018). https://analyticsindiamag.com/why-convolutional-neural-networks-are-the-go-to-models-in-deep-learning/. Accessed 13 Feb 2022
33. Chatterjee, H.S.: A basic introduction to convolutional neural network (2019). https://medium. com/@himadrisankarchatterjee/a-basic-introduction-to-convolutional-neural-network-8e39019b27c4, Accessed 19 May 2022
34. Sankesara, H.: UNet (2019). https://towardsdatascience.com/u-net-b229b32b4a71, Accessed 19 May 2022
35. Ronneberger, O., Fischer, P., Brox, T.: U-net: Convolutional networks for biomedical image segmentation. In: MICCAI (2015)
36. Xu, K., Ba, J., Kiros, R., Cho, K., Courville, A., Salakhudinov, R., Zemel, R., Bengio, Y.: Show, attend and tell: neural image caption generation with visual attention. In: Bach, F., Blei, D. (eds.) Proceedings of the 32nd International Conference on Machine Learning. Proceedings of Machine Learning Research, (Lille, France), vol. 37, pp. 2048–2057. PMLR (2015)
37. Oktay, O., et al.: Attention U-net: learning where to look for the. Pancreas. **arXiv [cs.CV]** (2018). https://doi.org/10.7937/K9/TCIA.2016.tNB1kqBU
38. He, K., Zhang, X., Ren, S., Sun, J.: Deep Residual Learning for Image Recognition, arXiv [cs.CV], (2015). [Online]. Available: http://arxiv.org/abs/1512.03385
39. Li, H., Xu, Z., Taylor, G., Studer, C., Goldstein, T.: Visualizing the Loss Landscape of Neural Nets. **arXiv [cs.LG]** (2017) [Online]. Available: http://arxiv.org/abs/1712.09913
40. Iakubovskii, P.: segmentation_models: Segmentation models with pretrained backbones. Keras and TensorFlow Keras. Github. [Online]. Available: https://github.com/qubvel/segmentation_models,. Accessed 06 May 2022
41. Wikipedia Contributors: ImageNet (2022). https://en.wikipedia.org/w/index.php?title= ImageNet&oldid=1083632180
42. Kingma D.P., Ba, J. A.: A method for stochastic optimization, arXiv [cs.LG], (2014). [Online]. Available: http://arxiv.org/abs/1412.6980
43. Godoy, D.: Understanding binary cross-entropy/log loss: a visual explanation (2018). https:// towardsdatascience.com/understanding-binary-cross-entropy-log-loss-a-visual-explanation-a3ac6025181a, Accessed 12 Feb 2022
44. Yaqub, M., Jinchao, F., Zia, M.S., Arshid, K., Jia, K., Rehman, Z.U., Mehmood, A.: State-of-the-Art CNN optimizer for brain tumor segmentation in magnetic resonance images. Brain Sci. **10** (2020)
45. Classification: Accuracy, Google Developers. https://developers.google.com/machine-learning/crash-course/classification/accuracy, Accessed 24 Apr 2023
46. Duque-Arias, D., Velasco-Forero, S., Deschaud, J.-E., Goulette, F., Serna, A., Decencière, E., Marcotegui, B.: On power jaccard losses for semantic segmentation. In: Proceedings of the 16th International Joint Conference on Computer Vision, Imaging and Computer Graphics Theory and Applications. SCITEPRESS - Science and Technology Publications (2021)

Machine Learning Based Perception Architecture Design for Semi-autonomous Vehicles

Joydeep Dey and Sudeep Pasricha

1 Introduction

In 2021, it was reported that an estimated 31,730 people died in motor vehicle traffic crashed in the United States, representing an estimated increase of about 12 percent compared to 2020 [1]. By eliminating the possibility of human driving errors through automation, advanced driver assistance systems (ADAS) are becoming a critical component in modern vehicles, to help save lives, improve fuel efficiency, and enhance driving comfort. ADAS systems typically involve a 4-stage pipeline involving sequential execution of functions related to *perception*, decision, control, and actuation. An incorrect understanding of the environment by the perception system can make the entire system prone to erroneous decision making, which can result in accidents due to imprecise real-time control and actuation. This motivates the need for a reliable *perception architecture* that can mitigate errors at the source of the pipeline and improve safety in emerging semi-autonomous vehicles.

The standard SAE-J3016 effectively classifies the capabilities of a perception architecture supported by a vehicle according to their targeted level of autonomy. In general, an optimal vehicle perception architecture should consist of carefully defined location and orientation of each sensor selected from a heterogeneous suite of sensors (e.g., cameras, radars) to maximize environmental coverage in the combined field of view obtained from the sensors. In addition to ensuring accurate sensing via appropriate sensor placement, a high object detection rate and low false

J. Dey (✉)
Department of Electrical and Computer Engineering, Colorado State University, Fort Collins, CO, USA
e-mail: Joydeep.Dey@colostate.edu

S. Pasricha
Colorado State University, Fort Collins, CO, USA
e-mail: sudeep@colostate.edu

© The Author(s), under exclusive license to Springer Nature Switzerland AG 2023   625
V. K. Kukkala, S. Pasricha (eds.), *Machine Learning and Optimization Techniques for Automotive Cyber-Physical Systems*, https://doi.org/10.1007/978-3-031-28016-0_22

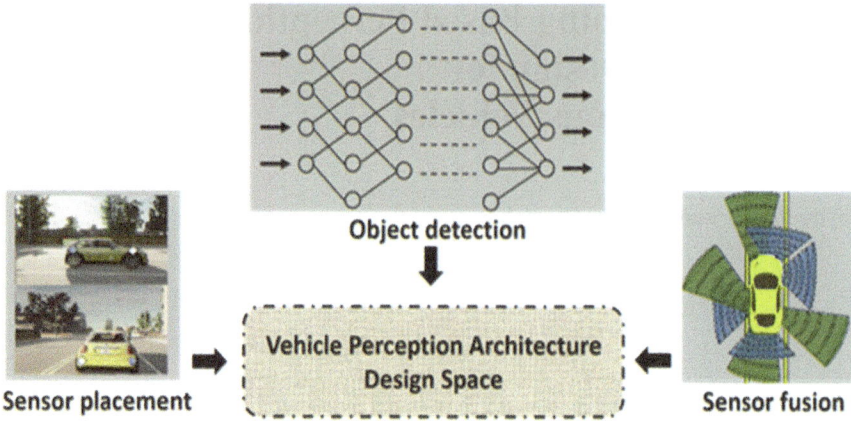

Fig. 1 Breakdown of perception architecture design space

positive detection rate needs to be maintained using efficient deep learning-based object detection and sensor fusion techniques.

State-of-the-art deep learning based object detection models are built with different network architectures, uncertainty modeling approaches, and test datasets over a wide range of evaluation metrics [2]. Object detectors that are capable of real time perception are resource-constrained by latency requirements, onboard memory capacity and computationally complexity. Optimizations performed to meet any one of these constraints often results in a trade-off with the performance of others [3]. As a result, comparison and selection from among the best set of deep learning based object detectors for perception applications remains a challenge.

In real-world driving scenarios, the position of obstacles and traffic are highly dynamic, so after detection of an object, tracking is necessary to predict its new position. Due to noise from various sources there is an inherent uncertainty associated with the measured position and velocity. This uncertainty is minimized by using sensor fusion algorithms [4]. An important challenge with sensor fusion algorithms is that the complexity of tracking objects increases as the objects get closer, due to a much lower margin for error (uncertainty) in the vicinity of the vehicle.

As summarized in Fig. 1, the design space of a vehicular perception architecture involves determining appropriate sensor selection and placement, object detection algorithms, and sensor fusion techniques. The possible configurations for each of these decisions is non-trivial and can easily lead to a combinatorial explosion of the design space, making exhaustive exploration impractical. Conversely, an optimization of each of these decisions individually before composing a final solution can lead to solutions that are sub-optimal and perform poorly in real environments. Perception architecture design depends heavily on the target features and use cases to be supported in the vehicle, making the already massive design space addressing the problem even larger and harder to traverse. Consequently, today there are no generalized rules for the synthesis of perception architectures for vehicles.

In this chapter, we describe a framework called PASTA (*P*erception *A*rchitecture *S*earch *T*echnique for *A*DAS), first introduced in [37], to perform perception architecture synthesis for emerging semi-autonomous vehicles. Our experimental results indicate that the proposed framework is able to optimize perception performance across multiple ADAS metrics, for different vehicle types.

The main contributions in this chapter include:

- A global co-optimization framework capable of synthesizing robust vehicle-specific perception architecture solutions that include heterogeneous sensor placement, deep learning based object detector design, and sensor fusion algorithm selection;
- An exploration of various design space search algorithms tuned for the vehicle perception architecture search problem;
- A fast and efficient method for co-exploration of the deep learning object detector hyperparameters, through adaptive and iterative environment- and vehicle-specific transfer learning;
- A comparative analysis of the framework efficiency across different vehicle models (Audi TT, BMW Minicooper).

2   Related Work

State-of-the-art semi-autonomous vehicles require robust perception of their environment, for which the choice of sensor placement, object detection algorithms, and sensor fusion techniques are the most important decisions. These decisions are carefully curated to support ADAS features (e.g., blindspot warning, lane keep assist) that characterize the autonomy level to be supported by a vehicle under design.

Many prior works have explored vehicle perception system design with different combinations of sensor types to overcome limitations that plague individual sensor types. The work in [5] used a single camera-radar pair for perception of headway distance using a Continental radar mounted on the geometric center of the front bumper and a Nextbase 512G monocular camera behind the windscreen. Vehicle detection was performed on the collected camera frames, by sorting potential candidates in a fixed trapezoidal region of interest in the horizontal plane. In [5] a camera-radar fusion based perception architecture was proposed for target acquisition with the well-known SSD (Single Shot Detection) object detector on consecutive camera frames. This allowed their perception system to differentiate vehicles from pedestrians in real time. The detection accuracy was optimized with the use of a Kalman filter and Bayesian estimation, which reduced computational complexity compared to [5]. In [6] a single neural network was used for fusion of all camera and radar detections. The proposed neural fusion model (CRF-Net) used an optimized training strategy similar to the 'Dropout' technique, where all input neurons for the camera data are simultaneously deactivated in random training steps, forcing the network to rely more on the radar data. The training focus towards radar overcame the bias introduced by starting with pre-trained weights from the feature extractor that was trained from the camera data. The work in [7] optimized merging camera detection with LiDAR processing. An efficient clustering technique inspired by the DBSCAN algorithm allowed for a better exploitation of features from the raw LiDAR point cloud. A fusion scheme was then used to sequentially merge the 2D detections made by a YOLOv3 object detector using cylindrical projection with the detections made from clustered LiDAR point cloud data. In [8], an approach to fuse LiDAR and stereo camera data was proposed, with a post-processing method for accurate depth estimation based on a patch-wise depth correction approach. In contrast to the cylindrical projection of 2D detections in [7], the work in [8] uses a projection of 3D LiDAR points into the camera image frame instead, which upsamples the projection image, creating a more dense depth map.

All of the prior works discussed above optimize vehicle perception performance for rigid combinations of sensors and object detectors, without any design space exploration. Only a few prior works have (partially) explored the design space of sensors and object detectors for vehicle perception. An approach for optimal positioning and calibration of a three LiDAR system was proposed in [9]. The approach used a neural network to learn and qualify the effectiveness of different LiDAR location and orientations. The work in [10] proposed a sensor selection and exploration approach based on factor graphs during multi-sensor fusion. The work in [11] heuristically explored a subset of backbone networks in the Faster R-CNN object detector for perception systems in vehicles. The work in [12] presented a framework that used a genetic algorithm to optimize sensor orientations and placements in vehicles.

The optimized perception techniques discussed in [5–12] provide highly accurate detections which enable design of efficient energy management strategies for ADAS. The work in [13] derives a prediction mechanism for optimal energy management for ADAS using a nonlinear autoregressive artificial neural network (NARX). Multiple sources are used as input to the neural network such as data from drive cycle information, current vehicle state, global positioning system, travel time data and detected obstacles. In addition, dynamic programming is used to derive an optimal energy management control strategy which shows significant fuel economy improvements compared to highly accurate predictive baseline models. The work in [14] proposes a predictive optimal energy management strategy that leverages sensor data aggregation and dynamic programming to achieve vehicle fuel economy improvement for ADAS compared to existing vehicle control strategies. The work discussed in [13, 14] leverage existing ADAS technology in modern vehicles to realize prediction based optimal energy management, which enables fuel economy improvements for ADAS with minor modifications.

Unlike prior works that fine-tune specific perception architectures, e.g., [5–8], or explore the sensing and object detector configurations separately, e.g [9–12]., this chapter proposes a holistic framework that jointly co-optimizes heterogeneous sensor placement, object detection algorithms, and sensor fusion techniques. To the best of our knowledge, this is the first effort that performs co-optimization across such a comprehensive decision space to optimize ADAS perception, with the ability to be tuned and deployed across multiple vehicle types.

3 Background

3.1 ADAS Level 2 Autonomy Features

In this chapter, our exploration of perception architectures on a vehicle, henceforth referred to as an *ego vehicle*, targets four ADAS features that have varying degrees of longitudinal (i.e., in the same lane as the ego vehicle) and lateral (i.e., in neighboring lanes to the ego vehicle lane) sensing requirements. The SAE-J3016 standard [15] defines adaptive cruise control (ACC) and lane keep assist (LKA) individually as level 1 features, as they only perform the dynamic driving task in either the latitudinal or longitudinal direction of the vehicle. Forward collision warning (FCW) and blindspot warning (BW) are defined in SAE-J3016 as level 0 active safety systems, as they only enhance the performance of the driver without performing any portion of the dynamic driving task. However, when all four features are combined, the system can be described as a level 2 autonomy system. Figure 2 shows an overview of the four features we focus on for level 2 autonomy, which are discussed next.

While modern ACC systems differ in their implementation and perception architectures, they take perform longitudinal control operations instead of the driver. The challenge in ACC is to maintain an accurate track of the lead vehicle (immediately ahead of the ego vehicle in the same lane) with a forward facing sensor and using longitudinal control to maintain the specified distance while maintaining driver comfort (e.g., avoiding sudden velocity changes). **LKA** (lane keep assist) systems determine whether the ego vehicle is drifting towards any lane boundaries and are an evolution of lane departure warning systems. LKA systems have been known to over-compensate, creating a "ping-pong" effect where the vehicle oscillates back and forth between the lane lines [16]. The main challenges in LKA are to reduce this ping-pong effect and the accurate detection of lane lines on obscured (e.g., snow covered) roads. **FCW** (forward collision warning) systems are used for real-time prediction of collisions with a lead vehicle. A critical requirement for FCW systems is that they avoid false positives and false negatives to improve driver comfort, safety and reduce rear end accidents [17]. Lastly, **BW** (blindspot warning) systems use lateral sensor data to determine whether there is a vehicle towards the rear on either side of the ego vehicle (Fig. 2) in a location the driver cannot see with their side mirrors. *A perception architecture designed to support Level 2 autonomy in a vehicle should support all four of these critical features.*

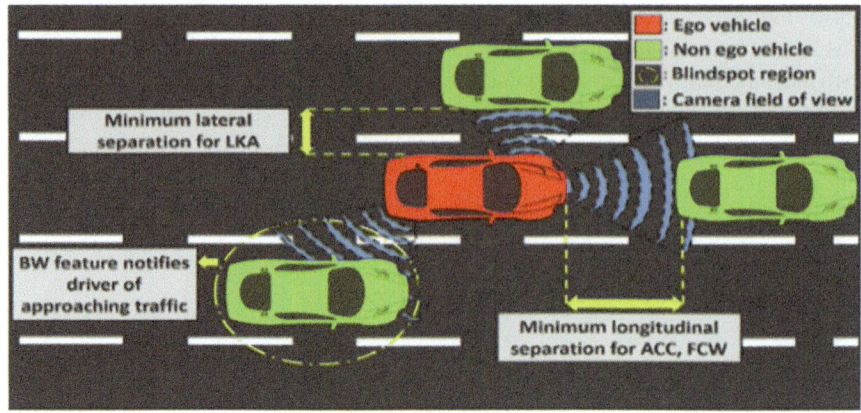

(a)

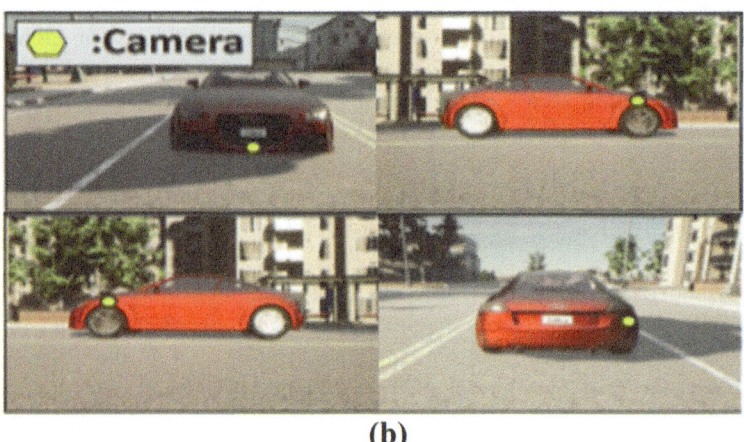

(b)

Fig. 2 Visualization of common scenarios in ACC, FCW, LKA, and BW

3.2   Sensor Placement and Orientation

In order to capture data most relevant to each feature, a strategic sensor placement strategy must be used on the ego vehicle such that the chosen position and orientation of selected sensors maximize coverage (of the vehicle environment). Figure 2 visualizes an example of field of view coverage (in blue) corresponding to three unique placements of camera sensors on the body of the ego vehicle (in yellow, lower images) to meet coverage goals. For the ACC and FCW features, the ego vehicle is responsible for slowing down to maintain a minimum separation between the ego and lead vehicle. The camera must be positioned somewhere on the front bumper to measure minimum longitudinal separation accurately while keeping the lead vehicle in the desired field of view. For LKA, there is a need to maintain a safe minimum lateral distance between non-ego vehicles in neighboring lanes. Here a front camera is needed to extract lane line information, while side cameras are required for tracking this minimum lateral separation. As BW requires information about a specific area near the rear of the vehicle, it is a challenge to find an optimal sensor placement that maximizes the view of the blind spot. If the sensor is too far forward or too far back, it will miss key portions of the blind spots. Beyond placement, the orientation of sensors can also significantly impact coverage for all features [17]. Thus sensor placement and orientation remains a challenging problem.

3.3 Object Detection for Vehicle Environment Perception

The two broad goals associated with deep learning based object detection are: determining spatial information (relative position of an object in the image) via *localization* followed by identifying which category that object instance belongs to via *classification* [18]. As an example, Fig. 3 shows object detection of multiple car instances (using the YOLOv3 deep learning based object detector [19]) by creating a bounding box around the 'car' object instances and predicting the object class as 'car'. The pipeline of traditional object detection models can be divided into informative region selection, feature extraction, and classification [20]. Depending on which subset of these steps are used to process an input image frame, object detectors are classified as single-stage or two-stage.

Modern single-stage detectors are typically composed of a feed-forward fully convolutional network that outputs object classification probabilities and box offsets (w.r.t. pre-defined anchor/bounding boxes) at each spatial position. The YOLO family of object detectors is a popular example of single-stage detectors [17]. SSD (single shot detection) is another example, based on the VGG-16 backbone [21]. An advantageous property of single-stage detectors is their very high detection throughput (e.g., ~40 frames per second with YOLO) that makes them suitable for real time scenarios. Two-stage detectors divide the detection process into separate region proposal and classification stages. The first stage involves identification of

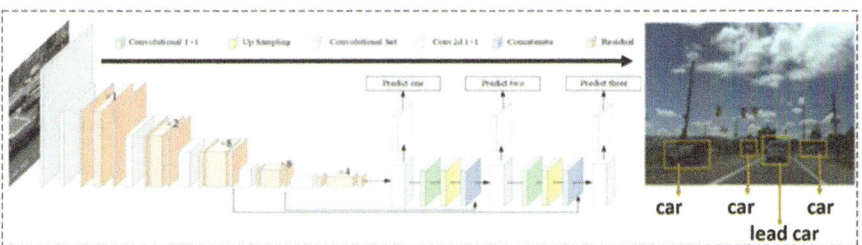

Fig. 3 Example of vehicle (object) detection with YOLOv3

several regions in an image that have a high probability to contain an object using a region proposal network (RPN). In the second stage, proposals of identified regions are fed into convolutional networks for classification. Region-based CNN (R-CNN) is an example of a two-stage detector [22]. R-CNN divides an input image into 2000 regions generated through a selective search algorithm, after which the selected regions are fed to a CNN for feature extraction followed by a Support Vector Machine (SVM) for classification. Fast R-CNN [23] and subsequently Faster R-CNN [24] improved the speed of training as well as detection accuracy compared to R-CNN by streamlining the stages.

Two-stage detectors have high localization and object recognition accuracy, whereas one-stage detectors achieve higher inference speed [25]. *In this chapter, we considered both types of object detectors to exploit the latency/accuracy tradeoffs during perception architecture synthesis.*

3.4 Sensor Fusion

Perception architectures that use multiple sensors in their sensing framework often must deal with errors due to imprecise measurements from one or more of the sensors. Conversely, errors can also arise when only a single sensor is used due to measurement uncertainties from insufficient spatial *(occlusion)* or temporal *(delayed sensor response time)* coverage of the environment. The Kalman filter is one of the most widely used sensor fusion state estimation algorithms that enables error-resilient tracking of targets [26]. The Kalman filter family is a set of recursive mathematical equations that provides an efficient computational solution of the least-squares method for estimation. The filters in this family have the ability to obtain optimal statistical estimations when the system state is described as a linear model and the error can be modeled as Gaussian noise. If the system state is represented as a nonlinear dynamic model as opposed to a linear model, a modified version of the Kalman filter known as the Extended Kalman Filter (EKF) can be used, which provides an optimal approach for implementing nonlinear recursive filters [27]. However, for real time ADAS operations the computation of the Jacobian (matrix describing the system state) in EKF can be computationally expensive and contribute to measurement latency. Further, any attempts to reduce the cost through techniques like linearization makes the performance unstable [28]. The unscented Kalman filter (UKF) is another alternative that has the desirable property of being more amenable to parallel implementation [29]. *In our design space exploration of perception architecture, we explore the family of Kalman filters as candidates for sensor fusion.*

4 PASTA Architecture

4.1 Overview

Figure 4 presents a high-level overview of our proposed *PASTA* framework. The heterogeneous sensors, object detection model library, sensor fusion algorithm library, and physical dimensions of the vehicle model are inputs to the framework. An algorithmic design space exploration is used to generate a perception architecture solution which is subsequently evaluated based on a cumulative score from performance metrics relevant to the ADAS autonomy level being targeted. As part of the framework, we evaluate the search efficacy of three design space search exploration algorithms: genetic algorithm (GA), differential evolution (DE), and the firefly algorthm (FA). The process of perception architecture generation and evaluation iterates until an algorithm-specific stopping criteria is met, at which point the best design points are output. The following subsections describe each component of our framework in detail.

4.2 Problem Formulation and Metrics

In our framework, for a given vehicle, a *design point* is defined as a perception architecture that is a combination of three components: a *sensor configuration* which involves the fixed deployment position and orientation of each sensor selected for the vehicle, an *object detector algorithm*, and a *sensor fusion algorithm*. The goal is to find an optimal design point for the given vehicle that minimizes the cumulative

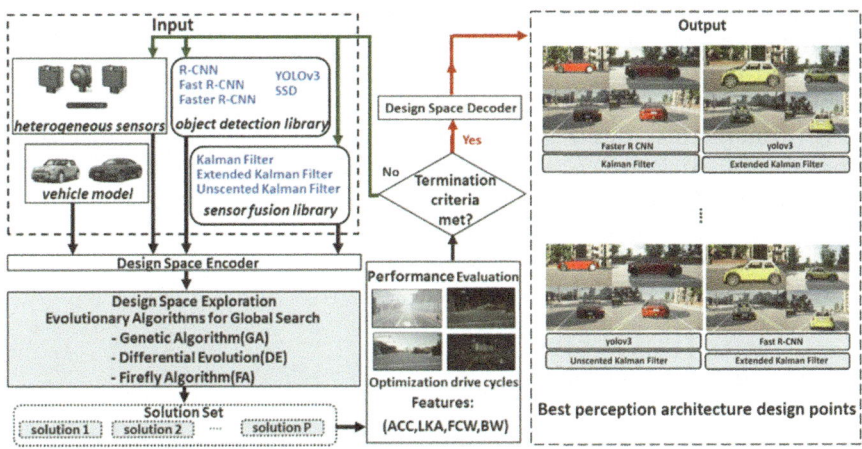

Fig. 4 An overview of the proposed PASTA framework

error across eight metrics that are characteristic of the ability to track and detect non-ego vehicles across road geometries and traffic scenarios.

The eight selected metrics are related to our goal of supporting level 2 autonomy with the perception architecture. In the descriptions of the metrics below, the ground truth refers to the actual position of the non-ego vehicles (traffic in the environment of the ego vehicle). The metrics can be summarized as: (1) *longitudinal position error and* (2) *lateral position error:* deviation of the detected positional data from the ground truth of non-ego vehicle positions along the y and x axes, respectively; (3) *object occlusion rate:* the fraction of passing non-ego vehicles that go undetected in the vicinity of the ego vehicle; (4) *velocity uncertainty:* the fraction of times that the velocity of a non-ego vehicle is measured incorrectly; (5) *rate of late detection:* the fraction of the number of 'late' non-ego vehicle detections made over the total number of non-ego vehicles. Late detection is one that occurs after a non-ego vehicle crosses the minimum safe longitudinal or lateral distance, as defined by Intel RSS safety models for pre-crash scenarios GA is a popular evolutionary algorithm that can solve optimization problems by mimicking the process of natural selection [30].2. This metric directly factors in the trade-off between latency and accuracy for object detector and fusion algorithms; (6) *false positive lane detection rate:* the fraction of instances when a lane marker is detected but there exists no ground truth lane; (7) *false negative lane detection rate:* the fraction of instances when a ground truth lane exists but is not detected; and (8) *false positive object detection rate:* the fraction of total vehicle detections which were classified as non-ego vehicle detections but did not actually exist.

4.3   Design Space Encoder/Decoder

The design space encoder receives a set of random initial design points as input which are expressed as a vector. This encoded format is best suited for various kinds of rearrangement and splitting operations during design space exploration. The encoder adapts the initial selection of inputs for our design space such that a design point is defined by the location and orientation of each sensor's configuration (consisting of six parameters: x, y, z, roll, pitch, and yaw), together with the object detector and fusion algorithm. The design space decoder converts the solutions into the same format as the input so that the output perception architecture solution(s) found can be visualized with respect to the real-world co-ordinate system.

4.4   Design Space Exploration

The goal of a design space exploration algorithm in our framework is to generate perception architectures (design points) which are aware of feature to field of view (FOV) zone correlations around an ego vehicle. Figure 5a shows the 10 primary

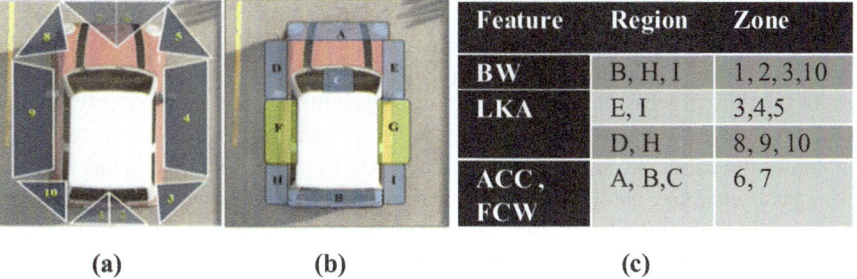

Feature	Region	Zone
BW	B, H, I	1, 2, 3, 10
LKA	E, I	3, 4, 5
	D, H	8, 9, 10
ACC, FCW	A, B, C	6, 7

(a)   (b)   (c)

Fig. 5 (**a**) Field of view (FOV) zones; (**b**) sensor placement regions; (**c**) feature, region, and zone relationship

FOV zones around the ego-vehicle. These zones of interest are defined as the most important perception areas in the environment for a particular ADAS feature. Figure 5b shows the regions on the vehicle on which sensors can be mounted (in blue). Regions F and G (in yellow) are exempt from sensor placement due to the mechanical instability of placing sensors on the door of a vehicle. The correlation between ADAS features, zones, and regions, is shown in Fig. 5c. For exploration of possible locations within a region, a fixed step size of 2 cm in two dimensions across the surface of the vehicle is considered, which generates a 2D grid of possible positions in each zone shown in Fig. 5b. The orientation exploration of each sensor involves rotation at a fixed step size of 1 degree between an upper and lower bounding limit for roll, pitch, and yaw respectively, at each of these possible positions within the 2D grid. The orientation exploration limits were chosen with caution with the caveat that some sensors, such as long range radars, have an elevated number of recorded false positives with extreme orientations.

To get a sense of the design space, consider four sensors (e.g., two cameras and two radars). Just the determination of the optimal placement and orientation of these sensors involves exploring $^{1.24e+26}C_4$ and $^{7.34e+25}C_4$ configurations for the Audi-TT and BMW-Minicooper vehicles, respectively. Coupled with the choice of different object detectors and sensor fusion algorithms, the resulting massive design space cannot be exhaustively traversed in a practical amount of time, necessitating the use of intelligent design space search algorithms that support hill climbing to escape local minima. In our framework, we explored three evolutionary algorithms: (1) *Genetic Algorithm (GA)*, (2) *Differential Evolution (DE)*, and the (3) *Firefly Algorithm (FA)*. As shown in Fig. 4, each algorithm generates a solution set of size *'P'* at every iteration until the termination criteria is met. The algorithms simultaneously co-optimize sensor configuration, object detection, and sensor fusion, and proceed to explore new regions of the design space when the termination (perception) criteria is not met. We briefly describe the three algorithms below.

4.4.1 Genetic Algorithm (GA)

GA is a popular evolutionary algorithm that can solve optimization problems by mimicking the process of natural selection [30]. Initially, the GA randomly selects a solution set of fixed size referred to as the population and then improves the quality of the candidate solutions in each iteration by modifying them using various GA operations. GA has the ability to optimize problems where the design space is discontinuous and also if the cost function is non differentiable. In our GA implementation, in the selection stage, the cost function values are computed for 50 design points at a time, and a roulette wheel selection method is used to select which set of chromosomes will be involved in the crossover step based on their cost function probability value (fraction of the cumulative cost function sum of all chromosomes considered in the selection). In the crossover stage, the crossover parameter is set to 0.5, allowing half of the 50 chromosomes to produce offspring. The mutation parameter is set to 0.2 which determines the new genes allowed for mutation in each iteration.

4.4.2 Differential Evolution (DE)

Differential Evolution (DE) [31] is another stochastic population-based evolutionary algorithm that takes a unique approach to mutation and recombination. An initial solution population of fixed size is selected randomly, and each solution undergoes mutation and then recombination operations. DE generates new parameter vectors by adding the weighted difference between two population vectors to a third vector to achieve difference vector-based mutation. Next, crossover is performed, where the mutated vector's parameters are mixed with the parameters of another predetermined vector, the target vector, to yield a trial vector. If the trial vector yields a lower cost function value than the target vector, the trial vector replaces the target vector in the next generation. To ensure that better solutions are selected only after generation of all trial vectors at every iteration, greedy selection is performed between the target vector and trial vector. Unlike GA where parents are selected based on fitness, every solution in DE takes turns to be one of the parents [30]. In our DE implementation, we set initial population size to 50 and use a crossover probability of 0.8 to select candidates participating in crossover.

4.4.3 Firefly Algorithm (FA)

FA is a swarm-based metaheuristic [32] that has shown superior performance compared to GA for certain problems [33]. In FA, a solution is referred to as a firefly. The algorithm mimics how fireflies interact using flashing lights (bioluminescence). The algorithm assumes that the attractiveness of a firefly is directly proportional to its brightness which depends on the fitness function value. Further, a given firefly can be attracted by any other firefly in the design space irrespective of the gender of both. Initially, a random solution set is generated and the fitness (brightness) of each candidate solution is measured. In the design space, a firefly is attracted to another with higher brightness (more fit solution), with brightness decreasing exponentially over distance. FA is significantly different from DE and GA, as both exploration of new solutions and exploitation of existing solutions to find better solutions is achieved using a single position update step.

4.5 Performance Evaluation

Each iteration of the design space exploration involves performance evaluation of the generated solution set where each design point undergoes multiple drive cycles. A drive cycle here refers to a virtual simulation involving an ego-vehicle (with a perception architecture under evaluation) following a fixed set of waypoint co-ordinates, while performing object detection and sensor fusion on the environment and other non-ego vehicles. A total of 20 different drive cycles were considered, with 5 drive cycles customized for each ADAS feature. As an example, drive cycles for ACC and FCW involve an ego vehicle following different lead vehicles at different distances, velocities, weather conditions, and traffic profiles. The fitness of the perception architectures generated by the framework are computed using the cumulative metric scores (Sect. 4.2) across the drive cycles.

5   Experiments

5.1 Experimental Setup

To evaluate the efficacy of the PASTA framework we performed experiments in the open-source simulator CARLA (Car Learning to Act) implemented as a layer on Unreal Engine 4 (UE4) [34]. The UE4 engine provides state-of-the-art physics rendering for highly realistic driving scenarios. We leveraged this tool to design a variety of drive cycles that are roughly 5 min long and contain scenarios that commonly arise in real driving environments, including adverse weather conditions (rain, fog) and a few overtly aggressive/conservative driving styles observed with vehicles. To ensure generalizability, we consider a separate set of test drive cycles to evaluate solution quality, which are different from the optimization drive cycles used iteratively by the framework to generate optimized perception architecture solutions.

We target generating perception architectures to meet level 2 autonomy goals for two vehicle models: Audi-TT and BMW-Minicooper (Fig. 6). A maximum of 4 mid-range radars and 4 RGB cameras are considered in the design space, where

BMW Minicooper

Length	Width	Height	Wheelbase
3.835	1.727	1.414	2.495

(a)

Audi TT

Length	Width	Height	Wheelbase
4.177	1.832	1.353	2.505

(b)

Fig. 6 BMW Minicooper (top) and Audi TT (bottom)

each sensor can be placed in any zone (Fig. 5a, b). Using a greater number of these sensors led to negligible improvements for the level 2 autonomy goal. The RGB cameras possess 90° field of view, 200 fps shutter speed, and image resolution of 800 × 600 pixels. The mid-range radars selected generate a maximum of 1500 measurements per second with a horizontal and vertical field of view of 30° and a maximum detection distance of 100 m. We considered 5 different object detectors (YOLOv3, SSD, R-CNN, Fast R-CNN, and Faster R-CNN) and 3 sensor fusion algorithms (Kalman filter, Extended Kalman filter, and Unscented Kalman filter). For the design space exploration algorithms, the cost function was a weighted sum across the eight metrics discussed in Sect. 4.2, with the weight factor for each metric chosen on the basis of their total feature-wise cardinality across all zones shown in Fig. 5c. During design space exploration, if the change in average cost function value was <5% over 250 iterations, the search was terminated. All algorithmic exploration was performed on an AMD Ryzen 7 3800X 8-Core CPU desktop with an NVIDIA GeForce RTX 2080 Ti GPU.

Table 1 Object detector latency and accuracy comparison

Object detector	Latency GPU (ms)	Latency CPU (ms)	mAP(%)
R-CNN	48956.18	66090.83	73.86
Fast R-CNN	1834.71	2365.86	76.81
Faster R-CNN	176.99	286.72	79.63
SSD	53.25	70.32	70.58
YOLOv3	24.03	32.92	71.86

5.2   Experimental Results

In the first experiment, we explored the inference latency and accuracy in terms of mean average precision (mAP) for the five different object detectors considered in this chapter. Table 1 summarizes the inference latency on a CPU and GPU, as well as the accuracy in mAP for the object detectors on images from our analyzed drive cycles, with all detectors trained on the MS-COCO dataset. It can be observed that the two-stage detectors (R-CNN, Fast R-CNN, and Faster R-CNN have a higher accuracy than the single stage detectors (SSD, YOLOv3). However, the inference time for the two-stage detector is significantly higher than for the single stage detectors. For real-time object detection in vehicles, it is crucial to be able to detect objects with low latency, typically less than 100 ms [35]. As a result, single stage detectors are preferable, with YOLOv3 achieving slightly better accuracy and lower inference time than SSD. However, in some scenarios, delayed detection can still be better than not detecting or wrongly detecting an object (e.g., slightly late blindspot warning is still better than receiving no warning) in which case the slower but more accurate two-stage detectors may still be preferable. Our PASTA framework is aware of this inherent trade-off and factors in the detection accuracy and rate of late detection in performance evaluation metrics (Sect. 4.2) to explore both single-stage and two-stage detectors. Also, detectors with a higher mAP value sometimes did not detect objects that other detectors with a lower mAP were able to; thus, we consider all five detectors in our exploration.

Next, we explored the importance of global co-optimization for our problem. We select the genetic algorithm (GA) variant of our framework to explore the entire design space (GA-PASTA) and compared it against five other frameworks. Frameworks GA-PO and GA-OP use the GA but perform a local (sequential) search for sensor design. In GA-PO, sensor position is explored before orientation, while in GA-OP the orientation for fixed sensor locations (based on industry best practices) is explored before adjusting sensor positions. For both frameworks, the object detector used was fixed to YOLOv3 due to its sub-100 ms inference latency and reasonable accuracy, while the extended Kalman filter (EKF) was used for sensor fusion due to its ability to efficiently track targets following linear or non-linear trajectories. The framework GA-VESPA is from prior work [12] and uses GA for exploration across sensor positions and orientations simultaneously, with the YOLOv3 object detector and EKF fusion algorithm. Frameworks GA-POD and GA-POF use GA for a more comprehensive exploration of the design space. GA-POD simultaneously explores the sensor positioning, orientation, and object detectors, with a fixed EKF fusion algorithm. GA-POF simultaneously explores the sensor positioning, orientation, and sensor fusion algorithm, with a fixed YOLOv3 fusion algorithm.

Figure 7a depicts the average cost of solution populations (lower is better) for the BMW-Minicooper across the different frameworks plotted against the number of iterations, with each exploration lasting between 80–100 h. It can be observed that GA-PO performs better than GA-OP, which confirms the intuitive importance of exploring sensor positioning before adjusting sensor orientations. GA-VESPA outperforms both GA-PO and GA-OP, highlighting the benefit of co-exploration of sensor position and orientation over a local sequential search approach used in GA-PO and GA-OP. GA-POD and GA-POF in turn outperform these frameworks, indicating that decisions related to object detection and sensor fusion can have a notable impact on perception quality. GA-POD terminates with its solution set having a lower average cost than GA-POF, which indicates that co-exploration of object detection and sensor placement/orientation is slightly more effective than co-exploration of sensor fusion and sensor placement/orientation. Our proposed GA-PASTA framework achieves the lowest average cost solution, highlighting the tremendous benefit that can be achieved from co-exploring sensor position/orientation, object detection, and sensor fusion algorithms. Figure 7b summarizes the objective function cost of the best solution found by each framework, which aligns with the population-level observations from Fig. 7a.

The comparative analysis for the BMW-Minicooper was repeated three times with different initializations for all six frameworks, and the results for the other two runs show a consistent trend with the one shown in Fig. 7. Note also that the relative trend across frameworks observed for the Audi-TT is similar to that observed for the BMW-Minicooper, and thus the results for the Audi TT are omitted for brevity.

In the next experiment, we explored the efficacy of different design space exploration algorithms (GA, DE, and FA; see Sect. 4.4) to determine which algorithm can provide optimal perception architecture solutions across varying vehicle models. Figure 8 shows the results for the three variants of the PASTA framework, for the Audi-TT and BMW-Minicooper vehicles. The best solution was selected across three runs of each algorithmic variant (variations for the best solution across runs are highlighted with confidence intervals, with bars indicating the median). It can be seen that for both considered vehicle models the FA algorithm outperforms the DE and GA algorithms. For Audi-TT, the best solution found by FA improves upon the best solution found with DE and GA by 18.34% and 14.84%, respectively. For the BMW-Minicooper the best solution found by FA outperforms the best solution found by DE and GA by 3.16% and 13.08%, respectively. Figure 9a depicts the specific sensor placement locations for each vehicle type, with a visualization of sensor coverage for the best solutions found by each algorithm shown in Fig. 9b.

Finally, in our quest to further improve perception architecture synthesis in PASTA, we focused on a more nuanced exploration of the object detector design space. We selected the FA search algorithm due to its superior performance over

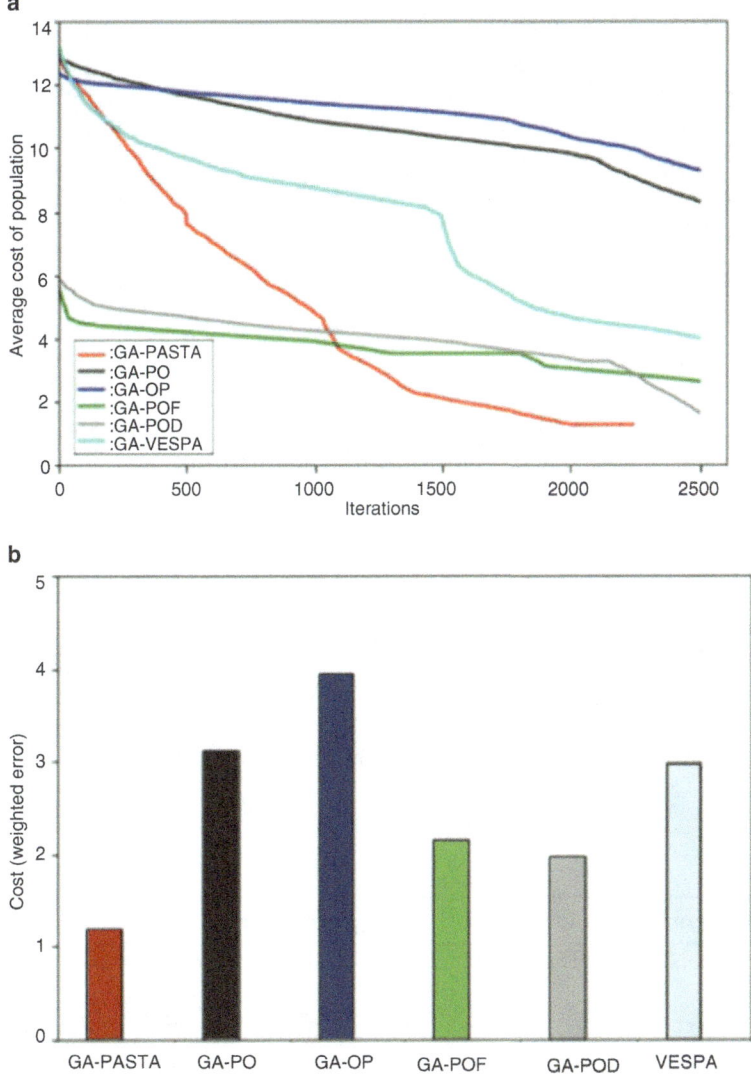

Fig. 7 (**a**) Comparison of perception architecture exploration frameworks; (**b**) Cost of best solution from each framework

GA and DE, and modified FA-PASTA to integrate a neural architecture search (NAS) for the YOLOv3 object detector, with the aim of further improving YOLOv3 accuracy across drive cycles while maintaining its low detection latency. Our NAS for YOLOv3 involved transfer learning to retrain network layers with a dataset consisting of 6000 images obtained from the KITTI dataset, using the open source tool CADET [36]. The NAS hyperparameters that were explored involved

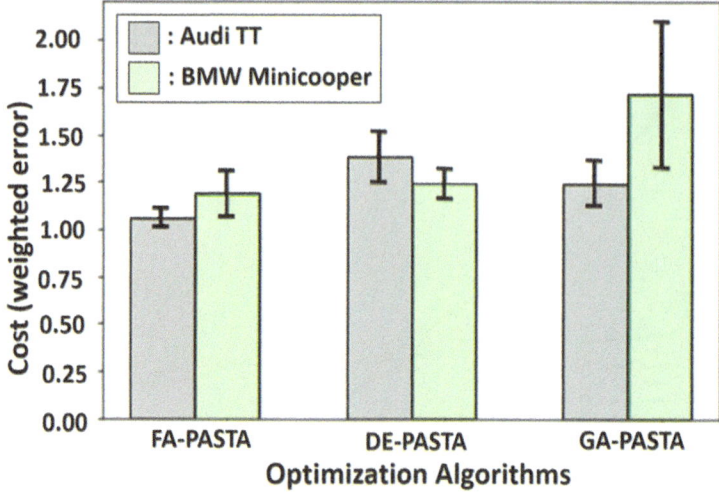

Fig. 8 Comparison of three variants of PASTA framework with genetic algorithm (GA), differential evolution (DE), and Firefly algorithm (FA)

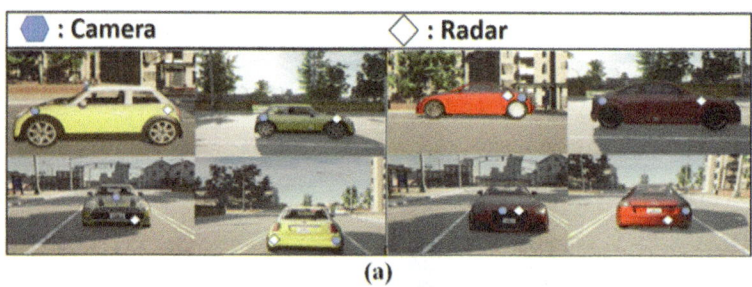

(a)

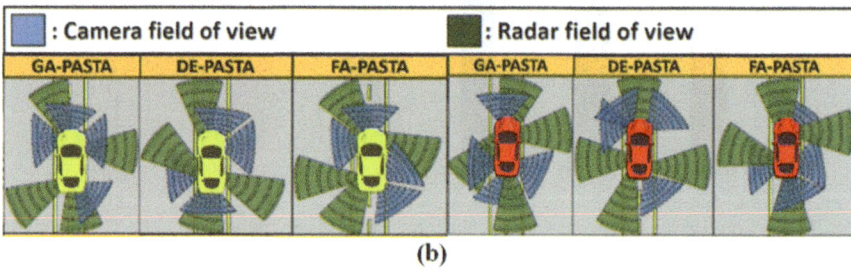

(b)

Fig. 9 (a) Sensor placement for best solution found with FA algorithm (top yellow vehicle: BMW-Minicooper, bottom red vehicle: Audi-TT) (top); (b) Sensor coverage for best solutions found by GA, DE, and FA search algorithms (bottom)

the number of layers to unfreeze and retrain (from a total of 53 layers in the Darknet-53 backbone used in YOLOv3; Fig. 10a), along with the optimizer learning rate, momentum, and decay. The updated variant of our framework, FA-NAS-

a

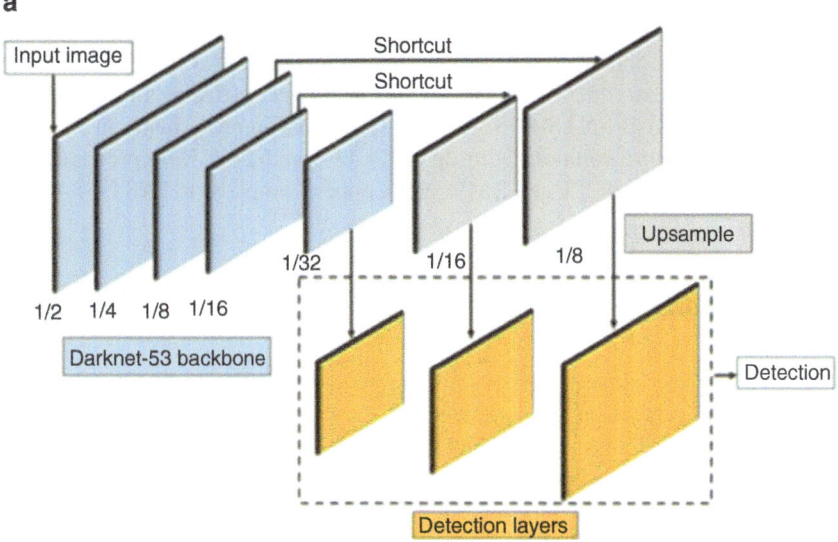

b

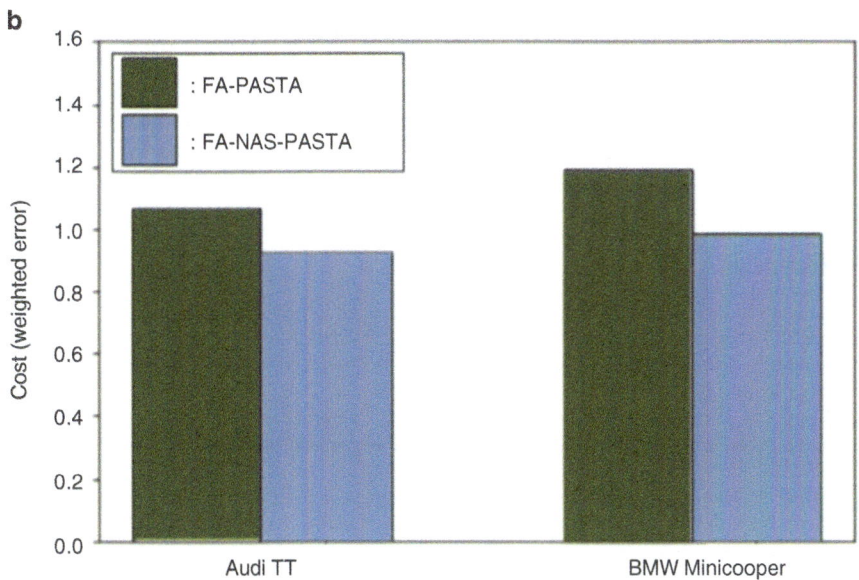

Fig. 10 (**a**) YOLOv3 object detector architecture with Darknet-53 backbone network that was fine-tuned using neural architecture search (NAS); (**b**) results of integrating object detector NAS with PASTA

PASTA, considered these YOLOv3 hyperparameters along with the sensor positions and orientations, and sensor fusion algorithms, during iterative evolution of the population of candidate solutions in the FA algorithm.

Figure 10b shows the results of this analysis for the two vehicles considered. FA-PASTA is the best performing variant of our framework (from Fig. 8), while FA-NAS-PASTA is the modified variant that integrates NAS for YOLOv3. It can be observed that fine tuning the YOLOv3 object detector during search space exploration in FA-NAS-PASTA leads to notable improvements in the best perception architecture solution, with up to 14.43% and 21.13% improvement in performance for the Audi-TT and BMW-Minicooper, compared to PASTA-FA.

6 Conclusions

In this chapter, we propose an automated framework called *PASTA* that is capable of generating perception architecture designs for modern semi-autonomous vehicles. *PASTA* has the ability to simultaneously co-optimize locations and orientations for sensors, optimize object detectors, and select sensor fusion algorithms for a given target vehicle. Our experimental analysis showed how PASTA can synthesize optimized perception architecture solutions for the Audi TT and BMW Minicooper vehicles, while outperforming multiple semi-global exploration techniques. Integrating neural architecture search for the object detector in PASTA shows further promising improvements in solution quality. Our future work will explore how to integrate PASTA with machine learning based techniques for anomaly detection [38–43] and robust vehicle network scheduling [42–44] in semi-autonomous vehicles.

References

1. NHTSA (National Highway Traffic Safety Administration), National Center for Statistics and Analysis, "Data Estimates Indicate Traffic Fatalities Continued to Rise at Record Pace in First Nine Months of 2021" (2022)
2. Gupta, A., Anpalagan, A., Guan, L., Khwaja, A.S.: Deep learning for object detection and scene perception in self-driving cars: Survey, challenges, and open issues. Array. **10**, 100057 (2021)
3. Feng, D., Harakeh, A., Waslander, S.L., Dietmayer, K.: A Review and Comparative Study on Probabilistic Object Detection in Autonomous Driving, vol. 23, pp. 9961–9980. IEEE Transactions on Intelligent Transportation Systems (2021)
4. Kukkala, V., Tunnell, J., Pasricha, S.: Advanced driver assistance systems: A path toward autonomous vehicles. IEEE Cons. Electron. **7**(5) (2018)
5. Zhexiang, Y., Jie, B., Sihan, C., Libo, H., Xin, B.: Camera-Radar Data Fusion for Target Detection via Kalman Filter and Bayesian Estimation. SAE Technical Paper (2018)
6. Nobis, F., Geisslinger, M., Weber, M., Betz, J., Lienkamp, M.: A Deep Learning-based Radar and Camera Sensor Fusion Architecture for Object Detection. IEEE Sensor Data Fusion: Trends, Solutions, Applications (SDF) (2020)
7. Verucchi, M., Bartoli, L., Bagni, F., Gatti, F., Burgio, P., Bertogna, M.: Real-time Clustering and LiDAR-Camera Fusion on Embedded Platforms for Self-Driving Cars. IEEE International Conference on Robotic Computing (IRC) (2020)

8. Meng, L., Yang, L., Tang, G., Ren, S., Yang, W.: An Optimization of Deep Sensor Fusion Based on Generalized Intersection Over Union. International Conference on Algorithms and Architectures for Parallel Processing, Springer (2020)

9. Meadows, W., Hudson, C., Goodin, C., Dabbiru, L., Powell, B., Doude, M., Carruth, D., Islam, M., Ball, J.E., Tang, B.: Multi-LIDAR Placement, Calibration, Co-registration, and Processing on a Subaru Forester for off-Road Autonomous Vehicles Operations. Autonomous Systems: Sensors, Processing and Security for Vehicles and Infrastructure (2019)

10. Chen, H., Ling, P., Danping, Z., Kun, L., Yexuan, L., Yu, C.: An Optimal Selection of Sensors in Multi-Sensor Fusion Navigation with Factor Graph. Ubiquitous Positioning, Indoor Navigation and Location-Based Services (UPINLBS) (2018)

11. Ji-Qing, L., Sheng Fang, H., Shao, F., Zhong, Y., Hua, X.: Multi-scale Traffic Vehicle Detection Based on Faster R–CNN with NAS Optimization and Feature Enrichment. Defence Technology (2021)

12. Dey, J., Taylor, W., Pasricha, S.: VESPA: A Framework for Optimizing Heterogeneous Sensor Placement and Orientation for Autonomous vehicles. IEEE Consumer Electronics Magazine (2020)

13. Asher, Z., Tunnell, J., Baker, D.A., Fitzgerald, R.J., Banaei-Kashani, F., Pasricha, S., Bradley, T.H.: Enabling Prediction for Optimal Fuel Economy Vehicle Control. SAE International (2018)

14. Tunnell, J., Asher, Z., Pasricha, S., Bradley, T.H.: Towards Improving Vehicle Fuel Economy with ADAS. SAE International (2018)

15. SAE International Standard J3016, "Taxonomy and definitions for terms related to driving automation systems for on-road motor vehicles" (2018)

16. C. Kirchner, "Lane keeping assist explained," Motor review, [Online]. Available: https://motorreview.com/lane-keeping-assist-explained (2014)

17. Li, H., Zhao, G., Qin, L., Aizeke, H., Zhao, X., Yang, Y.: A survey of safety warnings under connected vehicle environments. IEEE Trans. Intell. Transp. Syst. **22** (2020)

18. Zhao, Z.Q., Zheng, P., Xu, S.T., Wu, X.: Object Detection with Deep Learning: A Review. IEEE Transactions on Neural Networks and Learning Systems (2019)

19. Redmon, J., Divvala, S., Girshick, R., Farhadi, A.: You Only Look Once: Unified, Realtime Object Detection. Proceedings of the IEEE Conference on Computer Vision and Pattern Recognition (2016)

20. Han, J., Zhang, D., Cheng, G., Liu, N., Xu, D.: Advanced deep-learning techniques for salient and category-specific object detection: a survey. IEEE Signal Process. Mag. **35**, 84–100 (2018)

21. Liu, W., Anguelov, D., Erhan, D., Szegedy, C., Reed, S., Fu, C.Y., Berg, A.C.: "SSD: Single Shot Multibox Detector", European Conference on Computer Vision. Springer (2016)

22. Girshick, R., Donahue, J., Darrell, T., Malik, J.: Rich Feature Hierarchies for Accurate Object Detection and Semantic Segmentation. In Proceedings of the IEEE Conference on Computer Vision and Pattern Recognition (2014)

23. Girshick, R.: Fast R-CNN. Proceedings of the IEEE International Conference on Computer Vision (2015)

24. Ren, S., He, K., Girshick, R., Sun, J.: Faster R-CNN: towards Real-Time Object Detection with Region Proposal Networks. Advances in Neural Information Processing systems (NIPS) (2015)

25. Fayyad, J., Jaradat, M.A., Gruyer, D., Najjaran, H.: Deep learning sensor fusion for autonomous vehicle perception and localization: A review. Sensors. **20** (2020)

26. Kalman, R.E.: A New Approach to Linear Filtering and Prediction Problems. Transactions of the American Society of Mechanical Engineers (ASME) –Journal of Basic Engineering. (1960)

27. Simon, J., Uhlmann, J.: New Extension of the Kalman Filter to Nonlinear Systems. Signal Processing, Sensor Fusion, and Target Recognition International Society for Optics and Photonics (1997)

28. Yeong, D., Hernandez, G., Barry, J., Walsh, J.: Sensor and sensor fusion technology in autonomous vehicles: A review. Sensors. **21** (2021)

29. Wan, A., Merwe, R.: The Unscented Kalman Filter for Nonlinear Estimation. Proceedings of the IEEE Adaptive Systems for Signal Processing, Communications, and Control Symposium (2000)
30. Reeves, C.: Genetic Algorithms: Handbook of Metaheuristics. International Series in Operations Research & Management Science (2003)
31. Storn, R., Price, K.: Differential evolution–a simple and efficient heuristic for global optimization over continuous spaces. J. Glob. Optim. **11**, 341–359 (1997)
32. Yang, X.S.: Firefly algorithms for multimodal optimization. Stochastic Algorithms: Foundations and Applications (2009)
33. Zhou, G.D., Yi, T.H., Zhang, H., Li, H.N.: A Comparative Study of Genetic and Firefly Algorithms for Sensor Placement in Structural Health Monitoring. Shock and Vibration (2015)
34. Dosovitskiy, A., Ros, G., Codevilla, F., Lopez, A., Koltun, V.: CARLA: An Open Urban Driving Simulator. 1st Annual Conference on Robot Learning Conference on robot learning (2017)
35. Brekke, Å., Vatsendvik, F., Lindseth, F.: Multimodal 3d Object Detection from Simulated Pretraining. Symposium of the Norwegian Artificial Intelligence Society, Springer (2019)
36. Lin, S., Zhang, Y., Hsu, C., Skach, M., Haque, M., Tang, L., Mars, J.: The Architectural Implications of Autonomous Driving: Constraints and Acceleration. Proceedings of the Twenty-Third International Conference on Architectural Support for Programming Languages and Operating Systems (2018)
37. Dey, J., Pasricha, S.: Co-Optimizing Sensing and Deep Machine Learning in Automotive Cyber-Physical Systems. IEEE Euromicro Conference on Digital Systems Design (2022)
38. Thiruloga, S.V., Kukkala, V.K., Pasricha, S.: TENET: Temporal CNN with Attention for Anomaly Detection in Automotive Cyber-Physical Systems. IEEE/ACM Asia & South Pacific Design Automation Conference (ASPDAC) (2022)
39. Kukkala, V.K., Thiruloga, S.V., Pasricha, S.: LATTE: LSTM Self-Attention Based Anomaly Detection in Embedded Automotive Platforms. IEEE/ACM CODES+ISSS (ESWEEK) (2021)
40. Kukkala, V.K., Thiruloga, S.V., Pasricha, S.: INDRA: Intrusion Detection Using Recurrent Autoencoders in Automotive Embedded Systems. IEEE Transactions on Computer-Aided Design of Integrated Circuits and Systems, (TCAD). **39**(11) (2020)
41. Kukkala, V.K., Thiruloga, S.V., Pasricha, S.: "Roadmap for Cybersecurity in Autonomous Vehicles", to Appear, vol. 11, pp. 13–23. IEEE Consumer Electronics (2022)
42. Kukkala, V., Pasricha, S., Bradley, T.H.: SEDAN: Security-aware design of time-critical automotive networks. IEEE Transactions on Vehicular Technology (TVT). **69**(8) (2020)
43. Kukkala, V., Pasricha, S., Bradley, T.H.: JAMS-SG: A framework for jitter-aware message scheduling for time-triggered automotive networks. ACM Transactions on Design Automation of Electronic Systems (TODAES). **24**(6) (2019)
44. Kukkala, V.K., Pasricha, S., Bradley, T.: JAMS: Jitter-aware message scheduling for FlexRay automotive networks. IEEE/ACM International Symposium on Networks-on-Chip (NOCS) (2017)

Part V
Robust Control

Machine Learning and Optimization Techniques for Automotive Cyber-Physical Systems: Predictive Control During Acceleration Events to Improve Fuel Economy

Samantha White, Aaron Rabinowitz, Chon Chia Ang, David Trinko, and Thomas Bradley

Abbreviations

Abbreviations

AE	Acceleration Event. 650, 651, 656–658, 661–664, 666
APP	Accelerator Pedal Position. 663, 664
BMS	Battery Management System. 659, 661
BPPS	Brake Pedal Position Sensor. 661
BSFC	Brake Specific Fuel Consumption. 650, 651, 653, 666
CSU	Colorado State University. 664
DP	Dynamic Programming. 649–657, 664, 667
EM	Electric Motor. 650, 659, 661–664, 667
EPA	Environmental Protection Agency. 655, 657
FE	Fuel Economy. 649–651, 653–655, 657, 658, 666, 667
HEV	Hybrid Electric Vehicles. 649–652, 659, 663, 667
HSC	Hybrid Supervisory Controller. 659, 661, 664, 665
HV	High Voltage. 663
ICE	Internal Combustion Engine. 650, 661–664, 666, 667
IPO	Input-Processing-Output. 661
MPG	Miles Per Gallon. 666
NYCC	New York City Cycle. 657–659

S. White · A. Rabinowitz · C. C. Ang · D. Trinko · T. Bradley (✉)
Department of Systems Engineering, Colorado State University, Fort Collins, CO, USA
e-mail: samwhite@rams.colostate.edu; aaron.rabinowitz@colostate.edu; chon.ang@colostate.edu; dtrinko@colostate.edu; thomas.bradley@colostate.edu

© The Author(s), under exclusive license to Springer Nature Switzerland AG 2023
V. K. Kukkala, S. Pasricha (eds.), *Machine Learning and Optimization Techniques for Automotive Cyber-Physical Systems*, https://doi.org/10.1007/978-3-031-28016-0_23

1   Concept

The transportation sector has grown to become the leading contributor to greenhouse gas emissions, accounting for 36% of U.S. carbon emissions in 2021 [33]. Myriad emissions-reduction targets at all levels of government and industry are set to take effect in the coming decades and will require rapid reductions in transportation emissions, and therefore, to transportation fuel consumption. The emissions of vehicles powered by combustion can be reduced in the near term by improving their fuel efficiency, which is typically measured as Fuel Economy (FE). Common examples of technologies used to increase FE include engine sizing, advanced engine control, friction/mass/drag reduction, and powertrain electrification [13]. We focus on a category of controls-based FE improvement technologies, Optimal Energy Management Strategy (Optimal EMS), which can theoretically enable FE improvements of up to 30% for Hybrid Electric Vehicles (HEV) under conditions of ideal prediction and actuation [5].

An Optimal EMS is the application of optimal control to vehicle powertrain operation with the objective of minimizing fuel consumption (equivalently, maximizing FE). Computation of an Optimal EMS leverages predictions of future states of the vehicle, which are made based on information that may be, for example, gathered by sensors on the vehicle, obtained via communications with other vehicles and infrastructure, or learned based on historical driving data. The Optimal EMS technique was first published by Lin et al. [19], who derived the globally optimal control using Dynamic Programming (DP) for a hybrid electric truck. Since then, researchers have investigated stochastically robust strategies [20, 22, 24, 35, 36] as well as fast computation strategies [10, 11, 16, 23, 25] with the goal of progressing this technology toward commercial implementation. The technology has still not been realized commercially using such strategies, due in part to the computational cost of making predictions and calculating optimal control strategies in real time.

We conceive and test a novel method to realize Optimal EMS implementation. Instead of using a *real-time* computed *non-globally-optimal* EMS such as stochastic dynamic programming, equivalent consumption minimization strategy (ECMS), or a heuristic method, we use DP to compute *globally optimal* EMS *in advance*. In effect, this enables exchanging infeasible processing power requirements for potentially feasible memory requirements, improving the feasibility of commercial implementation. Furthermore, we target the strategy to one category of driving events, those in which the vehicle is accelerating from one speed (often zero) to another, or Acceleration Event (AE). We choose to target AE because they can be simpler to predict than general driving, and because they account for a high fraction of fuel consumption relative to their time durations. Thus, we refer to this strategy as Predictive Acceleration Event (PAE) control.

In this chapter, we summarize several years of research defining and testing the PAE method, which have resulted in a series of publications, theses, and patents [1, 4, 5, 7, 21, 26, 27, 31, 32]. The current section includes a brief summary of the simulation investigations that established the feasibility and potential of the PAE strategy. Details of the process for implementing the PAE strategy in a physical vehicle are described in Sect. 2, and results from physical testing are presented in Sect. 3.

1.1   Optimal EMS Mechanism

HEV achieve higher FE than conventional internal-combustion-engine-only vehicles in part because they enable the Internal Combustion Engine (ICE) to operate at high efficiency more of the time. This can be conveniently visualized using a Brake Specific Fuel Consumption (BSFC) map, which illustrates the ICE fuel consumption efficiency as a function of engine rotation speed and supplied torque (Fig. 1). HEV leverage their multiple degrees of freedom for power sourcing—power can be supplied by the ICE or the Electric Motor (EM)—to adjust the BSFC "operating points" in ways that pure ICE vehicles cannot. With a given power request, the only means available to ICE vehicles for adjusting torque and/or speed of the engine is to change the gear ratio of the transmission. In contrast, HEV enable much more flexibility because the EM can supply the difference, positive or negative, between the power supplied by the ICE and the requested propulsion power. For example, when operating at low powers/speeds, an HEV can avoid operating the engine at low efficiency either by using the EM to supply all the power, or by running the engine at high power and efficiency to supply propulsion power in addition to regeneration power through the EM.

Optimal EMS leverage predictions to further this FE improvement approach. In addition to modifying the BSFC operating points during each instant, an Optimal EMS aims to modify BSFC operating points over time. For the PAE strategy, the time horizon is on the order of tens of seconds (the duration of an AE). If the speed trajectory of the AE is known in advance, it is possible to use an optimization method such as DP to obtain a time series of BSFC operating points that supply the power needed to complete the AE while globally minimizing fuel consumption.

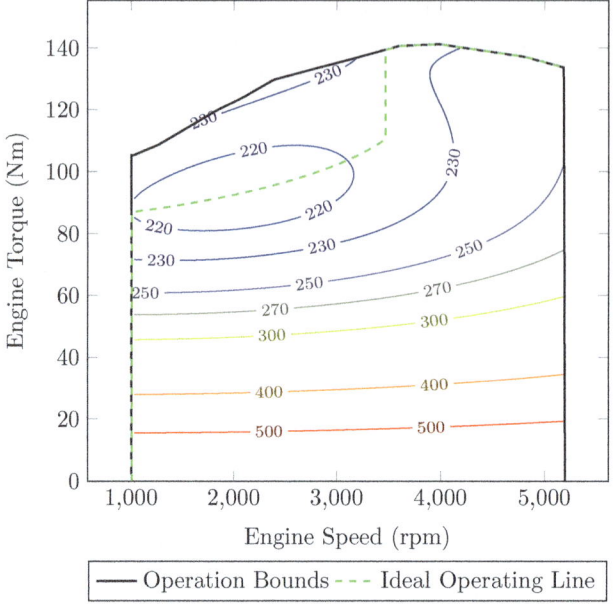

Fig. 1 BSFC map characterizes the fuel efficiency of an internal combustion engine (grams of fuel per kWh supplied) as a function of speed and torque

Table 1 Significant parameters defining the 2010 Toyota prius model

m	1380 kg	A_{front}	2.6005 m^2
$P_{\text{ICE,max}}$	73 kW	C_{rr}	0.008
m_{fuel}	$f(T_{\text{ICE}}, \omega_{\text{ICE}})$[17]	r_{fd}	3.27
$\omega_{\text{trac,max}}$	10,000 rpm	r_{wheel}	0.317 m
$\omega_{\text{gen,max}}$	13,500 rpm	C_d	0.250
$Q_{\text{batt,0}}$	6.5 Ah	V_{oc}	201.6 V
N_{sun}	30	N_{ring}	78
R_{int}	0.373 Ω		

1.2 Model Details

The Toyota Prius has consistently achieved the highest FE in its class [34], so it is an ideal vehicle to model for investigations of new HEV FE improvement techniques. The 2010 model was chosen for its commercial prevalence and publicly available parameter information. A model of the 2010 Toyota Prius, derived using the Autonomie modeling software, has been shown to correlate closely with real-world performance [18]. The referenced model is not publicly available, so a model was developed and validated by modifying a 2004 Toyota Prius model included with Autonomie with 2010 Prius parameters. Table 1 is a list of key parameters defining the model, where m = vehicle mass; $P_{\text{ICE,max}}$ = maximum engine power;

m_{fuel} is the fuel consumption model; T_{ICE} = engine torque; ω_{ICE} = engine speed; $\omega_{\text{trac,max}}$ is the maximum traction motor speed; $\omega_{\text{gen,max}}$ is the maximum generator motor speed; $Q_{\text{batt,0}}$ is the initial battery capacity; N_{sun} and N_{ring} are the number of teeth on the sun and ring gears in the planetary gearset; R_{int} is the battery's internal resistance; A_{front} is the frontal area of the vehicle; C_{rr} = coefficient of rolling resistance; r_{fd} = final drive ratio; r_{wheel} = wheel radius; C_d = drag coefficient; and V_{oc} = open circuit battery potential.

The Autonomie software produces high fidelity models that are useful for realistic modeling of a variety of vehicle signals, including power split control in a HEV, but are computationally expensive in simulation. Even if disregarding concerns about long computation times, it would be infeasible to use the Autonomie model with DP to derive the Optimal EMS, because states in Autonomie are dependent on preceding states, which is incompatible with the DP formulation (described in Sect. 1.2.2). Instead, the Autonomie model was used only to simulate the Baseline EMS engine control strategy, which was used as an input to a lower fidelity "power split" vehicle model for the remaining vehicle signal calculations. Details on the original development of the power split model are in a previous publication from the author's lab group [4] and reproduced briefly here.

The power split model is based on equations describing vehicle dynamics, a modeling approach that is well-defined in the literature [2, 12, 17, 28]. The power required to propel the vehicle at velocity v must be provided as a sum of engine power and electric propulsion system power:

$$P_{\text{prop}} = F_{\text{prop}} v = P_{\text{elec}} + P_{\text{ICE}} \tag{1}$$

P_{ICE} is an input to the power split model, so the equation is rearranged to solve for P_{elec}:

$$P_{\text{elec}} = F_{\text{prop}} v - P_{\text{ICE}} \tag{2}$$

F_{prop} effects vehicle acceleration and counteracts the forces opposing vehicle motion:

$$F_{\text{prop}} = m\dot{v} + C_{rr} mg + \frac{1}{2} C_d \rho_{\text{air}} v^2 A_{\text{front}} \tag{3}$$

where $\dot{v}$ is the acceleration of the vehicle, calculated using a numerical derivative; g is acceleration due to gravity $\left(9.81 \ \frac{\text{m}}{\text{sec}^2}\right)$; and ρ_{air} is the density of air $\left(1.1985 \ \frac{\text{kg}}{\text{m}^3}\right)$. For this research, grade angle is assumed to be zero.

P_{elec} is served by the battery, with an efficiency penalty modeled as a function of torque and speed of the generator and traction motors:

$$\eta_{\text{elec}} = f(\omega_{\text{gen}}, T_{\text{gen}}, \omega_{\text{trac}}, T_{\text{trac}}) \tag{4}$$

as defined by efficiency maps supplied with the Autonomie model. η_{elec} is enforced such that energy is always lost due to inefficiencies in the electric system, whether charging or discharging:

$$P_{batt} = \eta_{elec} P_{elec} \text{ if } P_{elec} \le 0 \tag{5}$$

$$P_{batt} = \frac{1}{\eta_{elec}} P_{elec} \text{ if } P_{elec} > 0 \tag{6}$$

where positive values of P_{batt} represent discharging. At timestep i, battery State of Charge (SOC) is calculated for the next timestep $i+1$ using the following equation:

$$SOC_{i+1} = SOC_i - \frac{V_{oc} - \sqrt{V_{oc}^2 - 4 P_{batt} R_{int}}}{2 R_{int} Q_{batt,o}} \Delta t \tag{7}$$

To enable fast computation when solving the DP formulation, fuel consumption is modeled using a cubic response surface [15] representation of a publicly available BSFC map for the Generation III Prius [17]:

$$BSFC \left(\frac{g}{kWh} \right) = A_1 + A_2 \omega_{ICE} + A_3 T_{ICE} + A_4 \omega_{ICE} T_{ICE} + A_5 \omega_{ICE}^2 +$$
$$A_6 T_{ICE}^2 + A_7 \omega_{ICE} T_{ICE}^2 + A_8 \omega_{ICE}^2 T_{ICE} + A_9 T_{ICE}^3 \tag{8}$$

where all A values are fitted constants. This BSFC surface has an ideal operating line [14] that represents the instantaneous optimal FE operating point (in terms of torque and speed) as a function of engine power. The fuel consumption during a timestep Δt is thus

$$m_{fuel} \text{ (grams)} = \left(BSFC \times \frac{1\,h}{3\,s} \right) P_{ICE} \Delta t \tag{9}$$

where P_{ICE} is in kW and Δt is in seconds.

The angular speeds of powertrain components are constrained by a planetary gearset:

$$\omega_{ICE} = \omega_{gen} \frac{\rho}{1+\rho} + \omega_{ring} \frac{1}{1+\rho} \tag{10}$$

where $\rho = \frac{N_{sun}}{N_{ring}}$. Speeds are also constrained by limits on the electric motors, given in Table 1. The ring gear speed is linearly related to vehicle speed:

$$\omega_{ring} = \frac{r_{fd}}{r_{wheel}} v \tag{11}$$

The model's powertrain is controlled by one of two different control strategies: a Baseline EMS, meant to simulate stock vehicle performance, and an Optimal EMS, derived via DP to optimize FE over a predicted driving schedule.

1.2.1 Baseline EMS

The Autonomie model is simulated over a drive cycle $v(t)$, defining the Baseline EMS $P_{ICE}(t)$, which also implicitly defines $P_{elec}(t)$ via Eq. (2). The power split model is used to calculate the remaining outputs, including m_{fuel}, SOC, and FE. This is illustrated in Fig. 2a.

To validate the Baseline EMS, the process in Fig. 2a was applied to three standard Environmental Protection Agency (EPA) FE test schedules and the FE results, corrected for change in SOC using a method standardized by Society of Automotive Engineers [30], were compared with experimental results for the 2010 Toyota Prius measured by Argonne National Laboratory [3] (Table 2). Additional validation steps included by comparing fuel consumption, SOC, and engine speed traces to actual data. These validations are documented in [31].

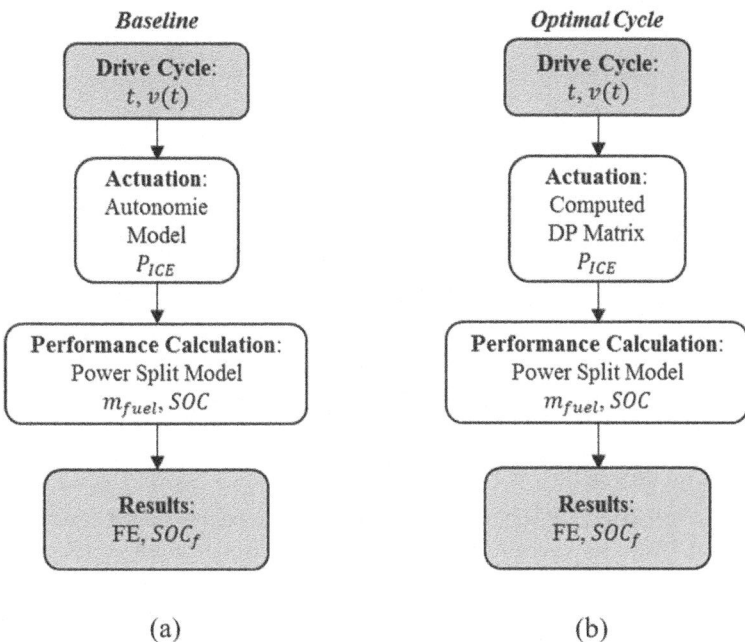

Fig. 2 FE simulation method for (**a**) Baseline EMS and (**b**) Optimal EMS (exact schedule prediction)

Table 2 FE results demonstrating validation of Baseline EMS model for FE investigations

EPA drive cycle	Simulated FE	Measured FE	% Difference
UDDS	76.4 mpg	75.6 mpg	+1.1%
US06	45.0 mpg	45.3 mpg	−0.6%
HWFET	69.1 mpg	69.9 mpg	−1.1%

1.2.2 Optimal EMS

The Optimal EMS is derived using deterministic DP, which uses backwards recursion to avoid solutions that are not optimal as defined by the Bellman Principle of Optimality [8, 9]. The DP scheme used for this study was detailed and validated in a previous publication [4] and will be described only briefly in this section.

In general, DP is used to compute optimal control as a function of system state by minimizing a cost function, subject to system constraints. For this study, the optimal control variable is engine power P_{ICE}, which also implicitly defines P_{elec} via Eq. (2); the state variable is battery SOC; and the cost function is fuel consumption m_{fuel}. For the purposes of the DP scheme, vehicle velocity trace $v(t)$ is an exogenous input upon which the state variable, SOC, partially depends. The state and cost are given by the following equations:

$$SOC(k + 1) = SOC(k) + f(SOC, P_{ICE}, v, k)\Delta t \tag{12}$$

$$Cost = \sum_{k=0}^{N-1} m_{fuel} + W \left(SOC_f - SOC(N)\right)^2 \tag{13}$$

where W is a penalty weight arbitrarily set at 10,000, k is the timestep index, N is the number of timesteps, and Δt is the size of a timestep. Equation (12) incorporates Eqs. (3)–(7) and (10)–(11), and Eq. (13) incorporates Eqs. (8)–(9). The allowable state and control spaces are

$$40\% \leq SOC(k) \leq 80\% \quad (k = 0, \dots N) \tag{14}$$

$$0\,kW \leq P_{ICE}(k) \leq 73\,kW \quad (k = 0, \dots N - 1) \tag{15}$$

To summarize, the DP scheme is used to calculate engine power (discretization $\Delta P_{ICE} = 0.1$ kW) for every feasible battery SOC (discretization $\Delta SOC = 0.02\%$) for every timestep in a drive cycle (discretization $\Delta t = 0.4$ s) to minimize fuel consumption for a velocity trace $v(t)$ and desired SOC_f. In future studies, other measurements (e.g. battery temperature) and cost variables (e.g. battery life impacts) may also merit inclusion but were not included in this research.

The output of DP for a velocity trace can be visualized as a two-dimensional matrix of engine power, where row indices represent values of SOC and column indices represent timesteps (see Fig. 3a). For any initial SOC (SOC_i), the DP matrix

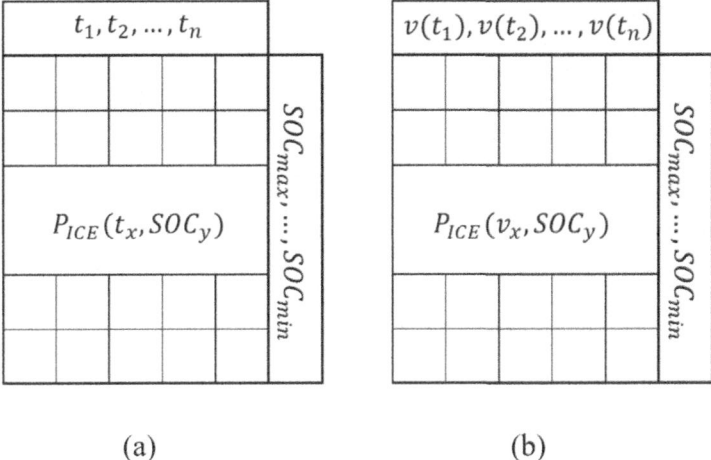

Fig. 3 (a) Illustration of matrix generated by dynamic programming algorithm; (b) Illustration of matrix with conversion of time index to velocity index

can be used as a lookup table to generate the optimal control solution $P_{ICE}(t)$ achieving the driving schedule $v(t)$ that results in a desired SOC_f.

1.3   Pre-Computing Optimal EMS for Approximate AE Prediction

Because the matrix generated via DP is a discrete array of optimal $P_{ICE}(k, SOC)$ for timesteps $k = 0...N$, it can be used as a lookup table for a different drive cycle with the same number of timesteps. This can yield a near-optimal solution if the new drive cycle is similar to the one to which DP was applied. However, the constraint of identical durations makes this method challenging to apply in practice.

If optimal control is only applied to AE, there is a way around the constraint of equal duration. AE are monotonically increasing segments of $v(t)$, so they can be indexed using velocity. This enables the DP matrix to be converted from a mapping with respect to time and SOC ($P_{ICE}(k, SOC)$) to a mapping with respect to *velocity* and SOC ($P_{ICE}(v_k, SOC)$), as shown in Fig. 3. With this conversion, it is possible to derive a DP matrix for one AE (the "expected AE") and apply it to any other AE ("actual AE") with the same velocity range as the expected AE, regardless of any difference in duration. Whereas drivers are not necessarily constrained to repeat AE with equal time durations, accelerator pedal traces, or other attributes, traffic laws encourage repetition of AE with equal velocity ranges (for example, 0–25 mph AE on neighborhood streets). Thus, a single DP matrix can serve as a lookup table for improved control of any AE with a similar velocity range to the AE for which it is computed.

In our simulations, a global upper limit to FE during a full drive cycle (containing one or more AE) is achieved by applying DP to compute optimal control for the full velocity profile of a drive cycle, a scenario we refer to as "optimal cycle control." When a DP matrix is computed for an AE and applied to control that same AE, we refer to this implementation as "optimal AE control." (By assuming knowledge of an exact velocity profile, these control scenarios simulate situations in which an AE or a full drive cycle is predicted exactly and DP is applied in real time. Due to sensing, predicting, and computing limitations, we assume these scenarios to be infeasible in practice.) When, instead, a DP matrix is computed for a category of AE (defined by its starting and ending speeds) and applied to another member of that category, this simulates a disturbance to the control loop. Thus, we refer to it as "disturbed AE control."

1.4 Drive Cycle Simulations

A variety of driving schedules, or drive cycles, were simulated to investigate the feasibility of the PAE strategy in real driving contexts. Because city driving is typically characterized by many low-speed AE, the PAE strategy demonstrates highest FE improvement potential when applied to city driving cycles. In this section, we present simulation results for a standard EPA city cycle, the New York City Cycle (NYCC). These simulations demonstrate that FE improvement potential is high for both Optimal Cycle and Optimal AE control, and Disturbed AE FE is nearly as high as Optimal AE FE. Most likely as a result of the low-aggression driving common to these cycles, effective fuel consumption reduction is achieved in the vast majority of AE, leading to high FE gains. The simulated FE results are plotted in Fig. 4 and selected simulation outputs are plotted in Fig. 5.

The Disturbed Optimal EMS achieves a significant portion (77%) of the FE improvement achieved by Optimal AE control and 35% of the FE improvement achieved by Optimal Cycle control. With the exception of some instances of high engine power for SOC correction, the Disturbed engine power trace appears to follow the Optimal AE engine power trace closely, indicating that the categorization scheme sufficiently limits prediction error to provide a close match between the Expected and Actual Optimal EMS.

The FE results for all seven drive cycles simulated, sorted in order of increasing Disturbed AE FE improvement, are given in Fig. 6. The cycle for which Disturbed AE control is least successful is US06, the aggressive cycle; the next three are the highway cycles; and the cycles with the best Disturbed AE performance are the city cycles. This is one indication that Disturbed AE control is most successful in city driving and less successful with increasing aggressiveness.

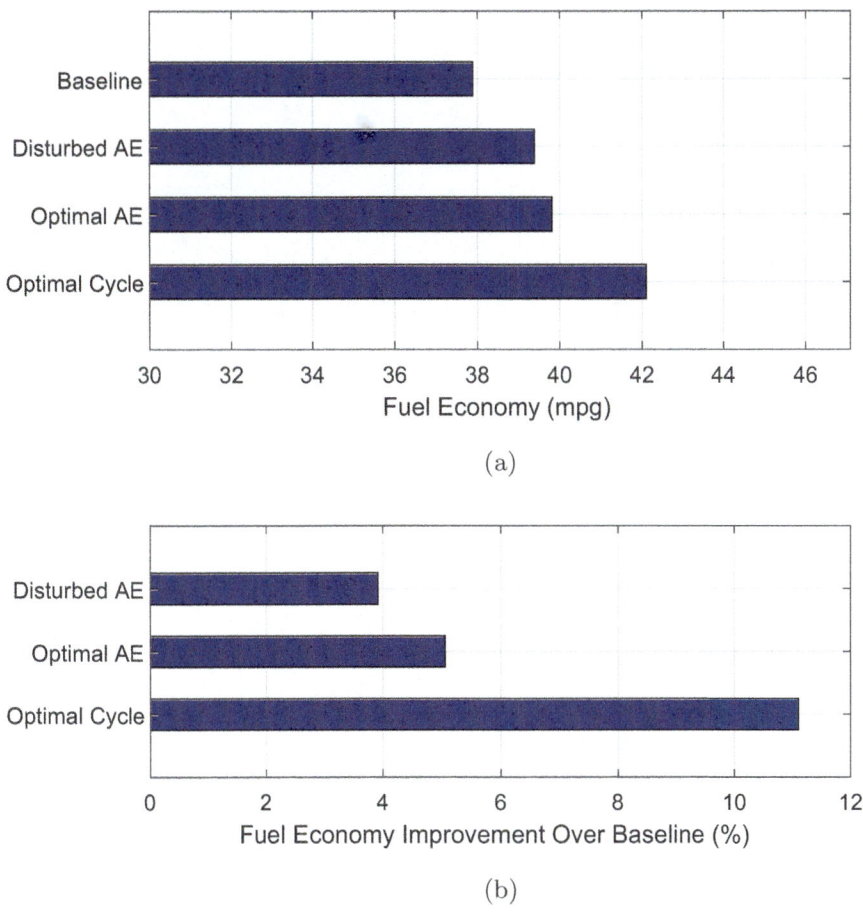

Fig. 4 FE results for NYCC cycle. (a) NYCC cycle FE results. (b) NYCC cycle FE improvement results

2 Implementation

In order to physically validate the effectiveness of PAE control, a Parallel-3 (P3) HEV was developed from a stock 2018 Toyota Tacoma [1, 21]. In a P3 HEV, the EM is located in between the transmission and the differential. This vehicle configuration was chosen for a number of reasons. The first was the relative ease of manufacturing in comparison to the other types of hybrid configurations. This type of powertrain was also comparatively easier to implement a supervisory controller in. The P3 configuration allowed the main structure of the vehicle to remain relatively unchanged.

This modification added an electric motor between the transmission and the differential as well as the necessary components to support the electrified powertrain.

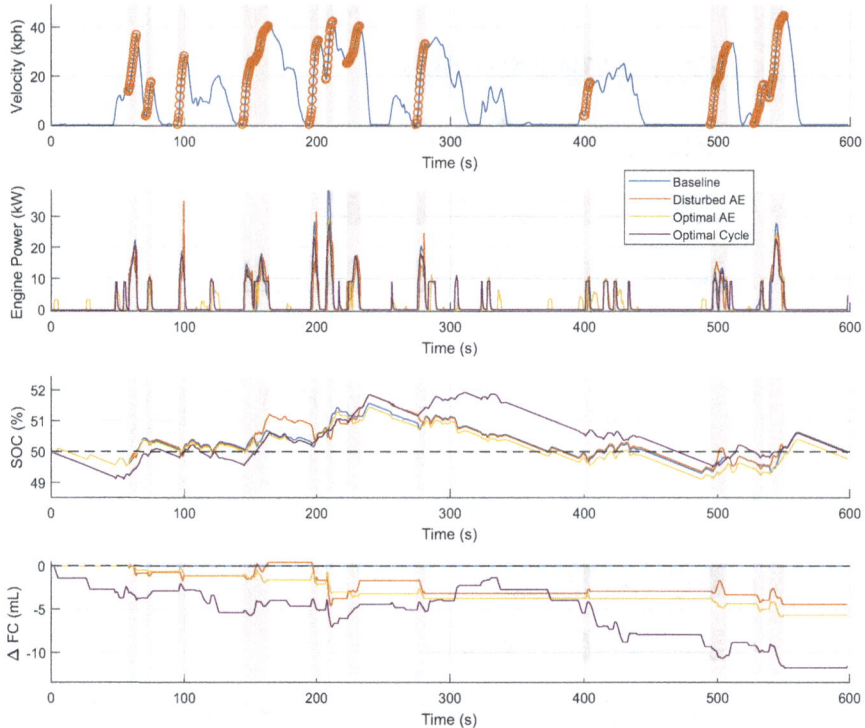

Fig. 5 Simulation outputs for the NYCC cycle

This included the Inverter, Battery, Battery Management System (BMS), Hybrid Supervisory Controller (HSC), Toyota Gateway, On-Board Charger (OBC), and the needed 12 volt powered components to control and provide thermal regulation of the components.

Central to this project was the desire to implement PAE control in a manner which fit industry norms. For this reason, in addition to the P3 conversion, the group elected to accomplish PAE control by leveraging the vehicle's existing distributed computing network and adding minimal computing load to the vehicle. For this reason, the group elected to control the vehicle using only one additional controller, the HSC. The HSC operated as an Input-Processing-Output (IPO) model where it converted input signals to output signals to control the vehicle's behavior. A 112 pin Woodward Motohawk (ECM-5644-112 SECM-112) was used as the hardware for the HSC. This controller is a typical firmware based automotive controller which might be used on a commercial vehicle. Matlab's Simulink software was the development environment for this program allowing for the use of Woodward's Motohawk rapid controller development software. This software was used to build the HSC code that was compatible with the Motohawk hardware. It also allowed for values within the controller to be viewed and calibrated in real time, on the vehicle.

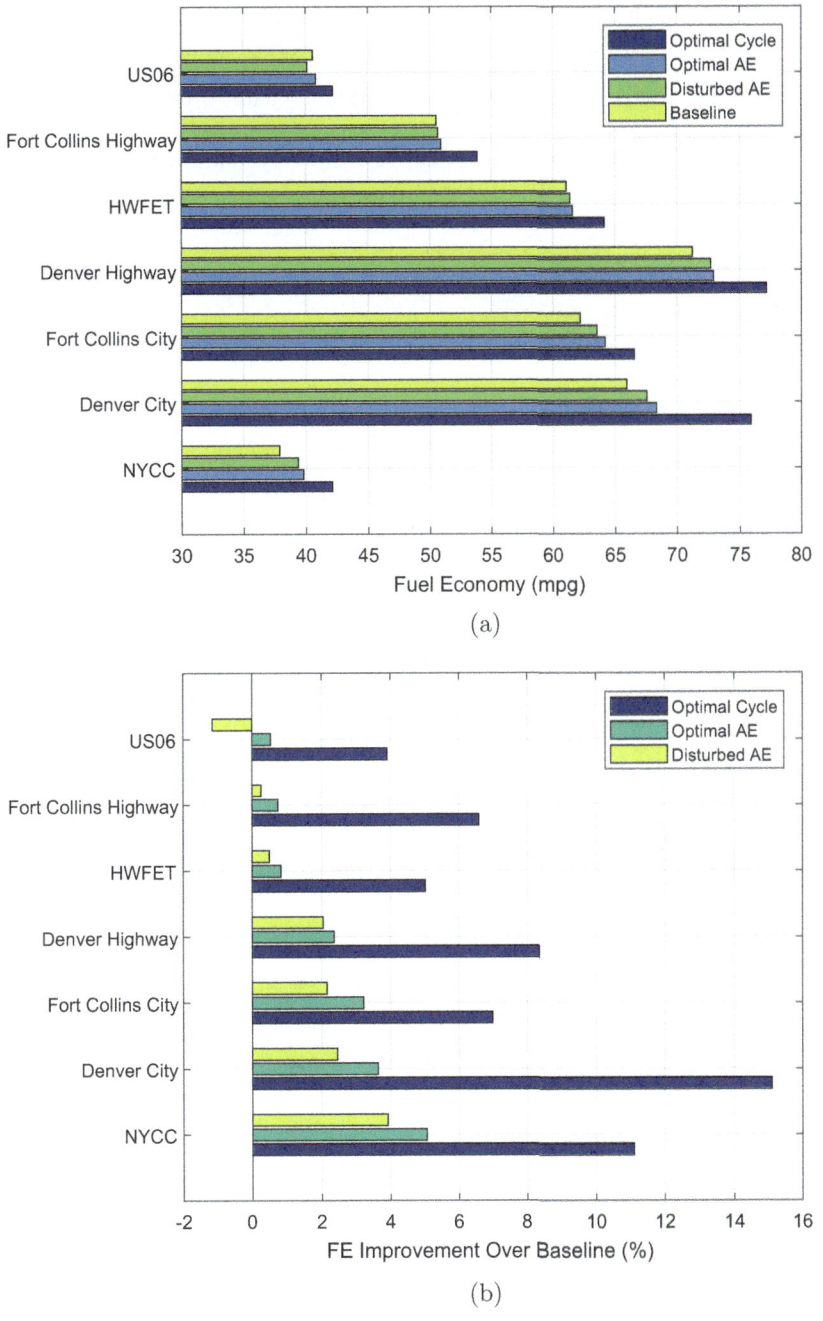

Fig. 6 FE results for all seven cycles. (**a**) FE for all cycles for all control strategies. (**b**) Improvement over baseline for each control strategy

The selected architecture imposed a series of limitations. When developing the HSC, the following constraints were observed and worked around:

- The HSC's storage capacity was too small to contain a full Optimal Control Matrix (OCM).
- The signals from the Toyota Gateway were the only values that could be used from the base vehicle.
- The BMS was an unreliable source of information, especially with producing SOC values.
- The driver could not consistently reproduce exact AEs under manual control using the accelerator pedal input.
- The HSC had no signal to differentiate between the vehicle's accessory mode and a fully on mode.
- The engine could not be controlled directly by torque requests.
- The Brake Pedal Position Sensor (BPPS) did not output non-binary values.

Figure 7 shows a simplified structure of the HSC and the flow of signals through the controller. It shows the signals that occurred when the vehicle was turned on and off as well as the signals that occurred continuously while the HSC was on. This flowchart illustrates the basic outline of the HSC code. It begins with the input signals and ends with the output signals. The flowchart shows the connection between power mode, vehicle state, pedal logic, torque split, ICE control, inverter and EM control, battery control, and SOC calculations.

With this architecture in mind, a baseline and PAE control were developed.

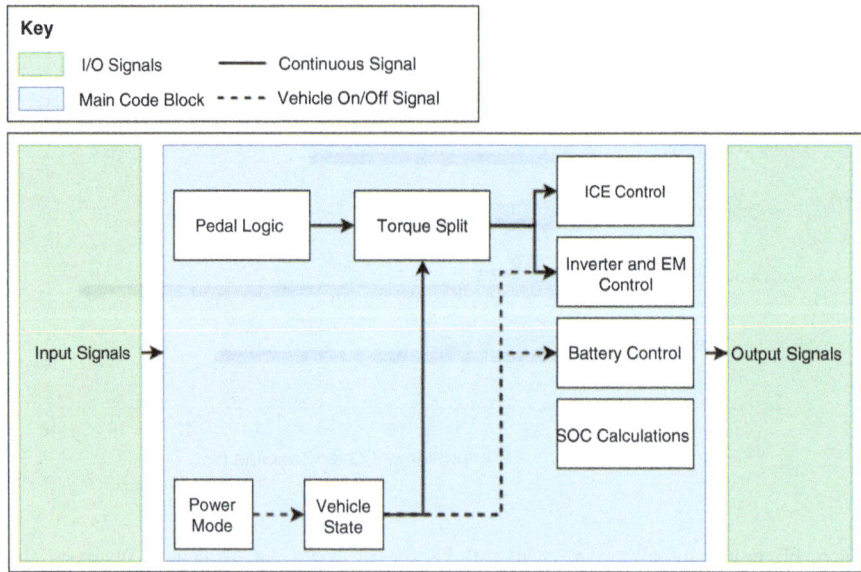

Fig. 7 Simplified controller flowchart

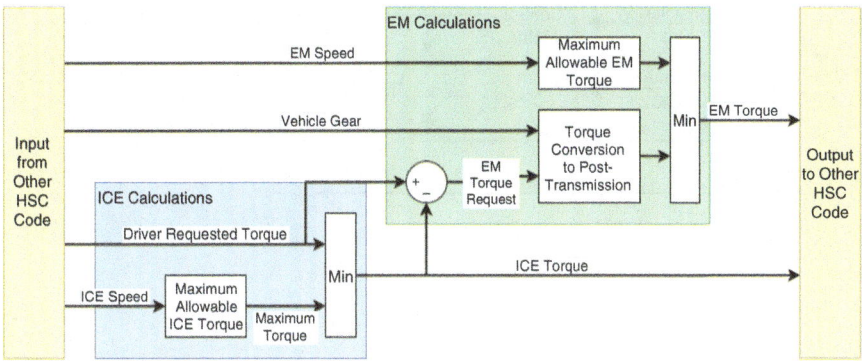

Fig. 8 Baseline torque split

2.1 Baseline Torque Split Control

A load following torque split strategy was chosen to act as the Baseline Torque Split Method. The load following method calculated the ICE torque first, compared that torque value to the driver's requested torque, and filled in the difference with EM torque. This method worked in all driving scenarios, and was calculated on the vehicle. No calculations needed to happen before the vehicle was driven. However, this strategy did not provide an optimal torque split for the AE. The flowchart for the Baseline strategy can be found in Fig. 8.

This strategy was composed of two main components: ICE calculations and EM calculations. The ICE calculations started with the minimum of the driver requested torque and maximum allowable torque for the given engine speed being selected for the ICE torque output. The ICE torque output was then subtracted from the driver requested torque to obtain the EM torque request value.

Since this was a P3 HEV, and the ICE and EM were located on opposite sides of the transmission, the EM torque request value must be multiplied by the vehicle gear ratio to obtain the post-transmission EM torque request value. The post-transmission value and maximum allowable EM torque are then compared, with the smaller value selected to be the EM torque output.

2.2 PAE Torque Split Control

Two PAE methods were used to calculate torque split values: on-vehicle torque computation through an optimal torque split matrix and pre-computed torque traces. The optimal torque split matrix, used for the on-vehicle torque split computations,

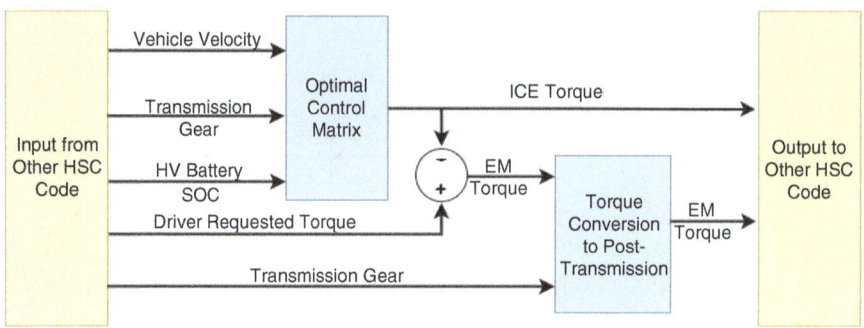

Fig. 9 OCM torque computation flowchart

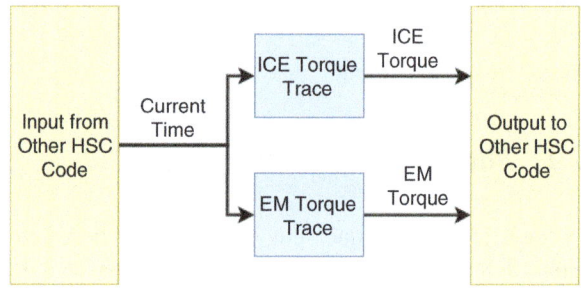

Fig. 10 Pre-computed torque trace flowchart

was obtained from the PAE model, as discussed in Sect. 1.3. The pre-computed torque traces were also obtained as an output from this model.

agraphOCM Torque Computation This torque split method required the AEs to be pre-computed to generate an OCM of torque split output values. This OCM was then added to the controller, so that the engine torque could be determined from the SOC and vehicle speed while performing an AE. Figure 9 shows the simplified flowchart of this method.

In this figure, it can be seen that the OCM requires three inputs: vehicle velocity, transmission gear, and the SOC of the High Voltage (HV) battery. This matrix then output the ICE torque which, when subtracted from the driver requested torque, produces the pre-transmission EM torque value. This value then was multiplied by the transmission gear to convert to a post-transmission EM torque value.

agraphPre-Computed Torque Trace The Pre-Computed Torque Trace involved generating the ICE and EM PAE torque traces in advance and using the traces to control the torque instead of the driver's Accelerator Pedal Position (APP) input. Figure 10 shows the flowchart for this PAE method.

As seen in the figure, the current time was input into lookup tables that contained ICE and EM torque traces which converted the time value to a torque output. After each AE, the driver reset the current time value back to zero via a manual switch, located on the vehicle's dashboard. A second dash switch was used to trigger the block that contains the torque trace logic.

3 Results

3.1 PAE Strategy Results

The bulk of the Baseline and PAE comparison testing was completed at the Christman Airfield, a 4000 foot long runway that was used as a Colorado State University (CSU) testing facility. This airfield was used for closed-course, straight-line testing. It runs in a north-south line, and every test that was completed with this vehicle was in the north direction. This reduced the effects of the slight slope of the runway.

The PAE control strategy was tested using the OCM method and the torque trace strategy. The OCM method, reduced in size and scope due to HSC memory limitations, was discovered to output torque split values that didn't prioritize recharging the battery to the starting SOC. Instead, it would command large, positive torque outputs from the EM.

Figure 11 shows one of the tests completed with the OCM generating the torque split between the EM and ICE for the AE. As seen in the figure, the EM torque

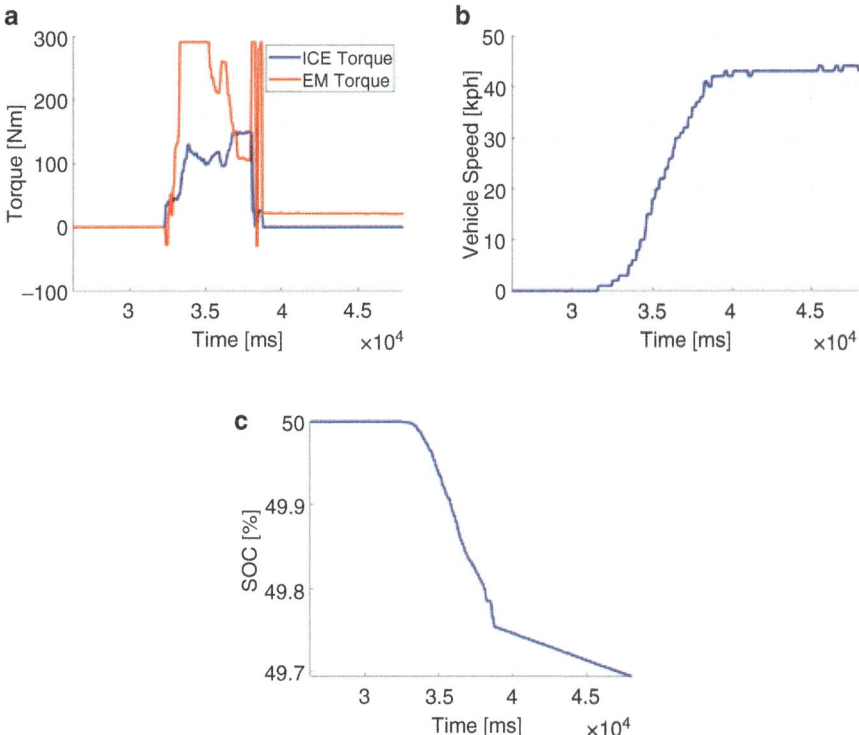

Fig. 11 Results of PAE using OCM method. (**a**) Torque output. (**b**) Vehicle speed trace. (**c**) SOC trace

Fig. 12 Vehicle speed trace used to generate pre-computed ICE and EM torque for baseline and PAE method

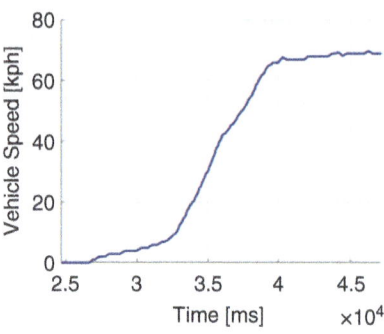

Table 3 PAE testing results

Mean baseline FE	Mean PAE FE	% Improvement
11.820	16.455	0.282

Table 4 PAE testing results statistics

Number of runs	St.D. of baseline FE	St.D. of PAE FE	Significance (p-value)
5	0.233	0.5267	6.0e-9

was applied at a much larger amount than the ICE torque. If the OCM was working correctly here, the EM torque output would have been negative during the AE to increase the SOC to ensure that the SOC at the beginning and end of the AE were equal. Ultimately, DP based control is subject to discretization and below a certain point, the control will no longer function optimally. The memory limitations on the controller were sufficient to make the matrix method infeasible.

In torque trace method, ICE and EM torque were pre-calculated using the speed trace from Fig. 12. For Baseline, APP signal that would result in the speed trace in Fig. 12 was generated and used as an input to control the vehicle behavior in test. For PAE, ICE and EM torque that would result Fig. 12 was generated and used as an input instead. In contrast, the torque trace method generated more favorable results as the vehicle was to perform as expected when pre-computed torque traces were fed into the HSC. for this reason the torque trace method was selected for data collection.

3.2 PAE vs Baseline Results

PAE control was tested for an AE of 0–30 kph. Results for these events are shown in Tables 3 and 4.

From the tests, it was found that the torque trace PAE method improved the FE by an average of 28.17%. To get to this number, the amount of energy used for the time that it took to complete the AE was converted into a gallon equivalent amount. This was then combined with the fuel usage of the ICE and was used to divide the

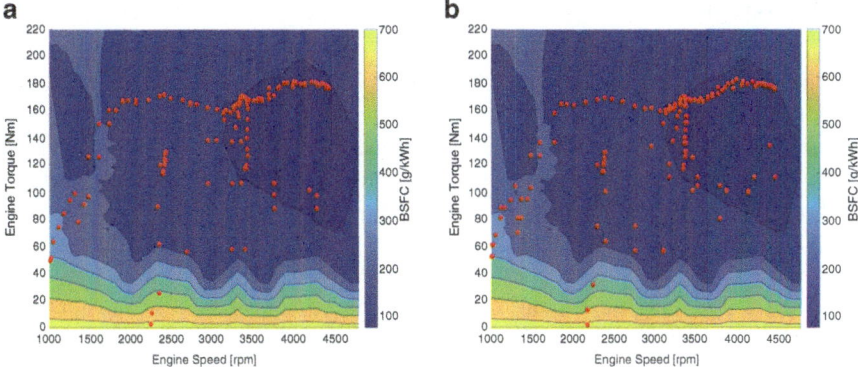

Fig. 13 Engine speed vs engine torque under PAE method. Red dots represent engine behavior during PAE operation. (**a**) PAE AE 1. (**b**) PAE AE 2

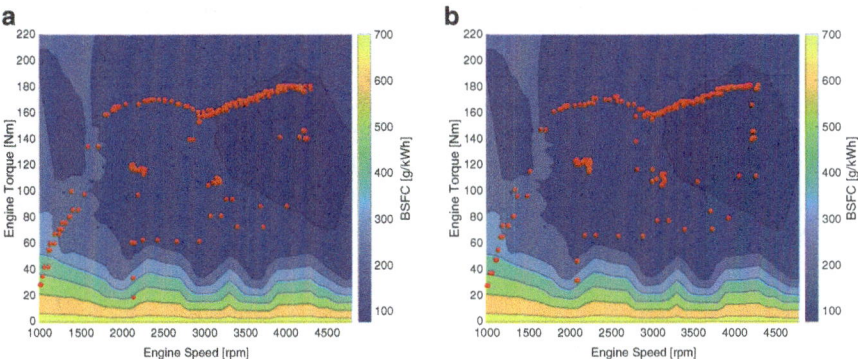

Fig. 14 Engine speed vs engine torque under baseline method. Red dots represent engine behavior during Baseline operation. (**a**) Baseline AE 1. (**b**) Baseline AE 2

distance driven to obtain the Miles Per Gallon (MPG) value. This value could then be used for comparison. For two PAE AEs and two Baseline AEs, the engine torque was plotted against the engine speed on the BSFC map developed previously. This can be seen in Figs. 13 and 14.

The PAE controlled AE shown in Fig. 13a spent 42.9% of the event in the darkest blue, or most fuel efficient zone. Figure 13b shows the next PAE AE which spent 43.7% of the time in that state. Figure 14a's AE, which used Baseline control, spent 29.6% of the event in the dark blue zone. The Baseline controlled AE in Fig. 14b spent 29.1% in the fuel efficient zone. Using this information with the BSFC plots, it can be seen that the AEs using PAE control spend more time with the ICE in the darkest blue, or most fuel efficient, section of the map than the Baseline AEs. The PAE control method was commanding the ICE into the more fuel efficient states during the AE allowing the vehicle to improve the FE by an average of 28.17%.

4   Conclusion

As outlined in Sect. 1, much research has been conducted into the theoretical use of optimal control for torque split in HEVs. Nearly every study of predictive powertrain control to date has used complicated and computationally expensive ways to optimize the operation of the powertrain [6, 29]. These studies include methods like DP, machine learning, and model predictive control to determine the optimal control for the vehicle.

In this study, a map-based control method, based on offline learning was used to realize a pseudo-optimal control that was robust to disturbances and realized measurable fuel economy gain. The control system was able to use the pre-computed PAE values as a lookup table to determine the optimal torque output for the ICE and EM. This meant that by pre-computing the PAE map, the PAE control could occur on the Motohawk control hardware with the rest of the supervisory control algorithm.

Testing this vehicle marked the first time in literature that a test vehicle was able to demonstrate the FE benefits of predictive control algorithms [6]. The previous literature has calculated, modeled, and simulated predictive control hypothesizing the fuel economy benefits while this paper demonstrated the actual improvements in FE for a PAE control when compared to a baseline control.

This real-world validation is significant as, despite research efforts, currently, all of the HEVs on the road currently use instantaneously optimized control to be able to control the powertrain of the vehicle [6]. PAE is an implementable way of improving the FE of HEVs, using information from the predicted vehicle trajectory. This research demonstrated that these strategies were feasible and could improve the FE of HEVs.

References

1. Adelman, D.: Post-transmission parallel hybrid vehicle design and validation for predictive acceleration event energy management strategies. Master's Thesis, Colorado State University, 2021
2. Arata, J.P. III.: Simulation and control strategy development of power-split hybrid-electric vehicles. Ph.D. Thesis, Georgia Institute of Technology, 2011
3. Argonne National Laboratory: Downloadable dynamometer database | argonne national laboratory (2017). https://www.anl.gov/energy-systems/group/downloadable-dynamometer-database, Accessed 16 Oct 2017
4. Asher, Z.D., Baker, D.A., Bradley, T.H.: Prediction error applied to hybrid electric vehicle optimal fuel economy. IEEE Trans. Control Syst. Technol. 26(6), 1–14 (2017)
5. Asher, Z.D., Trinko, D.A., Bradley, T.H.: Increasing the fuel economy of connected and autonomous Lithium-Ion electrified vehicles. In: Behaviour of Lithium-Ion Batteries in Electric Vehicles. Green Energy and Technology, pp. 129–151. Springer, Cham (2018)
6. Asher, Z.D., Patil, A.A., Wifvat, V.T., et al.: Identification and review of the research gaps preventing a realization of optimal energy management strategies in vehicles. SAE Int. J. Altern. Powertrains 8(2), 133–150 (2019)

7. Asher, Z.D., Trinko, D.A., Payne, J.D., et al.: Real-time implementation of optimal energy management in hybrid electric vehicles: globally optimal control of acceleration events. J. Dyn. Syst. Meas. Control **142**(8) (2020). https://doi.org/10.1115/1.4046477

8. Bellman, R.: Dynamic programming and lagrange multipliers. Natl. Acad. Sci. **42**(10), 767–769 (1956)

9. Bertsekas, D.P.: Dynamic Programming and Optimal Control, vol. I. Athena Scientific, Nashua (1995)

10. Bianchi, D., Rolando, L., Serrao, L., et al.: A rule-based strategy for a Series/Parallel hybrid electric vehicle: an approach based on dynamic programming. In: ASME 2010 Dynamic Systems and Control Conference. American Society of Mechanical Engineers, pp 507–514 (2010)

11. Borhan, H.A., Vahidi, A., Phillips, A.M., et al.: Predictive energy management of a power-split hybrid electric vehicle. In: 2009 American Control Conference, pp 3970–3976 (2009). ieeexplore.ieee.org

12. Burress, T.A., Campbell, S.L., Coomer, C., et al.: Evaluation of the 2010 toyota prius hybrid synergy drive system (2011). https://doi.org/10.2172/1007833

13. EPA, NHTSA, CARB: Draft technical assessment report: midterm evaluation of Light-Duty vehicle greenhouse gas emission standards and corporate average fuel economy standards for model years 2022-2025 (EPA-420-D-16-900, July 2016). Technical Report, EPA-420-D-16-900, Office of Transportation, Air Quality U.S. Environmental Protection Agency National Highway Traffic Safety Administration U.S. Department of Transportation, and the California Air Resources Board, 2016

14. Frank, A.A.: Control method and apparatus for internal combustion engine electric hybrid vehicles. US Patent 6054844 (2000)

15. Gunst, R.F.: Response surface methodology: process and product optimization using designed experiments. Technometrics **38**(3), 284–286 (1996)

16. HomChaudhuri, B., Vahidi, A., Pisu, P.: Fast model predictive control-based fuel efficient control strategy for a group of connected vehicles in urban road conditions. IEEE Trans. Control Syst. Technol. **25**(2), 760–767 (2017)

17. Kawamoto, N., Naiki, K., Kawai, T., et al.: Development of new 1.8-liter engine for hybrid vehicles. In: SAE Technical Paper. SAE International, 2009. https://doi.org/10.4271/2009-01-1061

18. Kim, N., Rousseau, A., Rask, E.: Autonomie model validation with test data for 2010 toyota prius. In: SAE Technical Paper, SAE International (2012). https://doi.org/10.4271/2012-01-1040

19. Lin, C.C., Kang, J.M., Grizzle, J.W., et al.: Energy management strategy for a parallel hybrid electric truck. In: Proceedings of the 2001 American Control Conference. (Cat. No.01CH37148), vol 4, pp 2878–2883 (2001). ieeexplore.ieee.org

20. Lin, C.C., Peng, H., Grizzle, J.W.: A stochastic control strategy for hybrid electric vehicles. In: Proceedings of the 2004 American Control Conference, vol 5, pp 4710–4715 (2004). ieeexplore.ieee.org

21. Mckenney, B.: Comparison of design and implementation of hybrid systems in prototype vehicles. Master's Thesis, Colorado State University, 2021

22. Onori, S., Serrao, L., Rizzoni, G.: 2010. Adaptive equivalent consumption minimization strategy for hybrid electric vehicles. In: ASME 2010 Dynamic Systems and Control Conference. American Society of Mechanical Engineers, pp. 499–505

23. Onori, S., Serrao, L., Rizzoni, G.: Equivalent consumption minimization strategy. In: Hybrid Electric Vehicles. SpringerBriefs in Electrical and Computer Engineering, pp. 65–77. Springer, London (2016)

24. Opila, D.F., Wang, X., McGee, R., et al.: Real-time implementation and hardware testing of a hybrid vehicle energy management controller based on stochastic dynamic programming. J. Dyn. Syst. Meas Control **135**(2), 021,002 (2013)

25. Paganelli, G., Delprat, S., Guerra, T.M., et al.: Equivalent consumption minimization strategy for parallel hybrid powertrains. In: Vehicular Technology Conference. IEEE 55th Vehicular Technology Conference, VTC Spring 2002 (Cat. No.02CH37367), vol. 4, pp 2076–2081 (2002). ieeexplore.ieee.org

26. Payne, J., Stefanon, H., Geller, B., Aoki, T., Bradley, T., Asher, Z. and Trinko, D.: Colorado State University Research Foundation, Toyota Motor Engineering and Manufacturing North America Inc, 2021. Systems and methods for determining engine start time during predicted acceleration events. U.S. Patent 10,946,852

27. Rabinowitz, A., Araghi, F.M., Gaikwad, T., et al.: Development and evaluation of velocity predictive optimal energy management strategies in intelligent and connected hybrid electric vehicles. Energies **14**(18), 5713 (2021)

28. Rajamani, R.: Vehicle Dynamics and Control. Springer Science & Business Media, New York (2011)

29. Sabri, M., Danapalasingam, K.A., Rahmat, M.F.: A review on hybrid electric vehicles architecture and energy management strategies. Renew. Sust. Energ. Rev. **53**, 1433–1442 (2016)

30. Society of Automotive Engineers: Recommended practice for measuring the exhaust emissions and fuel economy of Hybrid-Electric vehicles. Technical Report, J1711, SAE International, 2002

31. Trinko, D.A.: Predictive energy management strategies for hybrid electric vehicles applied during acceleration events. Master's Thesis, Colorado State University, 2019

32. Trinko, D.A., Asher, Z.D., Bradley, T.H.: Application of pre-computed acceleration event control to improve fuel economy in hybrid electric vehicles. Technical Report, SAE Technical Paper, 2018

33. U.S. Energy Information Administration: Annual Energy Outlook 2022. Technical Report, 2022

34. U.S. Environmental Protection Agency: 2017 best and worst fuel economy vehicles. https://www.fueleconomy.gov/feg/best-worst.shtml, Accessed 16 Oct 2017

35. Vagg, C., Akehurst, S., Brace, C.J., et al.: Stochastic dynamic programming in the Real-World control of hybrid electric vehicles. IEEE Trans. Control Syst. Technol. **24**(3), 853–866 (2016)

36. Yang, C., Du, S., Li, L., et al.: Adaptive real-time optimal energy management strategy based on equivalent factors optimization for plug-in hybrid electric vehicle. Appl. Energy **203**(Supplement C), 883–896 (2017)

Learning-Based Social Coordination to Improve Safety and Robustness of Cooperative Autonomous Vehicles in Mixed Traffic

Rodolfo Valiente, Behrad Toghi, Mahdi Razzaghpour, Ramtin Pedarsani, and Yaser P. Fallah

1 Introduction

Autonomous vehicles (AVs) have been an attractive research area for decades, as it offers the potential to generate more efficient and safer road networks [1]. The adoption of AVs will not become a reality until they can co-exist with humans, as part of a complex social system. In order to maximize the potential of AVs and optimize for safety and traffic efficiency of all the vehicles on the road, AVs have to coordinate and influence the other agents [1–3].

We recognize the importance of social interaction and behavior in safety and reliability and identify two important research directions. First, AVs must be social actors and behave predictably and safely. Driver behavior is shaped by habits and expectations in the traffic environment. The vehicle's interaction will be influenced by the way AV decisions are perceived. Therefore, the ability of AVs to drive in a socially obedient manner is critical for the safety of passengers and other vehicles because predictable behavior allows humans to comprehend and respond appropriately to the AV's actions. Second, AVs must be social-aware and learn to identify social cues of egoism or altruism, understand the behavior of human drivers and learn how to interact and coordinate with all agents in a mixed traffic

R. Valiente (✉) · Behrad Toghi · Mahdi Razzaghpour · Y. P. Fallah
Connected & Autonomous Vehicle Research Lab (CAVREL), University of Central Florida, Orlando, FL, USA
e-mail: rvalienter90@knights.ucf.edu; toghi@knights.ucf.edu; razzaghpour.mahdi@knights.ucf.edu; Yaser.Fallah@ucf.edu

R. Pedarsani
Department of Electrical and Computer Engineering, University of California, Santa Barbara, CA, USA
e-mail: ramtin@ece.ucsb.edu

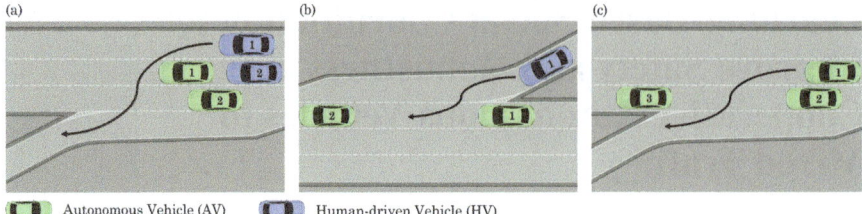

Fig. 1 (a) **Interaction of AV-HV to benefit a HV:** Altruistic agents create alliances and direct the behavior of HVs to improve traffic flow and prevent dangerous circumstances. AV1 and AV2 can create a formation to guide HV2 and provide a route for HV1, allowing the HV to change lanes and navigate to the exit ramp. (b) **Interaction of AV-AV to benefit a HV:** The goal of HV1 is to integrate onto the highway. Egoistic AVs disregard the merging vehicle and do not make room for it, possibly resulting in dangerous situations, however, if they exhibit sympathy for the merging HV, they can compromise on their own interest to create a safe path for HV1 to merge into the highway. (c) **Interaction of AV-AV to benefit another AV:** The goal of AV1 is to exit the highway. If AV2 acts selfishly, AV1 may miss the exit and be unable to complete its task. However, if AV2 and AV3 consider AV1's mission and act altruistically, they can free up space in the platoon by AV2 decelerating and AV3 accelerating, allowing AV1 to safely take the exit

environment, adapting and influencing the HVs behaviors to optimize for a social utility that improves traffic flow and safety.

In this chapter, we focus on social awareness challenges and seek a solution that can ensure the safety and robustness of AVs in the presence of human drivers with heterogeneous behavioral traits. Vehicle-to-vehicle (V2V) communication allows connected and autonomous vehicles (CAVs) to interact directly with their neighbors [4, 5]. By using V2V communication CAVs can create an extended perception that facilitates explicit cooperation among vehicles to overcome the limits of a non-cooperative agent [6–8]. While planning in a fully AV scenario is relatively easy to achieve, coordination in the presence of HVs is a significantly more challenging task, as the AVs not only need to react to road objects but also need to consider the behaviors of HVs [9–11].

In contrast to the individual non-cooperative approaches, we investigate the mixed-autonomy decision-making challenge from a multi-agent and social perspective. Rewarding AVs for adopting an altruistic behavior and taking into consideration the interests of other vehicles allows them to see the broad picture and find solutions that maximize the utility of the group. In addition to the potential benefits of altruistic decision-making in terms of safety and efficiency, altruism results in more societally advantageous outcomes [2]. Figure 1a demonstrates how a group of AVs can guide HV to increase safety and efficiency, while Fig. 1b and c show how AVs can collaborate to accomplish a social objective that benefits another HV or AV.

Currently, AVs lack an understanding of human behavior and frequently act extremely cautiously to avoid collisions. This conservative behavior not only leaves AVs unprotected from aggressive HVs, but also results in unexpected reactions that confuse HVs, creating bottlenecks in traffic flow and causing accidents. It's critical to distinguish between a human driver's individual traits, such as aggressiveness, conservativeness, and risk tolerance, and their social preferences, such as egoism and altruism. Even though the two categories are related, they have distinctive natures and so behave differently in mixed traffic. An aggressive driver, for example, is not inherently egoistic or selfish, but their aggression may hinder their ability to collaborate with other drivers and participate in a socially desirable coexistence with AVs [3, 12, 13]. In the field of behavior planning for AVs in mixed-autonomy traffic, we identify two fundamental problems. First, human drivers differ considerably in their individual traits and social preferences, making AV behavior planning exceedingly difficult because it is difficult for the AV to foresee the type of behavior it would encounter when dealing with a human driver. Furthermore, relying on a real-time inference of HV behaviors is not always feasible because vehicle interactions can be brief, such as when two vehicles meet at an intersection. Second, driving requires complex interactions of agents in a partially observable and non-stationary environment, as HVs do not follow a fixed policy and modify their policies in real-time in response to the actions of other vehicles.

The integration of AVs into the real world requires them to address those challenges. Due to the differences in maneuverability and reaction time between AVs and HVs, a road shared by both becomes a competitive situation. In contrast with the full-autonomy case, here the coordination between HVs and AVs is not as straightforward since AVs do not have an explicit means of harmonizing with humans and are therefore required to locally account for the other HVs and AVs in their proximity. This dilemma intensifies if AVs act egoistically and optimize solely for their local utility. As an illustration, Figs. 2 and 3 demonstrate a highway exiting and merging scenario in mixed traffic. We consider a general setup where AVs and HVs with different behaviors coexist. Vehicles need to efficiently merge onto the lane or exit the highway without collisions. In an ideal cooperative environment, AVs should proactively decelerate or accelerate to provide sufficient room for vehicles to safely exit/merge and prevent hazardous situations, while also being resilient to various situations and behaviors and assuring safety in decision-making [14]. For instance, in Fig. 2 (merging scenario) if AVs act egotistically, the merging vehicle must rely on the HV to slow down to allow it to merge. However, due to the unpredictability of HVs, relying solely on HVs might result in suboptimal or even dangerous circumstances. Therefore, if all AVs act egotistically, the merging vehicle would either be unable to join the highway or it will wait for an HV and risk cutting into the highway without knowing whether the HV will slow down. Nevertheless, if AVs act altruistically, they can coordinate to guide the traffic on the highway to allow for a seamless and safe merging. In particularly challenging driving scenarios, such altruistic AVs can achieve societally beneficial results without relying on or making assumptions about HVs behaviors.

To address these challenges, existing literature either depend on models of human behavior generated from pre-recorded driving datasets [15, 16] or define social utilities that can impose cooperative behavior among AVs and HVs [17]. Other works focus on rule-based methods that use heuristics and hand-coded rules to guide the AVs [18] or probabilistic driver modeling [19] learned from human driving data.

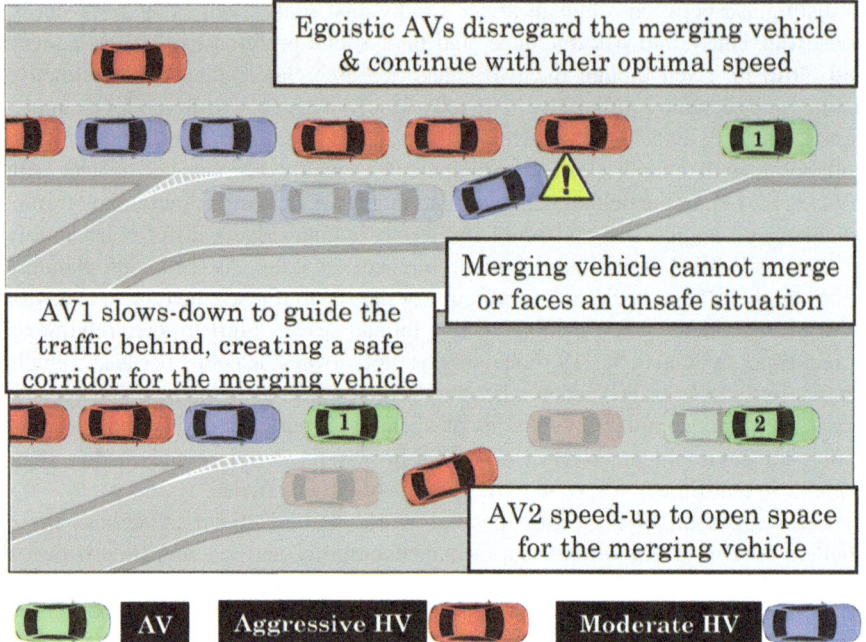

Fig. 2 For a seamless and safe highway merging, all AVs must coordinate and account for the utility of HVs. *(top)* Egoistic AVs optimize only for their own utility, *(bottom)* Altruistic AVs consider also the HV's utility

While this is feasible for simple situations, these methods become impractical in complex scenarios. Additionally, the human driver models learned in the absence of AVs, are not necessarily valid when humans confront AVs. This limits the application of the generated solutions, as they are frequently limited to the human behaviors with which AVs interacted during training. To account for this, several works in the literature adopt an excessively cautious approach when interacting with humans [20]. This strategy not only leaves the AVs vulnerable to other aggressive drivers, particularly in competitive situations, but it also causes traffic congestion and significant safety risks [1, 2].

On the other hand, data-driven methods such as reinforcement learning (RL) have received increased attention [21] as RL-based methods can learn decision-making and driving behaviors that are hard for traditional rule-based designs. However, the majority of the RL approaches are designed for a single AV, or try to handle the interaction between AVs and HVs either by predicting human behavior or by relying on the fact that humans are willing to collaborate or can be influenced to do so [15, 22], which could compromise safety or lead to sub-optimal performance. Recent works consider social interactions of AVs and train altruistic AVs that learn from experience and influence HVs to optimize a social utility function that benefits all vehicles on the road [10, 23].

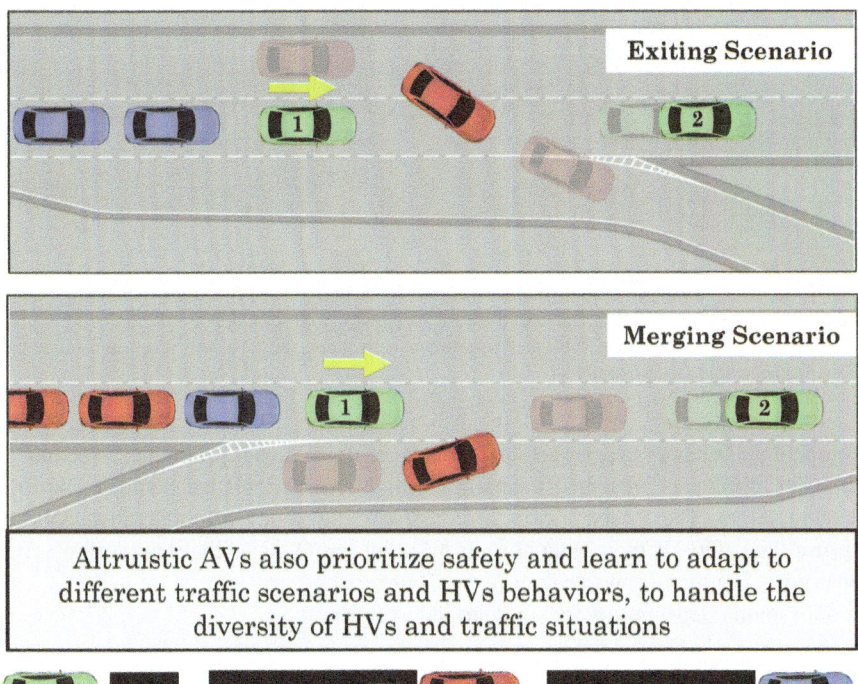

Fig. 3 Highway exiting and merging scenarios with AVs (green) and aggressive HVs (red) sharing the road. Altruistic AVs must learn to cooperate to exit/merge successfully and safely while being adaptable to a variety of scenarios and HV behaviors

In contrast, we consider a data-driven multi-agent reinforcement learning (MARL) approach and let the autonomous agents implicitly learn the decision-making process of human drivers only from experience, while optimizing for a social utility. By incorporating a cooperative reward structure into our MARL framework, we can train AVs that coordinate with each other, sympathize with HVs, and, as a result, demonstrate enhanced performance in competitive driving scenarios, such as highway exiting and merging. Despite not having access to an explicit model of the human drivers, the trained autonomous agents learn to implicitly model the environment dynamics, including the behavior of human drivers, which enables them to interact with HVs and guide their behavior.

This research aims to create a safe and robust training regimen that allows AVs to collaborate and influence the behavior of human drivers to achieve socially desirable outcomes, regardless of HV individual traits and social preferences. We based our work on the following insights. First, we rely on a decentralized reinforcement learning architecture that optimizes for a social utility that learns from experience and exposes the learning agents to a wide range of driving. As a result, the agents become more resistant to human driver behavior and can handle cooperative-

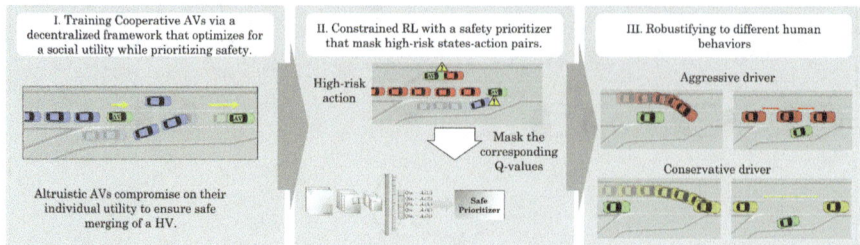

Fig. 4 An overview of our approach to leverage social awareness and coordination to improve the safety and reliability of CAVs. Our social-aware AVs learn from scratch not only to drive but also to understand the behavior of HVs and coordinate with them, they learn to adapt and influence HVs in a robust and safe manner

competitive behaviors regardless of HV's hostility or social preference. Second, a safety prioritizer is presented to minimize high-risk actions that could jeopardize driving safety. The safety prioritizer constrains the policy of cooperative AVs to ensure the safety of their behavior via masking the Q-states that lead to high-risk outcomes. Figure 4 shows an overview of our process.

Our main contributions are summarized as follows:

- We formulate the mixed-autonomy problem as a decentralized MARL problem and present an approach to training altruistic agents which utilizes a decentralized reward mechanism for achieving socially advantageous behaviors and takes advantage of a 3D convolutional deep reinforcement learning architecture to capture the temporal information in driving data.
- A training algorithm is proposed to make AVs robust to different drivers' behavior and situations while producing socially desirable outcomes. We investigate the effect of HVs behaviors on our altruistic AVs agents and especially conclude that the higher the traffic aggressiveness, the higher the importance of social coordination.
- We investigate the scenarios in which altruistic AVs can learn cooperative policies that are robust to diverse traffic scenarios and HV behaviors without compromising efficiency and safety, and present the results on transfer learning and domain adaptation in mixed-autonomy traffic.

The purpose of this chapter is to study the challenges of robust and safe AVs in mixed-autonomy traffic, especially in intrinsically competitive driving scenarios like those shown in Fig. 2, in which coordination is essential for safety and efficiency. The intention is to utilize the autonomous driving challenge as a case study to examine the use of social theories from psychology literature in the MARL domain. To apply these theories to real-world roads, more study is required. Nonetheless, the research on altruistic AVs that are robust, safe, and capable of learning to influence HVs in desirable ways, without the limitations of current solutions are promising.

2 Related Work

2.1 Multi-Agent Reinforcement Learning

The intrinsic non-stationarity of the environment is a key problem for MARL. To address those limitations a MARL derivation of importance sampling is proposed and used to remove the outdated samples from the replay buffer [24]. In [25] is presented another solution to address this issue by including latent representations of partner strategies allowing partner modeling and more scalable MARL.

To mitigate the problem of credit assignment in multi-agent systems, [26] proposed the *counterfactual multi-agent (COMA)* algorithm, which employs a centralized critic and decentralized actors. In [27] is proposed a deep RL algorithm with full environment observability and a centralized controller to govern the joint-actions of all the agents. Other current research on mixed-autonomy focuses on addressing cooperative and competitive challenges by assuming the nature of interactions between autonomous agents [28]. In [29] a variation of an actor-critic approach with a *centralized q-function* is proposed. The algorithm has access to local observations and the actions of all agents. In our work, in contrast, we consider a decentralized controller with partial observability, and train altruistic agents that optimize for a social utility.

2.2 Driver Behavior and Social Coordination

Existing works on driver behavior and social navigation approach agents coordination by either modeling driver behaviors [19, 30, 31] or simplifying and making assumptions about the nature of agent interactions [28, 32]. In [33] is presented a maneuver-based dataset and a model for classifying driving maneuvers is proposed. Other works on driver behavior modeling consider graph theory [34], data mining [35], driver attributes [36] or game theory [2]. In [31] is proposed a method for modeling and forecasting human behavior in circumstances that involve multi-human interactions in highly multi-modal situations.

Current research in social navigation has demonstrated the importance of AVs as social actors and the advantages of coordination between AVs and HVs [37]. Human driving patterns are learned from demonstration using inverse RL in [38] and [22]. Similarly, in [39] is presented a centralized game-theory model for cooperative inverse RL. The authors in [40] and [41] proposed a shared reward function to enable cooperative trajectory planning for robots and humans. Sadigh et al. presents a strategy based on imitation learning to learn a reward function for human drivers, demonstrating how AVs can influence human actors [15]. The importance of coordination and the advantage of using AVs to guide the traffic has been also investigated at the traffic level. Wu et al. [42] analyzes the capability of AVs to stabilize a system of HVs and presents the conditions in which when concurrently enforcing safety constraints on the AVs while stabilizing traffic improves traffic performance. Similar works have highlighted the potential of influencing HVs and how AVs can be used to stabilize and guide the traffic flow [42, 43]. Recent works focus on optimizing traffic networks in mixed autonomy to reduce traffic congestion and improve safety. In [44] is presented a model of vehicle flow and a model of how AV makes decisions among routes with various prices and latencies. The planner optimizes for a social objective and shows improvement in traffic efficiency. The vehicle routing problem is studied in [45] that proposed an innovative learning-augmented local search system to mitigate the problem by using a Transformer architecture. Cameron et al. explores how humans can supervise agents in order to attain an acceptable degree of safety [46]. In contrast to previous works, we do not rely on human cooperation and our AVs learn cooperative behaviors directly from experience, our focus is on the emerging altruistic behavior that allows agents to coordinate and optimize for a social utility.

2.3   Safe and Robust Driving

Safety is critical for AVs [47], and it is especially important for AVs that have been trained via RL. We must prioritize safety; because coordination is frequently associated with risk. In cooperative driving, there are often safe actions that have low rewards and riskier actions with higher rewards [48]; however, the risky action increases the likelihood of crashes when cooperation fails. Especially, AVs utilizing trained RL algorithms, may not always operate safely since the trained models may pick dangerous actions [20]. Several attempts in this direction use pure reward shaping to avoid collisions. While this is a frequent technique in RL, safety is not implicitly emphasized, and AVs implementing such RL methods may not behave properly in some cases due to function approximation.

To overcome this problem, the concept of safe RL is proposed in [20], which aims to increase safety in unobserved driving conditions when the RL algorithm performs dangerously. Wang et al. [49] proposes a rule-based decision-making system that evaluates the controller's decisions and substitutes collision-causing actions. A short-horizon safety supervisor is included in Nageshrao et al. [50] to replace unsafe actions with safer ones. A Q-masking strategy is presented in [51] to prevent collisions by deleting actions that might lead to a crash. Chen et al. proposes a novel priority-based safety supervisor that reduces collisions considerably [52].

We leverage these approaches in this work using a decentralized reward function, local actions, and assuming partial observability, to increase the altruistic agents' safety while also being adaptable to varied driver behaviors and circumstances. As shown in Fig. 2, we analyze a particular situation in which AVs and HVs with various characteristics coexist. The picture depicts two frequent traffic situations in which vehicles must either merge into a lane effectively or depart the highway without colliding with other vehicles. In an ideal cooperative context, vehicles should proactively decelerate or accelerate to provide enough room for vehicles to safely exit/merge and prevent stalemate situations, while also being resilient to various conditions and behaviors and assuring safety in decision-making.

3 Preliminaries and Formalism

We study safety and robustness in the maneuver-level decision-making problem for AVs to see what kinds of behaviors might lead to socially desirable results. We're interested in the question of how AVs can be trained from scratch to drive safely and reliably, while also taking into account the social aspects of their mission, i.e., optimizing for a social utility that takes into account the interests of other vehicles in the vicinity. Social awareness and coordination are essential to improve safety and reliability on the roads. In this work, we explore that insight. Thus, we continue this section by providing a quantitative description of an agent's level of altruism and formally defined our problem.

It is possible to define the MARL problem as a centralized or decentralized problem. It's simple to create a centralized controller that provides a central joint reward and joint action. However, in the real world, such assumptions are unfeasible. In this chapter, we focus on a decentralized controller with partial observability and formulate the problem as a partially observable stochastic game (POSG) defined by $\langle \mathcal{I}, \mathcal{S}, P, \gamma, \{\mathcal{A}_i\}_{i\in\{1,...,N\}}, \{O_i\}_{i\in\{1,...,N\}}, \{R_i\}_{i\in\{1,...,N\}}\rangle$ where

- $\mathcal{I}$: a finite set of agents $N \geq 2$.
- $\mathcal{S}$: a set of possible states that contains all configurations that N AVs can take (probably infinite).
- P: a state transition probability function from state $s \in \mathcal{S}$ to state $s' \in \mathcal{S}$, $P(S = s'|S = s, A = a)$.
- γ: a discount factor, $\gamma \in [0, 1]$.
- $\mathcal{A}_i$: a set of possible actions for agent i.
- O_i: a set of observations for agent i.
- R_i: a reward function for the ith agent, $R_i(s, a)$.

At a given time t the agent senses the environment and receives a local observation $o_i : \mathcal{S} \rightarrow O_i$, based on the observation o_i and its stochastic policy $\pi_i : O_i \times \mathcal{A}_i \rightarrow [0, 1]$, the agent takes an action within the action-space $a_i \in \mathcal{A}_i$. Consequently, the agent transits to the next state s' which is determined based on the state transition probability function $P(s'|s, a) : \mathcal{S} \times \mathcal{A}_1 \times ... \times \mathcal{A}_N \rightarrow \mathcal{S}$ and receives a decentralized reward $r_i : \mathcal{S} \times \mathcal{A}_i \rightarrow \mathbb{R}$. The goal of each agent i is to optimally solve the POSG by deriving a probability distribution over actions in $\mathcal{A}$ at a given state, that maximizes its cumulative discounted sum of future rewards over an infinite time horizon and find the corresponding optimal policy $\pi^* : \mathcal{S} \rightarrow \mathcal{A}$.

An optimal policy maximizes the action-value function, i.e.,

$$\pi^*(s) = \arg\max_a Q^*(s, a) \tag{1}$$

where,

$$Q^{\pi}(s, a) := \mathbb{E}_{\pi}[\sum_{k=0}^{\infty} \gamma^k R_k(s, a)|s_0 = s, a_0 = a]. \tag{2}$$

The optimal action-value function is determined by solving the Bellman equation,

$$Q^*(s, a) = \mathbb{E}\left[R(s, a) + \gamma \max_{a'} Q^*(s', a')|s_0 = s, a_0 = a\right] \tag{3}$$

3.1 Double Deep Q-Network

Deep Q-network (DQN) has been widely used in RL problems. DQN uses a deep neural network (NN) with weights $\mathbf{w}$ as the function approximator to estimate the state-action value function, i.e., $\tilde{Q}(.; \mathbf{w}) \cong Q(.)$. DDQN improves DQN by decomposing the max operation in the target into action selection and action evaluation, mitigating the over-estimation problem. The idea is to periodically sample data from a buffer and compute an estimate of the Bellman error or loss function, written as

$$\mathcal{L}(\mathbf{w}) = \mathbb{E}_{s,a,r,s' \sim \mathcal{RM}}[(Target - \tilde{Q}(s, a; \mathbf{w}))^2] \tag{4}$$

$$Target = R(s, a) + \gamma \tilde{Q}(s', \arg\max_{a'} \tilde{Q}(s', a'; \mathbf{w}); \hat{\mathbf{w}})) \tag{5}$$

The DDQN algorithm then performs mini-batch gradient descent steps as $\mathbf{w}_{i+1} = \mathbf{w}_i - \alpha_i \hat{\nabla}_{\mathbf{w}} \mathcal{L}(\mathbf{w})$, on the loss $\mathcal{L}$ to learn the approximation of the value function ($\tilde{Q}(.)$). The $\hat{\nabla}_{\mathbf{w}}$ operator denotes an estimate of the gradient at $\mathbf{w}_k$, $\mathbf{w}$ are the weights of the online network and $\hat{\mathbf{w}}$ are the weights of the target network which are updated at a lower frequency ($Target_{update}$) to stabilize training. The experience replay buffer (RM) is used to generate training samples (s, a, r, s'), which are randomly drawn to protect from correlated observations and non-stationary data distribution.

3.2 Driving Scenarios

Our objective is to investigate driving scenarios in which the lack of AV coordination hinders safety and efficiency. We also study adaptability among scenarios and driver behaviors. For this, we design a set of scenarios $\mathcal{F}$ with highway exiting and merging ramps as the main scenarios, as shown in Fig. 2, where a mission vehicle (in our case an exiting/merging vehicle) attempts to accomplish its task in a mixed-traffic environment.

The exiting and merging scenarios are designed in such a way that coordination is necessary for safety. AVs must coordinate, and neither can achieve a safe and smooth traffic flow on its own, i.e., exiting/merging will not be feasible without the coordination of the other AVs. To facilitate safe exiting/merging while also responding to varied traffic scenarios, altruistic AVs must learn to account for the interests of all vehicles, coordinate, make compromises, and influence human behavior. In Fig. 2, for example, the AV1 has to compromise its own utility and reduce speed to guide the traffic of aggressive HVs, creating space for the exiting/merging vehicle, while the other AVs have to increase speed to create room for the mission vehicle. The exiting and merging scenarios are defined as $f_e, f_m \in \mathcal{F}$ correspondingly. We particularly chose those scenarios as a case of study because of their intrinsic similarity and the need for coordination, as the exiting/merging vehicle's utility contrast with that of the HV highway vehicles.

3.3   Social Value Orientation for AVs

In this section, we introduce *Social Value Orientation (SVO)* to formally investigate the social conflicts between humans and agents in diverse environments. It is critical to quantify an individual's social preference to understand whether they would cooperate or not in a particular scenario, such as opening a gap in our highway merging example. For that purpose, SVO is a commonly used concept in the social psychology literature that has lately been applied in robotics research [2]. In our context, SVO defines the degree of an agent's egoism or altruism toward others. Based on the value placed on the utility of others, an HV or an AV's behavior can range from egoistic to completely altruistic. We rely on AVs to guide traffic toward more socially advantageous outcomes since the SVO of HV is unknown. In formal terms, an AV's SVO angle ϕ determines how the AV balances its own reward against that of others [10, 17, 53]. In terms of rewards, an AV's total reward R_i is defined as:

$$R_i = r_i \cos \phi_i + r_i^- \sin \phi_i \tag{6}$$

where r_i is the agent's individual utility, r_i^- is the total utility of other agents from the perspective of the ith agent which in general is a function $f(.)$ of their individual utilities,

$$r_i^- = f(r_j), \quad \text{where } j \neq i \tag{7}$$

The SVO angle can varied from $\phi = 0$ (entirely selfish) to $\phi = \pi/2$ (entirely altruistic). Nonetheless, none of the limits are optimal, and a point in the middle, known as the optimal SVO angle ϕ^* gives the most socially favorable outcome.

Fig. 5 The SVO angle ϕ quantifies the level of altruism of an agent. In the figure the diameter of the circles, represents the size of the human population that holds the associated SVO [55]

SVO allows us to understand the behaviors that make possible the socially desirable outcomes in Fig. 2.

Autonomous agents must be aware of human drivers' social preferences as well as their desire to collaborate. Humans, on the other hand, are known to be diverse in SVO, and so their preferences are uncertain [54]. Figure 5 depicts a range of altruism across individuals with varying SVO. As a result of the wide range of altruistic behavior seen in humans, is not safe to rely on humans to guide the traffic, instead, we should rely on AV to guide the traffic toward more socially advantageous goals. Therefore, our objective is that the AVs learn to create alliances and influence HV behavior to improve the global utility of the group.

3.4 Autonomous Vehicles as Social Actors

AVs in a mixed environment will be social actors in the traffic road that will react to HVs and influence and adapt to their behaviors. The traffic environment is rich with habits and expectations, that determine driver behaviors. The vehicle's interaction will be influenced by the way AV decisions are perceived [2, 56]. For instance, some human drivers may be grateful if the AV stops for them but frustrated if it does not perform as expected. Also, they might behave aggressively if they're stuck behind an overly cautious AV, which reduces speed constantly. Another example is the case that when crossing a street while a vehicle is waiting, pedestrians move faster (a gesture of respect for the driver). On the other hand, will pedestrians speed up for an AV, or will they behave differently? If an AV is understood as a social actor, the HVs will learn the individual and social traits of AVs and behave accordingly in mutual interactions. This would fit with current preconceptions that make assumptions about drivers based on the brand and type of vehicle they drive. Current AVs' driving is as conservative as possible to ensure safety. They will slow down in front of a crossing because they believe the other vehicle will want to go first, even though this is against the law. They wait for pedestrians when in doubt. It's not difficult to see how other agents and HVs could take advantage of and exploit these over-conservative behaviors. As AVs are going to be social actors in mixed autonomy traffic, the safety and reliability of AVs will be coupled with their social awareness and their ability to engage in complex social interactions. We consider risk awareness and social behavior as fundamental traits for decision-making.

Failure to identify social cues of selfishness or collaboration by an AV has ramifications for the general flow of the traffic network, as well as the safety of traffic participants. Current AVs ignore social signs and driver personality in favor of explicit communication or driver modeling. Because these methods can't handle complicated interactions, they tend to be conservative, restricting autonomy solutions to simple road interactions [2, 56]. The ability of AVs to drive in a socially obedient manner is critical for the safety of passengers and other vehicles because predictable behavior allows humans to comprehend and respond appropriately to the AV's actions.

3.5 Driving Behaviors

The problem of simulating varied behaviors may be defined as determining the appropriate range of parameters to produce heterogeneous behaviors within the simulator. Some works in social traffic psychology show that driving behavior falls between conservative and aggressive. Nevertheless, the specific definition is still under discussion and fluctuates across works [3]. The phrase "aggressive driving" refers to a wide range of unsafe driving practices, including running red lights and speeding. The root of aggressive driving has a variety of factors that aren't necessarily clear. Some are caused by hazardous road conditions, while others are caused by personal characteristics or mental states. Moreover, there is a correlation between aggressiveness and egoism, as egoistic drivers are less likely to yield and have a tendency to over-speeding and engage in unsafe actions. While there is a correlation between these concepts [12, 13], we distinguish aggressiveness from egoism in this study by describing individual traits and social preferences.

In this work, we discriminate between individual traits and social preferences because they result in different behaviors. We define altruism and egoism as social preferences; in that sense, an egoistic driver is a selfish driver who accounts for his personal utility irrespective of his aggression. We define conservatism and aggressiveness as individual traits, and we describe an aggressive driver as someone whose actions result in aggressive behavior. Individual traits such as aggressiveness are characterized by the outcomes of their actions, but social preferences such as egoism are distinguished by their social objectives and purposes. In this direction, an egoistic driver is a self-centered driver who lacks social motive, a driver who believes he controls the road and disregards the other drivers. Egoist drivers frequently engage in violent actions, and while ego defensiveness is not the primary source of aggression, it is a major contributor to aggressive driving [12, 13]. Despite their similarities, the two groups have different origins and result in different behaviors. A driver, for example, could be egoistic and conservative. We may envision a driver who drives cautiously to protect himself (selfish motivation/preference) and, as a result, is conservative in his behavior (outcome of his actions).

Properly, we described social preferences (altruism or egoism) by the AV's SVO angular phase ϕ; and individual traits (conservativeness and aggressiveness) by the HV driver model parameters ($\mathcal{P}$) as described in Sect. 5.5. Based on the values of these parameters, a driver will behave conservative or aggressive. In the simulations, the AVs have no access to HVs' SVO, we consider the SVO of HVs to be undetermined as they cannot communicate that directly. Finally, we define a set of behaviors $\mathcal{B}$, i.e, aggressive, moderate and conservative, $b_a, b_m, b_c \in \mathcal{B}$ based on the parameters ($\mathcal{P}$) obtained in Sect. 5.5.

4   Problem Formulation

We investigate the safety and robustness of the scenarios described in Fig. 2, an exiting/merging vehicle, which can be either HV or AV. This configuration contains a group of AVs that hold the same SVO, as well as a group of HVs which are heterogeneous in their SVO, making it unclear whether they are allies or opponents. Formally, the road is shared by a set of HVs $h_k \in \mathcal{H}$, with an undetermined SVO ϕ_k and heterogeneous behaviors $b_k \in \mathcal{B}$; a set of AVs $i_i \in \mathcal{I}$, that are connected together using V2V communication, controlled by a decentralized policy and sharing the same SVO, and a *mission vehicle*, $M \in \mathcal{I} \cup \mathcal{H}$ that is aiming to accomplish its mission (highway exiting/merging) and it can be either AV or HV. We focus on the multi-agent maneuver-level decision-making problem for AVs in mixed-autonomy environments and study the following problems: how AVs can learn in a mixed-autonomy environment optimal cooperative policies $\pi^*(s)$ that are robust to different scenarios $f \in \mathcal{F}$ and behaviors $b \in \mathcal{B}$ while ensuring safety on the decision-making, and how sensitive is the performance of the altruistic AVs to the HVs' behaviors.

As AVs are connected, we assume that they receive an accurate local observation of the environment $\tilde{o}_i \in \tilde{O}_i$, sensing all the vehicles within their perception range, i.e, a subgroup of HVs $\tilde{\mathcal{H}} \subset \mathcal{H}$ and a subgroup of AVs $\tilde{\mathcal{I}} \subset \mathcal{I}$. Nevertheless, AVs are unable to share their actions or rewards, and they take individual actions from a set of high-level actions $a_i \in \mathcal{A}_i (|\mathcal{A}_i| = 5)$. The goal of this work is to train social-aware AVs that learn how to drive in a mixed-autonomy scenario in a robust, efficient, and safe manner. We are interested in how to obtain a utility function that enables AVs to handle competitive driving scenarios (such as those in Fig. 2) and leads them into socially-desirable decisions that improve traffic efficiency, safety, and robustness.

5 Safe and Robust Social Driving

In this section, we present the safe and robust MARL approach. Our approach uses a general decentralized reward function that optimizes for social utility and induces altruism in the AVs; the general reward function accounts for any anticipated vehicle's mission, allowing it to be applied to a variety of environments; and collisions are reduced by the safety prioritizer. What we define as "driving" is the outcome of decades of human learning from experience. Consequently, we take the same approach and train AVs that learn from experience and define the optimization problem as the eventual desirable social outcome with adaptability, expecting AVs to learn how to drive safely during the process. We carefully design a decentralized general reward function, a suitable architecture, and a safety prioritizer to promote the desired safe altruistic behavior in AVs' decision-making process. The overview of our approach as presented in Figs. 4 and 2 helps us to create intuition on these points, by introducing driving scenarios in which altruistic AVs lead to socially advantageous results while adapting to different traffic scenarios.

Action Space The goal of this research is to look at inter-agent and agent-human interactions, as well as behavioral elements of mixed-autonomy driving. Thus, we choose a more abstract level and define the action-space as a set of discrete meta-actions $a_i \in \mathcal{A}_i$. In particular, we select a set of five high-level actions a_i as,

$$a_i \in \mathcal{A}_i = \begin{bmatrix} \texttt{LaneLeft} \\ \texttt{Idle} \\ \texttt{LaneRight} \\ \texttt{Accelerate} \\ \texttt{Decelerate} \end{bmatrix} \tag{8}$$

These meta-actions are then converted into trajectories and low-level control signals, which ultimately control the vehicle's movement.

Observation Space We use a *multi-channel VelocityMap* observation (o_i) that embeds the relative speed of the vehicle with respect to the ego vehicle in pixel values [17]. We represent the information in multiple semantic channels that embed: (1) an attention map to highlight the position of the ego vehicle, (2) the HVs, (3) the AVs, (4) the mission vehicle, and (5) the road layout. Figure 6 illustrates an example of this multi-channel representation. In order to map the relative speed of the vehicles into pixels, we use a clipped logarithmic function, which improves dynamic range and yields better results than a linear map, i.e.,

$$Z_j = 1 - \beta \log(\alpha |v_j^{(l)}|) \mathbb{1}(|v_j^{(l)}| - v_0) \tag{9}$$

where Z_j is the pixel value of the jth vehicle in the state representation, $v^{(l)}$ is its relative Frenet longitudinal speed from the kth vehicle's point-of-view, i.e.,

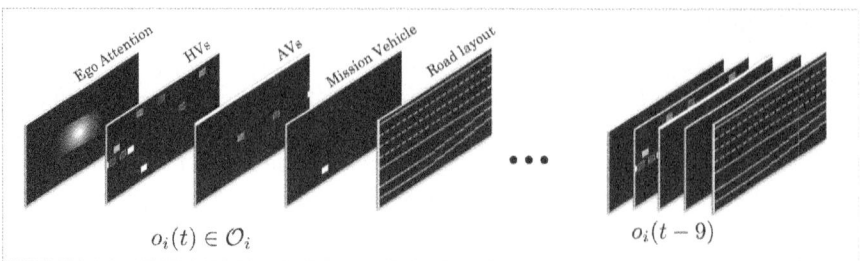

$$o_i(t) \in \mathcal{O}_i \qquad\qquad\qquad o_i(t-9)$$

Fig. 6 Multi-channel VelocityMap state representation embeds the speed of the vehicle in pixel values

$\dot{l}_j - \dot{l}_k$, v_0 is speed threshold, α and β are dimensionless coefficients, and $\mathbb{1}(.)$ is the Heaviside step function. Such non-linear mapping gives more importance to neighboring vehicles with smaller $|v^{(l)}|$ and almost disregards the ones that are moving either much faster or much slower than the ego vehicle. As temporal information is necessary for safe decision-making, we use a history of successive VelocityMaps observations to create the input state to the Q-network.

5.1 Distinguishing Sympathy from Cooperation

In our mixed-autonomy problem, we divide inter-agent relations into interactions between autonomous agents (AV-AV interactions) and interactions between autonomous agents and human drivers (HV-AV interactions). By decoupling the two, we can analyze the interactions between human drivers with unclear SVO and our autonomous agents in a methodical way. In that sense, we define *sympathy* as the autonomous agent's altruism toward a human, and *cooperation* as the altruistic behavior among autonomous agents. The fact that the components of altruism differ in nature is our reasoning for separating them. Sympathy, for example, may not be reciprocated since agents differ in their SVO, whereas cooperation among autonomous entities is fundamentally homogeneous if they share the same SVO. Following this concept, we can rewrite the AV reward in Eq. (6) as,

$$
\begin{aligned}
R_i &= r_i \cos \phi_i + (\sin \theta_i R_i^{\mathrm{AV}} + \cos \theta_i R_i^{\mathrm{HV}}) \sin \phi_i \\
&= \underbrace{r_i \cos \phi_i}_{\text{egoistic term}} + \\
&\quad \underbrace{\sin \theta_i \sin \phi_i R_i^{\mathrm{AV}}}_{\text{cooperation term}} + \underbrace{\cos \theta_i \sin \phi_i R_i^{\mathrm{HV}}}_{\text{sympathy term}}
\end{aligned}
\tag{10}
$$

where θ is the sympathy angular phase determining the cooperation-to-sympathy ratio. Parameters R_i^{AV} and R_i^{HV} denote the total utility of other AVs and HVs, respectively, as perceived from the ith agent's perspective. We expand on this topic in Sect. 5.2 where we introduce the distributed reward structure.

5.2 Decentralized Social Reward

The AVs are trained using the partial local observations and the decentralized reward function, and we anticipate them to learn how to drive in a variety of settings while taking into consideration the individual diver's missions. As a result, we create a well-engineered general reward function that considers social utility, traffic metrics, and individual diver's missions. Following the definition of sympathy and cooperation in equation (10) we decompose the decentralized reward received by agent $I_i \in I$ as,

$$
\begin{aligned}
R_i(s, a) &= R^{\text{ego}} + R^{\text{social}} \\
R^{\text{ego}} &= \cos \phi_i r_i(s, a) \\
R^{\text{social}} &= R^{\text{coop}} + R^{\text{symp}} \\
R^{\text{coop}} &= \sin \theta_i \sin \phi_i \left[\sum_j r_{i,j}^{AV}(s, a) + \sum_j r_{i,j}^{M}(s, a) \right] \\
R^{\text{symp}} &= \cos \theta_i \sin \phi_i \left[\sum_k r_{i,k}^{HV}(s, a) + \sum_k r_{i,k}^{M}(s, a) \right]
\end{aligned}
\tag{11}
$$

in which R^{ego}, R^{social} represents the egoistic and social reward, $i \in I$, $j \in (\widetilde{I} \setminus \{I_i\})$, $k \in \widetilde{H}$. The term r_i represents the ego vehicle's reward obtained from traffic metrics and the angle ϕ allows to adjust the level of egoism or altruism. R^{coop} is the cooperation term (the altruistic behavior among AVs, i.e, AV's altruism toward others AVs) and R^{symp} is the sympathy term (AV's altruism toward HVs). The sympathy reward term, $r_{i,k}^{HV}$ considers the individual reward of the HVs, while the cooperation reward term, $r_{i,j}^{AV}$ considers the individual reward of the other AVs, and are defined as

$$
r_{i,k}^{HV} = \frac{\mathcal{W}_k}{d_{i,k}^{\lambda}} \sum_m \omega_m x_m \qquad r_{i,j}^{AV} = \frac{\mathcal{W}_j}{d_{i,j}^{\lambda}} \sum_m \omega_m x_m
\tag{12}
$$

in which $d_{i,k}/d_{i,j}$ represents the distance between the agent and the corresponding HV/AV, λ is a dimensionless coefficient, $\mathcal{W}_k$ is a weight value for individual vehicle's importance, m is the set of traffic metrics that have been considered in the vehicle's utilities (speed, crashes, etc.), in which x_m is the m metric normalized value and w_m is the weight associated to that metric. The term r^M accounts for the reward of the vehicle's mission. A mission is defined as any desired specific outcome for a particular vehicle, as merging, exiting, etc.

$$r_{i,j}^{M} = \begin{cases} \frac{w_j}{(d_{i,j})^{\mu}}, & \text{if} g(j) \\ 0, & \text{o.w.} \end{cases} \quad r_{i,k}^{M} = \begin{cases} \frac{w_k}{(d_{i,k})^{\mu}}, & \text{if} g(k) \\ 0, & \text{o.w.} \end{cases} \tag{13}$$

The function $g(v)$ is an independent function to evaluate the mission; $g(v)$ returns true if the vehicle v has a mission defined and the mission has been accomplished in the recent time window. μ is a dimensionless coefficient, w_j/w_k are weights for an individual vehicle's mission (importance of the mission). This allows defining a general reward independent of the driving scenario and mission goals for different vehicles. In the experiments, a **HV** can be assigned a merging mission or a highway exiting mission, as referred to in Fig. 2.

5.3   Deep MARL Architecture for Social Driving

As shown in Fig. 8, we leverage a 3D Convolutional Neural Network (CNN) with a safety prioritizer for our MARL architecture. To account for the temporal information, the 3D CNN operates as a feature extractor and leverages a history of VelocityMap observations. The network receives a stack of 10 VelocityMap observations, i.e., a $10 \times (4 \times 512 \times 64)$ tensor that captures the latest 10 time-steps episodes. To mitigate the non-stationarity issue in MARL, agents are trained in a semi-sequential manner, as illustrated in Fig. 7. The agents are trained independently for $N_{iterations}$ iterations while freezing the policies of the remaining AVs, $\mathbf{w}^-$. Subsequently, the other agents' policies are updated with the new policy, $\mathbf{w}^+$.

To improve safety we train our agents using a safety prioritizer that, in the cases where the action selected by the agent policy is unsafe, selects a safe action and stores the unsafe action (a_t) and the related state in the RM with a suitable penalty on the reward (r_{unsafe}) for the unsafe state-action pair. The safety prioritizer reduces episode resets due to imminent collisions improving sample efficiency. The unsafe state-action pairs are not removed so the agent can also learn from unsafe experiences. The experience ($\psi(s_t), a_t, r_{unsafe}, \emptyset$) is stored in RM with a terminal next state $\emptyset$, the target for this unsafe pair (s_t, a_t) is $Target(s_t, a_t)^{DDQN} = r_{unsafe}$. The details of the safety prioritizer are given in the next Sect. 5.4.

The proposed deep MARL architecture is described in Algorithm 1. As part of the implementation, we start the learning process after the replay buffer has been filled with a sufficient number of sample simulations. Furthermore, we update the experience replay buffer to adjust for the extremely skewed training data [17]. Balancing skewed data is a frequent practice in machine learning, and it was effective in our MARL problem.

i) Repeat the weight update k times for agent I_i:

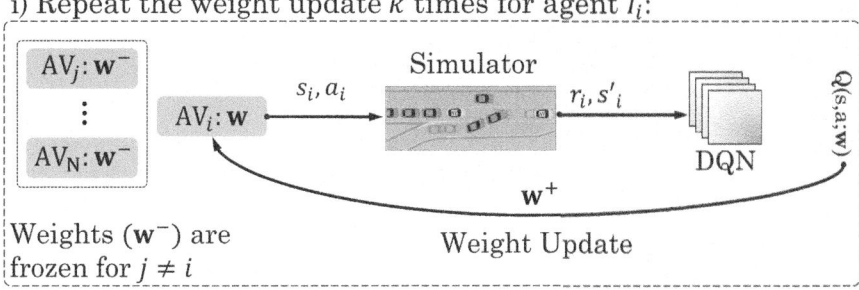

ii) Disseminate updated weights ($\mathbf{w}^+$):

Fig. 7 The multi-agent training and policy dissemination process

5.4 Safety Prioritizer

We include a safety prioritizer to the MARL algorithm that penalizes and reduce imminent crashes. This helps the agent to increase sample efficiency during training and avoid collisions when in deployment. If the agent comes into an unexpected situation and decides to perform a risky action, that action will be prevented. The safety prioritizer enhances simulation results and is crucial in real-world scenarios. The safety prioritizer included Algorithms 2 and 3.

Algorithm 2 During action selection of the agent I_i, once an action a_t is chosen, the safety prioritizer checks if the action is safe by computing a safety score for N_{steps} of planning. We utilize the time-to-collision (ttc) as a safety score. If $safety_{score} < safe_{th}$ the action is unsafe and we need to select a safe action. The selection of a safe action is presented in Algorithm 3.

Algorithm 3 The safe action selection is different in training and testing. During training, to encourage exploration, we remove the unsafe actions and keep the random action selection following the current exploration policy on the remaining actions. During testing, we follow the greedy policy in the subset of safe actions $a_t = \max_{a' \in \tilde{\mathcal{A}}_{safe}} Q(\psi(s_t), a'; \mathbf{w})$. It should be noted that the algorithm does not choose the safest of all possible actions, as that action may lead to particularly conservative behaviors that can compromise traffic efficiency; we instead remove the imminent unsafe actions and follow the priority given by the learned altruistic policy. If it happens that all possible actions are unsafe, we return the action $a_t \in \mathcal{A}$ with the highest safety score. In that way during training the constrained exploration will keep the agent from taking unsafe actions which will lead to efficient sampling

Algorithm 1 Safety Prioritized Multi-agent DDQN

Initialize *experience replay buffer RM*.
Initialize $\tilde{Q}(.; \mathbf{w}^-)$ with random weights $\mathbf{w}^- = \mathbf{w}_{ini}$
Initialize target network $\tilde{Q}(.; \hat{\mathbf{w}})$ with weights $\hat{\mathbf{w}} = \mathbf{w}^-$
Pre-store experience of first's 50 episodes in RM
for e $= 50$ to $N_{episode}$ **do**
    Initialize $s_1 = \{\tilde{\mathbf{o}}_1\}$ and compute $\psi_1 = \psi(s_1)$
    for t = 1 to T **do**
        for I_i in $\mathcal{I}$ **do**
            Freeze $\mathbf{w}^-$ for all I_j, $j \neq i$
            for $m = 1$ to $N_{iterations}$ **do**
                With probability ϵ select a random action a_t ,
                otherwise select $a_t = \max_{a' \in A} Q(\psi(s_t), a'; \mathbf{w}^+)$
                if a_t is unsafe (Algorithm 2) **then**
                    Store $(\psi_t, a_t, r_{unsafe}, \emptyset)$ in RM
                    a_t = Compute a safe action (Algorithm 3)
                end if
                Execute safe action a_t , and observe $r_t, \tilde{\mathbf{o}}_t$
                Set $s_{t+1} = \{s_t, \tilde{\mathbf{o}}_{t+1}\}$ and $\psi_{t+1} = \psi(s_{t+1})$
                Store experience $(\psi_t, a_t, r_t, \psi_{t+1})$ in RM
                Sample a mini-batch of size M from RM
                Compute $\mathcal{L}(\mathbf{w}^+)$
                Performs gradient descent
                $\mathbf{w}^+_{k+1} \leftarrow \mathbf{w}^+_k - \alpha \hat{\nabla}_{\mathbf{w}} \mathcal{L}(\mathbf{w}^+)$
            end for
            $\mathbf{w}^- = \mathbf{w}^+$ for all $I_i \in \mathcal{I}$
        end for
        Every $Target_{update}$ reset $\hat{\mathbf{w}} \leftarrow \mathbf{w}^-$
    end for
end for

Algorithm 2 Safety score

Simulate I_i taking the action a_t
for v in $(\tilde{\mathcal{I}} \cup \tilde{\mathcal{V}}) \setminus \{I_i\}$ **do**
    Compute safety score of I_i, v for N_{steps} planning
    if $safe_{score} < safe_{th}$ **then**
        Return unsafe
    end if
end for
Return safe

---

and more stable learning; and during testing, the decision-making is based on the prosocial learned policy with minimum intervention from the safety prioritizer, achieving higher traveled distance while avoiding collisions (Fig. 8).

Algorithm 3 Safe action

Initialize $\widetilde{\mathcal{A}} = \mathcal{A}$
while $\widetilde{\mathcal{A}}$ is not empty **do**
    if during training **then**
        Select a_t following the exploration policy on set $\widetilde{\mathcal{A}}$
    else if during test **then**
        Select $a_t = \max_{a' \in \widetilde{\mathcal{A}}_{safe}} Q(\psi(s_t), a'; \mathbf{w})$
    end if
    if a_t is safe (Algorithm 2) **then**
        Return a_t
    end if
end while
Return a_t with highest safe score in $\mathcal{A}$

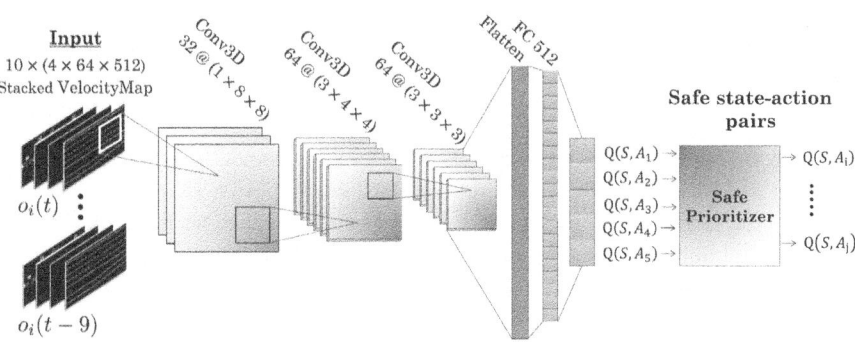

Fig. 8 Deep MARL architecture with the safety prioritizer

5.5 Modeling Driver Behaviors

We model the longitudinal movements of HVs using the *Intelligent Driver Model* (IDM) [57], while the lateral actions of HVs are based on the MOBIL model [58]. The MOBIL model considers two main criteria,

The safety criterion ensures that after the lane change, the deceleration of the new follower a_n in the target lane does not exceed a safe limit, i.e, $a_n > -b_{\text{safe}}$.

The incentive criterion determines the advantage of HV after the lane change, quantified by the total acceleration gain, given by

$$a'_{ego} - a_{ego} + \sin \phi_{ego} \left((a'_n - a_n) + (a'_o - a_o) \right) > \Delta a_{th} \qquad (14)$$

where a_o, a_n and a_{ego} represent the acceleration of the original follower in the current lane, the new follower in the target lane and the ego HV, correspondingly, and a'_o, a'_n, and a'_{ego} are the equivalent accelerations considering that the ego HV has changed the lane, $\sin \phi_{ego}$ is the politeness factor. Finally, the lane change is performed if the safety and incentive criteria are mutually satisfied.

The IDM Model determines the longitudinal acceleration of a HV $\dot{v}_k$ as follows,

$$\dot{v}_k = a_{max}\left[1 - \left(\frac{v_k}{v_k^0}\right)^\delta - \left(\frac{d^*(v_k, \Delta v_k)}{d_k}\right)^2\right] \tag{15}$$

in which v_k, d_k, δ, Δv_k, v_0^k denote the speed, the actual gap, the acceleration exponent, the approach rate, and the desired speed of the kth HV, respectively.

The desired minimum gap of the kth HV is given by,

$$d^*(v_k, \Delta v_k) = d_k^0 + v_k T_k^0 + \frac{v_k \Delta v_k}{(2\sqrt{a_{max} \cdot a_{des}})} \tag{16}$$

where T_k^0, d_k^0, a_{max}, and a_{des} are the safe time gap, the minimum distance, the comfortable maximum acceleration, and deceleration, correspondingly.

The typical parameters for the MOBIL model are $\sin \phi_{ego} = 0.5$, $\Delta a_{th} = 0.1\frac{m}{s^2}$ and $b_{safe} = 4\frac{m}{s^2}$. Table 1 shows typically used parameters of the IDM model [57].

Heterogeneous Driver Behaviors Although those parameters are typically used for IDM and MOBIL models, they simulate just one behavior. In order to generate diverse behaviors $\mathcal{B}$, we frame the task of simulating diverse behaviors as the problem of obtaining the appropriate range of parameters ($\mathcal{P}$) that can generate those behaviors. To achieve that, we leverage a behavior classifier and iteratively simulate the parameters and classify the behaviors, mapping parameters to behaviors. To classify the behaviors we represent traffic using a traffic-graph at each time step t, $\mathcal{G}_t$, with a set of edges $\mathcal{E}(t)$ and a set of vertices $\mathcal{V}(t)$ as functions of time, i.e, the positions of vehicles ($\tilde{\mathcal{H}} \cup \tilde{\mathcal{I}}$) represent the vertices. The adjacency matrix A_t is given by $A(k, m) = d(v_k, v_m), k \neq m$, in which $d(v_k, v_m)$ is the shortest travel distance between vertices k to m. Then we use centrality functions [34] to classify the behavior (level of aggressiveness) resulting from $\mathcal{P}$, and then use those simulation parameters $\mathcal{P}$ to model behaviors within the simulator with varying levels of aggressiveness. The centrality functions are defined as,

Closeness Centrality the discrete closeness centrality of the kth vehicle at time t is defined as,

$$C_C^k[t] = \frac{N-1}{\sum_{v_m \in \mathcal{V}(t)\setminus\{v_k\}} d_t(v_k, v_m)}, \tag{17}$$

Table 1 Common used parameters for the IDM model

Parameter	v^0	T^0	a_{max}	a_{des}	δ	d^0
Value	30 m/s	1.5 s	1 m/s^2	1.5 m/s^2	4	2 m

The more central the vehicle is located, the higher $C_C^k[t]$ and the closer it is to all other vehicles.

Degree Centrality the discrete degree centrality of the kth vehicle at time t is defined as,

$$C_D^k[t] = \left| \{v_m \in \mathcal{N}_k(t)\} \right| + C_D^k[t-1]$$
$$\text{such that } (v_k, v_m) \notin \mathcal{E}(\tau), \tau = 0, \ldots, t-1 \tag{18}$$

in which $\mathcal{N}_k(t) = \{v_m \in \mathcal{V}(t), A_t(k, m) \neq 0, v_m \leq v_k\}$ represents the set of vehicles in the proximity of the kth vehicle, given that $v_m \leq v_k$; and v_m, v_k denote the velocities of the mth and kth vehicles, $A_t(k, m)$ is the adjacency matrix. The more new vehicles seen by vehicle k that meet this condition, the higher $C_D^k[t]$.

With the centrality functions, we can measure the Style Likelihood Estimate (SLE) for different driver styles [34]. We consider two SLE measures. The SLE of overtaking and sudden lane changes (SLE_l) and the SLE of overspeeding (SLE_o). The SLE_l and SLE_o can be computed by measuring the first derivative of the centrality functions as,

$$\text{SLE}_l(t) = \left| \frac{\partial C_C(t)}{\partial t} \right| \quad \text{SLE}_o(t) = \left| \frac{\partial C_D(t)}{\partial t} \right| \tag{19}$$

The maximum likelihood $\text{SLE}_{\max}$ is calculated as $\text{SLE}_{\max} = \max_{t \in \Delta t} \text{SLE}(t)$.

Using those functions, we can approximately quantify and classify driver behaviors in our simulation. The intuition behind that is that an aggressive driver may frequently overspeed or perform sudden lane changes; while overspeeding the $C_D(t)$ monotonically increases (higher $\text{SLE}_o(t)$) and during sudden lane changes the slope and the extrema of $C_C(t)$ changes values. Thus higher values of $\text{SLE}_{\max}$ are related to increased levels of aggressiveness. Conversely, conservative drivers are not inclined toward those aggressive maneuvers, and the degree of centrality will be relatively flat, thus $\text{SLE}_o(t) \approx 0$ for conservative drivers.

We use these metrics as approximations of the driver's level of aggressiveness. In order to compute the suitable values for our simulation, we iteratively simulate the parameters from IDM and MOBIL models, and for each set of parameters, we quantify the resulting behavior in the simulation (using those metrics). A mapping of the parameters $\mathcal{P}$ to behaviors (quantified in the simulation for those parameters). The estimated simulation parameters that simulate conservative, moderate and aggressive behavior in our scenarios are presented in Table 2.

The desired velocity v^0 is set to $30m/s$ and the acceleration exponent $\delta = 4$.

Table 2 Estimated simulation parameters for conservative, moderate, and aggressive behaviors

Model	Parameter	Aggressive	Moderate	Conservative
MOBIL	$\sin \phi_e$	0	0.3	1
	Δa_{th}	$0\,\mathrm{m/s^2}$	$0.1\,\mathrm{m/s^2}$	$0.4\,\mathrm{m/s^2}$
	b_{safe}	$12.0\,\mathrm{m/s^2}$	$6.0\,\mathrm{m/s^2}$	$2.0\,\mathrm{m/s^2}$
IDM	T^0	$0.5\,\mathrm{s}$	$1\,\mathrm{s}$	$3\,\mathrm{s}$
	d^0	$1\,\mathrm{m}$	$2\,\mathrm{m}$	$6.0\,\mathrm{m}$
	$\mathrm{acc_{max}}$	$7.0\,\mathrm{m/s^2}$	$3.0\,\mathrm{m/s^2}$	$1.0\,\mathrm{m/s^2}$
	$\mathrm{acc_{des}}$	$12.0\,\mathrm{m/s^2}$	$7.0\,\mathrm{m/s^2}$	$2.0\,\mathrm{m/s^2}$

Table 3 Computation time for each agent

Computing platform	Online forward pass time
NVIDIA Tesla V100 GPU	3.7 ms
OnLogic Karbo 700 x2	65.2 ms
NVIDIA Jetson AGX Xavier GPU	32.9 ms
NVIDIA Jetson TX2 GPU	112.5 ms

5.6 Implementation and Computational Details

We customize the OpenAI Gym environment in [59] to suit our particular driving situation and MARL problem. We design a merging ramp and exiting highway scenario for our simulation running in python and used Pytorch for the implementation of our safety prioritized MARL DDQN algorithm. Our implementation on average uses 3.1GB of memory for 4 agents and 18 HVs using a GPU NVIDIA Tesla V100. The training process is repeated several times to ensure convergence of the experiments to a similar policy. The network is trained for $N_{episodes} = 10,000$ taking on average 8 h. While each round of $10,000$ training episodes in the Tesla V100 GPU takes around 8 h, a full forward pass during deployment for 4 simulated agents takes 15 ms (approximately 4 ms per agent).

In a real AV platform, each agent will receive a local observation of the environment that will be used by our algorithm to compute the safe optimal action based on the trained Q-network. The decision-making will take place on each AV's onboard computer; therefore, to verify the feasibility of the real-time operation of our decentralized algorithm we tested a forward pass of the Q-network during deployment in multiple hardware platforms. The results for the different platforms are presented in Table 3, for instance, an online forward pass of the network in the deployment phase using commodity GPU hardware, i.e, an NVIDIA Jetson AGX platform will be around 32.9 ms for each agent. We utilize 3200 GPU hours for all our simulation experiments. Table 4 lists our simulation and training hyper-parameters.

Table 4 List of hyper-parameters

Parameter	Value	Parameter	Value
N_{episode}	10,000	ϵ decay	Linear
RM buffer size	8000	Initial exploration ϵ_0	1.0
Batch size	32	Final exploration	0.05
Learning rate α_0	0.0005	Optimizer	ADAM
$Target_{update}$	300	Discount factor γ	0.95
$|\mathcal{H}|$	18	$|\mathcal{I}|$	4

6 Experiments and Results

6.1 Manipulated Variables

We study how the $safe_{th}$, the *level of aggressiveness*, the *traffic scenarios* (f_j) and the *HVs' behaviors* (b_k) impact the performance of AVs. We consider the case in which the mission vehicle (exiting/merging) in Fig. 2 is *human-driven*, $M \in \mathcal{H}$, and define the following terms:

- AV_S. Social AV ($\phi_i = \phi^*$) that act *altruistically* in the presence of diverse HVs behaviors $b \in \mathcal{B}$.
- AV_E. Egoistic AV ($\phi_i = 0$) that act *egoistically* in the presence of diverse HVs behaviors $b \in \mathcal{B}$.

with ϕ^* to be the optimal SVO angle tuned to reach the optimal level of altruism as in [17].

6.2 Performance Metrics

The performance of our system is measured based on safety, efficiency, altruistic performance gain (PG), and adaptation error A_{error}. To measure safety, we compute the percentage of episodes that encountered a crash ($C(\%)$). For efficiency, the average traveled distance ($DT(m)$) of the vehicles and the number of missions accomplished by the mission vehicle is used. The altruistic performance gain is measured by computing the difference in the safety/efficiency performance of AV_E and AV_S, as

$$PG_{safety}(\%) = \frac{(AV_E)_{C(\%)} - (AV_S)_{C(\%)}}{N_{Episodes}} \tag{20}$$

$$PG_{efficiency}(\%) = \frac{(AV_S)_{DT(m)} - (AV_E)_{DT(m)}}{(AV_E)_{DT(m)}} \tag{21}$$

Finally, the adaptation error is a weighted sum function of the safety ($C(\%)$) and efficiency ($DT(m)$) performance of the AV_S when trained and tested in different scenarios/behaviors. Defined as

$$A_{error}(\%) = w_s \times (C(\%)) + w_e \times 100(1 - \frac{DT}{DT_{max}}) \qquad (22)$$

such that an adaptation between different situations that result in 0% crash and $DT = DT_{max}$ will have $A_{error} = 0\%$.

6.3 Hypotheses

In this section we examine the following hypotheses

- **H1.** *In a mixed-autonomy scenario, the higher the level of aggressiveness, the bigger the impact of cooperation. We expect a higher performance gain (PG) when altruistic AVs face more aggressive environments.*
- **H2.** *Altruistic AVs agents using the decentralized framework can adapt to different driver behaviors and traffic scenarios without compromising the overall traffic metrics. However, the higher the similarity of testing scenarios to the ones seen during training ((f_{test}, b_{test}) $\approx$ (f_{train}, b_{train})), the lowest adaption error (A_{error}).*
- **H3.** *We anticipate an improvement in both safety and efficiency with the addition of the safety prioritizer. In the absence of a safety prioritizer ($safe_{th} = 0$) we expect that AVs will cause more crashes.*

6.4 Analysis and Results

Based on the hypotheses, we explore their correctness through the experiments in this section.

6.4.1 Sensitivity Analyses

To study the hypothesis **H1** we investigate the effect of HV behaviors on the altruistic AV agents. We focus on scenarios with a HV mission vehicle, with safe AVs that act *altruistically* (AV_S) or *egoistic* (AV_E), in environments with increasing levels of HVs aggressiveness. Figure 9 illustrates the altruistic performance gain for increasing levels of HVs' aggressiveness for 2 AVs (left) and 4 AVs (right). It demonstrates that the more aggressive the HVs are, the higher the impact of cooperation and thus confirms the **H1**. This is also observed in Fig. 10 where the level of aggressiveness is decomposed into lateral and longitudinal aggressiveness.

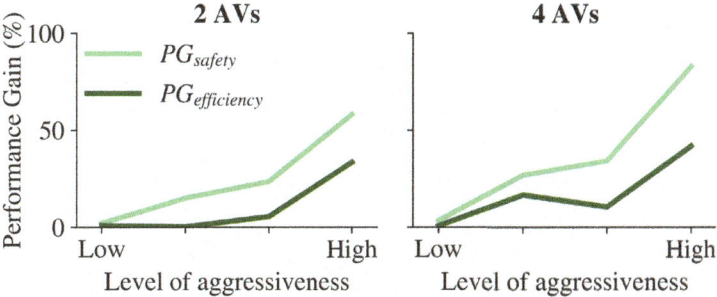

Fig. 9 Sensitivity analyses measured by altruistic performance gains (PGs) of AVs show that the more aggressive the HVs are, the more the impact/gain of cooperation

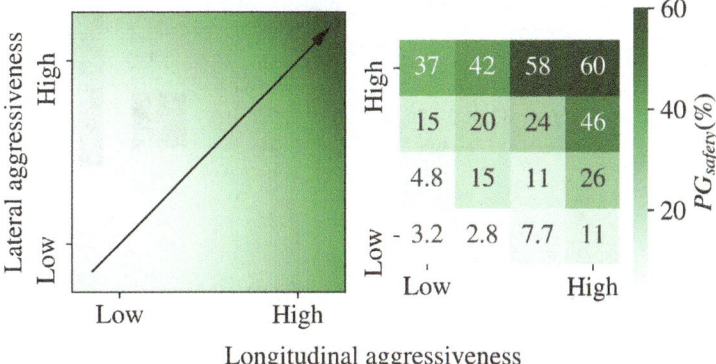

Fig. 10 Both lateral and longitudinal sensitivity analyses indicate an increase in altruistic performance gain (PG)

Lateral and longitudinal aggressiveness is varied by changing the MOBIL and IDM parameters (Table 2) from aggressive to conservative. Figure 10 shows that the altruistic gain increases in both directions, but is more pronounced in the longitudinal direction. That is probably due to the simulated scenarios having more longitudinal maneuvers.

6.4.2   Domain Adaptation

Following the sensitivity analysis, we investigate the domain adaptation of the AVs to validate the **H2**. Figures 11, 12 and 13 show how the altruistic AVs learn to adapt to different scenarios and behaviors by different performance metrics, i.e, crashed (**a**), distance traveled (**b**) and adaptation error (**c**). For the experiments, AV_S are trained in different scenarios $f_i \in \mathcal{F}$ in the presence of HVs with different behaviors $b_k \in \mathcal{B}$ and tested in other scenarios $f_j \in \mathcal{F}$ and behaviors $b_l \in \mathcal{B}$. In our experiments, we consider two case study scenarios $f_e, f_m \in \mathcal{F}$ (exiting/merging)

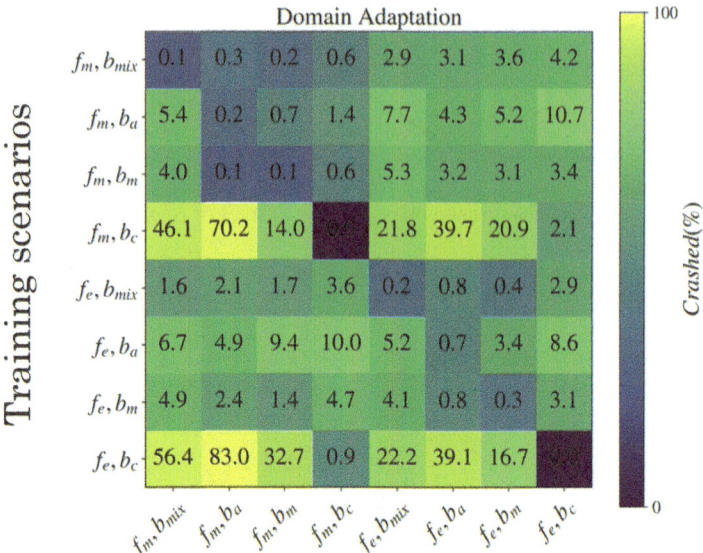

Crashed

Testing scenarios

Fig. 11 The domain adaptation matrix with crash percentage ($C(\%)$) between different traffic scenarios and behaviors. The lower $C(\%)$ the most suitable the adaptability in terms of safety (measured by $C(\%)$) between those domains. AV_S are trained (rows of the matrix) in different scenarios $f_i \in \mathcal{F}$ in the presence of HVs with different behaviors $b_k \in \mathcal{B}$ and tested (columns of the matrix) in other scenarios $f_j \in \mathcal{F}$ and behaviors $b_l \in \mathcal{B}$. Each pair (f_i, b_k) is a combination of scenario and behavior

in environments with three different HVs behaviors $b_a, b_m, b_c \in \mathcal{B}$ (aggressive, moderate, conservative) see Table 2; and a mixed behavior environment, in which HVs are created randomly and their behaviors are selected based on a uniform distribution over the behaviors in $\mathcal{B}$, given equal probability to the defined behaviors. In total, we have eight combinations of scenarios and behaviors, namely: (f_m, b_{mix}), $(f_m, b_a), (f_m, b_m), (f_m, b_c), (f_e, b_{mix}), (f_e, b_a), (f_e, b_m), (f_e, b_c)$.

The results are presented in Fig. 13 as an adaptation matrix, showing the A_{error} for different domains, the A_{error} is in percentage (%) and color-map in logarithmic scale to increase the perceived dynamic range for visualization. In our analyses, the weights used for $A_{error}(\%)$ are $w_s = \frac{2}{3}$ and $w_e = \frac{1}{3}$, which weighs the safety performance higher. DT_{max} is computed based on the maximum distance for each situation. Additionally, Figs. 11 and 12 illustrate how the AVs adapt in terms of safety (measured by $C(\%)$) and efficiency (measured by $DT(m)$), separately.

The matrix shows the best performances in its diagonal; where agents are trained and tested in the same environment $((f_i, b_k); (f_j, b_l)$ with $i = j$ and $k = l$); due to the fact that agents experience similar situations during testing as they do during

(b) Distance Traveled

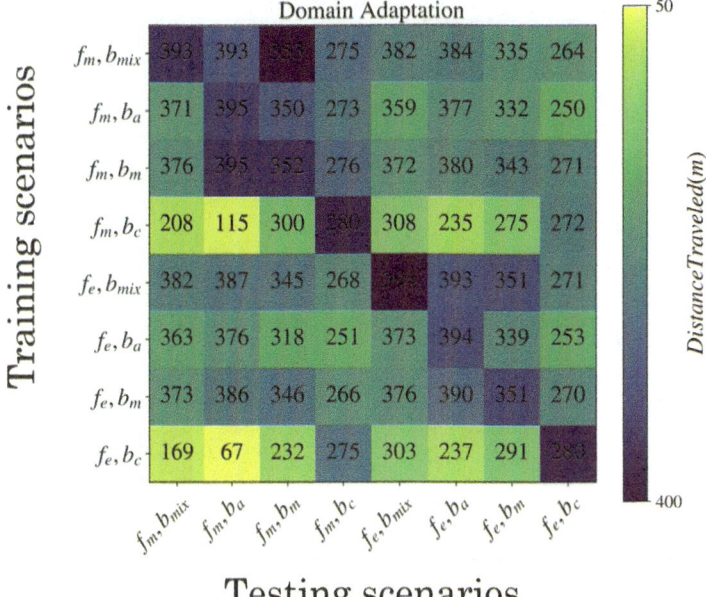

Fig. 12 The domain adaptation matrix with distance traveled ($DT(m)$). Illustrating how the AVs adapt to other situations in terms of efficiency (measured by $DT(m)$)

training. The vehicles trained in the merging environment can perform the exiting mission for different behaviors, and vice-versa. Interestingly, the performance of AVs trained in a conservative environment (b_c) is poor when tested in an aggressive environment (b_a). We believe that the reason is that in conservative environments, the HVs yield the mission vehicle, and the AVs learn to rely on HVs to guide the traffic. This learned policy is valid in a conservative environment where one can expect the HVs to always create a safe space for the mission vehicle. However, the same is not valid in more aggressive environments, in which AVs have to guide the traffic to avoid dangerous situations. As a result, the performance of vehicles trained in a conservative environment and tested in an aggressive one is the worse.

On the other hand, an adequate performance adaptation (lower A_{error}) is obtained when agents are trained in the presence of all moderate HVs (b_m) or a mixed behavior environment (b_{mix}), in which AVs face situations where the HVs yield, but also situations that require learning how to guide the traffic to optimize for the social utility. The results from the domain adaptation matrix indicate that a moderate or mixed environment is the most suitable for training robust AVs and show the adaptability of AVs to different situations, thereby confirming the **H2** hypothesis.

It can be concluded that the adaptation between the environments is not recipro-cal and environment and situations selection should be considered during training,

(c) Adaptation error

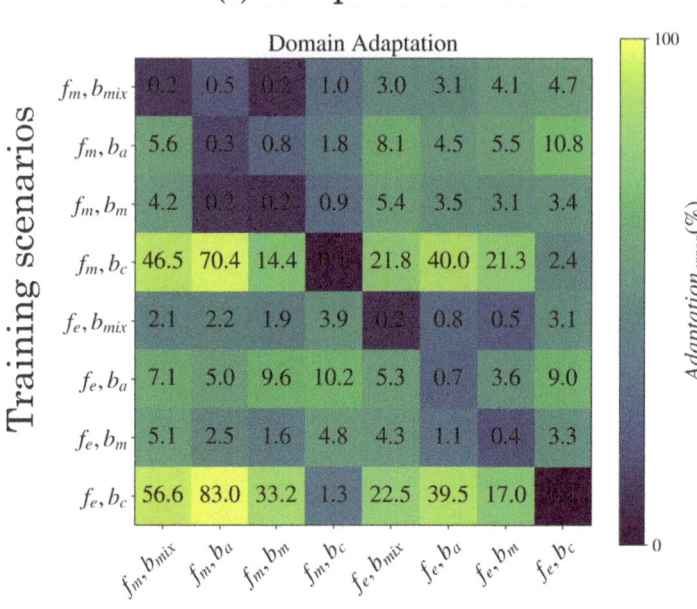

Fig. 13 The domain adaptation matrix with adaptation error (A_{error}) between different traffic scenarios and behaviors. The lower A_{error} the most suitable the adaptability between those domains

based on the application needs and target situations. The Domain adaptation matrix identifies the settings in which altruistic AVs can best learn cooperative policies that are robust to different traffic scenarios and human behaviors.

6.4.3 Transfer Learning

Through domain adaptation and transfer learning, we promote generalization while learning harder tasks efficiently from trained models and accelerate the learning process. We study how the policies learned during merging can be transferred to the exiting environment. For that, we train AVs agents from scratch for the mission/task of merging $AV_{merging}$ (T1), train AVs agents to drive on a highway, and then use that model as the starting point to learn the merging task $AV_{drive-to-merging}$ (T2), train AVs agents for the exiting task and then use that model as the starting point to learn the merging task $AV_{exiting-to-merging}$ (T3); and apply the same procedure for the exiting task, learning to exit from scratch $AV_{exiting}$ (T4), after learned how to drive $AV_{drive-to-exiting}$ (T5) and after learned how to merge $AV_{merging-to-exiting}$ (T6). The results of the experiments are presented in Fig. 14 and show that our transfer learning

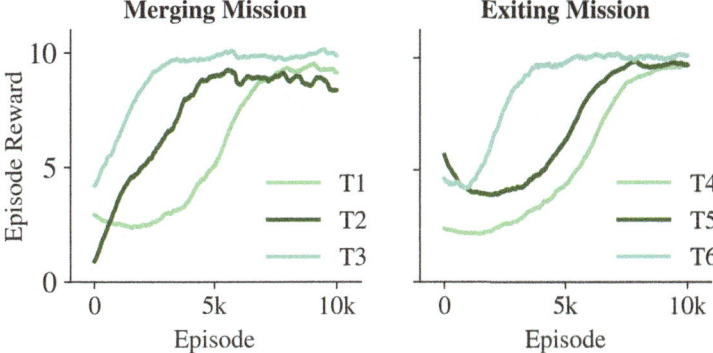

Fig. 14 The figure demonstrates how policies learned from merging can be transferred to the exiting environment to speed up the learning process while archiving similar performance to learning the task from scratch

approach speeds up the learning process while archiving similar performance as when learning the task from scratch.

6.4.4 Safety

Finally, we compared state of the art architectures related to our approach [10, 17, 23, 60] in terms of safety and efficiency to validate **H3**. We trained the different architectures in the same situations and examined their performance under different levels of HVs behaviors. As noted in Table 5 our safe altruistic agents consistently outperformed the other approaches (in bold is highlighted the best performance for each column), and the results are more notable when the level of aggressiveness is higher. We conclude that when using the safety prioritizer, immediate collisions are avoided reducing the overall number of crashes in the episodes. Our agents can learn from scratch not only how to drive, but also to understand the behavior of HVs and coordinate with them.

6.4.5 Importance of Social Coordination

We demonstrate that social awareness and coordination are essential to improve safety and reliability on the roads. Particularly in our sensitivity analyses (Fig. 9) we have shown that altruistic agents have a significant performance gain when compared to egoistic agents and the gain is more notable as the road becomes more aggressive. Additionally, to show that the performance gain vs. driver behaviors is not just because of a single altruistic agent but as a consequence of coordination among agents, we complement our results and conducted an experiment with the difference that only AV1 is altruistic and the others are egoistic AVs, we label this

Table 5 A comparison of the performance of related architectures. Our safe altruistic AVs outperform the other solutions, and performance improvements become more noticeable as the level of aggressiveness increases

	Aggressive HVs			Moderate HVs			Conservative HVs		
Approaches	C (%)	MF (%)	DT (m)	C (%)	MF (%)	DT (m)	C (%)	MF (%)	DT (m)
Conv2D+DQN [60]	31.2	28.9	316	25.4	20.3	302	14.0	7.9	274
Toghi et al. [17]	21.3	16.4	339	12.7	10.1	333	1.6	0.6	269
Conv3D+A2C [23]	14.8	12.6	341	9.4	8.8	328	1.1	0.1	267
Conv3D+DQN [10]	3.1	2.8	359	2.6	2.4	341	0.3	**0**	**284**
Ours	**0.2**	**0.1**	**397**	**0.1**	**0.1**	**354**	**0**	**0**	281

C Crashed, *MF* Mission Failed, *DT* Distance Traveled

Table 6 Importance of Social Coordination: AVs require to coordinate to enable a safe and seamless merging/exiting and none of them can achieve this goal if the others do not cooperate

	Aggressive HVs $C(\%)$	Moderate HVs $C(\%)$	Conservative HVs $C(\%)$
Multi-agent altruistic (MAA)	0.2%	0.1%	0%
Single altruistic agent (SAA)	24.1%	17.4%	2.3%

scenario as single altruistic agent **SAA**. Table 6 demonstrates the necessity of multi-agent coordination and the fact that a single altruistic AV, i.e., the Guide AV, is not able to achieve the mission of safe and seamless merging without help from the other AVs. Our results show that a non-cooperative SAA is not enough to guide the traffic and successful completion of the missions, as coordination is not guaranteed in a single-agent setting. All the AVs have to coordinate collectively to allow safe and efficient traffic, and this is unfeasible if the others do not collaborate. Table 6 complements our results in Fig. 9 and support the hypothesis **H1**.

6.5 Qualitative Analyses

We show a qualitative analysis of our altruistic AVs in the exit and merging scenarios. Figure 15 provides further intuition about the policies learned by altruistic AVs (*green*) in different situations, Figs. 15 and 16 show a set of snapshots for different policies learned in an exit/merging environment in the presence of HVs (*blue*) with different behaviors. In the presence of aggressive HVs, the guide AV has to slow down and guide the HVs in the platoon to allow a safe merging/exit of the mission vehicle; in this case, by slowing down the AV learn to compromise on their own utility for a more desirable social outcome. In the presence of moderate HVs

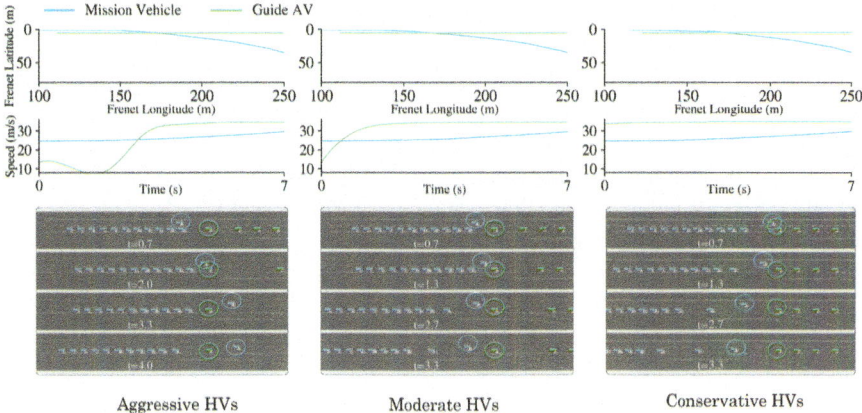

Fig. 15 Mission vehicle exiting the road under different HV behaviors (from left to right: aggressive, moderate and conservative HVs). AVs are shown in *green* and HVs are shown in *blue*

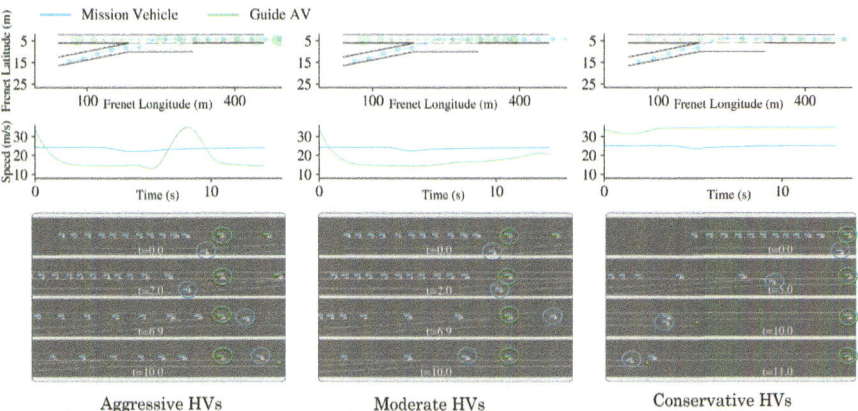

Fig. 16 Mission vehicle merging into the highway under different HV behaviors (from left to right: aggressive, moderate, and conservative HVs). AVs are shown in *green* and HVs are shown in *blue*. The diameter of the circles on the trajectory plot (first-row plot) shows the vehicles' speed

behaviors, the guide AV slows down (slowing down the vehicles in the platooning) to open a safe space for the mission vehicle and then quickly accelerates, the space created by the quick AV intervention is safe enough to allow the mission vehicle to exit/merge the road; in this case, the AV compromise in their own utility but does not need to compromise as much as in the aggressive traffic scenario, it learns to take sequences of actions to not only enable the mission vehicle to merge (by quick decelerating), but also manages to make the minimum compromise on its individual utility. Finally, in the conservative environment, the HVs are cautious enough to allow the mission vehicle to exit/merge safely, so the AVs learn to accelerate in those scenarios as the mission vehicle has enough space to merge, optimizing for their own utility (higher speed and longer distance travel), while also considering other vehicles utilities and safety; in this case the AV doe not need to compromise their own utility, it learns that HV will allow the exiting/merging so AVs does not need to guide the traffic. It is important to notice that the policies are learned by AVs from experience to optimize the social utility, AVs learn to adapt to different scenarios and behaviors. Is interesting to observe that our AVs develop some form of social awareness and learn the HVs' behaviors from experience, acting accordingly to optimize traffic efficiency while prioritizing safety.

7   Concluding Remarks

AVs need to learn to co-exist with HVs vehicles as deploying egoistic AVs that solely account for their individual interests on the road leads to sub-optimal and non-desirable social outcomes. Social awareness and coordination are essential to improve safety and reliability on the roads. We demonstrate how altruistic AVs learn the decision-making process from experience, considering the interests of all vehicles while prioritizing safety and optimizing a general decentralized social utility function. We expose the settings for our MARL problem in which transfer learning and domain adaptation are more feasible, and conducted a sensitivity analysis under different HVs' behaviors. Our experiments reveal that altruistic AVs learn to leverage social coordination to improve safety and reliability. Our social-aware AVs are robust to heterogeneous driver behaviors and can form alliances and affect the behavior of HVs to create socially-desirable outcomes that benefit the group of the vehicles.

Future Work Although we explored various elements of social navigation in a variety of settings and the presence of diverse HV behaviors, the HV models used are not from real human driver data, and the traffic scenarios are limited to merging and exiting. However, we believe that by leveraging and learning from actual human data and traffic circumstances, our approach might be beneficial in practical traffic conditions. For this strategy to be used in real-world circumstances, more attention to safety is necessary. We intend to investigate more sophisticated architectures and state representations in future work, as well as develop a more realistic simulation environment that incorporates data from real-world traffic and can handle more complex interactions between HVs and AVs, as well as diverse traffic agents like bicycles and pedestrians. Despite the drawbacks, we are excited to see safe and reliable social-aware AVs on the road that learns from experience. Beyond driving, we expect these principles to be applied to general multi-agent human-robot interactions in which agents influence humans and collaborate safely for a socially beneficial result.

References

1. Cosgun, A., Ma, L., Chiu, J., Huang, J., Demir, M., Anon, A.M., Lian, T., Tafish, H., Al-Stouhi, S.: Towards full automated drive in urban environments: a demonstration in gomentum station, California. In: 2017 IEEE Intelligent Vehicles Symposium (IV), pp. 1811–1818. IEEE (2017)
2. Schwarting, W., Pierson, A., Alonso-Mora, J., Karaman, S., Rus, D.: Social behavior for autonomous vehicles. Proc. Natl. Acad. Sci. **116**(50), 24972–24978 (2019)
3. Sagberg, F., Selpi, Piccinini, G.F.B., Engström, J.: A review of research on driving styles and road safety. Human Factors **57**(7), 1248–1275 (2015)
4. Toghi, B., Saifuddin, M., Mughal, M., Fallah, Y.P.: Spatio-temporal dynamics of cellular v2x communication in dense vehicular networks. In: 2019 IEEE 2nd Connected and Automated Vehicles Symposium (CAVS), pp. 1–5. IEEE (2019)
5. Shah, G., Valiente, R., Gupta, N., Gani S.O., Toghi, B., Fallah, Y.P., Gupta, S.D.: Real-time hardware-in-the-loop emulation framework for dsrc-based connected vehicle applications. In: 2019 IEEE 2nd Connected and Automated Vehicles Symposium (CAVS), pp. 1–6. IEEE (2019)
6. Valiente, R., Zaman, M., Ozer, S., Fallah, Y.P.: Controlling steering angle for cooperative self-driving vehicles utilizing cnn and lstm-based deep networks. In: 2019 IEEE Intelligent Vehicles Symposium (IV), pp. 2423–2428. IEEE (2019)
7. Razzaghpour, M., Shahram, S., Valiente, R., Fallah, Y.P.: Impact of communication loss on mpc based cooperative adaptive cruise control and platooning. Preprint (2021). arXiv:2106.09094
8. Valiente, R., Raftari, A., Zaman, M., Fallah Y.P., Mahmud, S.: Dynamic object map based architecture for robust cvs systems. SAE Technical Paper, Technical Report, 2020
9. Aoki, S., Higuchi, T., Altintas, O.: Cooperative perception with deep reinforcement learning for connected vehicles. In: 2020 IEEE Intelligent Vehicles Symposium (IV), pp. 328–334. IEEE (2020)
10. Toghi, B., Valiente, R., Sadigh, D., Pedarsani, R., Fallah, Y.P.: Cooperative autonomous vehicles that sympathize with human drivers. In: 2021 IEEE/RSJ International Conference on Intelligent Robots and Systems (IROS). IEEE (2021)
11. Jami, A., Razzaghpour, M., Alnuweiri, H., Fallah, Y.: Augmented driver behavior models for high-Fidelity simulation study of crash detection algorithms. https://arxiv.org/pdf/2208.05540.pdf
12. Harris P.B., Houston J.M., Vazquez J.A., Smither J.A., Harms, A., Dahlke, J.A., Sachau, D.A.: The prosocial and aggressive driving inventory (padi): a self-report measure of safe and unsafe driving behaviors. Accid. Anal. Prev. **72**, 1–8 (2014)
13. Vallières, E.F., Vallerand, R.J., Bergeron, J., McDuff, P.: Intentionality, anger, coping, and ego defensiveness in reactive aggressive driving. J. Appl. Soc. Psychol. **44**(5), 354–363 (2014)
14. Bouton, M., Nakhaei, A., Fujimura, K., Kochenderfer, M.J.: Cooperation-aware reinforcement learning for merging in dense traffic. In: 2019 IEEE Intelligent Transportation Systems Conference (ITSC), pp. 3441–3447. IEEE (2019)
15. Sadigh, D., Sastry, S., Seshia S.A., Dragan, A.D.: Planning for autonomous cars that leverage effects on human actions. In: Robotics: Science and Systems, vol. 2. Ann Arbor (2016)
16. Wu, C., Kreidieh, A., Vinitsky, E., Bayen, A.M.: Emergent behaviors in mixed-autonomy traffic. In: Conference on Robot Learning, pp. 398–407. PMLR (2017)
17. Toghi, B., Valiente, R., Sadigh, D., Pedarsani, R., Fallah, Y.P.: Social coordination and altruism in autonomous driving. Preprint (2021). arXiv:2107.00200
18. Rios-Torres, J., Malikopoulos, A.A.: A survey on the coordination of connected and automated vehicles at intersections and merging at highway on-ramps. IEEE Trans. Intell. Trans. Syst. **18**(5), 1066–1077 (2016)
19. Mahjoub, H.N., Raftari, A., Valiente, R., Fallah Y.P., Mahmud, S.K.: Representing realistic human driver behaviors using a finite size gaussian process kernel bank. In: 2019 IEEE Vehicular Networking Conference (VNC), pp. 1–8. IEEE (2019)
20. Li, Z., Kalabić, U., Chu, T.: Safe reinforcement learning: Learning with supervision using a constraint-admissible set. In: 2018 Annual American Control Conference (ACC), pp. 6390–6395. IEEE (2018)

21. Lin, Y., McPhee, J., Azad, N.L.: Anti-jerk on-ramp merging using deep reinforcement learning. In: 2020 IEEE Intelligent Vehicles Symposium (IV), pp. 7–14. IEEE (2020)
22. Sadigh, D., Landolfi, N., Sastry S.S., Seshia S.A., Dragan A.D.: Planning for cars that coordinate with people: leveraging effects on human actions for planning and active information gathering over human internal state. Auton. Robot. **42**(7), 1405–1426 (2018)
23. Toghi, B., Valiente, R., Sadigh, D., Pedarsani, R., Fallah, Y.P.: Altruistic maneuver planning for cooperative autonomous vehicles using multi-agent advantage actor-critic. In: Proceedings of the IEEE/CVF Conference on Computer Vision and Pattern Recognition (CVPR) Workshops (2021)
24. Foerster J.N., Chen R.Y., Al-Shedivat, M., Whiteson, S., Abbeel, P., Mordatch, I.: Learning with opponent-learning awareness. Preprint (2017). arXiv:1709.04326
25. Xie, A., Losey, D., Tolsma, R., Finn, C., Sadigh, D.: Learning latent representations to influence multi-agent interaction. In: Proceedings of the 4th Conference on Robot Learning (CoRL) (2020)
26. Foerster, J., Farquhar, G., Afouras, T., Nardelli, N., Whiteson, S.: Counterfactual multi-agent policy gradients. In: Proceedings of the AAAI Conference on Artificial Intelligence, vol. 32, no. 1 (2018)
27. Egorov, M.: Multi-agent deep reinforcement learning. CS231n: Convolutional Neural Networks for Visual Recognition, pp. 1–8 (2016)
28. Omidshafiei, S., Pazis, J., Amato, C., How J.P., Vian, J.: Deep decentralized multi-task multi-agent reinforcement learning under partial observability. In: International Conference on Machine Learning, pp. 2681–2690. PMLR (2017)
29. Lowe, R., Wu, Y., Tamar, A., Harb, J., Abbeel, P., Mordatch, I.: Multi-agent actor-critic for mixed cooperative-competitive environments. Preprint (2017). arXiv:1706.02275
30. Brown, K., Driggs-Campbell, K., Kochenderfer, M.J.: A taxonomy and review of algorithms for modeling and predicting human driver behavior. arxiv e-prints, article. Preprint (2020). arXiv:2006.08832
31. Ivanovic, B., Schmerling, E., Leung, K., Pavone, M.: Generative modeling of multimodal multi-human behavior. In: RSJ International Conference on Intelligent Robots and Systems, pp. 3088–3095. IEEE (2018)
32. Lauer, M., Riedmiller, M.: An algorithm for distributed reinforcement learning in cooperative multi-agent systems. In: In Proceedings of the Seventeenth International Conference on Machine Learning. Citeseer (2000)
33. Toghi, B., Grover, D., Razzaghpour, M., Jain, R., Valiente, R., Zaman, M., Shah, G., Fallah, Y.P.: A maneuver-based urban driving dataset and model for cooperative vehicle applications. In: 2020 IEEE 3rd Connected and Automated Vehicles Symposium (CAVS), pp. 1–6. IEEE (2020). https://ieeexplore.ieee.org/document/9334665
34. Chandra, R., Bhattacharya, U., Mittal, T., Bera, A., Manocha, D.: Cmetric: A driving behavior measure using centrality functions. In: 2020 IEEE/RSJ International Conference on Intelligent Robots and Systems (IROS), pp. 2035–2042. IEEE (2020)
35. Constantinescu, Z., Marinoiu, C., Vladoiu, M.: Driving style analysis using data mining techniques. Int. J. Comput. Commun. Control **5**(5), 654–663 (2010)
36. Beck K.H., Ali, B., Daughters, S.B.: Distress tolerance as a predictor of risky and aggressive driving. Traffic Inj. Prev. **15**(4), 349–354 (20140
37. Pokle, A., Martín-Martín, R., Goebel, P., Chow, V., Ewald, H.M., Yang, J., Wang, Z., Sadeghian, A., Sadigh, D., Savarese, S., et al.: Deep local trajectory replanning and control for robot navigation. In: 2019 International Conference on Robotics and Automation (ICRA), pp. 5815–5822. IEEE (2019)
38. Kuderer, M., Gulati, S., Burgard, W.: Learning driving styles for autonomous vehicles from demonstration. In: 2015 IEEE International Conference on Robotics and Automation (ICRA), pp. 2641–2646. IEEE (2015)
39. Hadfield-Menell, D., Russell S.J., Abbeel, P., Dragan, A.: Cooperative inverse reinforcement learning. Adv. Neural Inf. Proces. Syst. **29**, 3909–3917 (2016)

40. Trautman, P., Krause, A.: Unfreezing the robot: navigation in dense, interacting crowds. In: 2010 IEEE/RSJ International Conference on Intelligent Robots and Systems, pp. 797–803. IEEE (2010)
41. Nikolaidis, S., Ramakrishnan, R., Gu, K., Shah, J.: Efficient model learning from joint-action demonstrations for human-robot collaborative tasks. In: 2015 10th ACM/IEEE International Conference on Human-Robot Interaction (HRI), pp. 189–196. IEEE (2015)
42. Wu, C., Bayen A.M., Mehta, A.: Stabilizing traffic with autonomous vehicles. In: 2018 IEEE International Conference on Robotics and Automation (ICRA), pp. 6012–6018. IEEE (2018)
43. Lazar, D.A., Bıyık, E., Sadigh, D., Pedarsani, R.: Learning how to dynamically route autonomous vehicles on shared roads. Preprint (2019). arXiv:1909.03664
44. Bıyık, E., Lazar, D.A., Pedarsani, R., Sadigh, D.: Incentivizing efficient equilibria in traffic networks with mixed autonomy. IEEE Trans. Control Netw. Syst. **8**(4), 1717–1729 (2021)
45. Li, S., Yan, Z., Wu, C.: Learning to delegate for large-scale vehicle routing. Adv. Neural Inf. Process. Syst. **34** (2021)
46. Hickert, C., Li, S., Wu, C.: Cooperation for scalable supervision of autonomy in mixed traffic, pp. arXiv–2112. e-prints (2021)
47. Razzaghpour, M., Mosharafian, S., Raftari, A., Mohammadpour Velni, J., and Fallah, Y.P.: Impact of information flow topology on safety of tightly-coupled connected and automated vehicle platoons utilizing stochastic control. In: ECC (2022)
48. Wang W.Z., Beliaev, M., Biyik, E., Lazar D.A., Pedarsani, R., Sadigh, D.: Emergent prosocial-ity in multi-agent games through gifting. In 30th International Joint Conference on Artificial Intelligence (IJCAI) (2021)
49. Wang, J., Zhang, Q., Zhao, D., Chen, Y.: Lane change decision-making through deep reinforcement learning with rule-based constraints. In: 2019 International Joint Conference on Neural Networks (IJCNN), pp. 1–6. IEEE (2019)
50. Nageshrao, S., Tseng H.E., Filev, D.: Autonomous highway driving using deep reinforcement learning. In: 2019 IEEE International Conference on Systems, Man and Cybernetics (SMC), pp. 2326–2331. IEEE (2019)
51. Mohammadhasani, A., Mehrivash, H., Lynch, A., Shu, Z.: Reinforcement learning based safe decision making for highway autonomous driving. Preprint (2021). arXiv:2105.06517
52. Chen, D., Li, Z., Wang, Y., Jiang, L., Wang, Y.: Deep multi-agent reinforcement learning for highway on-ramp merging in mixed traffic. Preprint (2021). arXiv:2105.05701
53. Le, V.-A., Malikopoulos, A.A.: A cooperative optimal control framework for connected and automated vehicles in mixed traffic using social value orientation. Preprint (2022). arXiv:2203.17106
54. Murphy, R.O., Ackermann, K.A.: Social preferences, positive expectations, and trust based cooperation. J. Math. Psychol. **67**, 45–50 (2015).
55. Garapin, A., Muller, L., Rahali, B.: Does trust mean giving and not risking? Experimental evidence from the trust game. Rev. Econ. Polit. **125**(5), 701–716 (2015)
56. Müller, L., Risto, M., Emmenegger, C.: The social behavior of autonomous vehicles. In: Proceedings of the 2016 ACM International Joint Conference on Pervasive and Ubiquitous Computing: Adjunct, Ser. UbiComp '16, pp. 686–689. Association for Computing Machinery, New York (2016) [Online]. Available: https://doi.org/10.1145/2968219.2968561
57. Treiber, M., Hennecke, A., Helbing, D.: Congested traffic states in empirical observations and microscopic simulations. Phys. Rev. E **62**(2), 1805 (2000)
58. Kesting, A., Treiber, M., Helbing, D.: General lane-changing model mobil for car-following models. Transp. Res. Rec. **1999**(1), 86–94 (2007)
59. Leurent, E., Blanco, Y., Efimov, D., Maillard, O.-A.: Approximate robust control of uncertain dynamical systems. Preprint (2019). arXiv:1903.00220
60. Van Hasselt, H., Guez, A., Silver, D.: Deep reinforcement learning with double q-learning. In: Proceedings of the AAAI Conference on Artificial Intelligence, vol. 30, no. 1 (2016)

Evaluation of Autonomous Vehicle Control Strategies Using Resilience Engineering

Johan Fanas Rojas, Thomas Bradley, and Zachary D. Asher

1 Introduction

An autonomous vehicle (AV) is a system that can navigate through various driving scenarios and make judgments without the need for human intervention [1]. This technology is important because it can minimize traffic fatalities, potentially decrease traffic congestion, and offer transportation for the elderly and persons with disabilities [2–5]. Nonetheless, before these benefits can be realized, significant advancements in numerous facets of vehicle autonomy are needed including vehicle design, control, perception, planning, coordination, and human interaction [6].

Artificial Intelligence (AI) is a prerequisite to the development of AVs ability to learn how to perform tasks and improve itself based on collected data [7, 8]. AI has been used to perform computationally costly AV perception tasks such as object detection, Lidar processing, sensor fusion, and more [9–12]. Other applications of AI for AVs is the development of end-to-end networks to perform perception, planning and control in a single network instead of developing each subsystem separately which can be very beneficial in terms of engineering time [13]. The current drawbacks of AI are that it requires a large amount of data for training and that if the model does not perform as expected, determining the reason for the defect is difficult due to the model's complexity. Despite the potential of AI technology, there are still safety concerns and accidents have occurred in recent years [14].

J. F. Rojas (✉) · Z. D. Asher
Department of Mechanical and Aerospace Engineering, Western Michigan University, Kalamazoo, MI, USA
e-mail: johan.fanasrojas@wmich.edu; zach.asher@wmich.edu

T. Bradley
Department of Systems Engineering, Colorado State University, Fort Collins, CO, USA
e-mail: thomas.bradley@colostate.edu

Methods to predict potential accidents from the use of AI are largely nonexistent [15].

Development and evaluation of safe AVs is a challenging task. In general it is easy to argue that AV technology should be immediately implemented since at least 90% of current car accidents are attributable to human mistakes [16, 17]. It is anticipated that by eliminating people from vehicle control, hundreds of lives will be saved annually. Even while this technology has the potential to significantly improve ground transportation, society's acceptability may decrease as a result of the inherent dangers of autonomous cars [18]. Due to liability concerns in recent years, current research focuses on making AVs safer but it can be challenging to accurately measure an improvement in safety [19]. Numerous studies have been conducted to increase the safety of AVs when making decisions and to better estimate the risk associated with various maneuvers. However, there is no performance measurement for the overall autonomous system. Currently, the system's "safety" is measured by the death rate per mile driven. This measure is reactive since it analyzes the performance of the autonomous system after the event has transpired. The system may have failed during operation without causing an accident or disengagement which is not captured in the safety evaluation. In systems engineering, an accident is a series of events that result in an accident where each of those events must be understood and tracked [20]. Consequently, the system may have failed or exceeded the safety limits in a certain subsystem, yet current metrics establish that no accident occurred. This is an issue because, as automation technology becomes increasingly accessible to the public, fundamentally unsafe systems will only be exposed as such only when potentially fatal accidents accumulate on public roads. Consequently, this statistic cannot be utilized to evaluate the performance of an autonomous system.

Fortunately, methodology to assess the safety of complex systems has already been developed by systems engineering researchers. One technique that is applicable to AVs is Resilience engineering (RE) which is a subfield of systems engineering associated with safety management. Due to its potential to overcome constraints of conventional safety management systems, RE can enhance the operation and development of AVs [20]. RE offers methods for enhancing the operational resilience of complex systems. Resilience, from a systemic perspective, is the inherent ability of a system to modify its functioning prior to, during, or after changes and disturbances to the system, while maintaining needed operations under both anticipated and unanticipated conditions. The RE community sees AVs as a means of practical implementation, but the few extant studies are narrow in scope; they simulate AV effects on traffic rather than the functioning of the AV system [21, 22]. This is due to the fact that the engineering necessary to produce all driving features of an AV is extremely interdisciplinary, specialized, and a new skill set within the academic community [23]. RE examines the capacity of a system to continue functioning despite shocks, hence defining resilience. Unlike conventional safety management procedures, which simply investigate the source of the failure and its relationship with the other systems, a new approach is needed which examines the entire AV system. In contrast, conventional AV safety metrics are inherently reactive while RE can be regarded as a proactive safety methodology.

RE provides unique performance assessment methods and procedures for improvement, but it has not been utilized at the basic level of AV engineering despite the benefits that have been identified [20, 21]. Significant effort is necessary to understand and use resilience measurements, as well as to adapt the existing concepts for achieving resilience, to this complex application. Current research on AV operation focuses on safety through either a rigorous understanding of end-to-end neural network architecture and error outputs [24, 25] or a reintegration of subsystems and conventional control theory [26, 27]. Current research in this area is possible to demonstrate general autonomous control, but robust operation for safety reliability and safety certification is not yet unattainable [28, 29].

RE can also be applied to individual aspects of an AV such as various types of controllers. The development of robust lateral controllers and end-to-end Deep Learning (DL) is crucial to perform robust operation for safety reliability [30]. However, there is a gap between the evaluation metrics used for the design and development of these controllers since current metrics do not provide information on system performance. California's Department of Motor Vehicles (DMV) 2021s disengagement reports published that on several occasions the AV suffered from disengagement because it deviated from the reference path or did not perform properly the functions for which it was designed, which created a threat to the passenger [31]. This indicates the lack of robust metrics for the design and validation of these controllers for ADAS/Autonomous applications during deployment.

Overall this review shows that there is a significant research gap in the development of a proactive safety metric for AV system and component development such as vehicle controllers. RE theory provides these needed proactive evaluation metrics that can be used to assess the operational resilience of various AV controllers. There are studies in the literature describing resilience evaluation approaches for general engineering systems [32, 33], however evaluation applications for AVs are limited [34]. Currently, there is no overlap between systems engineering and AV engineering skill sets, resulting in AV RE studies that do not include perception or planning subsystem specifics [25, 35]. We are unaware of any existing RE evaluation of unmanned AVs that employs subsystem-level and algorithm-level examination of the technical engineering information that comprises these complex systems.

This chapter addresses this research gap by providing RE metrics as proactive safety metrics to evaluate the performance of a pure pursuit controller (PPC) and a Deep Learning controller (DLC) and comparing it versus traditional reactive metrics. In previous work, we applied RE metrics to evaluate a PPC using the resilience assessment grid (RAG) and measuring the resistance and recovery of the system [36]. This work extends upon the aforementioned study in order to provide some insight on DL methods used today for autonomous driving. In general it is widely regarded that end-to-end AV controllers do not perform as well as traditional controllers but quantitative proof is lacking [37]. A PPC and a DLC were developed in the CARLA simulator to perform path tracking and measure the resistance and recovery of the system. RAG was also used to show how these properties change as the speed increases. The resilience metrics were applied to each lateral controller and compared against traditional evaluation metrics such as average cross-track error, average heading error, and control effort. Overall this research demonstrates how to develop and implement proactive safety metrics for AV component performance evaluation which is the first step towards full AV safety evaluation.

2 Methodology

To evaluate the system's responsiveness, a PPC and a DLC were developed and tested at various speeds in the CARLA AV simulator. The following subsections detail the implementation of the PPC, the DLC, and a brief introduction of the AV simulator and its features. Then, an overview of RE and operational resilience performance of the AV controllers are presented. Lastly, we define the performance evaluation metrics used in this study.

2.1 Simulation Environment

CARLA is an open source simulator designed to train, validate, and test AV system algorithms [38]. This simulator provides the user with the ability to simulate environmental conditions (rainy, cloudy, sunny) and driving scenarios to test AV systems. Figure 1 shows an example of the CARLA simulator in different

Fig. 1 Example of a Tesla Model 3 vehicle performing path tracking in the CARLA simulator

environmental conditions. With this AV simulator you can test different algorithms of a subsystem. For example, you can evaluate the performance of computer vision in various environmental conditions, Lidar object detection, planning algorithms like A*, lateral controllers like Model Predictive Control (MPC) and more [39–41]. It can also be used to train neural networks and test them before deploying them to the real world [42, 43]. The use of simulated data greatly reduces the collection and labeling process for testing neural networks.

To interact with the simulator, the Robot Operating System (ROS) is used. ROS is a framework that provides a communication interface to create a level of abstraction for the user. So the CARLA simulator has a ROS bridge that allows the user to communicate with it and obtain a middleware suite that operates similar to a real-world vehicle equipped with AV sensors [44, 45].

2.2   Development of Autonomous Driving Controllers

2.2.1   Pure Pursuit Controller

A PPC is a lateral vehicle controller that geometrically determines the steering angle of the kinematic bicycle model to follow a prescribed reference path [46]. This controller was implemented in our previous foundational work applying RE to AVs and we are seeking to continue to utilize this strategy as a comparison case [36]. This previous study describes how the kinematic vehicle model has the origin on the rear axle and possesses a look ahead a distance (L_d) to determine the necessary steering angle to minimize the heading error between the vehicle and the path. This model assumes a fixed look ahead distance used; however, some derivations of this method have incorporated a dynamic look ahead distance that varies with speed. This allows the controller to adjust during cornering to have a better track performance.

2.2.2   Deep Learning Controller

A DLC was developed to perform path tracking using the front camera (raw image from the simulated camera can be seen in Fig. 2a) of the vehicle and the cross-track error of the vehicle with respect to the reference path. To perform path tracking using

Fig. 2 Raw image (**a**), Projected reference path to raw image (**b**), Raw image with drivable region (**c**), Masked image with drivable area (**d**)

a front camera, at each iteration we took the 30 waypoints ahead of the vehicle and created two parallel lines 2 m left/right from the reference path to create a drivable region. The right and left lines were then closed to create a drivable region. This drivable region was then projected from world coordinates to image coordinates using [47] (see Fig. 2b, c).

A masked image is obtained by creating an image with zeros (which in RGB channels means black) with the same dimensions as the original image and using the OpenCV function fillPolly to obtain a drivable region. The fillPoly function fills a contour with the specified color (in this case white) to the inputted image. Figure 2d shows the masked image with the drivable region.

This DL method applied to path tracking assumes that the waypoints are always in front of the vehicle; therefore, since we are only using one camera and the reference path isn't necessarily between two lanes, a large deviation from the path will cause the model to have a bad performance. An approach to correct this is to also use the cross-track error as a feature to the model since it possesses information of the deviation from the reference path. End-to-end DL models have been developed to predict the steering angle given the raw images from 3 frontal cameras [13, 48–50]. The purpose of these methods is to perform lane keeping by predicting the steering angle given the raw image. Others have used DL to perform behavioral cloning for lateral motion control [51, 52]. Our approach uses a drivable region and a cross-track error as features to perform path tracking.

2.2.2.1   Data Collection and Preprocessing

To collect the training data, the images of the front camera, the steering angle and the cross-track error were collected while performing path tracking using the PPC in the CARLA simulator. A total of 6077 samples were collected for training. When analyzing the recorded data, we noticed there is a bias towards small steering angles in the data due to many straight segments in the reference path. We observed that 4567 of the 6077 samples had small steering angles. Hence, the model will achieve a bad performance when cornering. Figure 3a shows the distribution of the samples of the dataset.

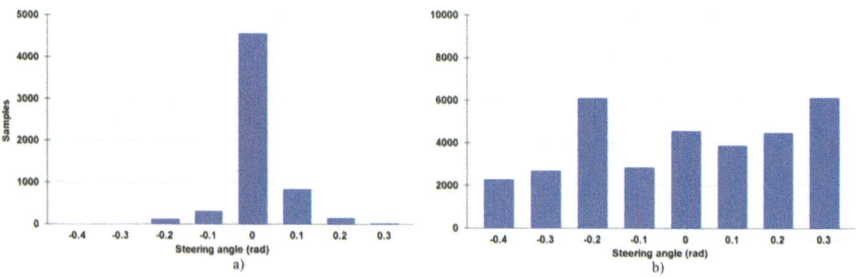

Fig. 3 Dataset with bias towards steering angles equal to zero (**a**), augmented dataset after removing bias towards small steering angles (**b**)

To correct this, the data was augmented to obtain a more balanced dataset. To augment the data and remove the bias, the data corresponding to the steering angles with low occurrence were repeated X/Y amount of times. Where X is the number of samples for the steering angles equal to zero and Y is the number of samples of each of the other steering angles, respectively. In other words the steering angles corresponding to 0.1 were repeated 4567/847 times. Where 4587 is the number of samples where the steering angle was equal to zero and 867 is the number of samples equal where the steering angle was 0.1 before the augmentation. Figure 3b shows the distribution of the number of samples of the augmented dataset.

To minimize the memory usage the images were resized from 800×600 to 32×32. Also, since the reference path possesses many left turns, the images were flipped with respect to the vertical axis and the steering angle was negated to account also for right turns. Which augmented even more the samples of the dataset. Obtaining a total of 66,308 training samples. The dataset was split to 80% for training and 20% for validation. Overall this approach is consistent with other end-to-end implementations of AV control used in research and development [13].

2.2.2.2   Model Development

A multi input and mixed data regression model was built using the Keras library in Python. The architecture consists of a Convolutional Neural Network (CNN) branch and a feedforward neural network (FNN) branch. The CNN branch uses the mask image as an input with the drivable region. The FNN branch uses the cross-track error of the vehicle from the reference path as the input. In the CNN branch, the masked image was fed through a 16, 32 and 64 convolutional layer. After each convolutional layer we perform max pooling with a 2×2 pool size and a 50% dropout to avoid overfitting. Each convolutional layer possesses a 3×3 kernel using valid padding. Then a flattening layer was applied followed by a dense layer. In the FNN branch the cross-track error was fed to a dense layer. The output of the dense layers from both branches were concatenated and then fed to another dense layer. This model uses mean squared error as its loss layer and the Adam optimized and root mean squared error as its metric. The output of this model is the predicted steering angle required to stay in the drivable region and minimize the cross-track error. Figure 4 shows the overall architecture.

The model was trained for 8 epochs using a batch size of 128. This model was trained on a Lenovo computer with 16GB of RAM, an Intel Xeon W-2123 processor, an Nvidia GeForce GTX 2070 graphics card, and Ubuntu 20.04 as the operating system. The respective CUDA drivers were properly installed to maximize the utility of the graphics card for the training phase. With these specifications, the DL model was trained in approximately 36 s and saved to later integrate it with the CARLA simulator.

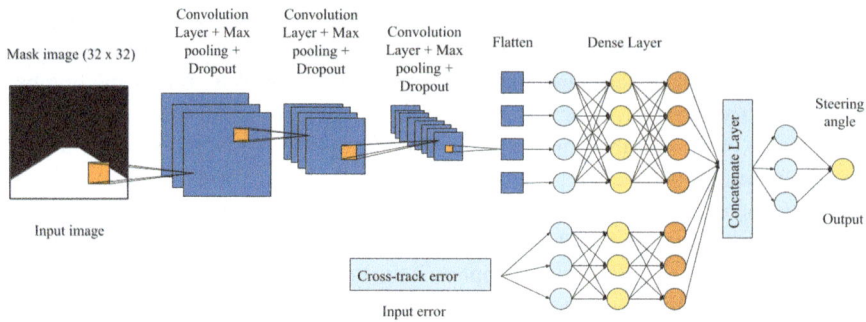

Fig. 4 DL architecture

2.3 Resilience Engineering Applied to AVs

2.3.1 Resilience Engineering Overview

RE is a safety management concept that focuses on socio-technical complexity [53]. An overview of the concepts and the foundation of RE were described in our previous study and will be used as a continuation to expand this study [36]. RE defines resilience as the ability of the system to achieve resilient performance rather than a property that a system possesses. This marks a difference between traditional safety management strategies (considered as Safety I concepts) and RE, since RE focuses on studying how the system works when it succeeds instead of studying why it fails (considered as Safety II concepts). Key RE concepts defined in our previous study are the ability of a system to learn, monitor, anticipate and respond. These are abilities that a resilient system must possess and improve system safety by taking a proactive approach.

2.3.2 Operational Resilience Performance on AVs

An AV is a complex system that is exposed to a dynamic environment and performs functions like sensing, planning and controls. There are studies on techniques to improve safety on AVs [54–56]. However, RE states that the resilient performance of a system is more concerned with how well the system performs, not so much how safe it maintains [57]. To improve the operational performance of a system, it is necessary to study how the system works and design operating rules so that the system is capable of continuing operation when it encounters threats or opportunities. This marks a difference between this new methodology and traditional methodologies. Since traditional methodologies study the system reactively. That is, they study how the system failed and then create operating rules so that it does not occur again. However, in very complex systems that are tightly coupled, they create nonlinear iterations [58]. In other words, creating more constraints on a complex system without studying how the system works will trigger more unexpected failures.

An AV requires a systematic study from the point of view of systems engineering to observe how the system works, the challenges of the system in certain conditions and thus develop a system with resilient performance. RE has several techniques to carry this out but it has not been applied due to the complexity of the system and the interdisciplinary expertise this methodology requires [57, 59, 60]. Nowadays, AVs possess a low operational resilience because when encountering unexpected scenarios, the system is unable to determine the correct decision or fails to sense features such as obstacles, curbs, lane exits, etc. California's DMV reported 2676 disengagements for AVs in 2021 due to operational failures and dangerous passenger decisions [31]. While the number of collisions reported between 2014 and 2021 [61] can be seen in Fig. 5.

Although these collisions do not mean that the cause of the accident was because of the AV, the system was unable to react to these circumstances. There is an increasing trend in Fig. 5, which means that current safety management strategies lack the techniques to improve the performance of the system. To drastically reduce the number of AV collisions as the technology becomes publicly available, a system capable of continuing its operation despite adversity and achieving operational resilience is required. RE has the potential to improve the operational resilience of AVs since this methodology focuses on coping with complex systems and how they succeed in this environment. Methods such as Functional Resonance Analysis Method (FRAM), RAG and others, can be used to recognize the limitations of the system in variable operations and their implications [62, 63].

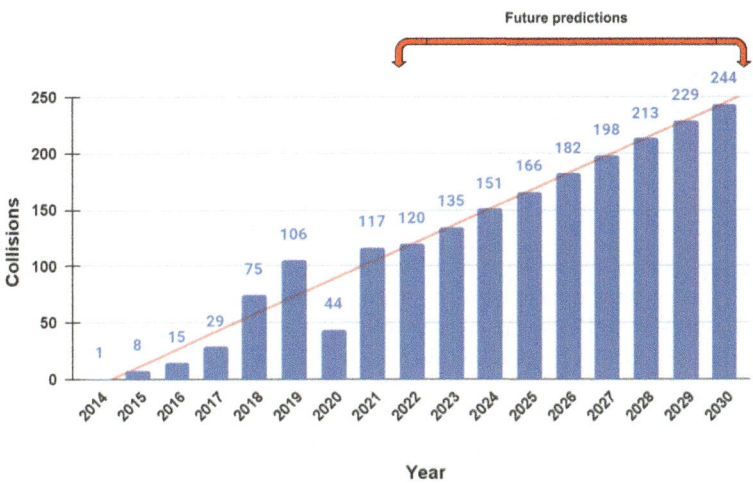

Fig. 5 California's DMV Collision Reports from 2014 to 2021 as well as a linear projection to 2030

2.4 Performance Evaluation Metrics

In this section, the traditional and resilience metrics will be presented. Also, a standard scaling technique will be introduced to provide some statistical information between the simulation results among all controllers.

2.4.1 Traditional Metrics

Evaluation methods for lateral controllers for AVs are used to measure how accurately the vehicle traveled the reference trajectory. So the typical metrics to evaluate lateral controllers when performing path tracking are average cross-track error, average heading error and control effort [64]. These metrics tell us the deviations of the vehicle with respect to the reference trajectory and the orientation of the vehicle in comparison to the reference yaw angle. We also introduce cross-track accumulation and heading error across the entire trajectory. These metrics will be described later.

The cross-track error is the lateral deviation of the vehicle with respect to the reference path. This tells us how much the vehicle deviated from the trajectory. The cross-track error can be determined by the following equation,

$$\varepsilon_{error} = \int_0^T \varepsilon_d(t) \cdot dt \tag{1}$$

Where ε_d is the cross-track error and ε_{error} is the sum of all the cross-track error throughout the whole path. The heading error is the difference between the vehicle heading and the reference heading. This tells us if the vehicle was headed in the same direction as the reference trajectory. The heading error can be determined by the following equation,

$$\psi_{error} = \int_0^T \left(\psi_{ego}(t) - \psi_{ref}(t) \right) \cdot dt \tag{2}$$

Where ψ_{ego} is the yaw angle of the vehicle, ψ_{ref} is the reference yaw of the path and ψ_{error} is the heading error. These two metrics give us the error accumulation over the entire trajectory. However, the average of these errors is commonly used and is represented by the Eqs. 1 and 2 divided by the number of samples. In other words,

$$\varepsilon_{avg} = \frac{1}{T} \int_0^T \varepsilon_d(t) \cdot dt \tag{3}$$

$$\psi_{avg} = \frac{1}{T} \int_0^T \left(\psi_{ego}(t) - \psi_{ref}(t) \right) \cdot dt \tag{4}$$

Where N is the number of samples, ε_{avg} is the average cross-track error and ψ_{avg} is the average heading error. The control effort is the amount of energy required by the lateral controller to minimize the error with respect to the reference trajectory. A higher control effort indicates that the controller has performed poorly and is unstable. The control effort is computed using the following equation,

$$e = \frac{1}{T} \int_0^T |\delta(t)| \cdot dt \qquad (5)$$

Where δ is the steering angle and e is the control effort.

2.4.2 Resilience Metrics

The measurement of RE metrics, such as "resistance" and "recovery," is enabled by exposing the AV system to a variety of driving environments and driving scenarios. Due to the low percentage of vehicle accidents and fatalities relative to the amount of kilometers driven, these measures cannot be relied solely on crash/accident situations. Instead, "resistance" will be computed as the divergence from the best trajectory, and "recovery" will be computed as the return time to the optimal trajectory.

The method of quantifying operational resilience based on robustness, speed, redundancy, and resourcefulness is referred to as the resilience triangle [32]. The resilience triangle describes how an incident affects the functionality of a system and the system's ability to respond after the event has occurred. Figure 6 depicts the resilience triangle and compares a system with low resilience performance

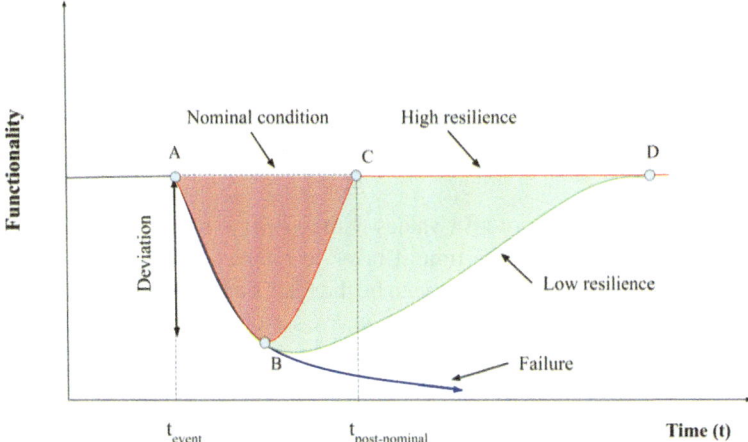

Fig. 6 Resilience triangle which shows the operation of a system with high resilience, low resilience and subjected to failure

to one with high resilience performance. Deviation is the magnitude (positive or negative) from the system's nominal condition. Recovery is the rate at which a system recovers to its normal state after an event has happened. This study applies resistance (robustness) and recovery (rapidity) as resilience indicators to AVs.

These are meaningful and significant real-world measurements since deviations are the situations that cause accidents, even though they do not always result in accidents. A failure recovery is conceivable if it is assessed that the AV system response resulted in collision with another object or deviance from the driveable path. A system with resilient performance can respond to an event to avoid significant repercussions and recover from accidents when they occur. The RAG, which measures the system's performance across the four abilities or pillars of RE, is another essential RE statistic. However, for the sake of this study, we will concentrate on the system's responsiveness in order to demonstrate the concept and execution of this statistic.

The resistance measures the system's susceptibility to external shocks, whereas the recovery measures the system's response speed. The system's resistance is its capacity to prevent any departure from the nominal condition, and its recovery is the amount of time it takes to return to the nominal condition.

To extend the definition of resistance to the sphere of AVs, resistance is defined here as the inverse of the cross-track error data point variance. The variance is defined as the average of the squared deviations from the mean. If the lateral controller is functioning properly, its mean should be close to zero; consequently, the variance of the cross-track error during the whole simulation should indicate the average amount by which each point deviates from the mean. Therefore, when resistance increases, deviations from the nominal state decrease. Using the following equation, we can calculate the controller's resistance,

$$Resistance = \frac{1}{\Theta} \tag{6}$$

Where ϑ can be defined as,

$$\Theta = \frac{\sum_{i=1}^{N} (x_i - \epsilon)^2}{N} \tag{7}$$

θ gives us the squared error of all x values with respect to the nominal condition ε.

The system's recovery is the time it takes to return to its normal state after a disturbance. We regard the disturbance to be the road's curvature and the cross-track error to be the divergence from the ideal state. We set a 5 cm (0.05 m) threshold since the cross-track error (deviation) can be incredibly small and never equal zero. If the system exceeds this threshold, we take the initial time (t_{event}). Similarly, after the system returns to its nominal condition (within the threshold $-0.05 \text{ m} \leq x \leq 0.05 \text{ m}$) we take the time post-event ($t_{post-nominal}$). This tendency is depicted graphically in Fig. 6. Where the return time is calculated by getting the time interval Δt between point A and point C by

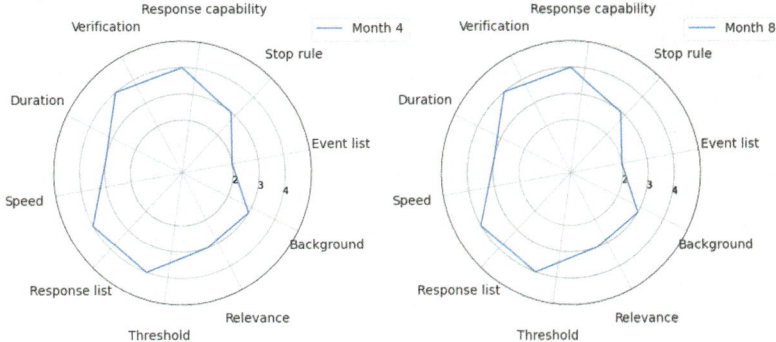

Fig. 7 An example of a traditional resilience assessment grid model which shows the evaluation of the ability to learn of a system [63]

$$RT = t_{post-nominal} - t_{event} \tag{8}$$

Another approach of evaluating RE is the resilience assessment grid. The RAG is a technique used to evaluate a system's performance on the four capabilities that a resilient system should possess [63]. RAG accomplishes this by asking a series of questions to determine the system's performance in a given ability. These four competencies are:

- **Ability to respond:** A system must be able to recognize when a change has occurred and respond to it effectively.
- **Ability to monitor:** A system should be able to monitor its own performance and detect external changes that could present an opportunity or a threat.
- **Ability to learn:** This ability provides the system with the means to learn from accidents and other occurrences that could have disturbed the system.
- **Ability to anticipate:** This skill is intended to predict possible risks and opportunities for the system.

Each skill is evaluated using a specific set of questions pertaining to the activities it should execute. For instance, if a system is capable of responding, it should do so in a timely and efficient manner. Therefore, some of the specific concerns could include speed, duration, and whether or not all occurrences were addressed, etc. Using a Likert-type scale or similar custom scale, each specific concern pertaining to the proper capacity is ranked. This study proposes a 1–10 scale since it provides a measurable evaluation. A radar chart is created with the issue's specifics and its rating. RAG is conducted numerous times to observe how the system's capabilities evolve after a predetermined amount of time. Figure 7 depicts an example of a system's ability to learn as measured by RAG. Both radar charts are very similar which means in four months, the performance of the system didn't change in the four capabilities.

Before measuring the capability of a system using RAG, the relevant variables must be defined. The relevant variables for measuring the system's responsiveness are deviations from nominal circumstances (cross-track error), recovery, and control effort. Additionally, the system's resistance was determined, but it was not presented using RAG. The factors of interest were utilized to generate the RAG radar chart. Typically, RAG is run multiple times under the same settings to assess the progression of a system's capabilities. However, we do not anticipate a substantial impact in response time for this application. We alter the speed in order to examine how the ability to respond changes as the pace increases. Due to the radius of curvature of the curves, we maintained a speed below 8 m/s for this study.

To establish a measurement standard between the three variables of interest, we devised a 1–10 scale. The scale determines the task's performance, with 1 indicating a low performance and 10 a high one. The scale was determined using a RAG score generated using the following formula,

$$RAG\ score = 10 \cdot \left(1 - \frac{z_\mu(x)}{z_{max}(x)}\right) \tag{9}$$

Where $z(x)$ is the z-score of all the data points and it is computed by the following equation

$$z(x) = \frac{x_i - \mu}{\sigma} \tag{10}$$

Where σ represents standard deviation, μ represents the mean, and x represents the observed value. The z-score indicates how many standard deviations the value x deviates from the data mean. However, as the average does not indicate the system's performance in many instances, we use ε as our nominal condition. A system with poor performance, an average error of 1, and minor deviations from the mean, for instance, can earn a low z-score and, thus, a high RAG score. However, we aim to eliminate the error. Therefore, the average does not indicate if the system fits our requirements. Therefore, we use ε as the nominal condition to determine the data's z-score. In addition, since it is irrelevant whether the observation point is to the right or left of the nominal condition, the absolute value of the numerator is obtained. Similarly, the standard deviation of the data indicates the proximity of the x value to the mean. However, as we have already shown, the average does not necessarily indicate the performance of our system. Instead, we are interested in the divergence of each x value from the nominal condition. Therefore, we calculate the pseudo standard deviation of each x value from the nominal condition using the square root of Θ. However, the z-score will be greater when the dispersion is modest. Consequently, the square root of $\sqrt{\Theta}$ is shifted to the numerator because the system obtains a greater RAG score when it achieves a smaller z-score. Hence, dividing by the pseudo standard deviation (ϑ) will produce a higher z-score and, ultimately, a lower RAG score. Therefore, Eq. 10 becomes

$$z(x) = |x_i - \epsilon| \cdot \sqrt{\Theta} \tag{11}$$

In Eq. 9, the average z-score (z_μ) of each test was divided by the highest z-score (z_{max}) across all tests. This is because, if we use the z_{max} of each test and the data has a small standard deviation, we will obtain a reasonably high z-score for each deviation. Alternatively, if we consider the z_{max} of all tests, the ratio z_μ/z_{max} will reveal the relationship between the average z-score of a test and the test with the highest z-score. If the z-score is very low, it indicates that the condition is closer to the mean. This ratio will yield a number between 0 and 1, thus by subtracting one and multiplying by ten, we can calculate the RAG score for the variable of interest.

2.4.3 Standardization

To evaluate the controllers developed in Sect. 2.2, we will use traditional metrics and RE metrics. To provide a standard among all the parameters to be measured, we use a standard scaling technique (also called standardization). This scaling technique centers all values around the mean with a unit standard deviation. For example, the standard scaling technique was used to standardize each output the control effort of all the controllers for a given test. The equation for this standardization is as follows,

$$S = \frac{x_i - \mu}{\sigma} \tag{12}$$

Where S is the standardized parameter, x are the outputs of each metrics of all controllers, μ is the mean and σ is the standard deviation.

3 Results

To do path tracking, we first gathered waypoints by subscribing to the odometry topic and manually driving the vehicle in the CARLA simulator's Town03 environment. The simulated vehicle was set to spawn in the starting position, and the Ackermann control package was utilized to provide the vehicle steering angle commands. At 5, 6, 7, and 8 meters per second, the PPC was employed to navigate the reference path. For clarity, we shall refer to them as Test No. 1, Test No. 2, Test No. 3, and Test No. 4, accordingly. The data for the DL model described in Sect. 2.2.2 was acquired by navigating the reference path at 5 m/s using the PPC. The preprocessed data was utilized to train the DL models. During runtime, the vehicle's position, cross-track error, heading error, and steering angle were captured for offline evaluation of both controllers. Figure 8 illustrates the reference path utilized in our investigation.

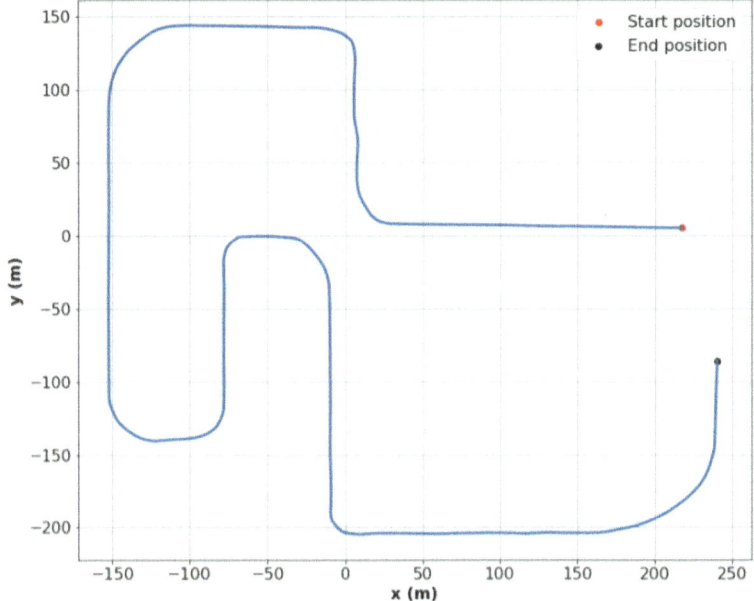

Fig. 8 Reference path for the path tracking algorithm used in this study generated in the Town03 of the CARLA Simulator

With the collected training data, the DL model was trained for 8 epochs. The training and validation loss were plotted to see the performance of our model. The model was trained for a few epochs to avoid overfitting since we used a large batch size. As we can see in Fig. 9, the training and validation loss gradually decreases which means that the model does not overfit the data. The training loss usually means how well the model is fitting the training data and the validation loss determines how well the model fits new data.

The PPC and DLC were developed and evaluated on the CARLA simulator at different speeds in order to assess their performance with traditional and resilience metrics as the speed increases.

The results of the traditional metrics were standardized using a standard scaling technique to provide more clarity between both controllers. As was mentioned in Sect. 2.4, the controller with the lowest value in each metric indicates that it obtained a better track performance. In other words, the controller with less deviations and control effort. The simulation results using the traditional metrics indicate that the PPC outperforms the DLC. From Fig. 10, we can see the PPC obtained a lower standard value than the DLC in all tests except in the control effort in Test No. 4. Test No. 4 of the PPC obtained the lowest standard value of -1.10, -0.99, -1.11 and -1.01 for the cross-track error, heading error, average cross-track error and average heading error, respectively. Test No. 4 of the DLC obtained a slightly lower value than the PPC for the control effort (which means a smoother response). Figure 10

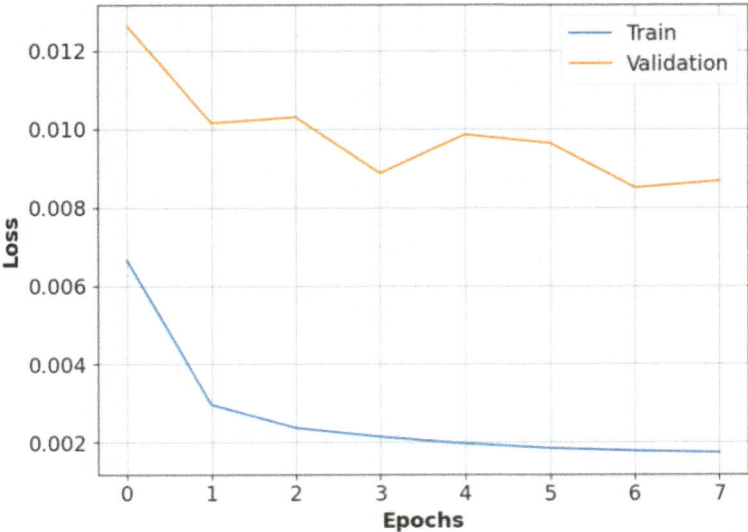

Fig. 9 Loss curves vs epochs of DL model. A gradually decaying training and validation loss means the model does well fitting the training data and fitting new data

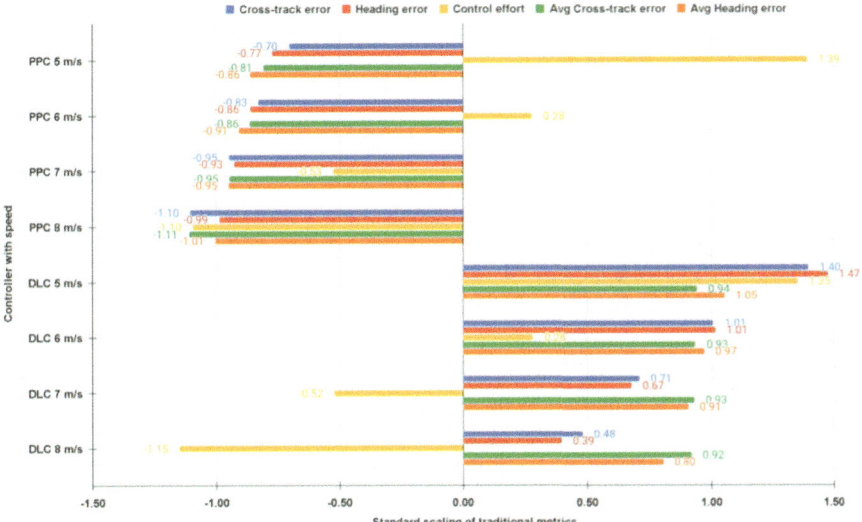

Fig. 10 Comparison of PPC and DLC using traditional metrics where a lower score indicates best performance according to these metrics

shows that as the speed increases, better track performance is achieved. Hence, the test with the highest speed in both controllers has better performance. This gives quantitative evidence that non AI-based controllers such as PPCs perform better

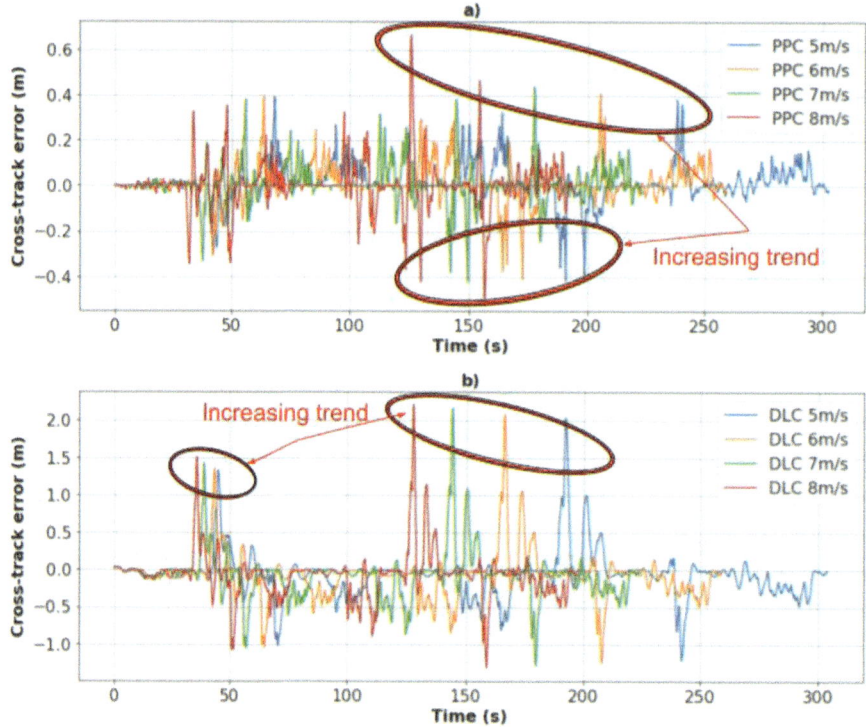

Fig. 11 Cross-track error of the PPC (**a**), Cross-track error of the DLC (**b**)

than DLCs which is likely the analysis that other researchers have conducted causing them to question AI-based control.

However, if we look at Fig. 11 we see in both controllers that the cross-track error increases as the speed increases. This contradicts our previous findings. This is because the metrics used are time dependent and when the same controller is compared at different speeds on the same trajectory, there is a discrepancy between the evaluation metrics and the operational performance of the system.

Similarly we can compare the heading error of both controllers throughout the whole trajectory. The PPC obtained less heading error than the DLC. This is to be expected since as we saw earlier, the DLC obtained more deviations than the PPC. Therefore, the heading error must also follow the same pattern. Figure 12 shows the error heading for both controllers for each test.

These metrics do not tell us about the operational performance of the system to respond to external disturbances. Additionally, the results of these tests are time dependent and cannot be used as an evaluation method to measure the operational performance of a system under the same trajectory at different speeds. For example, if the speed is increased while maintaining the same reference path, the vehicle will finish faster and therefore, the evaluation data possesses fewer samples. Therefore,

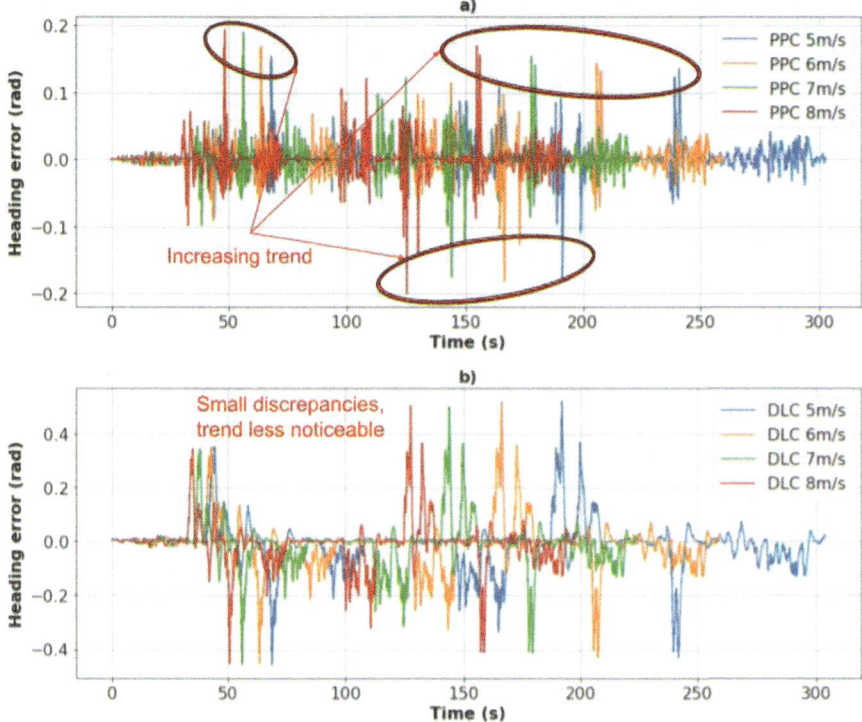

Fig. 12 Heading error of the PPC (**a**), Heading error of the DLC (**b**)

the sum of error will be lower at higher speed. For this reason, using the traditional metrics, the test with higher speed obtained better performance in all categories. Therefore, new evaluation metrics that are independent of time are needed that provides the user with another quantification of the operational performance of the system. Resistance and recovery are resilience metrics that indicate the system's ability to sustain in the face of a disturbance and how quickly it recovers after an incident. On the other hand, with the RAG score we evaluate the ability of the system to respond to disturbances as the speed increases. The RAG score provides us with a metric independent of the time.

Simulation results using the resilience metrics indicate that the PPC outperforms the DLC. However, unlike traditional metrics, we now see that the test with the highest speed performs the worst on most variables of interest. The simulation results were standardized using the same standard scaling technique described previously. However, since a RAG score of 10 means better performance and vice versa, now a higher standard value means better performance. As we can see in Fig. 13, the PPC obtained a higher performance in Test No. 2 among all tests. While the DLC obtained a higher performance in Test No. 1. This may be because the data collected for development of the DLC was at a speed of 5 m/s. As we saw in

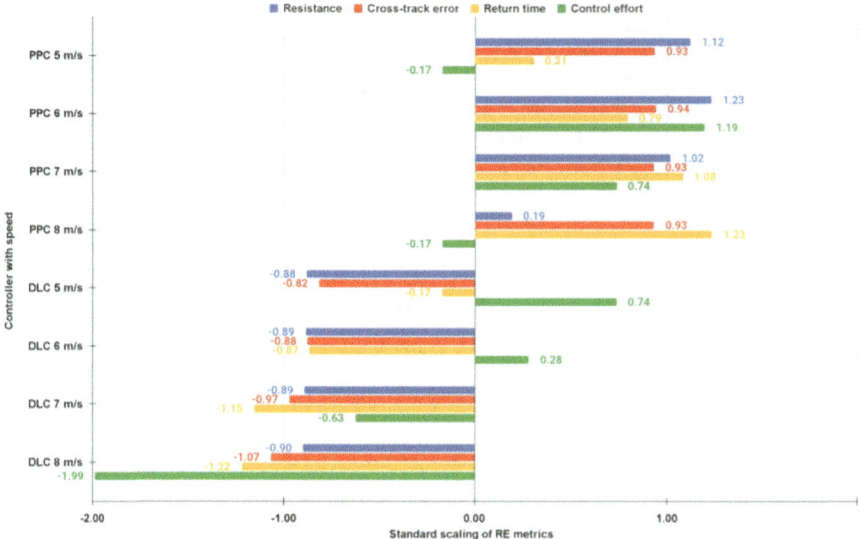

Fig. 13 Comparison of PPC and DLC using resilience metrics where a higher score is regarded as a better performance

Fig. 11, as the speed increases, the PPC has more deviations. Therefore, if the DLC was only trained with the controller that has the fewest deviations, then it will not perform well at other speeds. Comparing both controllers, we see that the PPC has higher RAG score than the DLC in all tests. It is likely that the PPC outperformed the DLC due to the training data and the features used for the DL model. The DL model was trained using the masked image with the drivable region and the cross-track error; however, feature engineering was not performed to see which parameters are more correlated for this path tracking problem. It is important to emphasize that in recovery, the PPC obtained a higher RAG score in test No. 4. This may be due to the nominal condition used for the RAG analysis. Overall we can see that evaluation using RE metrics provides direct insights into the system that match the logical development of the system. RE evaluation shows that AI-based control strategies may be more feasible that researchers realize and that the results shown here could be improved with more development of the AI-based controller.

It is important to note that the results presented using the RAG score depend on the nominal condition chosen for the study. For this study, zero was chosen as the nominal condition for cross-track error and control effort. For the return time, 1.5 s was used as the nominal condition. The nominal condition for cross-track error and control effort was chosen as zero since we want the controller with the least control effort and cross-track error. The nominal condition for the return time was chosen as 1.5 s because the data was observed to oscillate within this range. However, if we carry out a sensitivity study, we can observe that the RAG score varies depending on the nominal condition chosen by the user. This parameter must be chosen by the user depending on the requirements for the controller.

In this study, the RE metrics were used to evaluate the system's ability to respond. Given the definition of the system's ability to respond in Sect. 2.4, the ability to respond can be considered reactive since the system responds effectively after a change has occurred. However, that does not mean that RE metrics are reactive. The other abilities are considered proactive since they are capable of monitoring the state of the system, learning from past failures, anticipating future threats, and responding to these threats. However, this study serves as a path to demonstrate the potential that RE has to improve the operational performance of AVs. If this methodology is to be applied to assess the resilient performance of the overall system, all abilities must be considered. Nevertheless, it requires an interdisciplinary group with a level of expertise in systems engineering to perform a system level analysis and study how the system operates when it succeeds and develop metrics around its operational domain.

4 Conclusion

This chapter presented the overall methodology of RE and the potential it has to improve the operational performance of AVs. A PPC and DLC were developed to evaluate its performance using traditional and resilience metrics. The RE metrics used indicate how the system responds to disturbances and provides us with more information than traditional metrics since they are not time dependent. The resistance, recovery, and the RAG score were used to assess each lateral controller as the speed increases. Simulation results for both controllers using the traditional metrics show that the test with highest speed obtained a better performance. However, these results are time dependent and cannot be used to evaluate a controller on the same reference path while varying the speed. This is because if the vehicle travels faster, less samples will be collected and therefore less error accumulation. However, the resilience metrics are time independent. Therefore, we were able to conclude that Test No. 2 obtained a better performance for the chosen reference trajectory. Results may vary depending on the reference trajectory and the nominal conditions chosen for each variable of interest. It is important to mention that the PPC outperformed the DLC in all tests. This may be due to the training data collected and the features used to train the model, since feature engineering was not performed to see which features are more correlated to steering angle.

In this study the proposed resilience metrics provided genuine insights about future operational performance of a traditional lateral controller as well as a DL based lateral controller. However, resilience metrics can also be applied to evaluate an entire subsystem or specific aspects of an AV which can be challenging when new cutting-edge AI techniques are employed. Resilience is not an attribute of a system, rather it is the capacity that the system has in the operational domain. This means that this methodology can also be extended to evaluate the operational performance of the entire AV system. AVs are sophisticated systems that are subjected to a constantly changing environment and a high level of risk. For years, the rate of fatalities versus the amount of miles traveled has been used to assess these systems. This metric is considered as reactive since it tells you how the system performs after being deployed. Furthermore, the data from California's DMV collision reports tells us that underperforming or unsafe systems will come to light as they become publicly available. To assess the resilient performance of AVs under certain driving scenarios, a proactive evaluation metric is required. Additionally, RE demonstrated a clear and actionable path to improve performance and thus minimize crashes if this system were to go into commercial production. Future work in this area could focus on extending this study to a fusion sensor algorithm using traditional methods such as Kalman Filters and DL as well as full AV system evaluations.

References

1. Brodsky, J.S.: How an uncertain legal landscape may hit the brakes on self-driving cars. Berkeley Technol. Law J. **31**(2), 851–878 (2016)
2. Goberville, N., Zoardar, M.M., Rojas, J., Brown, N., Motallebiaraghi, F., Navarro, A., Asher, Z.: Techno-economic analysis of fixed-route autonomous and electric shuttles. In: SAE Technical Paper Series. SAE International, 400 Commonwealth Drive, Warrendale (2021). https://doi.org/10.4271/2021-01-0061
3. Motallebiaraghi, F., Yao, K., Rabinowitz, A., Hoehne, C., Garikapati, V., Holden, J., Wood, E., Chen, S., Asher, Z., Bradley, T.: Mobility Energy Productivity Evaluation of Prediction-Based Vehicle Powertrain Control Combined with Optimal Traffic Management. 2022-01-0141, SAE Technical Paper (2022)
4. Motallebiaraghi, F., Rabinowitz, A., Holden, J., Fong, A., Jathar, S., Bradley, T., Asher, Z.D.: High-fidelity Modeling of Light-Duty Vehicle Emission and Fuel Economy Using Deep Neural Networks. SAE Technical Paper 01–0181 (2021)
5. Rabinowitz, A., Araghi, F.M., Gaikwad, T., Asher, Z.D.: Development and evaluation of velocity predictive optimal energy management strategies in intelligent and connected hybrid electric vehicles. Energies. **14**, 18 (2021)
6. Schwarting, W., Alonso-Mora, J., Rus, D.: Planning and decision-making for autonomous vehicles. Annu. Rev. Control Robot. Auton. Syst. **1**(1), 187–210 (2018)
7. Xia, W., Li, H., Li, B.: A Control Strategy of Autonomous Vehicles Based on Deep Reinforcement Learning, pp. 198–201. 2016 9th International Symposium on Computational Intelligence and Design (ISCID), ieeexplore.ieee.org (2016)
8. Ren, J., Gaber, H., Al Jabar, S.S.: Applying Deep Learning to Autonomous Vehicles: A Survey, pp. 247–252. 2021 4th International Conference on Artificial Intelligence and Big Data (ICAIBD), ieeexplore.ieee.org (2021)
9. Zhou, Y., Tuzel, O.: Voxelnet: End-to-end Learning for Point Cloud Based 3d Object Detection, pp. 4490–4499. Proceedings of the IEEE Conference on Computer Vision and Pattern Recognition, openaccess.thecvf.com (2018)
10. Park, Y., Dang, L.M., Lee, S., Han, D., Moon, H.: Multiple object tracking in deep learning approaches: a survey. Electronics. **10**(19), 2406 (2021)
11. Goberville, N.A., Kadav, P., Asher, Z.D.: Tire track identification: A method for drivable region detection in conditions of snow-occluded Lane lines. In: SAE Technical Paper Series. SAE Technical Paper, 400 Commonwealth Drive, Warrendale (2022). https://doi.org/10.4271/2022-01-0083

12. Brown, N.E., Rojas, J.F., Alzu'bi, H., Alrousan, Q., Meyer, R., Asher, Z.: Higher accuracy and lower computational perception environment based upon a real-time dynamic region of interest. SAE Technical Paper (2022)
13. Chen, Z., Huang, X.: End-to-end learning for lane keeping of self-driving cars, pp. 1856–1860. 2017 IEEE Intelligent Vehicles Symposium (IV), ieeexplore.ieee.org (2017)
14. Nascimento, A.M., Vismari, L.F., Cugnasca, P.S., Camargo, J.B., Almeida, J.R.d., Inam, R., Fersman, E., Hata, A., Marquezini, M.V.: Concerns on the differences between AI and system safety mindsets impacting autonomous vehicles safety. In: Computer Safety, Reliability, and Security, pp. 481–486. Springer (2018)
15. Häring, I.: Technical safety and reliability methods for resilience engineering. In: Häring, I. (ed.) Technical Safety, Reliability and Resilience: Methods and Processes, pp. 9–26. Springer, Singapore., ISBN 9789813342729 (2021)
16. Singh, S., Critical Reasons for Crashes Investigated in the national motor vehicle crash causation survey (2015)
17. Walker Smith, B.: Human Error as a Cause of Vehicle Crashes. Center for Internet and Society. http://cyberlaw.Stanford.edu/blog/2013/12/human-Error-Cause-Vehicle-Crashes (2013)
18. Taeihagh, A., Lim, H.S.M.: Governing autonomous vehicles: emerging responses for safety, liability, privacy, cybersecurity, and industry risks. Null. **39**(1), 103–128 (2019)
19. Collingwood, L.: Privacy implications and liability issues of autonomous vehicles. Null. **26**(1), 32–45 (2017)
20. Hollnagel, E., Woods, D.D., Leveson, N.: Resilience Engineering: Concepts and Precepts. Ashgate Publishing, Ltd., ISBN 9780754681366 (2006)
21. Madni, A.M., Sievers, M.W., Humann, J., Ordoukhanian, E., D'Ambrosio, J., Sundaram, P.: Model-based approach for engineering resilient system-of-systems: application to autonomous vehicle networks. In: Disciplinary Convergence in Systems Engineering Research, pp. 365–380. Springer (2018)
22. Marshall, C., Roberts, B., Grenn, M.: Intelligent Control & Supervision for Autonomous System Resilience in Uncertain Worlds, pp. 438–443. 2017 3rd International Conference on Control, Automation and Robotics (ICCAR) (2017)
23. Koopman, P., Wagner, M.: Autonomous vehicle safety: an interdisciplinary challenge. IEEE Intell. Transp. Syst. Mag. **9**(1), 90–96 (2017)
24. Michelmore, R., Kwiatkowska, M., Gal, Y.: Evaluating Uncertainty Quantification in End-to-end Autonomous Driving Control. arXiv [cs.LG] (2018)
25. Zhang, F., Martinez, C.M., Clarke, D., Cao, D., Knoll, A.: Neural network based uncertainty prediction for autonomous vehicle application. Front. Neurorobot. **13**, 12 (2019)
26. Xu, S., Peng, H.: Design, analysis, and experiments of preview path tracking control for autonomous vehicles. IEEE Trans. Intell. Transp. Syst. **21**(1), 48–58 (2020)
27. Henaff, M., Canziani, A., LeCun, Y.: Model-Predictive Policy Learning with Uncertainty Regularization for Driving in Dense Traffic. arXiv [cs.LG] (2019)
28. Cosgun, A., Ma, L., Chiu, J., Huang, J., Demir, M., Añon, A.M., Lian, T., Tafish, H., Al-Stouhi, S.: Towards Full Automated Drive in Urban Environments: A Demonstration in GoMentum Station, California, pp. 1811–1818. 2017 IEEE Intelligent Vehicles Symposium (IV) (2017)
29. Koopman, P., Wagner, M.: Challenges in autonomous vehicle testing and validation. SAE Int. J. Transp. Saf. **4**(1), 15–24 (2016)
30. Muhammad, K., Ullah, A., Lloret, J., Ser, J.D., Albuquerque, V.H.C.d.: Deep learning for safe autonomous driving: current challenges and future directions. IEEE Trans. Intell. Transp. Syst. **22**(7), 4316–4336 (2021)
31. Disengagement Reports. https://www.dmv.ca.gov/portal/vehicle-industry-services/autonomous-vehicles/disengagement-reports/ (2020)
32. Furuta, K.: Resilience Engineering, pp. 435–454. Reflections on the Fukushima Daiichi Nuclear Accident (2014)
33. Woods, D.D.: Four concepts for resilience and the implications for the future of resilience engineering. Reliab. Eng. Syst. Saf. **141**, 5–9 (2015)

34. Liu, L., Lu, S., Zhong, R., Wu, B., Yao, Y., Zhang, Q., Shi, W.: Computing Systems for Autonomous Driving: State-of-the-art and Challenges. arXiv [cs.DC] (2020)
35. Kukkala, V.K., Tunnell, J., Pasricha, S., Bradley, T.: Advanced driver-assistance systems: a path toward autonomous vehicles. IEEE Ind. Electron. Mag. 7(5), 18–25 (2018)
36. Rojas, J.F., Brown, N., Rupp, J., Bradley, T., Asher, Z.D.: Performance Evaluation of an Autonomous Vehicle Using Resilience Engineering. SAE Technical Paper (2022)
37. Devineau, G., Polack, P., Altché, F., Moutarde, F.: Coupled Longitudinal and Lateral Control of a Vehicle Using Deep Learning, pp. 642–649. 2018 21st International Conference on Intelligent Transportation Systems (ITSC), ieeexplore.ieee.org (2018)
38. Dosovitskiy, A., Ros, G., Codevilla, F., Lopez, A., Koltun, V.: CARLA: An Open Urban Driving Simulator. arXiv [cs.LG] (2017)
39. Dworak, D., Ciepiela, F., Derbisz, J., Izzat, I., Komorkiewicz, M., Wójcik, M.: Performance of LiDAR object detection deep learning architectures based on artificially generated point cloud data from CARLA simulator, pp. 600–605. 2019 24th International Conference on Methods and Models in Automation and Robotics (MMAR) (2019)
40. Ebrahimpour, F.P., Ferdowsi, H.: Multi-constraint Predictive Control System with Auxiliary Emergency Controllers for Autonomous Vehicles, pp. 274–279. 2021 IEEE Intelligent Vehicles Symposium (IV), ieeexplore.ieee.org (2021)
41. Sakic, N., Krunic, M., Stevic, S., Dragojevic, M.: Camera-LIDAR Object Detection and Distance Estimation with Application in Collision Avoidance System, pp. 1–6. 2020 IEEE 10th International Conference on Consumer Electronics (ICCE-Berlin) (2020)
42. Gómez-Huélamo, C., Del Egido, J., Bergasa, L.M., Barea, R., López-Guillén, E., Arango, F., Araluce, J., López, J.: Train here, drive there: simulating real-world use cases with fully-autonomous driving architecture in CARLA simulator. In: Advances in Physical Agents II, pp. 44–59. Springer (2021)
43. Pérez-Gil, Ó., Barea, R., López-Guillén, E., Bergasa, L.M., Gómez-Huélamo, C., Gutiérrez, R., Díaz-Díaz, A.: Deep reinforcement learning based control for autonomous vehicles in CARLA. Multimed. Tools Appl. 81(3), 3553–3576 (2022)
44. Prescinotti Vivan, G., Goberville, N., Asher, Z., Brown, N., Rojas, J.: No cost autonomous vehicle advancements in CARLA through ROS. In: SAE Technical Paper Series. SAE International, 400 Commonwealth Drive, Warrendale (2021). https://doi.org/10.4271/2021-01-0106
45. Stević, S., Krunić, M., Dragojević, M., Kaprocki, N.: Development and Validation of ADAS Perception Application in ROS Environment Integrated with CARLA Simulator, pp. 1–4. 2019 27th Telecommunications Forum (TELFOR) (2019)
46. Wang, W.-J., Hsu, T.-M., Wu, T.-S.: The Improved Pure Pursuit Algorithm for Autonomous Driving Advanced System, pp. 33–38. 2017 IEEE 10th International Workshop on Computational Intelligence and Applications (IWCIA) (2017)
47. Liu, L., Cao, S., Liu, X., Li, T.: Camera Calibration Based on Computer Vision and Measurement Adjustment Theory, pp. 671–676. 2018 Eighth International Conference on Instrumentation Measurement, Computer, Communication and Control (IMCCC) (2018)
48. Jiang, H., Chang, L., Li, Q., Chen, D.: Deep Transfer Learning Enable End-to-end Steering Angles Prediction for Self-driving Car, pp. 405–412. 2020 IEEE Intelligent Vehicles Symposium (IV) (2020)
49. Valiente, R., Zaman, M., Ozer, S., Fallah, Y.P.: Controlling Steering Angle for Cooperative Self-driving Vehicles Utilizing CNN and LSTM-Based Deep Networks, pp. 2423–2428. 2019 IEEE Intelligent Vehicles Symposium (IV) (2019)
50. Eraqi, H.M., Moustafa, M.N., Honer, J.: End-to-end Deep Learning for Steering Autonomous Vehicles Considering Temporal Dependencies. arXiv [cs.LG] (2017)
51. Sharma, S., Tewolde, G., Kwon, J.: Behavioral Cloning for Lateral Motion Control of Autonomous Vehicles Using Deep Learning, pp. 0228–0233. 2018 IEEE International Conference on Electro/Information Technology (EIT) (2018)
52. Samak, T.V., Samak, C.V., Kandhasamy, S.: Robust Behavioral Cloning for Autonomous Vehicles Using End-to-end Imitation Learning. arXiv [cs.RO] (2020)

53. Righi, A.W., Saurin, T.A., Wachs, P.: A systematic literature review of resilience engineering: research areas and a research agenda proposal. Reliab. Eng. Syst. Saf. **141**, 142–152 (2015)
54. Yi, S., Worrall, S., Nebot, E.: Metrics for the Evaluation of Localisation Robustness, pp. 1247–1253. 2019 IEEE Intelligent Vehicles Symposium (IV) (2019)
55. Wang, J., Zhang, L., Huang, Y., Zhao, J.: Safety of autonomous vehicles. J. Adv. Trans. **2020**, 8867757 (2020). https://doi.org/10.1155/2020/8867757
56. Yang, Z., Huang, J., Yang, D., Zhong, Z.: Design and optimization of robust path tracking control for autonomous vehicles with fuzzy uncertainty. IEEE Trans. Fuzzy Syst. **30**, 1–10., undefined (2021)
57. Hollnagel, E., Nemeth, C.P.: From resilience engineering to resilient performance. In: Nemeth, C.P., Hollnagel, E. (eds.) Advancing Resilient Performance, pp. 1–9. Springer, Cham., ISBN: 9783030746896 (2022)
58. Cummings, L.L., Perrow, C.: Normal accidents: living with high-risk technologies. Adm. Sci. Q. **29**(4), 630 (1984)
59. Yarveisy, R., Gao, C., Khan, F.: A simple yet robust resilience assessment metrics. Reliab. Eng. Syst. Saf. **197**, 106810 (2020)
60. Azadeh, A., Salehi, V., Ashjari, B., Saberi, M.: Performance evaluation of integrated resilience engineering factors by data envelopment analysis: the case of a petrochemical plant. Process. Saf. Environ. Prot. **92**(3), 231–241 (2014)
61. Autonomous Vehicle Collision Reports. https://www.dmv.ca.gov/portal/vehicle-industry-services/autonomous-vehicles/autonomous-vehicle-collision-reports/ (2020)
62. Patriarca, R., Di Gravio, G., Woltjer, R., Costantino, F., Praetorius, G., Ferreira, P., Hollnagel, E.: Framing the FRAM: a literature review on the functional resonance analysis method. Saf. Sci. **129**, 104827 (2020)
63. Hollnagel, E.: RAG-Resilience Analysis Grid. Introduction to the Resilience Analysis Grid (RAG) (2015)
64. Rokonuzzaman, M., Mohajer, N., Nahavandi, S., Mohamed, S.: Review and performance evaluation of path tracking controllers of autonomous vehicles. IET Intell. Transp. Syst. **15**(5), 646–670 (2021)

Safety-Assured Design and Adaptation of Connected and Autonomous Vehicles

Xin Chen, Jiameng Fan, Chao Huang, Ruochen Jiao, Wenchao Li, Xiangguo Liu, Yixuan Wang, Zhilu Wang, Weichao Zhou, and Qi Zhu

1 Introduction

Connected and autonomous vehicles (CAVs) have the potential to transform the way we travel. They hold promise for increased mobility, reduced traffic congestion and better fuel efficiency with automated control, as well as the creation of a cooperative network that includes cars, traffic lights and other roadside infrastructures. Autonomous vehicles (AVs) typically employ a wide array of sensors to gather information about the road environment, and then use sophisticated techniques to fuse and process this data to come to a navigation decision in real time in an automated fashion. Many of the underlying components in an AV, such as perception, planning and control make use of deep learning or deep neural networks (DNNs) due to their superior performance. Moreover, greater benefits on safety and fuel economy can be achieved by enabling vehicles to exchange information with one another. In a connected vehicle (CV) system, vehicles are expected to exchange V2X (vehicle-to-everything) messages with surrounding vehicles and roadside units

X. Chen
University of Dayton, Dayton, OH, USA
e-mail: xchen4@udayton.edu

J. Fan · W. Li · W. Zhou
Boston University, Boston, MA, USA
e-mail: jmfan@bu.edu; wenchao@bu.edu; zwc662@bu.edu

C. Huang
University of Liverpool, Liverpool, UK
e-mail: chao.huang2@liverpool.ac.uk

R. Jiao · X. Liu · Y. Wang · Z. Wang · Q. Zhu (✉)
Northwestern University, Evanston, IL, USA
e-mail: ruochen.jiao@northwestern.edu; xiangguoliu2023@northwestern.edu;
yixuanwang2024@northwestern.edu; zhilu.wang@northwestern.edu; qzhu@northwestern.edu

© The Author(s), under exclusive license to Springer Nature Switzerland AG 2023
V. K. Kukkala, S. Pasricha (eds.), *Machine Learning and Optimization Techniques for Automotive Cyber-Physical Systems*, https://doi.org/10.1007/978-3-031-28016-0_26

(RSUs) for extended perception range, to learn about traffic status down the road, and to coordinate their planning and control decisions. Realizing these potentials of CAVs, however, require tackling the immense challenge of assuring their safety in uncontrolled, public road environments. Numerous recent accidents involving autonomous vehicles are reflective of the safety concerns that loom large in the rapid advancement of CAV technologies [38, 62, 65, 66, 78]. The U.S. Department of Transportation (USDOT) launched the Automated Vehicle Transparency and Engagement for Safe Testing (AV TEST) Initiative in June 2020 to improve the safety in the development and testing of automated driving systems [51]. The USDOT has also started deploying test sites for connected vehicle applications in Florida, New York, and Wyoming [70].

This book chapter will survey recent advances in designing and operating CAVs with safety assurance. Instead of reviewing existing safety standards and industry practices, it aims to bring into focus new methodologies and techniques that have the potential to reshape how we approach the problem of safety assurance of CAVs, paying special attention to two categories of problems—(1) safety verification of CAVs that employ neural network-based components and (2) system adaptation and design with safety guarantees. The chapter will end with a discussion of outstanding technical challenges, broader applications of the surveyed techniques, and the authors' outlook on this important topic of safety assurance of CAVs.

2 Safety Verification of Neural Network-Based Components in CAVs

In CAVs, neural network-based components have been widely used for sensing, perception and prediction, and increasingly being tried for planning and control as well. It is thus critical to conduct safety verification of these neural network-based components for ensuring overall system safety. In particular, this includes conducting robustness of individual neural networks, in particular those used for sensing, perception and prediction, and performing safety verification of a neural network controlled/planned system.

2.1 Robustness Analysis of Deep Neural Networks

Local Robustness Analysis of Neural Networks Robustness is one of the key metrics to measure how stable a neural network's outputs are under random noises, external perturbation, or adversarial attacks to its inputs. Recent studies have in particular highlighted the lack of robustness against adversarial perturbations for neural networks [21, 67]. These adversarial perturbations construct a local input region around each inputs. A neural network is verified to be robust if the neural network outputs are guaranteed to be correct for each local input region, i.e., verification of the **local robustness**.

Measurement of robustness can take the form of upper and lower bounds on certain key input and output parameters. For individual deep learning components, the robustness analysis problem can often be reduced to output range analysis of the neural networks. State-of-the-art methods for output range analysis mainly fall into two categories: constraint programming (CP) [14, 36] and abstract interpretation [63, 71]. CP-based methods can perform exact analysis of the neural networks. However, the scale of deep neural networks limits the usage of these methods because they require encoding an entire network into a large nonlinear programming problem (or an SMT problem) and then solving it. The main drawback with abstract interpretation, on the other hand, is that it is difficult to propagate the dependencies for nonlinear operations across layers [48]. While such methods can scale with the network's size, the performance degrades as the network becomes deeper.

In [29], we propose a layer-wise refinement method, *LayR* to compute a guaranteed and overapproximated range for the output of the neural network for a adversarially perturbed input region. By checking the overapproximated range, we can verify whether the neural network is robust against all possible adversarial perturbations within the input region. *LayR* bridges abstract interpretation with mixed integer linear programming (MILP) and iteratively improves approximation precision by systematically increasing the number of integer variables, as shown in Fig. 1.

Global Robustness Analysis of Neural Networks Most of the efforts in the literature focus on verifying/certifying the local robustness, which characterizes the robustness property for a small region of network input space. However, there are

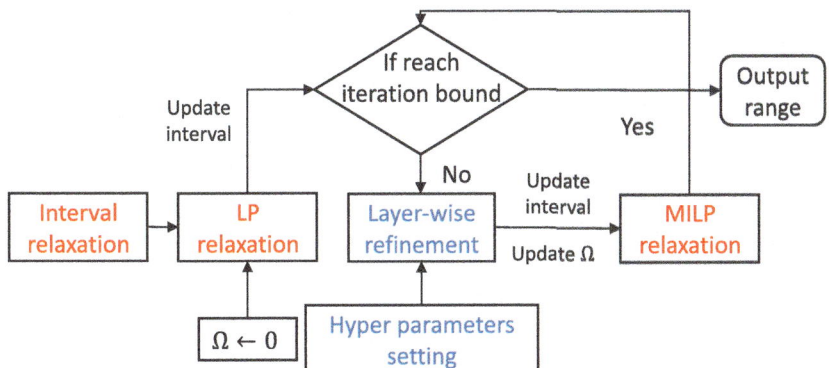

• Divide: For each neuron, divide the input space to refine the over-approximation;
• Slide: Layer-wise refinement by sliding-window based method.

Fig. 1 Divide-and-slide structure of LayR: Ω defines the number of slack integer variables of all the layers. In the refining process, Ω is monotonically increased to improve the output range estimation, until the iteration bound is reached

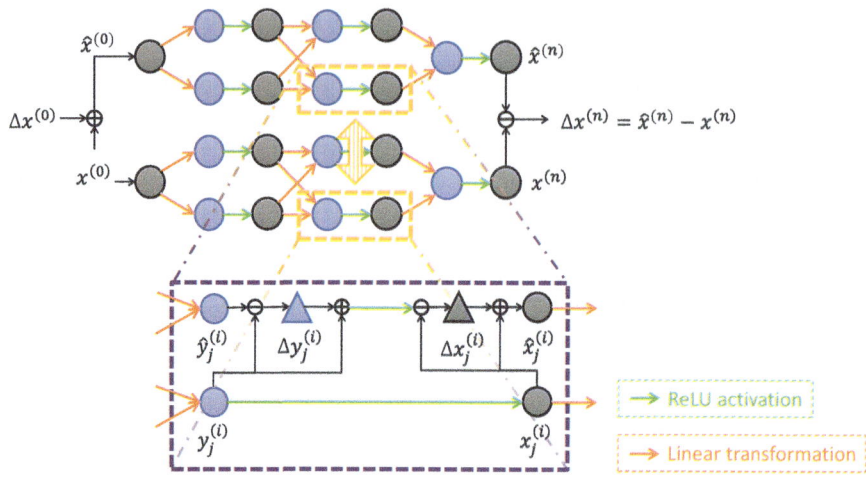

Fig. 2 Interleaving twin-network encoding (ITNE) for neural network global robustness certification. The hidden layer neurons are connected between the two copies of the neural network by distance variables $\Delta y_j^{(i)}$ and $\Delta x_j^{(i)}$

a lot of scenarios that need the robustness property over the entire network input domain, especially for cases that the network input samples cannot be obtained in advance. For instance, for image processing neural networks (like the perception modules in CAVs), the exact input samples during runtime are not always known at design time. In those cases, the global robustness property of the network should be considered, which can bound the worst-case output variation under perturbation for all possible network inputs. Directly conducting local robustness verification for all possible regions in the entire input domain by leveraging the divide-and-conquer techniques is not practical, especially for networks with high-dimension inputs, such as image inputs, as the complexity of divide-and-conquer is exponential to the input dimension. In [77], we developed an efficient global robustness certification algorithm that encoding two copies of the neural network side-by-side, as shown in Fig. 2. One network copy encodes the inference of a normal input while the other one encodes the inference of the disturbed input. Such encoding is formulated as an optimization problem that maximizes the output variation for all possible inputs and perturbations. The differences of hidden neurons between two networks are considered during the relaxation of the optimization problem to efficiently derive a tight over-approximation of the neural network output variation bound. Such over-approximated global robustness can be leveraged to enable the formal verification of the perception neural networks in CAV systems.

2.2 Safety Verification of Neural-Network Controlled Systems

An important class of CAVs can be described by a physical process such as the change of the velocities or distances of vehicles regulated by a learning-enabled controller which can be a neural network. We call such systems neural-network controlled systems. A *Neural-Network Controlled System (NNCS)* is a special sampled-data system which consists of a continuous-time physical process (plant) defined by an ordinary differential equation (ODE) and a feed-forward neural network (FNN) controller which works at discrete time moments. Figure 3 illustrates an execution of an NNCS. The physical process is defined by an ODE $\dot{x} = f(x, u)$ wherein x is the state variable and u is the control input. The FNN controller samples the system state every δ_c time and updates the control input value. Such a system is often safety-critical and it is significant formally verified the safety before implementation.

The safety verification problem asks whether a system can be in an unsafe situation or not. For example, it is crucial to know whether the distance between any of two connected vehicles could be too close at a near future time. Many safety verification problems can be reduced to *reachability problems*, that is, *determining whether the given state can be reached by the system*. Unfortunately, the reachability problem is not decidable even for linear hybrid systems [2, 24]. Hence, most of the existing reachability analysis techniques for hybrid dynamical systems seek to compute an overapproximation of the reachable set. If this overapproximation set does not contain any unsafe state, then the system is safe. Otherwise the safety is unknown, and either the reachable set overapproximation should be refined or an unsafe execution should be found.

NNCSs are particular hybrid dynamical systems such that only the dynamics is updated by the controller, while the system executions are still continuous. Therefore, regardless of the noises or uncertainties between the plant and the controller, an NNCS shows deterministic behavior from an initial state. In other words, a system execution, i.e., the reachable state and control input used at any time, is uniquely determined by the initial state, and we call the function that maps

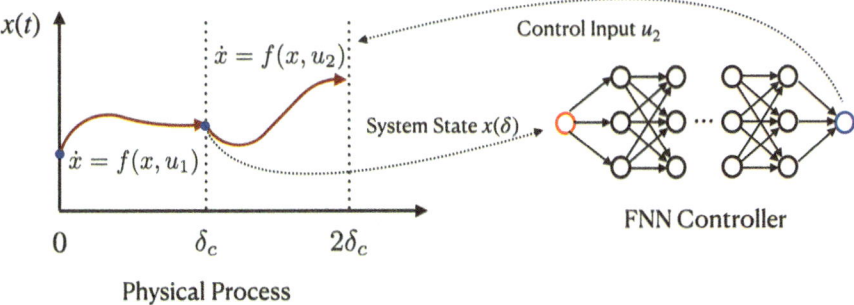

Fig. 3 State evolution of a neural-network controlled system

the initial state to its reachable state at a time *flowmap* which is essentially the solution of the piecewise ODE in Fig. 3. Hence, the reachability analysis task on an NNCS becomes computing the range of the flowmap w.r.t a given set of initial states.

The overapproximate reachability computation for NNCSs is at least as hard as that on general nonlinear sampled-data systems, and the main challenge is to accurately approximate the flowmap function which is a composition of a series of alternative neural network mappings and ODE evolution, and it often does not have a closed-form expression. *Set propagation* [9] is a popular scheme for computing time-bounded reachable sets under such dynamics. From a given initial state set, a set-propagation approach iteratively computes the reachable sets in small and consecutive time intervals the union of which is a cover of the time horizon. The reachable set segment which is also known as *flowpipe* computed in each iteration is propagated to the next time interval. For an NNCS, such an algorithm alternatively computes the flowpipes for the ODE and the output range of the controller until the upper bound of the time horizon is reached. A set-propagation approach for NNCS is often developed in the following two ways.

Pure Range Overapproximation A range overapproximation approach can be directly built by combining a neural network output range analysis method [15, 25, 36, 63, 69, 71, 73] and a reachability computation tool for ODEs [1, 8, 50]. It alternatively computes the reachable sets of the two components and propagates the result to the future time. Such a method mainly focuses on the range overapproximation and often cannot track the state dependency in a flowmap, therefore hard to control the accumulation of overapproximation error on highly nonlinear dynamics.

Functional Overapproximation A functional overapproximation approach seeks to compute an overapproximation for the flowmap function instead of only its range. Most of the existing methods [16, 19, 20, 26, 31–34] in this category uses *Taylor Models (TM)* [47] as the functional overapproximations. Unlike range overapproximations, a functional overapproximation is obtained by composing the functional overapproximations for the sub-components in a system, and it often requires more computational effort than computing a range overapproximation. However, functional overapproximations are able to keep the state dependency in flowmaps and effectively limit the accumulation of overapproximation error in reachability computation. Figure 4 illustrates an functional overapproximation represented by a TM for the output range of an FNN controller at the time $t = k\delta$. The actual flowmap that transforms an initial state x_0 to the control input $u_k = \kappa \circ \Phi(x_0, k\delta)$ used at $t = k\delta$ is overapproximated by a TM $p(x_0) + I$ wherein p is a polynomial and I is an interval remainder.

We briefly introduce the techniques we developed for computing functional overapproximations for the reachable sets of NNCSs.

ReachNN In [26], we present the ReachNN technique to compute reachable set overapproximations for NNCSs. The main contribution is an approach to obtain a TM-like overapproximation for the end-to-end relation of a neural network whose

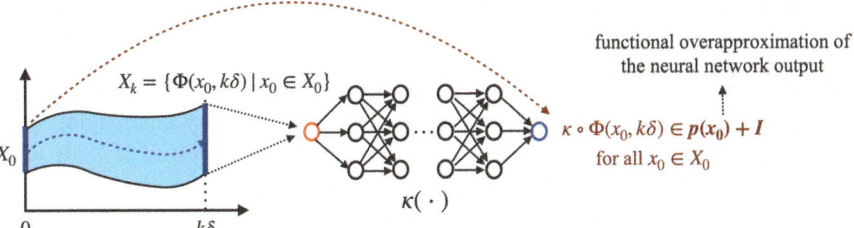

Fig. 4 Functional overapproximation of the control input range

activation functions are assumed to be all continuous. By Weierstrass approximation theorem [52] such a neural network over a compact input set can be uniformly approximated as closely as desired by a polynomial. The main method first computes a Bernstein interpolation for the input-output mapping of the neural network, and then a conservative interval remainder for it can be evaluated based on the adaptively selected samples from the input set, and an estimation of the Lipschitz constant of the neural network. We show that this method can be integrated with the reachability tool Flow* [8] which computes TM flowpipes for ODEs, and generate TM reachable sets which approximately keep the state dependency for NNCSs.

ReachNN* ReachNN* [20] leverages GPU-based parallel computing to compute the sampling-based error bound estimation in ReachNN. To further improve the runtime and error bound estimation, ReachNN* also features optional controller re-synthesis via a technique called *verification-aware knowledge distillation* [19] to reduce the Lipschitz constant of the neural network controller. ReachNN* demonstrated $7\times$ to $422\times$ efficiency improvement over ReachNN across a set of benchmarks.

The Polynomial Arithmetic (POLAR) Framework POLAR [31] is introduced for computing TM functional overapproximations for neural network outputs using layer-by-layer propagation. It is an extension of the standard TM arithmetic by introducing (A) Bernstein approximations for the activation functions in neural networks and (B) the symbolic representation of TM remainders in the layer-by-layer propagation framework for computing the output range of a neural network. It can be seamlessly integrated with the reachability tool Flow* to compute TM flowpipes for NNCSs. POLAR has the following main differences from ReachNN: (1) POLAR only uses Bernstein polynomials in approximating activating functions which are always univariate, but ReachNN needs to compute a multivariate Bernstein polynomial when the neural network has multiple inputs. It is much more time costly to compute multivariate Bernstein polynomials than the univariate ones. (2) POLAR uses layer-by-layer propagation framework to compute TM outputs for neural networks, however ReachNN performs an end-to-end overapproximation.

3 System Adaptation and Design with Safety Assurance

For safety-critical systems like CAVs, ensuring safety is a central focus during both the design stage and the runtime operation of them. It is a very challenging task, given the rapid increase of system functional complexity in terms of both scale and features, the usage of advanced architectural components such as multicore CPUs and GPUs, the stringent and contradicting requirements on various objectives such as performance, cost, fault tolerance, and reliability, the adoption of emerging machine learning components, particularly those based on deep neural networks, and the close interaction with a dynamic surrounding environment [60, 86]. In this section below, we will discuss these challenges in CAV design and adaptation, and introduce some of the proposed approaches to them, including those that leverage the methods from Sect. 2 as the underlying safety verification tools.

3.1 Safety-Assured Runtime Adaptation

The dynamic and uncertain environment of CAVs could put changing requirements on their objectives. For instance, a vehicle may need to enhance its planning, navigation and control performance in difficult-to-navigate terrains via more frequent sampling and processing [11–13] (especially for level 5 autonomy), to strengthen its security protection in an adversarial environment by adding monitoring tasks or authentication methods [40, 49], to improve its soft error tolerance in radioactive surroundings through task re-execution [41, 81], or to mitigate the impact under severe communication disturbance by running more computation locally. It is thus critical for those systems to be able to *adapt* to the dynamic environment and operation context.

Two major challenges in enabling runtime adaptation are to ensure that during and after the adaptation process, (1) functional safety is guaranteed, and (2) resource and timing constraints are met. To address the first requirement, we may leverage various verification/validation techniques, including those introduced in Sect. 2. To ensure both requirements, however, it is important to develop holistic approaches that span across functional, software, and hardware layers. Next, we will introduce our recent works in this area, along with some of the related works.

Opportunistic Intermittent Control with Safety Guarantees For safety-critical autonomous systems such as robots and automated vehicles, control schemes are often designed conservatively so that system safety can be maintained in a wide variety of situations [10, 43, 56]. During the operation of these systems, however, such schemes can be overly conservative and result in unnecessary resource and/or energy consumption. In [27, 28, 40, 41], we make the observation that certain control steps, even if they are skipped, do not impact either the performance or safety of the overall system. Armed with this observation, we propose an online scheme that opportunistically skips control computation and the corresponding actuation steps by learning specific characteristics of the system's operating environment. We further show that safety could be maintained with this more efficient control scheme.

Specifically, to address the safety, we first compute a *strengthened safe set* based on the notion of *robust control invariant* and *backward reachable set* of the underlying safe controller. Intuitively, the strengthened safety set represents the states at which the system can accept any control input at the current step and be able to stay within safe states, with the underlying safe controller applying input from the next step on. We then develop a monitor to check whether the system is within such strengthened safe set at each control step. Whenever it is found that the system state is out of the strengthened safe set, the monitor will require the system to apply the underlying safe controller for guaranteeing system safety. To efficiently leverage the characteristics of specific operation context and environment, we develop two approaches to leverage the characteristics of operation context and environment when the system is within the strengthened safe set, depending on the type of the underlying safe controller and whether the characteristics are known explicitly. In the simpler case where the safe controller has an analytic expression and the characteristics can be explicitly captured, we use a model-based approach to decide the skipping choices by solving a mixed integer programming (MIP) program. Otherwise, we use a deep reinforcement learning (DRL) approach to learn the mapping from the current state and the historical characteristics to the skipping choices, which implicitly reflects the impact of specific operation context and environment. Our approach is applied to a vehicle adaptive cruise control (ACC) example and shown to provide significant savings in actuation energy and computation load.

Switching Among Multiple Controllers with Safety Guarantees The work in [30] is our first attempt towards the safety adaptation and design for learning-enable systems, allowing a safe, efficient and intelligent switch between different system modes. Motivated by this work, we start considering a more general case, where *switching among multiple existing controllers*, including possibly both model-based ones and neural network-based ones, can be conducted to address system adaptation needs. This is show in Fig. 5. Note that the case where a control step is skipped can be viewed as a special case of switching to a trivial controller.

For safety-critical systems such as CAVs, the key to enable such switching among multiple controllers is to formally ensure safety. In [72], we extend the work from [30] to achieve energy-efficient control adaptation with safety guarantees by switching among multiple controllers (including neural network based ones) via *control invariant set* computation and reinforcement learning. Once a system starts from a control invariant set, it will never leave the set and therefore the safety can be guaranteed. However, it is a hard problem to compute the control invariant set for neural network controlled systems. To solve this problem, we first partition the system space into multiple regions, and on each small local region, we overly approximate the neural network controller by Bernstein polynomials with bounded error. After this transformation, we obtain a hybrid system with polynomial dynamics and compute the invariant set by solving a semi-definite programming

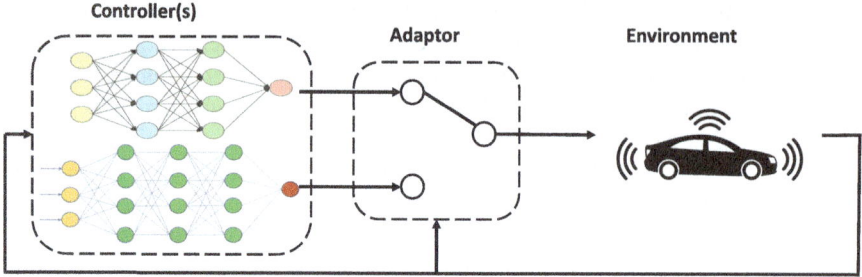

Fig. 5 Adaptation through switching among multiple controllers, which could include both model-based ones and neural network-based ones

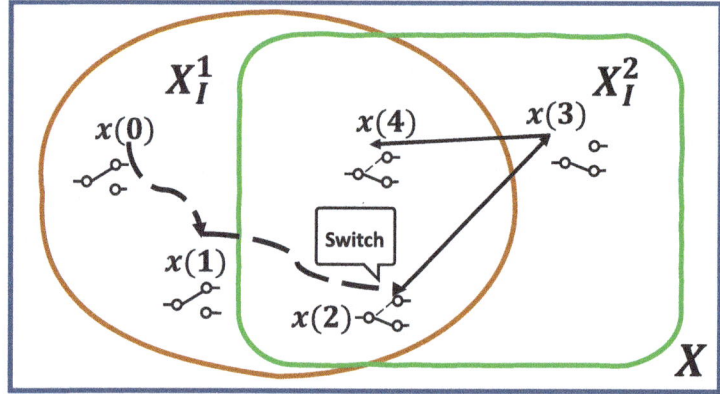

Fig. 6 An example illustrating the energy-efficient switching control with safety guarantees in [72]. We compute the control invariant sets X_I^1 and X_I^2 for controllers κ_1 and κ_2, respectively, and efficiently switch between them based on DRL when the system state is within the intersection of the two invariant sets. For example, in the figure, a control switching happens when the system is at $x(2)$, where both controllers can be safely chosen, and DRL picks κ_2 for energy efficiency

(SDP) problem. The union of all the invariant sets define the safe adaptation space, where we apply deep reinforcement learning (DRL) to learn an energy-efficient strategy. Figure 6 shows an example illustrating our framework. In two case studies, including an ACC example, our framework with invariant set and DRL achieves the best safety-energy consumption efforts when compared to baseline methods.

Cross-Layer Adaptation with Safety-Assured Job Skipping
For many practical systems such as CAVs, the ability to adapt to dynamic requirements is often limited by the tight resource constraints. Moreover, most safety-critical systems employ rigid timing requirements, such as periodic execution and hard deadlines, to guarantee the functionality under worst-case analysis, which further restricts the system adaptation ability. In these cases, it is important to address adaptation with cross-layer approaches.

In the literature, there are a number of methods that adapt task execution with cross-layer consideration. For instance, in [61], the simplex control architecture is proposed, where multiple controllers are being switched at runtime based on the system state and a safety controller keeps the system safe. In [11], an online adaptation approach is proposed for hard real-time systems to temporarily increase control sampling frequency under disturbances while maintaining schedulability. In [5–7], feedback schedulers assign new sampling periods to control tasks during runtime to optimize the control performance under earilest deadline first (EDF) scheduling. In [57], an approach is proposed to adaptively minimize tasks' usage of high quality-of-service resources while meeting control performance requirements.

In [75], different from the previous adaptation approaches that are based on traditional hard timing constraints, we propose an approach that explores proactive task job skippings based on the dynamic system state for state-aware tasks and static *weakly-hard constraints* for other state-unaware tasks. Note that with weakly-hard constraints [4, 55], occasional deadline misses are allowed in a bounded manner. Such paradigm provides more flexibility on the system design than traditional hard real-time constraints, while still allows the possibility of formally guaranteeing functional correctness that soft deadlines cannot provide, using formal analysis techniques such as those in [27, 28].

More specifically, we propose a cross-layer runtime adaptation framework in [75] that allows proactive skipping of task executions and re-allocate resources to the tasks that need performance improvement, as shown in Fig. 7. The system safety is guaranteed under the execution skipping, while the runtime task status is taken into account to maximize the freedom of resource re-allocation. This adaptation framework also involves an efficient runtime scheduler to ensure the timing property during the resource re-allocation. Based on the resource re-allocation, this adaptation framework achieves the dynamic adaptation goals in the best-effort manner. Case study on a robot car example demonstrates the effectiveness of this approach in meeting adaptation needs with safety assurance.

Runtime Safety-Guided Policy Repair For learning-based control systems, runtime safety assurance is particularly crucial and yet challenging. A common approach to providing such kind of assurance is to pair a learning-based controller with a safety controller at runtime. The learning-based controller is usually the primary controller. It learns control policy to attain high performance for the task through data-driven methods. However, it does not provide any safety guarantee especially in scenarios unseen during the training stage. The safety controller is tasked with predicting impending safety violation and taking over control when it deems necessary. It is often designed based on conservative models, has inferior performance compared with its learning-based counterpart, and may require significant computation resources if implemented online.

In order to mitigate the performance loss resulted from the undesirable alternations from the learning-based controller the safety controller while preserving safety, we propose to repair the learning-based controller's control policy by leveraging the interventions carried out by the safety controller in [85]. A naive repair

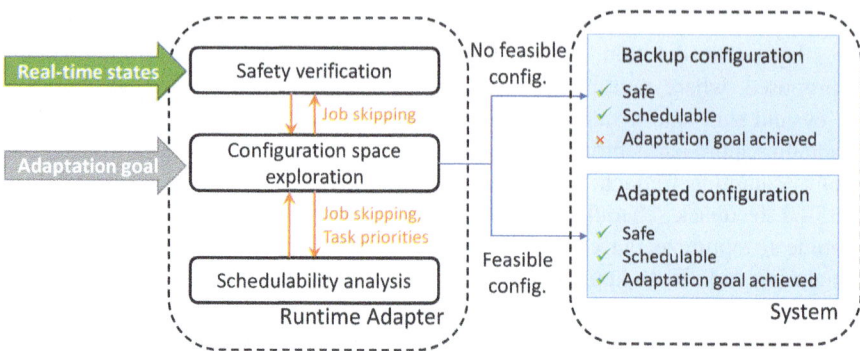

Fig. 7 An overview of the cross-layer runtime adaptation framework proposed in [72]. The system initially runs under a backup configuration that guarantees schedulability and safety. During runtime, an adaptation goal can be given by an external party. The adapter explores the configuration space to search for a feasible solution that achieves the adaptation goal, while ensuring schedulability and safety. If a solution is found, the system will run at this new configuration; otherwise, it will stay at the backup configuration

scheme is to have the learning-based controller learn from the safe control inputs generated by the safety controller until the policy no longer perform unsafe behavior. However, re-training the policy may undermine the policy's performance for the task. To address this, we introduce minimally deviating policy repair via trajectory synthesis. Basically, we synthesize safe trajectories such that by learning from those trajectories, the policy is safe and its parameters are minimally changed, as shown in Fig. 8. This policy repair scheme require naive policy repair as a precondition so that a safe policy is present. Then we formulate an optimization problem where the objective is to perturb the parameters of the safe policy to regress towards those of the original unsafe policy while the constraint is that the perturbation should not result in the policy generating unsafe trajectories. We use local linearization to transform this optimization problem into a trajectory optimization problem. The motivation is that, after applying the optimal perturbation to the policy parameters, the policy should be able to generate the trajectories solved from the trajectory optimization problem. This work can be viewed as data augmentation strategy where the data is optimized specifically for the learning model.

End-to-End Uncertainty-Based Adaptation for Mitigating Adversarial Attacks to CAVs Performing runtime adaptation for CAVs may significantly improve system safety, robustness and security in practice. For instance, in [35], we present an approach for runtime detection and mitigation of adversarial attacks. CAVs have been shown to be susceptible to adversarial attacks, where small perturbations in the input may cause significant errors in the perception results and lead to system failure. For instance, [84] designs a malicious billboard to attack end-to-end deep learning-based driving models. [59] generates a dirty road patch with carefully-designed adversarial patterns, which can appear as normal dirty patterns for human drivers while leading to significant perception errors and causing vehicles to deviate

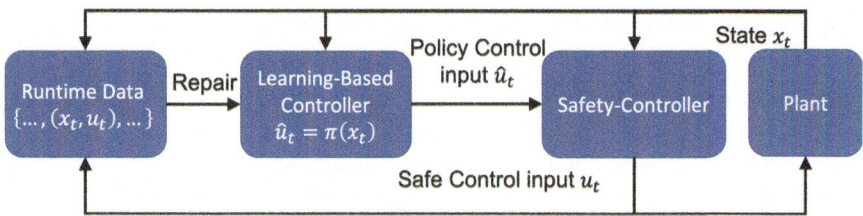

Fig. 8 The combination of a learning-based controller and a safety controller provides runtime safety assurance. Given the state x_t, the safety controller filters the control input $\hat{u}_t$ generated by the learning-based control policy π, and produces a safe control input u_t to the plant. The data (x_t, u_t) is collected at runtime to repair the policy π

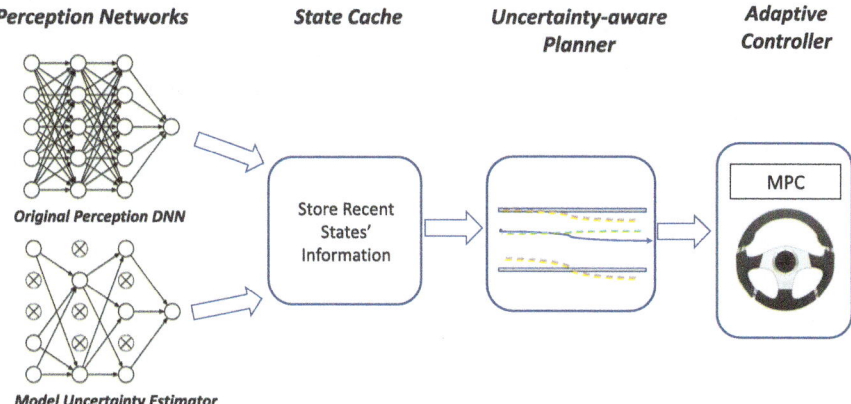

Fig. 9 An end-to-end detection and mitigation framework for adversarial attacks to CAVs [35]. In the perception module, the original neural network is to predict lane lines with confidence value and the data uncertainty while the other neural network is used to estimate the model uncertainty by Monte-Carlo dropout. The state cache will store recent predictions and then the planner will select one based on confidence values. The planner will calculate the center line in a safe region by considering both uncertainties and lane predictions. Finally, the controller will optimize the low-level control by an uncertainty-aware MPC

from their lanes within as short as 1 s. On the defense side, most previous works focus on detecting anomaly in the input data [39, 44] or making the perception neural networks themselves more robust against input perturbation [46].

In [35], instead of addressing adversarial attacks only on perception module, we develop *an uncertainty-based end-to-end approach* that detects and mitigates adversarial attacks throughout perception, planning, and control modules. In particular, we measure the confidence and uncertainty of perception modules, and conduct robust adaptation in the following modules accordingly based on the uncertainty analysis, as shown in Fig. 9. We apply the framework to the commercial automated lane centering system in OpenPilot and demonstrate that the impact of attacks can be reduced by up to 90%.

3.2 Safety-Driven Learning and System Design

Besides runtime adaptation, another critical aspect for CAV safety is to **design and learn** neural network-based components that can ensure system safety (i.e., not entering unsafe states) and robustness (i.e., being safe under disturbances from random noises or malicious attacks). Next, we will first introduce works that improve the robustness of neural networks, and then introduce techniques that try to learn safe neural network-based controllers from multiple experts, with verification in the loop, and based on physical information, respectively.

Learning Provably Robust Neural Networks Most of the current verification techniques for learning-enabled systems focus on analyzing trained systems, e.g., whether a trained neural network satisfies some specification. It is more desirable to have these systems "correct-by-construction". In fact, the same power of modern compute and data that has been fueling data-driven learning can be leveraged to scale up verification and enable provably-correct training of neural networks. We give such an example below.

For adversarial robustness problems in neural networks [3, 23, 45, 79, 83], given a model f_θ, loss function $\mathcal{L}$, and training data distribution $\mathcal{X}$, the training algorithm aims to minimize the loss whereas the adversary aims to maximize the loss within a neighborhood $\mathbb{S}(x, \epsilon)$ of each input data x as follows:

$$\min_\theta E_{(x,y)\in\mathcal{X}} \left[\max_{x'\in\mathbb{S}(x,\epsilon)} \mathcal{L}(f_\theta(x'), y) \right] \tag{1}$$

In general, the inner maximization is intractable. Most existing techniques focus on finding an approximate solution. There are two main approaches to approximate the inner loss (henceforth referred to as *robust loss*). One direction is to generate adversarial examples to compute a lower bound of robust loss. The other is to compute an upper bound of robust loss by over-approximating the model outputs.

Verification techniques [17, 36, 53, 54, 58] for neural networks can be used to compute a certified upper bound of *robust loss* (henceforth referred to as *abstract loss*). Given a neural network, a simple way to obtain this upper bound is to propagate value bounds across the network, also known as interval bound propagation (IBP) [23, 48]. Techniques such as CROWN [82], DeepZ [63], MIP [68] and RefineZono [64], can compute more precise bounds, but also incur much higher computational costs. Building upon these upper bound verification techniques, approaches such as DIFFAI [48] construct a differentiable *abstract loss* corresponding to the upper bound estimation and incorporate this loss function during training. However, [23] and [83] observe that a tighter approximation of the upper bound does not necessarily lead to a network with low robust loss. They show that IBP-based methods can produce networks with state-of-the-art certified robustness. More recently, COLT [3] proposed to combine adversarial training and zonotope propagation. Zonotopes are a collection of affine forms of the input variables and intermediate vector outputs in the neural network. The idea is to train the network with the so-called latent adversarial examples which are adversarial examples that lie inside these zonotopes. *AdvIBP* [18] proposed a principled framework for combining adversarial loss and abstract loss. Fan and Li [18] argues that minimizing *adversarial loss* and minimizing *abstract loss* can be viewed as bounding the true *robust loss* from two ends. From an optimization perspective, this amounts to an optimization problem with two objectives and can be solved using gradient descent methods if both objectives are semi-smooth. Inspired by the work on moment estimates [37], *AdvIBP* proposed a novel joint training scheme to compute the weights adaptively and minimize the joint objective with unbiased gradient estimates. For efficient training, *AdvIBP* uses FGSM and random initialization for computing the adversarial loss and IBP for computing the abstract loss. We summarize and compare the key features in Table 1.

Learning Neural Network Controllers from Multiple Experts In Sect. 3.1, we present an approach for switching among multiple controllers, including both model-based and neural network-based, with safety assurance [72]. After observing the benefit of such switching control, we then further propose a framework to automatically learn a better neural network-based controller from those multiple existing ones, by learning a system-level ensemble strategy and robust distillation via adversarial examples [74], as shown in Fig. 10. Specifically, we ensemble the multiple controllers by learning a linear combination weight for each expert through reinforcement learning optimization to enhance the control safety and efficiency. To achieve better verifiability based on the observation that smaller Lipschitz constant of the neural network leads to stronger robustness, we conduct teacher-student knowledge distillation with a novel probabilistic adversarial training to obtain the final controller. The final learned controller shows better control robustness when facing measurement noise and adversarial attacks, higher control energy efficiency, and better verifiability in terms of reachable set and invariant set computation.

Verification-in-the-Loop Control Learning with Safety Guarantees Traditionally, control synthesis/learning for a safety-critical system often follows the *design-then-verify* open-loop process, which could result in many iterations between design and verification, and may still fail to provide any safety guarantees. In [76], we instead propose a closed-loop process for control learning by integrating the verification results into the design module via propagating the feedback as an approximated gradient, i.e., a *design-while-verify* process. In particular, the verification results refer to the computed reachable set in this work. We establish two distance metrics, including the geometric distance and the Wasserstein distance, to measure how far the computed reachable set of the current controller is from the goal region and the unsafe region. We then add perturbations to the controller and approximate the gradient for it by a difference method for update until the final reach-avoid property is met.

Physics-Aware Safety-Assured Design of Hierarchical Neural Network Planner for CAVs In designing CAVs in practice, it is critical to consider the safety

Table 1 Comparison of different methods for training robust neural networks. We highlight the loss function used in each method. If there is an *abstract loss* used in training or post-training verification, we also list the corresponding verification method. We categorize the methods along five dimensions, with ✓ indicating a desirable property or an explicit consideration

Method	L-2ptoss	Abstract loss	Efficiency[a]	Empirical robustness	Provable robustness	No weight[b] tuning/ scheduling
Baseline	Regular loss	n/a	✓			n/a
FGSM [22]	Adversarial loss	n/a	✓	✓		n/a
FGSM+random init [80]	Adversarial loss	n/a	✓	✓		n/a
PGD [45]	Adversarial loss	n/a		✓		n/a
COLT [3]	Latent adversarial loss	RefineZono[c]		✓	✓	n/a
DIFFAI [48]	Abstract loss[d]	DeepZ	✗[e]		✓	n/a
CROWN-IBP [83]	Regular loss + abstract loss	CROWN + IBP	✓		✓	
IBP method [23]	Regular loss + abstract loss	IBP	✓		✓	
AdvIBP [18]	*Adversarial loss + abstract loss*	*IBP*	✓	✓	✓	✓

The bold text is the best performing method for training robust neural networks

[a] The efficiency baseline is the training time for each epoch during regular training. ✓ represents the training time is comparable to the baseline

[b] The weights here represent the weights for the different losses if there are multiple of them

[c] RefineZono is not used to construct an abstract loss. Instead, it is used to generate latent adversarial examples and for post-training verification

[d] In their experiments, DIFFAI shows that adding regular loss with a fixed weight can achieve better performance

[e] DIFFAI can also use IBP for training and verification for improved efficiency. However, the best robustness results are achieved using DeepZ

Fig. 10 Overview of the Cocktail framework to learn a better neural network controller from multiple existing control experts via system-level ensemble from reinforcement learning and robust distillation with probabilistic adversarial training

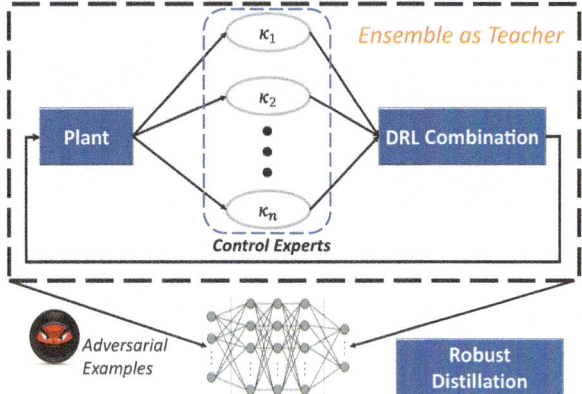

of the learning-based components. For instance, many recent neural network-based planners demonstrate significant performance improvement and accident rate reduction in average over traditional model-based methods. Some of those learn a single neural network for planning via reinforcement learning, imitation learning, supervised learning, etc., while others employ a hierarchical planner design, which usually consists of low-level planners for different modes and a high-level planner that is responsible for selecting the mode. However, even though safety improvement is often considered and demonstrated empirically through experiments in those works, formal system safety verification remains a challenging problem.

In [42], we propose a hierarchical neural network based planner that analyzes the underlying physical scenarios of the system and learns a system-level behavior planning scheme with multiple scenario-specific motion-planning strategies, as shown in Fig. 11. We develop an efficient verification method that incorporates overapproximation of the system state reachable set and novel partition and union techniques for formally ensuring system safety under our physics-aware planner. With theoretical analysis, we show that considering the different physical scenarios and building a hierarchical planner based on such analysis may improve system safety and verifiability. We also empirically demonstrate the effectiveness of our approach and its advantage over other baselines in practical case studies of unprotected left turn and highway merging, two common challenging safety-critical tasks in autonomous driving.

4 Conclusion and Future Directions

Safety is a critical challenge to the widespread adoption of CAVs. In this book chapter, we have outlined some specific technical problems and proposed solutions for verifying and improving the safety of CAVs, especially aiming at those challenges brought by the increasing usage of learning-based components. The road

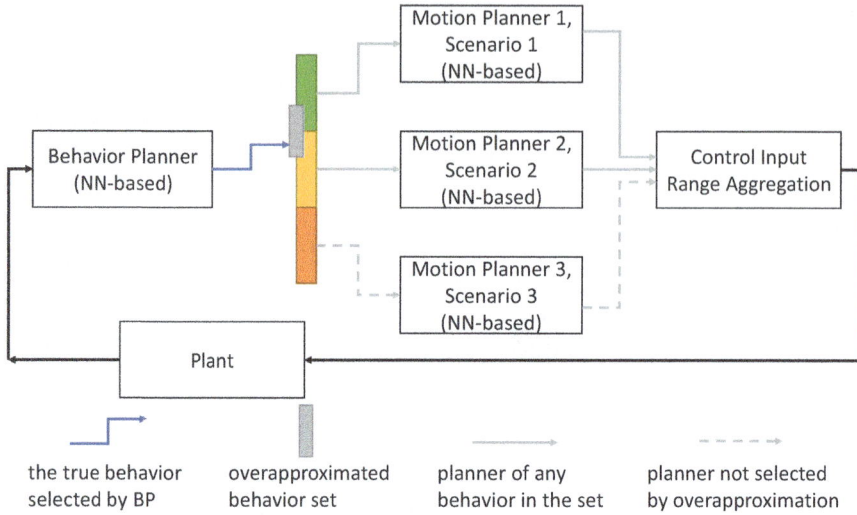

Fig. 11 Design of a hierarchical neural network-based planner that consists of one behavior planner μ and N motion planners $\{\kappa_1, \kappa_2, \ldots, \kappa_N\}$ [42]. In the figure, we have $N = 3$ for example. The behavior planner decides the most appropriate behavior given the system state x, and then the corresponding motion planner is enabled to control the system. To compute an overapproximation of the reachable set of the system under such hierarchical planner, we first compute an overapproximated behavior set, which is illustrated by the grey rectangle in the figure. Then for each behavior in the overapproximated behavior set, the corresponding motion planner's output range can be aggregated as the possible control input range, thus computing an overapproximation of the system state reachable set under all possible behaviors

to safe autonomy, however, still requires clearing major roadblocks in perception, control, and connectivity, and we discuss some of those below.

On the verification side, developing more efficient and rigorous techniques especially for CAVs with neural network-based perception modules will be a primary focus. The high dimensionality of the problem may necessitate sacrificing deterministic guarantees and adopting statistical or probabilistic analysis. In particular, for probabilistic safety verification of neural network-controlled systems, existing statistic model checking approach often requires a large number of system simulations and costs a lot of time. This may be relieved by approximately tracking the propagation of the probabilistic distributions of reachable states. Another possible direction is to perform property-directed reachability analysis for neural network-controlled systems. Existing reachability algorithms explore all state space that is possible to reach, and it is often unnecessary to do so when a safety property is simply defined by very few constraints. A property-directed reachability technique may exclude the state space that is not relevant to the safety condition and reduce a great amount of time in computing the reachable sets. For connected vehicles, abstract modeling of inter-vehicle information exchange and interactions and compositional analysis will be the key to leapfrogging the complexity challenge of verifying the safety of large-scale multi-agent systems.

On design and adaptation of CAVs, we believe that the key is to develop more end-to-end approaches that can address CAV safety across sensing, perception, planning and control stages, and more cross-layer approaches that can consider functional safety, software and hardware execution correctness, and even inter-vehicle communication reliability in a holistic manner. For instance, effectively addressing adversarial attacks to neural network-based perception modules will require quantitative analysis of their impact on downstream planning/control modules and ultimately on system-level safety, and will need end-to-end mitigation strategies that are developed based on such analysis. Runtime adaptation to mitigate component failures will need techniques to assess the impact of those failures across system layers, explore adaptation solutions that address the bottlenecks, and ensure the changed configurations meet various constraints across functional, software and hardware layers.

Acknowledgments We gratefully acknowledge the support from the US National Science Foundation (NSF) awards CCF-1646497, CCF-1834324, CNS-1834701, CNS-1839511, IIS-1724341, CNS-2038853, the US Office of Naval Research (ONR) grant N00014-19-1-2496, and the US Air Force Research Laboratory (AFRL) under contract number FA8650-16-C-2642.

References

1. Althoff, M.: An introduction to cora 2015. In: Proceedings of ARCH'15. EPiC Series in Computer Science, vol. 34, pp. 120–151. EasyChair (2015)
2. Alur, R., Courcoubetis, C., Halbwachs, N., Henzinger, T.A., Ho, P.-H., Nicollin, X., Olivero, A., Sifakis, J., Yovine, S.: The algorithmic analysis of hybrid systems. Theor. Comput. Sci. **138**(1), 3–34 (1995)
3. Balunovic, M., Vechev, M.: Adversarial training and provable defenses: Bridging the gap. In: International Conference on Learning Representations (2020)
4. Bernat, G., Cayssials, R.: Guaranteed on-line weakly-hard real-time systems. In: IEEE Real-Time Systems Symposium (RTSS) (2001)
5. Castane, R., Marti, P., Velasco, M., Cervin, A., Henriksson D.: Resource management for control tasks based on the transient dynamics of closed-loop systems. In: 18th Euromicro Conference on Real-Time Systems (ECRTS'06) (2006)
6. Cervin, A., Eker, J., Bernhardsson, B., Årzén, K.E.: Feedback–feedforward scheduling of control tasks. Real-Time Syst. **23**(1), 25–53 (2002)
7. Cervin, A., Velasco, M., Marti, P., Camacho, A.: Optimal online sampling period assignment: theory and experiments. IEEE Trans. Control Syst. Technol. **19**(4), 902–910 (2011)
8. Chen, X., Ábrahám, E., Sankaranarayanan, S.: Flow*: an analyzer for non-linear hybrid systems. In: Proceedings of CAV'13. LNCS, vol. 8044, pp. 258–263. Springer (2013)
9. Chen, X., Sankaranarayanan, S.: Reachability analysis for cyber-physical systems: are we there yet? In: Proceedings of NFM'22. LNCS, vol. 13260, pp. 109–130. Springer (2022)
10. Chisci, L., Rossiter, J.A., Zappa, G.: Systems with persistent disturbances: predictive control with restricted constraints. Automatica **37**(7) (2001)
11. Dai, X., Chang, W., Zhao, S., Burns, A.: A dual-mode strategy for performance-maximisation and resource-efficient cps design. ACM Trans. Embed. Comput. Syst. **18**(5s) (2019)

754   X. Chen et al.

12. Davare, A., Zhu, Q., Di Natale, M., Pinello, C., Kanajan, S., Sangiovanni-Vincentelli, A.: Period optimization for hard real-time distributed automotive systems. In: Design Automation Conference (DAC'07) (2007)
13. Deng, P., Zhu, Q., Davare, A., Mourikis, A., Liu, X., Natale, M.D.: An efficient control-driven period optimization algorithm for distributed real-time systems. IEEE Trans. Comput. **65**(12), 3552–3566 (2016)
14. Dutta, S., Jha, S., Sankaranarayanan, S., Tiwari, A.: Output range analysis for deep feedforward neural networks. In: NASA Formal Methods Symposium, pp. 121–138. Springer (2018)
15. Dutta, S., Jha, S., Sankaranarayanan, S., Tiwari, A.: Output range analysis for deep feedforward neural networks. In: Proceedings of NFM'18. LNCS, vol. 10811, pp. 121–138. Springer (2018)
16. Dutta, S., Chen, X., Sankaranarayanan, S.: Reachability analysis for neural feedback systems using regressive polynomial rule inference. In: 22nd ACM International Conference on Hybrid Systems: Computation and Control (HSCC), pp. 157–168 (2019)
17. Dvijotham, K., Stanforth, R., Gowal, S., Mann, T.A., Kohli, P.: A dual approach to scalable verification of deep networks. In: UAI, vol. 1, p. 2 (2018)
18. Fan, J., Li, W.: Adversarial training and provable robustness: a tale of two objectives. In: Proceedings of the AAAI Conference on Artificial Intelligence, vol. 35, pp. 7367–7376 (2021)
19. Fan, J., Huang, C., Li, W., Chen, X., Zhu, Q.: Towards verification-aware knowledge distillation for neural-network controlled systems. In: 2019 IEEE/ACM International Conference on Computer-Aided Design (ICCAD), pp. 1–8. IEEE (2019)
20. Fan, J., Huang, C., Chen, X., Li, W., Zhu, Q.: Reachnn*: a tool for reachability analysis of neural-network controlled systems. In: International Symposium on Automated Technology for Verification and Analysis (2020)
21. Fawzi, A., Moosavi-Dezfooli, S.-M., Frossard, P.: The robustness of deep networks: a geometrical perspective. IEEE Signal Process. Mag. **34**(6), 50–62 (2017)
22. Goodfellow, I.J., Shlens, J., Szegedy, C.: Explaining and harnessing adversarial examples. In: International Conferences on Learning Representations (2015)
23. Gowal, S., Dvijotham, K., Stanforth, R., Bunel, R., Qin, C., Uesato, J., Arandjelovic, R., Mann, T., Kohli, P.: On the effectiveness of interval bound propagation for training verifiably robust models. Preprint (2018). arXiv:1810.12715
24. Henzinger, T.A., Kopke, P.W., Puri, A., Varaiya, P.: What's decidable about hybrid automata? In: Proceedings of the 27th Annual ACM Symposium on Theory of Computing (STOC'95), pp. 373–382. ACM (1995)
25. Huang, X., Kwiatkowska, M., Wang, S., Wu, M.: Safety verification of deep neural networks. In: International Conference on Computer Aided Verification, pp. 3–29. Springer (2017)
26. Huang, C., Fan, J., Li, W., Chen, X., Zhu, Q.: Reachnn: reachability analysis of neural-network controlled systems. ACM Trans. Embedd. Comput. Syst. **18**(5s), 1–22 (2019)
27. Huang, C., Li, W., Zhu, Q.: Formal verification of weakly-hard systems. In: The 22nd ACM International Conference on Hybrid Systems: Computation and Control (HSCC) (2019)
28. Huang, C., Chang, K.-C., Lin, C.-W., Zhu, Q.: Saw: a tool for safety analysis of weakly-hard systems. In: 32nd International Conference on Computer-Aided Verification (CAV'20) (2020)
29. Huang, C., Fan, J., Chen, X., Li, W., Zhu, Q.: Divide and slide: layer-wise refinement for output range analysis of deep neural networks. In: International Conference on Embedded Software (EMSOFT) (2020)
30. Huang, C., Xu, S., Wang, Z., Lan, S., Li, W., Zhu, Q.: Opportunistic intermittent control with safety guarantees for autonomous systems. Proccedings of the Design Automation Conference (DAC'20) (2020)
31. Huang, C., Fan, J., Chen, X., Li, W., Zhu, Q.: Polar: a polynomial arithmetic framework for verifying neural-network controlled systems. Preprint (2021). arXiv:2106.13867
32. Ivanov, R., Weimer, J., Alur, R., Pappas, G.J., Lee, I.: Verisig: verifying safety properties of hybrid systems with neural network controllers. In: 22nd ACM International Conference on Hybrid Systems: Computation and Control (HSCC), pp. 169–178 (2019)
33. Ivanov, R., Carpenter, T.J., Weimer, J., Alur, R., Pappas, G.J., Lee, I.: Verifying the safety of autonomous systems with neural network controllers. ACM Trans. Embedd. Comput. Syst. (TECS) **20**(1), 1–26 (2020)

34. Ivanov, R., Carpenter, T., Weimer, J., Alur, R., Pappas, G., Lee, I.: Verisig 2.0: verification of neural network controllers using taylor model preconditioning. In: Silva, A., Rustan, K., Leino, M. (eds.) Computer Aided Verification, pp. 249–262. Springer International Publishing, Cham (2021)

35. Jiao, R., Liang, H., Sato, T., Shen, J., Chen, Q.A., Zhu, Q.: End-to-end uncertainty-based mitigation of adversarial attacks to automated lane centering. In: 2021 IEEE Intelligent Vehicles Symposium (IV), pp. 266–273 (2021)

36. Katz, G., Barrett, C., Dill, D.L., Julian, K., Kochenderfer, M.J.: Reluplex: an efficient smt solver for verifying deep neural networks. In: International Conference on Computer Aided Verification (CAV), pp. 97–117. Springer (2017)

37. Kingma, D.P., Ba, J.: Adam: a method for stochastic optimization. In: International Conference on Learning Representations (2016)

38. Lee, D., Hess, D.J.: Public concerns and connected and automated vehicles: safety, privacy, and data security. Hum. Soc. Sci. Commun. **9**(1), 1–13 (2022)

39. Lee, K., Lee, K., Lee, H., Shin, J.: A simple unified framework for detecting out-of-distribution samples and adversarial attacks. Adv. Neural Inf. Process. Syst. **31** (2018)

40. Liang, H., Wang, Z., Roy, D., Dey, S., Chakraborty, S., Zhu, Q.: Security-driven codesign with weakly-hard constraints for real-time embedded systems. In: 37th IEEE International Conference on Computer Design (ICCD'19) (2019)

41. Liang, H., Wang, Z., Jiao, R., Zhu, Q.: Leveraging weakly-hard constraints for improving system fault tolerance with functional and timing guarantees. In: 2020 IEEE/ACM International Conference On Computer Aided Design (ICCAD), pp. 1–9 (2020)

42. Liu, X., Huang, C., Wang, Y., Zheng, B., Zhu, Q.: Physics-aware safety-assured design of hierarchical neural network based planner. In: 2022 ACM/IEEE International Conference on Cyber-Physical Systems (ICCPS) (2022)

43. Löfberg, J: Minimax Approaches to Robust Model Predictive Control, vol. 812. University Electronic Press, Linköping (2003)

44. Lu, J., Issaranon, T., Forsyth, D.: Safetynet: detecting and rejecting adversarial examples robustly. In: Proceedings of the IEEE international conference on computer vision, pp. 446–454 (2017)

45. Madry, A., Makelov, A., Schmidt, L., Tsipras, D., Vladu, A.: Towards deep learning models resistant to adversarial attacks. Preprint (2017). arXiv:1706.06083

46. Madry, A., Makelov, A., Schmidt, L., Tsipras, D., Vladu, A.: Towards deep learning models resistant to adversarial attacks. In: International Conference on Learning Representations (2018)

47. Makino, K., Berz, M.: Taylor models and other validated functional inclusion methods. J. Pure Appl. Math. **4**(4), 379–456 (2003)

48. Mirman, M., Gehr, T., Vechev, M.: Differentiable abstract interpretation for provably robust neural networks. In: International Conference on Machine Learning, pp. 3578–3586 (2018)

49. Mundhenk, P., Paverd, A., Mrowca, A., Steinhorst, S., Lukasiewycz, M., Fahmy, S.A., Chakraborty, S.: Security in automotive networks: lightweight authentication and authorization. ACM Trans. Des. Autom. Electron. Syst. **22**(2), 25:1–25:27 (2017)

50. Nedialkov, N.S.: Implementing a rigorous ode solver through literate programming. In: Rauh, A., Auer, E. (eds.) Modeling, Design, and Simulation of Systems with Uncertainties. Mathematical Engineering, vol. 3, pp. 3–19. Springer, Berlin/Heidelberg (2011)

51. NHTSA Media.: U.S. transportation secretary elaine l. chao announces first participants in new automated vehicle initiative web pilot to improve safety, testing, public engagement. NHTSA (2020)

52. Phillips, G.M.: Interpolation and Approximation by Polynomials. Springer, Berlin (2003)

53. Prabhakar, P., Afzal, Z.R.: Abstraction based output range analysis for neural networks. In: Advances in Neural Information Processing Systems, pp. 15788–15798 (2019)

54. Raghunathan, A., Steinhardt, J., Liang, P.S.: Semidefinite relaxations for certifying robustness to adversarial examples. In: Advances in Neural Information Processing Systems, pp. 10877–10887 (2018)

55. Ramanathan, P.: Overload management in real-time control applications using (m, k)-firm guarantee. IEEE Trans. Parallel Distrib. Syst. **10**(6), 549–559 (1999)
56. Richards, A.G.: Robust constrained model predictive control. Ph.D Thesis, Massachusetts Institute of Technology, 2005
57. Roy, D., Chang, W., Mitter, S.K., Chakraborty, S.: Tighter dimensioning of heterogeneous multi-resource autonomous cps with control performance guarantees. In: ACM/IEEE Design Automation Conference (DAC), pp. 1–6 (2019)
58. Ruan, W., Huang, X., Kwiatkowska, M.: Reachability analysis of deep neural networks with provable guarantees. In: International Joint Conferences on Artificial Intelligence (2018)
59. Sato, T., Shen, J., Wang, N., Jia, Y., Lin, X., Chen, Q.A.: Dirty road can attack: Security of deep learning based automated lane centering under {Physical-World} attack. In: 30th USENIX Security Symposium (USENIX Security 21), pp. 3309–3326 (2021)
60. Seshia, S.A., Hu, S., Li, W., Zhu, Q.: Design automation of cyber-physical systems: challenges, advances, and opportunities. IEEE Trans. Comput. Aided Des. Integr. Circuits Syst. **36**(9), 1421–1434 (2017)
61. Seto, D., Krogh, B., Sha, L., Chutinan, A.: The simplex architecture for safe online control system upgrades. In: American Control Conference (ACC), vol. 6, pp. 3504–3508 (1998)
62. Siddiqui, F., Lerman, R., Merrill, J.B.: Teslas running autopilot involved in 273 crashes reported since last year. The Washington Post (2022)
63. Singh, G., Gehr, T., Mirman, M., Püschel, M., Vechev, M.: Fast and effective robustness certification. In: Advances in Neural Information Processing Systems, pp. 10802–10813 (2018)
64. Singh, G., Gehr, T., Püschel, M., Vechev, M.: Boosting robustness certification of neural networks. In: International Conference on Learning Representations (2019)
65. Summary Report: Standing general order on crash reporting for automated driving systems. Technical Report, NHTSA, 2022
66. Summary Report: Standing general order on crash reporting for level 2 advanced driver assistance systems. Technical Report, NHTSA, 2022
67. Szegedy, C., Zaremba, W., Sutskever, I., Bruna, J., Erhan, D., Goodfellow, I., Fergus, R.: Intriguing properties of neural networks. International Conferences on Learning Representations (2014)
68. Tjeng, V., Xiao, K.Y., Tedrake, R.: Evaluating robustness of neural networks with mixed integer programming. In: International Conference on Learning Representations (2019)
69. Tran, H.-D., Bak, S., Xiang, W., Johnson, T.T.: Verification of deep convolutional neural networks using imagestars. In: International Conference on Computer-Aided Verification (2020)
70. U.S. Department of Transportation: Using connected vehicle technologies to solve real-world operational problems. USDOT ITS Research - Connected Vehicle Pilot Deployment Program (2022)
71. Wang, S., Pei, K., Whitehouse, J., Yang, J., Jana, S.: Formal security analysis of neural networks using symbolic intervals. In: 27th {USENIX} Security Symposium ({USENIX} Security 18), pp. 1599–1614 (2018)
72. Wang, Y., Huang, C., Zhu, Q.: Energy-efficient control adaptation with safety guarantees for learning-enabled cyber-physical systems. In: Proceedings of the 39th International Conference on Computer-Aided Design, ICCAD '20, New York, NY, USA. Association for Computing Machinery (2020)
73. Wang, S., Zhang, H., Xu, K., Lin, X., Jana, S., Hsieh, C.-J., Kolter, J.Z.: Beta-crown: efficient bound propagation with per-neuron split constraints for neural network robustness verification. In: Proceedings of NeurIPS'21, vol. 34 (2021)
74. Wang, Y., Huang, C., Wang, Z., Xu, S., Wang, Z., Zhu, Q.: Cocktail: learn a better neural network controller from multiple experts via adaptive mixing and robust distillation. In: 2021 58th ACM/IEEE Design Automation Conference (DAC), pp. 397–402. IEEE (2021)
75. Wang, Z., Huang, C., Kim, H., Li, W., Zhu, Q.: Cross-layer adaptation with safety-assured proactive task job skipping. ACM Trans. Embed. Comput. Syst. **20**(5s) (2021)

76. Wang, Y., Huang, C., Wang, Z., Wang, Z., Zhu, Q.: Design-while-verify: correct-by-construction control learning with verification in the loop. In: 59th ACM/IEEE Design Automation Conference, DAC 2022, San Francisco, CA, USA, July 10–14 (2022)

77. Wang, Z., Huang, C., Zhu, Q.: Efficient global robustness certification of neural networks via interleaving twin-network encoding. In: DATE'22: Proceedings of the Conference on Design, Automation and Test in Europe (2022)

78. Wiggers, K.: Waymo's driverless cars were involved in 18 accidents over 20 months. VentureBeat (2020)

79. Wong, E., Kolter, Z.: Provable defenses against adversarial examples via the convex outer adversarial polytope. In: International Conference on Machine Learning, pp. 5286–5295 (2018)

80. Wong, E., Rice, L., Kolter, J.Z.: Fast is better than free: revisiting adversarial training. In: International Conferences on Learning Representations (2020)

81. Zheng, B., Gao, Y., Zhu, Q., Gupta, S.: Analysis and optimization of soft error tolerance strategies for real-time systems. In: 2015 International Conference on Hardware/Software Codesign and System Synthesis (CODES+ISSS), pp. 55–64 (2015)

82. Zhang, H., Weng, T.-W., Chen, P.-Y., Hsieh, C.-J., Daniel, L.: Efficient neural network robustness certification with general activation functions. In: Advances in Neural Information Processing Systems, pp. 4939–4948 (2018)

83. Zhang, H., Chen, H., Xiao, C., Li, B., Boning, D., Hsieh, C.-J.: Towards stable and efficient training of verifiably robust neural networks. In: International Conference on Learning Representations (2020)

84. Zhou, H., Li, W., Kong, Z., Guo, J., Zhang, Y., Yu, B., Zhang, L., Liu, C.: Deepbillboard: Systematic physical-world testing of autonomous driving systems. In: 2020 IEEE/ACM 42nd International Conference on Software Engineering (ICSE), pp. 347–358. IEEE (2020)

85. Zhou, W., Gao, R., Kim, B., Kang, E., Li, W.: Runtime-safety-guided policy repair. In: Deshmukh, J., Ničković, D. (eds.) Runtime Verification, pp. 131–150. Springer International Publishing, Cham (2020)

86. Zhu, Q., Sangiovanni-Vincentelli, A.: Codesign methodologies and tools for cyber–physical systems. In: Proceedings of the IEEE **106**(9), 1484–1500 (2018)

Identifying and Assessing Research Gaps for Energy Efficient Control of Electrified Autonomous Vehicle Eco-Driving

Farhang Motallebi Araghi, Aaron Rabinwoitz, Chon Chia Ang, Sachin Sharma, Parth Kadav, Richard T. Meyer, Thomas Bradley, and Zachary D. Asher

1 Introduction

Transportation's reliance on nonrenewable hydrocarbon fuels creates serious concerns about energy supply, cost, and environmental safety. In the pursuit for green, sustainable transportation systems, consideration of vehicle energy consumption is crucial [1]. Efficient alternative energy vehicles and advanced vehicle control technologies are two areas of research that might provide solutions to the need for increasing Fuel Economy (FE) and complying with current and upcoming environmental regulations [2, 3].

The need for energy efficient vehicles has facilitated the development of new vehicle technologies such as Hybrid Electric Vehicles (HEVs) and Plug-in Hybrid Electric Vehicles (PHEVs) and Battery Electric Vehicles (BEVs) [4]. Compared to vehicles powered with only an Internal Combustion Engine (ICE), HEVs provide significantly improved fuel efficiency [4]. The reason is due to their ability to recover braking energy and the fact that an extra powertrain degree of freedom is available to more cost-effectively meet the driver-required power. PHEVs exhibits even longer range and even further reduced need for hydrocarbon fuels [5]. This is owed to their enhanced battery capacity and their ability to be charged from wall power. BEVs are projected to further improve automotive transportation sustainability with a commercially viable and a readily accessible product [6].

F. M. Araghi (✉) · S. Sharma · P. Kadav · R. T. Meyer · Z. D. Asher
Department of Mechanical and Aerospace Engineering, Western Michigan University, Kalamazoo, MI, USA
e-mail: farhang.motallebiaraghi@wmich.edu; sachin.sharma@wmich.edu; parth.kadav@wmich.edu; richard.meyer@wmich.edu; zach.asher@wmich.edu

A. Rabinwoitz · C. C. Ang · T. Bradley
Department of Systems Engineering, Colorado State University, Fort Collins, CO, USA
e-mail: Aaron.rabinowitz@colostate.edu; ccang@colostate.edu; thomas.bradley@colostate.edu

© The Author(s), under exclusive license to Springer Nature Switzerland AG 2023
V. K. Kukkala, S. Pasricha (eds.), *Machine Learning and Optimization Techniques for Automotive Cyber-Physical Systems*, https://doi.org/10.1007/978-3-031-28016-0_27

The other major development facilitated by the need for energy efficient vehicles is advanced vehicle control technologies which is the focus of this research. Collectively these advanced control strategies are typically referred to as energy management strategies. These technologies are also becoming more implementable thanks to developments in AVs such as advanced perception subsystems, planning subsystems, and more. As AV technology continues to evolve commercially, it is crucial to ensure synergistic development with energy efficient controls to ensure transportation sustainability as well.

Further details regarding energy efficient vehicles and energy efficient control technologies are presented in the following subsections.

1.1   The Evolution of BEVs: The Modern Era

Since 2000, BEVs have experienced a substantial amount of progress and significant commercial milestones [7]. The first Tesla Roadster shipped in 2008 and it was the first highway-legal BEV to employ a lithium-ion battery and drive more than 200 miles on a single charge [8]. The Mitsubishi i-MiEV, which went into serial production in 2009, was the first modern highway-legal BEV [9]. The first Nissan Leaf was delivered to customers in 2010. Until 2011, Mitsubishi's i-MiEV had been the world's most popular BEV where between 2008–2012, 2450 were sold in 30 countries [10]. In 2016, more than one million BEVs were sold throughout the world. Tesla introduced the Model 3 in 2017 and not long after, sales of BEVs surpassed the one million mark for the first time and annual worldwide market share surpassed 1%. Annual worldwide sales surpassed two million units for the first time in 2018. The Tesla Model 3 was the first BEV to sell more over 100,000 units in a single year, a milestone it achieved in 2015. BEVs accounted for one out of every two new vehicles registered in Norway in 2019. The Tesla Model 3 overtook the Nissan Leaf as the best-selling BEV in history by 2020. More than 500,000 Tesla Model 3 s have been sold worldwide since its launch in 2013. Tesla became the first automaker to build more than one million BEVs. In Norway, 10% of the vehicles on the road are BEVs. Additionally, in 2020, the worldwide sales of the Nissan Leaf achieved the milestone of 500,000 units and global BEV sales crossed the ten million unit mark for the first time. The Tesla Model 3's worldwide sales surpassed one million units in 2021 and BEVs come in 27 distinct configurations, with 11 different manufacturers producing them. Table 1 shows the top five BEVs with the greatest ranges in 2020 model year [11]. The literature on BEVs is quite extensive, and it is constantly and fast evolving. BEVs were formerly seen to be a niche sector with an uncertain future, but that has since changed [12, 13]. The takeaway here is that BEVs are a commercially mature and desirable technology that has the potential to establish sustainable transportation [14, 15]. As we focus on a real-world implementation of energy efficient control technologies, the applicability to

Table 1 Model year 2020 BEV examples

Make and model	Vehicle type	Electric motor/battery-pack	City (mile/gge)	Electric range (miles)
Tesla model 3 long range	Sedan/wagon	211 KW/ 230 Ah	136	373
Chevrolet bolt BEV	Sedan/wagon	150 KW/ 188 Ah	127	259
Hyundai Kona electric	SUV	150 KW/ 180 Ah	132	258
Kia soul	Sedan/wagon	201 KW/ 180 Ah	127	243
Jaguar I-PACE	SUV	201 KW/ 223 Ah	80	234

BEVs must be a top consideration as they are growing and are an important part of the transportation sector.

Identifying and Assessing Research Gaps for Energy Efficient Control of Electrified Autonomous Vehicle Eco-driving.

As a vehicle, a BEV is quiet, simple to drive, and free of gasoline expenditures when compared to conventional vehicles [16]. Additionally, as a form of urban transportation, it has many benefits. It does not produce any emissions along urban corridors (reducing urban air pollution due to transportation), it easily handles frequent start-stop driving, it gives full torque from a stop, and eliminates the need for gas station stops provided that charging is available at in public or at home [17]. Additionally, the utility industry is evolving, with renewable energy sources gaining traction and the "smart grid" which is the next generation of the electricity grid, is now in the process of being built. BEVs are seen as a key component of this new power system, which includes renewable energy sources and high-tech grid technologies [18, 19]. All of this has resulted in increased interest and growth in this method of transportation.

As a system, BEVs can be modeled as a combination of several subsystems. Each of these subsystems interacts with the others to make the BEV function, and a variety of technologies may be used to run them. Figure 1 depicts major subsystem components and their contribution to the overall system. Some of these components must communicate significantly with others, while others have little or no interaction. Regardless of the situation, the operation of a BEV is dependent on such interaction of all these subsystems [20]. These subsystems are important to understand and utilize to develop energy efficient control strategies.

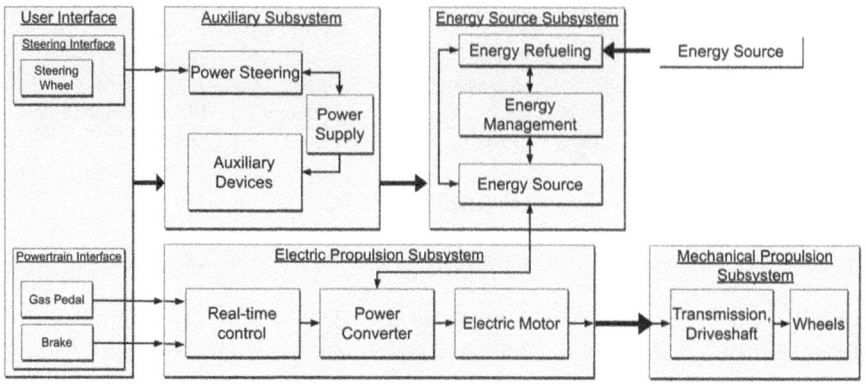

Fig. 1 A general systems-level viewpoint of BEV. (Adapted from Ref. [21])

1.2 Energy Management and Energy Efficient Strategies for Electrified Vehicles

Energy management strategies (EMS) in HEVs and PHEVs control the power/-torque split selection between the combustion engine and the electric motor, in which the amount of power/torque provided by each power source is combined to satisfy driver demand while reducing the amount of non-renewable fuel use and increasing powertrain utilization efficiency [22, 23]. Energy efficiency strategies seek to directly decrease the energy required to drive from one point to another by either modifying the second-by-second vehicle velocity or by choosing an alternative route.

This is particularly important for BEVs since energy efficiency strategies result in a direct increase in range thus enabling higher utility [24]. Overall, it has been shown that intelligent energy management and energy-efficient use of electrified vehicles increase vehicle FE and reduce global energy consumption, greenhouse gas emissions, and air pollution emissions.

Generally there are three types of vehicle control that reduce fuel consumption for a drive cycle with a fixed starting point and a fixed ending point: (1) powertrain EMS (P-EMS), (2) Eco-Routing (ER), and (3) Eco-driving (ED) [25, 26]. P-EMS decreases fuel consumption by increasing the efficiency of the vehicle powertrain operation without modification of the drive cycle [27]. However, ED and ER decrease fuel consumption by decreasing the energy output of the vehicle through modification of the drive cycle and route [28, 29].

1.2.1 Powertrain EMS (P-EMS)

As previously mentioned, electrified vehicles will benefit greatly from the development of P-EMSs. To meet driving demands, the primary goal of a P-EMS is to distribute the power request into multiple propulsion sources (specifically for HEVs and PHEVs) [30, 31]. If we take into account battery performance (i.e. the current rate and lifespan) and tailpipe emissions levels, an efficient P-EMS can improve fuel efficiency. However, it is challenging to devise P-EMSs due to the uncertainty of future driving conditions [32–35]. Furthermore, the P-EMS should have a sufficiently simple and fast real-time controller with a desired computational speed for the implementation of a global optimization algorithm. The performance of P-EMSs strongly depends on future vehicle velocity and power request which is influenced by external factors (e.g., traffic information and surrounding vehicles) [36]. Research groups all over the world have proposed various solutions which are briefly summarized:

1. *Rule-based P-EMS*: Here a P-EMS is implemented with either deterministic rules or with fuzzy rules.

    (a) *Deterministic*: The first application of deterministic rule-based techniques to the energy management of HEVs was in [37]. In place of the original electric assistance technique, Banvait et al. [38] described a charge depletion–charge sustaining (CD–CS) strategy. Following the cooperative control approach for the auxiliary power unit, the speed-switching power is compelled to acquire a proper curve together with the ideal brake specific fuel consumption [39, 40].

    (b) *Fuzzy Logic*: Fuzzy logic belongs to intelligent control strategies, but it dispenses with precise mathematical models of controlled systems. However, it contains self-learning capability, high flexibility, and resilience, and is thus commonly used to solve complicated nonlinear issues [41–43]. Denis et al. [44] developed a Sugeno-type fuzzy logic controller by using the moving average of the previous speed and the present global discharge rate as inputs in order to take use of the trip data. Li et al. has presented an adaptive-equivalent consumption reduction technique that combines a fuzzy inference system to increase self-adaptation [45].

2. *Optimization-based P-EMS*: In most cases, an optimization-based P-EMS is generated by formulating an optimal control problem. An Optimization-based P-EMS delivers FE improvements by explicitly or implicitly simulating vehicle operation and managing the vehicle powertrain components to reduce fuel consumption. An optimization- based P-EMS can accomplish FE improvements for conventional cars with ICE and BEVs, but the highest FE benefits are gained from vehicles with additional powertrain operating degrees of freedom such as HEVs and PHEVs [26, 46, 47]. The actual FE improvement from an Optimization-based P-EMS is significantly dependent on the chosen driving cycle and propulsion systems [48]. One of the first optimization-based P-EMS studies, for example, showed a 28% FE increase in a HEV by optimizing gear changing and battery charging/ discharging [49].

    (a) *Globally Optimal P-EMS*: Dynamic programming (DP) [48, 50], Pontryagin's minimum principle (PMP) [51–53], Stochastic Dynamic Programming

(SDP) [54–57], Genetic Algorithm (GA) [58, 59], and Particle Swarm Optimization (PSO) [60, 61] are among the key optimization techniques.

(b) *Real-time Optimal P-EMS*: Real-time optimization-based P-EMS is primarily composed of equivalent consumption-minimization strategies (ECMSs) and its variations such as adaptive ECMS [30, 62–64]. But predictive rules-based strategies can also be implemented in real time and DP methods can be implemented in real time through the use of a look-up table [65].

(c) *Prediction-based Optimal P-EMS*: The goal of a Prediction-based Optimal P-EMS is to discover the best control strategy for minimizing fuel consumption within the time frame when prediction data exists [3, 35, 66–70].

1.2.2   Eco-Routing (ER)

Classical vehicle routing algorithms seek the quickest or shortest routes [71, 72], while ER algorithms seek routes with the lowest energy consumption costs. When given a starting point and a destination, ER generates a route that minimizes the amount of energy required to complete the journey. Routing is often performed on a graph where intersections represent different junctions, connected by edges roads, and costs indicate the estimated energy required to go between two junctions that the road links. The route with the lowest overall energy for the journey may then be found using minimal path routing. The complicated time-variant functions that explain the expenses are often derived by researchers. For example, Dijkstra's routing method is a popular option among academics [73]. Users' route preferences, such as favoring highways or avoiding toll roads, might be considered while planning a path. Furthermore, it may utilize the number of passengers as an input to determine if the car is eligible to use high-occupancy vehicle lanes. Similar to other shortest route routing applications, a green routing service needs a server to handle diverse routing requests. However, operating and managing a routing server is costly and needs precise and thorough real-world traffic and network data which is difficult to access and analyze [74–77]. It should be noted that the ER navigation system may produce up to three routes for each journey depending on multiple minimization criteria, such as distance, travel time, and energy usage. For conventional ICE vehicles, there are currently various ER algorithms capable of generating energy-optimal routes based on historical and real-time traffic data [78–80], but there has been minimal study on PHEVs to date [81]. As shown in [82], the performance of ER algorithms is very sensitive to the energy model used to estimate the energy cost on each network connection. The most difficult component of solving the ER algorithm for PHEVs is locating an energy model capable of calculating both the electrical energy consumption and the gasoline consumption. Jurik et al. [83] addressed the ER challenge for HEVs using longitudinal dynamics. The eco-route for PHEVs was investigated using a charge-depleting first approach in [84, 85]. To address the ER of HEVs, De Nunzio et al. [86] recently developed a semi-analytical solution to the powertrain energy management based on Pontryagin's minimal principle. Houshmand et al. [87] conceived a combined routing and powertrain control algorithm that simultaneously identifies the energy-optimal route and the ideal energy management approach in terms of battery state of charge and fuel consumption. In [87], however, the option to recharge the battery on some portions of the trip was omitted, and either charge-sustaining or discharge-only operation was permitted.

1.2.3 Eco-Driving (ED)

ED decreases fuel consumption for all vehicle types by applying fuel-efficient driving behaviors along a predetermined route, which may affect travel duration [88]. Due to this increase in travel time, it is difficult to persuade drivers to adopt ED practices [89]. If the driving conditions along the route can be anticipated, ED may be treated as an optimum control question if the driver input is eliminated or disregarded. Current practical use of ED is realized through a set of heuristic goals, such as eliminating stops, traveling at a fuel-efficient speed (generally, this could be a higher or lower overall speed), and reducing acceleration and deceleration magnitudes, which can achieve FE improvements of approximately 10% for modern vehicles and 30% for fully AVs [90]. FE improvements realized through ED are the focus of this literature review because the energy savings is sufficiently large, and because ED can be directly implemented through AV technology.

Historically, ED implementation research has focused on the FE impact of a single intelligent vehicle technology, such as camera systems, radar systems, LiDAR, Vehicle to Vehicle (V2V), Vehicle to Infrastructure (V2I), or Vehicle to Everything (V2X). As an example of a typical ED study, researchers used projections of the traffic light Signal Phase and Timing (SPaT), a sort of V2I, to influence driving behavior and shown a FE improvement of 12–14% [91]. ED is difficult to adopt in reality since most drivers dislike giving up control [92]. Many studies of ED for AVs conclude that vehicle perception, sensor fusion, and planning must all be achieved for successful implementation. On the other hand, a comprehensive grasp of how each of these components should fit together at the system level is not as clearly defined.

1.2.4 Summary

To summarize, FE improvements realized using a fixed drive cycle are realized through a P-EMS which is a very active area of research but is most effective for HEVs and PHEVs [34, 93]. FE improvements from modifying the route is realized through ER which is highly applicable to BEVs but is a relatively mature technology. If FE is improved by modifying the drive cycle but keeping the route the same, then the technique is considered ED which is highly applicable to BEVs and has tremendous potential for further improvements once AV technology is also included [90].

1.3 Automated Cyber-Physical Vehicles

In addition to improvements in powertrain technology, embedded and cyber-physical systems have had a profound effect on the modern world [92, 94]. Embedded computer systems have been integrated with a variety of technological artifacts, such as the power grid, medical devices, automotive and transportation systems, and industrial control and production lines [95–97]. Modern engineering topics are often multidisciplinary and require significant interdisciplinary problem-solving capabilities. AVs are a kind of vehicular cyber-physical system that has experienced tremendous recent innovation and has garnered considerable interest from both industry and academia [98–100]. The strategy for establishing AVs as the primary mode of transportation on the road may have several advantages such as improvements of safety on the roads (e.g., collision avoidance); better mobility for young, elderly, and disabled; and individual improvements of energy efficiency [101]. But at the same time AV technology may increase travel demand and overall mileage due to new user groups, the reduced cost of driver's time, and potential for mode switching (walk, low speed shuttles, transit, regional air, etc.) [102–104]. While the full impact of AV technology remains unknown, it is certain that AV technology will begin to experience commercial adoption in the near future [105].

According to the projections shown in Table 2, in 2050, the reference case (all fleet vehicles will be Autonomous ICE) will have the lowest transportation energy consumption. At first glance, this may appear counter-intuitive; how could AVs with ICE consume less than BEVs? The reason for this according to Energy Information Administration (EAI) independent statistics and analysis projections data is that more people would prefer to use fleet services rather than their personal vehicles. Case 2, which assumes that all Autonomous Light-duty Vehicles (LDVs) will be BEVs. Another assumption is that AVs will enter both household and fleets, which means that more people will have access to AVs, making transportation easier for people who own vehicles. This, in turn, would have an impact on reliance on public transportation. As a result, an additional research focus is warranted to improve energy efficiency of Autonomous BEVs. In the next subsection, a derivation of the research gap for this type of technology based on its systematic readiness level is given.

Based on the uncertainty of the field there are four new contributions to the field in this article on the topic of ED in autonomous electrified vehicles (BEV and P/HEV), which builds on previous concepts and literature:

1. A holistic and systems-level understanding of the subsystems and integrations needed to implement ED in AVs allowing for comparison between all studies in the field.
2. Application of technology, integration, and system readiness analysis to ED realization in AVs.
3. A definition of the research gaps existing between the current state of the art and realization of ED usage in AVs.
4. A review of initial studies that have started to explore the identified research gaps.

Table 2 Reference and AV case description

Case name	Assumptions	Description
Reference	AVs enter fleet light-duty vehicles	1% of new light-duty passenger vehicles sales by 2050 and 100% are fleet sales
	AVs are used more intensively	Driven 65,000 miles per year and scrapped more quickly
	Autonomous LDV fuel type	100% conventional gasoline ICE
	Autonomous LDVs affect mass transit modes	Decreases use of transit bus by 12%, transit rail % by 2050
Autonomous BEV	AVs enter household and fleet LDVs	16% are new fleet sales and 84% are new household sales by 2050
	AVs are used more intensively	Driven 65,000 miles per year and scrapped more quickly (fleet) and + 10% more annual vehicle miles (household)
	Autonomous LDV fuel type	Increasing share of BEVs with 96% of fleet and 82% household by 2050
	Autonomous LDVs affect mass transit modes	Decreases use of transit bus by 17% by 2033, transit rail 35% by 2050 use of commuter rail 48% by 2050
Autonomous HEV	AVs enter household and fleet LDVs	16% are new fleet sales and 84% are new household sales by 2050
	AVs are used more intensively	Driven 65,000 miles per year and scrapped more quickly (fleet) and + 10% more annual vehicle miles (household)
	Autonomous LDV fuel type	Increasing share of HEVs with 96% of fleet and 71% household by 2050
	Autonomous LDVs affect mass transit modes	Decreases use of transit bus by 17% by 2033, transit rail 35% by 2050 use of commuter rail 48% by 2050

Adopted from Ref. [102]

2   Research Gap Derivation

One of the most important aspects of scientific advancement is the systematic identification and review of existing research gaps [106]. In order to identify research gaps, a systematic approach is applied to understand components of a general electric AV with ED implementation and the logical flow of operation. In this section, overall system architecture is introduced, and a holistic evaluation of system maturity based on the Department of Defense (DoD) approach is conducted.

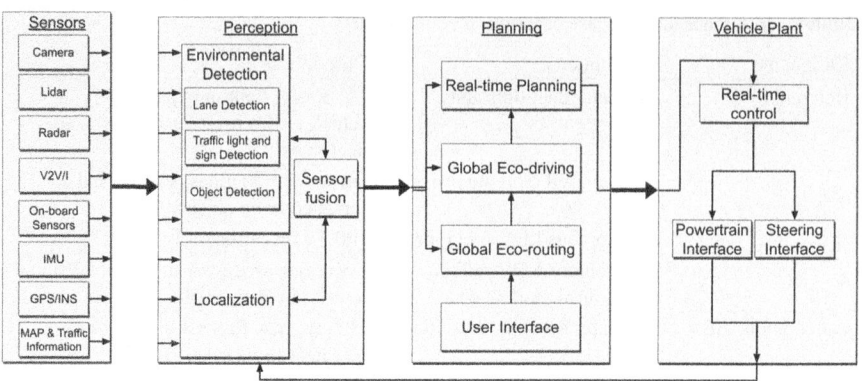

Fig. 2 A proposed systems-level viewpoint of ED implementation for an AV

2.1 AED System Architecture

A systems-level perspective of ED implementation for autonomous BEVs, represented in Fig. 2, is recommended to clarify communication between academic researchers, automotive sector manufacturers and suppliers, government officials, and other organizations.

The systems-level viewpoint is composed of four subsystems: a suite of sensors, a vehicle perception subsystem, a vehicle planning subsystem, and a vehicle plant subsystem which include a vehicle running controller. It is the goal of this systems-level perspective to remain closely aligned with the widely acknowledged systems-level perspective on autonomous BEV operation that use energy management strategies. This system receives input from a set of sensors that detect environmental information and also can be used to localize, therefore defining the vehicle's surroundings. An AV learns about its surroundings in two phases. The first step is to look down the road ahead to see if anything has changed, such as traffic lights and signs, a pedestrian crossing, or a barrier. The second phase is concerned with the perception of surrounding traffic. Camera, LiDAR, Radar, V2V and V2I, Inertial Measurement Unit (IMU), GPS and Inertial Navigation System (INS), and map and traffic information are the most typical sensors and data that comprise the sense and perception subsystems of AVs [107]. The real-time planning subsystem employs inputs from the perception subsystem to develop and solve both the long-range (such as Global ER and Global ED) and short- term planning strategies (such as maneuver planning and trajectory planning). It is worth noting that these subsystems also depend on the driving context, and their boundaries are quite blurred [108]. The real-time control subsystem tracks the longitudinal and lateral trajectories, and interfaces with the vehicle actuators. A control architecture is interfaced to the vehicle powertrain (e.g. controlling propulsion torque, braking torque and gear shifting) and to its steering system. The real-time planning subsystem require feedback from the vehicle, its position relative to the surrounding environment, and

Table 3 Technology readiness levels definition

Technology readiness level (TRL) and definition	
9	Actual System Proven Through Successful Mission Operations
8	Actual system completed and qualified through test and demonstration
7	System prototype demonstration in relevant environment
6	System/subsystem model or prototype demonstration in relevant environment
5	Component and/or breadboard validation in relevant environment
4	Component and/or breadboard validation in laboratory environment
3	Analytical and experimental critical function and/or characteristic proof-of-concept
2	Technology concept and/or application formulated
1	Basic principles observed and reported

predictions of moving obstacles [109], and finally, the powertrain operation from the running controller is actuated in the vehicle plant.

2.2 Holistic Evaluation of System Maturity

The National Aeronautics and Space Administration (NASA) developed a seven-level Technology Readiness Level (TRL) rating (shown in Table 3) in the 1980s to quantify the risk associated with technology development [110]. NASA now use this measure to assess the maturity of a specific technology and to compare the maturity of several technologies. Given its practical value, the DoD adopted a TRL model in 1999. While TRL is used similarly by NASA and the DoD, the understanding of TRL varies. For example, NASA requires TRL 6 technologies before a mission can be responsible for them [111], and the DoD requires TRL 7 technologies before they can be included in a weapons system program [112].

Further, the concept of a System Readiness Levels (SRL) was previously introduced by systems engineering researchers to address the problems applicable at the operating system level. This approach leverages the traditional TRL scale while also including the concept of Integration Readiness Levels (IRL) to produce an SRL index dynamically [112]. The definition of TRL, IRL and SRL and their corresponding levels are tabulated in Tables 3, 5 and 7 respectively.

2.2.1 Technology Readiness Levels (TRLs)

Table 4 provides a summary of the TRL for each of the subsystems shown in Fig. 2, as determined by the authors. These subsystems consist of (1) Sensors and (2) Vehicle Perception for Worldview Creation (3) vehicle planning and (4) application of a physical vehicle plant. These technologies are tabulated in the first column of Table 4. The perception subsystem takes in sensor and other pertinent inputs, defines the vehicle's environment, and computes future vehicle operation as an output. The

Table 4 TRL analysis of the individual technologies involved in ED in AVs implementation

Technology and TRL	Technology description	TRL definition	TRL justification
Sensors subsystem TRL:9	Detect environmental information	"Actual system proven through successful mission operations"	Cameras, radar, and even lidar have many commercial products
Perception subsystem TRL:6	Receives sensor/signal data and fuse data	"System/subsystem model or prototype demonstration in relevant environment"	Mobileye exists but it doesn't provide all of the functionality in our diagram, does not work in bad weather, etc. plus sensor fusion is not well developed.
Planning subsystem TRL:7	Solves several planning problems (maneuver planning, path planning, and trajectory planning)	"System prototype demonstration in relevant environment"	ER is mature. Derivation of ED is mature.
Vehicle plant subsystem TRL:9	Receives driver requests and component statuses and actuates vehicle operation	"Actual system proven through successful mission operations"	Vehicles by themselves are completely mature

vehicle perception is sent into the planning subsystem, which then computes the best control. The planning subsystem is simply responsible for computing the optimum control and issuing a control request; it is not responsible for attaining the goal.

2.2.2 Integration Readiness Levels (IRLs)

Table 6 summarizes the IRL for the three alternative integration sites in Fig. 2 as viewed by the authors. Table 6's column 1 contains descriptions of each integration scope. While the TRL is used to assess individual subsystems, the IRL assesses the readiness of each subsystem to integrate with others [112]. A more comprehensive assessment of each subsystem's integration is required than that of the individual subsystem, which normally consists of a basic input/output architecture. If the vehicle operating controller and the vehicle plant are viewed as one high IRL subsystem, there are three theoretically distinct integration points: (1), (2), and (3) and execution. Due to the little quantity of research that employs these integration scopes, each of these integration points was determined to have a poor technical maturity.

Table 5 Integration readiness levels definition

IRL	Definition
7	The integration of technologies has been **verified and validated** with sufficient detail to be actionable
6	The integrating technologies can **accept, translate, and structure information** for its intended application
5	There is sufficient **control** between technologies necessary to establish, manage, and terminate the integration
4	There is sufficient detail in the **quality and assurance** of the integration between technologies
3	There is **compatibility** (i.e. common language) between technologies to orderly and efficiently integrate and interact
2	There is some level of specificity to characterize the **interaction** (i.e. ability to influence) between technologies through their interface
1	An **interface** (i.e. physical connection) between technologies has been identified with sufficient detail to allow characterization of the relationship

Table 6 The IRL analysis demonstrates that the technology integrations involved in ED in AVs implementation require significant research

Integration and IRL	Integration description	IRL definition	IRL justification
Perception and planning integration: IRL 3	Detect environmental information	"There is compatibility (i.e. common language) between technologies to orderly and efficiently integrate and interact."	In some cases, the interface between vehicles and SPaT is converted to ED derivation constraints. Sensor fusion specifically has no commonality for ED
Planning when subjected to faulty inputs: IRL 2	Receives sensor/signal data and fuse data	"There is some level of specificity to characterize the interaction (i.e. ability to influence) between technologies through their interface."	Very limited literature
Planning and use of a vehicle plant: IRL 3	Solves several planning problems (maneuver planning, path planning, and trajectory planning)	"There is compatibility (i.e. common language) between technologies to orderly and efficiently integrate and interact."	Some researchers are starting to implement ED on a physical vehicle but progress is slow and there is a lot of work to be done

Table 7 System readiness levels definition

SRL	Name	Definition
5	Operations and support	Execute a support program that meets operational support performance requirements and sustains the system in the most cost-effective manner over its total life cycle
4	Production and development	Achieve operational capability that satisfies mission needs
3	System development and demonstration	Develop a system or increment of capability; reduce integration and manufacturing risk; ensure operational supportability; reduce logistics footprint; implement human systems integration; design for producibility; ensure affordability and protection of critical program information; and demonstrate system integration, interoperability, safety, and utility
2	Technology development	Reduce technology risks and determine appropriate sets of technologies to integrate into a full system
1	Concept refinement	Refine initial concept. Develop system/technology development strategy

Table 8 The SRL analysis demonstrates that the technology integrations involved in ED in AVs implementation require significant research

System and SRL	System description	SRL definition	SRL justification
Optimal ED implementation: SRL 1	Perception and planning subjected to errors and implemented in a vehicle plant	"Refine initial concept. Develop system/technology development strategy."	The sets of technologies are not defined and risks basically unknown, so this does not meet SRL 2

2.2.3 System Readiness Levels (SRLs)

The SRL analysis is the more appropriate method of assessment for the overall system of AED implementation, where the TRL analysis has been performed to individual subsystems and the IRL analysis has been used to subsystem integration. Table 5 shows that, despite the relatively high TRLs of each subsystem, the low IRLs result in a low total SRL. According to the SRL study, if the IRLs are improved, the total SRL will be improved, and optimal ED for AVs will be applicable to commercial production (Table 8).

2.2.4 Research Gap Analysis Summary

The SRL analysis has clearly indicated three research gaps that are inhibiting the implementation of AED, all of which are caused by subsystem integration. These gaps show that research should focus on advancing the following integrations:

1. Performance of integrated sensors and perception subsystems: The effect of Real-world AV perception on identifying the parameters associated with an ED problem.
2. Planning subsystem and noisy inputs: Effect of sparse or missing sensor data on global derivation of AED.
3. Planning and use of a vehicle plant: Performance of a planning subsystem integrated with a physical vehicle plant

3 Literature Review

There are a few important studies that have already begun to address these identified research gaps. While there are hundreds of articles that include ED, these integration-based research gaps must be addressed before an ED application of AVs can be commercialized. Each integration research requirement is addressed in the following subsections. Each subsection describes the scope of the research gap and critical studies that are beginning to bridge this research gap are identified and summarized. Studies that lack adequate integration to match the scope of the research gap are excluded.

3.1 Research Gap 1: Real-World AV Perception with Application to the AED Problem

The first research gap focuses on real-world AV perception using data from any real-world AV sensors to determine parameters for an ED problem; the scope is illustrated in Fig. 3. Many published ED studies exist that artificially create constraints for a mathematical optimization problem, but real world constraints derived from real world sensors are needed.

Researchers from University of Utah and San Diego State University proposed an ED algorithm for CAVs to improve fuel and operational efficiency of vehicles on the freeways [113]. The proposed algorithm optimizes CAV trajectories with three main objectives - travel time minimization, fuel time minimization, and traffic safety improvement. The first stage of two-state control logic proposed, provides optimal CAV trajectories that can simultaneously minimize freeway travel time and fuel consumption with traffic sensor data and trajectory information as inputs. The second stage of the control logic is focused on ensuring operational safety of CAVs by real time adaptive actions to adjust speeds in response to local driving conditions.

To achieve improved mobility and energy efficiency in mixed traffic conditions, researchers from University of California at Riverside proposed a combination of vision-perceptive technologies and V2I communications [114]. With a neural network extracting vision and V2I information; and a deep Reinforcement Learning

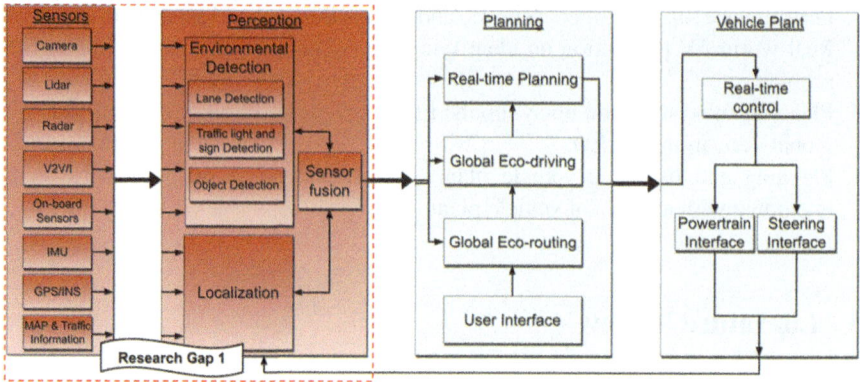

Fig. 3 The integration scope defined in research gap 1: Real-world AV perception with application to the AED problem

(RL) based policy network generates both longitudinal and lateral ED actions with a rule-based driving manager working to regulate the collaboration between rule-based policies and RL policies.

Fleming et al. [115] from Loughborough University outlined a system that uses real-time data from Global Positioning System (GPS) and automotive radar to predictably optimize a vehicle's speed profile and train a driver toward fuel-saving and CO_2-reducing behavior. Driving data was generated using STISIM Drive simulation software and validated on an instrumented vehicle equipped with radar and GPS sensor.

Table 9 summarizes the work in research gap 1. These papers are greatly advancing the commercial implementation of ED in AVs because real world sensors are being used to derive ED constraints. More research is needed in this area especially considering the possibility to utilize traditional AV sensors such as cameras, radar, and lidar.

3.2   Research Gap 2: Sparse or Missing Sensor Data on Global Derivation of AED

The second research gap focuses on the effect of sparse, missing, or incorrect sensor data which informs the ED problem's constraints. Figure 4 shows the integration scope associated with this research gap. This gap can include the failure of sensors and infrastructure signals in providing the necessary information for AVs to perform ED as well as studies investigating how an AV can execute an ED function without all necessary information being available to it. Despite this being a common occurrence in real-world applications, there are not many ED papers that address this issue.

Table 9 Summary of existing research that includes the integration scope of Real-world AV perception with application to the AED problem, thus addressing research gap 1

Research Group	Sensors/signals	Data collection technique	Planning techniques	Vehicle plant
University of Utah and San Diego State University [113]	V2V and V2I	Macroscopic traffic flow model by dividing vehicles into different classes	Travel time minimization, fuel consumption	Custom mathematical model
University of California at Riverside [114]	Front camera, radar, on-board diagnostics (OBD) and V2V-based SPaT (signal phase and timing) information	Intelligent driver model (IDM) for traffic environment	Hybrid reinforcement learning (HRL)	Unity-based simulator
Loughborough University and University of Southampton [115]	GPS-based localization and long-range radar	STIMSIM drive simulation software which simulated 21 km route around Southampton, UK	Fuel consumption and driver preference, and predictive optimization of vehicle speed	2004 fiat Stilo

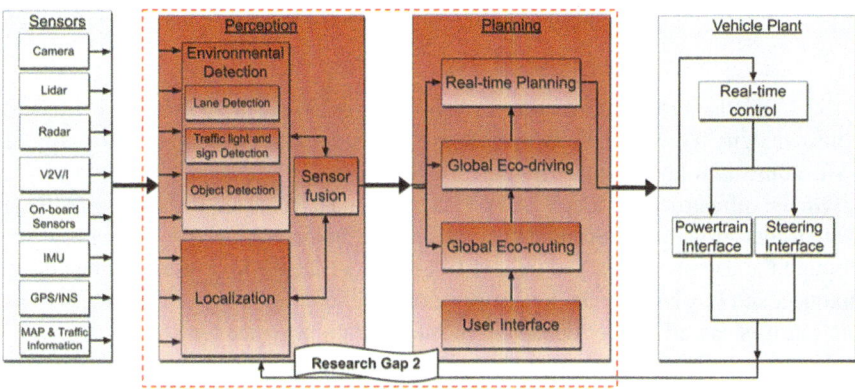

Fig. 4 The integration scope defined in research gap 2: Sparse or missing sensor data on global derivation of AED

On the vehicle side, researchers in University of California Berkeley have developed a stochastic approach with DP optimization to address scenarios in which limited SPaT data is available for AED vehicles [116]. Additionally, a two-layer receding horizon control framework has been proposed to address vehicle speed in scenarios where limited SPaT data is available with the control framework putting emphasis on safety control over velocity planning [117].

Table 10 Summary of existing research that includes the integration scope of sparse or missing sensor data on global derivation of AED problem, thus addressing research gap 2

Research Group	Sensors/signals	Perception model	Planning techniques	Faulty/Noisy data
University of California, Berkeley [116]	DSRC, camera, radar, LIDAR, GPS/INS	V2V/I perception and localization	Stochastic approach with DP optimization	Limited SPaT
VEDECOM [119]	Camera, LIDAR	Roadside infrastructure (RSI) central perception unit	None. This paper is more focused on external parties providing data for incoming AV	Object distance registered by camera and synchronization time for message transmission
University of California, Berkeley [117]	SPaT	ED and adaptive cruise control model	Two-layer receding horizon control framework (velocity planning and safety control)	Limited SPaT
SZTAKI [118]	Vehicle reference speed and following distance	Vehicle reference speed and following distance	Three layer control framework with driver safety having priority over vehicle cruise speed	Vehicle speed and acceleration

SZTAKI also proposed a similar framework to prioritize drive safety over vehicle cruise velocity but with a three layer control framework as opposed to University of California's two-layer control framework [118].

On the infrastructure side, VEDECOM proposed a Roadside Infrastructure (RSI) system that provides environmental information for incoming AV at intersections through the use of camera and lidar [119]. From VEDECOM's research consideration needs to be given for the height positioning and environment of RSI sensors as such factors can affect the robustness of information provided by RSI.

In reviewing the sources related to research gap 2, summarized in Table 10, few sources were available in directly addressing how AVs would perform autonomous driving features in faulty sensor and external infrastructure scenarios. While there are sources outlining the benefits and disadvantage of various sensors used in AV perception, such sources lack sufficient coverage on appropriate protocols in events where limited sensor and signal data are available [120, 121]. This suggests future research into research gap could focus on development of such protocols.

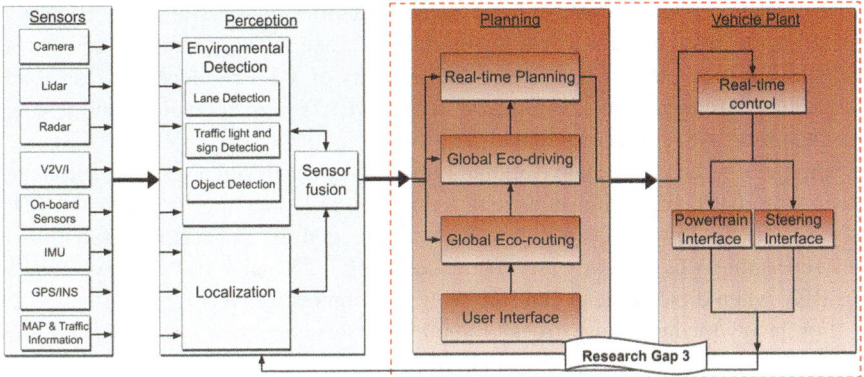

Fig. 5 The integration scope defined in research gap 3: Performance of a planning subsystem equipped with AED integrated with a physical vehicle plant

3.3 Research Gap 3: Performance of a Planning Subsystem Equipped with AED Integrated with a Physical Vehicle Plant

Research outlining the work done on physical AED implementation can be broken down into four distinct sections namely:

(i) what Drive Cycle was used to test AED control algorithm, (ii) what planning model was used to enable AV to generate a solution, (iii) what type of vehicle plant is used to validate performance of control algorithm and (iv) what physical vehicle is used to evaluate control algorithm in real time. Figure 5 provides the context of the research gap scope within the AV architecture.

Using a Rollout Algorithm (approximation of DP algorithm) with a multi-layer hierarchical Model Predictive Control (MPC) framework, researchers at Ohio State University evaluated the performance of AED through simulations and physical vehicle implementation [122]. Physical vehicle testing shows the vehicle consumed 22% less fuel compared to baseline scenario with 2.9% savings in trip time while maintaining State of Charge (SOC) at 50%. Results of physical testing were in line with findings from simulation.

University of Wisconsin-Madison developed a control system called Eco-Drive, used to optimize fuel efficiency for purely gasoline vehicles. Eco-Drive uses data available from ODB II port of gasoline vehicles to calculate an optimal vehicle speed to maximize fuel efficiency and implementation was done by automating accelerator pedal position via outputs from Eco-Drive [123]. Testing of Eco-Drive under 100 miles of driven road outline a fuel efficiency improvement of 10–40% depending on urban environments.

Leveraging NREL's Transportation Secure Data Center (TSDC) dataset, a joint effort between General Motors LLC, Carnegie Mellon University, and NREL was carried out to develop an AED vehicle that uses InfoRich Eco-Autonomous

Driving (iREAD) to generates optimal travel trajectories [124]. Evaluating iREAD's performance in large-scale, in-depth simulations along with physical evaluations in Vehicle-In-Loop, the research found fuel savings of 10–20% depending on road conditions. Although plans were made to test iREAD in road testing, such testing was not done by the time of publication.

On a similar note, Argonne National Labs (ANL) developed a set up to automate evaluation of ED algorithm in a Vehicle-In-Loop (VIL) setting for BEV and ICE vehicles [125]. Testing has shown ANL was successfully in creating a functional and repeatable VIL system with VIL test out-ling a 22% and 16% energy savings for BEV when driven in lead and following position respectively.

For heavy/medium duty trucks, Southwest Research Institute evaluated the performance of SwRI's ED control algorithm in class 8 trucks in accordance with J1321 test procedures [126]. Physical testing of class 8 trucks found SwRI's control algorithm resulted in 7% decrease in fuel consumption and 6% decrease in trip time.

Applying AED in a fleet-based setting, University of California and University of Cincinnati deployed a CAV fleet to evaluate the performance of AED in real time [127, 128]. Evaluating AED performance over 7 road segments and driven over 47 miles, University of Cincinnati's Relaxed Pontryagin's Minimum Principle (RPMP) based AED algorithm yielded fuel savings of 3.3 to 21.2% with variation depending on hill length and slope grade. Testing their control algorithm over 8 signalized intersections of Southern California, results of University of California's control algorithm outline a fueling savings of 30.98% for CAV fleet AED in exchange for an 8.51% increase in trip time compared to baseline.

Researchers at Colorado State University also applied predictive acceleration events control to the actual vehicle using customized 2019 Toyota Tacoma parallel-3 (P3) HEV. Their methodology combats long run time issues dynamic programming has for physical implementation by pre-computing the optimal solution for acceleration events. According to the findings of track-based testing using predictive acceleration event control in the real world 7% improvement in FE can be achieved. According to the author, this is the first time this sort of testing has ever been conducted on a real-world vehicle.

The parameters of interest are summarized in Table 11. In researching physical implementation of AED, we found that a majority of physical ED research was done on gasoline vehicles. This indicates that ED for physical BEV or Hybrids may be a potential avenue for future research.

4 Conclusions

This literature review provides an overview of automotive energy efficient control strategies and discusses that AED for BEVs should be a focus of future research efforts. A systems-level diagram of AED is proposed and an expansion of NASA's TRL analysis (SRL analysis) is performed which identifies three existing research gaps: real-world AV perception with application to the AED problem, sparse or

Table 11 Summary of existing research that includes the integration scope of Performance of a planning subsystem equipped with AED integrated with a physical vehicle plant, thus addressing research gap 3

Research group	Drive cycle	Planning model	Vehicle plant	Vehicle realization type
Ohio State University [122]	Custom route, Columbus, Ohio	Rollout algorithm (approximate DP) and model predictive control	P0 mild-HEV, 2016 VW Passat retrofitted with 48 V mild hybrid system	Actual vehicle
University of Wisconsin Madison [123]	Custom mid-size US city drive data	Eco-drive (DP)	Gasoline vehicle plant	Actual vehicle
National Renewable Energy lab oratory [124]	NREL's Transportation Secure Data Center (TSDC) drive cycle data	InfoRich Eco-Autonomous Driving (iREAD)	Cadilac CT6 (BEV)	GM internal model
Argonne National Labs [125]	Multiple custom drive cycle of varying speed limit and HWFET	Analytical closed form solutions	Chevrolet bolt (BEV)	Actual vehicle
Southwest Research Institute (SwRI) [126]	Modified NREL port drayage cycles	Control algorithm with objective of minimizing jerk and acceleration events	2017 Volvo VNL64T300	Actual vehicle
University of Cincinnati [127]	Rolling segments in Virginia and Maryland	Relaxed Pontryagin's minimum principle (RPMP)	2013 ICE Cadillac SRX	Actual vehicle
University of California, Berkeley [128]	Custom Route Model built using July 2019 Sensys Network data	ECO-ACC (Eco driving controller-adaptive cruise controller)	Unknown, PHEV is only stated to have 8.89kWh battery capacity	Actual vehicle
University of California, Riverside [129]	NA	NA	2015 Volvo VNL	Actual vehicle
University of Michigan [130]	Custom route, Ann Arbor, MI	Prediction of queuing profile using shock-wave profile model [131]	2017 Toyota Prius four Turing HEV	Actual vehicle
Colorado State University [132]	Custom route, Fort Collins, co	Predictive acceleration event model	2019 Toyota Tacoma	Actual vehicle

missing sensor data on global derivation of AED, and performance of a planning subsystem equipped with AED integrated with a physical vehicle plant. In other words, there are gaps in knowledge concerning

1. An understanding of critical sensors and signals for perception and sensor fusion that enable effective FE vehicle control through AED.
2. An in-depth comprehension of the sorts of fault or missing data from perception that might impact effective FE vehicle control.
3. The operational and real-world problems of effective AED control implementation.

Investigation of the AED literature revealed that, despite the availability of hundreds of papers addressing the idea of ED, there are few papers that provide insights into the AED research gaps which are currently slowing commercial realization. A summary of relevant papers that are beginning to address these gaps are provided and a summary of missing knowledge is given.

The overall conclusion of this research is that focused studies addressing AED research gaps are needed before AV technology and its associated infrastructure is rolled out and fully commercialized. ED considerations need to be a part of AV RD efforts to ensure that transportation sustainability is improved at the same rate as transportation safety. There are many inconclusive studies about the effect of widespread AV adoption on transportation energy use but some of these worst case scenarios could be alleviated with ED implementation. Focused studies are needed that utilize real-world AV sensors, that investigate the effects of sensor errors, and that include real world BEV implementation.

References

1. Mundorf, N., Redding, C.A., Bao, S.: Sustainable transportation and health. Int. J. Environ. Res. Public Health. **15**, 542 (2018)
2. Motallebiaraghi, F., Yao, K., Rabinowitz, A., Hoehne, C., Garikapati, V., Holden, J., Wood, E., Chen, S., Asher, Z., Bradley, T.: Mobility energy productivity evaluation of prediction-based vehicle powertrain control combined with optimal traffic management. Tech. Rep. 2022-01-0141, SAE Technical Paper (2022)
3. Rabinowitz, A., Araghi, F.M., Gaikwad, T., Asher, Z.D., Bradley, T.H.: Development and evaluation of velocity predictive optimal energy management strategies in intelligent and connected hybrid electric vehicles. Energies. **14**, 5713 (2021)
4. Electric Vehicle Benefits and Considerations. https://afdc.energy.gov/fuels/electricitybenefits.html
5. Jung, H.: Fuel economy of plug-in hybrid electric and hybrid electric vehicles: effects of vehicle weight, hybridization ratio and ambient temperature. World Electric Vehicle Journal. **11**, 31 (2020)
6. Bose, P., Mandal, D.K.: The future has arrived, are we ready for EV? IOP Conf. Ser.: Mater. Sci. Eng. **1080**, 012004 (2021)
7. Muratori, M., Alexander, M., Arent, D., Bazilian, M., Cazzola, P., Dede, E.M., Farrell, J., Gearhart, C., Greene, D., Jenn, A., Keyser, M., Lipman, T., Narumanchi, S., Pesaran, A., Sioshansi, R., Suomalainen, E., Tal, G., Walkowicz, K., Ward, J.: The rise of electric vehicles – 2020 status and future expectations. Prog. Energy Combust. Sci. **3**, 022002 (2021)

8. Ahmad, S., Khan, M., Others: Tesla: Disruptor or sustaining innovator. Journal of Case Research. **10**, 1 (2019)
9. Linde, A.: Electric Cars – the Future Is Now! Veloce Publishing Ltd (2010)
10. Zhou, Y., Wang, M., Hao, H., Johnson, L., Wang, H., Hao, H.: Plug-in electric vehicle market penetration and incentives: a global review. Mitig. Adapt. Strateg. Glob. Chang. **20**, 777–795 (2015)
11. Husain, I.: Electric and Hybrid Vehicles: Design Fundamentals, 2nd edn. CRC Press (2011)
12. Mahmoudzadeh Andwari, A., Pesiridis, A., Rajoo, S., Martinez-Botas, R., Esfahanian, V.: A review of battery electric vehicle technology and readiness levels. Renew. Sust. Energ. Rev. **78**, 414–430 (2017)
13. Kadav, P., Asher, Z.D.: Improving the Range of Electric Vehicles. Int. J. Electr. Hybrid Veh (2019)
14. Council on Future Mobility & Electrification, "2020 report," Tech. Rep. 1, Michigan Office of Future Mobility and Electrification, 2020
15. Wards Auto.: Powering up Electric Vehicles Key Part of Michigan Future Plans (2020)
16. Malmgren, I.: Quantifying the societal benefits of electric vehicles. World Electric Vehicle Journal. **8**, 996–1007 (2016)
17. Boulanger, A.G., Chu, A.C., Maxx, S., Waltz, D.L.: Vehicle electrification: status and issues. Proc. IEEE. **99**, 1116–1138 (2011)
18. Camacho, O.M.F., Norgard, P.B., Rao, N., Mihet-Popa, L.: Electrical Vehicle Batteries Testing in a Distribution Network Using Sustainable Energy, vol. 5, pp. 1033–1042 (2014)
19. Camacho, O.M.F., Mihet-Popa, L.: Fast charging and smart charging tests for electric vehicles batteries using renewable energy. Oil & Gas Science and Technology – Revue d'IFP Energies nouvelles. **71**(1), 13 (2016)
20. Un-Noor, F., Padmanaban, S., Mihet-Popa, L., Mollah, M.N., Hossain, E.: A comprehensive study of key electric vehicle (EV) components, technologies, challenges, impacts, and future direction of development. Energies. **10**, 1217 (2017)
21. Chan, C.C.: The State of the Art of Electric and Hybrid Vehicles, vol. 90, pp. 247–275 (2002)
22. Lin, C.-C., Peng, H., Grizzle, J.W., Kang, J.-M.: Power management strategy for a parallel hybrid electric truck. IEEE Trans. Control Syst. Technol. **11**, 839–849 (2003)
23. Wu, G., Zhang, X., Dong, Z.: Powertrain architectures of electrified vehicles: review, classification and comparison. J. Frankl. Inst. **352**, 425–448 (2015)
24. Rauh, N., Franke, T., Krems, J.F.: Understanding the impact of electric vehicle driving experience on range anxiety. Hum. Factors. **57**, 177–187 (2015)
25. Z. D. Asher, V. Wifvat, A. Navarro, S. Samuelsen, and T. Bradley, "The Importance of HEV Fuel Economy and Two Research Gaps Preventing Real World Implementation of Optimal Energy Management," 2017
26. Asher, Z.D., Patil, A.A., Wifvat, V.T., Frank, A.A., Samuelsen, S., Bradley, T.H.: Identification and review of the research gaps preventing a realization of optimal energy management strategies in vehicles. SAE Int. J. Alt. Power. **8** (2019)
27. Zeng, X., Wang, J.: A two-level stochastic approach to optimize the energy management strategy for fixed-route hybrid electric vehicles. Mechatronics. **38**, 93–102 (2016)
28. Hibberd, D.L., Jamson, A.H., Jamson, S.L.: The design of an in-vehicle assistance system to support eco-driving. Transp. Res. Part C: Emerg. Technol. **58**, 732–748 (2015)
29. Zhou, M., Jin, H., Wang, W.: A review of vehicle fuel consumption models to evaluate eco-driving and eco-routing. Transp. Res. Part D: Trans. Environ. **49**, 203–218 (2016)
30. Panday, A., Bansal, H.O.: A review of optimal energy management strategies for hybrid electric vehicle. Int. J. Veh. Technol. **2014**, 1–19 (2014)
31. Amjad, S., Neelakrishnan, S., Rudramoorthy, R.: Review of design considerations and technological challenges for successful development and deployment of plug-in hybrid electric vehicles. Renew. Sust. Energ. Rev. **14**, 1104–1110 (2010)
32. Zhang, F., Wang, L., Coskun, S., Pang, H., Cui, Y., Xi, J.: Energy management strategies for hybrid electric vehicles: review, classification, comparison, and outlook. Energies. **13**, 3352 (June 2020)

33. Yang, C., Zha, M., Wang, W., Liu, K., Xiang, C.: Efficient Energy Management Strategy for Hybrid Electric Vehicles/Plug-in Hybrid Electric Vehicles: Review and Recent Advances under Intelligent Transportation System. IET Intel. Transport Syst, Mar (2020)
34. Zhang, F., Hu, X., Langari, R., Cao, D.: Energy management strategies of connected HEVs and PHEVs: recent progress and outlook. Prog. Energy Combust. Sci. **73**, 235–256 (2019)
35. Asher, Z.D., Baker, D.A., Bradley, T.H.: Prediction Error Applied to Hybrid Electric Vehicle Optimal Fuel Economy. IEEE Trans. Control Syst. Technol (2017)
36. Sulaiman, N., Hannan, M.A., Mohamed, A., Majlan, E.H., Wan Daud, W.R.: A review on energy management system for fuel cell hybrid electric vehicle: issues and challenges. Renew. Sust. Energ. Rev. **52**, 802–814 (2015)
37. Burke, A.F., Smith, G.E.: Impacts of use-pattern on the design of electric and hybrid vehicles. In: SAE Technical Paper Series. no. 810265, (400 Commonwealth Drive, Warrendale, PA, United States), SAE International (1981)
38. Banvait, H., Anwar, S., Chen, Y.: A rule-based energy management strategy for plug-in hybrid electric vehicle (PHEV). 2009 American Control Conference, 3938–3943 (2009)
39. Shen, Y., Ge, G., Liu, A., Zheng, Z.: Operation of an ICE/PM/TTRB APU in a range extender electric vehicle Power-Train, pp. 3205–3210. 2019 IEEE Innovative Smart Grid Technologies – Asia (ISGT Asia) (2019)
40. Frank, A.A., Francisco, A.: Ideal Operating Line CVT Shifting Strategy for Hybrid Electric Vehicles. tech. rep (2002)
41. Li, Q., Chen, W., Li, Y., Liu, S., Huang, J.: Energy management strategy for fuel cell/battery/ultracapacitor hybrid vehicle based on fuzzy logic. Int. J. Electr. Power Energy Syst. **43**, 514–525 (2012)
42. Jun, L., Faming, Z., Xiong, T., Biao, L., Wenbin, W.: Simulation research on PHEV based on fuzzy logic control strategies. Journal of Chongqing Jiao Tong University (Natural Science). **32**(2), 329–333 (2013)
43. Mohd Sabri, M.F., Danapalasingam, K.A., Rahmat, M.F.: Improved fuel economy of through-the-road hybrid electric vehicle with fuzzy logic-based energy management strategy. Int. J. Fuzzy Syst. **20**, 2677–2692 (2018)
44. Denis, N., Dubois, M.R., Desrochers, A.: Fuzzy-based blended control for the energy management of a parallel plug-in hybrid electric vehicle. IET Intell. Transp. Syst. **9**, 30–37 (2015)
45. Li, P., Li, Y., Wang, Y., Jiao, X.: An intelligent logic Rule-Based energy management strategy for Power-Split plug-in hybrid electric vehicle, pp. 7668–7672. 2018 37th Chinese Control Conference (CCC) (2018)
46. Meyer, R.T.: Distributed switched optimal control of an electric vehicle. Energies. **13**, 3364 (2020)
47. Abotabik, M., Meyer, R.T.: Switched Optimal Control of a Heavy-Duty Hybrid Vehicle, vol. 14. Energies (2021)
48. Liu, J., Chen, Y., Li, W., Shang, F., Zhan, J.: Hybrid-trip-model-based energy management of a PHEV with computation-optimized dynamic programming. IEEE Trans. Veh. Technol. **67**, 338–353 (2018)
49. C.-C. Lin, J.-M. Kang, J. W. Grizzle, and H. Peng, "Energy Management Strategy for a Parallel Hybrid Electric Truck," 2001
50. Bellman, R.: Dynamic programming. Science. **153**, 34–37 (1966)
51. Rousseau, G., Sinoquet, D., Rouchon, P.: Constrained optimization of energy management for a mild-hybrid vehicle. Oil & Gas Science and Technology – Revue de l'IFP. **62**(4), 623–634 (2007)
52. Wei, X., Guzzella, L., Utkin, V.I., Rizzoni, G.: Model-Based Fuel Optimal Control of Hybrid Electric Vehicle Using Variable Structure Control Systems, vol. 129, pp. 13–19 (2007)
53. Serrao, L., Rizzoni, G.: Optimal control of power split for a hybrid electric refuse vehicle, pp. 4498–4503. 2008 American Control Conference (2008)
54. Du, Y., Zhao, Y., Wang, Q., Zhang, Y., Xia, H.: Trip-oriented stochastic optimal energy management strategy for plug-in hybrid electric bus. Energy. **115**, 1259–1271 (2016)

55. Liu, B., Li, L., Wang, X., Cheng, S.: Hybrid electric vehicle downshifting strategy based on stochastic dynamic programming during regenerative braking process. IEEE Trans. Veh. Technol. **67**, 4716–4727 (2018)

56. Moura, S.J., Fathy, H.K., Callaway, D.S., Stein, J.L.: A Stochastic Optimal Control Approach for Power Management in Plug-in Hybrid Electric Vehicles, vol. 19, pp. 545–555 (2011)

57. T. Leroy, J. Malaize, and G. Corde, Towards real-time optimal energy management of HEV powertrains using stochastic dynamic programming, in 2012 IEEE Vehicle Power and Propulsion Conference, pp. 383–388, ieeexplore.ieee.org, 2012

58. Lu, X., Wu, Y., Lian, J., Zhang, Y., Chen, C., Wang, P., Meng, L.: Energy management of hybrid electric vehicles: A review of energy optimization of fuel cell hybrid power system based on genetic algorithm. Energy Convers. Manage. **205**, 112474 (2020)

59. Wieczorek, M., Lewandowski, M.: A mathematical representation of an energy management strategy for hybrid energy storage system in electric vehicle and real time optimization using a genetic algorithm. Appl. Energy. **192**, 222–233 (2017)

60. Chen, Z., Xiong, R., Wang, K., Jiao, B.: Optimal energy management strategy of a plug-in hybrid electric vehicle based on a particle swarm optimization algorithm. Energies. **8**, 3661–3678 (2015)

61. Chen, Z., Xiong, R., Cao, J.: Particle swarm optimization-based optimal power management of plug-in hybrid electric vehicles considering uncertain driving conditions. Energy. **96**, 197–208 (2016)

62. Rezaei, A., Burl, J.B., Zhou, B.: Estimation of the ECMS Equivalent Factor Bounds for Hybrid Electric Vehicles. IEEE Trans. Control Syst. Technol (2017)

63. Zhang, Y., Chu, L., Fu, Z., Guo, C., Zhao, D., Li, Y., Ou, Y., Xu, L.: An improved adaptive equivalent consumption minimization strategy for parallel plug-in hybrid electric vehicle. Proc. Inst. Mech. Eng. Pt. D: J. Automobile Eng. **233**, 1649–1663 (2019)

64. Chen, Z., Liu, Y., Ye, M., Zhang, Y., Chen, Z., Li, G.: A survey on key techniques and development perspectives of equivalent consumption minimisation strategy for hybrid electric vehicles. Renew. Sust. Energ. Rev. **151**, 111607 (2021)

65. Asher, Z.D., Trinko, D.A., Payne, J.D., Geller, B.M., Bradley, T.H.: Real-Time implementation of optimal energy management in hybrid electric vehicles: Globally optimal control of acceleration events. J. Dyn. Syst. Meas. Control. **142** (2020)

66. Xie, S., Hu, X., Qi, S., Tang, X., Lang, K., Xin, Z., Brighton, J.: Model predictive energy management for plug-in hybrid electric vehicles considering optimal battery depth of discharge. Energy. **173**, 667–678 (2019)

67. T. Gaikwad, A. Rabinowitz, F. Motallebiaraghi, T. Bradley, and others, Vehicle Velocity Prediction Using Artificial Neural Network and Effect of Real World Signals on Prediction Window, 2020

68. Patil, A.A.: Comparison of Optimal Energy Management Strategies Using Dynamic Programming, Model Predictive Control, and Constant Velocity Prediction. PhD thesis, Western Michigan University (2020)

69. Meyer, R.T., DeCarlo, R.A., Pekarek, S.: Hybrid model predictive power management of a battery-supercapacitor electric vehicle. Asian J. Control. **18**, 150–165 (2016)

70. Meyer, R.T., Johnson, S.C., DeCarlo, R.A., Pekarek, S., Sudhoff, S.D.: Hybrid electric vehicle fault tolerant control. J. Dyn. Syst. Meas. Control. **140** (2017)

71. Bertsekas, D.P.: Dynamic Programming and Optimal Control. Athena Scientific (1995)

72. Braekers, K., Ramaekers, K., Van Nieuwenhuyse, I.: The vehicle routing problem: state of the art classification and review. Comput. Ind. Eng. **99**, 300–313 (2016)

73. Dijkstra, E.W.: A note on two problems in connexion with graphs. Numer. Math. **1**, 269–271 (1959)

74. Ericsson, E., Larsson, H., Brundell-Freij, K.: Optimizing route choice for lowest fuel consumption–potential effects of a new driver support tool. Transp. Res. Part C: Emerg. Technol. **14**(6), 369–383 (2006)

75. Boriboonsomsin, K., Barth, M.J., Zhu, W., Vu, A.: Eco-routing navigation system based on multisource historical and real-time traffic information. IEEE Trans. Intell. Transp. Syst. **13**, 1694–1704 (2012)

76. Zhu, L., Chiu, Y.-C.: Transportation routing map abstraction approach: algorithm and numerical analysis. Transp. Res. Rec. **2528**, 78–85 (2015)
77. Zhu, L.: Routing Map Topology Analysis and Application. PhD thesis (2014)
78. Barth, M., Boriboonsomsin, K., Vu, A.: Environmentally-Friendly Navigation, pp. 684–689. 2007 IEEE Intelligent Transportation Systems Conference (2007)
79. Andersen, O., Jensen, C.S., Torp, K., Yang, B.: EcoTour: Reducing the environmental footprint of vehicles using eco-routes, vol. 1, pp. 338–340. 2013 IEEE 14th International Conference on Mobile Data Management (2013)
80. Yang, B., Guo, C., Jensen, C.S., Kaul, M., Shang, S.: Stochastic skyline route planning under time-varying uncertainty, pp. 136–147. 2014 IEEE 30th International Conference on Data Engineering (2014)
81. Guanetti, J., Kim, Y., Borrelli, F.: Control of connected and automated vehicles: state of the art and future challenges. Annu. Rev. Control. **45**, 18–40 (2018)
82. M. Kubicka, J. Klusacek, A. Sciarretta, A. Cela, H. Mounier, L. Thibault, and S. I. Niculescu, "Performance of Current Eco-Routing Methods," 2016
83. Jurik, T., Cela, A., Hamouche, R., Natowicz, R., Reama, A., Niculescu, S.-I., Julien, J.: Energy Optimal Real-Time Navigation System, vol. 6, pp. 66–79 (2014)
84. Sun, Z., Zhou, X.: To save money or to save time: intelligent routing design for plug-in hybrid electric vehicle. Transp. Res. Part D: Trans. Environ. **43**, 238–250 (2016)
85. Z. Qiao and O. Karabasoglu, "Vehicle Powertrain Connected Route Optimization for Conventional, Hybrid and Plug-in Electric Vehicles," 2016
86. De Nunzio, G., Sciarretta, A., Ben Gharbia, I., Ojeda, L.L.: A constrained Eco-Routing strategy for hybrid electric vehicles based on Semi-Analytical energy management, pp. 355–361. 2018 21st International Conference on Intelligent Transportation Systems (ITSC) (2018)
87. Houshmand, A., Cassandras, C.G.: Eco-Routing of Plug-In Hybrid Electric Vehicles in Transportation Networks, pp. 1508–1513. 2018 21st International Conference on Intelligent Transportation Systems (ITSC) (2018)
88. Mahmoud, Y.H., Brown, N.E., Motallebiaraghi, F., Koelling, M., Meyer, R., Asher, Z.D., Dontchev, A., Kolmanovsky, I.: Autonomous eco-driving with traffic light and lead vehicle constraints: an application of best constrained interpolation. IFAC-PapersOnLine. **54**(10), 45–50 (2021)
89. Asher, Z.D., Trinko, D.A., Bradley, T.H.: Increasing the fuel economy of connected and autonomous lithium-ion electrified vehicles. In: Pistoia, G., Liaw, B. (eds.) Behaviour of Lithium-Ion Batteries in Electric Vehicles: Battery Health, Performance, Safety, and Cost, pp. 129–151. Springer International Publishing, Cham (2018)
90. Michel, P., Karbowski, D., Rousseau, A.: Impact of Connectivity and Automation on Vehicle Energy Use. tech. rep., SAE Technical Paper (2016)
91. Mandava, S., Boriboonsomsin, K., Barth, M.: Arterial Velocity Planning Based on Traffic Signal Information Under Light Traffic Conditions, pp. 1–6., ieeexplore.ieee.org. 2009 12th International IEEE Conference on Intelligent Transportation Systems (2009)
92. Potvin-Bernal, J., Hansma, B., Donmez, B., Lockwood, P., Shu, L.H.: Influencing greater adoption of Eco-Driving practices using an associative graphical display. J. Mech. Des. **142** (2020)
93. Zhang, P., Yan, F., Du, C.: A comprehensive analysis of energy management strategies for hybrid electric vehicles based on bibliometrics. Renew. Sust. Energ. Rev. **48**, 88–104 (2015)
94. Davis, K.R., Davis, C.M., Zonouz, S.A., Bobba, R.B., Berthier, R., Garcia, L., Sauer, P.W.: A cyber-physical modeling and assessment framework for power grid infrastructures. IEEE Trans. Smart Grid. **6**, 2464–2475 (2015)
95. Rajkumar, R., Lee, I., Sha, L., Stankovic, J.: Cyber-physical systems: The next computing revolution, pp. 731–736., ieeexplore.ieee.org. Design Automation Conference (2010)
96. Giraldo, J., Sarkar, E., Cardenas, A.A., Maniatakos, M., Kantarcioglu, M.: Security and privacy in cyber-physical systems: a survey of surveys. IEEE Design Test. **34**, 7–17 (2017)
97. Ebert, C., Jones, C.: Embedded software: facts, figures, and future. Computer. **42**, 42–52 (2009)

98. Chattopadhyay, A., Lam, K.-Y.: Security of Autonomous Vehicle as a Cyber-Physical System, pp. 1–6., ieeexplore.ieee.org. 2017 7th International Symposium on Embedded Computing and System Design (ISED) (2017)

99. Chen, B., Yang, Z., Huang, S., Du, X., Cui, Z., Bhimani, J., Xie, X., Mi, N.: Cyber-Physical System Enabled Nearby Traffic Flow Modelling for Autonomous Vehicles, pp. 1–6., ieeexplore.ieee.org. 2017 IEEE 36th International Performance Computing and Communications Conference (IPCCC) (2017)

100. Abid, H., Phuong, L.T.T., Wang, J., Lee, S., Qaisar, S.: V-Cloud: vehicular cyber-physical systems and cloud computing. In: Proceedings of the 4th International Symposium on Applied Sciences in Biomedical and Communication Technologies no. Article 165 in ISABEL '11, pp. 1–5. Association for Computing Machinery, New York, NY, USA (2011)

101. Harper, C.D., Hendrickson, C.T., Mangones, S., Samaras, C.: Estimating potential increases in travel with autonomous vehicles for the non-driving, elderly and people with travel-restrictive medical conditions. Transp. Res. Part C: Emerg. Technol. **72**, 1–9 (2016)

102. Chase, N., Maples, J., Schipper, M.: Autonomous vehicles: Uncertainties and energy implications. In: 2018 EIA Energy Conference, vol. 5, p. 2018. PowerPoint, Washington, DC

103. Mohan, A., Sripad, S., Vaishnav, P., Viswanathan, V.: Trade-Offs between Automation and Light Vehicle Electrification, vol. 5, pp. 543–549. Nature Energy (2020)

104. Goberville, N., Zoardar, M.M., Rojas, J., Brown, N., Motallebiaraghi, F., Navarro, A., Asher, Z.: Techno-economic analysis of fixed-route autonomous and electric shuttles. SAE Technical Paper, 01–0061 (2021)

105. Goberville, N.A., Kadav, P., Asher, Z.D.: Tire Track Identification: A Method for Drivable Region Detection in Conditions of Snow-Occluded Lane Lines. tech. rep., SAE Technical Paper (2022)

106. Robinson, K.A., Saldanha, I.J., Mckoy, N.A.: Development of a Framework to Identify Research Gaps from Systematic Reviews, vol. 64, pp. 1325–1330 (2011)

107. Rosique, F., Navarro, P.J., Ferna'ndez, C., Padilla, A.: A systematic review of perception system and simulators for autonomous vehicles research. Sensors. **19** (2019)

108. Paden, B., a'p, M.C., Yong, S.Z., Yershov, D., Frazzoli, E.: A survey of motion planning and control techniques for Self-Driving urban vehicles. IEEE Transactions on Intelligent Vehicles. **1**, 33–55 (2016)

109. Li, X., Sun, Z., Cao, D., He, Z., Zhu, Q.: Real-time trajectory planning for autonomous urban driving: framework, algorithms, and verifications. IEEE/ASME Trans. Mechatron. **21**, 740–753 (2016)

110. Mankins, J.C.: Technology readiness assessments: a retrospective. Acta Astronaut. **65**, 1216–1223 (2009)

111. Shishko, R.: Optimizing technology investments: A broad mission model approach. In: AIAA Space 2003 Conference & Exposition. AIAA SPACE Forum, American Institute of Aeronautics and Astronautics (2003)

112. Sauser, B., Verma, D., Ramirez-Marquez, J., Gove, R.: From TRL to SRL: The Concept of Systems Readiness Levels, pp. 1–10. Conference on Systems Engineering Research, Los Angeles (2006)

113. Yang, X.T., Huang, K., Zhang, Z., Zhang, Z.A., Lin, F.: Eco-driving system for connected automated vehicles: multi-objective trajectory optimization. IEEE Trans. Intell. Transp. Syst. **22**, 7837–7849 (2021)

114. Bai, Z., Hao, P., Shangguan, W., Cai, B., Barth, M.J.: Hybrid Reinforcement Learning-Based Eco-Driving Strategy for Connected and Automated Vehicles at Signalized Intersections, vol. 23, pp. 15850–15863 (2022)

115. Fleming, J., Yan, X., Allison, C., Stanton, N., others: Real-Time Predictive Eco-Driving Assistance Considering Road Geometry and Long-Range Radar Measurements, vol. 15, pp. 573–583. IET Intel. Transport Syst (2021)

116. Sun, C., Guanetti, J., Borrelli, F., Moura, S.J.: Optimal eco-driving control of connected and autonomous vehicles through signalized intersections. IEEE Internet Things J. **7**(5), 3759–3773 (2020)

117. Bae, S., Choi, Y., Kim, Y., Guanetti, J., Borrelli, F., Moura, S.J.: Real-time ecological velocity planning for plug-in hybrid vehicles with partial communication to traffic lights. CoRR. abs/1903.08784 (2019)
118. Nemeth, B., Miha'ly, A., Ga'spa'r, P.: Design of Fault-Tolerant Cruise Control in a Hierarchical Framework for Connected Automated Vehicles, pp. 1–6. 5th International Conference on Control and Fault-Tolerant Systems, 2021
119. Jandial, A., Merdrignac, P., Shagdar, O., Fevrier, L.: Implementation and evaluation of intelligent roadside infrastructure for automated vehicle with i2v communication. In: Laouiti, A., Qayyum, A., Saad, M.N.M. (eds.) Vehicular Ad-hoc Networks for Smart Cities, pp. 3–18. Springer, Singapore (2020)
120. Vargas, J., Alsweiss, S., Toker, O., Razdan, R., Santos, J.: An overview of autonomous vehicles sensors and their vulnerability to weather conditions. Sensors. **21** (2021)
121. Yurtsever, E., Lambert, J., Carballo, A., Takeda, K.: A survey of autonomous driving: Common practices and emerging technologies. CoRR. abs/1906.05113 (2019)
122. Deshpande, S.R., Gupta, S., Gupta, A., Canova, M.: Real-Time Ecodriving Control in Electrified Connected and Autonomous Vehicles Using Approximate Dynamic Programing. Journal of Dynamic Systems, Measurement, and Control. **144**, 011111 (2022)
123. Kang, L., Qi, B., Janecek, D., Banerjee, S.: Ecodrive: A mobile sensing and control system for fuel efficient driving. In: Proceedings of the 21st Annual International Conference on Mobile Computing and Networking, MobiCom '15, pp. 358–371. Association for Computing Machinery, New York, NY, USA (2015)
124. R. Rajkumar, J. Zhao, C. F. Chang, and J. Gonder, Corroborative evaluation of the real-world energy saving potentials of inforich eco-autonomous driving (iread) system
125. Jeong, J., Karbowski, D., Kim, N., Han, J., Stutenberg, K., Di Russo, M., Grave, J.: Vehicle-in-the-loop workflow for the evaluation of energy-efficient automated driving controls in real vehicles. Tech. Rep (2022)
126. Bhagdikar, P., Gankov, S., Frazier, C., Rengarajan, S. et al., Demonstration of Energy Consumption Reduction in Class 8 Trucks Using Eco-Driving Algorithm Based on On-Road Testing, SAE Technical Paper 2022-01-0139, 2022, https://doi.org/10.4271/2022-01-0139
127. Ma, J., Hu, J., Leslie, E., Zhou, F., Huang, P., Bared, J.: An eco-drive experiment on rolling terrains for fuel consumption optimization with connected automated vehicles. Transportation Research Part C: Emerging Technologies. **100**, 125–141 (2019)
128. Bae, S., Kim, Y., Choi, Y., Guanetti, J., Gill, P., Borrelli, F., Moura, S.J.: Ecological adaptive cruise control of plug-in hybrid electric vehicle with connected infrastructure and on-road experiments. Journal of Dynamic Systems, Measurement, and Control. **144**, 011109 (2022)
129. Wang, Z., Hsu, Y.-P., Vu, A., Caballero, F., Hao, P., Wu, G., Boriboonsomsin, K., Barth, M.J., Kailas, A., Amar, P., Garmon, E., Tanugula, S.: Early findings from field trials of Heavy-Duty truck connected Eco-Driving system, pp. 3037–3042. 2019 IEEE Intelligent Transportation Systems Conference (ITSC) (2019)
130. Amini, M.R., Hu, Q., Wang, H., Feng, Y., Kolmanovsky, I., Sun, J.: Experimental Validation of Eco-Driving and Eco-Heating Strategies for Connected and Automated HEVs. SAE International, Warrendale (2021)
131. Yang, Z., Feng, Y., Gong, X., Zhao, D., Sun, J.: Eco-trajectory planning with consideration of queue along congested corridor for hybrid electric vehicles. Transp. Res. Rec. **2673**, 277–286 (2019)
132. White, S.: Physical Validation of Predictive Acceleration Control on A parallel Hybrid Electric Vehicle. PhD Thesis

Index

SPRINGER NATURE

GPSR Compliance

The European Union's (EU) General Product Safety Regulation (GPSR) is a set of rules that requires consumer products to be safe and our obligations to ensure this.

If you have any concerns about our products, you can contact us on ProductSafety@springernature.com

In case Publisher is established outside the EU, the EU authorized representative is:

Springer Nature Customer Service Center GmbH
Europaplatz 3
69115 Heidelberg, Germany

The manufacturer's authorised representative in the EU is Springer
Nature Customer Service Centre GmbH, Europaplatz 3, 69115 Heidelberg,
Germany. If you have any concerns regarding our products, please
contact ProductSafety@springernature.com

Printed and bound by CPI Group (UK) Ltd, Croydon, CR0 4YY
27/04/2026
02097573-0012